# MICROBIAL PROTEOMICS

METHODS OF
BIOCHEMICAL ANALYSIS

Volume 49

# MICROBIAL PROTEOMICS
## FUNCTIONAL BIOLOGY OF WHOLE ORGANISMS

Ian Humphery-Smith
Biosystems Informatics Institute
Newcastle-Upon-Tyne
United Kingdom

Michael Hecker
Institute for Microbiology
Ernst-Moritz-Arndt-Universität
Greifswald, Germany

WILEY-LISS
A JOHN WILEY & SONS, INC., PUBLICATION

Published by John Wiley & Sons, Inc., Hoboken, New Jersey
Published simultaneously in Canada



***Library of Congress Cataloging-in-Publication Data:***

Humphery-Smith, Ian.
Microbial proteomics / Ian Humphery-Smith, Michael Hecker.

Includes bibliographical references and index.
ISBN-13 978-0-471-69975-0 (cloth)
ISBN-10 0-471-69975-6 (cloth)
1. Microbial biotechnology. 2. Proteomics. I. Hecker, M. (Michael) II. Title.

TP248.27.M53H86 2006
660.6′2–dc22 2005056954

Printed in the United States of America
10 9 8 7 6 5 4 3 2 1

# CONTENTS

# PREFACE

Here, we have attempted to regroup those authors best able to provide an overview of *microbial proteomics: whole-organism functional biology*. Never has the scientific community better been placed to assimilate and interpret whole-organism functionality, so this text advances quite some distance down this rarely, if ever, trodden path in much detail and focus. This tome has been designed to allow sector specialists and novices to delve beneath the surface of this fascinating field and start to meaningfully grapple with more esoteric concepts, such as a definition of life itself and life processes in the simplest living organisms. Schrödinger, in his seminal work *What Is Life* (Cambridge University Press, 1945), made significant in-roads by defining the properties of a necessary information-containing system, namely his *aperiodic crystal*. Not so long after, Alfred Hershey (1908–1997) and his assistant Martha Chase (1923–2003) at the Cold Spring Harbor Laboratory showed that deoxyribonucleic acid (DNA), and not proteins, of the phage virus contained the phage genes, that is, the code of life. Armed with access to a critical mass of experimentally acquired data akin to that presented in this textbook, we are now well placed to go further and refine the shared characters of living systems and to define features of life itself. Such definitions are capable of mathematically constraining our understanding of such complex phenomena within limited parameter sets.

As a consequence of our increased understanding over the last half century, we are gradually succeeding in placing these mathematical and observational constraints of what exactly constitutes a living system. In the case of microbes, their relative simplicity has made them attractive tragets for extensive experimental manipulation directed toward both dissecting their biology in a quest for improved disease prevention and treatment and an improved understanding of whole-organism functional biology. In this text, we shall see excellent examples of the reductionist approach directed at analyzing the multitude of parts and more holistic modeling exercises. Future endeavors will face the more daunting task of better linking these information sets between different resolution scales in time and space within living systems and the relations of the whole with the external environment—all in a concerted effort to survive and deal with a wide range of stress conditions.

Following decades of study dedicated to microbial physiology, environmental ecology, and biochemistry, our ability to examine holistic behavior in the light of a detailed knowledge base as to the molecular constituents has only been possible for at most a decade, namely since the sequencing of the first bacterial genome by Fleishman et al. in June 1995 in *Science*. By a strange coincidence, it was also June 1995, that Wasinger et al. first employed and defined the term *proteome* in the scientific literature in the journal *Electrophoresis*.

Over the last decade we have seen genomic, transcriptomic, proteomic, metabolomic, and systems modeling applied to microbial systems. The academic and commercial stimulus has undoubtedly been the attraction of tractability, that is, for the first time ever genomewide, gene-product-wide analyses could be contemplated and experimentally implemented under stringently controlled conditions, for example, chemostats.

Although the experimental tractability of microbes may prove attractive, the reality may be somewhat different whereby this knowledge base starts to demand further answers in an epigenic context in full cognizance of the high-dimensional complexity inherent in all living systems. The molecules and atoms involved in life processes must remain totally oblivious of any distinction between the interior of a living cell and the far reaches of our universe; that is, the same laws of the universe must apply no matter where a molecule finds itself. The dilemma and challenge to a more accurate understanding of these living systems are major: *How then is it that living cells undertake ordered molecular work*? Much of the answer can be attributed to both temporal and spatial segregation and maintenance of life processes by keeping reactants apart; that is, *the greatest art of life is very much stopping life from stopping*.

In reality, even when armed with increasingly complete experimental data sets, we are only beginning to scratch the surface of those life forces that are transmitted from one generation to the next in living cells and organisms. Nonetheless, as shown in this text, we can now accurately compute information linked to cellular half-lives (cellular doubling times), levels of ribonucleic acid (RNA) and protein expressions over time, RNA and protein synthesis and breakdown rates, spatial and temporal segregation within cells, metabolite production, and systems modeling dealing with biochemical complexity including both positive and negative feedback loops that influence physiological thresholds. The latter, in turn, commit whole cells to a given biological activity. From an initial starting point, the building blocks acquired from the previous generation, the progeny must manage intracellular molecular concentrations and exploit its genetic machinery to respond to the cell's needs in the face of environmental stress, shock of many descriptions, and on-going need to survive. This balance between constituents and molecular abundance intracellularly must then grapple with an astounding 9.6 trillion distinct possibilities for protein–protein interactions, as facilitated by the diversity of 5-*mer* binding sites expressed on the surface of the proteome of *Escherichia coli* (Humphery-Smith and Gestel, unpublished data).

We wish our readers a stimulating encounter with this fascinating discipline of microbial proteomics: whole-organism functional biology.

IAN HUMPHERY-SMITH and MICHAEL HECKER

# ACKNOWLEDGMENTS

We wish to thank most sincerely Lisa Shipley and Carol Mattacola for their assistance with the collation of chapters and the implementation of editorial changes to submitted manuscripts. Elsewhere, the editors wish to thank all authors for their contributions and particularly their collective patience while we awaited completion of all chapters. Without these efforts, this text would never have achieved the necessary level of overall discipline coverage so as to do justice to our chosen field.

## CONTRIBUTORS

**Tim Alefantis,** Vital Probes, Mayfield, Pennsylvania, *Identification of Protein Candidates for Developing Bacterial Ghost Vaccines against* Brucella.

**David J. Anderson,** Biological Sciences Division, Pacific Northwest National Laboratory, Richland, Washington, *AMT Tag Approach to Proteomic Characterization of* Deinococcus radiodurans *and* Shewanella oneidensis.

**Gordon A. Anderson,** Biological Sciences Division, Pacific Northwest National Laboratory, Richland, Washington, *AMT Tag Approach to Proteomic Characterization of* Deinococcus radiodurans *and* Shewanella oneidensis.

**Haike Antelman,** Institut für Mikrobiologie, Ernst-Moritz-Amdt-Universität, Greifswald, Germany, *A Proteomic Survey through the Secretome of* Bacillus subtilis.

**Rolf Apweiler,** European Bioinformatics Institute, Hinxton, Cambridge, United Kingdom, *Databases and Resources for* in silico *Proteome Analysis.*

**Robert Benyon,** Protein Function Group, Faculty of Veterinary Science, University of Liverpool, Liverpool, United Kingdom, *Strategies for Measuring Dynamics: The Temporal Component of Proteomics.*

**Sierd Bron,** Department of Genetics, Groningen Biomolecular Sciences and Biotechnology Institute, Haren, The Netherlands, *A Proteomic Survey through the Secretome of* Bacillus subtilis.

**Catherine R. Bruce,** Aberdeen Oomycete Group, College of Life Sciences and Medicine, University of Aberdeen, Aberdeen, Scotland, *Proteomic Studies of Plant-Pathogenic Oomycetes and Fungi.*

**Andreas Burkovski,** Lehrstuhl für Mikrobiologie, Friedrich-Alexander-Universität Erlangen-Nürnberg, Erlangen, Germany, *Protemics of* Corynebacterium glutamicum: *Essential Industrial Bacterium.*

**Phillip Cash,** Department of Medical Microbiology, University of Aberdeen, Aberdeen, Scotland, *Analyzing Bacterial Pathogenesis at the Level of the Proteome.*

**Ricardo Cavicchioli,** School of Biotechnology and Biomolecular Sciences, University of New South Wales, Sydney, New South Wales, Australia, *Proteomics of Archaea.*

**Ian N. Clarke,** Molecular Microbiology Group, University Medical School, Southampton General Hospital, Southampton, United Kingdom, *Quest for Complete Proteome Coverage.*

**Diego Comerci,** Instituto de Investigaciones Biotecnologicas, Universidad Nacional de General San Martin, Buenos Aires, Argentina, *Identification of Protein Candidates for Developing Bacterial Ghost Vaccines against* Brucella.

**Vito G. DelVecchio,** Vital Probes, Mayfield, Pennsylvania, *Identification of Protein Candidates for Developing Bacterial Ghost Vaccines against* Brucella.

**Oleg V. Demin,** A. N. Belozersky Institute of Physico-Chemical Biology, Moscow State University, Moscow, Russia, *Cellular Modeling of Biochemical Processes in Microorganisms.*

**Dwayne A. Elias,** Biological Sciences Division, Pacific Northwest National Laboratory, Richland, Washington, *AMT Tag Approach to Proteomic Characterization of* Deinococcus radiodurans *and* Shewanella oneidensis.

**Jim Frederickson,** Environmental Molecular Sciences Laboratory, Pacific Northwest National Laboratory, Richland, Washington, *AMT Tag Approach to Proteomic Characterization of* Deinococcus radiodurans *and* Shewanella oneidensis.

**Concha Gil,** Department of Microbiology II, Faculty of Pharmacy, Complutense University of Madrid, Madrid, Spain, Candida albicans *Biology and Pathogenicity: Insights from Proteomics; Contributions of Proeteomics to Diagnosis, Treatment, and Prevention of Candidiasis.*

**Carol S. Giometti,** Argonne National Laboratory, Argonne, Illinois, *A Tale of Two Metal Reducers: Comparative Proteome Analysis of* Geobacter sulferreducens *PCA and* Shewanella oneidensis *MR-1.*

**Amber Goodchild,** Johnson & Johnson Research, Eveleigh, New South Wales, Australia, *Proteomics of Archaea.*

**Igor Goryanin,** Edinburgh Centre for Bioinformatics, University of Edinburgh, Edinburgh, Scotland, *Cellular Modeling of Biochemical Processes in Microorganisms.*

**Celia W. Goulding,** UCLA-DOE Institute of Genomics and Proteomics, Los Angeles, California, *Structural Proteomics and Computational Analysis of a Deadly Pathogen: Combating* Mycobacterium tuberculosis *from Multiple Fronts.*

**Guido Grandi,** Chiron Vaccines Research, Sienna, Italy, *Genomics and Proteomics in Reverse Vaccines.*

**Laura J. Grenville-Briggs,** Aberdeen Oomycete Group, College of Life Sciences and Medicine, University of Aberdeen, Aberdeen, Scotland, *Proteomic Studies of Plant-Pathogenic Oomycetes and Fungi.*

**Michael Hecker,** Institut für Mikrobiologie, Ernst-Moritz-Amdt-Universität, Greifswald, Germany, *A Proteomic Survey through the Secretome of* Bacillus subtilis.

**Richard Herrmann,** ZMBH Universität Heidelberg, Heidelberg, Germany, *Proteome of* Mycoplasma pneumoniae.

**Kim K. Hixson,** Biological Sciences Division, Pacific Northwest National Laboratory, Richland, Washington, *AMT Tag Approach to Proteomic Characterization of* Deinococcus radiodurans *and* Shewanella oneidensis.

**Lothar Jänsch,** Deutsche Forschung Gesellschaft, Biologisches Zentrum, Braunschweig, Germany, *Elucidation of the Mechanisms of Acid Stress in* Listeria monocytogenes *by Proteomic Analysis.*

**Paul Kersey,** European Bioinformatics Institute, Hinxton, Cambridge, United Kingdom, *Databases and Resources for* in silico *Proteome Analysis.*

**Akbar Khan,** Chemical and Biological Defense Directorate, Defense Threat Reduction Agency, Alexandria, Virginia, *Identification of Protein Candidates for Developing Bacterial Ghost Vaccines against* Brucella.

**Mogens Kilstrup,** Microbial Physiology and Genetics Group, Bio-Centrum-DTU, Technical University of Denmark, Lyngby, Denmark, *Proteomics of* Lactococcus lactis: *Phenotypes for a Domestic Bacterium.*

**Tamara Kulikova,** European Bioinformatics Institute, Hinxton, Cambridge, United Kingdom, *Databases and Resources for* in silico *Proteome Analysis.*

**Bernard Labedan,** Évolution Moléculaire et Bioinformatique des Génomes, Institut de Génétique et Microbiologie, Universite Paris-Sud, Orsay, France, *Interspecies and Intraspecies Comparison of Microbial Proteins: Learning about Gene Ancestry, Protein Function, and Species Life Style.*

**Galina V. Lebedeva,** A. N. Belozersky Institute of Physico-Chemical Biology, Moscow State University, Moscow, Russia, *Cellular Modeling of Biochemical Processes in Microorganisms.*

**Olivier Lespinet,** Évolution Moléculaire et Bioinformatique des Génomes, Institut de Génétique et Microbiologie, Universite Paris-Sud, Orsay, France, *Interspecies and Intraspecies Comparisons of Microbial Proteins: Learning about Gene Ancestry, Protein Function, and Species Life Style.*

**Ka Yin Leung,** Department of Biological Sciences, Faculty of Science, National University of Singapore, *Unraveling* Edwardsiella tarda *Pathogenesis Using the Proteomics Approach.*

**Mary S. Lipton,** Biological Sciences Division, Pacific Northwest National Laboratory, Richland, Washington, *AMT Tag Approach to Proteomic Characterization of* Deinococcus radiodurans *and* Shewanella oneidensis.

**Werner Lubitz,** Biotech Innovation Research Development and Consulting GmbH & Co KEG, Wien, Austria, *Identification of Protein Candidates for Developing Bacterial Ghost Vaccines against* Brucella.

**Maria Ines Marchesini,** Instituto de Investigaciones Biotecnologicas, Universidad Nacional de General San Martin, Buenos Aires, Argentina, *Identification of Protein Candidates for Developing Bacterial Ghost Vaccines against* Brucella.

**Christophe Masselon,** Biological Sciences Division, Pacific Northwest National Laboratory, Richland, Washington, *AMT Tag Approach to Proteomic Characterization of* Deinococcus radiodurans *and* Shewanella oneidensis.

**Eugeniy A. Metelkin,** A. N. Belozersky Institute of Physico-Chemical Biology, Moscow State University, Moscow, Russia, *Cellular Modeling of Biochemical Processes in Microorganisms.*

**Matthew E. Monroe,** Biological Sciences Division, Pacific Northwest National Laboratory, Richland, Washington, *AMT Tag Approach to Proteomic Characterization of* Deinococcus radiodurans *and* Shewanella oneidensis.

**Heather Mottaz,** Environmental Molecular Sciences Laboratory, Pacific Northwest National Laboratory, Richland, Washington, *AMT Tag Approach to Proteomic Characterization of* Deinococcus radiodurans *and* Shewanella oneidensis.

**César Nombela,** Department of Microbiology II, Faculty of Pharmacy, Complutense University of Madrid, Madrid, Spain, Candida albicans *Biology and Pathogenicity: Insights from Proteomics; Contributions of Proteomics to Diagnosis, Treatment, and Prevention of Candidiasis.*

**Thomas Nyström,** Department of Cell Molecular Biology, Göteborg University, Göteborg, Sweden, *Oxidation of the Bacterial Proteome in Response to Starvation.*

**C. David O'Connor,** Centre for Proteomic Research and School of Biological Sciences, University of Southampton, Southampton, United Kingdom, *Quest for Complete Proteome Coverage.*

**Ljiljana Pasa-Tolic,** Biological Sciences Division, Pacific Northwest National Laboratory, Richland, Washington, *AMT Tag Approach to Proteomic Characterization of* Deinococcus radiodurans *and* Shewanella oneidensis.

**Luan Phan-Thanh,** Unité Pathologie Infectieuse et Immunologie, Institut National de la Recherche Agronomique, Nouzilly, France, *Elucidation of the Mechanisms of Acid Stress in* Listeria monocytogenes *by Proteomic Analysis.*

**Aida Pitarch,** Department of Microbiology II, Faculty of Pharmacy, Complutense University of Madrid, Madrid, Spain, Candida albicans *Biology and Pathogenicity: Insights from Proteomics; Contributions of Proteomics to Diagnosis, Treatment, and Prevention of Candidiasis.*

**Julie M. Pratt,** Protein Function Group, Faculty of Veterinary Science, University of Liverpool, Liverpool, United Kingdom, *Strategies for Measuring Dynamics: The Temporal Component of Proteomics.*

**Manuela Pruess,** European Bioinformatics Institute, Hinxton, Cambridge, United Kingdom, *Databases and Resources for* in silico *Proteome Analysis.*

**Mark Raftery,** Bioanalytical Mass Spectrometry Facility, University of New South Wales, Sydney, New South Wales, Australia, *Proteomics of Archaea.*

**P. S. Srinivasa Rao,** Department of Biological Sciences, Faculty of Science, National University of Singapore, *Unraveling* Edwardsiella tarda *Pathogenesis Using the Proteomics Approach.*

**Margaret F. Romine,** Environmental Molecular Sciences Laboratory, Pacific Northwest National Laboratory, Richland, Washington, *AMT Tag Approach to Proteomic Characterization of* Deinococcus radiodurans *and* Shewanella oneidensis.

**Thomas Ruppert,** ZMBH Universität Heidelberg, Heidelberg, Germany, *Proteome of* Mycoplasma pneumoniae.

**Paul Skipp,** Centre for Proteomic Research and School of Biological Sciences, University of Southampton, Southampton, United Kingdom, *Quest for Complete Proteome Coverage.*

**Richard D. Smith,** Biological Sciences Division, Pacific Northwest National Laboratory, Richland, Washington, *AMT Tag Approach to Proteomic Characterization of* Deinococcus radiodurans *and* Shewanella oneidensis.

**Michael Strong,** UCLA-DOE Institute of Genomics and Proteomics, Los Angeles, California, *Structural Proteomics and Computational Analysis of a Deadly Pathogen: Combating* Myobacterium tuberculosis *from Multiple Fronts.*

**Yuen Peng Tan,** Department of Biological Sciences, Faculty of Science, National University of Singapore, *Unraveling* Edwardsiella tarda *Pathogenesis Using the Proteomics Approach.*

**Nikola Tolic,** Biological Sciences Division, Pacific Northwest National Laboratory, Richland, Washington, *The AMT Tag Approach to Proteomic Characterization of* Deinococcus radiodurans *and* Shewanella oneidensis.

**Rodolfo A. Ugalde,** Instituto de Investigaciones Biotecnologicas, Universidad Nacional de General San Martin, Buenos Aires, Argentina, *Identification of Protein Candidates for Developing Bacterial Ghost Vaccines against* Brucella.

**Jan Maarten van Dijl,** Department of Molecular Bacteriology, University of Groningen, Groningen, The Netherlands, *A Proteomic Survey through the Secretome of* Bacillus subtilis.

**Pieter van West,** Aberdeen Oomycete Group, College of Life Sciences and Medicine, University of Aberdeen, Aberdeen, Scotland, *Proteomic Studies of Plant-Pathogenic Oomycetes and Fungi.*

**Valerie Wasinger,** Bioanalytical Mass Spectrometry Facility, University of New South Wales, Sydney, Australia, *Holistic Biology of Microorganisms: Genomics, Transcriptomics, and Proteomics.*

**Jun Zheng,** Department of Biological Sciences, Faculty of Science, National University of Singapore, *Unraveling* Edwardsiella tarda *Pathogenesis Using the Proteomics Approach.*

**Ekaterina A. Zobova,** A. N. Belozersky Institute of Physico-Chemical Biology, Moscow State University, Moscow, Russia, *Cellular Modeling of Biochemical Processes in Microorganisms.*

# PART I

# GENERAL PROTEOMICS OF MICROORGANISMS/MODEL ORGANISMS

CHAPTER 1

# Holistic Biology of Microorganisms: Genomics, Transcriptomics, and Proteomics

VALERIE WASINGER

University of New South Wales, Sydney, Australia

## 1.1 INTRODUCTION

An organism's phenotype is determined by the environment and its genetic content, and as such understanding the relationship between genome, transcriptome, and proteome is one of the fundamental goals of biology. Over the last decade quantitative biological analysis has been performed at the genomic, transcribed, and translated level. This has occurred through large-scale DNA sequencing, genomewide genetic analysis, DNA and protein chips, and two-dimensional gel electrophoresis (2DE) or multidimensional liquid chromatography integrated with mass spectrometry (LC-MS) for the fast and highly sensitive analysis of proteins. The rapid speed of progress in genomics has been impressive and unrivaled in proteomics, with more than 160 organisms having been sequenced, including the human genome. Despite this achievement, the information provided by any given genome has given little insight into the workings of a cell. A better understanding of the cellular machinery, biology, and disease processes will ultimately result from mining these genomes for their *cognate* proteins—proteomics. This chapter aims to highlight the contribution proteomics has made to our collective knowledge of whole-organism microbiology.

The expansion of the field of proteomics from the display of large numbers of proteins using 2DE to the incorporation of new technologies and a global perspective of protein expression, requisite to genomic sciences, has refined proteomics into a robust scientific field. A focus on quantitative measurement of proteins, patterns of changes in protein expression, and protein interactions in the context of a whole cell has given the field of proteomics a new level of maturity previously not seen.

## 1.2 PROTEOME IN PERSPECTIVE

Unlike genome sequencing, there has been no completion of a proteome, while a proteomic endpoint remains ill defined. Despite this, the study of the total protein

*Microbial Proteomics: Functional Biology of Whole Organisms*, Edited by Ian Humphery-Smith and Michael Hecker. 

complement of a genome is feasible [1–3]. What remains the biggest challenge to the identification and analysis of complex biological samples is the collective variability of all protein's physicochemical properties and a dependence on in vivo and ex vivo parameters. These parameters contribute to proteome complexity and are phenotypically manifested as a variation of relative protein abundance; modifications or truncations, for example enzymatic cleavage; altered molecular or protein interactions; complex formation or breakdown; and presence or absence of proteins. Proteome complexity can be explained from three diverse perspectives: evolutionary complexity, internal complexity, and sample complexity (addressed later).

### 1.2.1 Evolutionary Complexity

Complexity is often attributed to the number of base pairs in a sequence that give rise to functional genes [4]. We readily admit that as we move up the evolutionary ladder, the most complex organism of all is our own kind. Nonetheless, our genetic likeness to plants (*A. thaliana*) and worms (*C. elegans*) is striking [5]. However, annotation of a gene within a DNA sequence is not an indication that a gene is expressed or able to serve a useful metabolic or structural purpose. The "one gene–one protein" tenet has long been rejected [6, 7] because alternate gene-splicing and posttranslational modifications can result in multiple active forms of a protein. Gene numbers from theoretical proteomes of fully sequenced organisms are compared in Figure 1.1 and are also available from http://www.ebi.ac.uk/. This figure demonstrates evolutionary complexity is not governed by gene content, nor is sequence information sufficient to describe the individuality of an organism [8]. Complexity is more significantly a reflection of protein regulation [9, 10]. As we move up the evolutionary hierarchy there is a distinct increase in gene product modification: for *Mycoplasma genitalium* this has been quantified at a rate of 1.2 times more proteins than

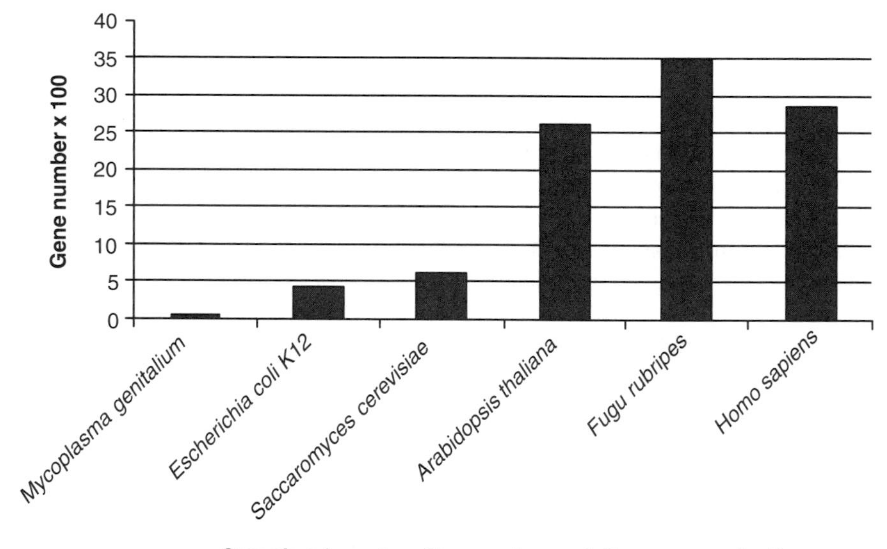


**Figure 1.1** Predicted proteome of fully sequenced organisms collated from nonredundant proteome sets from SwissProt and TREMBL entries (http://www.ebi.ac.uk/).

genes [1]; for *Escherichia coli* it is 1.3 proteins per gene; for yeast, 3 proteins per gene; and for human, as many as 10 proteins per gene [8]. This has been estimated from both two-dimensional (2D) sodium dodecyl sulfate polyacrylamide gel electrophoresis (SDS-PAGE) and tandem MS (MS/MS) work.

### 1.2.2 Internal Complexity

Internal complexity is a property of the network of interacting protein, transcript, and genes within a cell. As life is not just an assembly of these individual components, the study of gene function has been explored in terms of the relatively static localized gene and protein interaction networks and the massively parallel study of global networks embracing the higher order and collective behavior of genes and proteins as a proteome-scale network [11]. The study of the sum of all interacting parts in a biological system is called systems biology and addresses the network of interactions of biochemical/signaling pathways and their modulation, the influence of spatial and temporal differences in these networks, and how this knowledge can be applied to providing therapeutic targets in disease processes [12]. The recent interest in large-scale identification of functionally linked proteins can be affiliated to the development of high-throughput experimentation and computational procedures and in extensive database curation of this information. Computational approaches for finding gene and protein interactions complement and extend experimental approaches such as synthetic lethal and suppressor screens, yeast two-hybrid experiments, and high-throughput MS interaction assays [13]. Computational methods based on sequence do not assume knowledge about protein function and can therefore be used to assign function to uncharacterized proteins linked to network pathways of known function. Computational methods for studying protein associations include (i) phylogenetic profiling, based on the co-occurrence of proteins in different genomes [14]; (ii) domain fusion or "Rosetta stone" sequences [15, 16] where fused domains of a protein in one organism are used to predict interaction of these domains separated through evolution in another organism; and (iii) gene clustering in one organism where the genes have been separated by evolution in another organism [17].

The very nature of proteomics embodies all of these complexities. As a result, diverse approaches have and are being used to overcome some of the issues associated with these complexities (e.g., large protein numbers), thereby creating unique data sets. Furthermore, comparing multiple technologies then has the advantage of highlighting differences in measurements, thereby extending the proteomic coverage, but only if these approaches can be unified meaningfully.

## 1.3 AMALGAMATION OF TOOLS FOR A PROTEOMIC "TOOL BOX"

The field of proteomics has always been heavily reliant on protein characterization technologies. A powerful repertoire of tools has been implemented for the separation and identification of thousands of proteins simultaneously. In 1995, this consisted of Edman sequencing amino acid analysis, peptide mass fingerprinting, ladder sequencing, and expression of cloned inserts, to name a few, and these were approached in an hierarchical manner for protein identification. Success using these tools was contingent on the sensitivity of instruments for accurate mass determination [1] and available protein databases for protein identification. There was little scope for characterization of

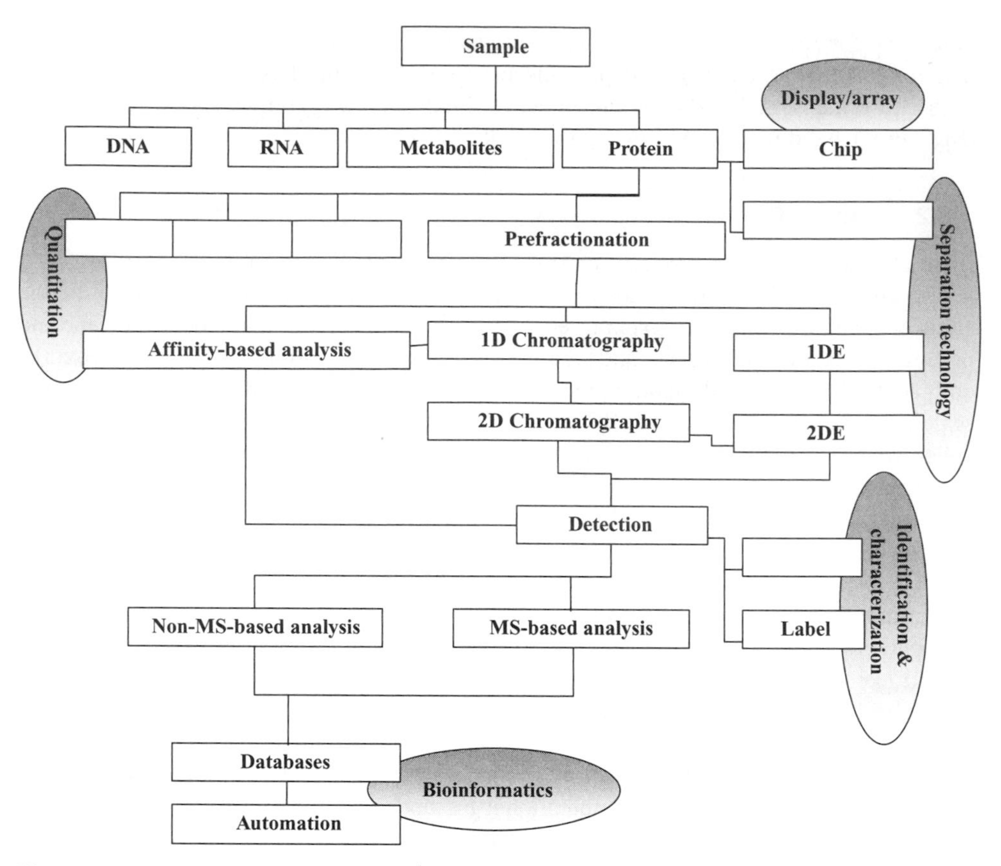


**Figure 1.2** Schematic of integration of technologies for proteomics. The unification of DNA, mRNA, and metabolite information is essential for a complete understanding of biological function [58].

"unknown" proteins on a large scale. Today, the proteomic tool box has assimilated genomic, metabolomic, transcriptomic, and proteomic tools, with MS technologies still providing the backbone for much of the proteomic analysis. A representation of an integrated proteomic tool box is given in Figure 1.2.

One of the major technological challenges facing protein analysis is the dynamic range of expression of proteins within massively complex samples [18]. This difference can be as great as 12 orders of magnitude, and this is currently beyond the range of detection capable for our existing technologies. Additionally, a few highly abundant proteins making up a large percentage of total proteins can mask the detection of lower abundance and biologically significant proteins [19]. A third technological challenge, alluded to earlier, is sample complexity. The expression of a myriad of different protein species with similar physicochemical properties is compounded by the numerous modifications that can take place posttranscriptionally. Over 300 different modifications have been described for proteins [20, 21].

These on-going challenges have been addressed in a number of ways and have benefited agriculture, industry, and biomedicine. Two-dimensional gel electrophoresis for proteomic applications is still widely accepted as the technology capable of delivering the greatest separation power for highly complex samples. The expansion of pH separation using a series of overlapping narrow-range gels to assemble a contiguous collage of protein expression and allow for a more meticulous analysis of proteomes was established in 1997 for *O. anthropi* [22] and *M. genitalium*. Close to two-thirds of *M. genitalium* open reading frames (ORFs) were shown to be expressed using this approach [2]. This approach is now commonly used and bacterial examples include *Spiroplasma melliferum* [23], *Mycobacterium tuberculosis* [24], *Corynebacterium glutamicum* [25], and *Streptococcus mutans* [26].

Proteomics is also providing practical applications through the identification of immunogenic proteins as potential vaccine targets. Many hypotheses generated in silico by genomics have been validated through functional studies of the transcriptome and proteome and have led to the identification of essential genes. Using 2DE [27] and DNA microarrays [28] for differential expression analysis has confirmed that the resistance of *M. tuberculosis* to the anti-tuberculosis cell envelope drug isoniazid was the result of overexpression of components of the fatty acid synthase system, in particular AcpM and a carrier protein synthase KasA. A compilation of *M. tuberculosis* [29] and *Mycobacterium bovis* [30] induced immunoresponses in splenocytes using 2D liquid-phase electrophoresis has also identified multiple antigens associated with posttranslationally modified proteins. In 1999, Jungblut et al. [31] compared the proteomes of two nonvirulent vaccine strains of *M. bovis* BCG with two virulent strains of *M. tuberculosis* to identify protein candidates for potential vaccine development as well as for diagnostic and therapeutic purposes. Over 30 differences were identified among the strains, including three cell envelope proteins, some antigenic proteins, and novel unannotated proteins. This work, as well as work from others, has contributed to a better understanding of the pathogenic and physiological mechanism available to this organism and also contributed to the development of new vaccine candidates. *Mycobacterium bovis* BCG (live attenuated) has been the only widely used vaccine available against tuberculosis until recently, with new tuberculosis vaccine trials begun in 2004.

Two-dimensional gel electrophoresis is a powerful protein separation technique in combination with MS, yet it does not always deliver the appropriate sensitivity for the discovery of low-level proteins [32], proteins with extreme isoelectric point (pI) or mass [11], or membrane-associated proteins because of solubility issues [33]. For these reasons, the development of alternative approaches to 2DE, such as multidimensional chromatography coupled to MS, has been vital to the success of *Bifidobacterium* proteomics. *Bifidobacterium* is a gram-positive prokaryote that naturally colonizes the human gut exerting health-promoting effects. Clinical studies have claimed that bifidobacterial probiotics promote gastrointestinal tract homeostasis and health because of antidiarrheal, immunomodulating, and possibly anticarcinogenic properties. Proteomics has contributed to the comprehensive understanding of the physiological mechanisms underlying these properties [34]. A predominant portion of *Bifidobacterium infantis* proteome consists of enzymes of the glycolytic and pentose-phosphate pathways, enzymes of anaerobic metabolism, transcriptional factors, shock proteins, ribosomal proteins, and proteases. The high level of these proteins during the exponential phase of growth underlines their central role in cell survival, replication, and energy metabolism, giving an indication of the basal functions, which are essential for the vitality of the bifidobacteria biological system.

A unification of genomics, transcriptomics, and proteomics will provide a comprehensive knowledge base of gene function and a powerful reference of protein properties. There are, however, significant challenges that are being addressed but may remain unresolved until technological advances can overcome them. The goal of deciphering the entire protein complement of an organism and making heuristic relationships is formidable and often limited by cell dynamics. Understanding biological events such as modification, complexity, environmental input, and technological constraints of detection thresholds and bioinformatics will be required to overcome the challenges for a complete proteome study [1, 35].

## 1.4 RELATIONSHIP BETWEEN GENOME, TRANSCRIPTOME, AND PROTEOME

All life is linked by a common genetic scaffold constrained to 4 nucleotides and 20 amino acids and as such phenotypic differences require far more investigation than was first anticipated. Genomics presents only one level of functional information. Both qualitative and quantitative unique levels of information are also given by the transcriptome and proteome as well as the interactome and metabolome.

### 1.4.1 Transcriptome and Proteomics

Transcriptome analysis involves messenger ribonucleic acid (mRNA), the relayer of information for protein synthesis. Several methods, including serial analysis of gene expression (SAGE), oligonucleotide and complementary deoxyribonucleic acid (cDNA) microarrays, and large-scale sequencing of expressed tags, are available to measure gene expression at the mRNA level globally and quantitatively. Measurements of protein expression and transcript do not always correlate. This is due to protein abundance also being influenced by protein stability, translation rate, modulation of transcript abundance by other proteins, posttranslational modifications, and half life; therefore, mRNA cannot always be a predictor of protein abundance [36].

The absolute range of transcript abundance is largely unknown in microorganisms; however, a significant amount of work has been done in yeast. A study by Futcher et al. [37] has revealed that for each mRNA there are approximately 4000 molecules of cognate protein produced and a small number of these proteins make up at least 50% of all cellular protein in yeast. For these abundant proteins, several statistical methods have shown a correlation between mRNA and protein abundances [3]. A study involving the use of SAGE [38] for mRNA measurements and 2DE for protein measurements [39] has also shown a close correlation between mRNA and protein levels for high-copy-number proteins but relatively poor correlation for proteins transcribed at 10 or less copies per cell. Seventy-five percent of genes are transcribed at one or fewer copies per cell [38] with some transcripts per cell as low as 0.001 per cell generation [40]. This has also been confirmed recently using immunodetection of high-affinity epitope-tagged ORFs known as tandem affinity purification (TAP) [41] for which $\sim$80% of the *Saccharomyces cerevisiae* proteome was analyzed [3]. Transcript numbers, protein molecules, and the positive correlation between mRNA and protein abundance for some proteins are shown in Figure 1.3, as is a comparison of the range of proteomic coverage for the techniques of TAP, multidimensional protein identification technology (MudPIT), and 2DE [3].

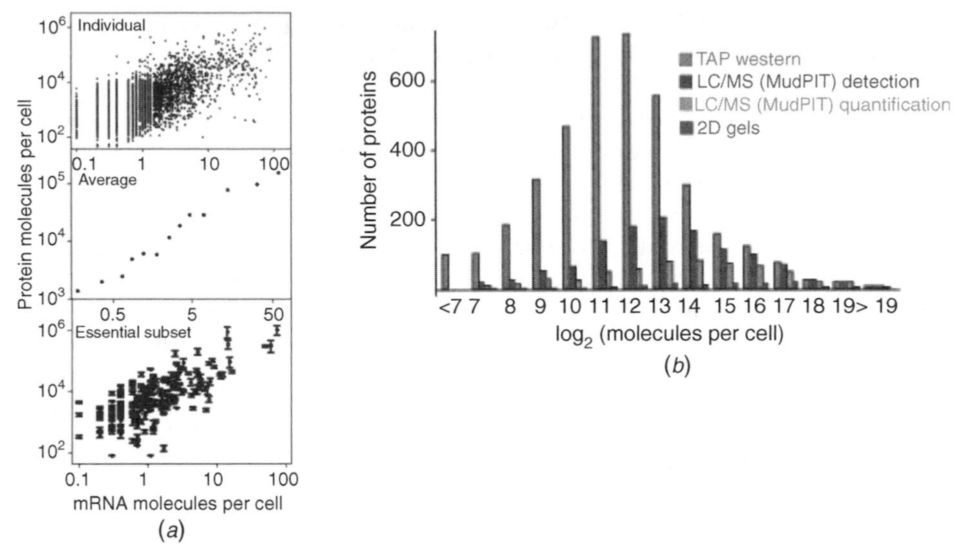


**Figure 1.3** Analysis of protein expression in yeast showing (*a*) correlation between protein abundance and transcripts: Top panel shows the relationship between steady-state mRNA and protein levels in yeast as determined by microarray analysis. Middle panel shows ORFs sorted into discrete mRNA levels and means plotted against mean protein abundance. Lower panel shows only a comparison of protein and mRNA levels for some essential soluble proteins. (*b*) Absolute range of expression possible using TAP/Western blot, LC-MS using a multidimensional chromatography approach, and 2DE [3]. (See color insert.)

### 1.4.2 Metabolome and Proteomics

There is a nonlinear relationship between the presence of a metabolite and a gene, as many genes may be involved in the synthesis or degradation of one metabolite. An example of metabolomics is provided by a well studied microbe, *C. glutamicum,* important in metabolic engineering as its production of glutamine and lysine is used extensively in the food industry [25]. This organism has been comprehensively studied in terms of genome [42, 43], transcriptome [44, 45], proteome [46–48], and metabolome [45]. However, until recently integration of these studies had not occurred. An in-depth study of glycolysis, pentose-phosphate pathway, tricarboxylic acid (TCA) cycle, and lysine synthesis by Krömer et al. [45] of this organism has revealed a dynamic relationship between transcriptome, metabolome, and fluxome (changes in metabolites). It was found that growth continued despite depletion of essential threonine and methionine from growth media (achieved by scavenging intracellular stores) with an increase in total soluble proteins, indicating protein synthesis was still active. Additionally, genes active in translation such as ribosomal proteins and other protein synthesis machinery were significantly expressed. A maximal flux for most enzymes correlated with maximal gene expression, and this could be affected by down regulation at the transcriptional level of enzymes. A conclusion of this work is that it is essential to measure gene, transcript, protein, and metabolite differences to gain insight into biological systems.

## 1.5 WHAT HAVE MODEL ORGANISMS CONTRIBUTED TO OUR UNDERSTANDING OF BIOLOGICAL SYSTEMS?

Organisms that are representative of more complex systems are amenable to experimental study and are associated with extensive accumulated information from many sources and can be defined as model organisms [49]. In the last 30 years, fundamental research involving model organisms has profoundly advanced our understanding of biological systems. As a result of considerable effort around the globe, organism-specific databases have been created as knowledge of an organism as well as comparative information of relevance to numerous research directions and perspectives becomes available. Many metabolic and developmental pathways are conserved in nature, irrespective of classification level. This feature invites the use of model organisms to minimize the effort required to understand complex biological systems. Table 1.1 summarizes some of the advances that three model organisms have afforded us and how this information has been applied to other organisms.

Life in its most minimal form is still biologically complex. The smallest bacterial genome capable of independent survival is that of *M. genitalium* (517 genes). Interest in this bacterium has focused on derivation of the minimal gene set or what is essentially required to make a cell alive. By selectively switching off each gene in turn, Hutchison et al. [50] were able to determine an essential requirement of 260–350 *M. genitalium* genes, 100 of which had unknown function. This has since been revised to 250 by comparison to other bacterial genomes [51].

On a similar line of thought, the simplest eukaryotic genome, *S. cerevisiae*, plays an important role as a model organism for understanding more complex genomes such as our own. The best example of the value of yeast as a model involves the study of some human-disease-causing genes and their orthologues in yeast, such as hereditary nonpolyposis colon cancer (*MSH2* and *MLH1*), neurofibromatosis type 1 (*IRA2*), ataxia telangiectasia (*TEL1*), and Werner's syndrome (*SGS1*) [32]. In humans, genetic inheritance of these

**TABLE 1.1 Summary of Three Model Organisms and How They Have Influenced Proteomics**

| Organism | Reason for Model Status | Celebrated Findings | Applications | References |
|---|---|---|---|---|
| *M. genitalium* | Smallest self-replicating bacterium; determine core genes that drive life | Minimal gene set of 300, 427 abundantly expressed proteins | Derive a synthetic minimal cell for genetic or biochemical manipulation, uses in nanotechnology, therapeutic vector delivery of DNA | [2, 50, 52] |
| *S. cerevisiae* | Simplest eukaryotic model for more complex organisms; best-studied eukaryotic system | Description of protein interaction map, first chromosome ever sequenced | Unknown function prediction in other organisms | [15]; [16]; [53]; [54]; [55] |
| *E. coli* | Biochemical, molecular, and metabolically best characterized system | 2D SDS-PAGE developed using *E. coli* | Standard protein separation technique in proteomics; virtual cell based on accumulated *E. coli* knowledge and applied across species | [56]; [57] |

genes results in disease. Initial insight as to the function of these genes was obtained because of their sequence homology to yeast genes and genes of other organisms.

*Escherichia coli* is the epitome of bacterial model organisms as it is one of the best-studied microbes in terms of genome sequence, metabolic and regulatory networks, proteome, and mutant phenotype studies. Both the pathogenic (O157:H7) and nonpathogenic (K12) strains have been sequenced and compared and databases exist enabling comparison of metabolic and genomic information (Ecocyc: http://ecocyc.org/; Kegg: http://www.genome.ad.jp/kegg/kegg2.html).

There are many more microbial model organisms, and some of their contributions to proteomics and science in general are discussed by contributing authors in this book.

## 1.6 CONCLUSION

There is an amazing lack of cross communication between the fields of genomics, transcriptomics, and proteomics. This dilemma can be attributed to the dynamic and open-ended nature of the proteome. It is compounded by the evolutionary, internal, and sample complexities of studied organisms and the rapid generation of volumes of data from diverse groups. However, it is apparent that the best understood organisms have had information contributed from all fields. This has most easily occurred for the microbes and will also occur with higher organisms as the inertia to propel research forward and unify knowledge is reached. This will result in a holistic understanding that will be beneficial for all biological systems.

## ACKNOWLEDGMENTS

I would like to acknowledge support in part from the Australian Government Systemic Infrastructure Initiative grant.

## REFERENCES

1. Wasinger, V. C., Cordwell, S. J., Cerpa-Poljak, A., Yan, J. X., Gooley, A. A., Wilkins, M. R., Duncan, M. W., Harris, R., Williams, K. L., and Humphery-Smith, I., Progress with gene-product mapping of the Mollicutes: *Mycoplasma genitalium, Electrophoresis* 1995, **16**, 1090–1094.
2. Wasinger, V. C., Pollack, J. D., and Humphery-Smith, I., The proteome of *Mycoplasma genitalium*; CHAPS soluble component, *Eur. J. Biochem.* 2000, **267**, 1571–1582.
3. Ghaemmaghami, S., Huh, W-K., Bower, K., Howson, R., Belle, A., Dephoure, N., O'Shea, E., and Weissman, J., Global analysis of protein expression in yeast, *Nat. Biotechnol.* 2003, **425**, 737–741.
4. Adami, C., Ofria, C., and Collier, T. C., Evolution of biological complexity, *PNAS* 2000, **97**(9), 4463–4468.
5. Southan, C., Has the yo-yo stopped? An assessment of human protein-coding gene number, *Proteomics* 2004, **4**(6), 1712–1726.
6. Jaenisch, R., and Bird, A., Epigenetic regulation of gene expression: How the genome integrates intrinsic and environmental signals, *Nat. Genet.* 2003, **33**, s245–254.

7. Strohman, R., Epigenesis: The missing beat in biotechnology? *Biotechnology (NY)* 1994, **12**(2), 156–164.
8. Kellner, R., Proteomics. Concepts and perspectives, *Frenius J. Anal. Chem.* 2000, **366**, 517–524.
9. Harrison, P. M., Kumar, A., Lang, N., Snyder, M., and Gerstein, M., A question of size: The eukaryotic proteome and the problems in defining it, *Nucleic Acids Res.* 2002, **30**(5), 1083–1090.
10. Graveley, B. R., Alternative splicing: Increasing diversity in the proteomic world, *Trends Genet.* 2001, **17**(2), 100–107.
11. Huang, S., Back to the biology in systems biology: What can we learn from biomolecular networks? *Brief Funct. Genomic Proteomic* 2004, **2**(4), 279–297.
12. Zhu, H., Huang, S., and Dhar, P., The next step in systems biology: Simulating the temporospatial dynamics of molecular network, *Bioessays* 2004, **26**(1), 68–72.
13. Date, S. V., and Marcotte, E. M., Discovery of uncharacterized cellular systems by genome-wide analysis of functional linkages, *Nat. Biotechnol.* 2003, **21**(9), 1055–1062.
14. Pellegrini, M., Marcotte, E. M., Thompson, M. J., Eisenberg, D., and Yeates, T. O., Assigning protein functions by comparative genome analysis: Protein phylogenetic profiles, *Proc. Natl. Acad. Sci. USA* 1999, **96**(8), 4285–4288.
15. Marcotte, E. M., Pellegrini, M., Thompson, M. J., Yeates, T. O., and Eisenberg, D., A combined algorithm for genome-wide prediction of protein function, *Nature* 1999, **402**(6757), 83–86.
16. Enright, A. J., Iliopoulos, I., Kyrpides, N. C., and Ouzounis, C. A., Protein interaction maps for complete genomes based on gene fusion events, *Nature* 1999, **402**(6757), 86–90.
17. Dandekar, T., Snel, B., Huynen, M., and Bork, P., Conservation of gene order: A fingerprint of proteins that physically interact, *Trends Biochem. Sci.* 1998, **23**(9), 324–328.
18. Corthals, G. L., Wasinger, V. C., Hochstrasser D. F., and Sanchez, J-C., The dynamic range of protein expression: A challenge for proteomic research, *Electrophoresis* 2000, **21**, 1104–1115.
19. Oda, Y., Nagasu, T., and Chait, B. T., Enrichment analysis of phosphorylated proteins as a tool for probing the phosphoproteome, *Nat. Biotechnol.* 2001, **19**, 379–382.
20. Krishna, R. G., and Wold, F., Post-translational modification of proteins, *Adv. Enzymol. Relat. Areas Mol. Biol.* 1993, **67**, 265–298.
21. James, P., Mass spectrometry and the proteome, in P. James (Ed.), *Proteome Research: Mass Spectrometry*, Springer, 2001, p. 6.
22. Wasinger, V. C., Bjellqvist, B., and Humphery-Smith, I., Proteomic "contigs" of *Ochrobactrum anthropi*, application of extensive pH gradients, *Electrophoresis* 1997, **18**, 1373–1383.
23. Cordwell, S. J., Basseal, D. J., Bjellqvist, B., Shaw, D. C., and Humphery-Smith, I., Characterisation of basic proteins from *Spiroplasma melliferum* using novel immobilised pH gradients, *Electrophoresis* 1997, **18**(8), 1393–1398.
24. Urquhart, B. L., Cordwell, S. J, and Humphery-Smith, I., Comparison of predicted and observed properties of proteins encoded in the genome of *Mycobacterium tuberculosis* H37Rv, *Biochem. Biophys. Res. Commun.* 1998, **253**(1), 70–79.
25. Schaffer, S., Weil, B., Nguyen, V. D., Dongmann, G., Gunther, K., Nickolaus, M., Hermann, T., and Bott, M., A high-resolution reference map for cytoplasmic and membrane-associated proteins of *Corynebacterium glutamicum*, *Electrophoresis* 2001, **22**(20), 4404–4422.
26. Len, A. C., Cordwell, S. J., Harty, D. W., and Jacques, N. A., Cellular and extracellular proteome analysis of *Streptococcus mutans* grown in a chemostat, *Proteomics* 2003, **3**(5), 627–646.
27. Mdluli, K., Slayden, R. A., Zhu, Y., Ramaswamy, S., Pan, X., Mead, D., Crane, D. D., Musser, J. M., and Barry, C. E. 3rd., Inhibition of a *Mycobacterium tuberculosis* beta-ketoacyl ACP synthase by isoniazid, *Science* 1998, **280**(5369), 1607–1610.

28. Wilson, M., DeRisi, J., Kristensen, H. H., Imboden, P., Rane, S., Brown, P. O., and Schoolnik, G. K., Exploring drug-induced alterations in gene expression in *Mycobacterium tuberculosis* by microarray hybridization, *Proc. Natl. Acad. Sci. USA*, 1999, **96**(22), 12833–12838.

29. Covert, B. A., Spencer, J. S., Orme, I. M., and Belisle, J. T., The application of proteomics in defining the T cell antigens of *Mycobacterium tuberculosis, Proteomics* 2001, **1**(4), 574–586.

30. Gulle, H., Fray, L. M., Gormley, E. P., Murray, A., and Moriarty, K. M., Responses of bovine T cells to fractionated lysate and culture filtrate proteins of *Mycobacterium bovis* BCG, *Vet. Immunol. Immunopathol.* 1995, **48**(1–2), 183–190.

31. Jungblut, P. R., Schaible, U. E., Mollenkopf, H. J., Zimny-Arndt, U., Raupach, B., Mattow, J., Halada, P., Lamer, S., Hagens, K., and Kaufmann, S. H., Comparative proteome analysis of *Mycobacterium tuberculosis* and *Mycobacterium bovis* BCG strains: Towards functional genomics of microbial pathogens, *Mol. Microbiol.* 1999, **33**(6), 1103–1117.

32. Botstein, D., Chervitz S. A., and Cherry J. M., Yeast as a model organism, *Science* 1997, **277**(5330), 1259–1260.

33. Anderson, N. L., Polanski, M., Pieper, R., Gatlin, T., Tirumalai, R. S., Conrads, T. P., Veenstra, T. D., Adkins, J. N., Pounds, J. G., Fagan, R., and Lobley, A., The human plasma proteome: A nonredundant list developed by combination of four separate sources, *Mol. Cell Proteomics* 2004, **3**(4), 311–326.

34. Vitali, B., Wasinger, V., Brigidi, P., and Guilhaus, M., A proteomic view of *Bifidobacterium infantis* generated by multi-dimensional chromatography coupled with tandem mass spectrometry, *Proteomics*, in press.

35. Wilkins, M. R., Sanchez, J. C., Gooley, A. A., Appel, R. D., Humphery-Smith, I., Hochstrasser, D. F., and Williams, K. L., Progress with proteome projects: Why all proteins expressed by a genome should be identified and how to do it, *Biotechnol. Genet. Eng. Rev.* 1996, **13**, 19–50.

36. Hatzimanikatis, V., and Lee, K. H., Dynamical analysis of gene networks requires both mRNA and protein expression information, *Metab. Eng.* 1999, **1**(4), 275–281.

37. Futcher, B., Latter, G. I., Monardo, P., McLaughlin, C. S., and Garrels, J. I., A sampling of the yeast proteome, *Mol. Cell. Biol.* 1999, **19**(11), 7357–7368.

38. Velculescu, V. E., Zhang, L, Vogelstein, B., and Kinzler, K. W., Serial analysis of gene expression, *Science* 1995, **270**(5235), 484–487.

39. Gygi, S. P., Corthals, G. L., Zhang, Y., Rochon, Y., and Aebersold, R., Evaluation of two-dimensional gel electrophoresis-based proteome analysis technology, *Proc. Natl. Acad. Sci. USA* 2000, **97**, 9390–9395.

40. Holland, M. J., Transcript abundance in yeast varies over six orders of magnitude, *J. Biol. Chem.* 2002, **277**(17), 14363–14366.

41. Rigaut, G., Shevchenko, A., Rutz, B., Wilm, M., Mann, M., and Seraphin, B., A generic protein purification method for protein complex characterization and proteome exploration, *Nat. Biotechnol.* 1999, **17**(10), 1030–1032.

42. Ikeda, M., and Nakagawa, S., The *Corynebacterium glutamicum* genome: Features and impacts on biotechnological processes, *Appl. Microbiol. Biotechnol.* 2003, **62**(2–3), 99–109.

43. Kalinowski, J., Bathe, B., Bartels, D., Bischoff, N., Bott, M., Burkovski, A., Dusch, N., Eggeling, L., Eikmanns, B. J., Gaigalat, L., Goesmann, A., Hartmann, M., Huthmacher, K., Kramer, R., Linke, B., McHardy, A. C., Meyer, F., Mockel, B., Pfefferle, W., Puhler, A., Rey, D. A., Ruckert, C., Rupp, O., Sahm, H., Wendisch, V. F., Wiegrabe, I., and Tauch, A., The complete *Corynebacterium glutamicum* ATCC 13032 genome sequence and its impact on the production of L-aspartate-derived amino acids and vitamins, *J. Biotechnol.* 2003, **104**(1–3), 5–25.

44. Hayashi, M., Mizoguchi, H., Shiraishi, N., Obayashi, M., Nakagawa, S., Imai, J., Watanabe, S., Ota, T., and Ikeda, M., Transcriptome analysis of acetate metabolism in *Corynebacterium*

*glutamicum* using a newly developed metabolic array, *Biosci. Biotechnol. Biochem.* 2002, **66**(6), 1337–1344.

45. Krmer, J. O., Sorgenfrei, O., Klopprogge, K., Heinzle, E., and Wittmann, C., In-depth profiling of lysine-producing *Corynebacterium glutamicum* by combined analysis of the transcriptome, metabolome, and fluxome, *J. Bacteriol.* 2004, **186**(6), 1769–1784.
46. Hermann, T., Pfefferle, W., Baumann, C., Busker, E., Schaffer, S., Bott, M., Sahm, H., Dusch, N., Kalinowski, J., Puhler, A., Bendt, A. K., Kramer, R., and Burkovski, A., Proteome analysis of *Corynebacterium glutamicum*, *Electrophoresis*, 2001, **22**(9), 1712–1723.
47. Bendt, A. K., Burkovski, A., Schaffer, S., Bott, M., and Farwick, M., and Hermann, T., Towards a phosphoproteome map of *Corynebacterium glutamicum*, *Proteomics* 2003, **3**(8), 1637–1646.
48. Hermann, T., Finkemeier, M., Pfefferle, W., Wersch, G., Kramer, R., and Burkovski, A., Two-dimensional electrophoretic analysis of *Corynebacterium glutamicum* membrane fraction and surface proteins. *Electrophoresis* 2000, **21**(3), 654–659.
49. Barr, M. M., Super models, *Physiol Genomics* 2003, **13**(1), 15–24.
50. Hutchison, C. A., Peterson, S. N., Gill, S. R., Cline, R. T., White, O., Fraser, C. M., Smith, H. O., and Venter, J. C., Global transposon mutagenesis and a minimal *Mycoplasma* genome. *Science* 1999, **286**(5447), 2089–2090.
51. Koonin, E. V., How many genes can make a cell: the minimal-gene-set concept, *Annu. Rev. Genomics Hum. Genet.* 2000, **1**, 99–116.
52. Zimmer, C., Genomics. Tinker, tailor: Can Venter stitch together a genome from scratch? *Science* 2003, **299**(5609), 1006–1007.
53. Schwikowski, B., Uetz P, and Fields, S., A network of protein-protein interactions in yeast, *Nat. Biotechnol.* 2000, **18**(12), 1257–1261.
54. Mewes, H. W., Hani, J., Pfeiffer, F., and Frishman, D., MIPS: A database for protein sequences and complete genomes, *Nucleic Acids Res.* 1998, **26**(1), 33–37.
55. Goffeau, A., Four years of post-genomic life with 6,000 yeast genes, *FEBS Lett.* 2000, **480**(1), 37–41.
56. O'Farrell, P. H., High resolution two-dimensional electrophoresis of proteins, *J. Biol. Chem.* 1975, **250**(10), 4007–4021.
57. Sundararaj, S., Guo, A., Habibi-Nazhad, B., Rouani, M., Stothard, P., Ellison, M., and Wishart, D. S., The CyberCell Database (CCDB): A comprehensive, self-updating, relational database to coordinate and facilitate in silico modeling of *Escherichia coli*, *Nucleic Acids Res.* 2004, **32**, D293–295.
58. Wasinger, V. C., and Corthals, G. L., Proteomic tools for biomedicine, *J. Chromatogr. B.* 2002, **2002**(771), 33–48.

CHAPTER 2

# Strategies for Measuring Dynamics: The Temporal Component of Proteomics

ROBERT J. BEYNON and JULIE M. PRATT

University of Liverpool, Liverpool, England

## 2.1 INTRODUCTION

The explosion of interest in proteomics reflects the ability, realizable for the first time, to measure a significant proportion of proteins that are expressed in a specific cell or cellular subsystem and also to compare two proteomes and assess the relative expression of specific proteins under two different conditions. Many such studies sit formally in the subdomain of comparative proteomics, with the express goal of discerning proteins that are differentially abundant, usually in a pairwise comparison. Unfortunately, "differential abundance" has sometimes been taken as synonymous with "differential expression" or, worse, "differential transcription". The latter assumption, that the only determinant of protein concentration in a cell is the rate of input into the protein pool (i.e., synthesis) is certainly not warranted and belies the roles that output from the pool (protein export or intracellular protein degradation) and posttranslational modifications can have on the concentration of a protein. More and more studies are now attempting to measure the levels (usually relative rather than absolute) of messenger ribonucleic acid (mRNA) and protein in the same experiment: the former usually by microarray experiment, the latter by two-dimensional gel electrophoresis (2DGE) quantitiative imaging or by stable isotope labeling. A recurrent theme in such studies is the weak correlation between the relative abundances of mRNA and protein (Gygi et al., 1999, 2000; Chen et al., 2002; Greenbaum et al., 2003; Griffin et al., 2002). However, there was no reason, a priori, why this correlation should ever have been expected to be high. The intracellular concentration of a protein is under the control of two opposing processes: protein synthesis and protein degradation. The former is the outcome of the concentration of cognate mRNA, the rate of initiation of translation, and the overall ribosomal activity. The latter is a function of the activity of the degradative process and the inherent susceptibility of the protein to the degradative processes. Rather than attempting to correlate the amount of protein to the amount of mRNA in the cell, we should therefore determine the relationship between the mRNA concentration and the rate of synthesis of that protein. Conversely, while measurements of the activity of the proteolytic machinery in the cell can give some indication of the overall degradative capacity, the intrinsic

*Microbial Proteomics: Functional Biology of Whole Organisms*, Edited by Ian Humphery-Smith and Michael Hecker. 

susceptibility of each protein varies, and thus, determination of the rate of degradation is required. This intrinsic susceptibility to the degradative apparatus can be modulated by ligands and posttranslational events such as phosphorylation, and therefore, while the transcriptome is directly linked to the rate of synthesis of a protein, the rate of degradation may in turn be linked more directly to the cellular manifestation of the metabolome.

It follows that a complete understanding of proteome dynamics requires that we deconvolute the single statistic defining the changing amount of a protein in the cell into the measurement of the rate of synthesis and the rate of degradation of each protein. For example, a protein can decline in amount in the cell through accelerated breakdown (a catabolic response) or by a diminution of the rate of synthesis (an anabolic response). Only the latter might be reflected in corresponding changes in the cognate transcript. In this chapter, we discuss some strategies for measurement of protein turnover in microbial systems. Although radiolabeling has been used with some success, predominantly with single proteins, we will restrict this discussion to stable isotopes, as these lend themselves to sensitive and high-throughput mass spectrometric analyses that underpin most proteomics studies (Beynon and Pratt, 2005). Radiolabeling has rarely been applied on a global, proteomewide scale (for a notable exception, see Grunenefelder et al., 2001).

## 2.2 MECHANISTIC PERSPECTIVE OF PROTEIN TURNOVER

The amount of protein in a cell is controlled by rates of input and of output. Even for intracellular, nonsecretory proteins, protein degradation ensures that for most proteins the pool is subject to turnover and there is continued synthesis of new protein molecules, matched in the steady state by an equal rate of replacement of protein molecules.

In the simplest of systems, protein synthesis can be considered to be a first-order process, in which, for a fixed mRNA concentration and a fixed ribosomal activity, the rate of synthesis is constant measurable in molecules over time and is insensitive to the cellular concentration of the protein. Of course, feedback mechanisms operate to regulate the rate of synthesis through transcriptional control. However, the process can still be considered to be zero order. By contrast, it is generally held that, to a reasonable approximation, protein degradation is a first-order process. The rate of degradation of proteins in a pool is expressed as a fractional rate constant, for example 0.1per hour, which is equivalent to the loss of 10% of the pool each hour. The flux through the degradative process, in molecules per unit time, is then expressed by size of protein pool multiplied by the fractional rate of degradation. This term now has the same dimensionality as protein synthesis, and the three terms are related such that $P = k_s/k_d$. It follows that measuring two of these three parameters permits calculation of the third. Since the amount of protein in the cell ($P$) is often the most readily determined, two common experimental strategies measure the rate of synthesis or the rate of degradation of the protein.

A first-order model of protein degradation brings with it a series of assumptions and consequences. First, a first-order process is intrinsically stochastic, and thus, a newly synthesized protein molecule has an equal chance of being degraded as a molecule synthesized some time ago. Thus, a protein does not "age" unless it undergoes irreversible covalent or noncovalent changes that change the intrinsic susceptibility to the degradative process. Second, we need to discriminate, at least conceptually, between the commitment phase of degradation and the completion phase. The commitment step, the point at which a protein molecule becomes destined for degradation, is the primary arbiter of clearance rate and is followed by a more rapid completion phase, in which the committed protein molecule is proteolyzed, ultimately to amino acids. It is clear that the first stage should be rate limiting and the second proteolytic phase should occur at a much higher rate than the commitment process. Otherwise, the cell would potentially accumulate substantial quantities of partially degraded proteolytic products that could retain partial functionality and interfere with normal cellular process, possibly as seen in senescence. At the same time, the proteolytic apparatus cannot be permitted to act in an uncontrolled fashion. Restrictions on the activity of the degradative machinery include (a) compartmentalization of the apparatus in vacuoles or tightly regulated macromolecular complexes, (b) the inherent resistance of most native protein structures to proteolysis, and (c) the requirement for a nonproteolytic commitment step.

## 2.3 MEASUREMENT OF PROTEIN TURNOVER IN MICROBIAL SYSTEMS

In all experimental designs to measure rates of protein turnover, the overall principles are the same but divided into two types of experiments. Proteins are metabolically labeled in an intact cell with either a stable or an unstable (radioactive) isotopically labeled precursor. Over time, the processes of protein metabolism cause a change in the abundance of a label in the protein pool. This can be due to either the incorporation of label, measuring the rate of synthesis of the protein, or the loss of label from prelabeled proteins, which is a measure of degradation. In both instances, there are complications that can compromise the experiments. First, during synthesis, some labeled proteins will be degraded and return label to the precursor pool. This results in an artifactual persistence of label in the protein pool if a single pulse label is administered. Alternatively, in a degradation study, reutilization of label can again manifest itself as an apparently lower rate of degradation. Fortunately, and particularly in the context of microbial systems, it is possible to control for these effects by appropriate design of labeling protocols.

Incorporation experiments can be used to generate partially labelled proteins, and the extent of labeling is then assessed to define the rate of synthesis. Alternatively, the protein pool can be fully prelabeled, followed by a "light" chase (the equivalent of a "cold" chase in radiolabeling experiments) and the loss of label from a protein can be determined as a measure of protein degradation. Inasmuch as it is possible to discriminate between the "heavy" and light pools by mass spectrometry, it can be argued that the two processes are formally equivalent. The decision as to which approach to take is more likely to be driven by other experimental considerations, such as the choice of precursor, the availability of an auxotroph for the particular amino acid, and the ability to deplete the precursor pools.

In proteomics, the parameter that is measured is that of the exact mass of a protein or a peptide. It follows that the incorporation of a stable-isotope-labeled amino acid, even if there is only one instance of that amino acid in the peptide, will lead to an increase in mass, and with a mass spectrometer of the appropriate performance (which means any instrument marketed today for proteomics), it ought to be possible to discern the stable isotope labeled from the unlabeled counterpart. It follows that the rate of incorporation or loss of a stable-isotope-labeled amino acid can be monitored by mass spectrometry. Given the exquisite sensitivity and selectivity of mass spectrometry, it should therefore be entirely feasible to use stable isotope labeling to monitor the dynamics of a proteome and, under appropriate conditions, be able to measure the rate of synthesis or degradation of any individual protein in that proteome. The only notable caveat here is of course low molecular weight proteins, wherein the likelihood of occurrence of individually labeled peptides decreases with reduction in mass.

The use of stable isotopes does, however, introduce a technical complexity. Radioisotopically labeled amino acids can be obtained at very high specific radioactivity, such that a trace level of incorporation can lead to a readily measurable signal, assessed, for example, by scintillation counting or fluorography. By contrast, the mass spectrometers used in proteomics are not capable of measuring trace levels of enrichment of stable isotopes. Typically, a minimum of 5% incorporation might be considered to be accurately measurable, and thus, there must be significant metabolic replacement (turnover) of a protein in order that the rate of turnover can be assessed. This requires an experimental design that can ensure that the labeling achieves this extent of incorporation.

There is one final consideration. In synthesis experiments, the protein of interest must not be allowed to become fully labeled, as a fully labeled protein is no longer informative with respect to its turnover. Fully labeled proteins are of greater value in comparative proteomics in what have become known as SILAC (stable isotope labeling in cell culture) experiments but are not appropriate for turnover studies. In fact, an important consideration in SILAC studies is to ensure that a protein pool has experienced sufficient half-lives, or the cells sufficient doubling times, so that complete labeling is assured.

## 2.4 STRATEGIES FOR MEASUREMENT OF PROTEOME DYNAMICS

### 2.4.1 Growth Conditions

Stable isotope incorporation into microbes requires the availability of media that sustain appropriate growth of the organism while being amenable to the facile substitution of the chosen precursor with labeled precursor. Commonly studied bacteria (e.g., *Escherichia coli, Salmonella typhimurium, Klebsiella pneumoniae, Pseudomonas aeruginosa)* (Nieweg and Bremer, 1997; Todar 2004) *or baker's yeast* (*Saccharomyces cerevisiae*) can grow on fully defined media in which the sole carbon source can be glucose, glycerol, or other appropriate sugar and the sole nitrogen source can be, for example, $NH_4Cl$. This allows the use of the less costly, universal labels $^{13}C$ and $^{15}N$ in the form of $^{13}C$-glycerol or $^{13}C$-glucose or $^{15}NH_4Cl$. Many yeasts can be grown with $^{15}NH_4Cl$ or glutamate as sole nitrogen source (*Candida utilis, Candida krusei, Pichia fermentans*, and *Kluyveromyces fragilis*; Large, 1986), allowing these precursors to be used to label cells with $^{15}N$. However, microbial growth under these conditions may be rather slow and therefore may not be appropriate for all studies. For studies in which the goal is to identify the effect of changing carbon source, stress (e.g., pH, osmotic stress, oxidative stress), the effect of antibiotics, or the effect of mutations on protein turnover within various pathways, a reasonable growth rate is more practicable. Growth rates can be substantially increased by including an excess of all the amino acids in the defined medium, for example, growth of coagulase-negative staphylococci (Hussain et al., 1991) and *Listeria monocytogenes* (Klarsfeld et al., 1994). The inclusion of amino acids extends investigations to additional organisms that are unable to synthesize many of the amino acids de novo (*Staphylococcus aureus, Staphylococcus epidermidis*) and also bacteria that require carbon- and nitrogen-containing supplements for growth. For example, *Lactobacillus* requires purines, pyrimidines, vitamins, and several amino acids in order to grow. *Haemophilus influenzae* is a fastidious bacterium which requires haemin and nicotinamide adenine dinucleotide (NAD) for growth. Valine, arginine, and cysteine are required for growth and enterotoxin production in *S. aureus* (Onoue and Mori, 1997). In all the above cases, it is possible to generate (often commercially available) a defined medium which may be supplemented with $^{13}$C-, $^{15}$N-, or $^{2}$H/$^{13}$C-labeled amino acids. Because the cells are unable to synthesize these amino acids de novo, they are incorporated into proteins with high efficiency, and the relative isotope abundance (RIA) of the incorporated amino acid will be the same as that for the precursor pool. Indeed, given such auxotrophy, the opportunity exists for complete labeling of the proteins using a precursor RIA of unity. The ability to include the labeled amino acid of choice in media permits the achievement of fully labelled cells after about seven generations, facilitating turnover studies of the type described by Pratt et al. (2002) for *S. cerevisiae*.

Many studies, however, are directed toward the changes that are occurring in gene expression when pathogens encounter the environment within the host, and in these cases considerable ingenuity is necessary to design a labeling strategy that will allow evaluation of changes in protein dynamics. Most pathogenic bacteria of animals, which have adapted themselves to growth in animal tissues, require complex media for their growth. Blood, serum, and tissue extracts are frequently added to culture media for the cultivation of pathogens. Prior dialysis to remove free amino acids will greatly increase the incorporation of the labeled amino acid introduced to the growth medium. However, the RIA is unlikely to be unity under these conditions, as proteins in the growth medium will be proteolyzed and contribute unlabeled amino acids to the precursor pool, diminishing the RIA. For pathogens that enter host cells, there will be access to all the precursors available within the specific cellular compartment targeted by the pathogen. If it is possible to grow host cells in tissue culture, these cells can be prelabeled with the chosen heavy-isotope precursor prior to infection to enhance the achievable levels of labeling. A particularly elegant solution to this problem was adopted in a study of the proteome of the malarial parasite, *Plasmodium falciparum*, during the cell cycle (Nirmalan et al., 2004). *Plasmodium falciparum* infects erythrocytes and obtains its amino acids for growth by uptake from the plasma by proteolysis of α- and β-hemoglobin and by de novo biosynthesis. However, human α- and β-hemoglobin lack isoleucine and *P. falciparum* is unable to synthesize this amino acid de novo. The parasite sustained growth in custom-made RPMI (Roswell Park Memorial Institute) medium, containing erythrocytes but lacking isoleucine other than $^{13}C_6{}^{15}N$-isoleucine and elicited over 90% labeling after 60 h. This method would be readily applicable to the determination of protein turnover in the malarial parasite. However, there are many organisms and specific experiments that would require more complex media and in these cases a more radical approach is required.

### 2.4.2 Culture Conditions

When measuring protein turnover rates in microbial organisms, the effect of changes in the concentration of nutrients, pH, and cell density on the growth rate of the culture should be considered. Usually cells are introduced into fresh medium from a stationary-phase, overnight culture. These cells will take a while to emerge from the lag phase, then will grow at their fastest rate in the log phase until nutrients become scarce and they will again enter the stationary phase. Thus the medium changes throughout growth in the concentration of nutrients, waste products, and protons (pH). Changes in pH can be minimized by including a buffer system at the optimal pH for growth, but there will be considerable fluctuations in nutrient availability. Further, changes in cell density (e.g., cell–cell interaction, quorum sensing) will certainly have an effect on gene expression and may affect protein turnover. For proteins that are rapidly turned over, the impact will be minimal as sampling can be completed during the log phase of growth, but for long-lived proteins, studies in batch culture will reflect only an average turnover rate. To overcome this problem, turnover can be studied by growth in a chemostat in which cell density and nutrient supply are kept constant.

### 2.4.3 Choice of Precursor

We suggest that the most appropriate precursor is a stable-isotope-labeled amino acid. There have been some studies of metabolic labeling using, for example, [$^{13}$C] glucose as sole carbon source (Cargile et al., 2004), but this approach brings specific problems. First, this approach seems to have been predominantly driven by the need to generate protein labeled for nuclear magnetic resonance (NMR) studies, and as such the goal is complete replacement of all $^{12}$C atoms with the $^{13}$C counterpart. However, in metabolic labeling for turnover, the one thing that cannot be allowed to occur is that a protein should become fully labelled within the time window of exposure to label. Under this circumstance, it is no longer possible to determine the rate of synthesis of the protein, as it maintains the same degree of labeling irrespective of the time of exposure to the precursor pool. Of course, if the goal is to label the protein completely in order to conduct a chase experiment, in which the loss of label is monitored (a direct measurement of degradation), then it is desirable to start with a fully labeled protein pool.

Accurate measurement of turnover rate requires that we measure the rate of incorporation of precursor into protein. To achieve this, we must determine the abundance of the stable isotope in the target protein or proteolytic peptides derived from a protein. However, the extent of incorporation still cannot be accessed unless we also know the extent of enrichment by the stable isotope of the precursor pool. This is defined as the precursor RIA. Stable-isotope-labeled amino acids can be purchased with RIA in excess of 98%, and thus it might be argued that since this is effectively unity, there are no experimental concerns. However, a number of factors conspire to reduce the precursor RIA to a lower, unknown value. First, preexisting biomass has the potential to dilute the precursor with unlabeled amino acid, both from the amino acid pool and the protein pool. This is less of a problem in microbial systems than in, for example, animal systems because a newly seeded culture contributes rather small quantities of biomass to the final expanded cell mass. Second, many microbial cells are fully competent in the biosynthesis de novo of all amino acids. Endogenous biosynthesis from unlabeled precursors will therefore also reduce the precursor RIA. However, microbes are often parsimonious in their investment in biosynthesis, and biosynthetic pathways for amino acids are usually tightly controlled. Provision of adequate concentrations of a specific amino acid in the medium will often repress endogenous biosynthesis and thus eliminate this dilution effect. If possible, use of a strain that is auxotrophic for the particular amino acid will eliminate this as a source of problem. An additional complication occurs when labeled amino acids are metabolized in such a way that stable isotope flows from one amino acid to another. This does not automatically invalidate turnover rate calculations but serves to complicate the mass offset of the stable-isotope-labeled variant of the protein and may hinder manual or automated interpretation of spectra. Ideally, there should be no metabolic flow of labeled atom centers either to or from the precursor amino acid. One modification that may be more informative is the loss of a single deuteron from an amino acid that is labeled at the $\alpha$-carbon atom. This deuteron, although chemically stable, is metabolically labile and is lost during the reversible reaction of transamination, for example (Fig. 2.1). Thus, the amino acid that is incorporated into protein has a labeled mass of one less than the amino acid added to the medium. This is not a particularly serious complication, although one

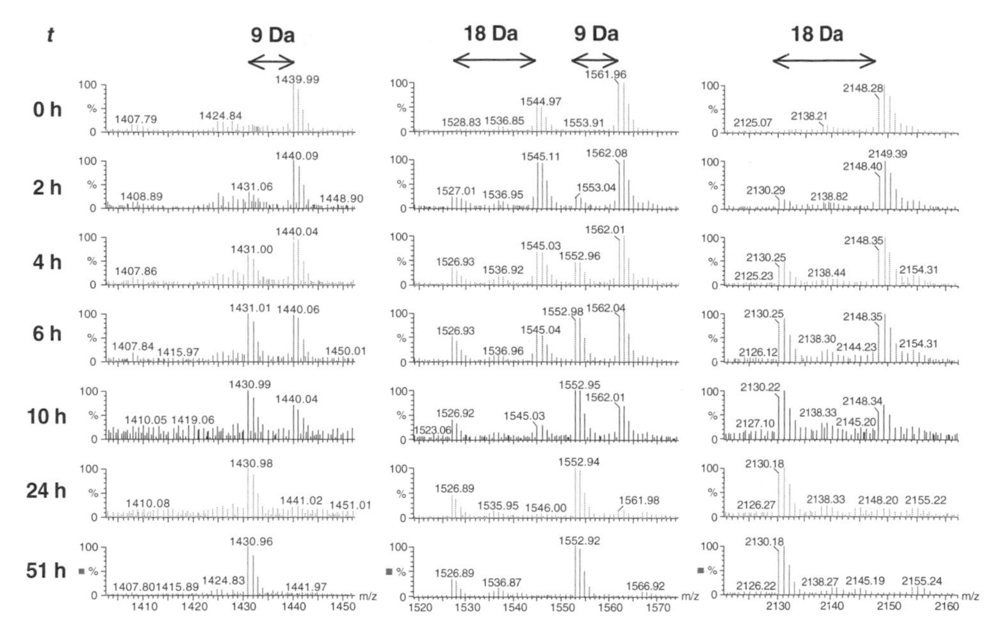


**Figure 2.1** Determination of protein turnover in *S. Scerevisiae. Saccharomyces cerevisiae* was grown in glucose-limited steady-state culture in a chemostat in the presence of a minimal medium containing [$^2H_{10}$]leucine. After seven doubling times, the proteins were virtually fully labeled with leucine. The medium feedstock was then changed to one containing unlabeled leucine, and at the same time, a large bolus of unlabeled leucine, equilvalent to at least a 20-fold excess over that present in the medium, was added. Because the cells were glucose limited, the addition of leucine had no effect on the growth rate. Subsequently, the culture was sampled, and proteins were separated by gel electrophoresis and analyzed by MALDI-TOF mass spectrometry. The loss of heavy label from leucine containing peptides is the result of a composite process, reflecting loss from the vessel, and intracellular degradation. Knowledge of the former permits calculation of the latter.

should be aware that this is a likely reaction for most amino acids and can serve to confirm rapid equilibration with the intracellular amino acid pool.

With microbial systems, the manipulation of the growth conditions should be able to achieve the goal of setting the precursor pool RIA to unity. Under these circumstances, there is no stochastic decision about incorporation of the labeled amino acid, and every instance of the amino acid in the protein is labeled (heavy $=$ H). Thus, a population of protein molecules pool shifts from a mass of $M$ to $M + nO$, where $n$ is the number of instances of the amino acid in the protein and $O$ is the mass offset attributable to the heavy amino acid. In turn, in a typical proteomics experiment, every peptide will also be fully labeled and, if the peptide contains at least one instance of the amino acid, will show the corresponding mass shift.

If the precursor RIA is significantly less than unity, additional complications arise in terms of patterns of labeling. To illustrate this, consider the situation where an amino acid has an RIA of 0.5. As the ribosomal apparatus encounters a codon for this amino acid, there is a 50% chance of incorporation of a light or a heavy variant—we can exclude isotope effects at this level. Moreover, the process is stochastic, so different protein molecules will have distinct distributions of the light and heavy amino acids. The labeling pattern follows a binomial distribution. For example, if we consider the situation where a tryptic peptide within a protein sequence contains a single instance of the amino acid, the options are simple; the peptide can either contain none or a single instance of the labeled amino acid. If the protein has two instances of the amino acid, then the three products, theoretically distinguishable by mass, are LL, [LH, HL], and HH in a ratio of 1 : 2 : 1. Extending to a third instance of the amino acid, the labeling patterns yield four distinguishable products, LLL, [LLH, LHL, HLL], [LHH, HLH, HHL], and HHH in the ratio 1 : 3 : 3 : 1. This progression will be recognized as a binomial series. The mass separation is readily discernible, and indeed, analysis of the distribution between the different species can yield valuable information about the precursor RIA. However, the net effect is to complicate the labeling pattern, and this situation can and should be avoided.

Such problems could also be reduced for tryptic peptides by incorporating stable-isotope-labeled lysine or arginine or both into proteins, as most tryptic peptides would then only contain a single instance of the label. If the labeled lysine and arginine showed the same mass offset, the analytical problem is even simpler, as the separation between heavy and light species is the same irrespective of whether the C-terminal amino acid is lysine or arginine. However, we would also make a case for a different mass separation of lysine- and arginine-terminated peptides, as the additional information obtained from knowledge of the C-terminal amino acid can bring about a significant reduction in search space, since an additional goal is identification of the protein (Pratt et al., 2002; Beynon, 2003).

## 2.5 EXPERIMENTAL STRATEGY

There are two fundamentally different approaches to the measurement of protein turnover by stable isotope labeling. Both require that the protein or a peptide derived therefrom be resolved from other molecules in order that the ratio of labeled to unlabeled variants may be determined. We will not dwell on the postlabeling methodologies, as these are routine to proteomics. For example, a protein might be isolated as a spot on a 2D gel or as an immunoprecipitate before being analyzed by matrix-assisted laser desorption ionization time-of-flight (MALDI-TOF) mass spectrometry. Alternatively, a protein or protein mixture can be proteolyzed and the peptides selectively enriched and/or resolved by one or two dimensions of chromatography prior to mass spectrometry. The objective is to measure the ratio of labeled peptide to its unlabeled cognate.

In synthesis experiments, the measured parameter is the rate of incorporation of labeled amino acid into a protein. Cells are exposed to the labeled precursor, and at time points subsequently, the proteins of interest are isolated and proteolyzed and the heavy and light variants are analyzed by mass spectrometry. If the cells are undergoing a transition from exponential to stationary phase or if the precursor is being depleted in the medium (such as could happen in batch culture), then the rate of synthesis or the precursor RIA could be changing through the labeling experiment. It is only by frequent sampling that this can be detected.

In degradation studies, a protein is prelabeled and the loss of label is then monitored. This is, of course, formally equivalent to the synthesis experiments described above, in which the roles of the light and heavy precursor variants are reversed but the overall design has notable differences. First, in batch culture, the prelabeled cells must be transferred to unlabeled medium or exposed to a large excess of unlabeled precursor to diminish reutilization of the label generated during degradation. This would normally require that cells in suspension culture be centrifuged and resuspended in fresh, unlabeled medium, bringing about consequential changes in growth rate, nutrient supply, and possibly cell density.

If there was significant labeled amino acid available for reutilization, the RIA of the precursor would almost certainly be greater than zero. Inspection of peptides containing multiple instances of the amino acid would be warranted, as the appearance of peptide variants consisting of the partially labelled peptide (e.g., the LLH or the LHH variants in a trilabelled peptide) would be suggestive of a significant degree of reutilization. For example, in our studies on yeast protein degradation, we preferred to establish the cells in steady state in continuous culture. The cells were auxotrophic for leucine, and the sole source of leucine in the medium was the [$^{2}H_{10}$] variant, which was incorporated into proteins as the non-adeuterated variant, due to the loss of the α-carbon deuteron. After seven doubling times, the cells were fully labeled with [$^{2}H_{9}$] leucine (without apparent detriment to growth rate or protein patterns on 2D gels). At this time, the medium feedstock was changed to an identical medium in which the labeled leucine was replaced with unlabeled leucine. At the same time, a 20-fold excess of unlabeled leucine was added to the chemostat to reduce the isotope abundance of the precursor pool such that the data would not be compromised by reutilization. Because the growth of the cells was glucose limited and the cells were auxotrophic for leucine, there was no effect from the perturbation of the medium by addition of the unlabeled leucine. Subsequently, the culture was sampled and proteins were resolved on 2D gels. The same protein spot was isolated, proteolyzed, and analyzed by MALDI-TOF mass spectrometry and the time course of "unlabeling" of the protein was assessed (Fig. 2.2). Noteworthy is the absence of

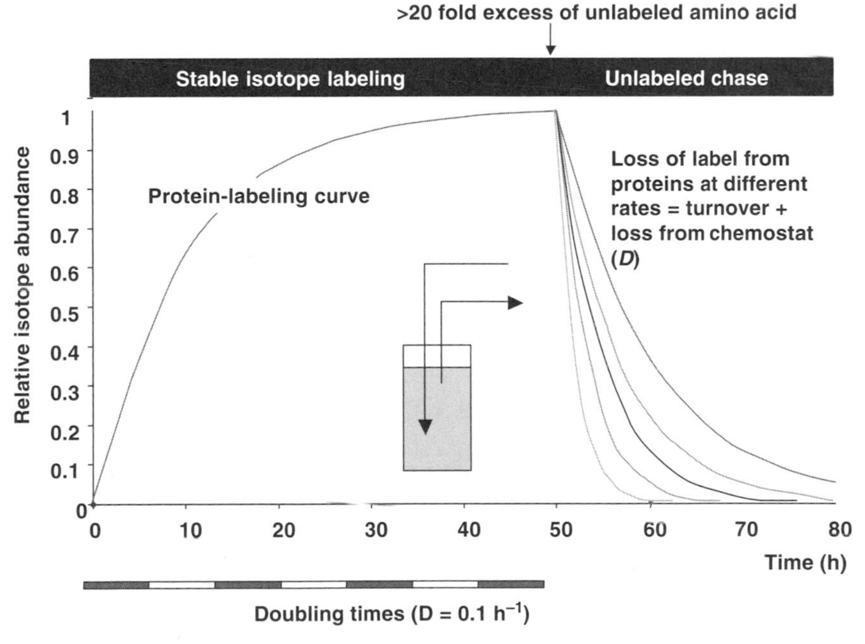


**Figure 2.2** Change in isotope labeling pattern in a turnover study. Protein recovered from *S. cerevisiae* cells grown as described in Figure 2.1 were resolved by 2DGE and analyzed by MALDI-TOF mass spectrometry. As the proteins lose label (through turnover and dilution) the [$^{2}H_{9}$]-labeled (heavy, H) peptides are gradually diminished and matched by the appearance of the light (L) counterparts, $9n$ daltons lower in mass, where $n$ is the number of leucine residues. The ion intensity ratio $I_H/(I_H + I_L)$ is the RIA of the isotope in the peptide and is therefore a concentration-independent term that reflects loss of labeled amino acid from the protein.

observable intermediate labeled forms, good presumptive evidence for an effective reduction of the precursor RIA to near zero by the unlabeled chase strategy.

## 2.6 CONCLUSIONS

The handful of publications outlined in this chapter that have described strategies for the measurement of protein turnover, either in batch culture or in steady state, have set the scene ofr determination of this overlooked dimension to proteomics investigations. With the increasing interest in systems biology, we should expect an explosion of studies in which RNA abundance, RNA turnover, protein abundance, and protein turnover together with metabolome-related parameters are measured in the same experiment. In addition, the feasibility of obtaining absolute quantification data across a proteome, yielding the numbers of molecules of each protein present in a cell (Beynon et al., 2005), will facilitate the development of computer models of the dynamics of cellular physiological processes. Appropriate experimental design and data standardization are necessary to ensure the coordination of such complex interrelated data and their accessibility to the scientific community.

## ACKNOWLEDGMENTS

The work described in this chapter has been supported by the Biotechnology and Biological Sciences Research Council.

## REFERENCES

Beynon, R. J. (2003). Stable isotope labelling with amino acids as an aid to protein Identification in peptide mass fingerprinting. In M. P. Conn (Ed.), *Handbook of Proteomic Methods.* Totowa, NJ: Humana, pp. 129–144.

Beynon, R. J., Doherty, M. K., Pratt, J. M., and Gaskell, S. J. (2005). Multiplexed absolute quantification in proteomics using artificial QCAT proteins of concatenated signature peptides. *Nature Methods* **2**:587–589.

Beynon, R. J., and Pratt, J. M. (2005). Metabolic labelling of proteins for proteomics. *Mol. Cell Proteomics*, in press.

Cargile, B. J., Bundy, J. L., Grunden, A. M., and Stephenson, J. L. Jr. (2004). Synthesis/degradation ratio mass spectrometry for measuring relative dynamic protein turnover. *Anal. Chem.* **76**(1):86–97.

Chen, G., Gharib, T. G., Huang, C. C., Taylor, J. M., Misek, D. E., Kardia, S. L., Giordano, T. J., Iannettoni, M. D., Orringer, M. B., Hanash, S. M., and Beer, D. G. (2002). Discordant protein and mRNA expression in lung adenocarcinomas. *Mol. Cell. Proteomics* **1**(4):304–313.

Greenbaum, D., Colangelo, C., Williams, K., and Gerstein, M. (2003). Comparing protein abundance and mRNA expression levels on a genomic scale. *Genome Biol.* **4**(9):117.

Griffin, T. J., Gygi, S. P., Ideker, T., Rist, B., Eng, J., Hood, L., and Aebersold, R. (2002). Complementary profiling of gene expression at the transcriptome and proteome levels in *Saccharomyces cerevisiae. Mol. Cell. Proteomics* **1**(4):323–333.

Grunenfelder, B., Rummel, G., Vohradsky, J., Roder, D., Langen, H., and Jenal, U. (2001). Proteomic analysis of the bacterial cell cycle. *Proc. Natl. Acad. Sci. USA* **98**(8):4681–4686.

Gygi, S. P., Rist, B., and Aebersold, R. (2000). Measuring gene expression by quantitative proteome analysis. *Curr. Opin. Biotechnol.* **11**(4):396–401.

Gygi, S. P., Rochon, Y., Franza, B. R., and Aebersold, R. (1999). Correlation between protein and mRNA abundance in yeast. *Mol. Cell. Biol.* **19**(3):1720–1730.

Hussain, M., Hastings, J. G., and White, P. J. (1991). A chemically defined medium for slime production by coagulase-negative staphylococci. *J. Med. Microbiol.* **34**(3):143–147.

Kelleher, N. L. (2004). Top-down proteomics. *Anal. Chem.* **76**(11):197A–203A.

Klarsfeld, A. D., Goossens, P. L., and Cossart, P. (1994). Five *Listeria monocytogenes* genes preferentially expressed in infected mammalian cells: *plcA, purH, purD, pyrE* and an arginine ABC transporter gene, *arpJ. Mol. Microbiol.* **13**(4):585–597.

Large, P. J. (1986). The degradation of organic nitrogen compounds by yeasts. *Yeast* **2**:1043.

Nieweg, A., and Bremer, E. (1997). The nucleoside-specific Tsx channel from the outer membrane of *Salmonella typhimurium, Klebsiella pneumoniae and Enterobacter aerognes:* functional characterization and DNA sequence analysis of the *tsx* genes. *Microbiology* **143** (Pt 2):603–615.

Nirmalan, N., Sims, P. F., and Hyde, J. E. (2004). Quantitative proteomics of the human malaria parasite *Plasmodium falciparum* and its application to studies of development and inhibition. *Mol. Microbiol.* **52**(4):1187–1199.

Onoue, Y., and Mori, M. (1997). Amino acid requirements for the growth and enterotoxin production by *Staphylococcus aureus* in chemically defined media. *Int. J. Food Mircobiol.* **36**(1):77–82.

Pratt, J. M., Petty, J., Riba-Garcia, I., Robertson, D. H., Gaskell, S. J., Oliver, S. G., and Beynon, R. J. (2002). Dynamics of protein turnover, a missing dimension in proteomics. *Mol. Cell. Proteomics* **1**(8):579–591.

Pratt, J. M., Robertson, D. H., Gaskell, S. J., Riba-Garcia, I., Hubbard, S. J., Sidhu, K., Oliver, S. G., Butler, P., Hayes, A., Petty, J., and Beynon, R. J. (2002). Stable isotope labelling *in vivo* as an aid to protein identification in peptide mass fingerprinting. *Proteomics* **2**(2):157–163.

Todar, K. (2004). *Todar's Online Textbook of Bacteriology.* http://textbookofbacteriology.net/nutgro.html.

CHAPTER 3

# Quest for Complete Proteome Coverage

C. DAVID O'CONNOR[1], IAN N. CLARKE[2], and PAUL SKIPP[1]

[1]University of Southampton, Bassett Crescent East, Southampton, United Kingdom
[2]University Medical School, Southampton General Hospital, Southampton, United Kingdom

## 3.1 INTRODUCTION

A decade ago almost all of the data sets gathered by molecular biologists were open ended and incomplete. For example, a search for members of a particular class of protein in an organism would necessarily be restricted to what could be uncovered in a reasonable time scale. The possibility that further members remained to be discovered could never be discounted even after exhaustive biochemical analyses. All this changed radically with the availability of complete genome sequences for a variety of living organisms, commencing with the report of the *Haemophilus influenzae* genome sequence in 1995 (Fleischmann et al., 1995). Because these data sets are comprehensive catalogues of all the possible components that could make up a living organism, they have had a major impact on the approaches used to study cellular processes. They have also triggered a fresh influx of mathematicians and engineers into the life sciences, primarily because genome data sets place strict limits on the number and type of components that can be synthesized in a cell. This in turn restricts the number of meaningful hypotheses that can be formulated about an organism's life processes and modus operandi and significantly simplifies attempts to model cellular processes.

Currently, the complete genomes for over 200 organisms are available (for an up-to-date total, refer to http://www.genomesonline.org/), and sequencing technologies have progressed to the extent that it is possible to obtain severalfold coverage of an entire microbial genome in a single day. These spectacular advances highlight the relative lack of progress in obtaining high coverage of components at the level of the proteome. To date, it has only proved possible to identify a fraction of the encoded proteins, even in organisms that are predicted to have very simple proteomes. For example, only about 30% of the ~1700 gene products of *H. influenzae* have been identified to date, despite the intensive and frequently ingenious efforts of several groups (Kolker et al., 2003; Langen et al., 2000). Clearly, our inability to obtain better proteome coverage is partially explicable by the dynamic nature of proteomes, whereby cells only express particular sets of proteins according to need, and by

*Microbial Proteomics: Functional Biology of Whole Organisms*, Edited by Ian Humphery-Smith and Michael Hecker. 

the extreme molecular diversity of proteins. However, the available evidence suggests that several other key factors contribute to the current impasse.

This chapter first considers the definition of a complete proteome before discussing in detail the current challenges in this area. It also offers some possible solutions to the current bottlenecks. As an illustration, we shall describe recent efforts to analyze the proteome of the bacterium *Chlamydia trachomatis*, an obligate intracellular pathogen with serovars that cause a broad spectrum of diseases. However, the concepts should apply to the proteomic analysis of many other organisms and, where it is appropriate, nonchlamydial studies are also considered.

## 3.2 WHAT DO WE MEAN BY COMPLETE PROTEOME COVERAGE?

At first sight, it would seem that defining complete coverage of a proteome is a trivial task. Thus, one simple aim might be to detect all the genes annotated in the genome sequence of the organism of interest. However, a moment's reflection shows that the concept of complete proteome coverage is complex. First, genome annotation is very much a case of "work in progress" and is subject to frequent revisions. For example, most software packages used to identify genes in genome sequences arbitrarily assume that the open reading frames for protein-coding genes are $\geq$150 bp, thereby potentially excluding small proteins from the predicted proteome of an organism. Although the extent of the problem has not yet been systematically studied, it is already clear that several gene products were initially missed because of this cutoff, for example in the pathogen *Mycobacterium tuberculosis* (Jungblut et al., 2001). Such software packages also tend to miss genes within genes. Although this phenomenon is best characterized for bacteriophage such as phiX174 (Sanger et al., 1978), it also occurs in microbial genomes, which tend to be highly streamlined (Ellis and Brown, 2003; Feltens et al., 2003).

A second complicating factor arises from the fact that genomes are not static. In particular, mobile genetic elements (e.g., bacteriophages, insertion sequences, and transposons) and recombination systems can bring about DNA rearrangements that significantly influence protein expression. Although the majority of these alterations will result in the abolition of the expression of one or more proteins, they can also activate previously cryptic genes. One dramatic example of alternative protein expression is provided by *Bacteroides fragilis*, which can readily modulate components at the cell surface through multiple DNA inversions (Krinos et al., 2001; Weinacht et al., 2004). Such reversible on–off phenotypes, which include the alternate expression of at least eight distinct capsular polysaccharides, clearly make it difficult to define the complete proteome of such organisms with any certainty. A related phase variation phenomenon, found in bacteria such as *Neisseria meningitidis* and *H. influenzae*, is associated with reversible changes within repeated simple sequence DNA motifs associated with so-called contingency genes. These microsatellite sequences, which are typically located in the promoter or coding regions of genes involved in the synthesis of surface antigens, can vary in length due to slipped strand mispairing during DNA replication (Saunders et al., 2000). Consequently, previously cryptic genes may sometimes be transcribed and translated following the addition or deletion of nucleotides in the repeated sequences, thereby producing proteins that may not have been predicted from the original genome sequence (Martin ct al., 2003). Because of the stochastic nature of the process, it is impossible to predict precisely which bacterial cells in a population will express which contingency genes at any given time. Moreover, the effect is combinatorial; as there are multiple contingency genes, cells can have thousands of possible combinations of these proteins.

Posttranslational modifications (PTMs) present a further confounding factor. It is increasingly clear that microbial proteins undergo an enormous range of chemical modifications, including some that were previously thought to be restricted to eukaryotes (see, e.g., Farris et al., 1998; Wacker et al., 2002). Ideally, a comprehensive description of a proteome should include information on such PTMs, as they frequently indicate the functional state of the protein and also give clues about its cellular location.

The presence of pseudogenes (i.e., DNA sequences that once coded for functional proteins but are no longer expressed) presents a fourth difficulty with regard to complete proteome coverage. Historically, bacteria were assumed to have very few pseudogenes because the energy expended in replicating (and in some cases transcribing) them would place the host cells at a selective disadvantage relative to their pseudogene-free counterparts. In keeping with this supposition, ~90% of a typical bacterial genome encodes proteins or structural RNAs. More recently, however, it has become apparent that certain bacteria have very high numbers of pseudogenes. For example, about one-quarter of the genome of *Rickettsia prowazekii* is noncoding and less than one-half of the 3.27-Mb genome of *Mycobacterium leprae* contains functional genes (Andersson et al., 1998; Cole et al., 2001). To account for these exceptions, it has been proposed that the "intracellular lifestyles" of these parasites shelter them from the influx of potentially deleterious genetic elements such as transposons and bacteriophage, which in other bacteria are normally limited by genetic deletion over evolutionary time (Lawrence et al., 2001). It should be noted, however, that other obligate intracellular parasites such as *C. trachomatis* do not seem to show evidence of genome decay, suggesting that additional explanations may need to be considered. Whatever the reason, the presence of pseudogenes may have a profound effect on the total number of proteins encoded by a genome. In many cases, it is relatively easy to detect such pseudogenes, for example due to the presence of premature stop codons relative to their positions in functional orthologues. However, other genes may have decayed more subtly and may even be expressed as full-length polypeptides even though they are nonfunctional. Interestingly, recent proteomic analyses of *Escherichia coli* suggest that a significant proportion of the genes that it has acquired by horizontal gene transfer are not detectable at the protein level and hence may fall into the pseudogene category. Following the identification of some 1480 (~35%) of the proteins predicted for the *E. coli* K-12 genome, Taoka and co-workers (2004) noticed that a disproportionately small number of proteins encoded by K loops, that is, DNA regions believed to be acquired by horizontal gene transfer, were present. Non detection at the protein level does not appear to be due to sensitivity problems as the liquid chromatography/mass spectrometry (LC-MS) system used was able to detect peptides down to $\sim 10^2$ copies and further investigation showed that the corresponding mRNAs of the K-loop genes were efficiently transcribed (Taoka et al., 2004). If these findings are confirmed and represent a more general trend, it may be necessary to radically revise our current concepts on what constitutes a functionally relevant gene and indeed a complete proteome.

## 3.3 CURRENT HURDLES TO BETTER PROTEOME COVERAGE

Despite the difficulties in precisely defining what is meant by a complete proteome, present experimental approaches appear to have a long way to go before significant coverage of any proteome can be achieved. It is convenient to divide the obstacles impeding better proteome coverage into those intrinsic to the microbe and those due to limitations in the technology. Chief among the problems in the former category is the fact that most microbes only express subsets of their total proteome under different growth conditions.

As unicellular organisms, almost all microbes encounter diverse environments during their life-cycles and hence need to appropriately adjust their expressed proteomes to maintain cellular homeostasis. However, the relevant environmental conditions encountered by microbes in nature are often exceedingly difficult to reproduce under laboratory conditions, and, although there has been some recent progress, it is still estimated that >99% of all soil and marine microbes are presently uncultivatable (Pace, 1997; Kaeberlein et al., 2002). Indeed, DNA array studies suggest that even for well-characterised organisms like *E. coli*, it has not yet proved possible to find in vitro conditions that detect the expression of all of its genes (Chang et al., 2004; Conway and Schoolnik, 2003; Tao et al., 1999). Since current microarray technology readily detects the transcription of genes that are expressed in <10 copies/cell (e.g., Lac repressor), it seems unlikely that the quiescent genes are simply being expressed below the limits of detection. Thus, one is forced to conclude either that investigators have not yet found the right environmental conditions to trigger their expression or that they are in fact pseudogenes.

The technical hurdles preventing better proteome coverage are also now evident. The inability of two-dimensional (2D) gel electrophoresis to detect low-abundance proteins or proteins with extreme physicochemical properties is well known (Gygi et al., 2000). However, recent direct comparisons suggest that 2D gels also underestimate the total number of induced components by two - to fourfold relative to DNA microarrays (Eymann et al., 2002; Hommais et al., 2001; Conway and Schoolnik, 2003). There has therefore been a major push toward proteomic approaches that circumvent these problems. One popular alternative to 2D gel analyses is the use of multidimensional LC (Link et al., 1999; Opiteck et al., 1997). For example, the multidimensional protein identification technology (MudPIT) strategy developed in the Yates laboratory uses strong cation exchange and reverse-phase nanocapillary columns arranged in series to provide a high-resolution separation and concentration of tryptic peptides derived from protein samples prior to their identification by tandem mass spectrometry (MS/MS) (Washburn et al., 2001; Wolters et al., 2001). Another method, termed GeLC-MS/MS, uses conventional sodium dodecyl sulfate polyacrylamide gel electrophoresis (SDS-PAGE) in conjunction with nanocapillary LC and MS/MS to similarly boost peak and load capacity, thereby increasing the number of peptides that can be identified (Schirle et al., 2003). Recently, we undertook a direct comparison of 2D gel, MudPIT, and GeLC-MS/MS approaches to determine which gave the best proteome coverage. Analysis of protein extracts from *C. trachomatis* serovar L2 showed that GeLC-MS/MS was markedly better than the other two approaches, identifying ≥35% of the predicted gene products (Skipp et al., in press). The approach identified the entire set of proteins previously identified by 2D electrophoresis and all but 26 of the 117 proteins identified by MudPIT. Importantly, GeLC-MS/MS, and to a lesser extent MudPIT, also sampled in a relatively unbiased manner low-abundance proteins, membrane proteins, high-molecular-weight proteins, and proteins with extreme isoelectric points (pIs).

Despite these improvements, it is clear that even relatively simple proteomes, such as that of *C. trachomatis*, generate challengingly complex mixtures of peptides when protein extracts from whole cells are digested with trypsin. Such mixtures present three major analytical problems. First, the complexity is such that the number of peptides eluting from current peptide separation systems into an online tandem mass spectrometer frequently exceeds the analytical capacity of the latter. This means that only a subset of the available peptides are analyzed during an experimental run. The challenge may be greater than is commonly appreciated as not all peptides ionize efficiently and hence are detected by the mass spectrometer. Thus, efforts to enhance peptide ionization (e.g., by the attachment of suitable chemical tags) may actually exacerbate the problem. A second (and related) major problem is that peptides from highly abundant proteins tend to saturate peptide separation media, thereby compromising the resolution and detection of their low-abundance counterparts. Third, the bioinformatics approaches currently available for matching peptides to specific proteins are pushed to their limits by complex mixtures and generate a significant number of false-positive identifications. This is particularly the case when the identification of a protein in a mixture is based on the detection of a single peptide. In part, the problem is due to the limitations of most tandem mass spectrometers, which are unable to dissociate peptide ions into complete fragment ion series and have imperfect mass accuracy and resolution. However, there remains scope for more rigorous statistical approaches to improve the assignment of peptides.

## 3.4 POTENTIAL STRATEGIES FOR IMPROVING PROTEOME COVERAGE

Given the formidable theoretical and practical challenges to achieving complete coverage of a proteome, what can be done to improve the situation? As indicated above, the fact that many microbes only express a limited portion of their proteome at any one time constitutes a major part of the difficulty. In principle, however, this can be circumvented by judicious choice of organism to be studied. For example, certain bacteria live in highly specialized niches and hence may be expected to express a larger proportion of their predicted proteomes than organisms that live in diverse environments. The obligate intracellular pathogen *C. trachomatis* is a good option in this respect as it appears to exist in only two forms—an inert but infectious extracellular form [the elementary body (EB)] and a metabolically active, replicating form [the reticulate body (RB)] (Ward, 1983). On infection of host cells, the primary differentiation process is initiated. This results in the conversion of the inactive EB form into the fully active RB form and the concomitant formation of an intracellular niche, known as an inclusion, in which the chlamydiae replicate. Importantly, at the midpoint in the infectious cycle (16–24 h postinfection), virtually every chromosomal and plasmid gene in *C. trachomatis* is transcribed (Belland et al., 2003). The remaining 28 genes, corresponding to ~3% of the genome, are transcribed at a late stage of the life cycle (40 h postinfection), when most of the RBs have developed into infectious EBs. These observations, together with the small size of the chlamydial genome (currently, 896 genes for serovar D), suggest that this organism is an excellent candidate for concerted efforts to achieve comprehensive proteome coverage.

Recent advances in proteomic technology should also enhance proteome coverage. While progress with 2D gel technology may now be subject to the law of diminishing returns, there have been significant recent improvements in other, predominantly nongel, approaches. By combining these approaches and deploying them strategically, it should be possible to achieve substantial improvements in proteome coverage. The first challenge in a "shotgun" proteomics project is to generate peptides that are amenable to MS analysis. Clearly, peptides need to be derived from as many proteins as possible if good proteomic coverage is to be achieved. Therefore, recent improvements in the efficiency of trypsin digestion of hitherto rather inaccessible proteins such as integral membrane proteins represent a significant advance (Blonder et al., 2004; Goshe et al., 2003). It is also important that such peptides are not too small or too large—ideally, they should be between 250 and 4000 *m/z* to fall within the analyzable mass range of most MS instruments—and they should have the right charge versus hydrophobicity characteristics (Pan et al., 2004). Recently, several groups have reported chemical derivitization strategies that either improve the separation characteristics of particular classes of peptides or improve their ionization (see, e.g., Brancia et al., 2001; Julka and Regnier, 2004; Pitteri et al., 2004).

Once suitable peptides have been produced, the next challenge is to separate them. Both GeLC-MS/MS and MudPIT approaches reduce complexity. In the first case, this is accomplished by only sampling the subsets of proteins present in gel slices. In contrast, MudPIT reduces the number of peptides entering the mass spectrometer by elution of subsets of peptides from a strong cation exchange column and subsequent further separation by reverse-phase chromatography. However, both approaches fail when highly abundant proteins are present in samples. To some extent, this problem is alleviated by repeating the analyses multiple times. This is because the initial runs tend to miss some of the peptides that coelute from a peptide separation system into the mass spectrometer. Eventually, however, the number of new peptides identified in this way dwindles even when MS software is written to exclude previously identified peptides from analysis. Accordingly, there remains a pressing need for better separation procedures. One promising strategy is the use of a peptide isolation procedure based on diagonal electrophoresis and diagonal chromatography. The combined fractional diagonal chromatography procedure, termed COFRADIC, led to the assignment of more than 800 proteins from *E. coli* and, importantly, can be modified to identify N-terminal peptides (Gevaert et al., 2002, 2003). Another promising approach is gas-phase fractionation of peptides in a tandem mass spectrometer (Utleg et al., 2003). In this procedure, mixtures of peptides eluting from an online microcapillary LC system are further separated in the gas phase by use of very narrow but overlapping *m/z* ranges. Data-dependent ion selection and fragmentation are then used to identify and assign the separated peptides. In principle, the combination of orthogonal procedures such as MudPIT, GeLC-MS/MS, or COFRADIC with gas-phase fractionation should result in very high resolution peptide separations, albeit at the cost of very long MS/MS runs.

Recent improvements in the accuracy (and affordability) of MS instrumentation are also helping to increase proteome coverage. For example, accurate mass measurements via a Fourier transform ion cyclotron resonance (FT-ICR) instrument coupled with precise measurement of peptide elution times after LC have been used both to confirm the peptide identification and to generate an accurate mass tag that uniquely defines the peptide of interest (Conrads et al., 2000; Page et al., 2004). In a dramatic illustration of the power of this approach, 1910 open reading frames in the ionizing radiation-resistant bacterium *Deinococcus radiodurans* were experimentally identified, corresponding to $\sim$61% of the predicted proteome (Lipton et al., 2002). While this study represents the current state of the art as far as microbial proteome coverage is concerned, it is highly probable that even better coverage will be achieved when the accurate mass tag approach is combined with improved methods for generating and separating peptides and/or ionizing them into the gas phase. Further, the recent development of hybrid FT-ICR instruments and improved software may reduce mass errors due to space charging, whereby overfilling of the ICR cell creates ion repulsion and consequent signal distortions (Bruce et al., 2000; Page et al.,

2004; Yates, 2004). Thus, FT-ICR approaches show considerable potential for even better proteome coverage in the near future.

Regardless of improvements in peptide production, separation, and assignment, it is inevitable that some proteins will prove refractory to identification via a high-throughput route due to their low abundance and/or physicochemical properties. There is therefore a need for additional highly sensitive proteomic methods that can be readily adapted to identify such proteins in a more focused and possibly more labor-intensive manner. The situation is analogous to genomic sequencing, where the fast, initial "shotgun" phase of data collection is followed by a slower gap closure and finishing phase. In principle, specific antibodies could be generated (e.g., using appropriate peptides as immunogens) and used to search for the missing proteins. However, this takes time (typically weeks to months) and they do not always have the desired selectivity. Additionally, they may not have a sufficiently high affinity for the protein of interest, thereby compromising the sensitivity of detection. For comprehensive proteome analysis, there is a need for a flexible, generic technique that is sufficiently sensitive to detect very low abundance proteins and that is able to discriminate between closely related proteins. Ideally, the method should also provide quantitative data on the level of expression of the protein.

Recently, Gerber and colleagues described a stable isotope dilution strategy that fulfills the above criteria and that has the potential to be applied to large-scale experiments. Their approach, termed AQUA (for the absolute quantification of proteins), uses [$^{13}$C]-labeled reference peptides and MS/MS to measure expression in terms of number of molecules per cell (Gerber et al., 2003). Peptides, corresponding to specific tryptic fragments of proteins of interest, are chemically synthesized with an incorporated [$^{13}$C] stable isotope (Fig. 3.1). Samples containing the protein of interest (e.g., slices excised from SDS-PAGE) are then

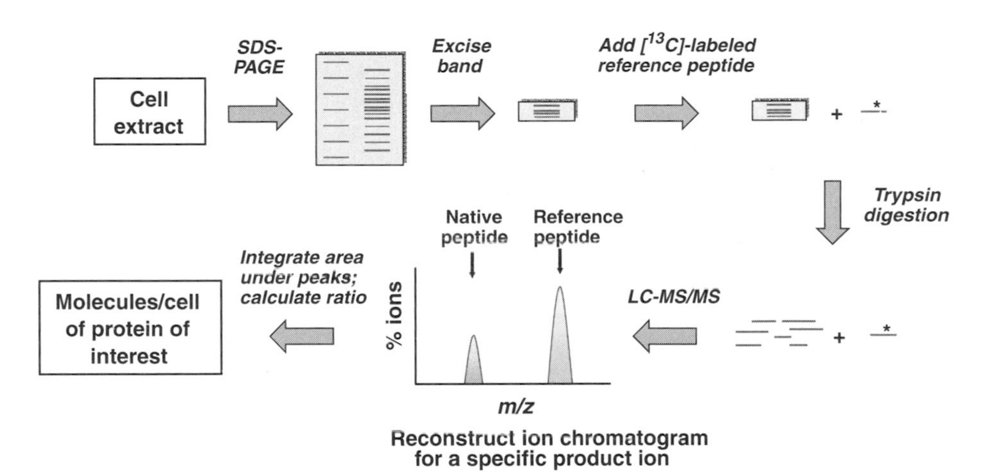


**Figure 3.1** Schematic representation of the AQUA strategy devised by Gerber and co-workers (2003). To quantify a protein of interest, it is necessary to know (i) its intact mass and (ii) the sequence of a peptide that can be reproducibly derived from it (e.g., by digestion with trypsin) and that is detectable by MS/MS. The peptide in question is chemically synthesized with an incorporated [$^{13}$C] stable isotope and added in known amounts to a gel slice containing the protein. Following digestion in situ with a site-specific protease such as trypsin, peptides are extracted from the slice and separated by nanocapillary LC. Online MS/MS is subsequently used to measure the level of the naturally occurring tryptic peptide from the protein of interest by comparison with the level of the corresponding internal standard.

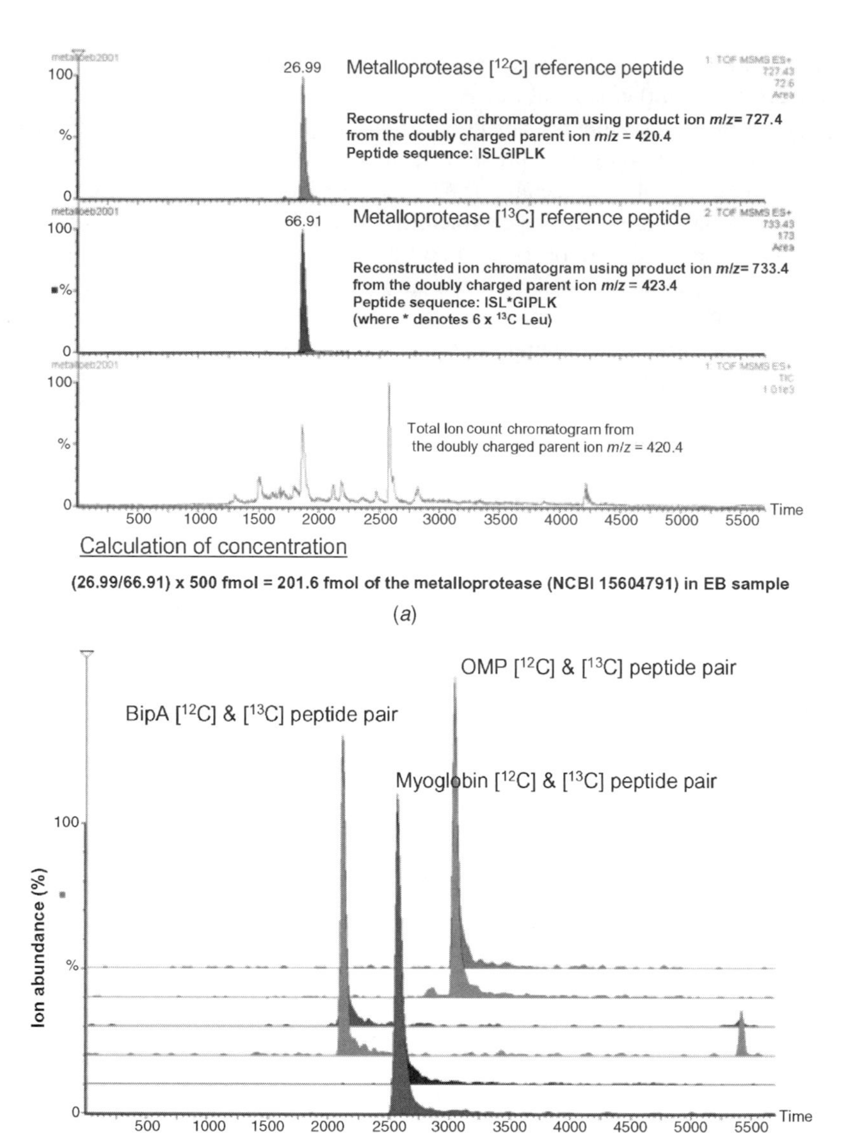


**Figure 3.2** (*a*) Quantification of hitherto hypothetical proteins by AQUA. The amounts of a previously hypothetical metalloprotease (NCBI 15604791) in EBs and RBs of *C. trachomatis* L2 were measured. The reference peptide used for quantification was ISL*GIPLK (where the asterisk denotes $6 \times [^{13}C]$-Leu). The first and second panels, showing the EB and RB results, respectively, indicate that 3.4-fold more of the protein (687.3 fmol) is present in the latter sample. (*b*) Multiplexing experiment. The figure above shows the results of an analysis where three gel slices and associated $[^{13}C]$ reference peptides were combined and processed for AQUA. Host cell extracts containing *C. trachomatis* L2 were spiked with known amounts of BipA protein and myoglobin prior to fractionation by SDS-PAGE. Product ions of the three peptide pairs [for the major outer membrane protein (OMP) of *C. trachomatis*, BipA protein and myoglobin] were monitored following collision-induced dissociation. Ion chromatograms for the product ions were constructed and the areas under the peaks were subsequently integrated using MassLynx 4 software.

subjected to digestion with excess trypsin, following the addition of the isotopically labeled peptide in known amount. Nanocapillary LC in conjunction with MS/MS is subsequently used to measure the level of the naturally occurring tryptic peptide from the protein of interest by comparison with the level of the corresponding internal standard. The AQUA method is precise and specific. For example, it can readily discriminate between phosphorylated and nonphosphorylated protein isoforms. It is also markedly more sensitive than many existing proteomic approaches; in the original paper, $< 20\ \mu g$ of HeLa cell lysates was required to quantify the protein separase and its phosphorylation state (Gerber et al., 2003). Recent studies have confirmed the sensitivity and precision of AQUA for studies with microbial proteomes (Owens et al., 2004). Moreover, it is now clear that it can be multiplexed so as to allow the simultaneous quantification of multiple components of interest (Fig. 3.2). Although currently used in conjunction with proteins separated by SDS-PAGE, it is probably best suited for use in nongel systems, with consequent further gains in quantification and sensitivity (Havlis and Shevchenko, 2004). The AQUA strategy therefore shows considerable potential for the "gap closure" stage in shotgun proteomics. Its application should result in markedly more comprehensive coverage of microbial proteomes.

## 3.5 CONCLUDING REMARKS

The goal of complete coverage of a microbial proteome is a highly attractive one, not least because such a complete data set would enormously facilitate efforts to model a living system. While the concept may be less tangible than originally thought, it remains a useful one insofar as it provides a powerful impetus for improving current proteomic technology and strategies. Moreover, the gap closure stage in shotgun proteomics has yet to be reached for any microbe as the detection and identification of genuine moderate- to low-abundance proteins remains the major bottleneck. In certain respects, such proteins resemble the "dark matter" postulated by astronomers—there is strong circumstantial evidence that they exist but they have yet to be observed directly.

Regardless of whether complete proteome coverage is ultimately a meaningful concept, recent advances in proteomic technology coupled with better strategies for protein identification, sensitivity, dynamic range, and throughput, promise to revolutionize our understanding of the number and type of components that make up a living microbe. This will not only enable the study of their systems biology and provide a test bed for still larger projects, but it will also have a significant impact on the development of better antimicrobial drugs and vaccines.

## ACKNOWLEDGMENTS

Studies in the authors' laboratories are supported by the BBSRC, Hope, the MRC, the National Institutes of Health (U.S.), and the Wellcome Trust.

## REFERENCES

Andersson, S. G. E., Zomorodipour, A., Andersson, J. O., Sicheritz-Ponten, T., Alsmark, U. C. M., Podowski, R. M., Naslund, A. K., Eriksson, A. S., Winkler, H. H., and Kurland, C. G. (1998).

The genome sequence of *Rickettsia prowazekii* and the origin of mitochondria. *Nature* **396**:133–140.

Belland, R. J., Zhong, G., Crane, D. D., Hogan, D., Sturdevant, D., Sharma, J., Beatty, W. L., and Caldwell, H. D. (2003). Genomic transcriptional profiling of the developmental cycle of *Chlamydia trachomatis*. *Proc. Natl. Acad. Sci. USA* **100**:8478–8483.

Blonder, J., Goshe, M. B., Xiao, W. Z., Camp, D. G., Wingerd, M., Davis, R. W., and Smith, R. D. (2004). Global analysis of the membrane subproteome of *Pseudomonas aeruginosa* using liquid chromatography-tandem mass. *J. Proteome Res.* **3**:434–444.

Brancia, F. L., Butt, A., Beynon, R. J., Hubbard, S. J., Gaskell, S. J., and Oliver, S. G. (2001). A combination of chemical derivatisation and improved bioinformatic tools optimises protein identification for proteomics. *Electrophoresis* **22**:552–559.

Bruce, J. E., Anderson, G. A., Brands, M. D., Pasa-Tolic, L., and Smith, R. D. (2000). Obtaining more accurate Fourier transform ion cyclotron resonance mass measurements without internal standards using multiply charged ions. *J. Am. Soc. Mass Spectrom.* **11**:416–421.

Chang, D. E., Smalley, D. J., Tucker, D. L., Leatham, M. P., Norris, W. E., Stevenson, S. J., Anderson, A. B., Grissom, J. E., Laux, D. C., Cohen, P. S., and Conway, T. (2004). Carbon nutrition of *Escherichia coli* in the mouse intestine. *Proc. Nat. Acad. Sci. USA* **101**:7427–7432.

Cole, S. T., Eiglmeier, K., Parkhill, J., James, K. D., Thomson, N. R., Wheeler, P. R., Honore, N., Garnier, T., Churcher, C., Harris, D., Mungall, K., Basham, D., Brown, D., Chillingworth, T., Connor, R., Davies, R. M., Devlin, K., Duthoy, S., Feltwell, T., Fraser, A., Hamlin, N., Holroyd, S., Hornsby, T., Jagels, K., Lacroix, C., Maclean, J., Moule, S., Murphy, L., Oliver, K., Quail, M. A., Rajandream, M. A., Rutherford, K. M., Rutter, S., Seeger, K., Simon, S., Simmonds, M., Skelton, J., Squares, R., Squares, S., Stevens, K., Taylor, K., Whitehead, S., Woodward, J. R., and Barrell, B. G. (2001). Massive gene decay in the leprosy bacillus. *Nature* **409**:1007–1011.

Conrads, T. P., Anderson, G. A., Veenstra, T. D., Pasa-Tolic, L., and Smith, R. D. (2000). Utility of accurate mass tags for proteome-wide protein identification. *Anal. Chem.* **72**:3349–3354.

Conway, T., and Schoolnik, G. K. (2003). Microarray expression profiling: Capturing a genome-wide portrait of the transcriptome. *Mol. Microbiol.* **47**:879–889.

Ellis, J. C., and Brown, J. W. (2003). Genes within genes within bacteria. *Trends Biochem. Sci.* **28**: 521–523.

Eymann, C., Homuth, G., Scharf, C., and Hecker, M. (2002). *Bacillus subtilism* functional genomics: Global characterization of the stringent response by proteome and transcriptome analysis. *J. Bacteriol.* **184**:2500–.

Farris, M., Grant, A., Richardson, T. B., and O'Connor, C. D. (1998). BipA: A tyrosine-phosphorylated GTPase that mediates interactions between enteropathogenic *Escherichia coli* (EPEC) amd epithelial cells. *Mol. Microbiol.* **28**:265–279.

Feltens, R., Gossringer, M., Willkomm, D. K., Urlaub, H., and Hartmann, R. K. (2003). An unusual mechanism of bacterial gene expression revealed for the RNase P protein of *Thermus* strains. *Proc. Nat. Acad. Sci. USA* **100**:5724–5729.

Fleischmann, R. D., Adams, M. D., White, O., Clayton, R. A., Kirkness, E. F., Kerlavage, A. R., Bult, C. J., Tomb, J. F., Dougherty, B. A., Merrick, J. M., Mckenney, K., Sutton, G., Fitzhugh, W., Fields, C., Gocayne, J. D., Scott, J., Shirley, R., Liu, L. I., Glodek, A., Kelley, J. M., Weidman, J. F., Phillips, C. A., Spriggs, T., Hedblom, E., Cotton, M. D., Utterback, T. R., Hanna, M. C., Nguyen, D. T., Saudek, D. M., Brandon, R. C., Fine, L. D., Fritchman, J. L., Fuhrmann, J. L., Geoghagen, N. S. M., Gnehm, C. L., McDonald, L. A., Small, K. V., Fraser, C. M., Smith, H. O., and Venter, J. C. (1995). Whole-genome random sequencing and assembly of *Haemophilus influenzae* Rd. *Science* **269**:496–512.

Gerber, S. A., Rush, J., Stemman, O., Kirschner, M. W., and Gygi, S. P. (2003). Absolute quantification of proteins and phosphoproteins from cell lysates by tandem MS. *Proc. Natl. Acad. Sci. USA* **100**:6940–6945.

Gevaert, K., Goethals, M., Martens, L., Van Damme, J., Staes, A., Thomas, G. R., and Vandekerckhove, J. (2003). Exploring proteomes and analyzing protein processing by mass spectrometric identification of sorted N-terminal peptides. *Nat. Biotechnol.* **21**:566–569.

Gevaert, K., Van Damme, J., Goethals, M., Thomas, G. R., Hoorelbeke, B., Demol, H., Martens, L., Puype, M., Staes, A., and Vandekerckhove, J. (2002). Chromatographic isolation of methionine-containing peptides for gel-free proteome analysis—Identification of more than 800 *Escherichia coli* proteins. *Mol. Cell.* Proteomics **1**:896–903.

Goshe, M. B., Blonder, J., and Smith, R. D. (2003). Affinity labeling of highly hydrophobic integral membrane proteins for proteome-wide analysis. *J. Proteome Res.* **2**:153–161.

Gygi, S. P., Corthals, G. L., Zhang, Y., Rochon, Y., and Aebersold, R. (2000). Evaluation of two-dimensional gel electrophoresis-based proteome analysis technology. *Proc. Natl. Acad. Sci. USA* **97**:9390–9395.

Havlis, J., and Shevchenko, A. (2004). Absolute quantification of proteins in solutions and in polyacrylamide gels by mass spectrometry. *Anal. Chem.* **76**:3029–3036.

Hommais, F., Krin, E., Laurent-Winter, C., Soutourina, O., Malpertuy, A., Le Caer, J. P., Danchin, A., and Bertin, P. (2001). Large-scale monitoring of pleiotropic regulation of gene expression by the prokaryotic nucleoid-associated protein, H-NS. *Mol. Microbiol.* **40**:20–36.

Julka, S., and Regnier, F. E. (2004). Benzoyl derivatization as a method to improve retention of hydrophilic peptides in tryptic peptide mapping. *Anal. Chem.* **76**:5799–5806.

Jungblut, P. R., Muller, E. C., Mattow, J., and Kaufmann, S. H. E. (2001). Proteomics reveals open reading frames in *Mycobacterium tuberculosis* H37Rv not predicted by genomics. *Infection and Immunity* **69**:5905–5907.

Kaeberlein, T., Lewis, K., and Epstein, S. S. (2002). Isolating "uncultivable" microorganisms in pure culture in a simulated natural environment. *Science* **296**:1127–1129.

Kolker, E., Purvine, S., Galperin, M. Y., Stolyar, S., Goodlett, D. R., Nesvizhskii, A. I., Keller, A., Xie, T., Eng, J. K., Yi, E., Hood, L., Picone, A. F., Cherny, T., Tjaden, B. C., Siegel, A. F., Reilly, T. J., Makarova, K. S., Palsson, B. O., and Smith, A. L. (2003). Initial proteome analysis of model microorganism *Haemophilus influenzae* strain Rd KW20. *J. Bacteriol.* **185**:4593–4602.

Krinos, C. M., Coyne, M. J., Weinacht, K. G., Tzianabos, A. O., Kasper, D. L., and Comstock, L. E. (2001). Extensive surface diversity of a commensal microorganism by multiple DNA inversions. *Nature* **414**:555–558.

Langen, H., Takacs, B., Evers, S., Berndt, P., Lahm, H. W., Wipf, B., Gray, C., and Fountoulakis, M. (2000). Two-dimensional map of the proteome of *Haemophilus influenzae. Electrophoresis* **21**:411–429.

Lawrence, J. G., Hendrix, R. W., and Casjens, S. (2001). Where are the pseudogenes in bacterial genomes? *Trends Microbiol.* **9**:535–540.

Link, A. J., Eng, J., Schieltz, D. M., Carmack, E., Mize, G. J., Morris, D. R., Garvik, B. M., and Yates, J. R. (1999). Direct analysis of protein complexes using mass spectrometry. *Nat. Biotechnol.* **17**:676–682.

Lipton, M. S., Pasa, T., Anderson, G. A., Anderson, D. J., Auberry, D. L., Battista, J. R., Daly, M. J., Fredrickson, J., Hixson, K. K., Kostandarithes, H., Masselon, C., Markillie, L. M., Moore, R. J., Romine, M. F., Shen, Y., Stritmatter, E., Tolic, N., Udseth, H. R., Venkateswaran, A., Wong, K. K., Zhao, R., and Smith, R. D. (2002). Global analysis of the *Deinococcus radiodurans* proteome by using accurate mass tags. *Proc. Natl. Acad. Sci. USA* **99**:11049–11054.

Martin, P., van de Ven, T., Mouchel, N., Jeffries, A. C., Hood, D. W., and Moxon, E. R. (2003). Experimentally revised repertoire of putative contingency loci in *Neisseria meningitidis* strain MC58: Evidence for a novel mechanism of phase variation. *Mol. Microbiol.* **50**:245–257.

Opiteck, G. J., Jorgenson, J. W., and Anderegg, R. J. (1997). Two-dimensional SEC/RPLC coupled to mass spectrometry for the analysis of peptides. *Anal. Chem.* **69**:2283–2291.

Owens, R. M., Pritchard, J. G., Skipp, P., Hodey, M., Connell, S. R., Nierhaus, K. H., and O'Connor, C. D. (2004). A dedicated translation factor controls the synthesis of the global regulator Fis. *EMBO J.* **23**:3375–3385.

Pace, N. R. (1997). A molecular view of microbial diversity and the biosphere. *Science* **276**:734–740.

Page, J. S., Masselon, C. D., and Smith, R. D. (2004). FTICR mass spectrometry for qualitative and quantitative bioanalyses. *Curr. Opin. Biotechnol.* **15**:3–11.

Pan, P., Gunawardena, H. P., Xia, Y., and McLuckey, S. A. (2004). Nanoelectrospray ionization of protein mixtures: Solution pH and protein pI. *Analy. Chem.* **76**:1165–1174.

Pitteri, S. J., Reid, G. E., and McLuckey, S. A. (2004). Affecting proton mobility in activated peptide and whole protein ions via lysine guanidination. *J. Proteome Res.* **3**:46–54.

Sanger, F., Coulson, A. R., Friedmann, T., Air, G. M., Barrell, B., Brown, N. L., Fiddes, J. C., Hutchison, C. A., Slocombe, P. M., and Smith, M. (1978). The nucleotide sequence of bacteriophage phiX174. *J. Mol. Biol.* **125**:225–246.

Saunders, N. J., Jeffries, A. C., Peden, P. F., Hood, D. W., Tettelin, H., Rappuoli, R., and Moxon, E. R. (2000). Repeat-associated phase variable genes in the complete genome sequence of *Neisseria meningitidis* strain MC58. *Mol. Microbiol.* **37**:207–215.

Schirle, M., Heurtier, A.-M., and Kuster, B. (2003). Profiling core proteomes of human cells by one-dimensional PAGE and liquid chromatography-tandem mass spectrometry. *Mol. Cell. Proteomics* **2.12**:1297–1305.

Skipp, P., Robinson, J., O'Connor, C. D., and Clarke, I. N. (2005). Shotgun proteomic analysis of *Chlamydia trachomatis*. *Proteomics* **5**:1558–1573.

Tao, H., Bausch, C., Richmond, C., Blattner, F. R., and Conway, T. (1999). Functional genomics: expression analysis of *Escherichia coli* growing on minimal and rich media. *J. Bacteriol.* **181**:6425–6440.

Taoka, M., Yamauchi, Y., Shinkawa, T., Kaji, H., Motohashi, W., Nakayama, H., Takahashi, N., and Isobe, T. (2004). Only a small subset of the horizontally transferred chromosomal genes in *Escherichia coli* are translated into proteins. *Mol. Cell. Proteomics* **3**:780–787.

Utleg, A. G., Yi, E. C., Xie, T., Shannon, P., White, J. T., Goodlett, D. R., Hood, L., and Lin, B. Y. (2003). Proteomic analysis of human prostasomes. *Prostate* **56**:150–161.

Wacker, M., Linton, D., Hitchen, P. G., Nita-Lazar, M., Haslam, S. M., North, S. J., Panico, M., Morris, H. R., Dell, A., Wren, B. W., and Aebi, M. (2002). N-linked glycosylation in *Campylobacter jejuni* and its functional transfer into *E. coli*. *Science* **298**:1790–1793.

Ward, M. E. (1983). Chlamydial classification, development and structure. *Br. Med. Bull.* **39**: 109–115.

Washburn, M. P., Wolters, D., and Yates, J. R. (2001). Large-scale analysis of the yeast proteome by multidimensional protein identification technology. *Nat. Biotechnol.* **19**:242–247.

Weinacht, K. G., Roche, H., Krinos, C. M., Coyne, M. J., Parkhill, J., and Comstock, L. E. (2004). Tyrosine site-specific recombinases mediate DNA inversions affecting the expression of outer surface proteins of *Bacteroides fragilis*. *Mol. Microbiol.* **53**:1319–1330.

Wolters, D. A., Washburn, M. P., and Yates, J. R. (2001). An automated multidimensional protein identification technology for shotgun proteomics. *Anal. Chem.* **73**:5683–5690.

Yates, J. R. (2004). Mass spectral analysis in proteomics. *Annu. Rev. of Biophys. Biomol. Struct.* **33**:297–316.

CHAPTER 4

# Proteome of *Mycoplasma pneumoniae*

RICHARD HERRMANN and THOMAS RUPPERT

ZMBH Universität Heidelberg, Heidelberg, Germany

## 4.1 INTRODUCTION

It is one of the major goals of molecular biology to know the functions of all genes encoded by a cell or organism. Obvious candidates for such comprehensive analyses are the classical prototypes of gram-negative and gram-positive bacteria, *Escherichia coli* and *Bacillus subtilis*. A huge amount of data on genetics, biochemistry, physiology, molecular biology, and cellular biology have been collected for both bacteria over the last 50 years, but despite all these achievements about one-third of the proposed 4288 genes of *E. coli* (Blattner et al., 1997) and 4100 genes of *B. subtilis* (Kunst et al., 1997) are functionally not assigned. Considering the tremendous efforts required for the functional analysis of a single gene, for the functional analysis of a complete cell we selected the relatively simple bacterium *Mycoplasma pneumoniae* M129, a human pathogenic bacterium causing upper respiratory diseases and an interstitial pneumonia (Jacobs, 1991; Taylor-Robinson, 1996; Waites and Talkington, 2004). It is characterized by a genome size of only 816 kb (Himmelreich et al., 1996), the lack of a cell wall, a polymorphic and sometimes flasklike cell shape (Fig. 4.1), a proteinous cytoskeleton-like structure (Krause and Balish, 2004; Meng and Pfister, 1980), motilitiy (Radestock and Bredt, 1977), and the requirement for cholesterol as a membrane component (Razin et al., 1998). The latest annotation predicted (Dandekar et al., 2000) 42 genes coding for RNAs only and 688 protein-encoding genes of which 458 were assigned to function. Therefore, 230 proteins/genes classified as hypothetical or conserved hypothetical are remaining for functional analyses. Before starting a comprehensive functional analysis it is important to confirm the open reading frame (ORF) predictions by expression data, since the functional analysis of a gene which is not expressed does not contribute useful information to the understanding of a cell. Gene expression can be monitored at the level of transcription (transcriptome) or translation (proteome). Both approaches have their merits. The microarray technology allows the identification of any RNA species synthesized in sufficient amounts in a cell independent of a genome annotation. In addition, species-specific RNA can be detected among a large excess of contaminating RNA, for instance, if the transcriptome of a bacterium infecting its host has to be established.

*Microbial Proteomics: Functional Biology of Whole Organisms*, Edited by Ian Humphery-Smith and Michael Hecker. 

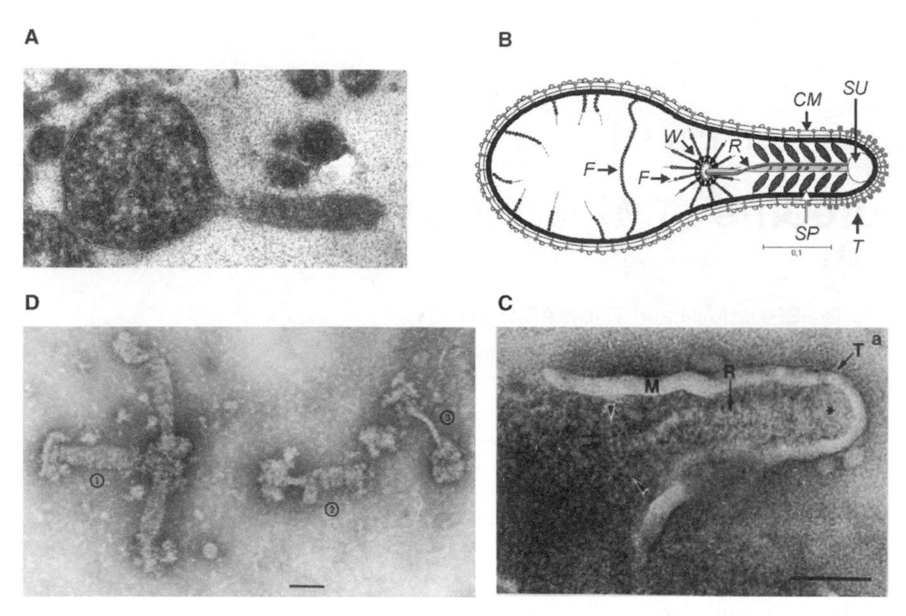


**Figure 4.1** *Mycoplasma pneumoniae*: structural organization. (A) Ultrathin section of *M. pneumoniae*. (B) Diagrammatic view of a longitudinal section through bacterium (CM, cytoplasmic membrane; F, fibrils; R, rod; SP, spokes; SU, undefined mass; T, tip; W, wheellike complex). (C) Cryo ultrathin section of untreated cell (T, tip region; R, rod; M, cytoplasmic membrane; arrowheads, row of protein subunits forming part of wheellike complex). (D) Examples of negative stained rods as seen in Triton X-100 insoluble fraction: (1) flat rods; (2) twisted rods; (3) rod seen edge-on. [(A) Courtesy of J. Hegeman. (B–D) From Hegemann et al., 2002, with permission.]

Although more difficult to achieve, the proteome analysis might be more important, since it represents the picture of the real concentration of the effective gene products in a cell. The standard proteome analysis, a combination of electrophoresis and mass spectrometry, is hampered by the necessity to separate complex protein extracts into individual proteins (spots) by two-dimensional (2D) gel electrophoresis before their identification by mass spectrometry. During this separation procedure, proteins with specific features like high isoelectric points or hydrophobic proteins were preferentially lost; therefore only a subset of a proteome is part of the analysis. Recently, this problem was solved to some extent by introducing a completely different approach in proteomics, that is, the multidimensional protein identification technology (MudPIT) for shotgun proteomics (Wolters et al., 2001) (Fig. 4.2). The extracted proteins are not separated by gel electrophoresis but directly digested by a protease, the latter process resulting in a very complex mixture of peptides. After digestion, the peptide mixture has to be separated before mass spectrometric analysis. Most commonly used is cation exchange liquid chromatography (IEX-LC) as a first dimension followed by reversed-phase chromatography (RP-LC) as a second dimension. These two orthogonal separation methods are used in an offline mode, where the two chromatographic steps are separated (Froehlich et al., 2003; Jaffe et al., 2004a; Peng et al., 2003) or in an online mode, in which the two separation columns are connected online (Mitulovic et al., 2004) or both separation matrices are within one column (Wolters et al., 2001). This method is not restricted by the limitations characteristic of the first dimension of 2D electrophoresis. Nonetheless, the solubility of membrane proteins still remains a major

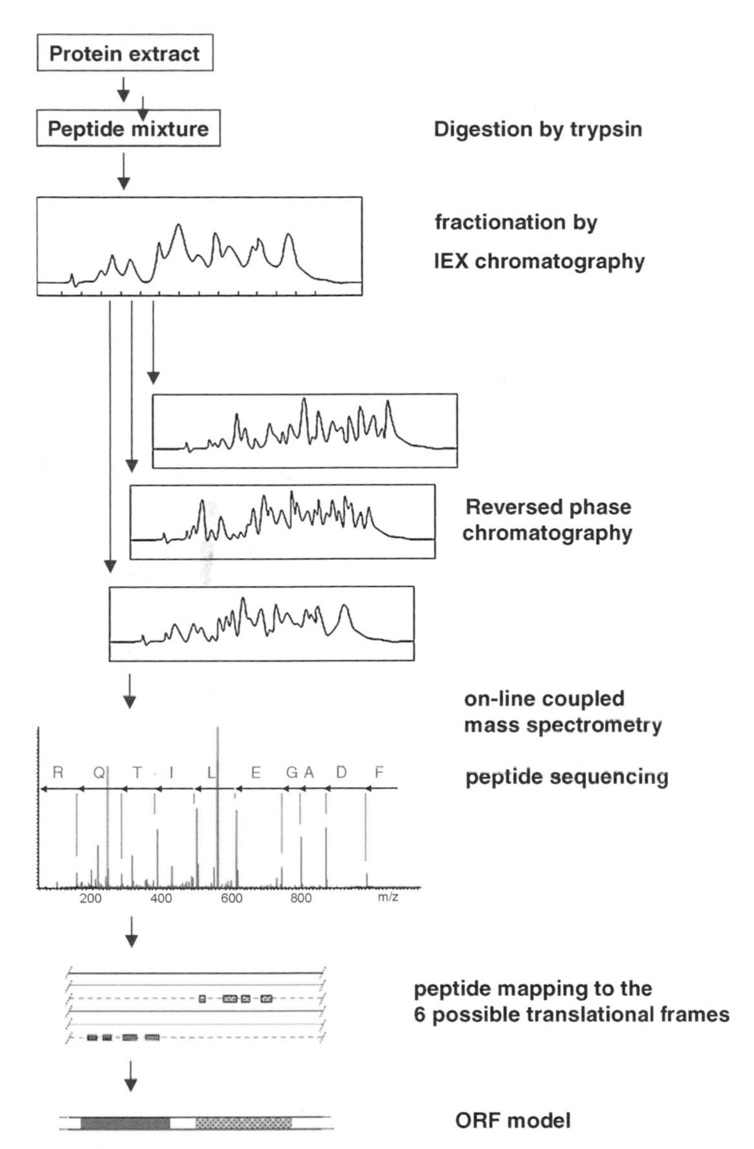


**Figure 4.2** Schematic drawing of proteogenomic mapping procedure as applied to *M. pneumoniae* (Jaffe et al., 2004a).

challenge for this nongel approach (Wu et al., 2003). However, there are more methods available to address this issue, such as the use of detergents (Han et al., 2001), organic solvents (Blonder et al., 2004), or organic acids (Washburn et al., 2001).

The data of two proteome analyses of *M. pneumoniae* are now available and permit a direct comparison of results obtained by these two different methods. In one study, a 2D proteome map was generated by the traditional 2D-gel-based method (Regula et al., 2000; Ueberle et al., 2002) and in another the multidimensional protein identification technology was applied (Jaffe et al., 2004a) for generating a "proteogenomic map" by aligning peptides to a genomic scaffold. Both methods were heavily dependent upon the nucleotide sequence of the complete genome.

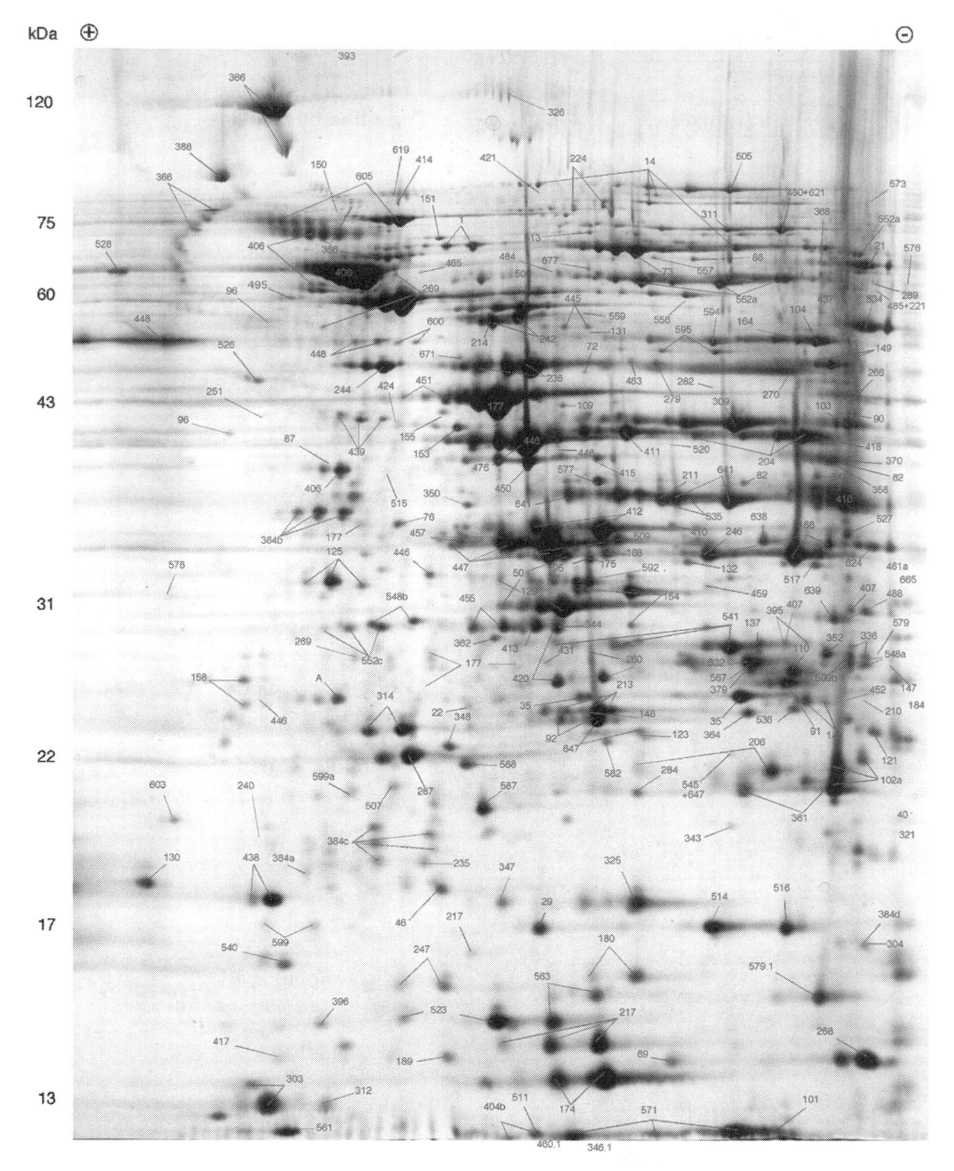


**Figure 4.3** Two-dimensional gel of soluble protein extract of *M. pneumoniae*. First dimension: immobilized pH gradient from 3 to 12; second dimension: vertical SDS polyacrylamide gel (12.5%). The gel was stained with silver. The identified spots were numbered according to original annotation (Himmelreich et al., 1996). The most prominent proteins are 406 (MPN 434, DnaK), 177 (MPN 665, EF-Tu), 446 (MPN 393, PdhA), 269 (MPN 573, GroEL), and 447 (MPN 392, PdhB). For details, see: *http://web.mpiib-berlin.mpg.de/cgi-bin/pdls/2d-page/extern/index.cgi*

## 4.2 TWO-DIMENSIONAL PROTEOME MAP OF *M. PNEUMONIAE*

The complete soluble protein extract of *M. pneumoniae* M129 was used as starting material for establishing a 2D proteome map. After separation by 2D gel electrophoresis and staining with colloidal Coomassie blue or silver staining, proteins from 225 genes were identified (Fig. 4.3). Since the concentration of individual

**TABLE 4.1 Selected Proteins of Triton X-100 Insoluble Fraction**

| MPN Number | Proposed Function/Annotation[a] |
|---|---|
| *Proposed Structural and Cytadherance-Associated Proteins* | |
| 452 | Cytadherence accessory protein (*HMW3*) |
| 309 | Protein P65 |
| 142 | Protein P40 and P90, cleavage products |
| 141 | Adhesin P1 |
| 567 | Protein P200 |
| 447 | Cytadherence accessory protein (*HMW1*) |
| 310 | Cytadherence accessory protein (*HMW2*) |
| *Unknown functions* | |
| 625 | Osmotic inducible protein-C-like family |
| 591 | Conserved hypothetical |
| 491 | MP-specific membrane nuclease |
| 474 | Hypothetical |
| 456 | Putative lipoprotein |
| 444 | Putative lipoprotein |
| 408 | Putative lipoprotein |
| 400 | Conserved hypothetical |
| 387 | Conserved hypothetical |
| 376 | Hypothetical |
| 323 | Probably Nrd1 |
| 314 | Conserved hypothetical |
| 297 | Conserved hypothetical |
| 295 | Hypothetical |
| 288 | Putative lipoprotein |
| 284 | Putative lipoprotein |
| 052 | Putative lipoprotein |

[a]Gene name is given in parentheses.

proteins in the soluble extract varied considerably and some proteins were just not sufficiently well expressed for mass spectrometric analysis, several enrichment procedures were tested with the aim of obtaining a larger quantity of under-represented proteins. Ion exchange chromatography, anion exchangers as well as cation exchangers, heat treatment by exposing a cytosolic protein extract to 85°C for 10 min, or heparin columns were used (Ueberle et al., 2002). In addition, complete bacteria were fractionated with Triton X-114 for enrichment of membrane and membrane-associated proteins and with Triton X-100 to get the Triton X-100-insoluble fraction, which was expected to contain components of the cytoskeleton-like structure (Regula et al., 2001).

In total, the number of proteins increased from 225 to 305 (Ueberle et al., 2002). Among these additional proteins were 5 proteins which were not predicted in the original annotation (Himmelreich et al., 1996). Three of these were probably excluded because of the cutoff point of 100 amino acids for ORFs without significant similarity to ORFs from other bacteria (MPN 377: 74aa; MPN 495: 80aa; MPN 272: 93aa; MPN 388: 142aa; MPN 254: 157aa).

The majority of the 80 proteins identified after enrichment procedures were separated on 1D sodium dodecyl sulfate (SDS) acrylamide gels only. The reduced complexity of the various subfractions made this possible. In some cases, specific subfractions were analyzed on 1D and 2D gels to gauge the comparative performance of methods of protein separation. The most convincing example for the limitations of 2D gels was the analysis of the enriched fraction of ribosomes. While all but 4 of the 52 ribosomal proteins were detected by 1D gels and mass spectrometry, only 12 ribosomal proteins could be identified from a 2D gel. This stresses that 2D electrophoresis is not well suited for separating proteins with isoelectric points above pH 10.5. Although some specialized laboratories achieve good separation of basic proteins, this general problem still exists (Wildgruber et al., 2002).

The other class of problematic proteins is membrane proteins. Specific programs (e.g., PRED-TMR2) predicted 55 proteins for *M. pneumoniae* with 5 or more (up to 14) transmembrane segments. Not a single one of these proteins could be identified on 2D gels, although proteins had been identified which build complexes or cooperate with membrane proteins. Typical examples were the ABC transporters. The expression of the characteristic adenosine triphosphate (ATP) binding proteins from several transporters could be shown (Dandekar et al., 2000; Himmelreich et al., 1996), but the predicted obligatory transmembrane proteins were not detected on 2D gels for oligopeptide or spermidine/putrescine ABC transporters. Another curiosity of the 2D gel analysis was the fragmented proteins. The gene products of 16 genes were only detected as fragments. Twelve of these genes coded for lipoproteins of the murein lipoprotein type of *E. coli* (Braun and Rehn, 1969), while the remaining 4 carried (proposed) transmembrane segments. The modified N-terminal fragments from the lipoproteins were never identified on 2D gels, indicating that the attached diacyl-glycerol moiety might prevent separation of not only the N-terminal fragment but also the complete protein in this gel-based system. To analyze the precursor–product relationship in more detail, the gene products of MPN 456 were separated by SDS polyacrylamide gel electrophoresis (SDS-PAGE) (Regula et al., 2000). An especially well suited example is MPN 456 because the protein starts with a putative lipoprotein signal sequence and was found only in four fragments. A rabbit antiserum generated against a fragment-specific polypeptide recognized only the specific fragment in 1D Western blots, but never the precursor protein. This suggested strongly that this lipoprotein was rapidly cleaved. It is still puzzling that 16 proteins, which are all probably surface exposed, appear only as fragments. The only lipoprotein so far found uncleaved was the subunit b of the $F_0F_1$-ATPase (Hilbert et al., 1996; Pyrowolakis et al., 1998). This was explained by the membrane topology of the subunit b. The precursor probably traverses the membrane twice—once by the signal peptide and again by a transmembrane segment located downstream of the processed N terminus ensuring that the C-terminal part of subunit b is oriented toward the cytosol (Pyrowolakis et al., 1998). It is still unresolved whether these fragmented proteins are the result of a *M. pneumoniae*–specific processing step or these fragments are artifacts caused during preparation of the cell extracts. The main argument against the "artifact theory" is the identification of many large, surface-exposed proteins as full-length products.

An important feature of a 2D gel is the visualization of the relative concentration of proteins. The precise determination of the concentration of individual proteins in total extracts of *M. pneumoniae* is rather difficult, because the standard method of labeling proteins with radioactive labeled amino acids of a well-defined specific activity is hampered by the lack of a defined minimal medium for mycoplasmas. Nevertheless, protein quantification in gels, either by determining the radioactivity of individual spots after incorporation of $^{14}C$-labeled amino acids or after staining by Coomassie blue or silver, always produced the same results for the most abundant proteins (Regula et al., 2000; Ueberle et al., 2002). By measuring the spot volume of individual proteins and comparing this to the spot volume of all proteins, it turned out that DnaK (MPN 434) and elongation factor Tu (EF-Tu, MPN 665) were by far the most abundant proteins followed by two subunits (α,β) of the pyruvate dehydrogenase complex (MPN 393, MPN 392) and heat-shock protein GroEL (MPN 573). The measurement of spot volumes is not very precise; for instance, the standardized spot volume for DnaK and EF-Tu varied between 9 and 18% (Fig. 4.3).

Nevertheless, the results from different staining methods or radioactive labeling were always comparable and the relative abundance of the spot volumes of the above-mentioned proteins remained unchanged. Removal of these few highly redundant proteins before running a 2D gel would allow more protein to be loaded onto the gel and to increase the number of detectable low-abundance proteins. Another possibility to identify low-copy-number proteins and to get some hints as to the function would be to enrich proteins in subfractions.

### 4.2.1 Triton X-100 Insoluble Fraction

The rationale for analyzing the components of the Triton X-100 insoluble fraction was based on the observation that *M. pneumoniae* has fiber- or rod like structures as seen by electron microscopy and that similar structures persist after treatment of *M. pneumoniae* with the detergent Triton X-100 (Biberfeld and Biberfeld, 1970; Göbel et al., 1981; Hegermann et al., 2002; Meng and Pfister, 1980; Regula et al., 2001; Stevens and Krause, 1991) (Fig. 4.1). In addition, indications from genetic and biochemical experiments concluded that *M. pneumoniae* possesses a cytoskeleton-like structure (Krause and Balish, 2004). Studies on the enrichment in the insoluble fraction of components of eukaryotic cytoskeletons from different cells that have been treated with Triton X-100 suggested that cytoskeleton-like structures of *M. pneumoniae* might also be concentrated in a Triton X-100 insoluble fraction. By silver staining of 2D gels, about 100 protein spots were visualized, while staining with colloidal Coomassie blue detected about 50 spots, of which 41 were identified by mass spectrometry. Among the known proteins, all the proposed structural cytoskeleton and cytoskeleton-associated proteins had been detected (Table 4.1). The most interesting set of proteins was the one encoded by genes not assigned to function, because these might be promising new candidates for proteins involved in the formation of the cytoskeleton-like structure (Table 4.1). The usefulness of this approach was shown by the recently reported finding that a mutation in MPN 387 caused loss of motility of *M. pneumoniae* (Hasselbring et al., 2004, 15th Congress of IOM, Athens, GA). Since motility should depend on an intact cytoskeleton, this protein could be a component of the cytoskeleton or a motility-specific protein interacting with it. It would be desirable to have mutants in all the genes encoding proteins of the Triton X-100 insoluble fraction. However, it remains a real handicap of *M. pneumoniae* as a model that specific mutants cannot be generated, for example, by homologous recombination. Consequently, one depends presently on transposon mutagenesis, which is unspecific and requires making every effort to isolate a specific mutant.

**TABLE 4.2 Functional categories of the National Center for Biotechnology Information (NCBI) Clusters of Orthologous Groups (COG) system**

| COG[a] | Proteins[b] | Description |
|---|---|---|
| 107 | 105 | Translation |
| 0 | 0 | RNA processing and modification |
| 18 | 17 | Transcription |
| 50 | 43 | Replication, recombination, and repair |
| 0 | 0 | Chromatin structure and dynamics |
| 25 | 19 | Cell cycle control, mitosis, and meiosis |
| 0 | 0 | Nuclear structure |
| 24 | 15 | Defense mechanisms |
| 6 | 6 | Signal transduction mechanisms |
| 13 | 12 | Cell wall/membrane biogenesis |
| 3 | 1 | Cell motility |
| 0 | 0 | Cytoskeleton |
| 0 | 0 | Extracellular structures |
| 9 | 8 | Intracellular trafficking and secretion |
| 21 | 21 | Posttranslational modification, protein turnover, chaperones |
| 22 | 22 | Energy production and conversion |
| 39 | 34 | Carbohydrate transport and metabolism |
| 29 | 21 | Amino acid transport and metabolism |
| 21 | 21 | Nucleotide transport and metabolism |
| 14 | 13 | Coenzyme transport and metabolism |
| 10 | 9 | Lipid transport and metabolism |
| 18 | 15 | Inorganic ion transport and metabolism |
| 0 | 0 | Secondary metabolites biosynthesis, transport and catabolism |
| 54 | 41 | General function prediction only |
| 16 | 16 | Function unknown |
| 190 | 126 | not in COGs |
| 689 | 565 | |

[a]The list was modified, so that each gene is only in one functional category
[b]Number of experimentally identified proteins

## 4.3 PROTEOGENOMIC MAPPING NOVEL APPROACH FOR WHOLE-PROTEOME ANALYSIS

Progress in mass spectrometry (MudPIT) technology and the development of improved software programs for the evaluation of data permitted a new approach for the analysis of the proteome of *M. pneumoniae* (Jaffe et al., 2004a). The authors named it *proteogenomic mapping*. The essential elements of these methods are shown in Figure 4.2. The crucial difference from the 2D gel approach is that first the total protein extract is digested with trypsin and then the complex peptide mixture separated into subfractions, first by strong cation exchange chromatography, then by reversed-phase chromatography. The reversed-phase column was directly interfaced to the mass spectrometer. The mass spectra were analyzed by SEQUEST and the peptide sequences mapped to the genome of *M. pneumoniae* M129, which had been translated in all six possible reading frames. In addition, the predicted ORFs were also displayed on this map. Therefore, it was straightforward to map the peptides identified and compare their location and reading frame to the ones predicted only from the nucleotide sequence.

This approach turned out to be much more sensitive and less biased than the 2D gel method. From 688 (Dandekar et al., 2000) predicted ORFs, the expression of 557 could be confirmed by 9709 unique corresponding peptides, providing an impressive 81% coverage. Further, 16 ORFs were newly proposed, of which 13 were shorter than 100 amino acids, while the N terminus was extended for 19 ORFs. The proteogenomic mapping confirmed 297 proteins and missed only 8 proteins that were identified for *M. pneumoniae* M128 (Ueberle et al., 2002). Combining the data from both analyses, 565 proteins from the 689 annotated genes were detected, bringing the proteome of *M. pneumoniae* close to completion. Most of the functionally assigned genes were detected (Jaffe et al., 2004a). There are several criteria for defining functional categories which all differ to some extent. To work with a comprehensible categorization, we used the classification in functional categories according to the NCBI COG system (http://www.ncbi.nlm.nih.gov/COG). The COG proteins were delineated by comparing protein sequences encoded by complete genomes. Therefore, genes listed in the COG system are very well conserved and widespread (Tatusov et al., 1997, 2003). The genes of *M. pneumoniae* were divided in 499 COG members, of which 439 (88%) were expressed, and 190 which are not in the COG system. From these, only 126 (66%) could be identified in proteome analyses (Table 4.3). These data support the notion of Jaffe et al. (2004a), using a slightly different functional classification, that genes designated as hypothetical or conserved hypothetical (Dandekar et al., 2000) were less frequently expressed. This result is not unexpected, since in the course of an annotation ORFs are defined by the rules established in molecular biology. If there are ORFs with no significant similarity to genes from other organisms, one has often to decide between several ORFs and these decisions might be wrong. Another possibility might be that some of these hypothetical genes are only expressed under very specific conditions.

The prediction on "not detected" ORFs (131), novel findings (16), N-terminal extensions (19), and deletions (6) have to be taken with some caution because the strain *M. pneumoniae* FH was used in this study but the genome of *M. pneumoniae* M129 was used as a genomic scaffold. These two strains are the prototypes for subtype 1 (*M. pneumoniae* M129) and subtype 2 (*M. pneumoniae* FH). This classification is mainly based of differences of the genes MPN 141 and 142, which code for the P1 protein, the main adhesin (MPN 141), and the proteins P40 and P90, cleavage products of a larger precursor coded for by MPN 142. The differences were caused by three repetitive DNA sequences, RepMP2/3, RepMP4, and RepMP5, of which one copy each is located in MPN 141 (RepMP2/3 and RepMP4) and in MPN 142 (RepMP5). In addition, 8–10 similar but not identical copies of each type of repetitive sequence are dispersed within the genome. A subtype switch could take place, at least on paper, by exchanging these copies. Multilocus sequencing analysis with selected genes (Dumke et al., 2003) and sequencing of about 200,000 bp of the genome of *M. pneumoniae* FH showed a high degree of sequence similarity to *M. pneumoniae* M129. For household genes the identity (amino acids) was about 98–100%. There were single mutations and also a few larger deletions in the FH strain (W. Reiser, unpublished), which eliminated, for example, MPN 137 and MPN 138, both of which were in the list of detected proteins but with very low sequence coverage. This shows that identifications of proteins with only one or two supporting peptides and a low percentage sequence coverage are questionable.

## 4.4 TWO-DIMENSIONAL ELECTROPHORESIS AND MUDPIT: COMPARISON

One aim of proteomics is the identification of all proteins expressed in a certain cell or organism. In this respect MudPIT offers clear advantages over 2D electrophoresis, mainly because there are no general limitations for certain protein classes. Proteins are digested at the beginning of a MudPIT analysis, giving rise to a high number of peptides with very different properties. In principle, identification of one of these peptides is sufficient to prove the expression of the protein. Therefore, a high percentage of the proposed ORFs of *M. pneumoniae* are identified at the protein level by this method. Nevertheless, not all proteins can be identified by this method, as exemplified for an extract of nuclear matrix proteins from Jurkat cells (Mitulovic et al., 2004); about 50% of the 94 proteins identified by 2D electrophoresis were not identified by MudPIT (174 proteins), despite the fact that the overall number of identified proteins was much lower. Identification of these proteins by MudPIT failed most likely because only a small fraction of detected peptide ions can be selected for peptide sequencing and the bias toward peptides of the more abundant proteins. This means that, in contrast to 2D electrophoresis, MudPIT is not limited by the amount of starting material or the detection range, but it is limited by the complexity of the peptide mixture itself, which leads in turn to the identification of abundant proteins by more and more peptides with decreased probability of sequencing peptides of less abundant proteins. Another disadvantage of MudPIT is that, for the statistical reasons described above, less abundant proteins are often identified by just a single peptide sequence, which causes two problems: (1) If the analysis is repeated, a number of these proteins will now be missed because that special peptide was by chance not selected for resequencing, whereas other proteins appear in the new protein list. Reproducible identification of such proteins from several analyses is therefore not high compared to 2D electrophoresis. (2) Protein identification by just one peptide sequence may lead to false-positive results because automated assignment of a measured tandem mass spectrometry (MS/MS) spectrum to a peptide sequence is based on the similarity to a theoretical MS/MS spectrum, but high similarity does not necessarily mean identity. True de novo sequencing is not possible simply because the number of such proteins is too high for a carefully manual interpretation of the MS/MS spectra. Therefore, the criteria for identification must be more stringent. One possibility is the prerequisite that there must be more than one peptide sequence to justify a significant score sufficient to identify a protein. In that case, however, many less abundant or small proteins may be missed.

Separation of proteins by 2D electrophoresis before identification selects against proteins with specific features such as high hydrophobicity and extreme isoelectric point or size (large and small). Additionally, less abundant proteins may be missed due to the limited protein-loading capacity of a gel. Protein separation, however, offers the possibility of separating modified and/or processed forms of the same protein. This is important for functional proteomics because alterations in the abundance of an activated form of a protein will have profound impact on cellular function even if the global abundance of this protein remains unchanged. This property of 2D electrophoresis can also be used to compare protein extracts of very similar but nevertheless different sources, such as strain FH and strain M129 of *M. pneumoniae*, because even very similar protein isoforms can be resolved. Furthermore, separated proteins can be stained and absolute quantification can be achieved to some extent after 2D electrophoresis using the intensity of the stained protein spots, whereas MudPIT offers no such possibility. Relative quantification, however, can be done by both systems in a very accurate way, namely, by the differential gel electrophoresis (DIGE) method combined with 2D electrophoresis as well as by stable isotope labeling combined with MudPIT (Gygi et al., 1999; Ong et al., 2003). Due to the very important advantages of these two methods, neither will replace the other. Both methods are complementary and should be used to overcome the shortcomings of the other.

## 4.5 POSTTRANSLATIONAL MODIFICATION

So far, four types of posttranslational modifications had been shown to take place in *M. pneumoniae*:

(i) Cleavage of signal peptides from putative exported/secreted proteins
(ii) Acylation of proteins combined with the cleavage of a signal peptide
(iii) Internal cleavage of proteins unrelated to a signal sequence
(iv) Phosphorylation of proteins

### 4.5.1 Cleavage of Signal Peptides

The final localization of a bacterial protein depends, among other factors, on the presence or absence of an N-terminal signal sequence. The typical features of a sequence for export or secretion comprise a positively charged amino terminus N-terminal end, a hydrophobic core region, and a neutral but polar and C-terminal cleavage domain (c-region). This c-region contains the recognition site for the enzyme signal peptidase I and, in the case of lipoproteins of the murein lipoprotein type of *E. coli,* the recognition site for signal peptidase II (Paetzel et al., 2002). The crucial difference is the obligatory cysteine, which will become the N-terminal amino acid in the mature lipoprotein after the signal peptide has been cleaved off. The processed protein remains associated with the membrane by the attachment of a diacyl-glycerol moiety to the $SH_2$ group of the cysteine before cleavage (Sankaran and Wu, 1994; Tokunaga et al., 1982).

Several programs exist which predict such signal peptides and their cleavage sites, for instance, signal P (Nielsen et al., 1997) or ExProt (Saleh et al., 2001). These programs were based and trained on experimentally determined signal sequences. Such experimental data are rare for mycoplasmas; therefore, the prediction should be taken with caution. In addition, a comparative analysis of (predicted) signal peptides from mycoplasmas, other gram-positive bacteria, and *E. coli* showed a different sequence pattern in signal peptides (Edman et al., 1999). The ExProt program predicted 248 exported/secreted proteins, including lipoproteins for *M. pneumoniae* (Saleh et al., 2001).

So far, the N termini of only two mature proteins with a predicted signal sequence from *M. pneumoniae* have been experimentally determined. The cleavage site of the precursor protein from MPN 142 was located between amino acid 25 and 26 (Catrein et al., 2002) and of the precursor protein of MPN 141 (Jacobs et al., 1987; Su et al., 1987) between amino acids 69 and 70.

Here, the experimentally determined cleavage site and that obtained by ExProt and signal P predicted cleavage site for MPN 142 were in agreement. However, there existed a clear discrepancy for MPN 141 where the cleavage was predicted to occur between amino acids, 27 and 28. The simplest explanation for this discrepancy would be an additional processing step, since a signal peptide of the 69 amino acid does not fit in any signal peptide pattern (Edman et al., 1999; Nielsen et al., 1997).

Most puzzling in this context is the absence of the conserved bacterial type I signal peptidase gene (SpaseI), which had not been found in annotations of the genome sequences of *M. pneumoniae* nor in the phylogenetically closely related *Mycoplasma genitalium* (Fraser et al., 1995). Since a signal peptidase I activity exists unambiguously in *M. pneumoniae*, a different protein has to take over this function. MPN 294 was proposed as a possible candidate based on several domains, which were in agreement with an intracellular protease (Dandekar et al., 2000). Another possibility would be that the signal peptidase II, which is normally lipoprotein specific, has a broader substrate range and possesses also signal peptidase I activity in *M. pneumoniae*. A signal peptidase II activity in *M. pneumoniae* was shown by processing the subunit b (AtpF) of the $F_0F_1$-type ATPase, namely via the inhibition of this process by the antibiotic globomycin (Pyrowolakis et al., 1998), which is a specific inhibitor of signal peptidase II (Inukai et al., 1978), and the labeling of the mature subunit b with $^{14}C$ palmitic acid. Altogether, the annotation of the genome sequence predicted 46 lipoproteins. The synthesis and processing of at least 20 of them could be proven by metabolic labeling of *M. pneumoniae* with $^{14}C$ palmitic acid and separation of the corresponding lipoproteins by SDS-PAGE (Pyrowolakis et al., 1998).

### 4.5.2 Cleavage of Proteins

Several gene products were only found as cleavage products. Most prominent is the example of MPN 142 (formally designated the ORF6 gene), which is involved in cytadherence. It codes for a protein of a molecular mass of about 130 kDa, but by SDS acrylamide gels and Western blotting only two subfragments were found with molecular masses of about 37 kDa (P40) and 85 kDa (P90) (Sperker et al., 1991). The precursor protein with a molecular mass of about 130 kD has not been unambiguously identified, but one has to consider that the classical pulse-labeling experiments with radioactively labeled amino acids, to prove such a precursor–product relationship, cannot be done properly in *M. pneumoniae* due to the lack of a defined minimal growth medium.

An unusual but well documented and analyzed family of lipoproteins also present in other *Mycoplasma* species has been best characterized in *Mycoplasma fermentans*. The surface-exposed lipoprotein MALP-404 appears in two versions, as a full-length protein with the N-terminal modified cysteine and as a small 14-amino-acid lipopeptide designated MALP-2. The latter corresponds to the N-terminal region of full-length mature MALP-404 lipoprotein (Calcutt et al., 1999). This lipoprotein MALP-2 exerts a distinct immunomodulatory activity (Muhlradt, 2002; Muhlradt et al., 1996, 1997). The protease cleaving the MALP-404 lipoprotein has not been detected. The orthologous protein of MALP-404 in *M. pneumoniae* is the lipoprotein MPN 052. Preliminary experiments indicated that *M. pneumoniae* also has a MALP-2 activity, but the MALP-2 peptide has not yet been isolated (P. F. Mühlradt, personal communication). The already described lipoproteins (see Section 4.2), which were only found as fragments, might also be the products of a specific internal cleavage, but so far there is no hint for an enzyme capable of doing this type of processing.

### 4.5.3 Phosphoproteins

Phosphorylation and deposphorylation of proteins are widespread posttranslational modifications which also occur in mycoplasmas. Although the almost ubiquitous two-component signal transduction system is absent in *M. pneumoniae*, other phosphorylation processes have been documented.

Recently a detailed study on phosphorylation and deposphorylation of the phosphocarrier protein HPr (MPN 053) by the serine kinase/phosphatase (MPN 223) has been reported. This kinase is the key regulator of carbon metabolism acting as a kinase or phosphotase depending on the growth conditions (Merzbacher et al., 2004). By analogy to the orthologous proteins in other bacteria, HPr was supposed to be phosphorylated at the serine in amino acid position 46 (Steinhauer et al., 2002). In the course of the proteogenomic mapping of *M. pneumoniae* (Jaffe et al., 2004a), this site of modification was confirmed by identification of a phosphopeptide containing this serine. Several years ago, the phosphorylation of the proteins HMW1 (MPN 447) and HMW2 (MPN 310) had been reported. HMW1 and HMW2 are cytadherence-associated proteins, which were phosophorylated at serine and threonine residues. This could be shown by labeling *M. pneumoniae* in vivo with $H_3\ ^{32}PO_4$ and in vitro (cell extracts) with [$\gamma$-$^{32}$P] ATP and by analyzing the protein profiles by SDS-PAGE and autoradiography (Dirksen et al., 1994; Krebes et al., 1995). Since this analysis was aiming to study only proteins with molecular masses above 90 kDa, only a part of the phosphoproteome could be seen on these gels. Nevertheless, it is obvious that even under these restricted conditions at least 10 different proteins were phosphorylated. So far, for none of these proteins have the function and consequences of phosphorylation/dephosphorylation been elucidated. In addition to these well-documented examples of phosphorylation, two meeting reports (U.B. Goebel, unpublished) (A. Borovsky and S. Rottem, unpublished) indicated that more proteins, including proteins of lower molecular masses, are phosphorylated. Of special interest is a 55-kDa protein (P55) which was present in *Mycoplasma penetrans, Mycoplasma gallisepticum*, and *M. pneumoniae* and which was suggested to be autophosphorylated. Applying the improved methods of mass spectrometry combined with enrichment procedures for phosphoproteins, an analysis of the complete phosphoproteome of *M. pneumoniae* is now feasible.

## 4.6 PROTEOME OF OTHER MOLLICUTES SPECIES

The pioneering work on proteome analyses of mollicutes was carried out by Humphery-Smith and his colleagues on *Spiroplasma melliferum* (Cordwell et al., 1997) and *M. genitaliium* (Cordwell et al., 1995). At that time complete genome sequences were not available and mass spectrometry was not as developed as today. Depending on cross-species identification and N-terminal sequencing of proteins, these analyses were very incomplete. A turning point was the publication of the genome sequence from *M. genitalium*, which was the second complete sequence to be published from a bacterial genome (Fraser et al., 1995). *Mycoplasma genitalium* was analyzed by 2D gel electroporesis and mass spectrometry from the logarithmic growth phase. From a total of 427 protein spots, 158 were identified which were encoded by 112 different genes (Wasinger et al., 2000), providing proof for the expression of 23% of the 480 proposed ORFs. Comparing this protein map with one established from bacteria of the stationary phase revealed that the number of proteins was lower and the intensity of spots relative to each other had changed. Recently, by a similar approach a proteome map of *M. penetrans* was established (Ferrer-Navarro et al., 2006) identifying the proteins from 153 genes of a genome encoding 1038 predicted ORFs (Sasaki et al., 2002).

New standards were set for genome sequencing projects by the publication of the genome of *Mycoplasma mobile* (Jaffe et al., 2004b). This is the first report of a genome sequence and its accompanying annotation being complemented by the proteogenomic mapping approach. This combined approach delivered previously unattained heights for the precision of ORF prediction. Unlike the case of *M. pneumoniae*, where the genome analysis and the proteogenomic mapping were done with two different although very similar strains, the identical *M. mobile* strain was used. Some 635 ORFs have been proposed for the 777-kb genome, of which 557 (88%) have been validated as expressed proteins. Interestingly, 26 genes were added to the DNA sequence based annotation through proteomics, thereby emphasizing the power of this combined approach. *Mycoplasma mobile* was isolated from a fish (Kirchhoff and Rosengarten, 1984). Its striking feature is the ability to glide on surfaces much faster than the other motile mycoplasmas: *M. pneumoniae, M. genitalium, M. galisepticum*, and *M. pulmonis* (Kirchhoff, 1992). Although the first genes involved in gliding motility have been identified for *M. mobile* (Uenoyama and Miyata, 2005) and *M. pneumoniae* (Hasselbring et al., 2005), the mechanisms in gliding have not been elucidated, but the genome sequences, the proteomes, and the successful generation of amotile mutants provide a promising basis for solving these problems.

So far, genomes of 12 species of the class Mollicutes have been sequenced, but the proteomes of only four species have been established. The new proteogenomic mapping approach should speed up the publication of new proteome analyses and provide very detailed information on gene expression of proposed hypothetical and conserved hypothetical genes.

## 4.7 FUTURE PERSPECTIVES

The next logical steps after genome sequencing and expression analyses are functional assignments of genes with unknown functions. The classical approach to generate mutants and to analyze the corresponding phenotype is hampered in *M. pneumoniae* by the lack of an efficient gene transfer system and the failure to get specific mutants by homologous recombination. A global transposon mutagenesis had been employed with *M. pneumoniae* (Hutchison et al., 1999), but this was only used to identify nonessential genes and did not lead to a collection of cloned mutants.

A second approach, known as structural genomics, is aiming to deduce functions of proteins from their three-dimensional folding patterns. For instance, the Berkeley Structural Genomic Center is conducting this approach with proteins of *M. genitalium* and *M. pneumoniae* (Kim et al., 2005). The first results of these efforts have been published recently. Based on the crystal structures of the proteins, it was proposed that MPN 625 has something to do with response to oxidative stress (Choi et al., 2003) and MPN 314 has a function in cell division (Chen et al., 2004). Although this approach is very promising, it will not be applicable to all proteins of *M. pneumoniae* (Grigoriev and Choi, 2002). Therefore, the genetic approach has to be pushed by developing tools for generating conditional mutants in *M. pneumoniae*. This would also permit the study of the function of essential genes and not only of nonessential genes as with transposon mutants.

## REFERENCES

Biberfeld, G, and Biberfeld, P. (1970). Ultrastructural features of *Mycoplasma pneumoniae*. *J. Bacteriol.* **102**:855–861.

Blattner, F. R., Plunkett, G., 3rd, Bloch, C. A., Perna, N. T., Burland, V., Riley, M., Collado-Vides, J., Glasner, J. D., Rode, C. K., Mayhew, G. F., Gregor, J., Davis, N. W., Kirkpatrick, H. A., Goeden, M. A., Rose, D. J., Mau, B., and Shao, Y. (1997). The complete genome sequence of *Escherichia coli* K-12. *Science* **277**:1453–1474.

Blonder, J., Conrads, T. P., Yu, L. R., Terunuma, A., Janini, G. M., Issaq, H. J., Vogel, J. C., and Veenstra, T. D. (2004). A detergent-and cyanogen bromide-free method for integral membrane proteomics: Application to *Halobacterium* purple membranes and the human epidermal membrane proteome. *Proteomics* **4**:31–45.

Braun, V., and Rehn, K. (1969). Chemical characterization, spacial distribution and function of a lipoprotein of the *E. coli* cell wall. *Eur. J. Biochem.* **10**:426–438.

Calcutt, M. J., Kim, M. F., Karpas, A. B., Muhlradt, P. F., and Wise, K. S. (1999). Differential posttranslational processing confers intraspecies variation of a major surface lipoprotein and a macrophage-activating lipopeptide of *Mycoplasma fermentans. Infect. Immun.* **67**: 760–771.

Catrein, I., Herrmann, R., Bosserhoff, A., and Ruppert, T. (in press). Absence of the conserved bacterial SPaseI gene but experimental proof for a signal peptidase I activity in *Mycoplasma pneumoniae. FEPS* **272**:2892–2900.

Chen, S., Jancrick, J., Yokota, H., Kim, R., and Kim, S. H. (2004). Crystal structure of a protein associated with cell division from *Mycoplasma pneumoniae* (GI: 13508053): A novel fold with a conserved sequence motif. *Proteins* **55**:785–791.

Choi, I. G., Shin, D. H., Brandsen, J., Jancarik, J., Busso, D., Yokota, H., Kim, R., and Kim, S. H. (2003). Crystal structure of a stress inducible protein from *Mycoplasma pneumoniae* at 2.85 Å resolution. *J. Struct. Funct. Genomics* **4**:31–34.

Cordwell, S. J., Basseal, D. J., and Humphery-Smith, I. (1997). Proteome analysis of *Spiroplasma melliferum* (A56) and protein characterisation across species boundaries. *Electrophoresis* **18**:1335–1346.

Cordwell, S. J., Wilkins, M. R., Cerpa-Poljak, A., Gooley, A. A., Duncan, M., Williams, K. L., and Humphery-Smith, I. (1995). Cross-species identification of proteins separated by two-dimensional gel electrophoresis using matrix-assisted laser desorption ionisation/time-of-flight mass spectrometry and amino acid composition. *Electrophoresis* **16**:438–443.

Dandekar, T., Huynen, M., Regula, J. T., Ueberle, B., Zimmermann, C. U., Andrade, M., Doerks, T., Sanchez-Pulido, L., Snel, B., Suyama, M., Yuan, Y. P., Herrmann, R., and Bork, P. (2000). Re-annotating the *Mycoplasma pneumoniae* genome sequence: Adding value, function and reading frames. *Nucleic Acids Res.* **28**:3278–3288.

Dirksen, L. B., Krebes, K. A., and Krause, D. C. (1994). Phosphorylation of cytadherence accessory proteins in *Mycoplasma pneumoniae. J. Bacteriol.* **176**:7499–7505.

Dumke, R., Catrein, I., Pirkl, E., Herrmann, R., and Jacobs, E. (2003). Subtyping of *Mycoplasma pneumoniae* isolates based on extended genome sequencing and on expression profiles. *Int. J. Med. Microbiol.* **292**:513–525.

Edman, M., Jarhede, T., Sjostrom, M., and Wieslander, A. (1999). Different sequence patterns in signal peptides from mycoplasmas, other gram-positive bacteria, and *Escherichia coli*: A multivariate data analysis. *Proteins* **35**:195–205.

Ferrer-Navarro, M., Gomez, A., Yanes, O., Planell, R., Aviles, F. X., Pinol, J., Perez Pons, J. A., and Querol, E. (2006). Proteome of the bacterium Mycoplasma penetrans. *J Proteome Res* **5**: 688–694.

Fraser, C. M., Gocayne, J. D., White, O., Adams, M. D., Clayton, R. A., Fleischmann, R. D., Bult, C. J., Kerlavage, A. R., Sutton, G., Kelley, J. M., Fritchman, R. D., Weidman, J. F., Small, K. V., Sandusky, M., Fuhrmann, J., Nguyen, D., Utterback, T. R., Saudek, D. M., Phillips, C. A., Merric, J. M., Tomb, J. F., Dougherty, B. A., Bott, K. F., Hu P. C., Lucier, T. S., Peterson, S. N., Smith, H. O., Hutchison, C. A. 3rd, and Venter, J. C. (1995). The minimal gene complement of *Mycoplasma genitalium. Science* **270**:397–403.

Froehlich, J. E., Wilkerson, C. G., Ray, W. K., McAndrew, R. S., Osteryoung, K. W., Gage, D. A., and Phinney, B. S. (2003). Proteomic study of the *Arabidopsis thaliana* chloroplastic envelope membrane utilizing alternatives to traditional two-dimensional electrophoresis. *J. Proteome Res.* **2**:413–425.

Göbel, U., Speth, V., and Bredt, W. (1981). Filamentous structures in adherent *Mycoplasma pneumoniae* cells treated with nonionic detergents. *J. Cell. Biol.* **91**:537–543.

Grigoriev, I. V., and Choi, I. G. (2002). Target selection for structural genomics: A single genome approach. *Omics* **6**:349–362.

Gygi, S. P., Rist, B., Gerber, S. A., Turecek, F., Gelb, M. H., and Aebersold, R. (1999). Quantitative analysis of complex protein mixtures using isotope-coded affinity tags. *Nat. Biotechnol.* **17**: 994–999.

Han, D. K., Eng, J., Zhou, H., and Aebersold, R. (2001). Quantitative profiling of differentiation-induced microsomal proteins using isotope-coded affinity tags and mass spectrometry. *Nat. Biotechnol.* **19**:946–951.

Hasselbring, B. M., Jordan, J. L., and Krause, D. C. (2005). Mutant analysis reveals a specific requirement for protein P30 in Mycoplasma pneumoniae gliding motility. *J Bacteriol* **187**:6281–6289.

Hegermann, J., Herrmann, R., and Mayer, F. (2002). Cytoskeletal elements in the bacterium *Mycoplasma pneumoniae. Naturwissenschaften* **89**:453–458.

Hilbert, H., Himmelreich, R., Plagens, H., and Herrmann, R. (1996). Sequence analysis of 56 kb from the genome of the bacterium *Mycoplasma pneumoniae* comprising the *dnaA* region, the *atp* operon and a cluster of ribosomal protein genes. *Nucleic Acids Res.* **24**: 628–639.

Himmelreich, R., Hilbert, H., Plagens, H., Pirkl, E., Li, B. C., and Herrmann, R. (1996). Complete sequence analysis of the genome of the bacterium *Mycoplasma pneumoniae. Nucleic Acids Res.* **24**:4420–4449.

Hutchison, C., Peterson, S., Gill, S., Cline, R., White, O., Fraser, C., Smith, H., and Venter, J. (1999). Global transposon mutagenesis and a minimal *Mycoplasma* genome. *Science* **286**:2165–2169.

Inukai, M., Takeuchi, M., Shimizu, K., and Arai, M. (1978). Mechanism of action of globomycin. *J. Antibiot. (Tokyo)* **31**:1203–1205.

Jacobs, E. (1991). *Mycoplasma pneumoniae* virulence factors and the immune response. *Rev. Med. Microbiol.* **2**:83–90.

Jacobs, E., Fuchte, K., and Bredt, W. (1987). Amino acid sequence and antigenicity of the amino-terminus of the 168 kDa adherence protein of *Mycoplasma pneumoniae. J. Gen. Microbiol.* **133(Pt. 8)**:2233–2236.

Jaffe, J. D., Berg, H. C., and Church, G. M. (2004a). Proteogenomic mapping as a complementary method to perform genome annotation. *Proteomics* **4**:59–77.

Jaffe, J. D., Stange-Thomann, N., Smith, C., DeCaprio, D., Fisher, S., Butler, J., Calvo, S., Elkins, T., FitzGerald, M. G., Hafez, N., Kodira, C. D., Major, J., Wang, S., Wilkinson, J., Nicol, R., Nusbaum, C., Birren, B., Berg, H. C., and Church, G. M. (2004b). The complete genome and proteome of *Mycoplasma mobile. Genome Res.* **14**:1447–1461.

Kim, S. H., Shin, D. H., Liu, J., Oganesyan, V., Chen, S., Xu, Q. S., Kim, J. S., Das, D., Schulze-Gahmen, U., Holbrook, S. R., Holbrook, E. L., Martinez, B. A., Oganesyan, N., DeGiovanni, A., Lou, Y., Henriquez, M., Huang, C., Jancarik, J., Pufan, R., Choi, I. G., Chandonia, J. M., Hou, J., Gold, B., Yokota, H., Brenner, S. E., Adams, P. D., and Kim, R. (2005). Structural genomics of minimal organisms and protein fold space. *J Struct Funct Genomics* **6**:63–70.

Kirchhoff, H. (1992). Motility. In J. Maniloff (Ed.), *Mycoplasmas. Molecular Biology and Pathogenesis.* American Society for Microbiology, Washington, D.C., pp. 289–306.

Kirchhoff, H., and Rosengarten, R. (1984). Isolation of a motile mycoplasma from fish. *J. Gen. Microbiol.* **130(Pt. 9)**:2439–2445.

Krause, D. C., and Balish, M. F. (2004). Cellular engineering in a minimal microbe: Structure and assembly of the terminal organelle of *Mycoplasma pneumoniae*. *Mol. Microbiol.* **51**:917–924.

Krebes, K. A., Dirksen, L. B., and Krause, D. C. (1995). Phosphorylation of *Mycoplasma pneumoniae* cytadherence-accessory proteins in cell extracts. *J. Bacteriol.* **177**:4571–4574.

Kunst, F., Ogasawara, N., Moszer, I., Albertini, A. M., Alloni, G., Azevedo, V., Bertero, M. G., Bessieres, P., Bolotin, A., Borchert, S., Borriss, R., Boursier, L., Brans, A., Braun, M., Brignell, S. C., Bron, S., Brouillet, S., Bruschi, C. V., Caldwell, B., Capuano, V., Carter, N. M., Choi, S. K., Codani, J. J., Connerton, I. F., Danchin, A., and et al. (1997). The complete genome sequence of the gram-positive bacterium *Bacillus subtilis* [see comments]. *Nature* **390**:249–256.

Meng, K. E., and Pfister, R. M. (1980). Intracellular structures of *Mycoplasma pneumoniae* revealed after membrane removal. *J. Bacteriol.* **144**:390–399.

Merzbacher, M., Detsch, C., Hillen, W., and Stulke, J. (2004). *Mycoplasma pneumoniae* HPr kinase/phosphorylase. *Eur. J. Biochem.* **271**:367–374.

Mitulovic, G., Stingl, C., Smoluch, M., Swart, R., Chervet, J. P., Steinmacher, I., Gerner, C., and Mechtler, K. (2004). Automated, on-line two-dimensional nano liquid chromatography tandem mass spectrometry for rapid analysis of complex protein digests. *Proteomics* **4**:2545–2557.

Muhlradt, P. F. (2002). Immunomodulation by mycoplasmas: Artifacts, facts and active molecules. In S. Razin and R. Herrmann (Eds.), *Molecular Biology and Pathogenicity of Mycoplasmas*, New York: Kluwer Academic/Plenum, pp. 445–472.

Muhlradt, P. F., Meyer, H., and Jansen, R. (1996). Identification of *S*-(2,3-dihydroxypropyl)cystein in a macrophage-activating lipopeptide from *Mycoplasma fermentans*. *Biochemistry* **35**:7781–7786.

Muhlradt, P. F., Kiess, M., Meyer, H., Sussmuth, R., and Jung, G. (1997). Isolation, structure elucidation, and synthesis of a macrophage stimulatory lipopeptide from *Mycoplasma fermentans* acting at picomolar concentration. *J. Exp. Med.* **185**:1951–1958.

Nielsen, H., Engelbrecht, J., Brunak, S., and von Heijne, G. (1997). Identification of prokaryotic and eukaryotic signal peptides and prediction of their cleavage sites. *Protein Eng.* **10**:1–6.

Ong, S. E., Kratchmarova, I., and Mann, M. (2003). Properties of 13C-substituted arginine in stable isotope labeling by amino acids in cell culture (SILAC). *J. Proteome Res.* **2**:173–181.

Paetzel, M., Karla, A., Strynadka, N. C., and Dalbey, R. E. (2002). Signal peptidases. *Chem. Rev.* **102**:4549–4580.

Peng, J., Elias, J. E., Thoreen, C. C., Licklider, L. J., and Gygi, S. P. (2003). Evaluation of multidimensional chromatography coupled with tandem mass spectrometry (LC/LC-MS/MS) for large-scale protein analysis: The yeast proteome. *J. Proteome Res.* **2**:43–50.

Pyrowolakis, G., Hofmann, D., and Herrmann, R. (1998). The subunit b of the F0F1-type ATPase of the bacterium *Mycoplasma pneumoniae* is a lipoprotein. *J. Biol. Chem.* **273**:24792–24796.

Radestock, U., and Bredt, W. (1977). Motility of *Mycoplasma pneumoniae*. *J. Bacteriol.* **129**:1495–1501.

Razin, S., Yogev, D., and Naot, Y. (1998). Molecular biology and pathogenicity of mycoplasmas. *Microbiol. Mol. Biol. Rev.* **62**:1094–1156.

Regula, J. T., Ueberle, B., Boguth, G., Görg, A., Schnölzer, M., Herrmann, R., and Frank, R. (2000). Towards a two-dimensional proteome map of *Mycoplasma pneumoniae*. *Electrophoresis* **21**:3765–3780.

Regula, J. T., Boguth, G., Görg, A., Hegermann, J., Mayer, F., Frank, R., and Herrmann, R. (2001). Defining the mycoplasma "cytoskeleton": The protein composition of the Triton X-100 insoluble fraction of the bacterium *Mycoplasma pneumoniae* determined by two-dimensional gel electrophoresis and mass spectrometry. *Microbiology* **147**:1045–1057.

Saleh, M. T., Fillon, M., Brennan, P. J., and Belisle, J. T. (2001). Identification of putative exported/secreted proteins in prokaryotic proteomes. *Gene* **269**:195–204.

Sankaran, K., and Wu, H. C. (1994). Lipid modification of bacterial prolipoprotein. Transfer of diacylglyceryl moiety from phosphatidylglycerol. *J. Biol. Chem.* **269**:19701–19706.

Sasaki, Y., Ishikawa, J., Yamashita, A., Oshima, K., Kenri, T., Furuya, K., Yoshino, C., Horino, A., Shiba, T., Sasaki, T., and Hattori, M. (2002). The complete genomic sequence of Mycoplasma penetrans, an intracellular bacterial pathogen in humans. *Nucleic Acids Res* **30**:5293–5300.

Sperker, B., Hu, P., and Herrmann, R. (1991). Identification of gene products of the P1 operon of *Mycoplasma pneumoniae*. *Mol. Microbiol.* **5**:299–306.

Steinhauer, K., Jepp, T., Hillen, W., and Stulke, J. (2002). A novel mode of control of *Mycoplasma pneumoniae* HPr kinase/phosphatase activity reflects its parasitic lifestyle. *Microbiology* **148**:3277–3284.

Stevens, M. K., and Krause, D. C. (1991). Localization of the *Mycoplasma pneumoniae* cytadherence-accessory proteins HMW1 and HMW4 in the cytoskeletonlike Triton shell. *J. Bacteriol.* **173**:1041–1050.

Su, C. J., Tryon, V. V., and Baseman, J. B. (1987). Cloning and sequence analysis of cytadhesin P1 gene from *Mycoplasma pneumoniae*. *Infect. Immun.* **55**:3023–3029.

Tatusov, R. L., Koonin, E. V., and Lipman, D. J. (1997). A genomic perspective on protein families. *Science* **278**:631–637.

Tatusov, R. L., Fedorova, N. D., Jackson, J. D., Jacobs, A. R., Kiryutin, B., Koonin, E. V., Krylov, D. M., Mazumder, R., Mekhedov, S. L., Nikolskaya, A. N., Rao, B. S., Smirnov, S., Sverdlov, A. V., Vasudevan, S., Wolf, Y. I., Yin, J. J., and Natale, D. A. (2003). The COG database: An updated version includes eukaryotes. *BMC Bioinformatics* **4**:41.

Taylor-Robinson, D. (1996). Infections due to species of *Mycoplasma* and *Ureaplasma*: An update. *Clin. Infect. Dis.* **23**:671–682; quiz 683–674.

Tokunaga, M., Tokunaga, H., and Wu, H. C. (1982). Post-translational modification and processing of *Escherichia coli* prolipoprotein in vitro. *Proc. Natl. Acad. Sci. USA* **79**:2255–2259.

Ueberle, B., Frank, R., and Herrmann, R. (2002). The proteome of the bacterium *Mycoplasma pneumoniae*: Comparing predicted open reading frames to identified gene products. *Proteomics* **2**:754–764.

Uenoyama, A., and Miyata, M. (2005). Identification of a 123-kilodalton protein (Gli123) involved in machinery for gliding motility of Mycoplasma mobile. *J Bacteriol* **187**:5578–5584.

Waites, K. B., and Talkington, D. F. (2004). *Mycoplasma pneumoniae* and its role as a human pathogen. *Clin. Microbiol. Rev.* **17**:697–728.

Washburn, M. P., Wolters, D., and Yates, J. R., 3rd (2001). Large-scale analysis of the yeast proteome by multidimensional protein identification technology. *Nat. Biotechnol.* **19**:242–247.

Wasinger, V. C., Pollack, J. D., and Humphery-Smith, I. (2000). The proteome of *Mycoplasma genitalium*. Chaps-soluble component. *Eur. J. Biochem.* **267**:1571–1582.

Wildgruber, R., Reil, G., Drews, O., Parlar, H., and Gorg, A. (2002). Web-based two-dimensional database of *Saccharomyces cerevisiae* proteins using immobilized pH gradients from pH 6 to pH 12 and matrix-assisted laser desorption/ionization time of flight mass spectrometry. *Proteomics* **2**:727–732.

Wolters, D. A., Washburn, M. P., and Yates, J. R., 3rd (2001). An automated multidimensional protein identification technology for shotgun proteomics. *Anal. Chem.* **73**:5683–5690.

Wu, C. C., MacCoss, M. J., Howell, K. E., and Yates, J. R., 3rd (2003). A method for the comprehensive proteomic analysis of membrane proteins. *Nat. Biotechnol.* **21**:532–538.

## CHAPTER 5

# Proteomics of Archaea

RICARDO CAVICCHIOLI, AMBER GOODCHILD*, and MARK RAFTERY

The University of New South Wales, Sydney, New South Wales, Australia

## 5.1 INTRODUCTION

Archaea represents one of the three primary domains of life, distinct from Bacteria and the Eucarya [1] (Fig. 5.1). While the ecology and physiology of archaea overlap with bacteria, at the genetic level, genes involved in information processing e.g., replication, transcription and translation) tend to be more similar to eukaryotes [2, 3]. Archaea also have a number of characteristics which distinguishes them from both bacteria and eucaryotes. They are unique in producing methane and in synthesizing a glycerol phosphate backbone for their phospholipids (G1P) that differs in stereospecificity from bacteria and eucaryotes (G3P) [4]. They also appear to be unique in not having pathogenic species (unlike bacteria and eucaryotes), despite their intimate association with human and other animal hosts [5]. Archaea are likely to have evolved early in biological history and therefore represent extant life forms with much to contribute to debates about the origin and evolution of life.

Archaea are often thought of as being extremophiles as they have been readily isolated from the environmentally sustainable limits of temperature, pH, and salt [6, 7]. Examples of their extreme abilities include an Fe(III)-reducing archaeon which was isolated from a hydrothermal vent and which grows well at 121°C in an autoclave [8] and *Picrophilus* spp. which have been isolated from volcanic soils in Japan and grow in 1.2 M $H_2SO_4$ with an optimal pH of 0.7 [9]. The capacity of the archaea to thrive in such extreme environments has led in recent years to an aggressive pursuit of extremophile ecology, biology, and biotechnology and fueled interest in astrobiology [10].

While it is clear that there are a seemingly disproportionate number of extremophiles that are members of the Archaea, ecological studies have proven that archaea are also abundant in nonextreme environments [11]. For example, in the oceans, archaea are estimated to be about one-third as abundant as bacteria, with numbers on the order of $10^{28}$ organisms [12]. Clearly archaea contribute a large amount of biomass to the world's biosphere in extreme and nonextreme environments.

*Present address: Johnson & Johnson Research, Eveleigh, New South Wales, Australia

*Microbial Proteomics: Functional Biology of Whole Organisms*, Edited by Ian Humphery-Smith and Michael Hecker. 

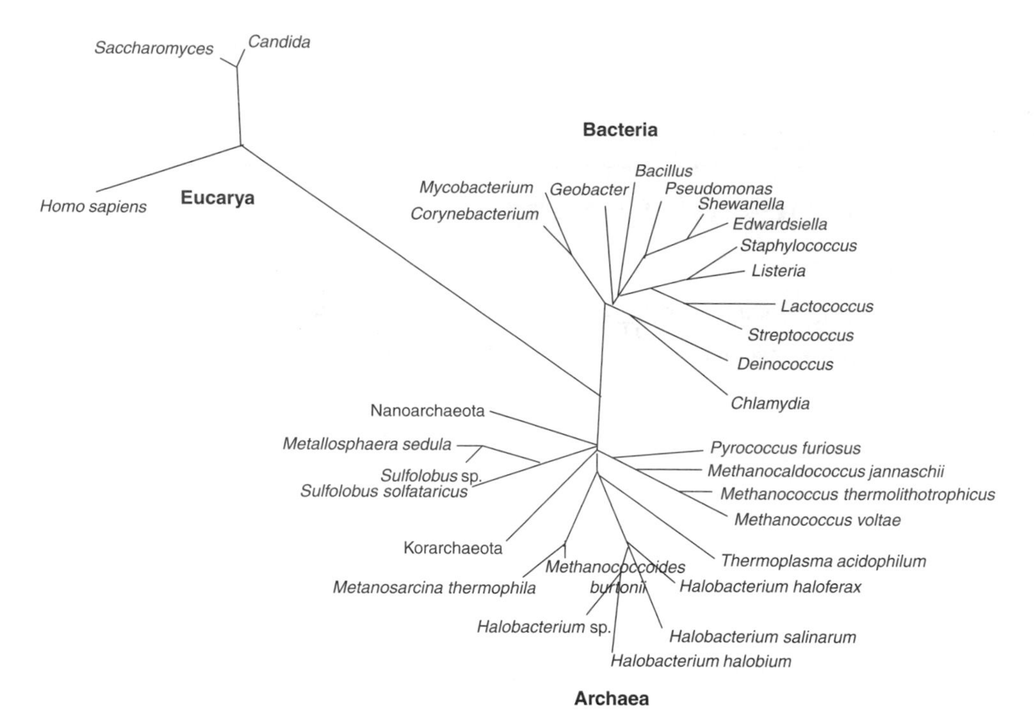


**Figure 5.1** Recombinant DNA (rDNA) phylogenetic tree of representatives of the three domains of life: Archaea, Bacteria, and Eucarya. The archaeal species used in proteomic studies described in this chapter are shown, and the bacteria which are represented in other chapters of the book are included. A phylogenetic tree was generated using ClustalW from 16S rDNA sequences obtained from GenBank for each archaea and bacteria and three 18S rDNA sequences for eucaryotic species. A representative of the archaeal kingdoms Nanoarchaeota and Korarchaeota are included. *Metallosphaera sedula, S. solfataricus*, and the *Sulfolobus* sp. are members of Crenarchaeota and the remaining archaea are members of Euryarchaeota.

Insight into the biology of archaea has rapidly advanced through the availability of genome sequences. A member of the Archaea, *Methanocaldococcus jannaschii*, was the third organism to have its genome completely sequenced [13]. Presently, 17 complete archaeal genome sequences and at least 27 draft genomes at various stages of completion have become available (http://www.genomesonline.org/). The availability of genome sequences underpins the ability to perform global functional studies, including proteomics. This is evidenced by the large number of in silico and expressed proteome studies which have included data from the genome sequence of *M. jannaschii*. Recently, the value of incomplete genome sequences for proteomics was highlighted in studies on the cold adaptation of an Antarctic archaeon, *Methanococcoides burtonii* [14].

The aim of this chapter is to provide an understanding of how proteomics has contributed to our understanding of archaea. While there is a significant scope for the expansion of the use of proteomics for studying archaea, reports of two-dimensional electrophoresis (2DE) in the literature date back to 1984 [15]. Since the success of contemporary proteomics has been driven by developments in technologies related to 2DE and liquid chromatography–mass spectrometry (LC-MS) methods, we have structured the chapter along these lines (1DE studies have not been addressed). Moreover, we separate studies that examine a snapshot of the expressed proteome from studies focused on differential expression. Posttranslational modifications are highlighted due to the importance they are likely to play in archaeal biology and to reflect the technical developments in this area. Studies combining proteomics with other functional genomics approaches are also highlighted. To balance the emphasis on technology-driven proteomics, Table 5.1 summarizes the archaeal proteomics literature according to organism. The phylogeny diagram (Fig. 5.1) also illustrates the relationship of all the archaea described in this chapter and places them on the tree relative to the other microbial species that are covered in this book.

## 5.2 PROTEOME SNAPSHOT

Proteome complexity for archaea may be considered on the basis of genome size and number of predicted open reading frames (ORFs). Based on the available genome sequences of archaea, theoretical proteomes would range in size from 552 proteins for *Nanoarchaeum symbiosum*, the organism with the smallest known genome of 490,885 bp [16], to 4528 for *Methanosarcina acetivorans*, the archaeon with the largest completely sequenced genome of 5,751,492 bp [17].

Expressed proteomes are however not a simple manifestation of their genetic capacity as they reflect the physiological state of the cell at the time of sampling. Proteomes are necessarily a product of gene expression, protein turnover, and protein stability, with the activity and function of all proteins being affected by translational and posttranslational modifications (collectively referred to as PTMs). The fact that one ORF can produce numerous forms of a protein particularly complicates archaeal 2DE profiles, where proteins matching a single ORF can be found at multiple positions of isoelectric point (pI) and molecular weight [14].

These features of archaeal proteomes are characteristics of microbial proteomes in general, and as a result, developments in proteomic technologies directly benefit archaeal proteomics. However, while proteomic technology has even been developed using archaeal biomass as the biological material, the field of archaeal proteomics is in its infancy. As a result, it is not clear what features of archaeal proteomics may be specific to archaea, and will require dedicated attention. For example, archaea possess metabolic pathways with novel cofactors which are unique to this domain of life (e.g., methanogenesis). Recently, the 22nd amino acid, pyrrolysine, was discovered in one of the methanogenesis enzymes, monomethylamine methyltransferase [18]. While pyrrolysine has subsequently been reported to be present in bacteria [19], the methyltransferases have been found to be highly abundant in methanogen proteomes [14, 20]. Moreover, their cellular levels are regulated by growth conditions (e.g., temperature) manifesting in multiple forms being observed by 2DE [14]. Despite the abundance of methyltransferases, pyrrolysine has not been able to be detected using MS methods [14, 18]. With the awareness and biological significance of this discovery, it will only be a matter of time before methods, such as FTMS, are developed to be able to detect pyrrolysine. However, it illustrates the potentially unique discoveries which may be realized through archaeal proteomics and the need to carefully consider the unique character of archaeal proteomes.

### 5.2.1 2DE Studies

***Methanogens*** The majority of proteomics studies performed on archaea (Table 5.1) have been on methanogens, in particular on *M. jannaschii*. This is due to the inherent

**Table 5.1 Proteomic Studies Performed on Archaea**

| Organism | Snapshot vs. Differential | 2DE vs. LC-MS | Main Outcomes | Posttranslational Modification | Reference |
|---|---|---|---|---|---|
| *Halobacterium* | Snapshot | 2DE | Halophilic archaea exhibit a general stress response | Yes | Daniels et al., 1984 |
| *Halobacterium halobium* | | 2DE | Structural modifications of flagellin proteins | Yes | Giometti et al., 2001 |
| | Snapshot | LC-MS | Identified >40 membrane proteins with 1–16 TM domains | | Blonder et al., 2004 |
| *Halobacterium salinarium* | Snapshot | 2DE | Modified sample preparation and electrophoresis for acidic proteins | | Cho et al., 2003 |
| *Haloferax volcanii* | Differential | 2DE | Proteins involved in salt adaptation | | Mojica et al., 1997 |
| *Metallosphaera sedula* | Differential | 2DE | Overexpression of a 66-kDa protein with increasing growth temperature | | Han et al., 1997 |
| *Methanocaldococcus jannaschii* | Differential | 2DE | Hydrogen partial pressure controls synthesis of flagella and chemotaxis regulates motility | | Mukhopadhyay et al., 2000 |
| | Differential | 2DE | Hydrogen partial pressure and ammonium concentration control flagella synthesis | Yes | Giometti et al., 2001 |
| | Snapshot | 2DE | Identified 170 proteins of which 34% have no known function | Yes | Giometti et al., 2002 |
| | Snapshot | 2DE | MALDI-TOF and MS/MS were equally useful for the identification of 100 proteins | Yes | Lim et al., 2003 |

| | | | | | |
|---|---|---|---|---|---|
| | Snapshot | LC-MS | Identified 72 proteins with 100% sequence coverage and mapped PTMs using FTMS | Yes | Forbes et al., 2004 |
| *Methanococcoides butonii* | Differential | 2DE | Identified 33 proteins involved in thermal adaptation | Yes | Goodchild et al., 2004; |
| *Methanococcus thermolitho autotrophicus* | Differential | 2DE | Expression of basic proteins dependent on growth pressure | | Jaenicke et al., 1988 |
| *Methanosarcina thermophila* | Differential | 2DE | Large numbers of proteins are specific to growth substrate | Yes | Jablonski et al., 1990; |
| | Differential | 2DE | Multiple homologues of proteins associated with substrate utilization | | Ding et al., 2002 |
| *Pyrococcus furiosus* | Differential | 2DE | MALDI-TOF and MS/MS were equally successful in the identification of 62 protein spots | | Lim et al., 2003 |
| *Sulfolobus* | Differential | 2DE | Three proteins likely to be involved in thermotolerance | Yes | Trent et al., 1990 |
| *Sulfolobus acidocaldaricus* | Differential | 2DE | Detected HSPs, phosphorylated proteins, and the presence of a two component regulatory system | Yes | Osorio and Jerez, 1990 |
| *Thermoplasma acidophilum* | Snapshot | 2DE | Archaeal proteasomes similar in size to eukaryotic proteasomes, indicating common ancestry | Yes | Zwickl et al., 1992 |
| | Snapshot | 2DE | Archaea contain ubiquitin | | Wolf et al., 1993 |

*Note*: MALDI-TDF, matrix-assisted laser desorption ionization time of flight; MS/MS, tandem mas spectrometry; PTM, posttranslational modification; FTMS, Fourier transform mass spectrometry.

stability of *M. jannaschii* proteins, resulting from it being a hyperthermophile, and the fact that it was the first archaeon to have its genome completely sequenced. As early as 1997 the *M. jannaschii* genome sequence was used, in a purely bioinformatic study, for testing algorithm design for cross-species matching [21].

Giometti et al. [20] identified 170 abundant cytosolic proteins by employing a range of separation techniques coupled with MALDI-TOF and LC-MS/MS. The total number of proteins detected was increased by independently using Coomassie and silver. While silver stain is clearly recognized for providing more sensitive protein detection, in this study a number of proteins were detected by one method but not by the other. In one case, an abundant S-layer protein was detected using Coomassie, but not by silver.

The study also highlighted the large number of hypothetical proteins which are not only encoded by the genome sequence (34%) but are expressed during growth (32%), some of which represented very abundant proteins. Little emphasis was placed on the biological implications of the identified proteins, although it was noted that 21% of the proteins were predicted to be involved in energy metabolism, 12% in protein synthesis, and the remainder in other cellular processes such as cell division and cell structure.

In a separate study, a comparison of MALDI-TOF peptide mass mapping and μLC–electrospray ionization (ESI) MS/MS was performed on 2DE separated proteins from *M. jannaschii* and *Pyrococcus furiosus* [22]. By optimizing protein preparation and database search parameters, the identities of proteins were determined from 100 spots from *M. jannaschii* and 62 spots from *P. furiosus*, with a success rate of essentially 100% for both methods. An important fundamental finding of this study was that up to seven proteins could be identified from a single 2DE spot using LC-MS/MS. This clearly has important implications for the use of 2DE for determining differential expression levels of proteins based on spot intensities.

Although the less abundant spots were not analyzed in these two studies [20, 22], the findings illustrate there are no inherent impediments in high-throughput analyses of archaeal proteomes. Moreover, the development of 2DE reference sites such as GELBANK [23] ensures the data will remain accessible and usable.

Most theoretical archaeal proteomes have a bimodal distribution of pIs [24, 25]. While 2DE has been effectively used for resolving most proteins in the neutral–acidic range, methods have required development for the analysis of basic proteins. Nonequilibrium pH gradient electrophoresis and LC-MS were used to successfully analyze basic proteins from *M. jannaschii* with pIs between 5.5 and 9.7 [20]. Immobilized pH gradient (IPG) electrophoresis has also been used to resolve around 70 spots with a pI above 7 from *M. burtonii* (Goodchild et al., unpublished results).

***Halophiles*** In contrast to nonhalophilic members of the Archaea, *Halobacterium* has a highly acidic proteome with no basic peak [26]. An acidic proteome has evolved in halophiles due to the need for a high positive surface charge of proteins that is required to counterbalance the high intracellular salt (e.g., 3–5 M potassium). The presence of high concentrations of salts interferes with first-dimension separation of proteins. Protocols for sample preparation and electrophoresis of proteins from the halophilic archaeon *Halobacterium salinarum* have been specifically developed for 2DE [27]. It is however noteworthy that 2DE has been used to examine the heat-shock response in *Halobacterium* [15] and osmotically induced proteins from *Haloferax volcanii* [28] using more standard techniques.

***Thermoplasma*** Research on various archaeal subproteomes has occurred since the early 1990s. A good example of this involves the 2DE analysis of the proteasome of

*Thermoplasma acidophilum* [29]. The study demonstrated that members of the Archaea express a proteasome that is essentially identical in size and shape to a eucaryotic proteasome, despite the fact that it only contains 2 subunits compared with 15–20 eucaryotic subunits. The $\alpha$ subunit was found to have PTMs involving phosphorylation and/or glycosylation. A regulatory role was assigned to the $\alpha$ subunit and a catalytic role to the $\beta$ subunit. The amino acid sequences of the archaeal $\alpha$ and $\beta$ subunits and the eucaryotic subunits are in fact sufficiently similar to invoke a single common ancestor gene.

The presence of ubiquitin in the *T. acidophilum* proteome provides a further link between archaeal and eucaryotic proteasome systems [30]. In eucaryotes, ubiquitin plays a role in directing proteolysis of misfolded proteins. To identify ubiquitin in *T. acidophilum*, Wolf et al. [30] trialed the use of a commercial antibody to bovine ubiquitin in a 2DE Western blot. When this proved unsuccessful, a brute-force approach was taken using N-terminal Edman sequencing of proteins below a molecular weight of 10 kDa and between a pI of 5 and 8. The archaeal ubiquitin was identified as a match to eucaryotic ubiquitin with only one amino acid difference in the first 19 amino acid residues. At the time, this study was the first to demonstrate that, unlike bacteria, Archaea contained ubiquitin homologues.

### 5.2.2 LC-MS Studies

Nano-LC-MS methods involve the separation of peptides using LC linked to MS/MS. Fragmentation spectra of protonated peptide ions are normally characteristic of the precursor's amino acid sequence, and proteins may be identified from this information using database search programs (e.g., Mascot or SEQEST) or manually with de novo sequencing strategies. For gel bands, gel spots, or relatively simple mixtures of proteins, one-demensional (1D) LC is sufficient to separate peptides for MS/MS analysis. When the sample contains complex mixtures of proteins (e.g., total cell lysates), separation and dynamic range may be improved using 2D liquid chromatography (LC/LC), sometimes referred to as "shotgun proteomics" [31]. Dynamic ranges of three to four orders of magnitude have been reported, indicating less abundant proteins and peptides may be identified in the presence of abundant proteins [32]. We have found that up to 10,000 individual MS/MS spectra can be obtained from a single LC/LC experiment and that in excess of 500 proteins may be identified from *M. burtonii* lysates using these methods (Goodchild et al., unpublished results). Proteins identified include those that are traditionally difficult to analyze with 2D SDS/PAGE, such as membrane proteins.

A novel approach identifying PTMs using ESI FTMS has recently been described [33, 34]. The approach, called "top-down MS," uses FTMS to isolate protein ions and fragment these using a novel ion activation technique called electron capture dissociation (ECD). Amino acid sequences may be obtained after ECD of proteins, and from subsequent database searches proteins and potential PTMs may be determined. Forbes et al. [34] used these methods with proteins from *M. jannaschii*. A sample enriched for ribosomes contained 26 proteins, including 8 proteins not predicted to be in the ribosome, and no PTMs were observed apart from truncation of the start Met [34].

***Membrane Proteins*** The ability to use proteomics to identify, characterize, and quantify membrane proteins is presently challenging [35]. However, due to the abundance of predicted membrane proteins in all forms of life [36] and their wide-ranging and important biological roles, there is a compelling reason to develop appropriate methods. A large number of proteins with predicted transmembrane domains were identified in a high-throughput proteome analysis of *M. burtonii*, several of which were hypothetical proteins (Goodchild et al., unpublished). This highlights the importance of being able to detect and analyze membrane proteins in archaea and the need to develop accompanying in silico methods for inferring their function.

Recently a simple and rapid method was developed to prepare membrane proteins without using detergents using a 60% methanol extraction and analysis with μLC-MS/MS [37]. This method contrasts with a detergent-based procedure that has also been developed for similar proteins [38].

The non-detergent-based method was developed using commercially available purple membranes from *Halobacterium halobium*, and the general usefulness of the method was evaluated using human epidermal tissue from newborn foreskins [37]. The purple membranes are present as distinct patches in lipid membranes of archaeal halophiles, similar to lipid rafts found in eucaryotic cells [39, 40]. Blonder et al. [37] identified all the tryptic peptides from bacteriorhodopsin, the light-driven proton pump known to be present in purple membranes. Forty additional proteins were identified, including 32 which were predicted to contain between 1 and 16 transmembrane domains, of which 8 were hypothetical proteins. Proteins involved in photo- and aerotactic signal transduction responses were identified. In addition, proteins involved in the Sec pathway which mediates protein transport across the lipid membrane were identified. While Sec proteins have been identified in all domains of life, little is known about protein translocation out of the cytoplasm in archaea [41, 42]. This study provides methods which may find general application in the proteomic analysis of membrane proteins and provides the specific foundation for differential expression studies of purple membrane proteins in *H. halobium*.

## 5.3 DIFFERENTIAL EXPRESSION

One of the key uses of proteomics is to examine changes in gene expression that are linked to specific growth and survival conditions. This is particularly relevant for probing mechanisms of adaptation in extremophiles. Reflecting the course of technology development, 2DE has been the method primarily used for differential expression studies in archaea. It is likely however that the rapid developments in quantitative LC-MS/MS techniques, such as isotope-coded affinity tagging (ICAT), will be adopted as a complementary approach to 2DE.

### 5.3.1 2DE studies

***Responses to Temperature*** Archaea are able to grow at temperatures ranging from around 0°C to above 121°C [8, 43]. In some environments archaea are exposed to perennially stable temperatures; however, in others they may be exposed to dramatic and sudden changes in temperature. A sudden cold shock may occur at a hydrothermal vent where temperatures may vary from greater than 350°C at the mouth of the vent to 1–2°C within a few meters from the vent. Conversely, archaea exposed to desert conditions in hypersaline pools may be subjected to a severe heat shock during peak heat conditions of the day.

Several studies have used proteomics to examine the heat-shock response in several genera of archaea. In 1984, Daniels et al. [15] provided the first demonstration that the heat-shock response occurred in seven strains of *Halobacteria*, thereby establishing the ubiquity of the heat-shock response in all domains of life. They probably also provided the first indication of PTM in archaea by observing heat-shock proteins with the same molecular weight and marginal differences in pI in 2DE. Another stress that is relevant to halophiles in their natural environment is salt dilution (e.g., after rain), and Daniels et al. [15] reported that the salt dilution stress increased the abundance of two protein spots that also appeared after heat-shock, thereby demonstrating a general stress response in archaeal halophiles.

In *Sulfolobus*, thermotolerance was shown to be acquired when cells were heat-shocked from 70°C to a sublethal temperature of 88°C, prior to exposure to the lethal temperature of 92°C [44]. Using 2DE and pulse labeling with $^{35}$S-methionine, it was discovered that only three proteins (55, 35, and 28 kDa) could be detected during the heat-shock (88°C) preincubation. The 55-kDa protein was by far the major protein, which appeared as a train of spots on gels differing only in pI, and therefore indicative of PTM.

In a separate study on *Sulfolobus acidocaldarius*, $^{35}$S-methionine/cysteine was used to detect heat-shock proteins, and $^{32}$P-phosphate was used to detect phosphorylated proteins [45]. This proteomic study clearly demonstrated the occurrence of a specific PTM (phosphorylation) in archaea. The study also had implications for phosphorylation pathways in archaea. The phosphorylation of proteins was reminiscent of bacterial responses to phosphate limitation, and the mechanism was confirmed by identifying homologues of the *Escherichia coli* genes *phoB* and *phoR* in the chromosome of *S. acidocaldarius*. The study provided the first experimental evidence that two-component regulatory systems appeared to be involved in signal transduction pathways in Crenarchaeota.

Similar heat-shock experiments were performed on a different member of the thermoacidophilic Crenarchaeota, *Metallosphaera sedula*, which grows optimally at 79°C and pH 2 [46]. Cells in continuous culture were subjected to abrupt or gradual increases in temperature and the ability of the cells to grow correlated with 2DE profiles. The main finding was the overexpression of numerous isomeric forms of a 66-kDa protein. This was shown by Western blot to be the chaperonin protein [46], analogous to the 55-kDa heat-shock protein expressed in *Sulfolobus* [44].

For organisms growing in thermally stable environments it is likely that exposure to temperatures even moderately outside what they naturally encounter will be stressful. Goodchild et al. [14] compared 2DE profiles for 4 and 23°C ($T_{opt}$) grown cultures of *M. burtonii*. Forty-three differentially expressed proteins were identified using μLC-ESI-MS/MS. Cold adaptation was linked to changes in fundamental cellular processes, including metabolism, transcription, and protein folding, with key roles identified for the pyrrolysine-containing trimethylamine methyltransferase, RNA polymerase subunit E, a response regulator from a two-component regulatory system, and peptidyl prolyl cis/trans isomerase. It was striking that increased levels of the heat-shock protein DnaK were observed during growth at 23°C. This indicates that growth of cold-adapted organisms at apparently optimum temperatures is stressful. This is an important finding that addresses a misconception that psychrophiles are not well adapted to their environmental temperatures because they tend to grow faster at temperatures exceeding the natural environment [47].

***Methanogens*** The methanogenic archaea are a metabolically diverse, strictly anaerobic group of organisms that can grow in environments ranging from Antarctic temperatures close to 0°C (e.g., *M. burtonii*) to hydrothermal vent temperatures up to 110°C (e.g., *Methanopyrus kandleri*). A variety of substrates can be used by methanogens for energy production, including acetate, methyl-containing compounds, and $H_2 + CO_2$. Decades of research have been devoted to understanding the ecology, physiology, biochemistry, and genetics of methanogenesis [48], in which 2DE has played a role since the late 1980s.

Many members of the Archaea live in the ocean under combinations of high or low pressure and temperature. *Methanococcus thermolithoautotrophicus* was isolated from a near-shore submarine hydrothermal vent [49], and its growth rate is enhanced by hydrostatic pressure up to 50 MPa but inhibited by higher pressure [50]. Two-dimensional electrophoresis was used to compare logarithmic- and stationary-phase cultures of cells grown at atmospheric pressure and 50 MPa [51]. No changes in protein profiles were detected after silver staining. To radiolabel the cells, $^{14}$C-sodium carbonate was used as a carbon source (while growth with $^{14}$C-actetate, -formate, -sulfate, or -cysteine was not found to be suitable). While the overall characteristics of the protein profiles for radiolabeled cells were found to be similar, a number of changes in basic proteins were identified in the molecular weight range 38–70 kDa. Analysis of differential expression was performed by manual inspection of X-ray film autoradiograms.

An elegant example of 2DE analysis reported in 1990 was a study performed on *Methanosarcina thermophila* to evaluate gene expression regulated by growth on methanol versus acetate [52]. In a time-honored process of manual inspection (no 2DE analysis programs available), up to 454 spots were identified on silver-stained gels with around 140 of those specific to the carbon source used. Many examples of charge trains were observed, indicative of PTM. In the absence of genome data and modern MS methods of analysis, spots corresponding to acetate kinase, phosphotransacetylase, and the five subunits of carbon monoxide dehydrogenase were identified by comparing 2DE profiles with those from each of the pure enzymes.

With the availability of genome sequence data for *M. thermophila* becoming available in the late 1990s, in a subsequent study, N-terminal Edman sequencing was used to identify proteins from 17 spots that were regulated by growth on the two substrates [53]. All three homologues of the two subunits of the methanol-specific methyltransferase I and one of the homologues of methyltransferase II were present during growth on methanol. In addition, one homologue of each of the methanol-specific methyltransferase I subunits was detected during growth on acetate. Likewise, two of the homologues of the trimethylamine-(TMA-) specific methyltransferase I (MttC-2 and MttC-3) were expressed when TMA was the growth substrate and one of these TMA-specific homologues (MttC-1) was expressed when methanol was the substrate for growth. The expression of multiple proteins that are not linked to the metabolism of that substrate would appear to be a wasteful process for the cell. However, it would enable *M. thermophila* to rapidly utilize the substrate when it becomes available or when the carbon flux changes sufficiently in the environment.

Two-dimensional electrophoresis has been used to examine the ability of archaea to respond to substrate availability that causes energy limitation. *M. jannaschii* is a strictly hydrogenotrophic methanogen that was isolated from a deep-sea hydrothermal vent where hydrogen is its only energy source [54]. The hydrogen partial pressure at sea vents is often low and also highly variable [55]. Moreover, hydrogenotrophic methanogenesis is a relatively poor energy-yielding pathway [56]. Using 2DE performed on cells grown at ecologically relevant cell densities and hydrogen partial pressures, Mukhopadhyay et al. [57] showed that the hydrogen partial pressure controlled the synthesis of flagellin proteins. Prior to this study it was suggested that *M. jannaschii* was thermotactic but probably not chemotactic [58]. The proteomics study provided the first evidence that *M. jannaschii* used chemical signals to regulate genes involved in motility. This was particularly striking as genes known to be involved in chemotaxis in bacteria had not been identified in the *M. jannaschii* genome, indicating that this archaeon possessed an entirely novel system of chemotaxis. Subsequent studies by others confirmed this finding and further demonstrated that growth phase and ammonium concentration also regulated flagellin proteins [59].

### 5.3.2 LC-MS Studies

Isotope-coded affinity tagging and MS/MS may be used in combination to selectively identify and quantify proteins that are differentially expressed from cells in two different states [60]. The ICAT reagents comprise a thiol-reactive group, a linker that occurs in isotopically heavy ($^{13}C_9$-labeled) or isotopically normal (unlabeled) form and a biotin group. Reduced Cys within proteins in the two samples to be compared are labeled with either the heavy or normal reagent, then combined, digested with trypsin, and the tagged peptides isolated by avidin affinity chromatography. The enriched peptides are separated by nano-LC and analyzed by MS. The relative abundances of each peptide and its amino acid sequence are determined and proteins identified by database searching. A difference in the ratio of peptide masses indicates changes in protein expression. Our initial comparison of protein expression of *M. burtonii* at 4 and 23°C with ICAT reagents indicates approximately 15% of the proteins identified had changes in protein expression (Goodchild et al., unpublished results).

## 5.4 POSTTRANSLATIONAL MODIFICATIONS

The presence of proteins with PTM can been inferred from most 2DE studies of archaeal proteomes (Table 5.1), and reference to these has been extensively made in the preceding sections of this chapter. The evidence from 2DE is based on the presence of the same protein from multiple spots on 2D gels. A good example of PTM occurring is for the flagella proteins from *H. halobium*, *Methanococcus voltae*, and *M. jannaschii*, which are encoded by multiple genes in each organism [59]. Expression of the flagellin genes appears to involve signal sequence processing, glycosylation, and other, as-yet-undefined form(s) of PTM [59]. In *M. jannaschii*, multiple forms of the flagellin proteins FlaB1, FlaB2, FlaB3, FlaD, and FlaE were identified based on differences between predicted and observed molecular weight and pI. This includes growth condition regulation of flagellin subunits that resulted in oligomeric forms remaining intact under denaturing conditions of electrophoresis [59]. Regulation of PTM appears to be a complex process dependent on growth conditions.

Recently, FTMS was used to detect methylation, acetylation, disulfide bonds, and N-terminal processing in proteins from *M. jannaschii* [34]. The ability to quantitate the level of methylation (50%) was demonstrated for protein MJ0556. This protein is predicted to encode a domain involved in donating methyl groups to a wide variety of substrates [34] and may therefore play a role in regulating its own methylation. The ability to effectively detect PTMs was also instrumental in establishing the lack of PTM. Similar to eucaryotes, histones are expressed in archaea; however, unlike archaea, there is no evidence that histones are posttranslationally modified. Forbes et al. [34] were able to show that the histones expressed in *M. jannaschii* were not acetylated. They extended this approach to fractionate proteins from *M. acetivorans* and confirm that they also lacked detectable PTMs. The implication of this study is that, unlike eucaryotes, archaea do not appear to modify histones in order to package their DNA or promote transcription.

## 5.5 COMBINING PROTEOMICS WITH OTHER ASSESSMENTS OF GENE FUNCTION

A more comprehensive view of the biology of cold adaptation of *M. burtonii* was achieved by determining the mRNA levels for all genes identified by proteomics and performing specific enzyme assays to compare to protein and mRNA levels [14]. The mRNA abundance for glutamate dehydrogenase (GDH) and glyceraldehyde-3-phosphate dehydrogenase (GAPDH) genes was compared to the protein abundance and enzyme activities. At 4°C, GDH had 5.9-, 2.0-, and 2.1-fold higher levels of protein, enzyme activity, and mRNA, respectively. For GDH, 2.4- and 1.8-fold higher levels of protein and enzyme activity were observed, with no change in mRNA levels. It was concluded that cold regulation of GDH activity appeared to involve protein and mRNA levels, whereas control of GAPDH activity appeared to be exclusively at the protein level.

The value of combining transcript and proteomic data was further highlighted in this study by examining proteins that appeared in multiple spots on 2D gels [14]. For example, the methyl-CoM-reductase α, β, and γ-subunits appeared in rows and clusters of spots, some of which changed as a function of growth temperature. The β and γ subunits were in spots with both increased and decreased intensities at 4°C, while the α subunit was in a spot that increased at 4°C. The mRNA abundance was approximately threefold higher for the α,-β-, and γ-subunit genes. This provided a good indication that the expression and activity of methyl-CoM-reductase were higher during growth at 4°C. Moreover, the equivalent levels of mRNA abundance for the three methyl-CoM-reductase genes and their organization on the genome in a five-gene cluster indicated the genes were likely to be of an operon structure. Integrating bioinformatics with proteomics and mRNA analysis helped to determine that thermal regulation in this archaeon was achieved through complex gene expression events involving gene clusters and operons through to protein modifications.

## 5.6. CONCLUDING REMARKS

The future looks bright for archaeal proteomics. The platforms established through 2DE and LC-MS highlight the opportunities for using proteomics for targeted and high-throughput analyses. Importantly, studies to date have been for a broad range of archaea, including methanogens, extreme halophiles, and thermoacidophiles (Table 5.1; Fig. 5.1). Moreover, the inherent biochemical properties of proteins from the hyperthermophilic archaea make them good practical candidates for proteomic development.

The rapid expansion of archaeal genomics is ensuring that the biology of these organisms can be approached through proteomics. We recently used LC-MS/MS and LC/LC-MS/MS methods to identify more than 500 *M. burtonii* proteins (Goodchild et al. unpublished results). This not only enabled a better understanding of the biology of the archaeon to be determined but also demonstrated the capacity to perform high throughput analyses of proteins from cold-adapted organisms, an area of extremophile studies that is rapidly expanding [61, 62].

Technological advances clearly provide new avenues for proteomic development. Excellent examples of this are the use of FTMS for detection of PTM and more sophisticated software for automated interpretation of MS/MS spectra for searching for possible PTMs. A synergy will also become increasingly possible by combining a proteomics approach with other complementary functional genomics approaches (e.g., transcriptomics, metabolomics) and by careful consideration of genome sequence information. A good example of this is the combining of 2DE, mRNA quantification, and specific enzyme assays for studies on *M. burtonii* [14]. More recently, lipid analysis using ESI/high-performance liquid chromatography–MS/MS was combined with LC/LC-MS/MS for proteomics to examine the role that lipid unsaturation plays in cold adaptation in *M. burtonii* (Nichols et al., unpublished results). This is a specific example illustrating how the use of global methods of analysis for lipids and proteomics linked to a draft genome sequence can be effectively combined to infer specific mechanisms of key biological processes.

## 5.7. ADDENDUM

A number of significant large-scale archaeal proteomics studies have appeared in the last half of 2004 and first half of 2005 that have made important contributions to the field. Proteomes for four archaea have been published, with proteins identified using 2DE and/or LC/LC-MS/MS technology. These include three methanogenic [63–65] and one halophilic archaea [66,67]. The coverage of the predicted proteomes was approximately 50% for *M. jannaschii* [63] and 25% for *M. burtonii* [64] using LC/LC-MS/MS methods, and 10% for *M. acetivorans* [65] and 34% for *H. salinarum* [66, 67] using 2DE and LC-MS/MS. The diversity of proteins identified not only highlighted the important technical developments that have occurred in the field, but provided valuable information regarding a range of cellular processes that occur in archaea. An important finding from all the studies was the significant number of expressed proteins that have no known predicted function (up to 44% of the proteins identified). This particular issue was addressed by Saunders et al. [68] who used a range of computational tools to infer functions for 55 of the hypothetical proteins from *M. burtonii*.

Two studies were published that examined differential protein abundance. The first study used 2DE and identified 34 proteins with differential abundance from *M. acetivorans* grown on two different substrates [69]. The proteomic analysis highlighted the intricate, and sometimes non-intuitive regulation that occurs in this methanogen. The second study used ICAT to examine the abundance of proteins in *M. burtonii* grown at high and low growth temperatures [70]. The study compared the outcomes of this work with a previous 2DE study that used the same growth conditions [14]. An important conclusion was that the 2DE and ICAT approaches were generally complementary, and pooling the data provided synergistic interpretations about archaeal adaptation.

One study used proteomics to identify proteins involved in lipid biosynthesis [71]. The study reported that cold adaptation in *M. burtonii* involves membrane lipid unsaturation and was achieved through membrane lipid analysis using ES/HPLC-MS/MS, proteomics using LC/LC-MS/MS to identify enzymes involved in lipid biosynthesis, and reconstruction of the pathway for lipid biosynthesis using the genome sequence. These recent examples highlight the ways in which proteomics can be used to empower our understanding of the biology of archaea.

We also identified two important proteomic studies published in 2002 that were missed when performing our literature review. Baliga et al. [72] performed proteomics and transcriptomics in wild-type and mutant strains of *Halobacterium* NRC-1 to study the regulation of energy production pathways. The large scale, integrated study identified differences in protein expression that were not reflected at the mRNA level, thereby indicating that cellular protein levels were mainly regulated by posttranscriptional events. This was also the first study to report the use of ICAT for differential expression in archaea. Luo et al. [73] performed 2DE and Northern blots to study differential expression of methanogenesis genes. Their study identified differences in the expression levels of specific methanogenesis genes when *Methanothermobacter thermoautotrophicus* was grown in a pure culture with $H_2$ plus $CO_2$ compared to growth in a syntrophic culture with a bacterial species under low hydrogen partial pressure. The authors reported that the switch in use of methyl coenzyme M reductase genes indicated that one form of the enzyme was more important for growth in natural methanogenic ecosystems. This chapter may not be exhaustive and we apologize if we have unintentionally missed relevant published literature.

## REFERENCES


1. Woese, C. R., Kandler, O., and Wheelis, M. L., *Proc. Natl. Acad. Sci. USA* 1990, **87**, 4576–4579.
2. Madigan, M. T., Martinko, J. M., and Parker, J., *Brock Biology of Microorganisms*, 10th ed., Prentice-Hall, Englewood Cliffs, NJ, 2003, pp. 445–471.
3. Blum, P., and Dixit, V., in M. Schaechter (Eds.), *The Desk Encyclopedia of Microbiology*, Elsevier Academic, London, 2004, pp. 108–116.
4. Koga, Y., Kyuragi, T., Nishihara, M., and Sone, N., *J. Mol. Evol.* 1998, **46**, 54–63.
5. Cavicchioli, R., Curmi, P.M., Saunders, N., and Thomas, T., *Bioessays* 2003, **25**, 1119–1128.
6. Cavicchioli, R., and Thomas, T., in M. Schaechter (Ed.), *The Desk Encyclopedia of Microbiology*, Elsevier Academic, London, 2004, pp. 436–453.
7. Cavicchioli, R., and Thomas, T., in J. Lederberg, M. Alexander, B. R. Bloom, D. Hopwood, et al., (Eds.), *Encyclopedia of Microbiology*, Academic, San Diego, 2000, pp. 317–337.
8. Kashefi, K., and Lovley, D. R., *Science* 2003, **301**, 934.
9. Schleper, C., Puhler, G., Kuhlmorgen, B., and Zillig, W., *Nature* 1995, **375**, 741–742.
10. Cavicchioli, R., *Astrobiology* 2002, **2**, 281–292.
11. Keller, M., and Zengler, K., *Nat. Rev. Microbiol.* 2004, **2**, 141–150.
12. Karner, M. B., DeLong, E. F., and Karl, D. M., *Nature* 2001, **409**, 507–510.
13. Bult, C. J., White, O., Olsen, G. J., Zhou, L., et al., *Science*, 1996, **273**, 1058–1073.
14. Goodchild, A., Saunders, N. F. W., Ertan, H., Raftery, M., et al., *Mol. Microbiol.*, 2004, **53**, 309–321.
15. Daniels, C. J., McKee, A. H. Z., and Doolittle, W. F., *EMBO J.* 1984, **3**, 745–749.
16. Waters, E., Hohn, M. J., Ahel, I., Graham, D. E., et al., *Proc. Natl. Acad. Sci. USA* 2003, **100**, 12984–12988.
17. Galagan, J. E., Nusbaum, C., Roy, A., Endrizzi, M. G., et al., *Genome Res.* 2002, **12**, 532–542.
18. Hao, B., Gong, W., Ferguson, T. K., James, C. M., et al., *Science*, 2002, **296**, 1462–1466.
19. Srinivasan, G., James, C. M., and Krzycki, J. A., *Science* 2002, **296**, 1459–1462.
20. Giometti, C. S., Reich, C., Tollaksen, S., Babnigg, G., et al., *J. Chromatogr. B: Analyt. Technol. Biomed. Life Sci.* 2002, **782**, 227–243.
21. Cordwell, S. J., Humphery-Smith, I., *Electrophoresis* 1997, **18**, 1410–1417.

22. Lim, H., Eng, J., Yates, J. R., 3rd, Tollaksen, S. L., et al., *J. Am. Soc. Mass Spectrom.* 2003, **14**, 957–970.
23. Babnigg, G., Giometti, C. S., *Nucleic Acids Res.* 2004, **32**, D582–585.
24. VanBogelen, R. A, Schiller, E. E., Thomas, J. D., and Neidhardt, F. C., *Electrophoresis* 1999, **20**, 2149–2159.
25. Schwartz, R., Ting, C. S., and King, J., *Genome Res.* 2001, **11**, 703–709.
26. Kennedy, S. P., Ng, W. V., Salzberg, S. L., Hood, L., and DasSarma, S., *Genome Res.* 2001, **11**, 1641–1650.
27. Cho, C., Lee, S., Choi, J., Park, S., et al., *Proteomics* 2003, **3**, 2325–2329.
28. Mojica, F. J., Cisneros, E., Ferrer, C., Rodriguez-Valera, F., and Juez, G., *J. Bacteriol.* 1997, **179**, 5471–5481.
29. Zwickl, P., Grziwa, A., Puhler, G., Dahlmann, B., et al., *Biochemistry* 1992, **31**, 964–972.
30. Wolf, S., Lottspeich, and F., Baumeister, W., *FEBS Lett.* 1993, **326**, 42–44.
31. Wolters, D. A., Washburn, M. P., and Yates, J. R., 3rd, *Anal. Chem.* 2001, **73**, 5683–5690.
32. Wu, S. L., Choudhary, G., Ramstrom, M., Bergquist, J., and Hancock, W. S., *J. Proteome Res.* 2003, **2**, 383–393.
33. Sze, S. K., Ge, Y., Oh, H., and McLafferty, F. W., *Proc. Natl. Acad. Sci. USA* 2002, **99**, 1774–1779.
34. Forbes, A. J., Patrie, S. M., Taylor, G. K., Kim, Y., et al., *Proc. Natl. Acad. Sci. USA* 2004, **101**, 2678–2683.
35. Wu, C. C., and Yates, J. R., 3rd, *Nat. Biotechnol.* 2003, **21**, 262–267.
36. Wallin, E., and von Heijne, G., *Protein Sci.* 1998, **7**, 1029–1038.
37. Blonder, J., Conrads, T. P., Yu, L.-R., Terunuma, A., et al., *Proteomics* 2004, **4**, 31–45.
38. Barnidge, D. R., Dratz, E. A., Jesaitis, A. J., and Sunner, J., *Anal. Biochem.* 1999, **269**, 1–9.
39. Corcelli, A., Colella, M., Mascolo, G., Fanizzi, F. P., and Kates, M., *Biochemistry* 2000, **39**, 3318–3326.
40. Weik, M., Patzelt, H., Zaccai, G., and Oesterhelt, D., *Mol. Cell* 1998, **1**, 411–419.
41. Bolhuis, A., *Microbiology* 2002, **148**, 3335–3346.
42. Eichler, J., *Eur. J. Biochem.* 2000, **267**, 3402–3412.
43. Franzmann, P. D., Liu, Y., Balkwill, D. L., Aldrich, H. C., et al., *Int. J. Syst. Bacteriol.* 1997, **47**, 1068–1072.
44. Trent, J. D., Osipiuk, J., and Pinkau, T., *J. Bacteriol.* 1990, **172**, 1478–1484.
45. Osorio, G., and Jerez, C. A., *Microbiology* 1996, **142**, 1531–1536.
46. Han, C. J., Park, S. H., and Kelly, R. M., *Appl. Environ. Microbiol.* 1997, **63**, 2391–2396.
47. Feller, G., and Gerday, C., *Nat. Rev. Microbiol.* 2003, **1**, 200–208.
48. Ferry, J. G., Chapmans Hall, New York, 1993.
49. Huber, H., Thomm, M., Konig, H., Thies, G., and Stetter, K. O., *Arch. Microbiol.* 1982, **132**, 47–50.
50. Bernhardt, G., Jaenicke, R., Ludemann, H.–D., Konig, H., and Stetter, K. O., *Appl. Environ. Microbiol.* 1988, **54**, 1258–1261.
51. Jaenicke, R., Bernhardt, G., Ludemann, H.-D., and Stetter, K. O., *Appl. Environ. Microbiol.* 1988, **54**, 2375–2380.
52. Jablonski, P. E., DiMarco, A. A., Bobik, T. A., Cabell, M. C., and Ferry, J. G., *J. Bacteriol.* 1990, **172**, 1271–1275.
53. Ding, Y. H., Zhang, S. P., Tomb, J. F., and Ferry, J. G., *FEMS Microbiol. Lett.* 2002, **215**, 127–132.

54. Jones, W., Leigh, J. A., Mayer, F., Woese, C. R., and Wolfe, R. S., *Arch. Microbiol.* 1983, **136**, 254–261.
55. Jannasch, H. W., and Mottl, M. J., *Science* 1985, **229**, 717–725.
56. Thauer, R. K., *Biochim. Biophys. Acta* 1990, **1018**, 256–259.
57. Mukhopadhyay, B., Johnson, E. F., and Wolfe, R. S., *Proc. Natl. Acad. Sci. USA* 2000, **97**, 11522–11527.
58. Faguy, D. M., and Jarrell, K. F., *Microbiology* 1999, **145( Pt. 2)**, 279–281.
59. Giometti, C. S., Reich, C. I., Tollaksen, S. L., Babnigg, G., et al., *Proteomics* 2001, **1**, 1033–1042.
60. Gygi, S. P., Rist, B., Gerber, S. A., Turecek, F., et al., *Nat. Biotechnol.* 1999, **17**, 994–999.
61. Cavicchioli, R., Siddiqui, K. S., Andrews, D., and Sowers, K. R., *Curr. Opin. Biotechnol.* 2002, **13**, 253–261.
62. Feller, G., and Gerday, C., *Nat. Rev. Microbiol.* 2003, **1**, 200–208.
63. Zhu, W., Reich, C., Olsen, G., Giometti, C., and Yates III, J., *J. Prot. Res.* 2004, **3**, 538–548.
64. Goodchild, A., Raftery, M., Saunders, N., Guilhaus, M., and Cavicchioli, R., *J. Prot. Res.* 2004, **3**, 1164–1176.
65. Li, Q., Li, L., Rejtar, T., Karger, B., and Ferry, J., *J. Prot. Res.* 2005, **4**, 112–128.
66. Tebbe, A., Klein, C., Bisle, B., Siedler, F., Scheffer, B., Garcia-Rizo, C., Wolfertz, J., Hickmann, V., Pfieffer, F., and Oesterhelt, D., *Proteomics* 2005, **5**, 168–179.
67. Klein, C., Garcia-Rizo, C., Bisle, B., Scheffer, B., Zischka, H., Pfeiffer, F., Siedler, F., and Oesterhelt, D., *Proteomics* 2005, **5**, 180–197.
68. Saunders, N., Goodchild, A., Raftery, M., Guilhaus, M., Curmi, P., and Cavicchioli, R., *J. Prot. Res.* 2005, **4**, 464–472.
69. Li, Q., Li, L., Rejtar, T., Karger, B., and Ferry, J., *J. Prot. Res.* 2005, **4**, 129–135.
70. Goodchild, A., Raftery, M., Saunders, N., Guilhaus, M., and Cavicchioli, R., *J. Prot. Res.* 2005, **4**, 473–480.
71. Nichols, D., Miller, M. R., Davies, N. W., Goodchild, A., Raftery, M., and Cavicchioli, R., *J. Bacteriol.* 2004, **186**, 8508–8515.
72. Baliga, N. S., Pan, M., Goo, Y. A., Yi, E. C., Goodlett, D. R., Dimitrov, K., Shannon, P., Aebersold, R., Ng, W. V., and Hood., L., *Proc. Natl. Acad. Sci. USA* 2002, **99**, 14913–14918.
73. Luo, H. W., Zhang, H., Suzuki, T., Hattori, S., and Kamagata, Y., *Appl. Environ. Microbiol.* 2002, **68**, 1173–1179.

PART II

# PROTEOMICS AND CELL PHYSIOLOGY

CHAPTER 6

# Elucidation of Mechanisms of Acid Stress in *Listeria monocytogenes* by Proteomic Analysis

LUU PHAN-THANH[1] and LOTHAR JÄNSCH[2]

[1]Unité Pathologie Infectieuse et Immunologie, Institut National de la Recherche Agronomique, Nouzilly, France
[2]Deutsche Forschung Gesellschaft, Biologisches Zentrum, Braunschweig, Germany

## 6.1 INTRODUCTION

In spite of ever-improving hygienic and sanitary conditions in the agricultural sector and food industry the food-borne pathogen *Listeria monocytogenes* remains a cause for concern for public health in industrialized countries. Much effort in the past decade has been deployed to delineate the genes involved in the virulence of this gram-positive bacterium [1]. Its genome has also been recently sequenced [2]. However, the study of its capabilities to survive in hostile conditions has not much advanced, especially under the mechanistic aspect. This kind of knowledge is of primary importance since the pathogenecity of a bacterium and its propagation largely depend on its ability to survive and express its virulence in defined environments. Among the environmental conditions unfavorable for the growth of *Listeria*, acidity is perhaps the most encountered, both in natural habitats and in infected host systems (e.g., acid rain, industrial and agricultural wastes, fermented food and feed products, gastric secretions, phagocytosomal vacuoles of the host cells). When facing such adverse conditions, *L. monocytogenes*, like other organisms, evolves strategies which help them to resist the adversity and eventually to adapt for survival. The majority of works on acid tolerance in microorganisms were done in gram-negative bacteria, especially in *Escherichia coli* and *Salmonella typhimurium* [3–5], much less in gram-positive microorganisms where *Lactobacillus* [6–8] occupied a preponderant place due to its industrial interest.

### 6.1.1 Acquired Acid Tolerance and Natural Acid Tolerance

The optimal pH for growth of *L. monocytogenes* is in the neutral range. In minimal medium when the pH goes down to 4 the bacterial growth stops [9]. The lowest pH *L. monocytogenes* can resist depends on the strain, the medium, and the kind of acid being present. When pH descends under 4 its lethal effect begins to be felt. At pH 3.5

*Microbial Proteomics: Functional Biology of Whole Organisms*, Edited by Ian Humphery-Smith and Michael Hecker. 

(hydrochloric acid) 27% of *L. monocytogenes* strain L028 survived 60 min of acid challenge. If before being stressed at pH 3.5 the bacteria were treated with a nonlethal acidic solution (pH 5.0–5.5) for 2 h, the survival percentage increased to 66% [9]. This acid tolerance was acquired by adaptation (habituation) to an intermediary pH. The extent of the acquired acid tolerance depends however on the duration of acid adaptation. Maximal tolerance was obtained with 2–3 h of adaptation for midexponential-phase bacteria. An extended adaptation time resulted in diminished tolerance and even rendered *Listeria* cells more fragile to posterior acid stress [9]. Apparently, the damage caused by long-term contact with the acid outdid the effort deployed by the bacteria to resist the acid. The acquired acid tolerance persisted for several weeks if the adapted bacteria were conserved at 4°C. This acid tolerance response (ATR) has been found in many other microorganisms belonging to both gram-positive and gram-negative bacteria [10, 11]. In the stationary phase of growth *L. monocytogenes* shows moreover a certain "natural" acid tolerance as compared to the exponential phase. A tolerance acquired by adaptation can eventually come to add itself to the tolerance of the stationary phase. That is why the acid-adapted stationary-phase cells are more resistant than the acid-adapted exponential-phase cells.

All the acids do not have the same effect on *Listeria*. At the same pH of the medium, weak organic acids exert a more deleterious effect than strong inorganic acids. The measuring of intracytoplasmic pH ($pH_i$) showed an inferior value with volatile organic acids (acetic acid, propionic acid, butyric acid) than with hydrochloric acid at the same external pH ($pH_e$) [9]. Strong acids completely dissociate in aqueous solutions and the cell membranes are semipermeable to protons. Protons can only pass the cell membranes with the help of energy-dependent systems such as $H^+$adenosine triphosphatases (ATPases), $Na^+/H^+$ antiporters, or electron transport systems. Weak organic acids, on the contrary, permeate the cell membranes more freely as undissociated molecules. Once inside the cell they dissociate and lower the $pH_i$, often to dramatic values that come to deregulate the metabolic machinery of the cell [12]. The low $pH_i$, however, is not the sole cause of lethality; accumulation of organic acid anions or whole molecules may also be toxic to the cell.

What we know about the mechanisms underlying the phenomenon of acid resistance in *L. monocytogenes* is scarce and fragmentary [13–15]. To study the mechanistic problem of stress response, two general approaches can be envisaged. One is genetic and consists of looking in stress-resistant mutants for the genes responsible for stress resistance. The other is proteomic and consists of identifying the proteins induced by stress directly in the wild-type strain. The identification of stress-induced proteins will inform us of the functions these proteins may fulfill in the process. The proteomic approach has been rendered feasible in recent years by improvements in the two-dimensional electrophoresis (2DE) technique and especially by progresses in mass spectrometry technology and also by the availability of the sequenced genome of *L. monocytogenes*. The characterization and identification of stress proteins, which have been until recently quite laborious and frustrating tasks, now can be done much more straightforward at the amount level of protein spots on a 2DE gel.

## 6.2 PROTEOMICS OF ACID STRESS IN *L. MONOCYTOGENES*

The differential analysis of the proteomes of *L. monocytogenes* (strain EGDe) stressed at pH 5.5 and pH 3.5 versus pH 7.2 was reported in a previous paper [16]. The acid stress proteins (ASPs), which are taken up in Table 6.1 and Figure 6.1, comprised

**TABLE 6.1 Proteins Induced in *L. monocytogenes* by HCl at pH 5.5 and pH 3.5**

| Identification Number | Induction Ratio[a] | Molecular Mass (kDa) | Isoelectric Point (pI) | Induced at pH |
|---|---|---|---|---|
| 52 | 2.6 | 43.8 | 5.76 | 3.5 |
| 152 | 3.2 | 18.2 | 5.70 | 3.5, 5.5 |
| 223 | 3.4 | 37.1 | 6.28 | 3.5 |
| 262 | 4.1 | 30.2 | 6.31 | 5.5 |
| 271 | 2.7, 2.2 | 28.2 | 6.79 | 3.5, 5.5 |
| 284 | 2.4 | 53.1 | 6.12 | 3.5 |
| 304 | 4.8 | 38.0 | 5.40 | 5.5 |
| 307 | 2 | 37.8 | 6.03 | 5.5 |
| 315 | 2.5, 8.5 | 12.6 | 5.65 | 3.5, 5.5 |
| 353 | 4.4, 8.3 | 18.2 | 5.47 | 3.5, 5.5 |
| 374 | 6.6 | 25.0 | 5.40 | 3.5 |
| 375 | 3.9 | 48.5 | 6.54 | 3.5 |
| 380 | 3.5 | 31.0 | 6.60 | 3.5 |
| 422 | 3.8 | 14.1 | 5.46 | 5.5 |
| 424 | 3.6 | 22.4 | 6.08 | 3.5 |
| 475 | 2.8 | 50.0 | 6.44 | 5.5 |
| 479 | 6.5 | 75.8 | 5.49 | 3.5 |
| 500 | 13.3, 15.4 | 54.9 | 4.95 | 3.5, 5.5 |
| 520 | 3.1 | 37.0 | 6.50 | 5.5 |
| 524 | 2.8 | 33.1 | 5.88 | 3.5 |
| 525 | 2.9, 6.3 | 31.6 | 6.11 | 3.5, 5.5 |
| 527 | 14.6, 7.7 | 36.3 | 6.84 | 3.5, 5.5 |
| 552 | 2.6 | 27.8 | 6.17 | 3.5 |
| 570 | 3.1 | 35.7 | 6.94 | 3.5 |
| 586 | 4.4 | 47.9 | 6.37 | 5.5 |
| 604 | 2.8, 5.6 | 30.2 | 5.90 | 3.5, 5.5 |
| 609 | 11 | 56.0 | 6.28 | 5.5 |
| 618 | 3.3 | 26.9 | 6.58 | 3.5 |
| 632 | 2.4 | 16.8 | 6.43 | 5.5 |
| 636 | 4.8 | 29.5 | 4.96 | 5.5 |
| 640 | 12.8 | 50.2 | 6.22 | 5.5 |
| 651 | 8.2, 9.3 | 26.1 | 6.48 | 3.5, 5.5 |
| 652 | 3.9 | 42.2 | 6.77 | 3.5 |
| 657 | 3.0 | 29.8 | 5.84 | 3.5 |
| 672 | 19.6 | 39.6 | 5.60 | 5.5 |
| 695 | 2.7 | 37.1 | 6.15 | 3.5 |
| 702 | 3.8, 8.6 | 50.3 | 6.65 | 3.5, 5.5 |
| 705 | 9.1, 3.6 | 23.5 | 6.26 | 3.5, 5.5 |
| 708 | 6.0 | 47.9 | 6.76 | 3.5 |
| 719 | 6.6, 11 | 41.7 | 6.29 | 3.5, 5.5 |
| 720 | 7.5, 8.5 | 50.9 | 6.76 | 3.5, 5.5 |
| 722 | 4.8, 11.4 | 27.0 | 5.85 | 3.5, 5.5 |
| 727 | 6.0 | 47.1 | 5.84 | 3.5 |
| 728 | 3.2 | 54.6 | 6.65 | 5.5 |
| 731 | 4.4, 5.4 | 45.7 | 6.55 | 3.5, 5.5 |
| 732 | 7.0, 6.7 | 48.9 | 6.54 | 3.5, 5.5 |
| 754 | 22.8, 9.5 | 42.1 | 5.59 | 3.5, 5.5 |

(*continued*)

**TABLE 6.1** (*Continued*)

| Identification Number | Induction Ratio[a] | Molecular Mass (kDa) | Isoelectric Point (pI) | Induced at pH |
|---|---|---|---|---|
| 773 | 5.7 | 27.4 | 6.21 | 3.5 |
| 783 | 2.7 | 35.5 | 5.30 | 3.5 |
| 800 | 7.5 | 23.7 | 6.81 | 3.5 |
| 802 | 8.7, 3.8 | 32.0 | 6.88 | 3.5, 5.5 |
| 822 | Novel | 13.7 | 6.72 | 5.5 |
| 825 | Novel | 22.1 | 5.60 | 3.5, 5.5 |
| 827 | Novel | 25.3 | 4.95 | 3.5, 5.5 |
| 828 | Novel | 25.3 | 5.10 | 3.5, 5.5 |
| 836 | Novel | 22.3 | 6.87 | 5.5 |
| 838 | Novel | 22.6 | 6.53 | 3.5, 5.5 |
| 839 | 4.5 | 16.0 | 6.20 | 3.5 |
| 840 | Novel | 17.8 | 6.75 | 3.5 |
| 843 | Novel | 15.9 | 6.29 | 3.5, 5.5 |
| 846 | Novel | 19.7 | 5.44 | 3.5 |
| 854 | Novel | 44.7 | 5.15 | 3.5 |
| 874 | 5.2 | 56.9 | 4.98 | 3.5 |
| 875 | Novel | 51.2 | 4.86 | 3.5, 5.5 |

[a]The induction ratios given are means of the values from several gels. When two ratios are given, they correspond to induction at pH 3.5 and 5.5, respectively.

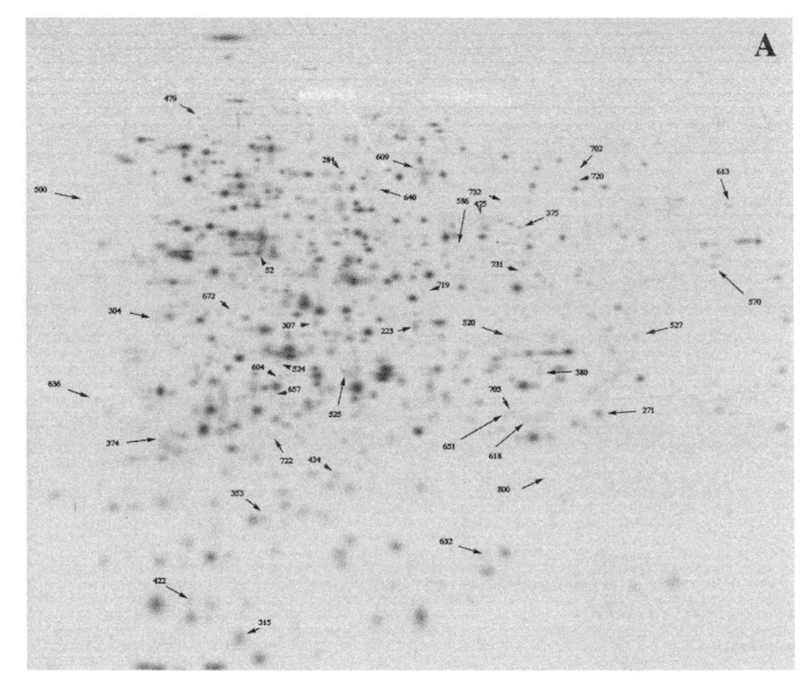


**Figure 6.1** 2DE autoradiograms of cellular proteomes of *L. monocytogenes* at (*a*) pH 7.2, (*b*) pH 3.5, and (*c*) pH 5.5. The acid-induced proteins are indicated by an arrow and an identification number with pI and molecular weight (MW) given in Table 6.1.

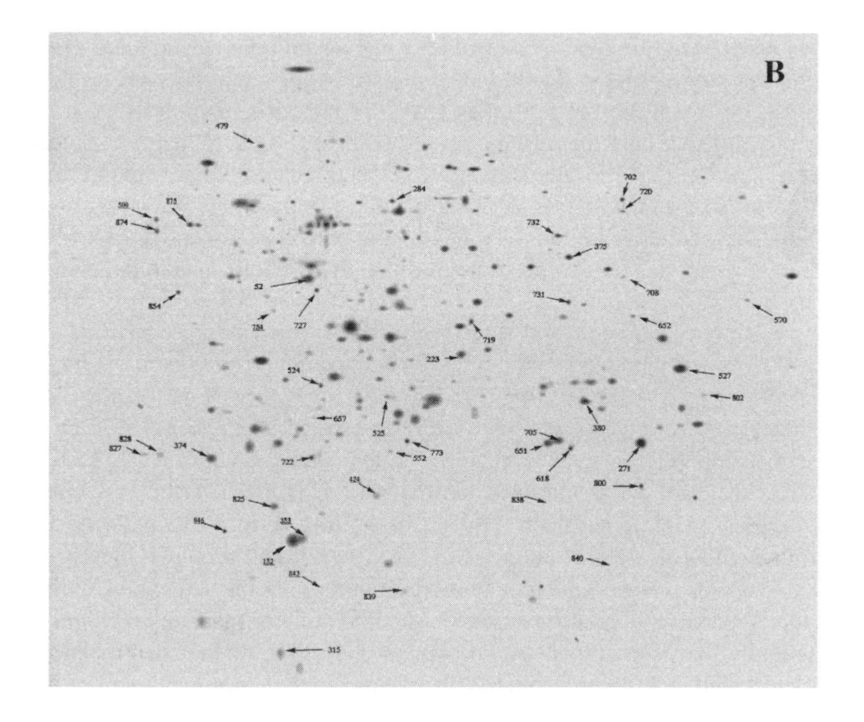


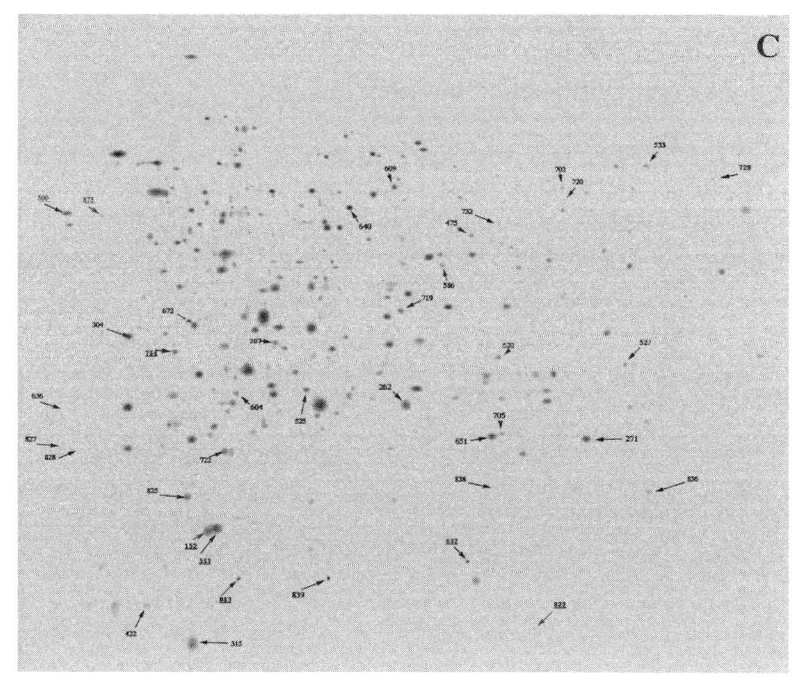


**Figure 6.1** (*continued*)

37 proteins which were up-regulated at pH 5.5 and 47 proteins which were up-regulated at pH 3.5. The two groups shared 23 proteins, suggesting that when the bacteria encounter an acid medium (pH 5.5), they resist by increasing synthesis of a number of necessary proteins that help them to mount a tolerance to acid. When the bacteria face a more severe acidity (pH 3.5), they require synthesis of additional proteins in order to mount a fiercer resistance. Thus, ASPs are either constitutive (their expressions are upgraded by the acidic conditions) or novel (not existing at neutral pH). To combat stresses, bacteria evolve overlapping strategies that involve both constitutive and inducible defense.

Using preparative 2DE gels and referring to the corresponding analytical 2DE gels, spots of ASPs (visible with Coomassie blue or silver staining) were excised and submitted to mass spectrometry (MS) analysis [16]. The identification was performed using either peptide fingerprinting [matrix-assisted laser desorption ionization (MALDI)] or tag peptide sequencing [electrospray ionization tandem mass spectrometry (ESI-MS/MS)] or both when the MALDI results were ambiguous. It is note-worthy that although the *L. monocytogenes* genome has been sequenced, the functions of the majority of its open reading frames (ORFs) are still unknown. Hence, beside the proteins directly identified using the *L. monocytogenes* genome, a number of other proteins had to be identified using the sequenced genome of *Bacillus subtilis* or other microorganisms, *B. subtilis* being phylogenetically the nearest to *L. monocytogenes*. In light of the identification results, which are presented in Tables 6.2 and 6.3, we attempted to determine the roles these ASPs may play in the process of acid resistance.

### 6.2.1 Biochemical significance of ASPs

***Proteins 353, 722, 728, 732, and 754 (Table 6.2)*** When facing external low pH, the bacteria attempt to maintain their intracytoplasmic $pH_i$. This can be done by several ways: decreasing membrane permeability to $H^+$, buffering their cytoplasm, and

**TABLE 6.2 Acid Stress Proteins in *L. monocytogenes* Identified by MALDI**

| Spot | Database | | | |
|---|---|---|---|---|
| Number | MW | pI | Species | Protein Name/Function |
| 152 | 19.2 | 5.12 | *B. subtilis* | ATP synthase (subunit b) |
| 353 | 19.7 | 5.67 | *B. subtilis* | Homologue of thioredoxin redutase |
| 424 | 25.3 | 5.62 | *B. subtilis* | Sigma H factor |
| 500 | 60.4 | 4.73 | *Methanobacterium thermoautotrophicum* | Chaperonin |
| 632 | 17.8 | 6.19 | *Klebsiella pneumonia* | Ferric uptake regulator |
| 754 | 43.5 | 6.45 | *B. subtilis* | Alkohol dehydrogenase |
| 822 | 18.2 | 6.45 | *B. subtilis* | Yfiv, similar to transcriptional regulator |
| 843 | 16.8 | 6.42 | *B. subtilis* | Similar to transcriptional regulator |
| 846 | 19.7 | 4.87 | *B. subtilis* | Transcriptional regulator |
| 875 | 56.7 | 4.84 | *Erwinia herbicola* | Similar to GroEL |

**TABLE 6.3 Acid Stress Proteins in *L. monocytogenes* Identified by ESI MS/MS**

| Protein Number | MW | pI | Peptide Sequence | Species | Function |
|---|---|---|---|---|---|
| 632 | 17.9 | 6.1 | IIVISDDVAKDEVR | *L. monocytogenes* | Mannose-specific phosphotransferase system (PTS), component IIb |
| 702 | 44 | 6.10 | NAAESLALVG | *L. monocytogenes* | Glycine betaine ABC transporter[a] (ATP binding protein) |
| 719 | 41 | 6.4 | LGFYAPTAGQE | *L. monocytogenes* | Pyruvate dehydrogenase[a] (E1 alpha subunit) |
| 722 | 26.5 | 5.71 | GNIYQGQTPQSFFNMK | *L. monocytogenes* | Short-chain dehygrogenase/ reductase family[a] |
| 728 | 53 | 6.5 | DVVIVAGNVATAEGAR | *L. monocytogenes* | Inosine monophosphate dehydrogensase[a] |
| 732 | 49 | 5.7 | VQFNSAIGPYK QKHIGPDTDVP | *L. monocytogenes* | NADP-specific glutamate dehydrogenase[a] |
| 825 | 20.8 | 5.3 | FILNDRFEK SDLGLTPNNDGSVL | *L. monocytogenes* | Ribosome recycling factors[a] |
| 846 | 19.2 | 5.2 | YGVPTSDYDFDSVR | Other Microorganisms | Intracellular protease |
| 854 | 41.5 | 4.9 | TPNPDCMCNK | *L. monocytogenes* | Putative tRNA-methyltransferase[a] |
| B1 | 13.8 | 9.6 | EDLNTQVDTIK LVDSAGGFVDVVK | *L. monocytogenes* | General stress protein[a] |
| B2 | 20 | 9.3 | EQLIFPEIDYDQVSK | *L. monocytogenes* | Ribosomal protein L5 (rplE) |
| B5 | 14.6 | 9.5 | VMTDPIADFLTRDVEYIE DDNAGTIR | *L. monocytogenes* | Ribosomal protein S8 (rpsH) |
| B8 | 24.5 | 9.6 | VYTAEEAVELAKFDATV EVAFR | *L. monocytogenes* | Ribosomal protein L1 (rplA) |

[a]Similar to this function.

equilibrating the external $pH_e$ through catabolism. The phenomenon of $pH_i$ homeostasis is largely recognized but not very well understood in detail. Some general views, however, can be outlined. For the $pH_i$ homeostasis, the permeability of the cytoplasmic membrane to protons has to be maintained at a low level. Environmentally induced modifications of the lipid composition may be one of the ways the cell strives to limit changes to proton permeability. But when the proton pressure exerted on the cell is too high, the effort deployed by the cell cannot be sufficient to render the membrane impermeable and protons pass into the cytoplasm. To evacuate protons out of the cytoplasm, the cell enhances its oxidation–reduction systems and accelerates electron-transferring reactions, since this process expel protons (respiration in aerobic bacteria [17]). Here the enzymes dehydrogenases and reductases found in proteins 353, 722, 728, 732, and 754 (Table 6.2) may intervene. These enzymes are substrate specific (pyruvate, NADP, or inosine monophosphate specific). The respiratory chain of aerobic bacteria contains a large number of electron-carrying proteins that act in sequence to transfer electrons from the substrates to oxygen (Fig. 6.2). The electron-carrying groups associated with these proteins include

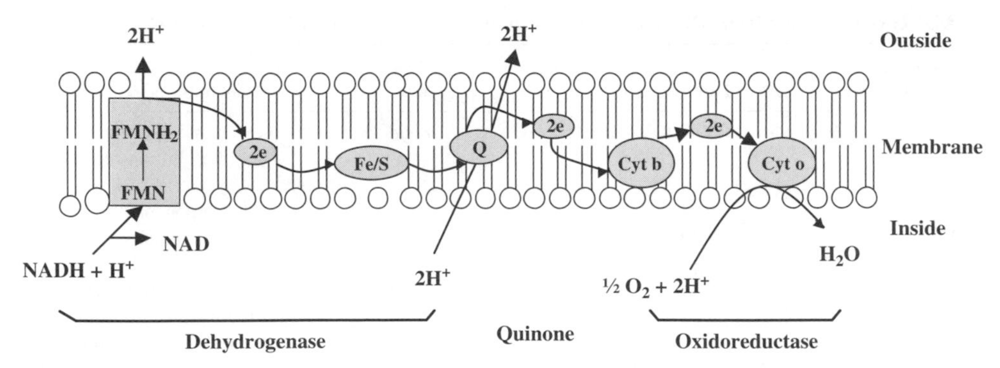


**Figure 6.2** Schematic model of bacterial respiratory chain. Electrons are transferred from NADH to oxygen thanks to dehydronenases, quinones (Q), and oxidoreductases. This electron flow is coupled to expulsion of $H^+$.

nicotinamide adenine dinucleotide (NAD), flavin mononucleotide (FMN), ubiquinone or coenzyme Q, and iron-containing proteins such as iron–sulfur centers or clusters (Fe–S) and cytochromes. The iron atoms in the iron–sulfur centers by undergoing Fe(II)–Fe(III) cycles transfer reducing equivalents from the $FMNH_2$ prosthetic group of NADH dehydrogenase to ubiquinone, the next electron carrier in the respiratory chain. The complexation of NADH–dehydrogenase with the iron–sulfur proteins (resulting in NADH–ubiquinone reductase) contains two kinds of electron-carrying structures, FMN and iron–sulfur centers, which appear to function in sequence [17]. Here may intervene Fur, protein 632 (Table 6.2) The role of the ferric uptake regulator is to mobilize iron for the iron–sulfur centers. Foster and Hall [18] also found Fur as an ASP in *S. typhimurium*, while other authors found Fur to be an important global regulator that controls sidephore-mediated iron assimilation [19, 20] or modulates the expression of alternative sigma factor genes [21], oxidative stress-protective genes [22, 23], and virulence genes [24, 25]. Furthermore, Thompson et al. [26] reported that in addition to the expected derepression of iron siderophore biosynthesis and receptor genes in FUR1, the *fur* mutation in *Shewanella oneidensis* affected the transcription of a number of genes involved in electron transport systems and energy metabolism. Thus, upon acid stress and following acidification of the cytoplasm, the respiratory chain of aerobic bacteria helps to evacuate $H^+$ from the cytoplasm to the outside of the cell. This transfer of $H^+$ coupled with the action of dehydrogenases and oxidoreductases does not necessitate energy; on the contrary, it supplies energy due to the flow of electrons from substrates to oxygen. The free energy of electron flow is served, among other purposes, as the driving force to transport $H^+$ outward against a gradient of $H^+$ through the so-called '$H^+$ pumps' and through such antiporters as $Na^+/H^+$and $K^+/H^+$, which exchange intracellular protons with extracellular $K^+$ or $Na^+$. (There are also other specific antiporters which couple the transfer of $H^+$ with that of other compounds.) In addition, the modulation of cation ($K^+$ and $Na^+$) movement across the membrane is intrinsic to generating and maintaining a transmembrane pH gradient [27]. In fact, the major components of the pH homeostatic mechanism are expressed constitutively and are quiescent until challenged. But the cytoplasmic buffering capacity is a finite resource that can be used by the cell to limit perturbation of intracellular $pH_i$ in response to acidification. Furthermore, pH homeostasis

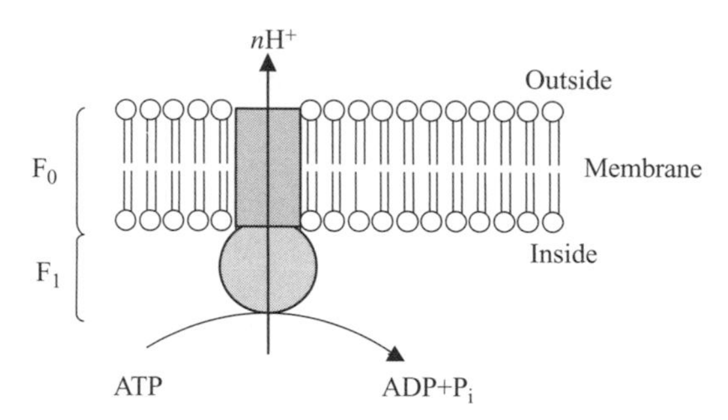


**Figure 6.3** Functioning of $F_0F_1$–$H^+$–ATPase system in fermentative bacteria. Energy supplied by hydrolyse of one ATP molecule is served to expel three or four $H^+$.

can only be achieved when tight control is exerted over the channels in the cytoplasmic membranes that possess the ability to overwhelm the capacity of the proton pumps and ion transport systems.

***Protein 152*** According to chemiosmotic theory [28], the $H^+$–ATPase system acts as a reversible proton-translocating pump which catalyzes the movement of $H^+$ across the cell membrane as a consequence of the hydrolysis or synthesis of ATP. In respiring (aerobic) bacteria, the energy of electron flow can be in part conserved under the form of ATP in the process of oxidative phosphorylation. In fermentative (anaerobic) bacteria, on the contrary, the membrane-bound $H^+$–ATPase system ($H^+$ pump) is believed to function in the reverse direction (Fig. 6.3) and expel the protons accumulated by the cell as a result of lactic acid production [29, 30]. This action necessitates energy, which is supplied by ATP. The $F_0F_1$–$H^+$–ATPase enzyme system (also called ATP synthase or ATP synthetase) is a multi-subunit complex constituted by two domains: a cytoplasmic domain $F_1$ (comprising nine subunits $\alpha,\beta,\gamma,\delta$, and $\varepsilon$) with catalytic sites permitting the synthesis/degradation of ATP and a trans-membrane domain $F_0$ (comprising a, b, and c subunits) serving as a proton channel. The activity of $F_0F_1$–$H^+$–ATPase enables the bacteria possessing no respiratory chain to maintain $pH_i$ homeostasis. *Listeria monocytogenes*, a facultatively anaerobic bacterium, can use either mechanism to resist acidity depending on the situation. Protein 152, which was up-regulated upon acid stress, was identified as the b subunit of the ATPase complex. Cotter et al. [31], using a mutagenesis technique, attempted to confirm the role of $F_0F_1$–ATPase in the acid tolerance response of *L. monocytogenes* but could not assess the acid sensitivity of the mutant because of its slow growth. Jordan et al. [5], using a proteomic approach, found in *E. coli* an acid-induced membrane protein identified as an $\varepsilon$ subunit of ATP synthase. Van der Heyden and Docampo [32] demonstrated in *Trypasonoma cruzi* the presence of a plasma membrane $H^+$–ATPase that regulated $pH_i$ and that was supported by $K^+$ channels.

***Protein 732 (Glutamate Dehydrogenase)*** Glutamate in the bacterial metabolism undergoes the oxidative deamination under the action of glutamate dehydrogenase, which requires $NAD^+$ as acceptor of reducing equivalents:

$$\text{Glutamate} + NAD^+ + H_2O \rightleftharpoons \alpha\text{-ketoglutarate} + NH_4^+ + NADH$$

The bacteria use the ammonia released to neutralize acidity, while α-ketoglutarate serves as fuel for the citric acid cycle to form ATP. Glutamate dehydrogenase is also responsible for most of the ammonia formed in animal tissues, and glutamate is the only amino acid whose α-amino group can be directly removed at a high rate in this manner [17]. The studies of acid resistance mechanisms in *E. coli* led to the identification of three systems that protect the cell against acidic pH [33–35]. All three systems were identified in stationary-phase cells. The first is a glucose-repressed system controlled by the alternative sigma factor (RpoS), adenosine-3′,5′-cyclic monophosphate (cAMP), and its receptor protein (CRP). It is an oxidative system whose mechanism is not yet well known. (It is partly regulated by Rpo). The three compounds RpoS, cAMP, and CRP are also involved in the regulation of starvation protein synthesis [36]. The second system requires glutamate to be activated during adaptation and utilizes an inducible glutamate decarboxylase. The third system requires arginine and an inducible arginine decarboxylase (gene *adiA*). The two decarboxylase systems are believed to consume protons during decarboxylation of glutamate or arginine, thus preventing $pH_i$ from decreasing to lethal levels. The end products γ-aminobutyric acid and agmatine are transported out of the cell in exchange for new substrates through specific antiporter systems (GadC for glutamate and unknown antiporter for agmatine). The glutamate decarboxylase system (GAD) encompasses three genes: *gadA*, *gadB*, and *gadC*. The genes *gadA* and *gadB* encode two homologous and independently regulated glutamate decarboxylases, and the gene *gadC* encodes a putative glutamate/γ-aminobutyric antiporter. The genes *gadB* and *gadC* form an operon. Both decarboxylase systems are induced by acidic conditions. However, expression of *gadA* is predominantly induced by acidic pH, whereas expression of *gadB* is primarily induced by entry into the stationary phase. In *L. monocytogenes* similar GAD systems were also found and proved to play a role in acid resistance in the stomach [37]. We wonder whether there is a mechanistic link between glutamate decarboxylase GadB and glutamate dehydrogenase (protein 732) described above, both being induced by mildly acidic pH 5.5. The two enzymes may conceivably participate in glutamate metabolism in the process of $pH_i$ homeostasis as a response to acid stress, concomitantly or alternatively.

***Protein 424*** This protein was identified as similar to sigma H factor, a secondary sigma factor in *B. subtilis*, a factor required for transcription of competence and early sporulation genes [38]. In general, the role of sigma factors is control of gene expression. Each sigma subunit of RNA polymerase is required for binding of the enzyme to a specific promoter. The precise function of the sigma factor H in the acid tolerance of *Listeria* is unclear. On the contrary, the role of another sigma factor, sigma B, has been well documented [39–41]. Sigma B is known to protect the bacteria against many environmental stresses (acid stress, oxidative stress, salt stress, cold stress, and carbon starvation). In *L. monocytogenes*, while $\sigma^B$ contributes to adaptive acid response with the greatest $\sigma^B$-specific effects observed in exponential-phase cells, $\sigma^B$ is mostly induced by entry into the stationary phase [42]. Sigma B protein did not appear in our proteomic analysis, probably because the bacterial samples we used for the analysis originated from exponential-phase cells.

***Protein 702 (GbuA)*** Protein 702 was identified as GbuA, an ATP-dependent, DNA-binding glycine betaine transporter which is known to play a role in osmotolerance and cryotolerance of *L. monocytogenes* [43]. It is thought to allow accumulation of compatible solutes as a common response to osmotic upshift in many bacteria species, and these solutes presumably alleviate the consequences of osmotic stress without altering the structure of intracellular proteins and other cellular machinery [44]. In *L. monocytogenes*, we found it to be up regulated upon acid stress at both pH 5.5 and 3.5. It may play a similar role in acid tolerance as in osmotolerance.

***Protein 500 (Chaperonin) and Protein 875 (GroEL)*** Although the damage caused to the cell by the lowering of intracytoplasmic $pH_i$ is not known in detail, this damage can be considered the noxious consequence of acid insult suffered by two main cellular compound classes. On the one hand, the damage at the DNA level can provoke selective depurination or lesions in double-stranded DNA [17], resulting in unrepaired DNA and mismatches in repaired sequences, which ultimately can be lethal to the cell. On the other hand, the damage at the protein level can lead to denaturation of many proteins and inactivation of their physiological activity. To survive the acidity, the bacteria have to up regulate synthesis of the necessary molecules that permit them to undo the adverse processes and/or to repair the damage done. Protein 500 (chaperonin) and 875 (GroEL) belong to such molecular classes. The chaperones (GroES, GroEL, DnaK) and chaperonins known as heat-shock proteins in many organisms may fulfill these functions. They participate in either the prevention of protein denaturation or the repair of proteins that have been damaged by stress. When the damage is irreparable, the chaperones help the cell to degrade and evacuate the damaged proteins [45]. The chaperones may conceivably play the same roles in acid stress as in heat stress, the two stresses being known to have cross-protective effects [46]. Using a genetic approach, Hanawa et al. [47] demonstrated that chaperone DnaK was essential in the heat tolerance and acid tolerance of *L. monocytogenes*. Similarly, a DNA-binding protein (Dps) was reported to have a pivotal role in the acid tolerance and oxidative stress tolerance of *E. coli* [48].

***Protein 846*** This protein was identified as a intracellular protease. This enzyme participates in the degradation of proteins, a necessary step prior to their evacuation out of the cell by opsonization. This protein was only induced at pH 3.5 and not at pH 5.5, that is at a severe acid pH where some proteins are damaged beyond repair by the stress.

***Basic ASPs B1, B2, B3, B4 (Table 6.3) (Basic 2DE Gels Not Shown)*** Four basic ASPs were identified, one as a general stress protein (which appeared in several stress conditions and whose function is unknown) and three others, ribosomal proteins L5 (rplE), S8 (rpsH), and L1 (rplA). Beside the fact that new and urgent protein synthesis in response to acid stress requires increased supply of ribosomal proteins, which in turn builds up the foundation for protein synthesis, ribosomes may be themselves sensors for stress [49]. Zang et al. [50] demonstrated that ribosomes provide an essential input for stress activation of sigma B factor, since the absence of ribosomal protein 11 blocks the stress activation of sigma B.

## 6.3 CONCLUSION

Although we still are far from an overall mechanistic elucidation of acid resistance, the proteomic analysis of the phenomenon provided us with a better understanding of the functions of many acid-induced proteins involved in this process. The number of proteins identified is still limited and many ASPs remain to be identified, especially those not detectable by Coomassie blue and silver nitrate. More acid-induced proteins in the basic pH zone also need to be identified. Furthermore, for a complete mechanistic elucidation, we need to identify not only the proteins induced but also those repressed by acid stress. It is also noteworthy that the ASPs we presented above were induced by hydrochloric acid. The response to organic acids may be different, as we have seen a more deleterious effect of organic acids at the same external pH. In resistance to organic acids, other mechanisms may be superposed onto those that maintain $pH_i$ homeostasis. With the advanced state of the mass-spectrometric technology and availability of the sequenced genome, the proteomic analysis now relies primarily on the quality of 2DE gels. Despite its inherent shortcomings in reproducibility, 2DE remains the sole methodology capable of separating thousands of proteins from a complex mixture of cellular proteins. Still, the 2DE technique we are using is essentially that of O'Farrel, as employed three decades ago. However, efforts have been made in recent years to bring about improvements from protein extraction to electrophoretic separation to gel image analysis, all aimed at increasing the quality and reproducibility of 2DE gels. Improvements include thiourea used in combination with urea to increase dissociation and solubility of membrane proteins. Noncharged tributylphosphine now routinely replaces charged dithiotreitol or dithioerythritol as reducing agent to avoid precipitation of proteins as consequence of migration of reducing agents off the gels during isoelectrofocusing (first dimension) and to spare the acetylation step with iodoacetamide before the second dimension [16]. Longer (24 cm) and more stable Immobiline Drystrips are now available commercially, rendering the making of bigger and more resolutive 2DE gels easier. Separation of basic proteins, however, remains a problem since gels in this pH zone are much less stable than those in neutral pH zone.

## ACKNOWLEDGMENT

We thank Jean-Pierre Le Caer of Ecole Supérieure de Physique et de Chimie Industrielle de Paris for his technical assistance in MALDI analysis.

## REFERENCES

1. Vazquez-Boland, J. A., Kuhn, M., Berche, P., Chakraborty, T., Dominguez-Bernal, G., Goebel, W., Gonzalez-Zorn, B., Wehland, J., and Kreft, J., *Clin. Microbiol. Rev.* 2001, 584–640.
2. Glaser, P., Frangeul, L., Buchrieser, C., Rusniok, C, et al., *Science* 2001, **294**, 849–852.
3. Hickey, E. W., and Hirshfield, I. N., *Environ. Microbiol.* 1990, **56**, 1038–1045.
4. Bearso, S., Bearson, B., and Foster, J., *FEMS Microbiol. Lett.* 1997, **147**, 187–216.
5. Jordan, K. N., Oxford, L., and O'Byrne, C. P., *Appl. Environ. Microbiol.* 1999, 3048–3055.
6. Bender, G. R., Sutton, S. V., and Marsuis, R. E., *Infect. Immun.* 1986, **53**, 331–338.
7. Yokota, A., Amachi, S., Ishii, S., and Tomata, F., *Biosci. Biotechnol. Biochem.* 1995, **59**, 2004–2007.
8. Sander, J. W., Lenhouts, K., Burghum, J., Brand, J. R., Venema, G., and Kok, J., *Mol. Microbiol.* 1998, **27**, 299–310.
9. Phan-Thanh, L., and Montagne, A., *J. Gen. Appl. Microbiol.* 1998, **44**, 183–191.
10. Foster, J. W., Microbial responses to acid stress, in G. Storz and R. Hengg-Aronis (Eds.), *Bacterial Stress Responses*, ASM Press, Washington DC, 2000, pp. 99–115.
11. Merell, D. S., and Camilli, A., *Curr. Opin. Microbiol.* 2002, **5**, 51–55.

12. Olson, E. R., *Mol. Microbiol.* 1993, **8**, 5–14.
13. Davis, M. J., Coote, P. J., and O'Byrne, C. P., *Microbiology* 1996, **142**, 2975–2982.
14. O'Driscoll, B., Gahan, C. G., and Hill, C., *Appl. Environ. Microbiol.* 1996, **62**, 1693–1698.
15. Marron, L., Emerson, N., Gahan, and C. G., Hill, C., *Appl. Environ. Microbiol.* 1997, **63**, 4945–4947.
16. Phan-Thanh, L., and Mahouin, F., *Electrophoresis* 1999, **20**, 2214–2224.
17. Lehninger, A. L., Nelson, D. L., and Cox, M. M., *Principles of Biochemistry*, 2nd ed, Worth, New York, 1993.
18. Foster J. W., and Hall, H. K., *J. Bacteriol.* 1990, **173**, 6896–6902.
19. Brickman, T. J., Ozenberger, B. A., and McIntosh, M. A., *J. Mol. Biol.* 1990, **212**, 669–682.
20. Griggs, D. W., and Konisky, J., *J. Bacteriol.* 1989, **171**, 1048–1054.
21. Ochsner, U. A., and Vasil, M. L., *Proc. Natl. Acad. Sci. USA* 1996, **93**, 4406–4414.
22. Hassan, H. M., and Sun, H. C. H., *Proc. Natl. Acad. Sci. USA* 1992, **89**, 3217–3221.
23. Nierderhoffer, E. C., Naranjo, C. M., Bradley, K. L., and Fee, J. A., *J. Bacteriol.* 1990, **172**, 1930–1938.
24. Goldberg, M. B., Boyko, S. A., and Calderwood, S. B., *J. Bacteriol.* 1991, **172**, 6863–6870.
25. Litwin, C. M., and Calderwood, S. B., *Clin. Microbiol. Rev.* 1993, **6**, 137–149.
26. Thompson, D. K., Beliaev, A. S., Giometti, C. S., Tollaksen, S. L., Khare, T., Lies, D. P., Nealson, K. H., Lim, H., Yates III, J., Brandt, C. C., Tiedje, J. M., and Zhou, J., *Appl. Environ. Microbiol.* 2002, **68**, 881–892.
27. Booth, I. R., *Novartis Found. Symp.* 1999, **221**, 19–28.
28. Mitchell, P., *Science* 1979, **206**, 1148–1159.
29. Kobayashi, H. T., Suzuki, T., and Unemoto, T., *J. Biol. Chem.* 1986, **261**, 627–636.
30. Nannen, N., and Hutkins, R. W., *J. Dairy Sci.* 1991, **74**, 747–751.
31. Cotter, P. D., Gahan, G. M., and Hill, C., *Int. J. Food Microbiol.* 2000, **60**, 137–146.
32. Van der Heyden, N., and Docampo, R., *Mol. Biochem. Parasitol.* 2000, **105**, 237–251.
33. Hersh, B. M., Farouq, F. T., Barstad, D. N., Blankenshorn, D. L., and Slonczewski, J. L., *J. Bacteriol.* 1996, **178**, 3978–3981.
34. Lin, J., Smith, Chaplin, K. C., Balk, H. S., Bennett, G. N., and Foster, J. W., *Appl. Environ. Microbiol.* 1996, **62**, 3094–3100.
35. Catanie-Cornet, M. P., Penfound, T. A., Smith, D. J., Elliot, J. F., and Foster, J. W., *J. Bacteriol.* 1999, **181**, 3525–3535.
36. Schultz, J. E., Latter, G. I., and Matin, A., *J. Bacteriol.* 1988, **170**, 3903–3909.
37. Cotter, P. D., Gahan, C. G., and Hill, C., *Mol. Microbiol.* 2001, **40**, 465–475.
38. Tatti, K. M., Carter III, H. L., Moir, A., and Moran, Jr., C. P., *J. Bacteriol.* 1989, **171**, 5928–5932.
39. Becker, L. A., Cetin, M. S., Hutkins, R. W., and Benson, A. K., *J. Bacteriol.* 1998, **180**, 4547–4554.
40. Wiedmann, M., Arvik, T. J., Hurley, R. J., and Boor, K. J., *J. Bacteriol.* 1998, **180**, 3650–3656.
41. Ferreira, A., O'Byrne, C. P., and Boor, K. J., *Appl. Environ. Microbiol.* 2001, **67**, 4454–4457.
42. Ferreira, A., Sue, D., O'Byrne, C. P., and Boor, K. J., *Appl. Environ. Microbiol.* 2003, 2692–2698.
43. Ko, R., and Smith, L. T., *Appl. Environ. Microbiol.* 1999, **65**, 4040–4048.
44. Yancey, P. H., Clark, M. E., Hand, S. C., Bowlus, R. D., and Somero, G. N., *Science* 1982, **217**, 1214–1222.
45. Hartl, F. U., Hlodan, R., and Langer, T., *TIBS* 1994, **19**, 20–25.
46. Phan-Thanh, L., Mahouin, F., and Aligé, S., *J. Food Microbiol.* 2000, **55**, 121–126.

47. Hanawa, T., Fukuda, M., Kawakami, H., Hirano, H., Kamiya, S., and Yamamoto, T., *Cell Stress Chaperones* 1999, **4**, 118–128.
48. Choi, S. H., Baumbler, D. J., and Kaspar, C. W., *Appl. Environ. Microbiol.* 2000, **66**, 3911–3916.
49. VanBogelen, R. A., and Neidhardt, F. C., *Proc. Natl. Acad. Sci.* 1990, **87**, 5589–5593.
50. Zang, S., Scott, J. M., and Haldenwang, W. G., *J. Bacteriol.* 2001, **183**, 2316–2321.

CHAPTER 7

# Oxidation of Bacterial Proteome in Response to Starvation

THOMAS NYSTRÖM

Göteborg University, Göteborg, Sweden

## 7.1 INTRODUCTION

Bacteria, such as *E. coli*, display a remarkable capacity to defend themselves against the vicissitudes of their environment. This self-defense encompasses a paradigm of global regulatory networks, such as the heat-shock and cold-shock regulons, oxidative stress defenses, and the SOS response to DNA damaging agents. These regulatory networks allow the bacterial cell to survive, or even to reproduce, under a large variety of potentially harmful conditions. The regulatory networks involved in defending the cells against a specific stress are usually induced by that specific stress condition (VanBogelen et al., 1990). Yet, global analysis of gene expression using two-dimensional (2D) gel electrophoresis has shown that many proteins produced during specific stresses are induced also during starvation for different essential nutrients when no other stress is imposed on the cells (Groat et al., 1986; Jenkins et al., 1988; VanBogelen et al., 1990; Matin, 1991).

Several of these "general" stress and starvation proteins have been identified using either N-terminal sequencing or mass spectrometry, demonstrating that many display functions related to oxidative management (Jenkins et al., 1988; Matin, 1991; Kolter et al., 1993; Hengge-Aronis, 1993, 2000; Nyström 2002, 2003). For example, primary oxidative defence proteins, such as the superoxide dismutases (SodA and SodB), and catalases (KatE and KatG) as well as proteins involved in the reduction, repair, and proteolysis of oxidized proteins are also increasingly produced in starved cells (Dukan and Nyström, 1999). The latter proteins include peptide methionine sulfoxide reductase, glutathione reductase, thioredoxin, glutaredoxin, and heat-shock chaperones (e.g., Eisenstark et al., 1996; Potamitou et al., 2002a,b; Dukan and Nyström 1998; Nyström, 2002). This suggests that there is an increased demand for oxidation management in cells subjected to nutrient starvation and that protein oxidation might be an intrinsic problem in growth-arrested cells.

*Microbial Proteomics: Functional Biology of Whole Organisms*, Edited by Ian Humphery-Smith and Michael Hecker. 

## 7.2 METABOLIC REARRANGEMENTS TO AVOID OXIDATIVE DAMAGE

Starvation of *E. coli* cells is accompanied by an increased synthesis of glycolysis enzymes and pyruvate formate lyase, phospho-transacetylase, and acetate kinase concomitantly with a reduced production of tricarboxylic acid (TCA) cycle enzymes (Nyström, 1994; Nyström et al., 1996). The two-component response regulator ArcA is required for the decreased synthesis of NADH/FADH$_2$-(reduced form of nicotinamide dinucleotide/flavin adenine dinucleotide) generating TCA cycle enzymes in carbon-starved stationary phase cells; an *arcA* deletion mutant fails to decrease the synthesis of malate dehydrogenase, isocitrate dehydrogenase, lipoamide dehydrogenase E3, and succinate dehydrogenase during stasis, while the increased production of the glycolysis enzymes phosphoglycerate mutase and pyruvate kinase is unaffected (Nyström et al., 1996). The ArcA-dependent response of glucose-starved cells results in a reduced respiratory activity. This modulation in catabolic activities appears important since cells lacking ArcA fail to perform reductive division early during starvation and lose viability at an accelerated rate after a few days in the stationary phase (Nyström et al., 1996).

Gene expression profiling analysis from Conway's laboratory demonstrated that the repression of aerobic metabolism upon the stationary phase correlates with up regulation of *arcA* (Chang et al., 2002). Further it was suggested that ArcA, via its sensor component ArcB, may be activated in the stationary phase by a mechanism encompassing redox control of the quinone pool. ArcB monitors the oxidation–reduction status of the cell by interacting with the membrane quinone pool. Under conditions suitable for aerobic respiration, the quinone pool is primarily oxidized, and this exerts a negative effect on ArcB activity. Interestingly, the gene encoding quinone oxidoreductase, *qor*, is markedly up regulated in stationary-phase cells and it was argued that this may result in an increased ratio of reduced to oxidized quinines. In turn this could activate the ArcB/ArcA system, resulting in repression of the ArcA regulon (Chang et al., 2002).

The role of an activated ArcB/ArcA system in stasis survival may be severalfold. First, the reduced production and activity of the aerobic respiratory apparatus during starvation may prevent an uncontrolled drainage of endogenous reserves. The rate of degradation of endogenous carbon energy sources, such as membrane lipids, is most likely feedback regulated by the activity of the catabolic apparatus. Thus, an uncontrolled respiratory activity, as seen in the Δ*arcA* mutant, may drain these sources and seriously debilitate the membrane, resulting in loss of cell integrity.

Second, the reduced production of respiratory substrate and components of the aerobic respiratory apparatus during starvation may be a defense mechanism mustered by the cell to protect itself against potentially damaging effects of reactive oxygen species produced by the electron transport chain. This notion is supported by data demonstrating that the accelerated die-off of Δ*arcA* mutants starved for glucose could be suppressed by overproducing the superoxide dismutase SodA (Nyström et al., 1996). Thus, ArcA-dependent alterations in the proteome/metabolome of starved *E. coli* cells might be part of an oxidative response—a system that together with the primary oxidative defence and repair proteins mitigates self-inflicted oxidative damage during growth arrest.

## 7.3 OXIDATIVE MODIFICATIONS OF PROTEINS

Proteins can become oxidized through a number of different chemical reactions, and different amino acids are targets for such oxidation. For example, cysteine residues can be oxidatively modified to form sulfydryl groups (Cy–SH) and sulfenic (Cy–SOH), sulfinic (Cy–$SO_2H$), or sulfonic (C–$SO_3H$) derivatives. In addition, disulfide bonds (Cy–S–S–Cy) between two nearby cystein residues within a protein (intramolecular crosslinking) or between two proteins (intermolecular crosslinking) and formation of mixed disulfide (Cy–S–SG) between sulfydryl group and GSH (S-glutathionylation) can result from oxidation of cysteins. Other oxidation reactions include the formation of 2-oxo-histidine from histidine, methionine sulfoxide from methionine, 3,4-dihydroxy-phenylalanine (DOPA) from tyrosine, *o*-tyrosine and *m*-tyrosine from phenylalanine, and *N*-formylkynureine, kynureine, 5-hydroxytryptophan, and 7-hydroxytryptophan from tryptophan (Dalle-Donne et al., 2003). Furthermore, lysine, arginine, proline, and threonine can become oxidized by a carbonylation reaction (Levine, 2002). The carbonyl derivatives are formed by direct oxidative attack on the amino acid side chains ($\alpha$-aminoadipic semialdehyde from Lys, glutamic semialdehyde from Arg, 2-pyrrolidone from Pro, and 2-amino-3-ketobutyric acid from Thr). Carbonyl derivatives on lysine, cysteine, and histidine can also be formed by secondary reactions with reactive carbonyl compounds derived from oxidation of carbohydrates (glycoxidation products), lipids (MDA, HNE, acrolein), and advanced glycation/lipoxidation end products (AGEs/ALEs)

Protein carbonylation is perhaps the most commonly used biomarker of severe oxidative protein damage, and diseases associated with protein carbonylation include Parkinson's disease, Alzheimers disease, cataractogenesis, diabetes, and sepsis (Dalle-Donne et al., 2003; Levine, 2002). Compared to Met sulfoxide and cysteinyl derivatives, carbonyls are relatively difficult to induce and are regarded as markers of severe oxidative stress. Thus, an elevated level of carbonyls is usually a sign not only of oxidative stress but also of a disease-derived protein dysfunction (Dalle-Donne et al., 2003). Moreover, carbonylation appears to be an irreversible process and the cell must rid itself of such protein derivatives by proteolysis. Different sensitive methods have been developed for the detection and quantification of protein carbonyl groups, and most of these involve derivatization of the carbonyl group with 2,4-dinitrophenylhydrazine and subsequent immunodetection of such derivatized groups using monoclonal or polyclonal antibodies (Levine, 2002). The method lends itself nicely to techniques such as Eliza, Western blot detection, and in situ detection in single cells (Ghezzi and Bonetto, 2003; Aguilaniu et al., 2003).

## 7.4 PROTEIN CARBONYLATION IN STARVING *E. COLI* CELLS

Oxidative carbonylation increase during stasis in wild-type *E. coli* cells (Dukan and Nyström, 1998), and proteome analysis demonstrates that this oxidation affects specific proteins, for example, the Hsp-70 chaperone DnaK, the histonelike protein H–NS, the universal stress protein UspA, elongation factors EF-Tu and EF-G, glutamine synthase, glutamate synthase, aconitase, malate dehydrogenase, and pyruvate kinase (Tamarit et al., 1998; Dukan and Nyström, 1998, 1999). Interestingly, some of these proteins have been demonstrated to be specifically carbonylated also in oxidation stressed yeast cells (Cabiscol et al., 2000), aging flies (Yan et al., 1997; Sohal, 2002), and Alzheimer's disease brain (Castegna et al., 2002).

In some cases, the levels of oxidatively damaged proteins have been shown to be associated with the physiological age or life expectancy of an organism rather than with its chronological age. For example, carbonyl levels are higher in crawlers (low life expectancy) than fliers in a cohort of houseflies of the same chronological age (Sohal et al.,

1993). This is also true using in situ detection of protein carbonylation in single *E. coli* cells and a density gradient centrifugation technique to separate culturable and nonculturable, senescing cells of the same chronological age. It was demonstrated that the proteins of nonculturable cells exhibit increased and irreversible oxidative damage (Desnues et al., 2003).

A question of interest is how the asymmetry in population damage is generated. Many genes are differentially expressed during progression through the bacterial division cycle (Weitao et al., 2000; Laub et al., 2000), and a sudden arrest of growth at a time in the cycle when specific gene products (e.g., superoxide dismutases A and B) are present at low levels could generate a subpopulation of cells experiencing increased damage during prolonged stasis. It is interesting to note that the abundance of Sod is much lower in starved *E. coli* cells that have become nonculturable, and the pattern of protein carbonylation is similar in these nonculturable cells and cells totally lacking cytoplasmic Sod activity. Thus, self-inflicted oxidation of proteins is enhanced in both *sod* mutants (Dukan and Nyström, 1999) and nonculturable cells (Desnues et al., 2003). Furthermore, the oxidized proteome of *sod*-deficient cells and nonculturable wild-type cells is very similar. Thus, it is possible that the loss of reproductive ability of some cells entering the stationary phase is linked to the abundance of Sod in individual cells.

## 7.5 MISTRANSLATION AND OXIDATION OF THE *E. COLI* PROTEOME

A possible explanation for the failure of cells to combat stasis-induced oxidation comes from recent results demonstrating that this oxidation occurs by a route that eludes the classical oxidative defense pathways. The levels of oxidized proteins increase upon treatment of cells with antibiotics and mutations causing increased mistranslation (Dukan et al., 2000). Interestingly, during these treatments, the rate of superoxide production and the activity of the superoxide dismutases and catalases are unchanged and the expression of oxidative stress defense genes does not increase (Dukan et al., 2000). In addition, it was demonstrated that increased oxidation during these treatments is primarily the result of aberrant protein isoforms being oxidized (Dukan et al., 2000). Thus, increased protein oxidation can be the result of increased production of aberrant proteins, and this does not require increased generation of reactive oxygen species (Dukan et al., 2000). Moreover, 2D gel electrophoresis of proteins demonstrated that the sudden increase in protein oxidation during the early stages of stasis in *E. coli* is strongly associated with the production of aberrant protein isoforms that appear to be specific targets for oxidative modifications (Ballesteros et al., 2001). In other words, proteomic analysis indicated that carbonylation was strongly associated with protein stuttering. The phenomenon called protein stuttering has been shown to be the result of erroneous incorporation of amino acids into proteins and can be detected on autoradiograms of 2D gels as satellite spots with similar molecular masses to those of the authentic protein but separated from it in the dimension corresponding to isoelectric focusing dimension (O'Farell, 1978; Parker et al., 1978). Frame shifting (Barak et al., 1996), missense errors (O'Farrell, 1978), and stop-codon read-through (Wenthzel et al., 1998; Ballcsteros et al., 2001) increase in response to stasis in *E. coli* cells, suggesting that protein oxidation in nonproliferating cells might be caused by an increase in mistranslation. Indeed, protein carbonylation and stuttering are drastically attenuated in the early stages of stasis in *E. coli* cells harboring intrinsically hyperaccurate ribosomes (Ballesteros et al., 2001). Thus, the elevated oxidation of proteins in nonproliferating cells might be due to the abundance of substrates (aberrant proteins) available for oxidative attack rather than an increased production of reactive oxygen species. The classical oxidative stress defence proteins might be ineffective in counteracting such mechanisms of oxidation.

## 7.6 PROTEOME OXIDATION AND ERROR PROPAGATION

Orgel (1963) has presented a conceptual and mathematical model explaining how an error feedback loop in macromolecular synthesis may cause an irreversible and exponential increase in error levels leading to an "error catastrophe." The feedback loop in Orgel's original model concerned ribosomes and translational accuracy such that errors in the sequences of proteins, which themselves functioned in protein synthesis (e.g., ribosomal proteins, elongation factors), might lead to additional errors. Such a positive-feedback loop was argued to lead to an inexorable decay of translational accuracy and, as a result, aging. The hypothesis is thus based on the assumption that mistranslated proteins can escape degradation and be incorporated into functional (but less accurate) ribosomes. However, several experimental and theoretical approaches, primarily using *E. coli* as a model system, have indicated that increased mistranslation does not cause a progressive decay in the proof-reading capacity of the ribosomes (see Gallant et al., 1997). The susceptibility of mistranslated proteins to oxidation may provide a molecular explanation for this. It has been shown that oxidized proteins are more susceptible to proteolytic degradation than their nonoxidized counterparts (Bota and Davies, 2002; Dukan et al., 2000; Starke et al., 1987). Thus, the rapid oxidation of an erroneous protein may ensure that such a polypeptide is directed to the proteolysis apparatus. This will effectively reduce incorporation of mistranslated proteins into mature machines (e.g., ribosomes and RNA and DNA polymerases) involved in information transfer. In this context, it should be pointed out that the reduced translation fidelity of growth-arrested cells is most likely the result of ribosomes being increasingly starved for charged transfer RNAs (tRNAs) (empty A sites are known to be slippery) rather than being intrinsically error prone.

## ACKNOWLEDGMENTS

Research on protein carbonylation in my laboratory is supported by the Swedish Research Council, VR, an award from the Göran Gustafsson foundation, and the Swedish Strategic Research Foundation, SSF.

## REFERENCES

Aguilaniu, H., Gustafsson, L., Rigoulet, M., and Nyström, T. (2001). Protein oxidation depends on the state rather than rate of respiration in *Sacchormyces cerevisae* cells in the $G_0$ phase. *J. Biol. Chem.* **276**:35396–35404.

Ballesteros, M., Fredriksson, Å., Henriksson, J., and Nyström, T. (2001). Bacterial senescence: Protein oxidation in non-proliferating cells is dictated by the accuracy of the ribosomes. *EMBO J.* **18**:5280–5289.

Barak, Z., Gallant, J., Lindsley, D., Kwieciszewki, B., and Heidel, D. (1996). Enhanced ribosome frame-shifting in stationary phase cells. *J. Mol. Biol.* **263**:140–148.

Bota, D. A., and Davies, K. J. (2002). Lon protease preferentially degrades oxidized mitochondrial aconitase by an ATP-stimulated mechanism. *Nat. Cell. Biol.* **4**:674–680.

Cabiscol, E., Piulats, E., Echave, P., Herrero, E., and Ros, J. (2000). Oxidative stress promotes specific protein damage in *Saccharomyces cerevisiae*. *J. Biol. Chem.* **275**:27393–27398.

Castegna, A., Aksenov, M., Aksenova, M., Thongboonkerd, V., Klein, J. B., Pierce, W. M., Booze, R., Markesbery, W. R., and Butterfield, D. A. (2002). Proteomic identification of oxidatively modified proteins in Alzheimer's disease brain. Part I: Creatine kinase BB, glutamine synthase, and ubiquitin carboxyl-terminal hydrolase L-1. *Free Radical Biol. Med.* **33**:562–571.

Chang, D. E., Smalley, D. J., and Conway, T. (2002). Gene expression profiling of *Escherichia coli* growth transitions: An expanded stringent response model. *Mol. Microbiol.* **45**:289–306.

Dalle-Donne, I., Giustarini, D., Colombo, R., Rossi, R., and Milzani, A. (2003). Protein carbonylation in human diseases. *Trends Mol. Med.* **9**:169–176.

Desnues, B., Gregori, G., Dukan, S., Aguilaniu, H., and Nyström, T. (2003). Differential oxidative damage and expression of stress regulons in culturable and nonculturable cells of *Escherichia coli. EMBO Rep.*, in press.

Dukan, S., Farewell, A., Ballestreros, M., Taddei, F., Radman, M., and Nyström, T. (2000). Proteins are oxidatively carbonylated in response to reduced transcriptional or translational fidelity. *Proc. Natl. Acad. Sci. USA* **97**:5746–5749.

Dukan, S., and Nyström, T. (1998). Bacterial senescence: Stasis results in increased and differential oxidation of cytoplasmic proteins leading to developmental induction of the heat shock regulon. *Genes Devel.* **12**:3431–3441.

Dukan, S., and Nyström, T. (1999). Oxidative stress defense and deterioration of growth-arrested *Escherichia coli* cells. *J. Biol. Chem.* **274**:26027–26032.

Eisenstark, A., Calcutt, M. J., Becker-Hapak, M., and Ivanova, A. (1996). Role of *Escherichia coli rpoS* and associated genes in defense against oxidative damage. *Free Radical Biol. Med.* **21**:975–993.

Gallant, J., Kurland, C., Parker, J., VERSUS Holliday, R., and Rosenberger, R. (1997). The error catastrophe theory of aging: Point counterpoint. *Exp. Gerontol.* **32**:333–346.

Ghezzi, P., and Bonetto, V. (2003). Redox proteomics: Identification of oxidatively modified proteins. *Proteomics* **3**:1145–1153.

Groat, R. G., Schultz, J. E., Zychlinsky, E., Bockman, A., and Matin, A. (1986). Starvation proteins in *Escherichia coli:* Kinetics of synthesis and role in starvation survival. *J. Bacteriol.* **168**:486–493.

Hengge-Aronis, R. (1993). Survival of hunger and stress: The role of *rpoS* in early stationary phase gene regulation in *E. coli. Cell* **72**:165–168.

Hengge-Aronis, R. (2000). The general stress response in *Escherichia coli*. In G. Storz and R. Hengge-Aronis (Eds.), *Bacterial Stress Responses.* Washington DC: ASM Press. pp. 161–179.

Jenkins, D. E., Schultz, J. E., and Matin, A. (1988). Starvation-induced cross protection against heat or $H_2O_2$ challenge in *Escherichia coli. J Bacteriol.* **170**:3910–3914.

Kolter, R., Siegele, D. A., and Tormo, A. (1993). The stationary phase of the bacterial life cycle. *Annu. Rev. Microbiol.* **47**:855–874.

Laub, M. T., McAdams, H. H., Feldblyum, T., Fraser, C. M., and Shapiro, L. (2000). Global analysis of the genetic network controlling a bacterial cell cycle. *Science* **290**:2144–2148.

Levine, R. L. (2002). Carbonyl modified proteins in cellular regulation, aging, and disease. *Free Radical Biol. Med.* **32**:790–796.

Matin, A. (1991). The molecular basis of carbon-starvation-induced general resistance in *Escherichia coli. Mol. Microbiol.* **5**:3–10.

Nyström, T. (1994). The glucose starvation stimulon of *Escherichia coli*: Induced and repressed synthesis of enzymes of central metabolic pathways and the role of acetyl phosphate in gene expression and starvation survival. *Mol. Microbiol.* **12**:833–843.

Nyström, T. (2002). Aging in bacteria. *Curr. Opin. Microbiol.* **5**:596–601.

Nyström, T. (2003). Viable but non-culturable bacteria: Programmed survival forms or cells at deaths door? *BioEssays* **25**:204–211.

Nyström, T., Larsson, C., and Gustafsson, L. (1996). Bacterial defense against aging: Role of the *Escherichia coli* ArcA regulator in gene expression, readjusted energy flux, and survival during stasis. *EMBO J* **15**:3219–3228

O'Farrell, P. H. (1978). The suppression of defective translation by ppGpp and its role in the stringent response. *Cell* **14**:545–557.

Orgel, L. E. (1963). The maintenance of the accuracy of protein synthesis and its relevance to ageing. *Proc. Natl. Acad. Sci. USA* **49**:517–521.

Parker, J., Pollard, J. W., Friesen, J. D., and Stanners, C. P. (1978). Stuttering: High-level mistranslation in animal and bacterial cells. *Proc. Natl. Acad. Sci. USA* **75**:1091–1095.

Potamitou, A., Holmgren, A., and Vlamis-Gardikas, A. (2002a). Protein levels of *Escherichia coli* thioredoxins and glutaredoxins and their relation to null mutants, growth phase, and function. *J. Biol. Chem.* **277**:18561–18567.

Potamitou, A., Neubauer, P., Holmgren, A., and Vlamis-Gardikas, A. (2002b). Expression of *Escherichia coli* glutaredoxin 2 is mainly regulated by ppGpp and sigmaS. *J. Biol. Chem.* **277**:17775–17780.

Sohal, R. S. (2002). Role of oxidative stress and protein oxidation in the aging process. *Free Radical Biol. Med.* **33**:37–44.

Sohal, R. S., Agarwal. S., Dubey, A., and Orr, W. C. (1993). Protein oxidative damage is associated with life expectancy of houseflies. *Proc. Natl. Acad. Sci. USA* **90**:7255–7259.

Starke, P. E., Oliver, C. N., and Stadtman, E. R. (1987). Modification of hepatic proteins in rats exposed to high oxygen concentration. *FASEB J.* **1**:36–39.

Tamarit, J., Cabiscol, E., and Ros, J. (1998). Identification of the major oxidatively damaged proteins in *Escherichia coli* cells exposed to oxidative stress. *J. Biol. Chem.* **273**:3027–3032.

VanBogelen, R. A., Hutton, M. E., and Neidhardt, F. C. (1990). Gene-protein database of *Escherichia coli* K-12: Edition 3. *Electrophoresis* **11**:1131–1166.

Weitao, T., Nordström, K., and Dasgupta, S. (2000). *Escherichia coli* cell cycle control genes affect chromosome superhelicity. *EMBO Rep.* **1**:494–499.

Wenthzel, A. M., Stancek, M., and Isaksson L. A. (1998). Growth phase dependent stop codon read-through and shift of translation reading frame in *Escherichia coli. FEBS Lett.* **421**:237–242.

Yan, L. J., Levine, R. L., and Sohal, R. S. (1997). Oxidative damage during aging targets mitochondrial aconitase. *Proc. Natl. Acad. Sci. USA* **94**:11168–11172.

CHAPTER 8

# Tale of Two Metal Reducers: Comparative Proteome Analysis of *Geobacter sulferreducens* PCA and *Shewanella oneidensis* MR-1

CAROL S. GIOMETTI

Argonne National Laboratory, Argonne, Illinois

## 8.1 INTRODUCTION

*Geobacter sulfurrenducens* PCA and *Shewanella oneidensis* MR-1 both have the capability to use oxidized metals as terminal electron acceptors in their respiratory processes. The reduction of metals as part of their natural electron transport mechanisms makes these microbes candidates for environmental bioremediation. The key to the development of efficient bioremediation protocols, however, lies in understanding the natural metal reduction processes performed by these organisms. Characterization of their proteomes is one component of such an understanding, since knowledge of the proteins they express and how that expression is regulated provides a view of their metabolic potential.

A gram-negative δ-proteobacteria generally considered to be a strict anaerobe, *G. sulfurreducens* was originally isolated from surface sediments at a hydrocarbon-contaminated site and is considered to be a model organism for subsurface soil remediation (Table 8.1) [1]. *Geobacter* spp. have been found as a predominant microbial component of diverse subsurface habitats, including aquatic sediments and pristine deep aquifer, in addition to petroleum-contaminated shallow aquifer [2]. While Fe(III) oxides are the most common terminal electron acceptors in this natural environment, *G. sulfurreducens* has also been observed to reduce elemental sulfur and fumarate. In addition, *G. sulfurreducens* has been shown to facilitate the reduction of soluble U(VI) to insoluble U(IV) in conjunction with iron oxide reduction [3]. In field studies, an increase in the *Geobacter* spp. population was observed to correlate with increased uranium precipitation when acetate (the natural carbon source and electron donor in the subsurface) was pumped into the soil [4]. The potential for the use of endogenous *Geobacter* spp. for waste site bioremediation was the motivation for the U.S. Department of Energy to fund the sequencing of the complete genome of *G. sulfurreducens* PCA [5]. The complete

*Microbial Proteomics: Functional Biology of Whole Organisms*, Edited by Ian Humphery-Smith and Michael Hecker. 

**TABLE 8.1 General Characteristics of *G. sulfurreducens* and *S. oneidensis* MR-1**

| *G. sulfurreducens* | *S. oneidensis* |
|---|---|
| Gram negative δ-proteobacterium | Gram negative γ-proteobacterium |
| Soil subsurface | Freshwater lakes |
| Obligate[a] anaerobe | Facultative anaerobe |
| 3466 ORFs | 4758 ORFs |
| >70 *c*-type cytochromes | 39 *c*-type cytochromes |
| 1011 hypothetical proteins[b] | 1957 hypothetical proteins[b] |
| No plasmid DNA detected | Contains plasmid |

*Note*: ORF = open reading frame.

[a] Prior to sequencing the genome *G. sulfurreducens* was characterized as a strict anaerobe [1].

[b] Total number of conserved hypothetical proteins (those with sequence similarity to a translation of coding sequence from other genomes but with no evidence of actual protein expression) and hypothetical proteins (those for which the genome sequence is consistent with a protein-coding region but for which no comparable genome sequence was found and for which no evidence of expression exists).

genome sequence of *G. sulfurreducens* consists of 3.8 Mb encoding a predicted 3466 proteins. The predicted proteome includes an unusually large number of *c*-type cytochromes, including 23 with sequence similarity to *c*-type cytochromes encoded in the *S. oneidensis* genome, 10 with similarity to those in *Desulfovibrio vulgaris*, and 43 apparently unique to *G. sulfurreducens*.

A gram-negative γ-proteobacteria first isolated from Lake Oneida, *S. oneidensis* MR-1 is found prevalently in aquatic environments and grows aerobically as well as anaerobically (Table 8.1) [6]. In the natural environment, this microbe thrives equally well near the surface of the water using oxygen for respiration or at the bottom of the pond or lake using iron or manganese oxides as electron acceptors. In the laboratory, the diverse respiratory capabilities of this microbe allow use of a broad range of terminal electron acceptors, including oxygen, numerous metal oxides such as Fe(III), Mn(III) and (IV), Cr(VI), and U(VI), fumarate, nitrate, trimethylamine *N*-oxide, dimethyl sulfoxide, sulfite, thiosulfate, and elemental sulfur [7]. The respiratory versatility and metal reduction capability of this microbe have made it a widely used model system for analysis of microbial respiratory pathways and mechanisms and, more recently, as a model system for environmental bioremediation with a focus on groundwater decontamination. The U.S. Department of Energy also funded the complete sequencing of the *S. oneidensis* MR-1 genome, which is comprised of approximately 5.0 Mb encoding a predicted 4758 proteins [8]. The respiratory versatility of *S. oneidensis* is believed to result from a multicomponent branched electron transport system containing a variety of *c*-type cytochromes, reductases, iron–sulfur proteins, and quinines [9]. In fact, the report describing the completed genome sequence of *S. oneidensis* MR-1 stated that this microbe had more *c*-type cytochromes than any other organism for which sequence was available [8]. However, with a total of only 42 predicted *c*-type cytochromes [10], we now know that *S. oneidensis* has the potential to express fewer predicted *c*-type cytochromes than *G. sulfurreducens*, which encodes more than 70.

With the complete genome sequences of both of these microbes publicly available, identification of the proteins they express is now possible, providing the opportunity to determine which proteins are produced under steady-state conditions and which are synthesized only under a specific set of environmental conditions. The proteomes of both *G. sulfurreducens* and *S. oneidensis* are being studied extensively using a variety of methods, including traditional [11, 12] and nondenaturing two-dimensional gel electrophoresis (2DE) [13,14] coupled with tandem mass spectrometry (MS/MS), liquid chromatography (LC) coupled with MS/MS [15], and LC coupled with Fourier transform ion cyclotron resonance (FTICR) MS [15]. Although the number of protein identifications is currently small, comparison of the proteomes expressed by each of these metal-reducing microbes provides a view of the proteins essential to their steady-state metabolism and the opportunity to assess the similarities and differences in their molecular mechanisms for growth, survival, and, perhaps most important, metal reduction. As proteome information accumulates through further research and new proteome methods, construction of a detailed model of the metal reduction pathways, including regulation, will be possible, thus allowing the design of bioremediation protocols that can capitalize on the natural capabilities of these microbes.

## 8.2 THEORETICAL PROTEOMES

The availability of complete genome sequences for *G. sulfurreducens* and *S. oneidensis* allows the prediction of the complete proteomes for each of these microbes (http://GelBank.anl.gov [16]). The theoretical two-dimensional (2D) distribution of these proteomes based on the isolectric points and molecular masses predicted from the ORFs found in the completed genome sequences produces bimodal patterns similar to those observed for many other predicted proteomes (Fig. 8.1). A majority of the proteins in the predicted proteomes for both of these metal-reducing microbes have isolectric points and molecular masses in the range of pH 4–7.5 and 10,000–100,000, respectively. Interestingly, most of the predicted proteins with isoelectric points above pH 10 in both genomes are annotated as either hypothetical or conserved hypothetical, indicating the sequences code for proteins with as yet undetermined function.

The theoretical 2D distributions provide a useful perspective for experimental proteome analyses, since each of the laboratory methods used for proteomics has certain limitations on the isoelectric point and molecular mass ranges attainable. Different classes of proteins require different approaches for detection and characterization, and knowledge of the predicted physical characteristics allows for optimization of the approaches used. For example, resolution of very alkaline proteins remains problematic for traditional 2DE separations and for some LC separations. Nonequilibrium pH gradient electrophoresis (NEPHGE) in the first dimension of 2DE allows the detection of some proteins with isoelectric points above pH 7 [17], and the analysis of peptide fragments by LC followed by MS methods provides more complete analysis of the basic protein constituents [15, 18]. The theoretical 2D maps indicate that over 50% of the *G. sulfurreducens* proteins annotated as cytochrome *c* have isoelectric points above pH 7 while approximately 90% of the *S. oneidensis* proteins annotated as cytochrome *c* have isoelectric points at or below pH 7. For *G. sulfurreducens* and *S. oneidensis*, the *c*-type cytochromes are a particularly important class of proteins to understand and monitor, since they are central to the electron transport chemistry required for metal reduction. With the insight provided by the theoretical 2D maps, strategies for the detection of the more alkaline *G. sulfurreducens* *c*-type cytochromes are obviously needed.

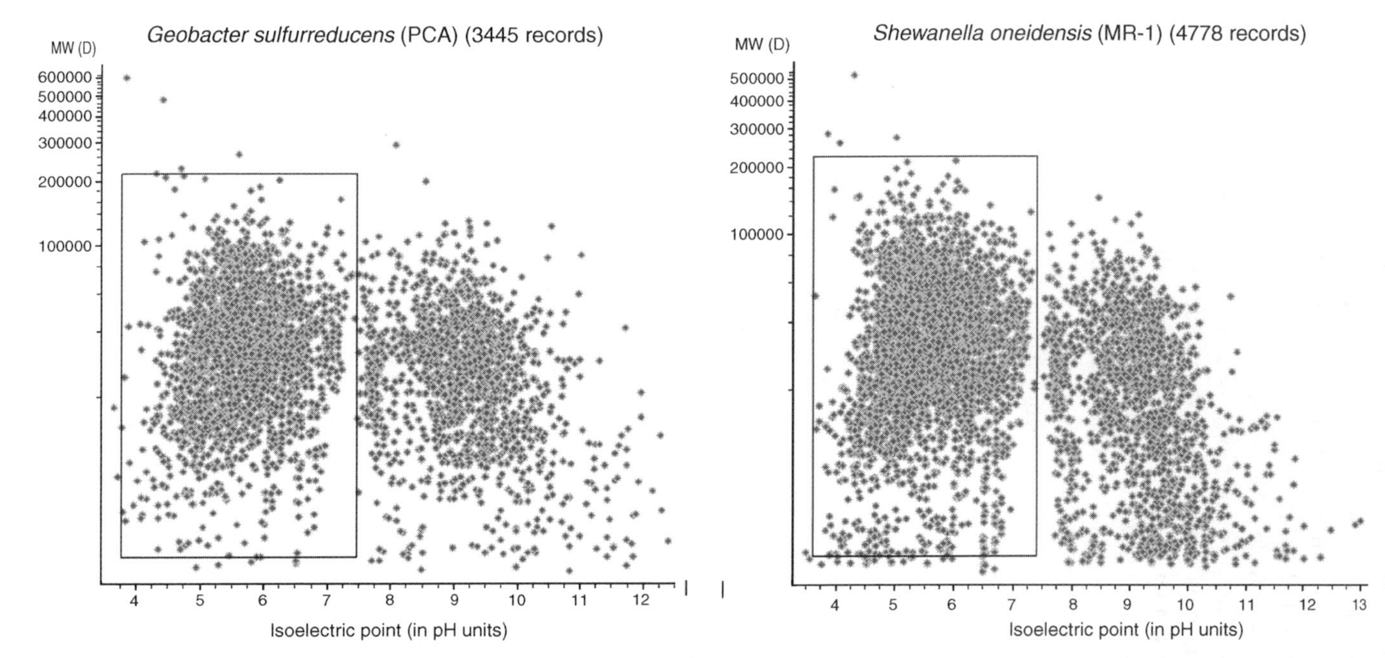


**Figure 8.1** Theoretical proteomes. Using the amino acid sequences predicted from the *G. sulfurreducens* and *S. oneidensis* genome sequences, theoretical 2D protein patterns can be generated (ref. GelBank). A rectangle has been drawn around the region of the graph that represents the isoelectric point and molecular weight range of the proteins resolved by traditional 2DE.

## 8.3 THE *C*-TYPE CYTOCHROMES: METAL REDUCER'S SPECIALIZED PROTEOME

An understanding of the electron transport systems in both *G. sulfurreducens* and *S. oneidensis* is essential to our understanding of their metal reduction capabilities. Even before the availability of the complete genome sequences for these microbes, biochemical methods were being used to isolate and characterize the *S. oneidensis c*-type cytochromes. Since the genome sequence has become available, modern molecular biology methods such as complementary deoxyribonucleic acid (cDNA) microarrays and targeted knock-out mutations are being used to further explore the *c*-type cytochromes in both *G. sulfurreducens* and *S. oneidensis*, although *c*-type cytochromes represent only a small fraction of either of their respective theoretical proteomes. A majority of the characterization of these *c*-type cytochromes has been done using 1D electrophoresis, thereby avoiding the complications associated with the alkaline isoelectric points.

Studies of *G. sulfurreducens c*-type cytochromes have focused on their expression in the context of electron transport to Fe(II), since this metal ion is relevant to the natural soil habitat of this microbe. In addition, Fe(III) reduction has been linked to the reduction of soluble U(VI) to insoluble U(IV), a reaction that is pertinent to the bioremediation goal of immobilizing uranium contamination at waste sites to avoid the contamination of groundwater [19, 20]. The *G. sulfurreducens* genome encodes over 70 *c*-type cytochromes, most of which are predicted to contain more than one heme moiety [5]. For example, structural analysis of one of these polyheme cytochromes, a member of the cytochrome *c*-type protein family, has revealed a three-heme core [21]. Comparison of the genome sequence for this particular cytochrome with the entire *G. sulfurreducens* genome indicated that four additional cytochromes with similar functions also exist. Functional studies of the *G. sulfurreducens c*-type cytochromes using knock-out mutations combined with cDNA microarray and protein electrophoresis analyses have revealed the differential expression of these proteins in response to the presence of Fe(III) in the growth medium. One of these *c*-type polyheme cytochromes, OmcB, has been found associated with the outer membrane [22]. Deletion of the gene for OmcB was observed to decrease the ability of *G. sulfurreducens* to reduce Fe(III), relative to wild-type cells, suggesting a direct role for this protein in the electron transport required for reduction of Fe(III).

In contrast to studies of *G. sulfurreducens c*-type cytochromes in which Fe(III) reduction has been the primary focus, studies of *S. oneidensis c*-type cytochromes have centered on their differential expression under a wide variety of different growth conditions (e.g., aerobic, suboxic, anaerobic with fumarate, iron, nitrate, manganese, chromate, or uranium). The largest number of different *c*-type cytochromes expressed, a total of 19 distinct heme positive proteins, has been observed in *S. oneidensis* grown suboxically without the addition of an alternative electron acceptor [23]. One of the most abundant of the *c*-type cytochromes expressed under both suboxic conditions and anaerobic conditions with fumarate or Fe(III) provided is the periplasmic tetraheme fumarate reductase [11, 24]. In addition to fumarate reductase, *S. oneidensis* has been shown to express six soluble cytochromes [10] and CymA, a tetraheme cytochrome localized in the cytoplasmic membrane, as well as OmcA, OmcB, and the decaheme cytochrome MtrC, all of which are associated with the outer membrane [25, 26]. As is the case with *G. sulfurreducens*, the *S. oneidensis c*-type cytochromes required for electron transport to metal oxides appear to reside on the surface of the cell, as demonstrated by Myers and Myers [27]. In nature, both in the soil environments where *G. sulfurreducens* is found and in the aquatic environments inhabited by *S. oneidensis,* insoluble metal oxides (e.g., goethite, an iron oxide) are the primary terminal electron acceptors in the respiratory process. Hence, location of the *c*-type cytochromes essential to the terminal metal reduction reactions is, most logically, going to be on the surface of the cell.

## 8.4 GLOBAL PROTEOMICS AND METAL REDUCERS

Whereas a majority of the information regarding the expression of *c*-type cytochromes has been obtained using classical biochemical and molecular biology methods such as subcellular fractionation, sodium dodecyl sulfate polyacrylamide gel electrophoresis, and mutation analysis, the availability of the complete genome sequences has prompted global proteome analyses of both *G. sulfurreducens* and *S. oneidensis*. Comprehensive characterization of these proteomes in the context of the completed genomes is providing inventories of expressed proteins and observations of differential expression that can be linked to specific metabolic pathways. As the catalogs of expressed proteins become more complete, the regulatory mechanisms controlling the metal reduction processes in *G. sulfurreducens* and *S. oneidensis* will be elucidated. Volumes of literature have been written describing the strengths and weaknesses of different technological approaches to the global analyses of proteomes. At the time of this manuscript preparation, 2DE and LC, both coupled with either matrix-assisted laser desorption ionization (MALDI) or electrospray ionization (ESI) MS methods, are the most frequently used methods for global proteomics. The use of LC, either one or two dimensional, coupled with MS/MS methods is the most efficient method to obtain an inventory of the proteins expressed by cells grown under a specific set of conditions, but for a relatively rapid assessment of the relative abundance of the major protein components in a proteome and the occurrence of posttransitional modifications, gel-based methods are advantageous. These methods have only recently been applied to the characterization of the proteomes of *G. sulfurreducens* and *S. oneidensis*, providing a superficial view into their respective protein constitution.

### 8.4.1 Expressed Proteins Detected in *G. sulfurreducens* and *S. oneidensis*

Using traditional 2DE with silver stain for protein detection, *S. oneidensis* and *G. sulfurreducens*, both grown anaerobically with ferric citrate as the electron acceptor, have been found to express approximately 600 proteins with isoelectric points between pH 3.5 and 7.5 (Fig. 8.2), compared to the 200–2900 (*G. sulfurrenducens* and *S oneidensis*, respectively) proteins predicted from the respective genome sequences (i.e., proteins with the rectangle shown in Fig. 8.1). Differences between the predicted and actual number of proteins detected can be attributed to (a) protein expression only under specific growth conditions, (b) the detection limit and specificity of the protein-staining method used, (c) the limited dynamic range of the 2DE method, and (d) posttranslational modifications that shift the isoelectric point away from the predicted value.

Using the proteins detected in 2DE patterns of whole-cell lysates, cytosol preparations, and crude membrane fractions (2DE patterns such as those shown in Fig. 8.2 as well as those generated using the nonequilibrium pH gradient electrophoresis method for first-dimension separation [17]), 575 *S. oneidensis* proteins have been identified on the basis of peptide mass and amino acid sequence (http://GelBank.anl.gov). These protein identifications represent the most abundant protein spots detected in the 2DE patterns

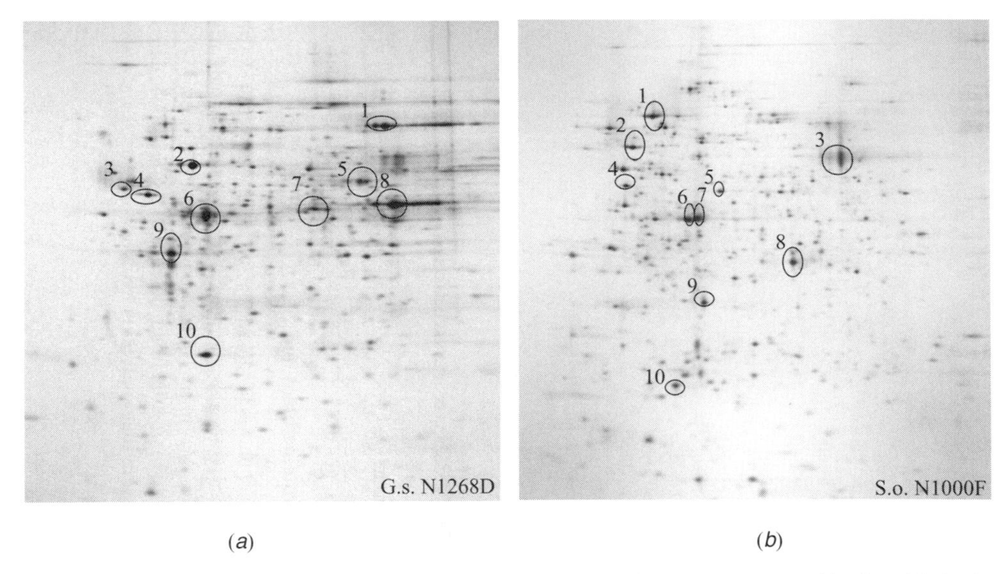


**Figure 8.2** Actual proteomes of *G. sulfurreducens* and *S. oneidensis* grown anaerobically with ferric citrate. Proteins from whole-cell lysates were separated by 2DE and detected by silver stain. Patterns are oriented with acidic proteins to the left and basic proteins to the right, high-molecular-weight proteins at the top and small molecular-weight proteins at the bottom. Protein standards (β-galactosidase, serum albumin, actin, carbonic anhydrase, and soybean trypsin inhibitor) were run within the same sets of gels to provide approximate molecular weights and isoelectric points. The 10 most abundant proteins in each pattern are shown with numbers corresponding to the spot numbers listed in Table 8.2.

stained with either Coomassie blue R250 [28] or silver nitrate [29] and include proteins expressed in *G. sulfurreducens* cells grown with either fumarate or Fe(III) and *S. oneidensis* cells grown with 50% dissolved oxygen (i.e., aerobic conditions), 0% dissolved oxygen (i.e., suboxic), Fe(III), nitrate, or fumarate. For most of the functional categories, the relative number of identified proteins in both microbes is quite similar (Fig. 8.3). Exceptions are found in the categories of cellular processes and protein fate with greater percentage of *S. oneidensis* proteins identified than *G. sulfurreducens* proteins. More proteins associated with regulatory functions, on the other hand, have been identified among the *G. sulfurreducens* proteins detected in 2DE than among the *S. oneidensis* proteins. Also, a larger number of *G. sulfurreducens* proteins with no annotation (NA) or annotated as having unknown function have been identified, whereas more proteins annotated as hypothetical proteins have been identified in the 2DE patterns of *S. oneidensis*. The latter observation could reflect the less mature annotation of the *G. sulfurreducens* genome compared to the *S. oneidensis* genome; that is, proteins currently having no functional annotation could be reannotated as proteins with known function as improved methods for functional annotations become available.

Comparison of the 2DE patterns of whole-cell lysates from *G. sulfurreducens* and *S. oneidensis* grown with Fe(III) shows both similarities and differences among the most abundant proteins detected by silver nitrate staining. Comparing 10 of the most intensely stained protein spots in 2DE patterns as a representative sampling of the identified proteins (Fig. 8.2), translation elongation factor Tu and trigger factor are expressed in relatively high abundance by both *G. sulfurreducens* and *S. oneidensis* (Table 8.2). Subunits of adenosine

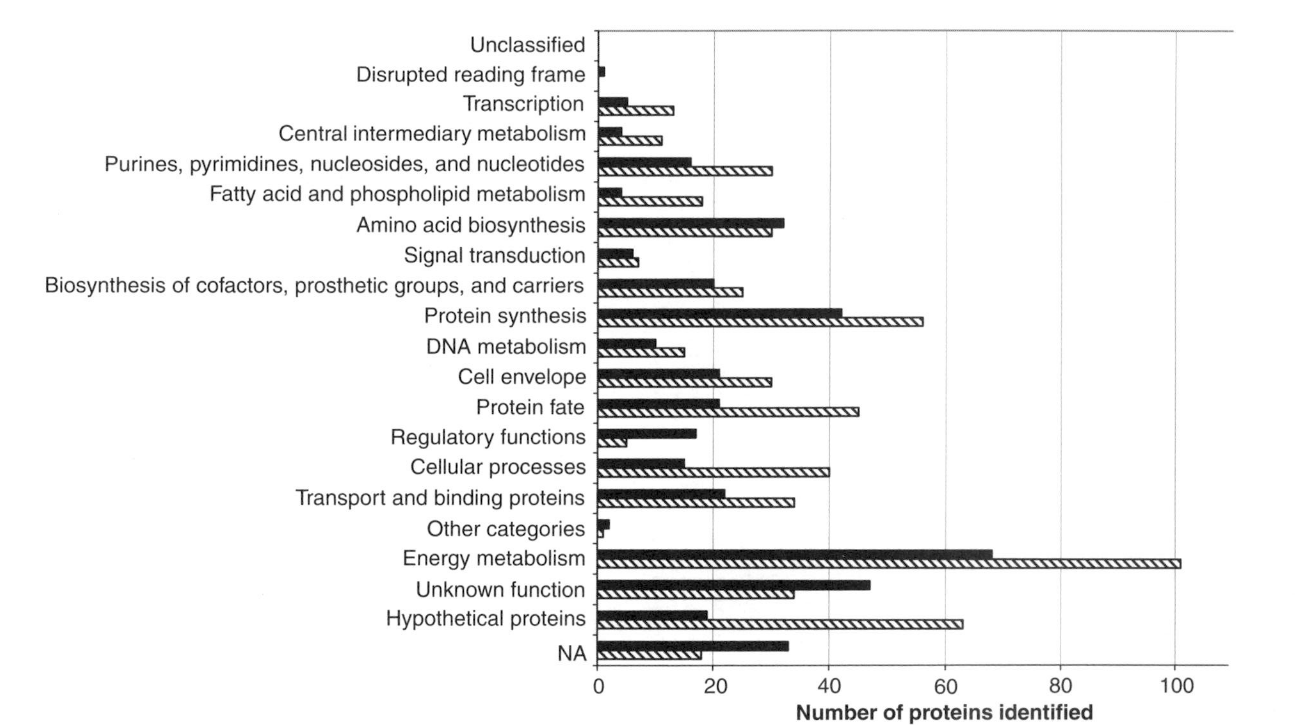


**Figure 8.3** The *G. sulfurreducens* and *S. oneidensis* proteins identified from 2DE gel patterns sorted by functional category. Proteins spots were cut from silver or Coomassie blue stained 2DE gels, the proteins were digested with trypsin, peptides were eluted from the gel pieces, and peptide masses were determined by LC-MS/MS. The peptide masses and amino acid sequences were used to search the ORF databases (http://NCBI website) using Sequest [34] for the annotation which best matched the observed peptide mass. The protein identifications obtained were sorted into functional categories using the Clusters of Orthologous Groups (COG) classification system [35, 36]. Solid bars indicate *G. sulfurreducens* proteins while diagonally hatched bars indicate *S. oneidensis* proteins.

**TABLE 8.2 Comparison of Ten Most Abundant Proteins Expressed by *G. sulfurreducens* and *S. oneidensis***

| Spot Number | *G. sulfurreducens* Protein | *S. oneidensis* Protein |
|---|---|---|
| 1 | Isocitrate dehydrogenase, NADP dependent (GI# 39996565) | Translation elongation factor G (GI# 24372431) |
| 2 | GroEL/HSP60 homolog (GI# 39998429) | Ribosomal protein S1 (GI# 24373949) |
| 3 | Trigger factor (GI# 39996892) | Fumarate reductase (GI# 24372557) |
| 4 | ATP synthase F1, beta subunit (GI# 39995224) | Trigger factor (GI# 24373359) |
| 5 | Prismane protein (GI# 39995780) | ATP synthase F1, alpha (GI# 24376221) |
| 6 | Translation elongation factor Tu (GI# 39997951 / 39997962) | Translation elongation factor Tu (GI# 24371827) |
| 7 | *O*-acetyl-L-homoserine sulfhydrylase (GI# 39983165) | Translation elongation factor Tu (GI# 24371815) |
| 8 | Glu/Leu/Phe/Val dehydrogenase family protein (GI# 39983287) | Alcohol dehydrogenase (GI# 24373064) |
| 9 | Malate dehydrogenase (GI# 39996566) | Translation elongation factor Tu (GI# 24373198) |
| 10 | Keto/oxoacid ferredoxin oxidoreductase, gamma subunit (GI# 39996570) | Alkyl hydroperoxide reductase, C subunit (GI# 24372545) |

*Note*: The identifications of 10 of the most intensely stained proteins shown in Fig. 8.2 are compared. These proteins were identified using the peptide masses obtained from μLC-MS/MS analysis of tryptic peptides to search the appropriate ORF databases using Sequest [31]. The GI# indicates the gi number assigned in the National Center for Biotechnology Information GenBank database.

triphosphate (ATP) synthase (alpha in *S. oneidensis* and beta in *G. sulfurreducens*), a major membrane protein in these microbes, are also among the most abundant proteins in these patterns. The remainder of the 10 most abundant proteins is indicative of the different habitats and metabolic features of these two metal reducers. Whereas *G. sulfurreducens* is naturally found in an anaerobic soil environment rich in iron oxides, *S. oneidensis* inhabits aquatic environments where iron is used as the terminal electron acceptor only in the absence of oxygen. Thus the facultative anaerobe *S. oneidensis* produces a significant amount of fumarate reductase when grown anaerobically with Fe(III), as well as protein-synthesis-related proteins, including translation elongation factors G and S and ribosomal protein S1 to support the needed de novo synthesis of the enzymes and transport proteins needed to survive in the absence of oxygen. Expression of alcohol dehydrogenase and alkyl hydroperoxide reductase is also relatively high in anaerobically grown *S. oneidensis*, indicating the metabolic adjustment to the lack of oxygen.

*Geobacter sulfurrenducens,* in contrast, shows a relatively high steady-state abundance of proteins involved in amino acid and energy metabolism (i.e., isocitrate dehydrogenase, malate dehydrogenase, and keto/oxoacid ferredoxin oxidoreductase), the protein components required for the normal anaerobic metabolism of the electron donor acetate. The function of prismane, also a predominant protein in *G. sulfurreducens* grown with Fe(III), is unknown, but the structure has been shown to include Fe–S clusters [30, 31].

When a subset of the proteins identified thus far from 2DE patterns of *G. sulfurreducens* and *S. oneidensis*, that is, those that have identifications based on greater than 25% amino acid sequence coverage, are compared, additional insight into the similarities and differences between the microbes is gained (Table 8.3). Numerous ribosomal proteins are among the most abundant common proteins detected in both microbes, as are the translation elongation factors. Several outer membrane proteins, including Omp W, K, and H, as well as TolC and MtrB are among the most well represented *S. oneidensis* proteins, while metabolic enzymes such as keto/oxoacid ferredoxin oxidoreductase and 3-isopropylmalate dehydrogenase and the oxidative proteins rubredoxin and catalase are more well represented in *G. sulfurreducens*. In the context of cell motility, flagellin protein is among the more abundant proteins identified in the 2DE patterns of *S. oneidensis*, while a pilus biogenesis protein is among the proteins identified in *G. sulfurreducens*. The *c*-type cytochromes are found in more abundance in the *S. oneidensis* than in the *G. sulfurreducens* patterns. The better representation of outer membrane proteins in the *S. oneidensis* than the *G. sulfurreducens* 2DE patterns suggests that the sample preparation methods used for these studies are more efficient at extracting the *S. oneidensis* proteins than the *G. sulfurreducens* proteins. Since the samples for traditional 2DE were all prepared using 9 M urea, 2% 2-mercaptoethanol, and 4% nonionic detergent [11], these observations indicate that the *G. sulfurreducens* membrane proteins are more difficult to extract than those of *S. oneidensis*. Such a difference in the solubility of the *G. sulfurreducens* membranes could explain the lower efficiency of *c*-type cytochrome extraction relative to that achieved with *S. oneidensis*. Given the obvious importance of a complete inventory of this component of the *G. sulfurreducens* proteome, the use of different solubilization approaches that optimize membrane protein solubility should be a focus for future studies of the global proteome of *G. sulfurreducens*.

Using a combination of "top-down" and "bottom-up" MS approaches for the analysis of the *S. oneidensis* proteins, VerBerkmoes and co-workers were able to identify 868 proteins [15]. This study utilized LC fractionation procedures to reduce the complexity of the protein samples prior to the top-down MS by FTICR, providing the necessary resolution for intact protein analysis. With the top-down approach, evidence of posttranslational processing was detected for several of the proteins identified, including fumarate reductase. A significant number of the MS/MS spectra obtained in this study identified ribosomal proteins, which have also been consistently found as relatively abundant components in both *G. sulfurreducens* and *S. oneidensis* 2DE patterns. Fumarate reductase was also identified, along with three decaheme cytochromes. In general, the proteins identified using the bottom-up LC-MS/MS method corresponded well with those identified using 2DE coupled with LC-MS/MS.

### 8.4.2 Differential Protein Expression in Response to Growth Conditions

The characterization of differential protein expression observed when *G. sulfurreducens* and *S. oneidensis* are grown with different electron acceptors is a major component in determining the mechanisms involved in their metal reduction capabilities. In *G. sulfurreducens*, fumarate respiration, measured as the amount of succinate accumulated, has been shown to be reduced in the presence of Fe(III) [32]. Using 2DE, the abundance of fumarate reductase was observed to be reduced in cultures growing with fumarate when Fe(III) was added and then increased again once the Fe(III) had been reduced to Fe(II). The change in the abundance of fumarate reductase observed using 2DE for global proteome analysis correlated with the messenger ribonucleic acid (mRNA) and fumarate reductase activity analyses [32]. In *S. oneidensis*, 2DE has shown that a significant up

**TABLE 8.3** *G. sulfurreducens* **and** *S. oneidensis* **Proteins Identified from 2DE Gels on Basis of Greater than 25% Matching Amino Acid Sequence Coverage**

| Common to Both *G. sulfurreducens* and *S. oneidensis* | | Unique to *G. sulfurreducens* | Unique to *S. oneidensis* |
|---|---|---|---|
| Trigger factor | Glyceraldehyde 3-phosphate dehydrogenase | Type IV pilus biogenesis protein PilB | Outer membrane protein TolC |
| Translation elongation factor Tu | Fumarate reductase | Type III secretion chaperone | Outer membrane protein precursor MtrB |
| Translation elongation factor Ts | Fructose-1,6-biphosphatase | Thymidylate kinase | Outer membrane protein OmpW |
| Translation elongation factor G | Cysteine synthase A | Rubrerythrin/rubredoxin protein | Outer membrane protein OmpK |
| Thioredoxin reductase | Cold-shock domain family protein | Rubredoxin–oxygen oxidoreductase | Outer membrane protein OmpH |
| Thioredoxin 1 | Clp B Protein | Pyruvate phosphate dikinase | Outer membrane porin, putative |
| Thiol:disulfide interchange protein | Gro Es | Nitroreductase family protection | Flagellin |
| Superoxide dismutase | Gro El | NifU-related protein | Flagellin hook protein |
| Sulfate ABC transporter | Dna K | N utilization substance protein A | Decaheme cytochrome *c* |
| Ribosomal proteins S1, 2, 4, 5, 6, 8,10, 13 | ATP synthase: alpha, beta, delta subunits | *O*-acetyl-L-homoserine sulfhydrylase | Cytochrome *c*551 peroxidase |
| Ribosomal proteins L1, 3, 5, 6, 7/12, 9, 10, 11, 13, 23, 25 | Aspartate semialdehyde dehydrogenase | Keto/oxoacid ferredoxin oxidoreductase | Cytochrome *c* oxidase, cbb3-type subunit III |
| Prismane protein | Aspartate aminotransferase | Glu/Leu/Phe/Val dehydrogenase family protein | Cytochrome $c'$ |
| Polyribonucleotidyltransferase | Adenylosuccinate synthetase | Cytochrome *c* (hsc) | Bacterial surface antigen |
| OmpA family (domain) protein | Aconitate hydratase 2 | DNA repair protein RecN | Antioxidant, AhpC/Tsa family |
| Nucleoside diphosphate kinase | 3,4-Dihydroxy-2-butanone 4-phosphate synthase | Catalase/peroxidase | Alkyl hydroperoxide reductase, C subunit |
| Malate dehydrogenase | Enolase | Arginosuccinate lyase | Alcohol dehydrogenase II |
| Isocitrate dehydrogenase, NADP dependent | DNA-directed RNA polymerase | Aminotransferase, Class B and Class I | Agglutination protein |
| Inosine-5′–monophosphate dehydrogenase | DNA-binding proteins, HU family | Aminopeptidase A/I | Aerobic respiration control protein ArcA |

*Note*: The identified proteins common to both microbes as well as those unique to either *G. sulfurreducens* and *S. oneidensis* are listed. All proteins included in this list were identified on the basis of more than five peptides. These proteins represent some of the most highly expressed proteins from 2DE gel patterns stained with either Coomassie blue R250 or silver nitrate, including proteins that are altered in expression as a result of specific growth conditions.

regulation of fumarate reductase expression occurs whenever the cells are deprived of oxygen, with the exception of anaerobic cultures provided with nitrate [11]. In contrast to *G. sulfurreducens*, the fumarate reductase abundance in *S. oneidensis* is comparable in cells grown with fumarate of Fe(III), suggesting significant differences in the utilization of this enzyme in the metabolic processes of the two microbes. The difference in subcellular location of fumarate reductase (membrane in *G. sulfurreducens* and periplasm in *S. oneidensis*) further supports a fundamental difference in the roles played by this enzyme.

In addition to differences in the abundance of fumarate reductase observed in *G. sulfurreducens* grown with fumarate or Fe(II), a nondenaturing 2DE method developed to allow the analysis of intact proteins [13] has demonstrated that key enzymes in carbohydrate metabolism as well as markers of oxidative stress are expressed more in *G. sulfurreducens* grown with Fe(III) than in cells grown with fumarate [14]. In *G. sulfurreducens*, fumarate is used as both a carbon source for growth and a terminal electron acceptor, whereas Fe(III) serves solely as an electron acceptor. Thus, in the absence of fumarate, an increase in the abundance or activity of the enzymes required to sustain the level of metabolism to support growth is required. In support of this, increases in key enzymes of the tricarboxylic acid (TCA) cycle, such as malate dehydrogenase and pyruvate phosphate dikinase, have been observed in *G. sulfurreducens* grown with Fe(III) relative to cells grown with fumarate.

Although fumarate reductase is, by far, the major difference visible in the 2DE patterns of *S. oneidensis* grown with or without oxygen, numerous other proteins have been identified as differing in abundance when cells grown with different electron acceptors have been analyzed. In a study designed to compare the relative abundance of mRNA and protein products from *S. oneidensis* grown with oxygen, fumarate, Fe(III), or nitrate, a subset of proteins was identified as being significantly more abundant (i.e., $P < 0.05$ with $N = 3$ using a two-tailed student $t$-test) in cells grown under particular conditions [11]. Aerobic cells were found to consistently express more DNA gyrase subunit β, alcaligin receptor protein, and agglutination protein as well as the α subunit of an electron transfer flavoprotein and a specific conserved hypothetical protein than found in cells grown anaerobically with any of the three alternate terminal electron acceptors. Compared to cells grown aerobically, cells grown anaerobically with any of the three alternate electron acceptors showed a dramatic increase in the abundance of formate acetyltransferase, phosphomannomutase, and dihydroxy-2-butanol 4-phosphate synthase. Subsequent studies have revealed the increased abundance of the aerobic respiration control proteins (ArcA) in *S. oneidensis* grown anaerobically with fumarate or Fe(III), a finding confirmed, at least for growth with Fe(III), by Vanrobaeys and co-workers [33]. In addition to these generalized increases in protein expression in response to anaerobic growth with all of the electron acceptors used, *S. oneidensis* grown anaerobically with Fe(III) or nitrate showed a dramatic increase in prismane expression (17-fold in cells grown with Fe(III) and 12-fold in cells grown in nitrate). No increase above the aerobic level was observed in cells grown with fumarate. The increase in *S. oneidensis* prismane abundance observed in response to Fe(III), together with the expression of prismane as one of the more abundant proteins in *G. sulfurreducens* grown with Fe(III) (discussed above), further indicates the involvement of this Fe–S protein in the iron reduction process together with its previously suggested role in aerobic–anaerobic respiration [12]. One unique protein specific to growth anaerobically with Fe(III) has been observed, a conserved hypothetical protein (Gi24375784) with unknown function. The specific expression of this hypothetical protein when cells are grown with Fe(III) suggested that it is associated with the transport or the reduction of the metal ion. Further characterization of this protein, using a combination of bioinformatics, molecular biology, and biochemistry approaches, should elucidate its role.

## 8.5 SUMMARY

*Geobacter sulfurreducens* and *S. oneidensis* are the subjects of intense research efforts due to their potential applications to bioremediation. The characterization of their proteomes, being done in parallel with the analysis of their genome sequences, transcriptomes, and metabolomes, is providing valuable insights to both their similarities and their differences. A primary target of interest in the proteomes of both of these metal-reducing microbes is the characterization of their *c*-type cytochromes. The discovery of their full compliment of *c*-type cytochromes and the description of what growth conditions trigger their expression is central to harnessing their bioremediation potential.

Proteome analyses thus far show that both *G. sulfurreducens* and *S. oneidensis* share the common location of a majority of their *c*-type cytochromes in their outer membranes. The *c*-type cytochromes of *G. sulfurreducens*, however, appear to be less soluble and therefore more difficult to isolate from the membranes than those expressed by *S. oneidensis*. The majority of the *G. sulfurreducens* *c*-type cytochromes also differ from those of *S. oneidensis* in that they have higher isoelectric points, most higher than pH 8.0. These characteristics of solubility and isoelectric point could be related and could indicate an underlying functional difference in the strategy for metal reduction between these two microbes.

The global proteome results available for *G. sulfurreducens* and *S. oneidensis* at the time of this writing are primarily the result of 2DE analysis coupled to protein identification by LC-MS/MS of tryptic peptides from in-gel digests and represent the most abundant proteins detected by Coomassie blue or silver nitrate staining. Currently, several complimentary efforts utilising the 2D-LC-MS/MS approaches are in progress, promising a more complete protein inventory for these microbes in the near future. As these data are added to those already available, the intricate network of metabolic processes, regulation of protein synthesis and protein function, transport of nutrients, and signal transduction will be elucidated. The existing tools of proteomics will be complimented with newer methods such as protein chips and phage display to further characterize these microbial systems. The end result, in the not too distant future, will be predictive models of *G. sulfurreducens* and *S. oneidensis* behavior in their natural habitats under a variety of environmental conditions.

## ACKNOWLEDGMENTS

I would like to acknowledge Dr. Derek Lovley and Dr. Ken Nealson for their willingness to collaborate with my laboratory group in the characterization of the *G. sulfurreducens* PCA and the *S. oneidensis* MR-1 proteomes, respectively. I also want to thank Sandra Tollaksen, Tripti Khare, Angela Ahrendt, and Gyorgy Babnigg for assistance with the preparation of this manuscript. This work was funded by the U.S. Department of Energy, Office of Biological and Environmental Research, through Microbial Genome, Natural and Accelerated Bioremediation Research, and Genomics:GTL Programs under contract W-31-109-ENG-38.

## REFERENCES

1. Caccavo, Jr., F., Lonegan, D. J. Lovley, D. R. Davis, M., et al., *Appl. Environ. Microbiol.* 1994, **60**, 3752–3759.
2. Coates, J. D., Phillips, E. J. P, Lonergan, D. J., Jenter, H., and Lovley, D. R. *Appl. Environ. Microbiol.* 1996, **62**, 1531–1536.
3. Holmes, D. E., Finneran, K. T., O'Neil, R. A., and Lovley, D. R., *Appl. Environ Microbiol.* 2002, **68**, 2300–2306.
4. Anderson, R. T., Vrionis, H. A., Ortiz-Bernad, I., Resch, C. T., et al., *Appl. Environ. Microbiol.* 2003, **69**, 5884–5891.
5. Methe, B., Nelson, K. E., Eisen, J. A., Paulsen, I. T., et al., *Science* 2003, **302**, 1967–1969.
6. Myers, C., and Nealson, K., *Science* 1988, **240**, 1319–1321.
7. Nealson, K. H., and Saffarini, D. A., *Ann. Rev. Microbiol.* 1994, **48**, 311–343.
8. Heidelberg, H. F., Paulsen, I. T., Nelson, K. E., Gaidos, E. J., et al., *Nat. Biotechnol.* 2002, **20**, 1118–1123.
9. Richardson, D. J., *Microbiology* 2000, **146**, 551–571.
10. Meyer, T. E., Tsapin, A. I., Vandenberghe I., de Smet, L., et al., *OMICS* 2004, **8**, 57–77.
11. Beliaev, A. S., Thompson, D. K., Khare, T., Lim, et al., *OMICS* 2002, **6**, 39–60.
12. Thompson, D. K., Beliaev, A. S., Giometti, C. S., Tollaksen, S. L., et al., *Appl. Environ. Microbiol.* 2002, **68**, 881–892.
13. Giometti, C. S., Khare, T., Tollaksen, S. L. Tsapin, A., et al., *Proteomics* 2003, **3**, 777–785.
14. Khare, T., Esteve-Núñez, A., Nevin, K. P., Zhu, W., et al., *Proteomics* 2006, **6**, 632–640.
15. VerBerkmoes, N. C., Bundy, J. J., Hauser, L., Asano, G., et al., *J. Proteome Res.* 2002, **1**, 239–252.
16. Babnigg, G., and Giometti, C. S., *Nucleic Acids Res.* 2003, **32**, D582–585.
17. O'Farrell, P. Z., Goodman, H. M., and O'Farrell, P. H., *Cell* 1977, **12**, 1133–1142.
18. Smith, R. D., *Nat. Biotechnol.* 2000, **18**, 1041–1042.
19. Lovley, D. R., Phillips, E. J. P., Gorby, Y. A., and Landa, E. R., *Nature* 1991, **350**, 413–416.
20. Lovley, D. R., and Anderson, R. R., *J. Hydrol.* 2000, **8**, 77–88.
21. Pokkuluri, P. R., Londer, Y. Y., Duke, N. E., Long, W. C., et al., *Biochemistry* 2004, **43**, 849–859.
22. Leang, C., and Coppi, M. C., *J. Bacteriol.* 2003, **185**, 2096–2103.
23. Blakeney, M. D., Moulaei, T., and DiChistina, T. J., *Microbiol. Res.* 2000, **155**, 87–94.
24. Tsapin, A., Neilson, K., Meyers. T, et al., *J. Bacteriol.* 1996, **178**, 6386–6388.
25. Myers, J. M., and Myers, C. R., *Appl. Environ. Microbiol.* 2001, **67**, 260–269.
26. Beliaev, A. S., Saffarini, D. A., McLaughlin, J. L., and Hunnicutt, D., *Mol. Microbiol.* 2001, **39**, 722–730.
27. Myers, C. R., and Myers, J. M., *Lett. Appl. Microbiol.* 2003, **37**, 254–259.
28. Anderson, N. L., Nance, S. L., Tollaksen, S. L., Giere, F. A., and Anderson, N. A., *Electrophoresis* 1985, **6**, 592–599.
29. Giometti, C. S., Gemmell, M. A., Tollaksen, S. L., and Taylor, J., *Electrophoresis* 1991, **12**, 536–543.
30. Pierik, A. J., Wolbert, R. B., Mutsers, P. H., Hagen, W. R., and Veeger, C., *Eur. J. Biochem.* 1992, **206**, 697–704.
31. Stokkermans, J. P., van den Berg, W. A., van dongen, W. M., and Veeger, C., *Biochim. Biophys. Acta* 1992, **1132**, 83–87.

32. Esteve-Nunez, A., Nunez, C., and Lovley, D. R., *J. Bacteriol.* 2004, **186**, 2897–2899.
33. Vanrobaeys, F., Devreese, B., Lecocq, E. Rychlewshi, L., et al., *Proteomics* 2003, **3**, 2249–2257.
34. Eng, J. K, McCormack, A. L., and Yates, J. R. III., *J. Am. Soc. Mass Spectrom.* 1994, **5**, 976–989.
35. Tatusov, R. L., Koonin, E. V., and Lipman, D. J., Science 1997, **278**, 631–637.
36. Tatusov, R. L., Natale, D. A., Garkavtsev, I. V., Tatusova, T. A., et al., *Nucleic Acid Res.* 2001, **29**, 22–28.

CHAPTER 9

# AMT Tag Approach to Proteomic Characterization of *Deinococcus radiodurans* and *Shewanella oneidensis*

MARY S. LIPTON,[1] MARGARET F. ROMINE,[2] MATTHEW E. MONROE,[1] DWAYNE A. ELIAS,[1] LJILJANA PASA-TOLIC,[1] GORDON A. ANDERSON,[1] DAVID J. ANDERSON,[1] JIM FREDRICKSON,[2] KIM K. HIXSON,[1] CHRISTOPHE MASSELON,[1] HEATHER MOTTAZ,[2] NIKOLA TOLIC,[1] and RICHARD D. SMITH[1]

[1]Biological Sciences Division and [2]Environmental and Molecular Sciences Laboratory, Pacific Northwest National Laboratory, Richland, Washington

## 9.1 INTRODUCTION

Microbes can respond, adapt, and change their external environment through differential expression of genes and proteins that lead to alterations in the structure, behavior, and metabolism of the organism. While microarray technology is now extensively used to characterize global messenger ribonucleic acid (mRNA) content, global proteome (the protein complement of an organism) analysis methodologies are still more immature and have been applied to fewer biological systems.

The advent of global analysis of both mRNA and proteins has been facilitated by the availability of complete genome sequences. The rate at which genomes can be sequenced currently far surpasses the rate at which they can be functionally characterized. While DNA microarrays are especially useful for unraveling cellular regulatory networks, other methodologies are necessary to validate predictions that a cell expressed a particular protein and that they are present under various growth conditions. It is only by looking at the protein expression patterns and levels of protein abundance directly that a more complete understanding of a microbial organism can be obtained. The maturation of mass spectrometry (MS) based global proteomics technologies opens the door for interpreting these biological systems.

Contemporary proteomic analyses have predominantly been based on MS coupled with either classical two-dimensional polyacrylamide gel electrophoresis (2D PAGE) separation of intact proteins followed by in gel digestion or more recently developed liquid-based chromatographic separations developed for the enzymatically digested protein complement of a cell, otherwise known as "gel-free" or "shotgun" proteomics

*Microbial Proteomics: Functional Biology of Whole Organisms*, Edited by Ian Humphery-Smith and Michael Hecker. 

[1–4]. Although, 2D gels are still widely utilized to determine the changes in protein expression patterns and to resolve protein isoforms, 2D PAGE-based proteomics relies on visualizing spots on the gel and therefore is severely limited in proteomic coverage, sensitivity, and dynamic range [5, 6]. Additionally, since protein identification requires significant postgel processing of the sample, throughput for protein identification is severely limited.

In recent years, efforts to develop gel-free approaches that employ alternative separation methodologies combined with MS have intensified to address these limitations [7–9]. These approaches bypass the need to visualize spots on 2D gels by analyzing peptides from enzymatically digested proteins without prior separation of the individual proteins from whole cells or cell fractions. Traditionally, peptides generated from a proteolytic digest of lysed cells (or a fraction therein) are separated by reversed-phase chromatography, eluted into a mass spectrometer, and analyzed by tandem mass spectrometry (MS/MS) techniques. If the eluting peptide has a high enough abundance in the mass spectrometer (as measured by an MS scan), the peptide is isolated within the instrument and fragmented, resulting in breaking the peptide, usually along the peptide backbone. The acquired mass spectra of the fragments serve as the basis for identification of the peptides by the application of specifically designed algorithms (i.e., SEQUEST or MASCOT). These algorithms match information from the MS/MS spectra to the sequenced genome of the organism [10].

The extent to which a proteome can be characterized using these gel-free approaches is limited by the separation efficiency of the single-dimension separation. Washburn et al. developed an online 2D separation that utilizes an ion exchange fractionation step in line with the reversed-phase separation [8, 11]. While this method results in an increased characterization of the proteome, it does so with a significantly increased experimental analysis time and sample consumption. Each experiment can take 20–36 h, which is over 5–10 times longer than a normal 1D experiment, but without a corresponding increase in identified peptides. With this method, peptide identification relies on the MS/MS data for the peptide in that particular analysis, and little data can be carried from run to run, resulting in the need to reanalyze each different culture condition separately. Additionally, since the eluting peptide is fragmented during the MS/MS experiment, information about the relative abundance of the peptide is lost, eliminating the potential for quantitative experiments. While practical for analyzing a few cellular conditions, this limitation hinders the application of this method for high-throughput analysis of many cellular conditions.

We, and others have developed approaches and the supporting technology that allows large numbers of proteins to be rapidly characterized directly from cell lysates without the need for separation by gel electrophoresis [12, 13]. In our case, the technology relies on creating peptide biomarkers referred to as mass-and-time (MT) tags for each of the proteins expressed in a particular organism, which are then used to identify the same expressed proteins in later experiments. Because the approach uses separate liquid chromatography (LC) MS/MS analysis to identify unique peptide tags, the approach enables higher throughput analysis and also facilitates the use of peptide abundance information for targeting quantitative changes in the proteome. This technology has recently been demonstrated in proteomc characterization studies of the intensely radiation-resistant microbe *Deinococcus radiodurans* [13]. Importantly, the approach can and is being applied to a variety of other biological systems.

The technology [14, 15] blends protein identifications obtained from LC-MS/MS experiments with accurate mass measurements obtained from LC Fourier transform ion cyclotron resonance (FTICR) MS experiments. By comparing calculated masses and elution times of the initial assignments or MT tags with those measured in the FTICR experiments, the initial identifications are validated as accurate mass-and-time (AMT) tags for future quantitative experiments. This chapter will explore the use of the AMT tag technology for surveying microbial proteomes. Specific examples from two microbial systems, *D. radiodurans* and *Shewanella oneidensis* MR-1, will be used to illustrate the types of studies that can be addressed with such a method.

*Deinococcus radiodurans* has the extraordinary ability to withstand 50–100 times more ionizing radiation than *Escherichia coli.* The level of radiation resistance within *D. radiodurans* depends on the culture conditions and the growth state of the organism. For example, *D. radiodurans* is much more radiation resistant in the stationary phase compared to the early log phase and in rich media compared to minimal media. Addition of amino acids, nicotinamide adenine dinucleotide (NAD), and/or Mn(II) has been shown to increase resistance [16, 17]. Recent investigations indicate that relative intracellular concentrations of Fe and Mn play an important role in resistance and recovery from ionizing radiation, exposure, or desiccation.

*Shewanella oneidensis* MR-1 is a gram-negative, facultative anaerobe and a respiratory generalist [18] that can oxidize organic compounds or $H_2$ using solid-phase metals such as Fe(III) or Mn(III,IV) oxides as electron acceptors and gain energy for maintenance and growth from such reactions [19–21]. As a dissimilatory metal-reducing bacterium (DMRB), *Shewanella* can also reduce relatively soluble and mobile contaminants, including Tc(VII), U(VI), and Cr(VI), to lower oxidation states that are much less soluble and therefore less mobile in the environment [18]. *Shewanella oneidensis* MR-1 exhibits remarkable metabolic versatility in regards to electron acceptor utilization [22–24]. This versatility is thought to allow MR-1 to efficiently compete for resources in environments where electron acceptor type and concentration fluctuate in space and time.

In the following sections, examples from both *D. radiodurans* and *S. oneidensis* MR-1 will be used to illustrate both organism-centric and protein-centric methods for interpreting proteomic data. The organism-centric method utilizes a systemswide evaluation of the data and incorporates information about culture condition, localization, and functional characterization of large subsets of proteins for biological characterization. The protein-centric method delves deeply into the aspects of a single protein and utilizes information about expression, localization, posttranslational modifications, and peptide coverage for protein characterization. Ultimately, the essential quantitative determination of changes in protein expression corresponding to change in culture conditions will require the blending of both sets of information.

## 9.2 MATERIALS AND METHODS

### 9.2.1 Bacterial Cultures

Methods for culturing *D. radiodurans* were described previously [13]. Samples analyzed included cells grown at 32°C in defined media [17] to midlog [$OD_{600}$ (optical density) 0.3–0.4], late-log ($OD_{600}$ 0.6), and stationary phase ($OD_{600}$ 0.9) or TGY [17] (rich) media to midlog ($OD_{600}$ 0.3–0.4), late-log ($OD_{600}$ 0.6), and stationary phase ($OD_{600}$ 0.9). For stress studies, cells were grown to midlog phase ($OD_{600}$ 0.3–0.4) at 32°C in rich media prior to transitioning to stress conditions. Heat-shock stress was induced by raising the temperature to 42°C and cold-shock stress by lowering the temperature to 0°C for 1 h prior to sampling. Hydrogen peroxide stress was induced for 2 h by adding $H_2O_2$ to a final concentration of 60 μM. Prolonged starvation (one week and four weeks) was applied to post-stationary-phase ($OD_{600}$ 0.9) cells without addition of fresh medium. Chemical shock was induced by 2 h exposure to 0.05% (v/v) TCE or *o*-xylene. Alkaline shock was induced for 1 h by raising the culture pH from 6.5 to 8.5 with 1 N NaOH. Cells were routinely collected by centrifugation at 10,000 rpm at 4°C, washed three times with phosphate-buffered saline (PBS), and aliquots quick frozen with Liquid $N_2$ for storage at −80°C.

*Shewanella oneidensis* MR-1 [American Type Culture Collection (ATCC) 700550] cultures were routinely grown at 30°C in defined or rich media, in shake flask, batch, or under chemostat control. Shake flask cultures were grown with complex media, either Luria broth (LB) or tryptic soy broth (TSB) (Becton Dickenson). Alternatively, Bioflow model 110 fermentors (New Brunswick), in either fed-batch or steady-state mode, were used to grow cells in a modified defined medium that was carbon limited. Defined medium was prepared as described elsewhere [23, 25] with the following modifications: the $K_2HPO_4$ was replaced with 0.6 g/L $NaH_2PO_4$ and $(NH_4)_2SO_4$, ethylenediaminetetraacetic acid (EDTA), and $NaHCO_3$ were excluded. D,L-Lactic acid was added to a final concentration of 1.123 g/L. Electron acceptors were either $O_2$ at 50, 20, or 0.1% dissolved oxygen (DO) (controlled by a DO probe during culturing) or fumaric acid (4.8 g/L). For steady-state conditions, fresh medium was pumped into the reactor at a dilution rate of 0.1 $h^{-1}$. Fermentors were maintained at 30°C and medium pH at 7.0 ± 0.1 by addition of 2 N KOH or 2 N NaOH. The stirring rate was 450 rpm. Cells from cultures were pelleted by centrifugation at 27,000 *g* for 8 min at 4°C. The resulting pellets and approximately 2–3 mL of the supernatant were transferred to a 1.0-mL cryovial and centrifuged at 9000 *g* for 4 min at 4°C. After removal of the supernatant, the cryovials containing packed cell pellets were immediately stored at −80°C for later analysis.

### 9.2.2 Cell Lysis and Tryptic Digestion

Cell lysis was achieved by bead beating (3 min) in a Biospec Minibeadbeater with 0.1-mm zirconia/silica beads in 0.5-mL sterile siliconized microcentrifuge tubes. Beads were immediately removed by centrifugation (16,000 rpm) at 4°C for 5 min and remaining lysates placed on ice to inhibit proteolysis. The protein concentration of the lysates was determined by the Pierce BCA assay kit. Cell lysates were treated by the addition of dry urea, thiourea, and DTT to final concentrations of 7 M, 2 M, and 5 mM, respectively, and incubated at 60°C for 30 min. The sample was diluted 10-fold with 100 mM ammonium bicarbonate (pH 8.4) and $CaCl_2$ was added to a final concentration of 1 mM. Sequencing-grade, modified trypsin (Promega) was prepared by adding 20 μL of 100 mM ammonium bicarbonate (pH 8.4) to a vial containing 20 μg trypsin and then, after 10 min incubation at 37°C, this mixture was added to the cell sample. Tryptic digestion was performed for 5 h at 37°C using a 1 : 50 (w/w) trypsin-to-protein ratio. Sample clean-up was achieved using a 1-mL SPE C18 column (Supelclean SPE LC-18, Supelco). The sample was eluted using 80% acetonitrile, 20% water with 0.1% trifluoroacetic acid (TFA) and concentrated via speed-vac. The final peptide concentration was determined as above. The samples were then quick frozen in liquid nitrogen and stored at −80°C until analyzed.

### 9.2.3 Capillary LC Separations

The capillary LC system consisted of a pair of syringe pumps (100-mL ISCO model 100DM) and controller (ISCO, series D), and in-house manufactured mixer, capillary column selector, and sample loop for manual injections. Most separations were achieved with 5000 psi reversed-phase packed capillaries (150 μm i.d. × 360 μm o.d.; Polymicro Technologies) [26] using two mobile-phase solvents consisting of 0.2% acetic acid and 0.05% TFA in water (A) and 0.1% TFA in 90% acetonitrile–10% water (B). More recently, separations using higher pressures and smaller inner diameters have also been implemented to provide higher separation peak capacities and higher sensitivity. In all cases, the mobile-phase selection valve was switched from position A to B 10 min after injection, creating an exponential gradient as mobile phase B displaced A in the mixer. Flow through the capillary high-performance liquid chromatography (HPLC) column was ~1.8 μL/min when equilibrated to 100% mobile phase A.

### 9.2.4 Mass Spectrometric Analysis

All samples were analyzed as previously described [13]. Eluent from the HPLC was infused into a conventional ion trap MS (LCQ, ThermoFinnigan) operating in a data-dependent MS/MS mode over a series of segmented *m/z* ranges and also with prior fractionation using ion exchange chromatography. For each cycle, the three most abundant ions from MS analysis were selected for MS/MS analysis using a collision energy setting of 35%. Dynamic exclusion was used to discriminate against previously analyzed ions. The collision-induced disassociation (CID) spectra from the conventional ion trap mass spectrometer were analyzed using SEQUEST [10] and the protein sequences deduced from the *S. oneidensis* MR-1 genome sequence. Additionally, all samples were analyzed by a 9.4-tesla (Bruker Daltonics) or in-house-built 11-tesla FTICR mass spectrometer. Both systems used an electrospray ionization interfaced to an electrodynamic ion funnel assembly coupled to a radio frequency (RF) quadruple electronic ion guide for collisional ion focusing and highly efficient ion accumulation and transport ions into the cylindrical ion cyclotron resonance (ICR) cell for analysis [27].

### 9.2.5 Peptide Verification and Mapping to Open Reading Frames (ORFs)

A variety of peptide identification cutoff filters were used for this work for comparison. Briefly, the most liberal filter allowed peptides with $X_{corr} \geq 2$ regardless of the charge state. Charge-state-based filters used a combination of both $X_{corr}$ and charge state for filtering, where, for +1 peptides, $X_{corr} \geq 2$; for +2 peptides, $X_{corr} \geq 2.5$; and for +3 peptides, $X_{corr} \geq 3.2$. The criteria for the filters based on both frequency and charge state are shown in Table 9.1. The standard filter used in this chapter for those studies not used for comparison was composed of preliminary peptide filters based on both frequency and charge state (Table 9.1). Additionally, peptides were required to be fully or partially tryptic, and each peptide was required to be detected with high accuracy in the FTICR analysis of the same sample. Since the separation systems for both the FTICR and the LCQ analyses were identical, the confirmation of the peptides was based on both the calculated mass (from the MT tag database) and measured mass (from the FTICR analysis) of the peptide matching to within 5 ppm and the "normalized" elution time matching to within 5%.

**TABLE 9.1** ***Shewanella oneidensis*** **ORF Identification as Function of Protein Localization**

| Location | Total | 2AMT[a] | 4AMT[a] |
|---|---|---|---|
| Cytosol | 2304 | 1891 (82%) | 1594 (69%) |
| Inner membrane | 1659 | 1463 (88%) | 1267 (76%) |
| Periplasm | 351 | 308 (87%) | 268 (76%) |
| Outer membrane | 122 | 111 (91%) | 101 (82%) |
| Outer membrane or periplasm | 136 | 119 (87%) | 101 (74%) |

[a]Designates the least number of mass tags for protein identification. Numbers in parentheses represent, precentage of the total ORFS predicted to be contained in each location that were identified with at least the designated number of AMT tags.

## 9.3 RESULTS AND DISCUSSION

### 9.3.1 Annotation

High-throughput methodologies for characterizing microbial proteomes is enabled by the availability of complete genome sequences and predictions of the sequences of proteins encoded therein. A comprehensive annotation of the functions for these proteins facilitates interpretation of comparative proteome analyses. Typically, only half of the proteins predictions are based on sequence similarity, resulting in populating protein databases with a rapidly growing number of "hypothetical" proteins predicted with low confidence. Proteome analysis is vital to validate that these proteins are indeed expressed. While there is more evidence to support predictions in the "conserved hypothetical" category, they too require validation by proteome analysis. This category is defined by The Institute for Genome Research (TIGR) to include proteins that have significant sequence similarity to the conceptual translation of a gene from another species for which there is no experimental evidence that a protein is actually expressed (http://www.tigr.org/CMR2/db_assignmentextver2.shtml).

For a protein to be validated, its sequence must be included in the databases used during SEQUEST analysis. Because there is frequently more than one version of the genome annotation, it is important that these databases include a compilation of all of protein sequences predicted in each annotation. For *D. radiodurans* and *S. oneidensis*, our databases are comprised of proteins associated with two distinct annotations of the genome available in Genbank, additional proteins found at TIGRs Comprehensive Microbial Repository (CMR), and predictions made by researchers focused on analyses of these organisms. In addition we include sequences deduced from genes with internal stop codons predicted. To account for possible sequencing mistakes, genome variability, and programmed frame shifting, we include both the truncated form of the protein and the "repaired" protein sequence. The appropriate "repair" is derived through sequence comparison to orthologous proteins or identification of programmed frame shifting signals [28].

### 9.3.2 AMT Tag Method

The AMT tag approach, a "bottom-up" comprehensive proteomics method, was developed to circumvent the need for solid-phase 2D gels and provide greater sensitivity, dynamic range, and throughput [29]. The AMT tag approach involves (1) generating an

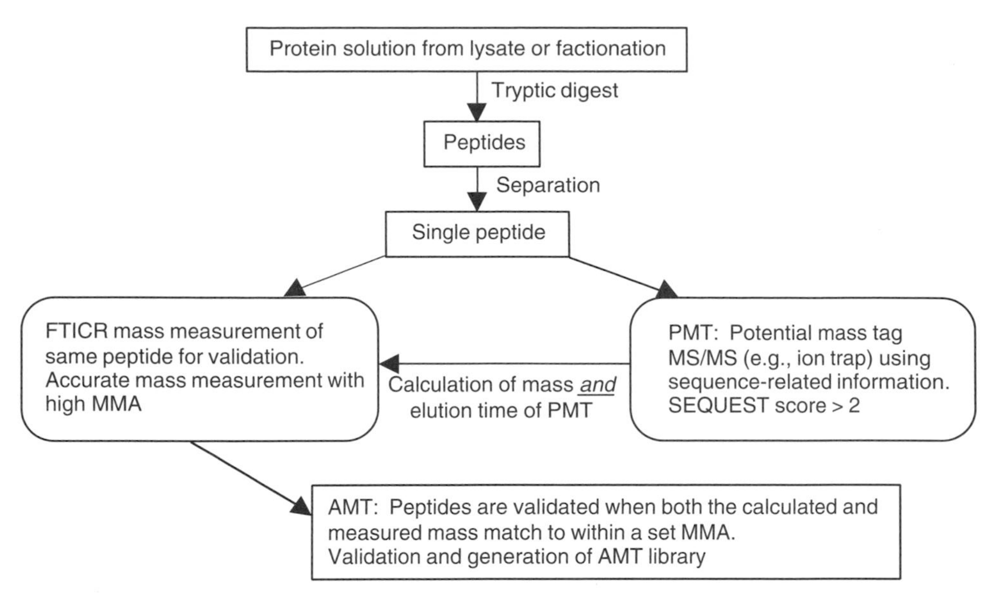


**Figure 9.1** Overview of process for creation of AMT tag database. Proteins are digested with trypsin and analyzed by both LC-MS/MS and LC-FTICR MS. The data from both types of mass spectrometers is used together to create a library of validated mass tags whose identity has been verified by both methods.

MT (i.e., peptide biomarker) database and (2) performing high-throughput LC-FTICR experiments that use this MT database to identify and quantify peptides and proteins [14]. An AMT tag is defined as a single peptide whose "measured molecular mass" is sufficiently accurate that it is unique among all possible peptides predicted from a genome [4].

Similar to other approaches, the AMT tag approach uses a multidimensional LC separation strategy coupled with conventional LC-MS/MS for initial peptide/protein identification (Fig. 9.1). An MT database is first populated as follows. Initial peptide identification is based on the comparison of the MS/MS fragmentation pattern of the peptide to sequenced genomes using SEQUEST [10]; after identification, the peptide's exact mass can be calculated and its LC elution time is now known. Final confirmation of the identification is based on (1) correlation of calculated mass, (2) LC elution time, and (3) observation of one or more peptides by LC-FTICR MS with high mass accuracy. Once the peptide has been confirmed, the parent protein containing the given peptide is verified. A major advantage to this approach is that once the MT tag database for an organism is created, subsequent analyses rely solely on LC-FTICR MS measurements enabling high-throughput analysis of any new samples provided for analysis. Although the mass tag database is continuously evolving with the addition of new experiments, the database is usable after the first sets of peptides are deposited. Additional LC-MS/MS experiments are needed only when the experimental culture conditions differ significantly from those already represented in the database. Thus, the AMT tag approach reduces the time required for analysis from days to hours, reducing the sample consumption without sacrificing sensitivity and dynamic range and, as such, is amenable for development into a high-throughput proteomics platform. Additionally, unlike LC-MS/MS experiments, the

LC-FTICR measurements are more quantitative, especially when coupled with isotope labeling, since all the data are used.

### 9.3.3 Global Protein Identification

***Filtering Approaches*** Most proteomic technologies are based on initial identification of the peptides by matching MS/MS spectra against a known genome. Such matching usually produces a score ($X_{corr}$ with SEQUEST or probability score with MASCOT). The challenge is to distinguish the correct identifications from the incorrect ones. For scores that lie at the extremes, this process is straightforward; however, most peptides fall in the gray area of false positives (incorrect identifications that are labeled correctly) or false negatives (correct identifications that are not identified correctly). Early proteomic work with SEQUEST set the $X_{corr}$ cutoff at 2 for a positive identification. This resulted in the inclusion of many false identifications in the data set. The use of accurate mass information provided a measure of the false-positive rate as a function of SEQUEST score [14]. While this second accurate mass filter was effective, large numbers of peptides identified incorrectly from the use of low scores ultimately increase the error rate even after application of the accurate mass criterion [14]. More recently, the analyses have incorporated a charge-dependent scoring system to increase confidence in peptide identifications [11]. Other groups have used a blend of two searching algorithms for peptide identification, taking those that are identified by both methods [30]. Still others have designed new algorithms for peptide identification [31].

The level at which the filters are set should take into account the goal of the analysis. For example, lower levels of false-positive identifications would be appropriate for studies aimed at validating protein predictions. However, higher levels of false positives may be tolerated in studies of induced changes in protein expression when complementary approaches (e.g., DNA microarray analysis, traditional 2D gel-based proteome analysis) are used to characterize samples and the data are carefully evaluated by a knowledgeable biologist.

An example of the use of these filters using different cutoff scores and criteria for SEQUEST identification of peptides from *S. oneidensis* samples is shown in Table 9.2. A filter that uses an $X_{corr}$ value cutoff of $>2$ yields 360,309 peptide identifications, which map to 4875 of a possible 5109 proteins present in the database. This high number of protein identifications is unexpected given the limited number of culture conditions examined. Incorporating a set of charge-state-dependent rules, only 3497 proteins were identified through 27,750 unique peptides characterized. The average number of peptides per protein decreased as well.

As the number of MS/MS spectra increases for the characterization of a proteome, so too does the likelihood that spectra include peaks that result incorrectly in a false identification. Stated differently, the number of incorrect assignments will likely increase nearly in proportion to the total number of spectra generated. By relating identification frequency to the scoring algorithm (i.e., the more spectra that have been assigned to a particular peptide, the more the confidence increases), the required score can be relaxed. For *S. oneidensis*, the proteome was characterized with more than 640,000 MS/MS spectra yielding 360,309 total peptide identifications where 292,908 were only identified by a single analysis. By raising the required score for these peptides, it was possible to remove many false positives while still retaining the correct peptide identifications. To date, we have used these types of filters to identify 30,605 unique peptides that were assigned to 3812 proteins in *S. oneidensis* (Table 9.2).

Ultimately, a separate analysis technique that employs different metrics is desirable for validating peptides. The AMT tag method incorporates such a mechanism by first identifying peptides using SEQUEST and then validating each peptide by matching masses and elution times with FTICR measurements. Such a method allows "further filtering" of the data and added confidence in peptide identifications. For *S. oneidensis*, preliminary peptides were filtered by a charge-state- and frequency-dependent method and validated by FTICR measurements, producing 27,654 peptides that corresponded to 3588 proteins (Table 9.2). These parameters will be used in discussions of peptide identification that follow.

As more LC-MS/MS experiments are used to characterize the proteome of an organism, the stringency of the filters should be increased. For example, while the early work on *D. radiodurans* was based on 17 different culture conditions [13], only about 500 LC-MS/MS runs were used to form the initial mass tag database. Using the liberal filters that were accepted as standard, the work yielded unique tags for slightly over 63% of the predicted proteome (summarized in Table 9.3). While the use of ion exchange chromatography for the fractionation of the peptides increased the dynamic range of the LC-MS/MS analyses and provided greater coverage of the proteome, the process also produced many more MS/MS spectra. As mentioned previously, an increase in the number of MS/MS spectra increases the false identifications. If one simply applied the original filters to this much larger data set, unique tags for all the predicted proteins would be identified (many incorrectly). However, applying more stringent criteria for peptide identification that is based on a combination of $X_{corr}$, charge state, and frequency of identification to the data from the fractionation experiments confidently identified unique tags for at least 2349 of the 3118 predicted proteins (Table 9.3). *Deinococcus radiodurans* has multiple genetic elements, two chromosomes predicted to encode 2633 and 365 proteins and two plasmids encoding 145 and 40 proteins. We verified the expression of 1959, 268, 99, and 23 proteins from the large chromosome, small chromosome, megaplasmid, and small plasmid, respectively. Therefore, while the application of ion exchange chromatography for the initial peptide identification yielded more potential MT tags and necessitated the use of more stringent filters, the confidence in the resulting peptide identifications as well as the proteome coverage is greater.

***Fully Tryptic versus Partially Tryptic Peptides*** Another debate in the proteomic community centers on the use of only full, complete or exact tryptic peptides (where both ends of the peptide correspond to predicted tryptic cut sites) versus also including partial tryptic peptides (where only one end corresponds to a tryptic cut site) or nontryptic peptides (a nonspecific peptide). While trypsin digestion should result in full tryptic peptides, some contend that trypsin has some nonspecific cleavage activity that will result in partial or nontryptic peptides. In the *S. oneidensis* database, 11,562 peptides are fully tryptic, corresponding to 3575 proteins. Adding the partial tryptic peptides, the protein count is raised to 3845. Nontryptic peptides are not added to the database and, as a result, were not used in the studies exemplified here. Since it is unlikely that many of the additional 270 proteins identified using partial tryptic peptides are valid, we settled on criteria requiring at least one fully tryptic peptide for a protein. However, once a protein has been identified with a tryptic peptide, partial tryptic peptides can be used for further confirmation. For instance, because many proteins may have *incorrectly assigned start* *sites* or *postprocessing events that change the N -terminus*, these events would be invisible if only full tryptic peptides were used.

***Confidence of Protein-Level Identifications*** An additional issue revolves around the level of confidence required for protein identifications, that is, how many peptides are required to confidently identify a protein. For example, using only AMT tags (i.e., peptides that have passed the frequency-based filters and were validated with the FTICR), the 30,605 AMT tags translate into 3812 proteins expressed in the *S. oneidensis* cultures studied to date (Table 9.4). Since the initial protein database used for these studies contained 4875 proteins, the level of proteome coverage corresponded to 78%. However, with the extra level of validation afforded by the AMT tag approach, the presence of a single peptide should be sufficient to identify the presence of expressed proteins in a cell culture. Of course, the more peptides identified as originating from a protein, the greater the likelihood that the protein is present. Raising the criteria to two peptides detected per protein, the level of proteomic coverage drops to almost 60%, indicating that 859 proteins were identified from the detection of a single peptide. Similarly, if the criterion is raised to 3 or 4 peptides required to identify an protein, the levels of proteome coverage drop to 48 and 39%, respectively. The average molecular mass of all predicted proteins to be expressed in *S. oneidensis* (Table 9.2) is about 32 kDa. Of the proteins observed whose identifications are based on the observation of one or more peptides, the average molecular mass increases to 38 kDa, indicating a slight bias for the identification of larger proteins. Not surprisingly, as the number of AMT tags needed for identification increases, the average molecular mass of the observed proteins also increases. This point illustrates that escalating the number of peptides required to identify an expressed protein discriminates against proteins that produce fewer tryptic peptides in the molecular mass range applicable for identification. Therefore, it is desirable to choose a criterion that maintains the increased confidence of protein identifications afforded by observing multiple peptides from a given protein while minimizing the discrimination against smaller proteins. Thus, all future analyses described here are based on the observation of at least two AMT tags for protein identification.

A better parameter for determining the confidence in peptide identification is a ratio of either peptides identified per protein to the number of predicted peptides possible or of the number of amino acids represented in the identified peptides to the total number of amino acids predicted in the protein. Both types of ratios take the size of the protein into account, but the former would take the portions of the protein not amenable to tryptic digestion into account. Take, for example, three proteins from *S. oneidensis*, SO0249 encoding ribosomal protein L30, SO0009 encoding DNA polymerase III, and SO0998 encoding formate dehydrogenase, each of which had been identified with 16 peptides. Accordingly, using the single peptide parameter count, each would be afforded the same confidence. However, because these proteins differ drastically in size, (SO0249, SO009, and SO0988 possess 60, 366, and 1428 amino acids, respectively), our amino acid coverage of these proteins varies greatly, producing 85, 62, and 23%, respectively.

***Functional Categorization*** Creating an overview of all the proteins that are expressed is just the first step in the characterization of the microbe.

Through annotation of the genome, each protein can be assigned to a functional category. The distribution of predicted proteins among the functional categories for both *D. radiodurans* and *S. oneidensis* is shown in Table 9.5. Not surprisingly, an overview of the coverage of proteins validated in each category for either organism revealed that we identified more of the proteins in categories such as protein and nucleotide synthesis and energy metabolism. The representation for most categories was higher in *D. radiodurans* than in *S. oneidensis*, with a notable exception in the category corresponding to biosynthesis of cofactors. The extraordinarily high abundance of *c*-type cytochromes (42 in all) in *S. oneidensis* requires greater levels of heme than most other bacteria. Consequently, we expect the proteins involved in synthesis of the heme cofactor to be highly abundant under many of our growth conditions and thus easier to detect by the AMT tag approach.

Over 80% of the *D. radiodurans* proteins that were assigned to the hypothetical category were identified, while a smaller representation of this category was observed for *S. oneidensis*. Due to the limited culture conditions that were used for *S. oneidensis* compared to *D. radiodurans*, this result is not unexpected. The representation of conserved hypothetical proteins was almost 80% for both organisms. Proteome validation of their expression can be used to update their annotation to unique proteins or conserved proteins, respectively.

***Culture Condition Specific to Protein Expression*** While an overview of all samples analyzed is useful for validating protein identifications, analysis of culture-specific expression can be used to develop and understanding of cellular response to the environment and to develop more precise predictions of protein function. Under a given condition, microbes are predicted to only express detectable levels of only a fraction of their proteome, as it is energetically unfavorable to do otherwise. Differential expression patterns can be used to identify proteins that are constitutively expressed (typically housekeeping or essential proteins) versus those that exhibit condition-specific expression. For example, *D. radiodurans* protein expression patterns derived from each of the 17 culture conditions studies were evaluated to determine which proteins were constitutive versus those that were induced. This study has been described in detail elsewhere [13] with a few representative proteins shown in Table 9.6 to demonstrate the value of this comparison. Proteins such as elongation factors, ribosomes, and S-layer protein serve housekeeping functions of the cell and are present under every condition. Similarly, some hypothetical and conserved hypothetical proteins also demonstrate this expression pattern potentially indicating that they also exhibit housekeeping functions. Conversely, sets of proteins are only expressed under stressed conditions. While the explanation for the expression patterns of some proteins in this set remain elusive, the hypothetical and conserved hypothetical proteins with this expression pattern may warrant further investigation as to their potential role in stress response.

Finally, the glycerol-3-phosphate regulon repressor is turned on during the stationary phase, possibly indicating that the organism is heading for stasis. The characterization of protein expression patterns is especially useful for developing subsequent experiments to functionally characterize proteins in the hypothetical and conserved hypothetical category. They also are useful for identifying which proteins play important roles in condition-specific functions of special interest to the biologist (e.g., proteins involved in dissimilatory metal respiration of *S. oneidensis* or radiation-induced DNA repair by *D. radiodurans*).

***Characterization of Membrane Proteins*** The detection and characterization of membrane proteins represent a challenge for proteomic methodologies. A prediction of protein localization can be derived by computational analyses with tools such as PSORT [32], PsortB [33] (predicts both protein localization and cleavage sites for signal peptidase I and II), and LipoP [34]. As with the annotation process, these tools are only predictive and must be validated by experimental data. PSORT of *S. oneidensis* proteins predicts that approximately 58% of the proteins are soluble, with 51% residing in the cytoplasm and 7% in the periplasm. Membrane proteins can be difficult to release during sample preparation and consequently are more difficult to detect. PSORT predicted that 42% of the *S. oneidensis* proteins are membrane associated, with 35% predicted to localize to the inner membrane and 7% to the outer membrane.

The routine sample preparation method used in the AMT tag approach separates soluble (cytoplasmic or periplasmic) and insoluble (membrane-bound) proteins for analysis. For *S. oneidensis*, procedures for release of proteins from the insoluble portion of a cell lysate or from outer membrane vesicle samples resulted in identification of between 88 and 91% (Table 9.7) of the proteins predicted to localize to the inner and outer membrane, respectively, using a filter that required at least two AMT tags per protein, while 76 and 82% were identified by filtering for identifications with a minimum of AMT tags. Only 82% of the predicted cytosolic proteins (based on two or more AMT tags per protein) were identified in this study. This lower number may reflect a greater number of proteins that are expressed at levels below detection (e.g., regulatory proteins) and a larger proportion of incorrect protein predictions (e.g., small hypothetical proteins under 100 aa in length).

PSORT and other tools such as TmPred [35] and Signal P [36] provide estimations of the number of transmembrane spanning regions for inner membrane proteins. *Shewanella oneidensis* possesses proteins predicted to have up to as many as 28 transmembrane spans. There was no clear correlation between the number of predicted transmembrane regions to the number of proteins identified using AMT tags. Indeed, the level of coverage for proteins predicted to have four or more transmembrane regions was slightly higher (91%) than those proteins predicted to possess three or less of these regions (88%). Not surprisingly, the average number of observed AMT tags was 21 for proteins with less than three transmembrane regions, while those with more averaged only 9 AMT tags, which we attribute to the transmembrane region being less accessible to trypsin digestion.

### 9.3.4 Proteomic Data for Alternate Coding Patterns and Gene Prediction

Besides functional predictions, the annotation process also involves predictions of the protein termini. While sequence homology and occurrence of conserved Shine–Dalgarno sites assists in predictions of the N-terminus, there will inevitably be mistakes. Post translational processing of either the C- or N-terminal ends of proteins is typically not included in the annotation. Data from proteome analysis can validate the sequence of the mature protein termini and provide evidence that supports posttranslational processing. For example, analysis of the *S. oneidensis* proteome identified N-terminal AMT tags for 338 proteins and C-terminal AMT tags for 644 proteins.

If the predicted protein N terminus is internal to the true N-terminus, it can be validated by reanalyzing the proteome spectra with a database that is comprised of proteins deduced from the longest stretch of amino acids between every start and stop codon in the genome sequence (start-to-stop database). For example, the *D. radiodurans* cold-shock protein encoded by DR0907 was predicted to be 133 amino acids in length. Additional orthologous bacterial cold-shock proteins deposited in public databases since the original

**Figure 9.2** Map of identified peptides that correspond to regions of predicted protein DR0907 whose alternate start site is designated by the arrow. All but two of the peptides (designated A and B) show a high level of protein coverage and overlap. The outlying peptides A and B correspond to regions of the protein whose expression is questionable and show minimal overlap with the other peptides.

annotation of *D. radiodurans* suggested that an additional 47 residues exist at the N-terminus. Reanalysis of the *D. radiodurans* raw data sets support this hypothesis. Twenty-nine AMT tags, illustrated in Figure 9.2, were found for the protein and overlapped throughout the additional N-terminal sequences predicted. Only two AMT tags (designated by A and B) mapped to the region outside the gene start-and-stop positions. Each peptide was identified in only three MS/MS analyses, was partially tryptic, and had charge states of +3 and $X_{corr}$ values of 2.4 and 2.4, respectively. These two peptides are probably incorrectly assigned and the other peptides validate the later start side.

While most proteomics analyses to date have in effect focused on validating annotation predictions, few studies have used proteomics methods to predict new or questionable ORFs. Recently, Jaffe and VeerannaPant have shown the usefulness of proteomic methodologies for gene prediction in *Mycoplasma pneumonia* [37] where the annotation was disregarded and the MS/MS data were searched against the entire genome sequence. We have used the AMT approach similarly with *D. radiodurans*, where the genome was translated in all six reading frames and possible ORFs were assigned at the start codon proceeding a stop codon and end at the next codon. This "stop-to-stop" database can be used to discover ORFs missed by the annotation process. One challenge in doing this is the large number of possible peptides, which increases the probability of false positive identifications. For this reason, the peptides need to pass higher stringency filter, and flexible tools for data visualization are helpful. One such tool is illustrated in Figure 9.3, which illustrates all six ORFs of the genome for the large plasmid of *D. radiodurans* as lines. The gray portion of the line represents the predicted annotation and the red squares represent the peptides that were observed. Region A corresponds to ORFs for which no peptides were identified, region B corresponds to sections where peptides did not align with annotated ORFs, region C corresponds to an annotated ORF that does have peptides that align well with the prediction, and region D corresponds to a potential frame shift. Using this stop-to-stop database, 12 new ORFs were discovered for *D. radiodurans* and are in the process of being validated by other methods.

Various mechanisms have been discovered that enable recoding of genomic DNA during the translation process [28]. Consequently, protein sequences deduced directly from the genome sequence may inaccurately describe the true sequence of the protein expressed. Alternatively, there can be mistakes in the genome sequence or growth-condition-specific variability in the genome sequence leading to incorrect predictions of

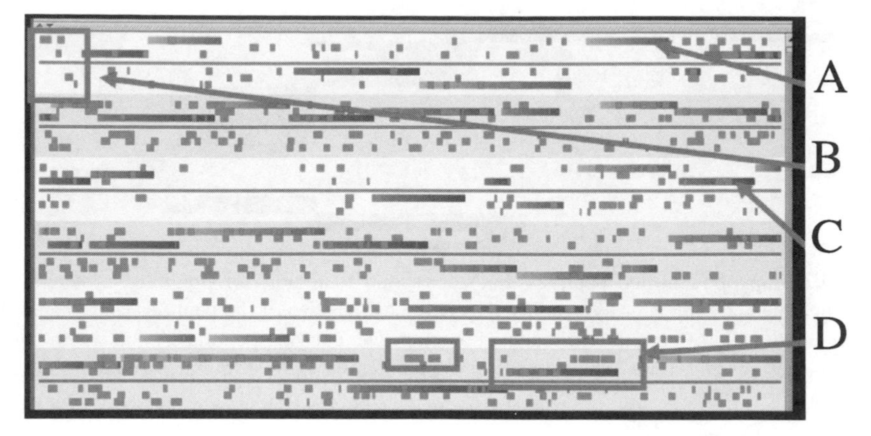


**Figure 9.3** Illustration of all six reading frames of genome for large plasmid of *D. radiodurans* as lines. The gray portion of the line represents the predicted annotation and the red squares represent the peptides that were observed. Region A corresponds to ORFs for which no peptides were identified, region B corresponds to sections where peptides did not align with annotated ORFs, region C corresponds to an annotated ORF that does have peptides that align well with the prediction, and region D corresponds to a potential frame shift.

protein sequences. A stop-to-stop database (deduced by translating sequences between all stop codons in the genome) can also be used to validate the occurrence of these events. For *D. radiodurans*, the Comprehensive Microbial Resource database (http://www.tigr.org/tigrscripts/CMR2/GenomePage3.spl?database=gdr) lists an additional 71 genes that are predicted to contain point mutations that result in premature termination of a protein or frame shifting. Using this approach, we identified peptides for asparaginyl transfer ribonucleic acid (tRNA) synthetase and several other proteins derived from genes believed to be mutated. Some of these peptides indicate that full-length proteins are expressed and consequently are likely functional. For example, we detected peptides that map to the DR0100 gene (putative single-stranded DNA-binding protein; SSB) [38], which is predicted to encode three frame shifts. In addition, Figure 9.4 shows that AMT tags were detected in all three reading frames of gene DR2477, suggesting that this gene encodes a full-length protein. Additional experimentation is necessary to determine whether a sequence mistake or specific cellular recoding function explains this finding.

### 9.3.5 Quantitative Measurements

Methods for relative or absolute quantitation of proteins include counting numbers of protein-specific peptides in a sample [39], modification of specific amino acids with isotope labels [isotope coded affinity tagging (ICAT) [40], and isotope labeling of culture medium [41], each having both advantages and disadvantages. While the number of peptides is roughly correlated to protein abundance, quantitation of relatively small or large proteins by the number of observed peptides is problematic. For example, small proteins have fewer peptides so small changes in numbers of peptides result in large changes in abundance. Tryptic peptides result from the cleavage of a protein at the N-terminal side of either lysine or arginine. The digestion procedure, however, is not always complete, resulting in intact cleavage sites. Further complicating this issue is the use of

Frame 1

**MKIKKAAVIGAGVMGAAIAAQLANAGIPVLLLDIVLPDKPDRNFLAKAGVERALKARPA**
**AFMDNDRARLIEVGNLEDDLKK**LKDVDWVLEAIIEK**LDAKHDLWEKVEKVVKK**TAIISS
NSSGIPMHLQIEGR**SEDFQRRFVGAHFFNPPR**YLHLLEVIPTDKTDPQVVKDFSEFAEH
TLGK**GVVVANDVPGFVANRIGVYGIVRAMQHMEKYGLTPAEVDQLTGPALGRASSATFR**
TADLSGLDIISHVATDIGGVTTADLSGLDIISHVATDIGGVTTADLSGLDIISHVATDI
GGVT

Frame 2

**KTKILNLNLQTGEYEDQGKVRVPAVDAVKGKPLAERVNALYTLEGKEGDFLRATMNDGF**
**WYAAKMAGTVSGRLQDIDNALK**WGFGWEQGPFETMDTIGVQQVIK**NLEAEAA**

Frame 3

**EPGSRSRTLPPLLQKMKESGADKFYNGNETVTPSGEKTEFKAPYFIVADLKKDATKVVK**
KRPGASVVDLGDGVLLVEWHAK**MNALGEDQLRAVQDAHKLVQDMGYAGLVVGNQGEHFS**
**AGANLPLILAQAQADEWDELDDQIKQFQQTTTSMRFSPHPTVSAPFNMTLGGGCEFSLH**
**ADR**VVASAETYMGLVEVGVGLIPGGGGTK**EMLLR**FTDLQQPGQQLGMTLLPAVQR**AFEL**
**IATAKVSTSALDARKLGFLKDHDTVVMNKNHVIEEAKRTVLALAPDYVQPVMR**TDIPVM
GDAAIGAIKSALYGMHEGGYITDYDLVVSNELAR**VLSGGTGNNRTAKVSEQHLLDLERE**
**AFLTLAGKKGTQQRIEHMLKTGKPLRN**

**Figure 9.4** One-letter assignment for protein DR2477. Peptides were detected in all three reading frames starting with frame 1 (designated in brown) shifting to frame 2 (designated in purple) and ending in frame 3 (designated in blue). The amino acids colored green are those for which no peptides were detected.

partial tryptic peptides where trypsin has cleaved at a site other than lysine or arginine. Conversely, for large proteins, a significant increase in the number of peptides corresponds to a relatively small change in abundance.

Other quantitation methods involve the addition of a labeled amino acid to the culture medium. While the stable isotope labeling with amino acids (SILAC) method has been shown to be useful in eukaryotic systems [42] that need a media supplemented with amino acids, most prokaryotic systems have biosynthetic pathways for these amino acids, thus creating a pool of unlabeled amino acids. The activity of such pathways in nonauxotrophic organisms results in the incomplete labeling of proteins and an inability to determine the relative abundance of the peptide. A related approach involves labeling of specific amino acids, for example, ICAT [43]. While broadly applicable and advantageous due to reduction in sample complexity, a drawback of such approaches can be artifactual reaction side products (which lead to misidentifications) and the failure to detect some proteins (e.g., significant fraction of prokaryotic proteins do not contain cysteine).

The most accurate methods for MS-based quantitation used with MS utilize stable isotope labeling strategies where one labeled species serves as an internal calibrant for the other unlabeled species [44–47]. One such strategy uses the metabolic incorporation of a stable isotope, usually $^{15}N$, in the growth media of the cell, thus ensuring uniform labeling. In this approach, the two distinct cultures of cells are created and mixed after harvest. As all subsequent processing of the sample is in one "pot," processing differences between samples are eliminated.

For example, *S. oneidensis* was cultured suboxically and aerobically under chemostat conditions utilizing natural and $^{15}N$-labeled defined media. After mixing cultures and

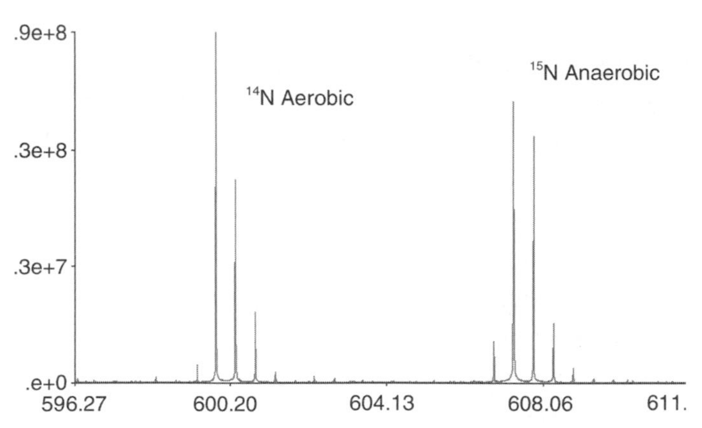


**Figure 9.5** Peptide mass spectra. Mass spectra of two peptides, one unlabeled and the other labeled with $^{15}N$. Two distinct isotopic distributions are detected for each peptide.

sample processing, tryptic digestion mixtures were analyzed by capillary LC-FTICR. For each peptide, two representative isotopic distributions were detected, as illustrated in Figure 9.5. Each distribution is separated according to the number of nitrogens in the peptide, and their intensities can be compared to determine abundance. However, the two different species do not always coelute, and a more effective manner of calculating abundance is to sum the intensity as measured by LC-FTICR of the mass spectral peaks across the elution time window. Figure 9.6 shows the comparison of abundance between aerobic and anaerobic conditions for peptides from three representative proteins: fumarate reductase, isocitrate lyase, and ribosomal protein L3. The elution profiles show that the abundance of the ribosomal protein does not change between the two conditions: fumarate reductase is significantly more abundant in anaerobic conditions and isocitrate lyase is more abundant in aerobic conditions. The protein abundances were determined by averaging the abundances of all the peptides identified for that protein.

A representative sample of these results is shown in Figure 9.7, where rows 1 and 2 are the control comparisons ($^{14}N$ aerobic to $^{15}N$ aerobic and $^{14}N$ suboxic to $^{15}N$ suboxic) and rows 3 and 4 are the experimental samples ($^{14}N$ aerobic to $^{15}N$ suboxic and $^{14}N$ suboxic to $^{15}N$ aerobic). Each abundance measurement represents an average of all the peptides detected for that protein. A black box designates no change detected between the two cultures, yellow boxes designates increase in abundance of the $^{14}N$ peptide over the $^{15}N$ peptide, and blue boxes indicates a decrease in abundance of the $^{14}N$ peptide over the $^{15}N$ peptide. Rows A and B represent the comparison of aerobic-to-aerobic and suboxic-to-suboxic samples. Not surprisingly, the boxes for the proteins are black, indicating that the peptide abundances for the proteins were unchanged. Column C represents triplicate samples mixing $^{14}N$ aerobic and $^{15}N$ suboxic mixtures, while column D represents triplicate samples mixing $^{14}N$ suboxic and $^{15}N$ aerobic mixtures. Proteins associated with housekeeping functions such as elongation factors and ribosomes were unchanged in each comparison. However, for proteins such as fumarate reductase, the box in column C is blue (indicating a decrease of the $^{14}N$ peptide compared to the $^{15}N$ peptide) and the box for column D is yellow (indicating an increase of the $^{14}N$ peptide compared to the $^{15}N$ peptide). As column C compares $^{14}N$ aerobic to $^{15}N$ suboxic and column D represents the comparison of $^{14}N$ suboxic to $^{15}N$ aerobic, these results indicate a higher level of

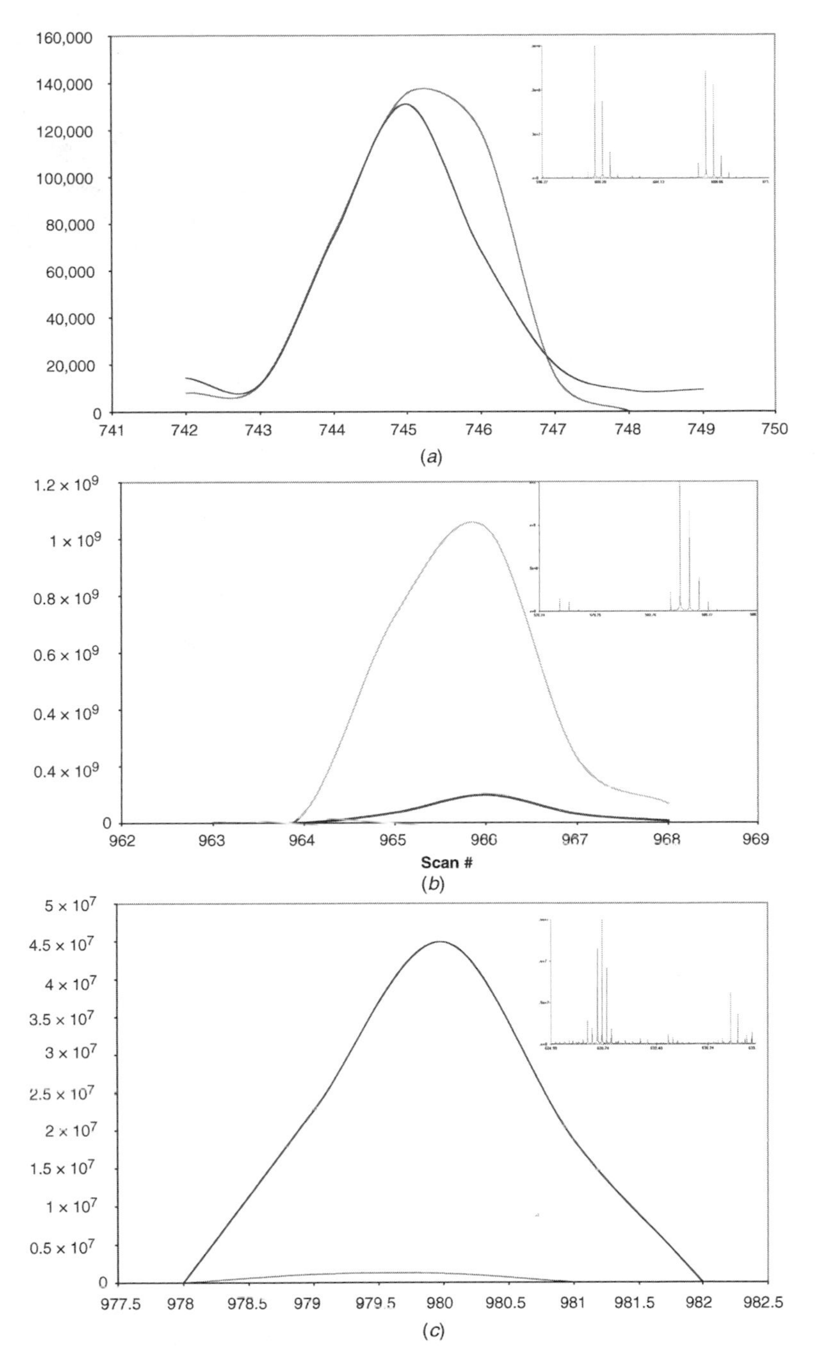


**Figure 9.6** Peptide elution profiles. Elution profiles of a representative peptide from three different proteins expressed in *Shewanella* grown on both unlabeled media under aerobic conditions (blue line) and $^{15}N$ labeled media grown under suboxic conditions (red line). The insets in each panel illustrate a representative spectra for each of the eluting peptides. For ribosomal protein L3 (*a*), the elution profiles are similar size, indicating the abundances of the peptide were similar in the two cultures. For fumarate reductase (*b*), the abundance is greater for the peptide isolated from the suboxic culture (red line) compared to the peptide isolated from the aerobic culture (blue line). The opposite trend is seen for isocitrate lyase (*c*), where the peptide isolated from the aerobic cultures (blue line) is in greater abundance than the peptide isolated from the suboxic culture (red line).

| Reference | Description | A | B | C | D |
|---|---|---|---|---|---|
| SO0217 | Translation elongation factor Tu | | | | |
| SO0220 | Ribosomal protein L11 | | | | |
| SO0221 | Ribosomal protein L1 | | | | |
| SO0222 | Ribosomal protein L10 | | | | |
| SO0673 | Hypothetical protein | | | | |
| SO0970 | Fumarate reductase flavoprotein subunit precursor | | | | |
| SO1484 | Isocitrate lyase | | | | |
| SO1776 | Outer membrane protein precursor MtrB | | | | |
| SO1777 | Decaheme cytochrome *c* MtrA | | | | |
| SO1778 | Decaheme cytochrome *c* | | | | |
| SO1779 | Decaheme cytochrome *c* | | | | |
| SO2345 | Glyceraldehyde 3-phosphate dehydrogenase | | | | |
| SO2361 | Cytochrome *c* oxidase cbb3-type, subunit III | | | | |
| SO2602 | Conserved hypothetical protein | | | | |
| SO2893 | Conserved hypothetical protein | | | | |
| SO4068 | Hypothetical protein | | | | |

**Figure 9.7** Illustration of differential abundances determined for selected proteins from *S. oneidensis* MR-1 as measured using $^{14}N/^{15}N$ metabolic labeling. Columns A and B represent triplicate samples in which the $^{14}N$ and $^{15}N$ from either aerobic or suboxic cultures were mixed with like culture. Column C represents triplicate samples mixing $^{14}N$ aerobic and $^{15}N$ suboxic mixtures, while column D represents triplicate samples mixing $^{14}N$ suboxic and $^{15}N$ aerobic mixtures. Black designates no change detected between the two cultures, yellow designates increase in abundance of the $^{14}N$ peptide over the $^{15}N$ peptide, and blue indicates a decrease in abundance of the $^{14}N$ peptide over the $^{15}N$ peptide. The representative housekeeping proteins do no change in abundance, while changes are demonstrated for proteins known to be involved with respiration or carbon metabolism.

abundance in the suboxic cultures and represent a reciprocity of the experimental conditions that further validates the result. Similarly, the results for isocitrate lyase and glyceraldehyde-3-phosphate dehydrogenase indicate a higher level of abundance in aerobic-versus-suboxic conditions.

While comparative measurements based on isotopic labeling are the best option, these approaches require detection of both labeled and unlabeled versions of the peptide. However, large changes in relative protein abundances between two labeled samples can result in the detection of only one of the peptides. Approaches based upon direct use of peak intensities are attractive for overcoming this limitation. In a peak-intensity approach, samples are analyzed in replicates by LC-FTICR. Similar to the stable isotope labeling approach, once the peptides are validated against the MT tag database, peak intensity can be calculated by summing the intensity as measured by LC-FTICR of the mass spectral peaks across the elution time window. The relative abundance of a particular protein is determined on the basis of the abundances of the peptides observed for that protein in replicate analyses. Recent reports have also described the use of peptide peak intensity information for determining changes in relative protein abundances based upon different normalization techniques. However, measurements of peptide abundances using MS signal intensities can vary significantly for reasons that include variations in MS ionization efficiencies and losses during sample preparation and separations. While useful for large differences in abundances, they have been less effective for studying more subtle variations.

| Reference | Description | A | B | C | D |
|---|---|---|---|---|---|
| SO0217 | Translation elongation factor Tu | | | | |
| SO0220 | Ribosomal protein L11 | | | | |
| SO0221 | Ribosomal protein L1 | | | | |
| SO0222 | Ribosomal protein L10 | | | | |
| SO0770 | Malate dehydrogenase | | | | |
| SO0970 | Fumarate reductase flavoprotein | | | | |
| SO1137 | Conserved hypothetical protein | | | | |
| SO1484 | Isocitrate lyase | | | | |
| SO1538 | Isocitrate dehydrogenase NAD dependent | | | | |
| SO1673 | Outer membrane protein OmpW putative | | | | |
| SO1755 | Phosphoglucomutase/phosphomannomutase family protein | | | | |
| SO1777 | Decaheme cytochrome *c* | | | | |
| SO1778 | Decaheme cytochrome *c* | | | | |
| SO1779 | Decaheme cytochrome *c* | | | | |
| SO2345 | Glyceraldehyde 3-phosphate dehydrogenase | | | | |
| SO2361 | Cytochrome *c* oxidase cbb3-type, subunit III | | | | |
| SO2363 | Cytochrome *c* oxidase cbb3-type, subunit II | | | | |
| SO2629 | Isocitrate dehydrogenase NADP-dependent | | | | |
| SO2691 | Hypothetical protein | | | | |
| SO3409 | OsmC/Ohr family protein | | | | |
| SO3502 | Hypothetical protein | | | | |
| SO3855 | Malate oxidoreductase | | | | |

**Figure 9.8** Illustration of absolute abundances determined for selected proteins from *S. oneidensis* MR-1 as calculated from direct abundance measurements. Columns A and B represent triplicaate comparisons of aerobic versus aerobic (A) and suboxic versus suboxic (B). Column C represents the triplicate comparision of suboxic to aerobic and column D represents the triplicate comparision of aerobic to suboxic in which the $^{14}N$ and $^{15}N$ from either aerobic or suboxic cultures were mixed with like culture. Black designates no change detected between the two cultures, yellow indicates and increase in abundance, and blue indicates a decrease in abundance. Similar to the $^{14}N/^{15}N$ experiments, the representative housekeeping proteins do no show a change in abundance, while changes are demonstrated for proteins known to be involved with respiration or carbon metabolism.

The same unlabeled *S. oneidensis* (aerobic-versus-suboxic) samples are analyzed by the absolute abundance method, and the results are shown in Figure 9.8. Similar to the isotopic labeling experiment shown in Figure 9.7, rows A and B represent the control experiments and rows C and D represent the comparison of suboxic to aerobic and aerobic to suboxic, respectively. In a similar labeling strategy to the labeled experiments, black represents no change, blue represents a decrease in abundance, and yellow represents an increase in abundance. For ribosomal proteins and some transcription factors, no change in the protein abundance was detected between suboxic and aerobic conditions. Among the proteins exhibiting significant changes, fumarate reductase increased in relative abundance by an order of magnitude. Also, while decaheme *c*-type cytochromes that participate in anaerobic electron transport increased in abundance, several key enzymes involved in the tricarboxylic acid (TCA) cycle decreased. In addition, several new proteins (including conserved hypotheticals) were observed to change significantly. These findings provide new insights into the energy production, electron transport chain, and regulatory mechanisms for this organism under anaerobic respiration. *At present, no single method is broadly effective for a comprehensive quantitative analysis of cellular proteins*. However, the combination of absolute and relative abundance approaches can provide confirmation and complementary information.

## 9.4 CONCLUSIONS

Although the concept of holistically understanding biological organisms at a systems level is not new [48], the experimental measurements and computational capabilities needed to develop such understandings have only recently begun to emerge. Key to enabling systems-wide cellular characterization is the ability to make quantitative global measurements of biological macromolecules; achieving a system-level understanding requires insight into several fundamental properties, including system structures and dynamics [49, 50]. With the huge number of microbial genome sequencing efforts completed or underway, such systems-level characterization of proteins has become possible using high-throughput proteomics methodology.

Application of proteomics to a microbial system yields a significant amount of information about the organism, particularly which proteins are expressed and under which conditions. However, presently, proteomic data must be interpreted with the understanding that peptide identifications may be imperfect and ideally supported by a statistically sound level of confidence. Such sets of expressed proteins, when they are expressed, where they are localized, and whether they are posttranslationally processed represent the compilation of information for the organism. A subset of all possible proteins is expressed constitutively within cells and change only modestly in abundance with changes in culture conditions. The quantitative characterization of the changes in these proteins with change in culture conditions is critical for an understanding how the cell works. For instance, fumarate reductase is expressed in *S. oneidensis* under aerobic and anaerobic conditions, but the level is increased over 10-fold under anaerobic conditions.

In addition, quantitative studies on systems have to date focused on steady-state end points, where systems are at equilibrium with their environmental conditions. The next step in the fundamental understanding of how these microbes adapt to their environment is the determination of the transitional response of the cells adapting to the change. Extensive quantitative studies that follow proteomic variations over time will be needed to gain this level of microbial characterization.

The important scientific aspect ultimately is not the presence of the protein but its activity, localization, modification state, and presence of cofactors. Future developments aimed at augmenting these types of proteomics studies with metabolomics and other measurements will be critical for unraveling microbial response and function. With this future goal in mind, proteomic technologies and studies must aim to achieve the systems level of understanding of microbial entities that are necessary for more comprehensive microbial studies.

## REFERENCES

1. Wilkins, M. R., Williams, K. L., Appel, R. D., and Hochstrasser, D. F., *Proteome Research: New Frontiers in Functional Genomics*, Springer Verlag, Berlin, 1997.
2. Figeys, D., *Anal. Chem.* 2002, **36**, 413A–419A.
3. Goodlett, D. R., and Yi, E. C., *Funct. Integr. Genomics* 2002, **2**, 138–153.
4. Conrads, T. P., Anderson, G. A., Veenstra, T. D., Pasa-Tolic, L., et al., *Anal. Chem.* 2000, **72**, 3349–3354.
5. Gygi, S. P., Corthals, G. L., Zhang, Y., Rochon, Y., et al., *Proc. Nat. Acad. Sci. USA* 2000, **97**, 9390–9395.

6. Westbrook, J. A., Wait, R., Welson, S. Y., and Dunn, M. J., *Electrophoresis* 2001, **22**, 2865–2871.
7. Link, A. J., Eng, J., Schieltz, D. M., Carmack, E., et al., *Nat. Biotechnol.* 1999, **17**, 676–682.
8. Washburn, M. P., Ulaszek, R., Deciu, C., Schieltz, D., et al., *Anal. Chem.* 2002, **74**, 1650–1657.
9. Davis, M. T., Beierle, J., Bures, E. T., McGinley, M. D., et al., *J. Chromatogr. B* 2001, **752**, 281–291.
10. Eng, J. K., McCormack, A. L., and Yates, J. R., *J. Am. Soc. Mass Spectrom.* 1994, **5**, 976–989.
11. Washburn, M. P., Wolters, D., and Yates, J. R., *Nat. Biotechnol.* 2001, **19**, 242–247.
12. Angerer, J., Mannschreck, C., and Gundel, J., *Int. Arch. Occupati. Environ. Health* 1997, **70**, 365–377.
13. Lipton, M. S., Pasa-Tolic, L., Anderson, G. A., Anderson, D. J., et al., *Proc. Natl. Acad. Sci. USA* 2002, **99**, 11049–11054.
14. Smith, R. D., Anderson, G. A., Lipton, M. S., Pasa-Tolic, L., et al., *Proteomics* 2002, **2**, 513–523.
15. Pasa-Tolic, L., Lipton, M. S., Masselon, C., Anderson, G. A., et al., *J. Mass Spectrom.* 2002, **37**, 1185–1198.
16. Vasilenko, A. G., E. K., Matrosova, V. Y., Ghosal, D. et al., in preparation.
17. Venkateswaran, A., McFarlan, S. C., Ghosal, D., Minton, K. W., et al., *Appl. Environ. Microbiol.* 2000, **66**, 2620–2626.
18. Venkateswaran, K., Moser, D. P., Dollhopf, M. E., Lies, D. P., et al., *Int. J. Syst. Bacteriol.* 1999, **49**, 705–724.
19. Liu, C., Zachara, J. M., Gorby, Y. A., Szecsody, J. E., et al., *Environ. Sci. Technol.* 2001, **35**, 1385–1393.
20. Kostka, J. E., Stucki, J. W., Nealson, K. H., and Wu, J., *Clays and Clay Minerals* 1996, **44**, 522–529.
21. Myers, C. R., and Myers, J. M., *J. Appl. Bacteriol.* 1994, **76**, 253–258.
22. Nealson, K. H., Moser, D. P., and Saffarini, D. A., *Appl. Environ. Microbiol.* 1995, **61**, 1551–1554.
23. Myers, C. R., and Nealson, K. H., *J. Bacteriol.* 1990, **172**, 6232–6238.
24. Moser, D., and Nealson, K. H., *Appl. Environ. Microbiol.* 1996, **62**, 2100–2105.
25. Zachara, J. M., Fredrickson, J. K., Li, S., Kennedy, D. W., et al., *Am. Mineral.* 1998, **83**, 1426–1443.
26. Shen, Y., Zhao, R., Rodriguez, N., Berger, S. J., et al., *Anal. Chem.*, submitted for publication.
27. Harkewicz, R., Anderson, G. A., Pasa–Tolic, L., Masselon, C., et al., *J. Am. Soc. Mass Spectrom.*, in press.
28. Baranov, P. V., Gurvich, O. L., Hammer, A. W., Gesteland, R. F., et al., *Nucleic Acids Res.* 2003, **31**, 87–89.
29. Smith, R. D., Anderson, G. A., Lipton, M. S., Masselon, C., et al., *OMICS* 2002, **6**, 61–90.
30. Resing, K. A., Meyer–Arendt, K., Mendoza, A. M., Aveline–Wolf, L. D., et al., *Anal. Chem.* 2004, **76**, 3556–3568.
31. Searle, B. C., Dasari, S., Turner, M., Reddy, A. P., et al., *Anal. Chem.* 2004, **76**, 2220–2230.
32. Nakai, K., and Horton, P., *Trends Biochem. Sci.* 1999, **24**, 34–36.
33. Gardy, J. L., Spencer, C., Wang, K., Ester, M., et al., *Nucleic Acids Res.* 2003, **31**, 3613–3617.
34. Juncker, A. S., Willenbrock, H., Von Heijne, G., Brunak, S., et al., *Protein Sci.* 2003, **12**, 1652–1662.
35. Hofmann, K., and Stoffel, W., *Biol. Chem. Hoppe-Seyler* 1993, **374**, 166.
36. Nielsen, H., Engelbrecht, J., Brunak, S., von Heijne, G., et al., *Protein Eng.* 1997, **10**, 1–6.

37. Jaffe, H., and VeerannaPant, H. C., *Biochemisty* 1998, **37**, 16211–16224.
38. White, O., Eisen, J. A., Heidelberg, J. F., Hickey, E. K., et al., *Science* 1999, **286**, 1571–1577.
39. Florens, L., Washburn, M. P., Raine, J. D., Anthony, R. M., et al., *Nature* 2002, **419**, 520–526.
40. Smolka, M. B., Zhou, H., Purkayastha, S., and Aebersold, R., *Anal. Biochem.* 2001, **297**, 25–31.
41. Jensen, P. K., Pasa-Tolic, L., Peden, K. K., Martinovic, S., et al., *Electrophoresis* 2000, **21**, 1372–1380.
42. Ong, S. E., Foster, L. J. and Mann, M., *Methods* 2003, **29**, 124–130.
43. Zhou, H. L., Ranish, J. A., Watts, J. D., and Aebersold, R., *Nat. Biotechnol.* 2002, **19**, 512–515.
44. Goshe, M. B., and Smith, R. D., *Curr. Opin. Biotechnol.* 2003, **14**, 101–109.
45. Berger, S. J., Lee, S.-W., Anderson, G. A., Pasa-Tolic, L., et al., *Anal. Chem.*, in press.
46. Regnier, F. E., Riggs, L., Zhang, R. J., Xiong, L., et al., *J. Mass Spectrom.* 2002, **37**, 133–145.
47. Smith, R. D., Shen, Y., and Tang, K., *Acc. Chem. Res.* 2004, **37**, 269–278.
48. Konopka, A., *ASM News* 2004, **70**, 163–168.
49. Kitano, H., *Science* 2002, **295**, 1662–1664.
50. Kitano, H., *Nature* 2002, **420**, 206–210.

# PART III

# PHYSIOLOGICAL PROTEOMICS OF INDUSTRIAL BACTERIA

CHAPTER 10

# Proteomics of *Corynebacterium glutamicum*: Essential Industrial Bacterium

ANDREAS BURKOVSKI

Friedrich-Alexander-Universität Erlangen-Nürnberg, Erlangen, Germany

## 10.1 *CORYNEBACTERIUM GLUTAMICUM*, A BIOTECHNOLOGY WORKHORSE

The first member of the genus *Corynebacterium*, namely *Corynebacterium diphtheriae*, was already described in 1896 (Lehmann and Neumann, 1896); today 60 different *Corynebacterium* species are known (for a present list, see http://www.dsmz.de/bactnom/nam0942.htm#0955: bacterial nomenclature up-to-date, genus *Corynebacterium*. Deutsche Sammlung für Mikroorganismen und Zellkulturen, Braunschweig, Germany). Typically, corynebacteria form straight or slightly curved rods with tapered or sometimes clubbed ends (*coryne*: Greek for "club"), and due to the typical snapping mechanism of cell division, often a V formation or even palisades of several interconnected cells can be observed. In the newest hierachic classification system, the suborder Corynebacterinae (order Actinomycetales) comprises six families, Corynebacteriaceae (genus *Corynebacterium* and *Turicella*), Mycobacteriaceae (genus *Mycobacterium*), Nocardiaceae (genus *Nocardia* and *Rhodococcus*), Gordoniaceae (genus *Gordonia*), Tsukumurellaceae (genus *Tsukumurella*), and Dietziaceae (genus *Dietzia*) (Stackebrandt et al., 1997).

The discovery and investigation of the biotechnologically most important *Corynebacterium* species, *Corynebacterium glutamicum*, is closely connected with the development of biotechnological processes for the industrial production of amino acids. In 1957, *C. glutamicum* was isolated in parallel by Kinoshita and co-workers (1959) and by Udaka (1960) in a screening program for L-glutamate-producing bacteria from a soil sample collected at Ueno Zoo in Tokyo. At that time it was named *Micrococcus glutamicus* (Kinoshita et al., 1959; Udaka, 1960; Eggeling and Sahm, 2001) due to the remarkable ability of even the wild type to excrete considerable amounts of L-glutamate into its surrounding. This property led to the almost immediate use of this bacterium in fermentation industry. Nowadays, more than 1,000,000 tons of L-glutamate and more than 560,000 tons of L-lysine are produced annually by different *C. glutamicum* strains. Due to

*Microbial Proteomics: Functional Biology of Whole Organisms*, Edited by Ian Humphery-Smith and Michael Hecker. 

its unique flavor, which is called *umami* in Japanese, the main product L-glutamate, is almost exclusively used as a flavor enhancer in Asian cuisine and in convenience food (instant soups, meals ready-to-eat, etc.) in the form of its monosodium salt (monosodium glutamate, or MSG). The daily per-capita consumption of MSG lies between about 0.8 g for European and 4 g for Asian populations. L-Lysine is mainly used as a feed additive for cattle and poultry to improve the nutritive value of the typically protein - and lysine-poor vegetable diet. The world market for this feed amino acid shows tremendous increase rates, with a factor of 20 in the last two decades and this is expected to grow further. Besides these bulk products, smaller amounts of some industrially less important amino acids (L-alanine, L-isoleucine, and L-proline) are produced with *C. glutamicum* strains.

Other important products of *C. glutamicum* strains and closely related species with a production scale of several 100,000 tons annually are different nucleotides, which are also used mainly as flavor enhancers.

Due to this industrial importance, starting in the 1960s research on *C. glutamicum* was strongly focused on amino acid synthesis pathways and central metabolism. However, with the development of new techniques and the rising demand for a rational design of production strains, *C. glutamicum* has been studied intensively with a great variety of different techniques. The subject of research broadened to basic topics such as cell wall structure, transport processes, osmoregulation, and so on, and all major global analysis approaches such as metabolic flux analysis and genome, transcriptome, proteome, and metabolome analysis have been established for this bacterium. Today, *C. glutamicum* is one of the most thoroughly investigated bacteria and might serve as a model organism for other biotechnologically important microorganisms. Additionally, it can be applied as a model for less accessible mycolic acid-containing actinomycetes (e.g., different pathogens such as *C. diphtheriae*, *Mycobacterium leprae*, or *Mycobacterium tuberculosis*).

In this chapter, the established methods, reported studies, and recent developments in proteome analysis of *C. glutamicum* are summarized.

## 10.2 PROTEOMES AND SUBPROTEOMES

As the standard method for the analysis of protein profiles of cells, the two-dimensional polyacrylamide gel electrophoresis (2D PAGE) of proteins has been established for *C. glutamicum* (Hermann et al., 1998, 2000, 2001; Schaffer et al., 2001). From the beginning, and in contrast to other proteome projects, a fractionation protocol according to different functional entities of the cell was established. *C. glutamicum* exhibits a very rigid and complex cell wall (Peyret et al., 1993; Marienfeld et al., 1997; Puech et al., 2001) which has to be opened by mechanic stress. After disruption using either a French press, a bead mill, or ultrasonic treatment, cell debris and cell extract are separated by low-speed centrifugation. The crude extract is further fractionated in cytoplasmic proteins and proteins of the membrane fraction by ultracentrifugation. Alternatively, whole cells can be treated with different detergents (Peyret et al., 1993; Marienfeld et al., 1997; Hermann et al., 2000). Due to the rigidity of the cell wall, the addition of detergents does not result in cell lyses, but cell surface proteins and cell-wall-associated proteins are released. In addition, secreted proteins can be isolated from the culture supernatant by acetone precipitation (Hermann et al., 2001). As a result, submaps of cytoplasmic proteins, proteins of the membrane fraction, cell-wall-associated proteins, and secreted proteins are available now (Hermann et al., 2001).

Besides the progress made in *C. glutamicum* proteomics in the last years, some problems still need to be solved. For example, membrane proteins are underrepresented on current 2D gels. In fact, only membrane-associated proteins were identified on such gels, while membrane proteins with more than one transmembrane helix or grand average of hydropathy (GRAVY) values higher than 0.013 could not be identified (Schaffer et al., 2001). Moreover, a graphical representation of the *C. glutamicum* proteome calculated from the genomic DNA sequence shows a bimodal isoelectric distribution, but when the actual 2D protein pattern obtained for *C. glutamicum* is compared with this calculated map, it is characterized by an almost complete absence of basic proteins (Hermann et al., 2001; Schaffer and Burkovski, 2005). These limitations of the established 2D PAGE protocols are currently being addressed in different laboratories. Efforts are being made to develop methods for the analysis of basic, membrane, and low-abundance proteins, including alternative protein separation techniques and the development of more sensitive staining methods (Herick et al., 2001).

## 10.3 WHO'S WHO?

Since *C. glutamicum* proteomics was started before the genome sequence of this bacterium was determined, initially protein spots detected on 2D gels had to be identified by amino-terminal (N-terminal) microsequencing and subsequent comparison of the amino acid sequences obtained with public databases (Hermann et al., 1998, 2000). This unfavorable, time-consuming, and often unsuccessful approach was replaced when the *C. glutamicum* genome was sequenced in parallel by several biotechnology companies (BASF, Degussa AG, and Kyowa Hakko) and made available to the public (Ikeda and Nakagawa, 2003; Kalinowski et al., 2003). This sequence information made the identification of proteins by matrix-assisted laser desorption ionization time-of-flight mass spectrometry (MALDI-TOF MS) fingerprint analyses possible (Hermann et al., 2001, Schaffer et al., 2001) and significantly improved the speed, sensitivity, and reliability of proteome analyses. Based on the established techniques, a "high-resolution" reference map of cytoplasmic and membrane-associated proteins from *C. glutamicum* cells grown in minimal medium with glucose as carbon source has been published (Schaffer et al., 2001). Moreover, all proteome data available have been recently integrated into a public database. Master gel images which show *C. glutamicum* protein profiles after fractionation of cytoplasmic, membrane-associated, and secreted proteins and a corresponding table of proteins are available online at http://www.fzjuelich.de/ibt/biochem/biochem.html. The master gel images published at this site show about 970, 660, and 40 spots, respectively. Protein table and images are clickable and allow easy cross-referencing for a given spot or protein entry. Currently, the database contains entries concerning 197 spots representing 164 different proteins (Schaffer and Burkovski, 2005).

## 10.4 FROM THE BEGINNING: N-TERMINAL PROCESSING OF *C. GLUTAMICUM* PROTEINS

A number of *C. glutamicum* proteins, mainly from 2D gels but also isolated by other approaches, were analyzed by N-terminal microsequencing (Garbe et al., 2000; Hermann et al., 1998, 2000, 2001; Lichtinger et al., 2001; Matsushita et al., 2001; Reinscheid, 1994;

Reinscheid et al., 1994). Additionally, N-terminal peptides of proteins were determined by MALDI-TOF-MS-based postsource decay (PSD) analysis (Schaffer et al., 2001). When these sequences were used for database searches, 33 proteins could be identified (summarized in Table 10.1). An astonishing number of six database entries (18%) suggested translation initiation at start codons different from those determined experimentally, indicating incorrect annotations or alternative translational start sites. These results emphasize the need for experimental verification of in silico data.

Twenty-four proteins showed methionine-aminopeptidase-dependent processing of their N termini (73%). The N-terminal methionine was always removed when the following amino acid residue is L-glutamine (one case), glycine (one protein), L-leucine (one case), L-proline (one case), or L-serine (four cases) and cleaved off in most cases when it was followed by L-alanine (eight out of nine proteins) or L-threonine (five out of seven cases). Other amino acid residues revealed a different behavior. In the case of L-lysine following the initiator methionine, one protein was processed and one N terminus was unprocessed, while L-arginine (two cases), L-aspartate (one protein), L-glutamate (one protein), and L-histidine residues (one protein) never promoted processing of the N-terminal methionine. From these data it can be concluded that N-terminal processing of *C. glutamicum* proteins is quite similar to protein processing in *Escherichia coli* (Link et al., 1997). Both in *E. coli* and in *C. glutamicum* the N-terminal methionine is always cleaved when the penultimate amino acid residue is either L-serine or L-alanine, while cleavage is variable with L-threonine. In contrast to *E. coli*, methionine-aminopeptidase-dependent processing appears to occur in *C. glutamicum* also in proteins with L-glutamine, L-lysine, and L-leucine following the initiator methionine.

## 10.5 ANALYSIS OF PROTEIN MODIFICATIONS

Phosphorylation is the most abundant covalent modification of proteins (Kaufmann et al., 2001) and plays a crucial role in the regulation of protein activity, especially in signal transduction processes in "eukaryotes". Recently, a global analysis of phosphorylation in *C. glutamicum* was published which represents the first study of this kind for bacteria (Bendt et al., 2003). In this communication, phosphorus-containing *C. glutamicum* proteins were identified by two independent approaches, only by immunostaining with phosphoamino-acid-specific monoclonal antibodies and by in vivo radiolabeling using [$^{33}$P]-phosphoric acid and subsequent autoradiography (Bendt et al., 2003). The first method is "specific" for distinct phosphorylated amino acid residues (e.g., phosphoryl serine or phosphoryl threonine) while detection of protein spots with the second method is independent of the phosphorylated amino acid but may also include proteins modified by other phosphorus-containing groups like adenylyl and uridylyl residues. After 2D PAGE approximately 90 immunostained protein spots and approximately 60 [$^{33}$P]-labeled protein spots were detected, 31 of these protein spots detected with both methods (Fig. 10.1). Forty-one different proteins were identified by peptide mass fingerprinting and a first phosphoproteome map was established based on the combined results of the two methods (Bendt et al., 2003). The unexpected high number of phosphorylated proteins observed even in this preliminary study indicates that phosphorylation is more widespread in bacteria than previously thought.

Also another modification involved in the regulation of protein activity was detected in a proteomics study: adenylylation of GlnK, the central signal transduction protein in the

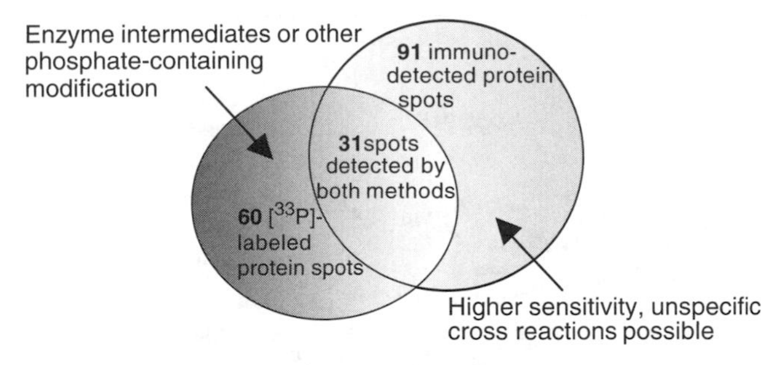


**Figure 10.1** Detection of *C. glutamicum* phosphoproteins. Two different methods for the detection of phosphorproteins were used, in vivo labeling and Western blotting (Bendt et al., 2003). Advantages and drawbacks of the methods are indicated.

nitrogen control of *C. glutamicum.* This was detected when the 2D patterns of cells grown with different nitrogen supply were compared (Silberbach et al., submitted). It was observed that the isoelectric point (pI) of GlnK shifted depending on the cellular nitrogen supply. Cryptic in-gel digests and MALDI-TOF MS subsequently revealed the modification of a distinct peptide with an adenosine monophosphate (AMP) residue, which causes the pI shift of the protein due to two additional negative charges (Silberbach et al., 2005). Additionally, a pI shift of glutamine synthetase from approximately 4.6 to 5.1 was observed in this study when cells were grown under nitrogen limitation. This is in accordance with a deadenylylation of the GS enzyme in response to nitrogen deprivation reported previously (Jakoby et al., 1997, 1999).

## 10.6 *C. GLUTAMICUM* PROTEOMICS AND PHYSIOLOGY

Proteome analyses are especially suitable for the comparison of protein profiles of cells facing different physiological conditions. The published studies using proteome techniques for the investigation of *C. glutamicum* physiology are summarized in the following paragraphs.

Most studies in this respect were carried out to investigate the cellular response of *C. glutamicum* to nitrogen limitation (Schmid et al., 2000; Nolden et al., 2001; Bendt et al., 2003, 2004; Beckers et al., 2004; Silberbach et al., 2005). In first studies a combination of [$^{35}$S]methionine in vivo labeling, 2D PAGE, and autoradiography was used (Schmid et al., 2000; Nolden et al., 2001). These experiments revealed initial indications for a crosstalk between nitrogen control and energy metabolism (Schmid et al., 2000) as well as evidence for a proposed regulation of part of the L-lysine biosynthesis pathway on the protein level (Nolden et al., 2001). The *dapD* gene product tetrahydrodipicolinate-*N*-succinyltransferase was found to be increasingly synthesized in increasing amounts during nitrogen limitation. This enzyme is part of the split diaminopimelate pathway (Schrumpf et al., 1991; Wehrmann et al., 1998), which is important for cell wall and L-lysine biosynthesis. Up regulation of tetrahydrodipicolinate-*N*-succinyltransferase activity increases the metabolite flux via the high-ammonium-affinity branch of this pathway, allowing synthesis of the cell wall building block diaminopimelate even under nitrogen limitation. The results obtained by shake flask experiments and in vivo labeling were confirmed recently, when cells grown under nitrogen limitation in a chemostat were

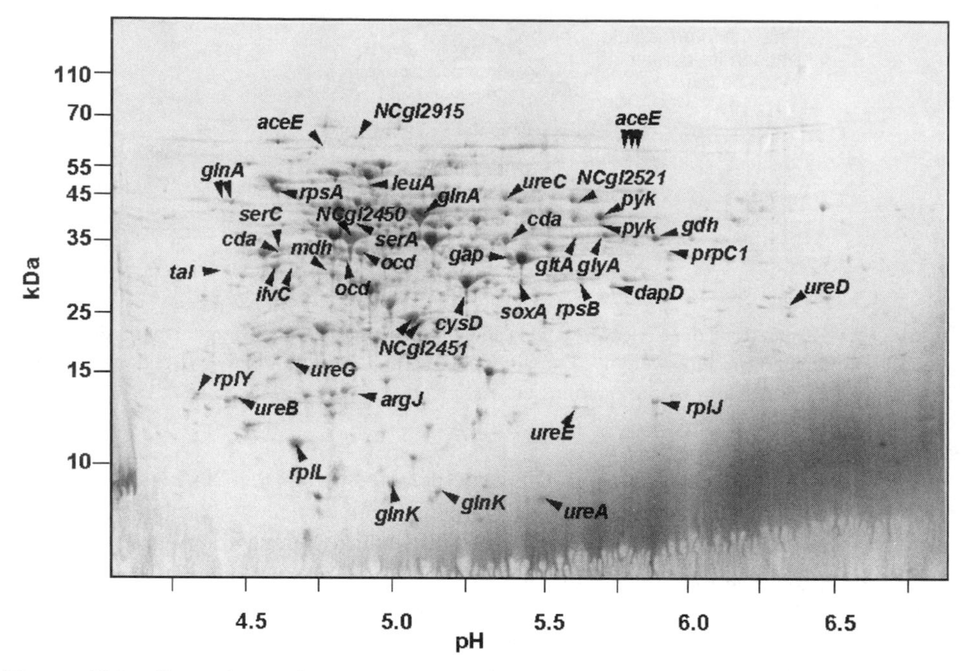


**Figure 10.2** Comparison of *C. glutamicum* protein profiles from cells grown with different nitrogen supply. One hundred micrograms of cytoplasmic protein fractions was separated by 2D PAGE and stained with Coomassie brilliant blue. For comparison, a false color overlay was produced using the DECODON Delta2D 3.1 software package (red, nitrogen-rich batch phase; green, nitrogen-limited continuous phase). Spots are designated according to the annotation of the corresponding genes (see also Silberbach et al., submitted). (Kindly provided by M. Silberbach, Institute of Biochemistry, University of Cologne.) (See color insert.)

analyzed by a combined proteomics and transcriptomics investigation (Silberbach et al., 2005). This study showed that *C. glutamicum* reacts in response to nitrogen starvation with a rearrangement of its cellular transport capacity, changes in metabolic pathways crucial for nitrogen assimilation and amino acid biosynthesis, an increased energy generation, and a decreased capacity for protein synthesis and cell division. In fact, more than 100 protein spots with different size and intensity were analyzed (Fig. 10.2). The abundances of two enzymes, creatinine deaminase and urease, which are indispensable for the utilization of creatinine and urea as alternative nitrogen sources, were additionally characterized by 2D PAGE in two independent studies (Beckers et al., 2004; Bendt et al., 2004). Increased abundances of creatinine deaminase and of five out of seven urease subunits could be shown in nitrogen-starved cells.

Recently, the effect of nitrogen starvation on the phosphorylation pattern of *C. glutamicum* proteins was also investigated (Bendt et al., 2003). However, differences between cultures grown with high nitrogen supply and nitrogen starved were not observed.

Carbon metabolism has been another topic of interest in *C. glutamicum* proteome research. To study differences in the protein profile of glucose—and acetate-grown *C. glutamicum* cells, [$^{35}$S]methionine in vivo labeling in combination with 2D PAGE and peptide mass fingerprint analysis was used (Gerstmeir et al., 2003). Of about 500 protein spots detected after 2D PAGE and autoradiography of cell extracts from glucose—and acetate-grown cultures, 54 were present in higher amounts and 26 in lower amounts in the lysate of acetate-grown cells. Ten acetate-induced proteins are identified, namely butyryl-CoA transferase, citrate synthase, cysteine synthase, fumarase, glycine-tRNA ligase, isocitrate lyase, malate:chinone oxidoreductase, malate synthase, a putative ABC transporter, and a putative aminotransferase. The enhanced synthesis of citrate synthase, fumarase, isocitrate lyase, and malate synthase is in agreement with data obtained by transcriptome analyses and also in accordance with increased activities of Krebs cycle and glyoxylate cycle enzymes in acetate-grown cells; "the physiological connection of the other proteins identified to acetate metabolism is not clear" (Gerstmeir et al., 2003). The investigation of propionate metabolism provides another example of the application of proteomics techniques to study carbon metabolism in *C. glutamicum* (Claes et al., 2002). Two-dimensional PAGE combined with MS revealed a strong induction of the *prpD2B2C2* gene products 2-methylcitrate dehydratase, 2-methylisocitrate lyase, and 2-methylcitrate synthase when propionate was added as an additional carbon source to acetate-grown cells. These results were confirmed by genetic studies (Claes et al., 2002).

Interestingly, the 2-methylcitrate dehydratase protein was also eightfold more abundant in the *C. glutamicum* wild type when this was grown in the presence of 300 mM L-valine (Lange et al., 2003). Other proteins with increased abundance under these conditions are the arginine repressor ArgR and *N*-acetyl glutamate semialdehyde dehydrogenase. In all cases the increased abundance of proteins correlates with increased mRNA levels of the respective genes (Lange et al., 2003). The physiological significance of the observed L-valine-induced changes in the protein expression profile is, however, not clear.

Engels et al. (2004) made use of 2D PAGE and MALDI-TOF MS to characterize *C. glutamicum* mutants with deletions in the genes coding for accessory subunits of the Clp protease complex. Deletion of *clpC* led to dramatically increased abundance of the ClpP1 and ClpP2 proteins, which represent the proteolytic subunits of the Clp protease. Subsequent studies led to the identification of the transcriptional activator ClgR which controls *clpC* and *clpP1P2* gene expression in *C. glutamicum*. Moreover, using 2D PAGE it could be demonstrated that the *M. tuberculosis* ortholog of ClgR is able to functionally replace *C. glutamicum* ClgR, supporting the hypothesis that the identified autoregulatory loop controlling *clp* gene expression and involving ClgR as well as the ClpCP protease is conserved in the order Actinomycetales.

When setting up a proteome map of the *C. glutamicum* cytoplasmic protein fraction, the presence of a highly abundant protein with similarity to GlpX-like fructose-1,6-bisphosphatases from other organisms was observed (Schaffer et al., 2001). The constitutive high abundance of this protein in *C. glutamicum* was in contrast to the tight regulation of the corresponding gene in other organisms, suggesting a housekeeping function of GlpX in *C. glutamicum*. In fact, GlpX was shown to have fructose-1,6-bisphosphatase activity and is required for growth of *C. glutamicum* on gluconeogenic substrates, thus probably representing the only fructose-1,6-bisphosphatase in *C. glutamicum* (Rittmann et al., 2003). This result is an example of how proteome mapping by itself can provide valuable information on individual proteins, stimulating subsequent research and ultimately leading to a functional characterization of these proteins.

## 10.7 TOOLS FOR INDUSTRY

In the last four years, production strains for amino acids, but also for other products, were very successfully generated by successive rounds of random mutagenesis and screening

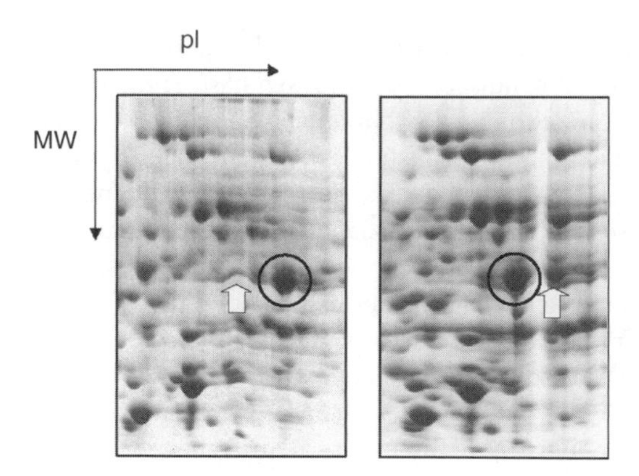


**Figure 10.3** Proteome analysis of *C. glutamicum* amino acid production strain. The sections shown are part of a comparison of the wild type (left) and an L-lysine production strain (right). A key enzyme of the L-lysine biosynthesis pathway (circled) is shifted in its position to a more basic isoelectric point (pI) due to an amino acid exchange. The corresponding positions of the wild type and mutant protein are marked by arrows. (Photographs kindly provided by T. Hermann, Degussa AG, Halle-Künsebeck.)

for the desired product. However, this classical approach has the disadvantage of accumulation of uncharacterized secondary mutations that might be detrimental for strain performance. Recently, Ohnishi and coworkers (2002) described the generation of an excellent L-lysine producer strain by the introduction of only three site-directed mutations in the wild-type genome. Based on this success, the identification of mutations in classically generated production strains as a basis for the rational design of strains is of great interest and the proteomics approach provides a valuable tool for the characterization of these strains. Unfortunately, due to obvious reasons companies are not too interested in publishing their research results and only very limited information is available on the 2D PAGE analysis of production strains. As an example for this kind of study, 2D profiles of the *C. glutamicum* wild type and an L-lysine production strain are shown (Fig. 10.3). The comparison of protein patterns led to the detection of a protein spot corresponding to a key enzyme of the L-lysine biosynthesis pathway, which shifted in its position in the production strain. Subsequent analyses revealed the presence of an amino acid exchange in this enzyme.

In principle, an analysis of cells from production cultures is also possible by 2D PAGE, for example, for monitoring cell extracts for the presence of stress proteins to evaluate the status of the fermentation process. However, the need for several time-consuming steps such as protein preparation, isoelectric focusing, gel electrophoresis, and staining is a drawback of this method. As an alternative, first attempts have been made to establish protein arrays. These arrays might be more suitable for a fast and efficient detection of distinct proteins for fermentation control.

## 10.8 PROSPECTS

The main challenge in *C. glutamicum* proteomics in the next years will be to combine this approach with other global approaches established for this bacterium. A combination of proteome and transcriptome techniques was already successful in a variety of different physiological studies (Gerstmeir et al., 2003; Lange et al., 2003; Beckers et al., 2004, 2005; Silberbach et al., 2005), while a combined transcriptome, metabolome, and fluxome analysis of a *C. glutamicum* L-lysine production strain was published recently (Krömer et al., 2004). The task will be to integrate all these global approaches to gain a holistic view on the physiology of *C. glutamicum*.

## ACKNOWLEDGMENTS

The author wishes to thank all previous and recent members of the group for their commitment and excellent work, Reinhard Krämer for his support and interest, and the Bundesministerium für Bildung und Forschung, the Degussa AG, and the Deutsche Forschungsgemeinschaft for financial support.

## REFERENCES

Beckers, G., Bendt, A. K., Krämer, R., and Burkovski, A. (2004). Molecular identification of the urea uptake system and transcriptional analysis of urea transporter- and urease-encoding genes in *Corynebacterium glutamicum*. *J. Bacteriol.* **186**:7645–7652.

Beckers, G., Strösser, J., Hildebrandt, U., Kalinowski, J., Farwick, M., Krämer, R., and Burkovski, A. (2005). Regulation of AmtR-controlled gene expression in *Corynebacterium glutamicum*: mechanism and characterization of the AmtR regulon. *Mol. Microbiol.* **58**:580–595.

Bendt, A. K., Krämer, R., Burkovski, A., Schaffer, S., Bott, M., Busker, E., Hermann, T., Pfefferle, W., and Farwick, M. (2003). Towards a phospho-proteome map of *Corynebacterium glutamicum*. *Proteomics* **3**:1637–1646.

Bendt, A. K., Beckers, G., Silberbach, M., Wittmann, A., and Burkovski, A. (2004). Utilization of creatinine as an alternative nitrogen source in *Corynebacterium glutamicum*. *Arch. Microbiol.* **181**:443–450.

Claes, W. A., Pühler, A., and Kalinowski, J. (2002). Identification of two *prpDBC* gene clusters in *Corynebacterium glutamicum* and their involvement in propionate degradation via the 2-methylcitrate cycle. *J. Bacteriol.* **184**:2728–2739.

Eggeling, L., and Sahm, H. (2001). The cell wall barrier of *Corynebacterium glutamicum* and amino acid efflux. *J. Biosci. Bioeng.* **92**:201–213.

Engels, S., Schweitzer, J., Ludwig, C., Bott, M., and Schaffer, S. (2004). *clpC* and *clpP1P2* gene expression in *Corynebacterium glutamicum* is controlled by a regulatory network involving the transcriptional regulators ClgR and HspR as well as the ECF sigma factor $\sigma^H$. *Mol. Microbiol.* **52**:285–302.

Garbe, T. R., Kobayashi, M., and Yukawa, H. (2000). Indole-inducible proteins in bacteria suggest membrane and oxidant toxicity. *Arch. Microbiol.* **173**:78–82.

Gerstmeir, R., Wendisch, V. F., Schnicke, S., Ruan, H., Farwick, M., Reinscheid, D., and Eikmanns, B. J. (2003). Acetate metabolism and its regulation in *Corynebacterium glutamicum*. *J. Biotechnol.* **104**:99–122.

Herick, K., Jackson, P. Wersch, G., and Burkovski, A. (2001). Detection of fluorescence dye-labeled proteins in 2-D gels using an Arthur 1442 Multiwavelength Fluoroimager. *BioTechniques* **31**:146–149.

Hermann, T., Wersch, G., Uhlemann, E.-M., Schmid, R., and Burkovski, A. (1998). Mapping and identification of *Corynebacterium glutamicum* proteins by two-dimensional gel electrophoresis and micro sequencing. *Electrophoresis* **19**:3217–3221.

Hermann, T., Finkemeier, M., Pfefferle, W., Wersch, G., Krämer, R., and Burkovski, A. (2000). Two-dimensional electrophoretic analysis of *Corynebacterium glutamicum* membrane fraction and surface proteins. *Electrophoresis* **21**:654–659.

Hermann, T., Pfefferle, W., Baumann, C., Busker, E., Schaffer, S., Bott, M., Sahm, H., Dusch, N., Kalinowski, J., Pühler, A., Bendt, A. K., Krämer, R., and Burkovski, A. (2001). Proteome analysis of *Corynebacterium glutamicum*. *Electrophoresis* **22**:1712–1723.

Ikeda, M., and Nakagawa, S. (2003). The *Corynebacterium glutamicum* genome: Features and impacts on biotechnological processes. *Appl. Microbiol. Biotechnol.* **62**:99–109.

Jakoby, M., Krämer, R., and Burkovski, A. (1999). Nitrogen regulation in *Corynebacterium glutamicum*: Isolation of genes involved and biochemical characterization of corresponding proteins. *FEMS Microbiol. Lett.* **173**:303–310.

Jakoby, M., Tesch, M., Sahm, H., Krämer, R., and Burkovski, A. (1997). Isolation of the *Corynebacterium glutamicum glnA* gene encoding glutamine synthetase I. *FEMS Microbiol. Lett.* **154**:81–88.

Kalinowski, J., Bathe, B., Bischoff, N., Bott, M., Burkovski, A., Dusch, N., Eggeling, L., Eikmanns, B. J., Gaigalat, L., Goesmann, A., Hartmann, M., Huthmacher, K., Krämer, R., Linke, B., McHardy, A. C., Meyer, F., Möckel, B., Pfefferle, W., Pühler, A., Rey, D., Rückert, C., Sahm, H., Wendisch, V. F., Wiegräbe, I., and Tauch, A. (2003). The complete *Corynebacterium glutamicum* ATCC 13032 genome sequence and its impact on the production of L-aspartate-derived amino acids and vitamins. *J. Biotechnol.* **104**:5–25.

Kaufmann, H., Bailey, J. E., and Fussenegger, M. (2001). Use of antibodies for detection of phosphorylated proteins separated by two-dimensional gel electrophoresis. *Proteomics* **1**:194–199.

Kinoshita, S., Udaka, S., and Shimono, M. (1959). Amino acid fermentation. I. Production of L-glutamic acid by various microorganisms. *J. Gen. Appl. Microbiol.* **3**:193–205.

Krömer, J. O., Sorgenfrei, O., Klopprogge, K., Heinzle, E., and Wittmann, C. (2004). In-depth profiling of lysine-producing *Corynebacterium glutamicum* by combined analysis of the transcriptome, metabolome, and fluxome. *J. Bacteriol.* **186**:1769–1784.

Lange, C., Rittmann, D., Wendisch, V. F., Bott, M., and Sahm, H. (2003). Global expression profiling and physiological characterization of *Corynebacterium glutamicum* grown in the presence of L-valine. *Appl. Environ. Microbiol.* **69**:2521–2532.

Lehmann, K. B., and Neumann, R. (1896). *Atlas und Grundriss der Bakteriologie und Lehrbuch der speziellen bakteriologischen Diagnostik.* J. F. Lehmann, Munich, Germany.

Lichtinger, T., Riess, F. G., Burkovski, A., Engelbrecht, F., Hesse, D., Kratzin, H. D., Krämer, R., and Benz, R. (2001). The low-molecular-mass subunit of the cell wall channel of the Gram-positive *Corynebacterium glutamicum*. Immunological localization, cloning and sequencing of its gene *porA*. *Eur. J. Biochem.* **268**:462–469.

Link, A. J., Robison, K., and Church, G. M. (1997). Comparing the predicted and observed properties of proteins encoded in the genome of *Escherichia coli* K-12. *Electrophoresis* **18**:1259–1313.

Marienfeld, S., Uhlemann, E.-M., Schmid, R., Krämer, R., and Burkovski, A. (1997). Ultrastructure of the *Corynebacterium glutamicum* cell wall. *Ant. Leeuwenhoek* **72**:291–297.

Matsushita, K., Otofuji, A., Iwahashi, M., Toyama, H., and Adachi, O. (2001). NADH dehydrogenase of *Corynebacterium glutamicum*. Purification of an NADH dehydrogenase II homolog able to oxidize NADPH. *FEMS Microbiol. Lett.* **204**:271–276.

Ohnishi, J., Mitsuhashi, S., Hayashi, M., Ando, S., Yokoi, H., Ochiai, K., and Ikeda, M. (2002). A novel methodology employing *Corynebacterium glutamicum* genome information to generate a new L-lysine-producing mutant. *Appl. Microbiol. Biotechnol.* **58**:217–223.

Peyret, J. L., Bayan, N., Joliff, G., Gulik-Krzywicki, T., Mathieu, L., Schechter, E., and Leblon, G. (1993). Characterization of the *cspB* gene encoding PS2, an ordered surface-layer protein in *Corynebacterium glutamicum*. *Mol. Microbiol.* **9**:97–109.

Puech, V., Chami, M., Lemassu, A., Laneelle, M. A., Schiffler, B., Gounon, P., Bayan, N., Benz, R., and Daffé, M. (2001). Structure of the cell envelope of corynebacteria: Importance of the non-covalently bound lipids in the formation of the cell wall permeability barrier and fracture plane. *Microbiology* **147**:1365–1382.

Reinscheid, D. J. (1994). Physiologische und genetische Untersuchungen des Acetat-Stoffwechsels in *Corynebacterium glutamicum*. Ph.D. thesis, Universität Düsseldorf.

Reinscheid, D. J., Eikmanns, B. J., and Sahm, H. (1994). Malate synthase from *Corynebacterium glutamicum*: Sequence analysis of the gene and biochemical characterization of the enzyme. *Microbiology* **140**:3099–3108.

Rittmann, D., Schaffer, S., Wendisch, V. F., and Sahm, H. (2003). Fructose-1,6-bisphosphatase from *Corynebacterium glutamicum*: Expression and inactivation of the *fbp* gene and biochemical characterization of the enzyme. *Arch. Microbiol.* **180**:285–292.

Schaffer, S., and Burkovski, A. (2005). Genome-based approaches: Proteomics. In M. Bott and L. Eggeling (Eds.), *Handbook of Corynebacterium glutamicum*. Boca Raton, FL: CRC Press, pp. 99–118.

Schaffer, S., Weil, B., Nguyen, V. D., Dongmann, G., Günther, K., Nickolaus, M., Hermann, T., and Bott, M. (2001). A high-resolution reference map for cytoplasmic and membrane-associated proteins of *Corynebacterium glutamicum*. *Electrophoresis* **22**:4404–4422.

Schmid, R., Uhlemann, E.-M., Nolden, L., Wersch, G., Hecker, R., Hermann, T., Marx, A., and Burkovski, A. (2000). Response to nitrogen starvation in *Corynebacterium glutamicum*. *FEMS Microbiol. Lett.* **187**:83–88.

Schrumpf, B., Schwarzer, A., Kalinowski, J., Pühler, A., Eggeling, L., and Sahm, H. (1991). A functionally split pathway for lysine synthesis in *Corynebacterium glutamicum*. *J. Bacteriol.* **173**:4510–4516.

Silberbach, M., Schäfer, M., Hüser, A., Kalinowski, J., Pühler, A., Krämer, R., and Burkovski, A. (2005). Adaptation of *Corynebacterium glutamicum* to ammonium limitation: A global analysis using transcriptome and proteome techniques. *Appl. Environ. Microbiol.* **71**:2391–2402.

Stackebrandt, E., Rainey, F. A., and Ward-Rainey, N. L. (1997). Proposal for a new hierachic classification system, *Actinobacteria* classis nov. *Int. J. Syst. Bacteriol.* **47**:97–102.

Wehrmann, A., Phillipp, B., Sahm, H., and Eggeling, L. (1998). Different modes of diaminopimelate synthesis and their role in cell wall integrity: A study with *Corynebacterium glutamicum*. *J. Bacteriol.* **180**:3159–3165.

Udaka, S. (1960). Screening method for microorganisms accumulating metabolites and its use in the isolation of *Micrococcus glutamicus*. *J. Bacteriol.* **79**:745–755.

CHAPTER 11

# Proteomics of *Lactococcus lactis*: Phenotypes for a Domestic Bacterium

MOGENS KILSTRUP

Technical University of Denmark, Lyngby, Denmark

## 11.1 INTRODUCTION

Lactic acid bacteria are used extensively in the dairy industry for acidification of milk, because of their efficient conversion of lactose into lactate. The lactic acid bacteria are phylogenetically diverse, but they all cluster in the *Bacillus–Lactobacillus–Streptococcus* subdivision of the gram-positive bacteria [1]. The *Lactococcus* and *Streptococcus* families are closely related and together they form a separate branch in lactic acid bacteria. Due to the loss of the heme biosynthesis pathway, all lactic acid bacteria are unable to synthesize catalase [2] and functional cytochromes, and they obtain their energy from fermentation of sugar to lactic acid. The metabolic capacities of the lactic acid bacteria are highly reduced compared to most free-living bacteria. They are auxothrophic for varying numbers of amino acids and vitamins, and some require nucleotide precursors. *Lactococcus lactis* has a functional nucleotide biosynthesis pathway [3]. From the genome sequence [4], it appears that in addition to an intact glycolysis pathway, *L. lactis* contains the first two reactions in the tricarboxylic acid (TCA) cycle, leading to 2-oxo-glutarate as well as the succinate dehydrogenase enzyme (http://www.genome.ad.jp/kegg/pathway.html). All other TCA enzymes are absent.

Interestingly, both *L. lactis* and *Streptococcus agalactiae* (as well as *Lactobacillus plantarum* and *Enterococcus faecalis*) contain a rudimentary respiration chain, including a cytochrome *bd* complex, although they are incapable of heme biosynthesis (http://www.genome.ad.jp/kegg/pathway.html). It has been found recently that *L. lactis* will respire when supplemented with a source of heme [5].

The natural habitat for *L. lactis* is not known, but due to the intact heme-dependent respiration pathway, plants and animals has been proposed as niches where heme could be obtained [5]. A prototrophic strain of *L. lactis* has been found in frozen peas [6], which suggests that there are habitats in nature where free-living lactococci require an intact metabolism. The gut of various wood- and soil-eating termites has recently been found to be a habitat for *L. lactis*. [7]

*Microbial Proteomics: Functional Biology of Whole Organisms*, Edited by Ian Humphery-Smith and Michael Hecker. 

From the *Lactococcus–Streprococcus* phylogenetic branch, only *L. lactis* and *Streptococcus thermophilus* are used in the dairy industry for the production of cheeses and other fermented milk products. *Streptococcus thermophilus* is used in the production of yoghurt and *L. lactis* is used in the production of buttermilk. The economically most important cheeses made by *L. lactis* and *S. thermophilus* are cheddar and mozzarella, respectively. The species *L. lactis* is divided into two subspecies, *L. lactis* subsp. *lactis* and *L. lactis* subsp. *cremoris*, which are distinguished both by 16S ribosomal ribonucleic acid (rRNA) phylogeny and by their different stress responses [8, 9]. Strains of subsp. *lactis* are generally more stress resistant than subsp. *cremoris*. The differences between the two subspecies are so extensive that they might actually be typed as separate species. Before the exclusion of *Lactococcus, Enterococcus*, and *Vagococcus* from the *Streptococcus* family, the two *L. lactis* subsp. were typed as *S. lactis* and *S. cremoris* [8]. Dairy strains of *L. lactis* are able to grow in milk by virtue of plasmid-encoded enzymes for both lactose uptake and conversion to glycolytic intermediaries as well as casein breakdown and utilization. In the two *L. lactis* model organisms IL1403 (subsp. *lactis*) [4] and MG1363 (subsp. *cremoris*) [10], these plasmids have been aborted. In the construction of MG1363 from the dairy strain NCDO712, both plasmids and dormant prophages were deleted [10], while IL1403 still contains prophages [4].

### 11.1.1 Multiple Approaches in Proteomic Analysis

Proteomics provides a very broad range of analysis, and for the sake of clarity, the field has been subdivided into branches by the different experimental approaches employed. This review will only deal with gel-based proteomics, but other methods are being developed. All of the branches described below rely on the identification of proteins by peptide mass spectrometry (MS) but may in special cases be identified by N-terminal amino acid sequencing.

- The first approach, *reference map proteomics*, deals with pure description and identification of the entire proteome, that is, the full protein complement of the genome of the cell in a reference physiological state [11]. The reference map may be divided into an acid and an alkaline proteome [12] or into a cytoplasmic [13] and a membrane-associated proteome [14]. The mapping of modified proteins, such as the phosphorylated proteome [15] or the glycosylated proteome [16], may also be considered as a part of reference map proteomics.
- A second approach, *expression proteomics* [17], deals with changes in expression patterns between proteomes from different physiological situations or from cells with different genotypes. Expression proteomics can be performed by two techniques. By in vitro (post mortem) labeling of proteins [18] or staining of the proteins in the gel by Coomassie or silver staining, the output of a density determination of the protein spots is the level of the individual proteins in the extract. If the cells are labeled in vivo using incorporation of radioactive amino acids [19], the amount of radioactivity corresponds to the synthesis rate of the protein. The two approaches reach similar results if the growths of the bacterial cultures are balanced (see below). During a shift situation, the synthesis rate responds much faster to changes in expression than the protein level, because the latter detects the cumulated protein load. The difference between the two responses has been exploited using false color

overlays of images from silver staining and autoradiograms in analysis of *B. subtilis* stress responses [20]. Because of the different approaches, it could be advantageous to differentiate between the two by naming the technique using pulse labeling as expression rate proteomics.

- A third branch follows a *cell map approach* [17], in which all physical interactions between proteins in a cell are mapped. So far no cell map proteomics studies have been reported for *Lactococcus* or *Streptococcus* species.

The scope of the reference map proteomics can be compared to the genome-sequencing projects where the main objective is to provide descriptive information as a stepping stone for the broader scientific community. It is generally agreed that the description of the entire proteome is a perquisite for detailed proteomic research [21] and that the location of the individual proteins on standardized two-dimensional (2D) gels is the key to deciphering the proteome. Ideally, the two first branches of proteomics should complement each other, so that the maps provided by reference map proteomics are readily used as a source of protein identification for the pragmatic expression researchers. Due to the elusive art of protein handling, however, not all proteins can be matched with a sufficient degree of certainty between 2D gels from different laboratories, even if very similar procedures are followed (compare gels in [22] and [23]). These problems have to be solved if 2D-gel-based reference maps shall function as identification keys for the larger scientific community. Identical growth conditions, protein purification techniques, identical protein load, and identical 2D gel electrophoresis (2DGE) techniques need to be applied. Furthermore, web-based comparison services should also be provided by proteome centers to compare 2D gels from external users to the annotated averaged master gels. A probability score should also accompany the identification for the subsequent evaluation by referees in the publishing procedure. Interlaboratory comparison will hopefully become very useful in the future, but as of today the domestic reference maps with cumulated information from the individual proteome laboratories are still of the greatest importance in the study of *Lactococcus* and other bacteria.

### 11.1.2 Proteomics Delivers Detailed Phenotypes

Whereas specialized proteomics laboratories usually perform reference map proteomics [12, 24], expression proteomics is often performed by individual research groups that take up proteomic analysis to answer specific questions in their research [25–27]. Phenotypic traits are scarce in bacteria, and where behavior, color, and morphological structure have been employed as phenotypic markers in higher organisms, bacteriology has resorted to physiological criteria, such as the ability to grow on or without specific substances. Due to a lack of visible phenotypes, the bacterial proteome structure has been used extensively as an extension of the cellular phenotype under particular environmental conditions [28, 29]. The phenotypic concept may appear to some to be stretched too far so as to cover proteomic structures, but it clearly fits Richard Lewontin's definition: "the descriptor of the phenome, the manifest physical properties of the organism, its physiology, morphology and behavior" (http://plato.stanford.edu/entries/genotype-phenotype/). That is, just as the genotype is the descriptor of the physical genome, the phenotype is the descriptor of a physical "phenome" that is the collection of all the distinctive features of the organism except its genome sequence. The distinction between genotype and phenotype was introduced in 1908 by the Danish biologist Wilhelm Johannesen, decades before DNA was discovered, to emphasize the differences between evolutionary and developmental pathways. Phenotypes can thus be extremely complex and include morphological, physiological, chemotactic, metabolomic, transcriptomic, and proteomic characteristics. A stress response can accordingly be defined as the transition from a defined reference phenotype to a stress phenotype.

### 11.1.3 Is Proteomics Purely Descriptive?

In bacterial stress research, a proteomic or transcriptomic phenotype is not the ultimate goal. Rather it is a starting point for elucidation of the underlying regulatory networks. A relevant question would be whether the field of proteomics is able to go beyond providing detailed phenotypes. Is proteomics exclusively descriptive or is it an experimental science that sets up hypotheses and tests these with the use of variables and positive versus negative controls? Reference map proteomics in its simplest form may be strictly descriptive with its output of annotated 2D gels. *In silico* proteomics deals with, among other topics, the bioinfomatic rendering of the calculated molecular masses and isoelectric points (pIs) of genome-encoded proteins as dots in scatter-plot diagrams [30]. These theoretical maps resemble 2D gels and may be used in comparison with real gels in reference map proteomics to test various predictions. Among the interesting topics could be the concordance between predicted and actual expression levels. This will be elaborated a little further under the discussion of the IL1403 *in silico* proteome. Through the identification of isoforms of proteins, reference map proteomics may also offer important clues for the analysis of posttranscriptional regulation [31]. Through the direct analysis of protein isoforms for the presence of chemical modifications using tandem mass spectrometry (MS/MS), both the type and site of modification can be identified. In itself this goes beyond pure description, but its highest potential is reached when expression proteomics is able to utilize this information in the quantification of differential expression ratios and thereby calculate the degree of protein modification under different conditions [31]. So even though proteomics in itself is a descriptive technique, the application of its various aspects in biological research may place proteomics within the experimental sciences.

### 11.1.4 Bacterial Growth Physiology as Prerequisite for Proteomics

Three decades ago the Copenhagen school showed, under the auspices of Ole Maaløe, the descriptive power of careful bacterial physiological analysis [32, 33]. Central to the school's idea was the importance of balanced growth. Under balanced growth, all cellular components are in a steady state and all concentrations of DNA, RNA, protein, and metabolites in the two daughter cells after a division are identical to that of the parental cell. This is contrasted by the cellular components after a metabolic shift. When a bacterium enters the stationary growth phase or any other shift situation, the metabolism is rearranged. If a bacterial culture is inoculated from a starter culture in the stationary phase, such as an outgrown overnight culture, it will experience a lag phase before the exponential growth phase sets in. Because the culture has experienced a shift, the subsequent entry into the exponential growth phase does not signify that the culture is uniform or stable. If a particular protein was overexpressed in the stationary phase to, say,

20-fold elevated levels, this would still be 10-fold elevated in the daughter cells, 5-fold in the next generation, and so on. The Copenhagen school demanded that a culture should have been in the exponential growth phase for eight generations before the bacteria were considered balanced. Any differences between the two physiological conditions would then be leveled out by a $2^8$-fold (256-fold) dilution.

Bacterial proteomics started under the influence of the Copenhagen school, and some the earliest studies were collaborations between Steen Pedersen [34] and Frederick Neidhard, who adopted the 2D GE method from O'Farrell [35] almost immediately after its invention. In a pioneering study from 1978, the synthesis rate of 140 proteins was determined in *Escherichia coli* under balanced growth with a wide span of different growth rates [34]. The synthesis rates of the proteins were determined by pulse labeling with [$^{35}$S]-methionine and with internal standardization using long-term [$^{14}$C] labeling. The significance of the individual protein synthesis rates in this study obviously relied upon balanced growth of the cultures. Yet, balanced growth is even more important in proteomics where the 2D gels are labeled post mortem for estimation of protein levels, because the excess stress proteins have to be diluted by cellular division in twofold steps. Protein synthesis rates, as measured by [$^{35}$S]-methionine pulse labeling, on the other hand, are primarily determined by the messenger RNA (mRNA) levels, which may change extremely fast due to the short half-lives. Controlled bacterial growth physiology is thus a prerequisite for expression proteomics for the definition of the reference state and the stress conditions.

Proteomics has inherited a useful vocabulary of regulatory typing from the earliest investigations: A collection of proteins or mRNA species (targets) that respond to a specific stress situation is termed a *stimulon*. If a collection of these targets within a stimulon share the same regulatory mechanism, that is, are controlled by the same regulatory protein, this collection is termed a *regulon*. The two concepts regulon and stimulon were introduced already in 1964 [36] and 1983 [37], respectively. To give an example of the use of the regulon and stimulon concepts, the heat-shock stimulon from *L. lactis* has more than 16 targets [22] and is comprised of two known heat-shock regulons, the HrcA regulon with the GroEL, GroES, HrcA, DnaK, [22] DnaJ, and GrpE targets and the CtsR regulon with the CtsR, ClpP, ClpB, ClpC, ClpE, [25] GroEL, and GroES, targets. The remaining heat-shock targets may belong to other regulons.

### 11.1.5 Synergy between Genetics and Proteomics in *L. lactis*

Proteomics, and especially expression rate proteomics, is a powerful means of detecting minor changes in the proteome induced by altered environmental conditions. Especially in combination with mutant selections, either site directed [25, 27] or by random mutagenesis, proteomics has been used as a powerful strategy in screening for proteins that are important for a given stress response. Overexpression of specific proteins by placing the respective genes under control of a foreign promoter has also brought interesting and unexpected changes in the bacterial proteome [26]. A new technique uses random artificial promoters as a means of obtaining promoter libraries where the expression of specific genes can be modulated around their normal cellular level [38, 39]. These subtle changes are needed for the analysis of protein levels by metabolic control analysis [40, 41]. The synergy between genetics and proteomics is enormous and will without doubt be exploited extensively in the future.

### 11.1.6 Synergy between Transcriptomics and Proteomics

Global analysis of gene expression at both the transcriptomics and proteomics levels has profound advantages. Primarily, it is possible to check expression patterns for the significance of the regulation identified at either level. A low level of regulation for the expression of a gene is more convincing if it was observed by both transcriptomics and expression rate proteomics. Differences in expression between the two levels of analysis may however be even more interesting. Posttranscriptional regulation has traditionally been identified by comparing the regulation of transcriptional versus translational reporter gene fusions [42]. Comparison of the regulation at the transcriptomics and proteomics levels offers the same type of information at a global scale. Whereas many transcriptional data from Northern blots have complemented proteomics data, few transcriptomics data have yet been published for *L. lactis*, so the value of a combined approach for lactococci will have to be seen in the future. One report, however, in which the purine deficiency stimulon was analyzed by both proteomic and transcriptomics techniques was presented in 2002 as a poster at the seventh Conference on Lactic Acid Bacteria in Edmond am Zee, The Nederlands (M. Kilstrup, N. H. Beyer, A. K. Nielsen, M. D. Pulka-Amin, M. D. Rasmussen, P. Roepstorff, and K. Hammer: Analysis of the purine nucleotide stress stimulon on both the mRNA and protein levels in *Lactococcus lactis*: Use of proteomics and custom designed DNA microarrays). A general lowering in expression level was detected by both approaches as well as up regulation of mRNA and proteins from genes belonging to the PurR regulon, confirming that PurR acts at the transcriptional level.

## 11.2 *LACTOCOCCUS* PROTEOME REFERENCE MAPS

The first reference map provided for the proteome of *L. lactis* had a very modest number of annotated protein spots [22]. The map consisted of a reference 2D gel of [$^{35}$S]-methionine-labeled proteins from strain MG1363 grown exponentially in the chemically defined synthetic amino acid (SA) medium [43] supplemented with glucose (GSA). The gel had a pH range from 4 to 7, where the authors claimed that more than 95% of all detectable proteins were situated. Following the lead from the influential *E. coli* proteome map [44], the gel was overlaid by a coordinate system, which permitted precise mapping and reference of protein spots (see Fig. 11.1). Due to the application of protein from very few bacteria in each gel ($\sim$1.5 $\times$ $10^7$ bacteria, $\sim$1.5 µg cytoplasmic protein), a special technique with lyophilization and grinding of the cellular pellet with glass beads in Eppendorf tubes was developed for extraction of the full protein complement [22]. By a combination of Western blotting and N-terminal sequencing of protein spots after blotting of 2D gels onto polyvinylidene diflouride (PDVF) membrane, the heat-shock proteins GroEL and DnaK were identified together with the abundant glycolytic enzymes pyruvate kinase (PYK), phosphoglucokinase (PGK), and two isoforms of glyceraldehyde-3-phosphate dehydrogenase (Gap). When proteins in the following are identified by coordinates (e.g., {47.0, 67.5} for GapB), their location on the MG1363 reference map can be found using the scale on the bottom and left axis in Figure 11.1.

### 11.2.1 Reference Map of *L. lactis* subsp. *cremoris*

A step further in the establishment of the *L. lactis* subsp. *cremoris* proteome reference map was taken in the identification of 17 protein spots from a 2D gel of strain NCDO763

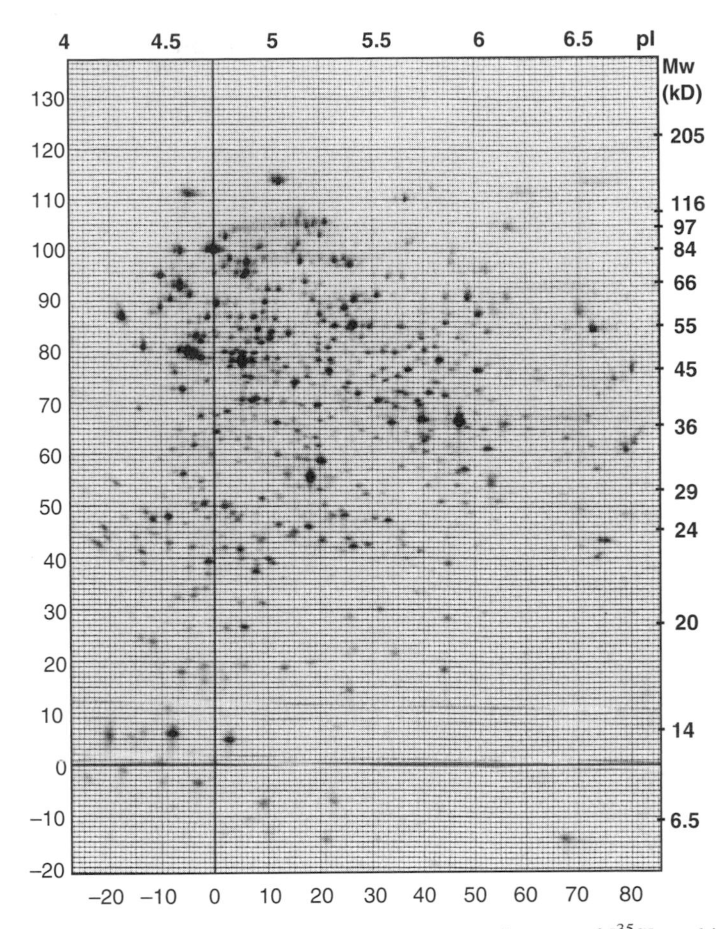


**Figure 11.1** Reference map of *L. lactis* subsp. *lactis*. Autoradiogram of [$^{35}$S]-methionine-labeled proteins from MG1363 grown in chemically defined SA medium supplemented with glucose. (From [22.])

proteins in the pH range 4–7 [45]. NCDO763 is identical to the plasmid and prophage containing dairy strain NCDO712, from which the model strain MG1363 was derived [10]. The strain was cultured in the chemically defined medium (CDM) [45] and harvested in the midexponential phase, and the 100-μg cytoplasmic protein fraction analyzed was obtained by sonic disruption and removal of the cellular envelope. The identified proteins were a mix of abundant proteins and faint spots, which were chosen for various reasons. Interestingly, three isoforms of the glycolytic GAPDH enzyme were identified, of which the two most alkaline were identical to the isoforms previously detected [22]. Both peptidases (PepO, PepN) and plasmid-encoded enzymes for lactose utilization (LacA, LacB, LacX) were detected as well as the HPr protein involved in the PTS sugar uptake system.

The NCDO763 proteome was later analyzed during early exponential growth in rich M17 broth medium supplemented with glucose [24]. The work, which was part of an expression proteomics study, included a Coomassie-stained reference gel from a 300-μg cytoplasmic protein fraction. In this study 161 protein spots were identified, including all glycolytic enzymes and medium to highly expressed proteins from a wide variety of functional classes. Plasmid-encoded proteins were identified, including enzymes involved in lactose (LacA, LacB, LacD, LacG, and LacX) and casein (PepF2) utilization.

### 11.2.2 In Silico Proteome of *L. lactis* subsp. *lactis* IL1403

When the predicted proteome of IL1403 was analyzed *in silico* [24], it was found that the apparent clustering of 95% of the *L. lactis* proteins between pH 4 and 7 [22] was far from the truth. The *in silico* proteome of *L. lactis* is separated into two dense areas [30] as observed for most bacteria [29]. One is most dense around pI 5 and the other around 9. Only 60% of the encoded proteins have pI values between pH 4 and 7. The claim that most proteins were found between pH 4 and 7 originated partly from the fact that only the most abundantly expressed proteins are seen under normal conditions and that these seem to cluster in the region. A correlation has been found to exist between the codon bias in a gene and the expression level of the encoded protein under normal unstressed growth conditions [46]. Highly expressed genes appear, by genetic drift, to have their codons fine tuned to match the most abundant transfer RNA (tRNA) species. A codon adaptation index (CAI) between 0 and 1 can be calculated for each gene in the genome using the overall codon bias of the genome as a guide [47]. This has previously been used as an index for the expression level of the specified protein [48]. The glycolytic genes of *L. lactis* have CAIs above 0.7, while those for 87% of all genes are below 0.7. The highest number of proteins (71%) have correspondinhg CAI values between 4 and 6 [24]. In Figure 11.2 the IL1403 in silico map from pI 3–13 is rendered in a way so that the sizes of the protein spots are determined by their corresponding CAI values. By selecting a steep gradient between CAI

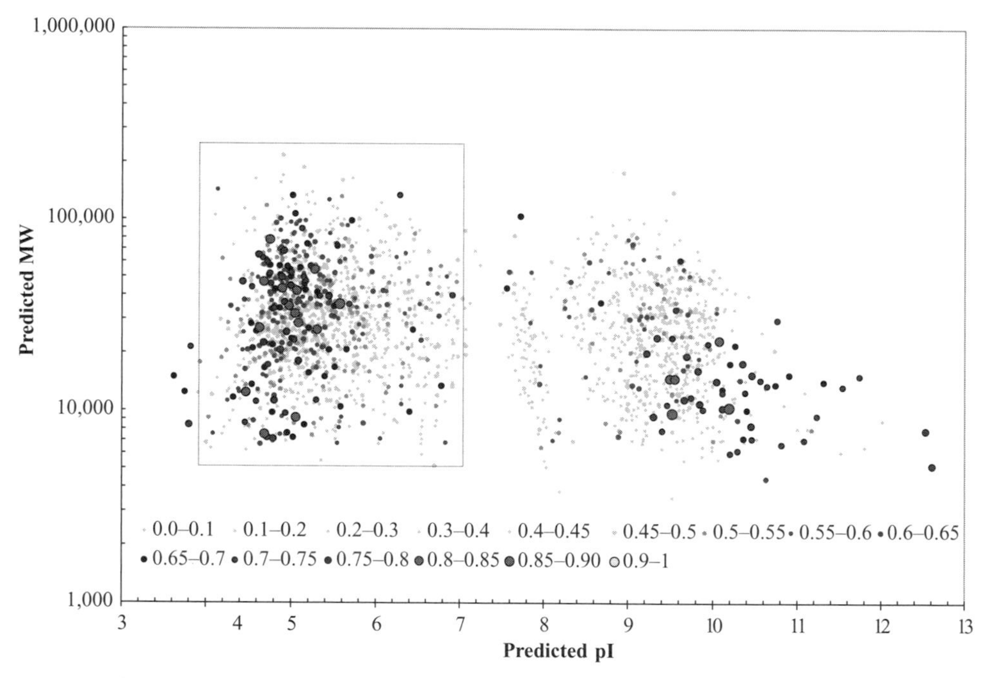


**Figure 11.2** In silico 2D gel for predicted proteome of *L. lactic* IL1403. Spots represent the predicted pI and molecular weight (MW) values for the proteins encoded by the IL1403 genome [4]. The size and color of the spots represent the CAI for the corresponding gene, as indicated.

value and size, the 976 proteins with highest CAI values (>0.45) are clearly visible while the rest are shown as small faint spots. It is interesting to see that this rendering gives a general pattern which can be recognized in the real 2D gels of *L. lactis*. It is possible to see the location of all the glycolytic enzymes, the HPr protein from the PTS system, as well as the translation elongation factors Tuf, Tsf, and FusA. This map can almost be superimposed upon the real gel. Membrane proteins usually require special procedures to be solubilized from the bacterial material. Most proteomics studies have thus analyzed only the cytoplasmic fraction, by removal of the cell envelope fraction prior to solubilization. All integral membrane proteins require hydrophobic membrane-spanning domains and therefore have a high percentage of hydrophobic amino acids. To identify those proteins *in silico*, a grand average of hydrophobicity (GRAVY) score can be calculated for each polypeptide as the sum of hydropathicity values divided by the number of residues. The extremes of the GRAVY scale are polylysine at –3.9 and polyisoleucine at 4.5. A hydrophobic protein will generally have a GRAVY score above zero. The *L. lactis in silico* proteome has more than 22% proteins with positive GRAVY values, which are therefore not expected to be members of the cytoplasmic proteome [24]. Interestingly, in the alkaline pI range above pH 7 almost 40% of the proteins have positive GRAVY scores [24]. If, however, all hydrophobic proteins from the *in silico* proteome map in Figure 11.2 were removed, the overall pattern would not change. This is because most hydrophobic proteins have CAI values below 0.45 and are only shown as faint spots. A correlation between GRAVY and CAI values are shown in Figure 11.3 for the IL1403 proteome. Interestingly, the highest CAI value for a hydrophobic protein is found for the mannose specific enzyme II (PtnC) for the PTS sugar uptake system, which is known to be the main transporter of glucose in the cell. It will be very important in the future to analyze if a

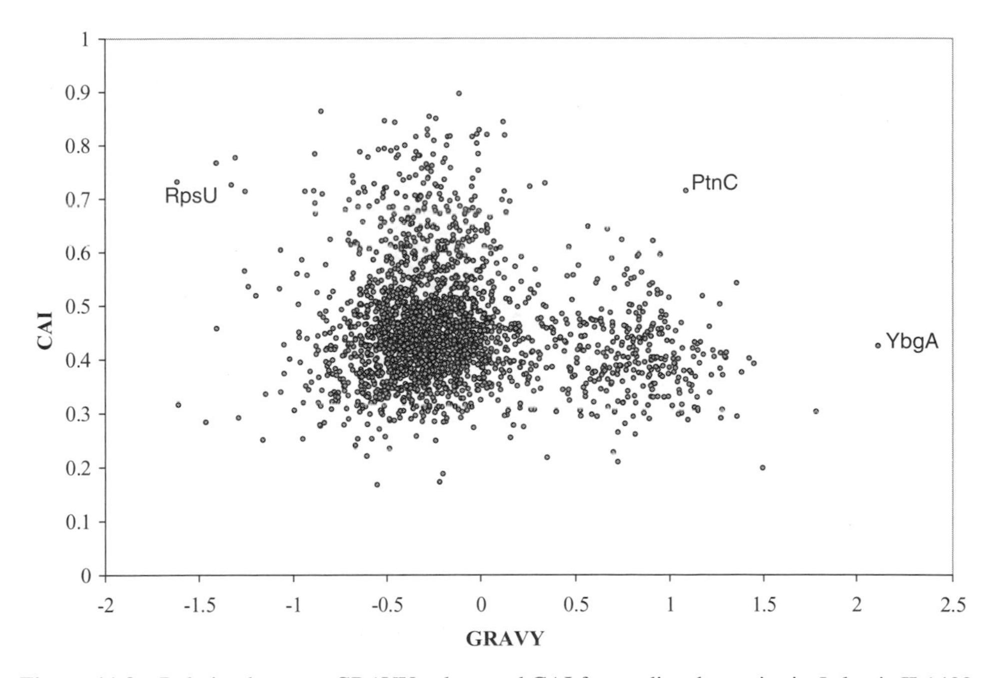


**Figure 11.3** Relation between GRAVY values and CAI for predicted proteins in *L.* lactis IL1403. Protein annotations [4] have been added to outliers.

strong correlation exists between actual expression levels (spot intensities) and the CAI values in the unstressed *L. lactis* proteome.

### 11.2.3 Reference Map of *L. lactis* subsp. *lactis*

In a proteome map of IL1403 published in 2003 by Guillot et al. [24], more than 230 spots were identified. All glycolytic enzymes were identified as highly abundant proteins. Interestingly, many of the proteins were detected as multible isoforms. GAPDH (Gap B) was previously detected as three isoforms in subsp. *cremoris*, but the present study identified a fourth minor spot located at the more alkaline pH. All four isoforms (two major flanked by two minor) apparently have identical molecular masses and are equidistantly spanned by pI values of approximately 0.12 pH unit. Pyruvate kinase (Pyk) was detected as three isoforms of identical molecular mass, with one major spot flanked by two minor, all spanned by approximately 0.07 pH unit. Several other glycolysis enzymes (PgiA, FbaA, Pgk, and Pgm) and enzymes involved in fermentation pathways (Ldh and Pfl) appeared as a major spot flanked by minor satellites of equal molecular mass. The Pfl was found as two series of spots, one series representing the full-length protein and the other a processed form. Most proteins are migrating according to their molecular weights and isoelectric points. Figure 11.4 shows the correspondence between determined and theoretical pI values for the identified spots based upon the values presented on the 2D reference map from http://genome.jouy.inra.fr/2dlactis/eng/gels/IL1403_glucose_M17.html. Only 13 proteins migrate far from their expected pI positions. In Figure 11.4, all proteins with pI isoforms are rendered with large circles. Although there is generally a very fine match between the predicted and actual pI values, it is not possible to determine, on the basis of the pI, which of the isoforms are unmodified and which are native.

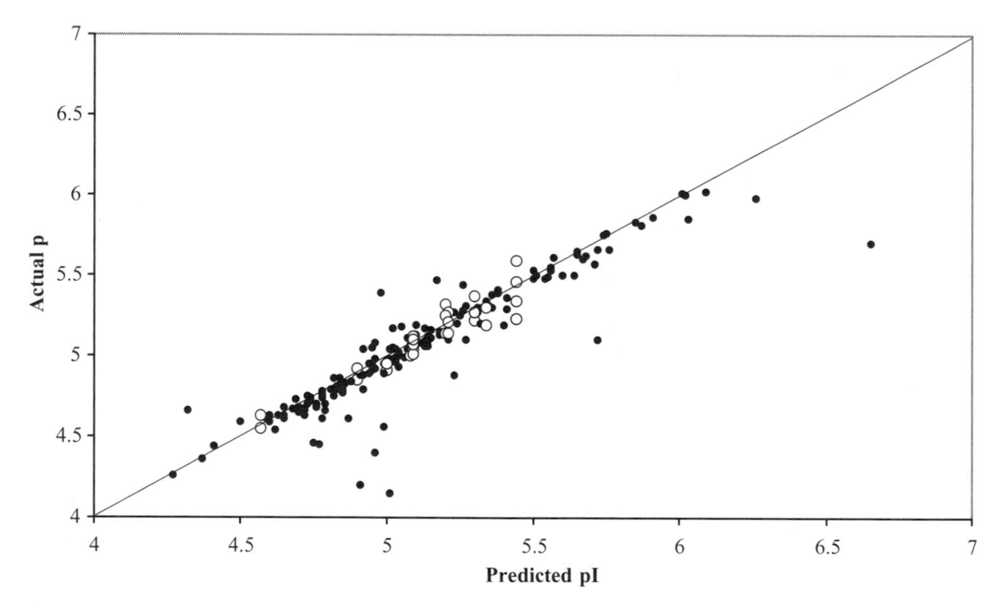


**Figure 11.4** Comparison between predicted and detected pI values for proteins from *L. lactis* IL1403 identified by Guillot et al. [24] and data from complementary web page at http://genome.jouy.inra.fr/2dlactis/eng/gels/IL1403_glucose_M17.html. Proteins identified as more than one isoform are shown as large circles.

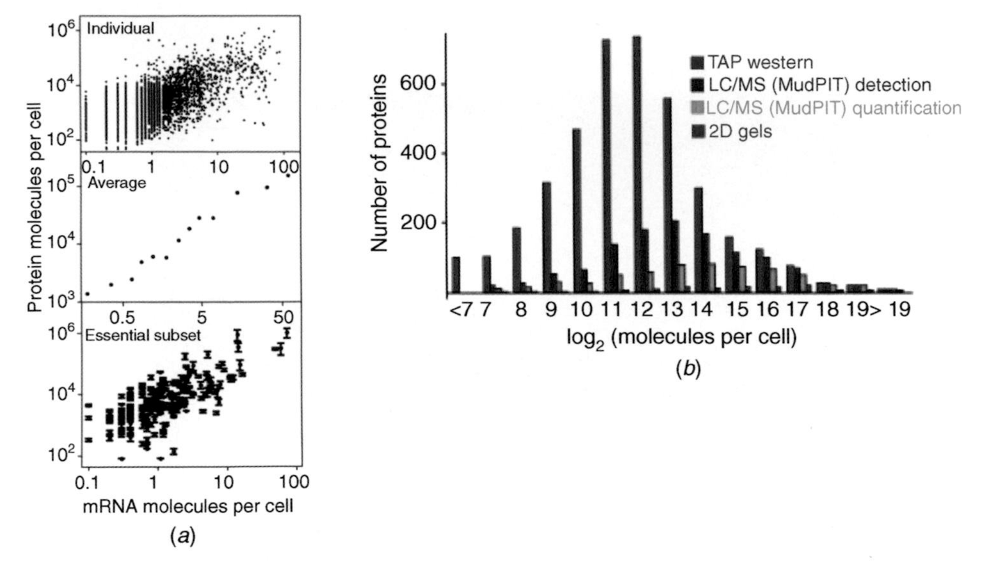


**Figure 1.3** Analysis of protein expression in yeast showing (*a*) correlation between protein abundance and transcripts: Top panel shows the relationship between steady-state mRNA and protein levels in yeast as determined by microarray analysis. Middle panel shows ORFs sorted into discrete mRNA levels and means plotted against mean protein abundance. Lower panel shows only a comparison of protein and mRNA levels for some essential soluble proteins. (*b*) Absolute range of expression possible using TAP/Western blot, LC-MS using a multidimensional chromatography approach, and 2DE [3].

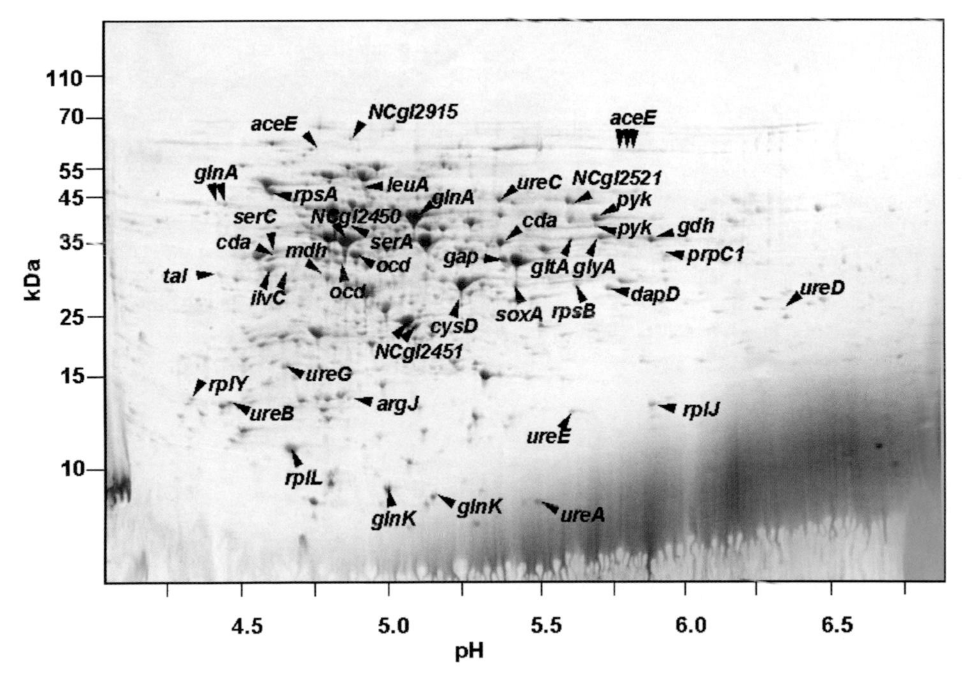


**Figure 10.2** Comparison of *C. glutamicum* protein profiles from cells grown with different nitrogen supply. One hundred micrograms of cytoplasmic protein fractions was separated by 2D PAGE and stained with Coomassie brilliant blue. For comparison, a false color overlay was produced using the DECODON Delta2D 3.1 software package (red, nitrogen-rich batch phase; green, nitrogen-limited continuous phase). Spots are designated according to the annotation of the corresponding genes (see also Silberbach et al., submitted). (Kindly provided by M. Silberbach, Institute of Biochemistry, University of Cologne.)

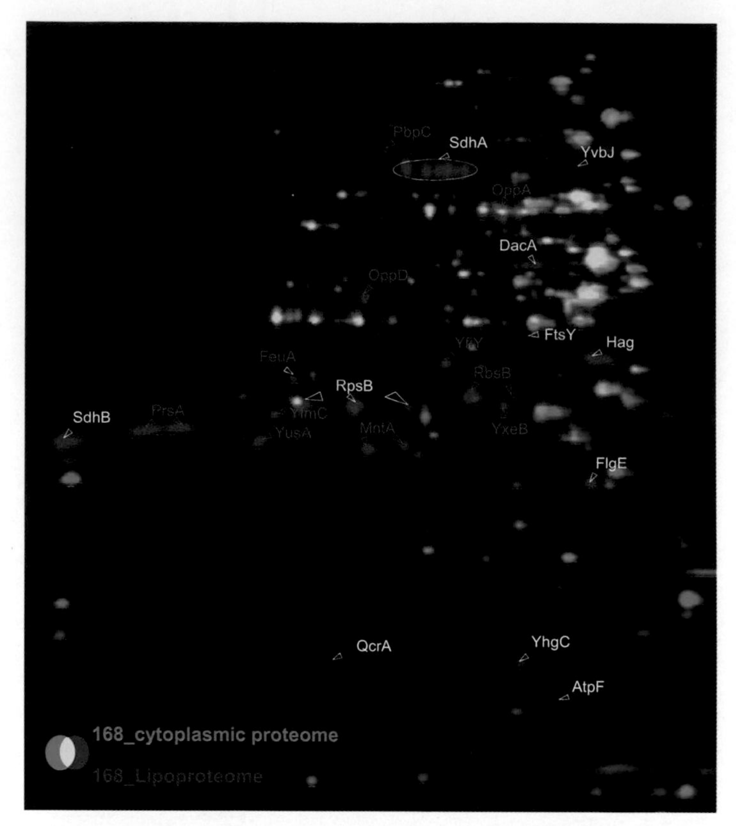


**Figure 12.5** Comparison of membrane-attached lipoproteome (red image) and cytoplasmic proteome (green image) of *B. subtilis*. Cells of *B. subtilis* 168 were grown in LB broth and harvested 1 h after entry into the stationary phase. Cytoplasmic protein extracts were compared to the nondetergent sulfobetaine- (NDSB-) washed membrane fraction obtained after stepwise extraction with $\beta$-D-maltoside (corresponding to fraction 4 according to Bunai et al., 2004). Image analysis was performed using the Decodon Delta 2D software that is based on the dual-channel imaging (Bernhardt et al., 1999). Proteins enriched in the membrane fraction that are present at lower amounts in the cytoplasmic fraction were identified. Known or predicted lipoproteins which could be extracted from the membrane fraction are indicated by red labels.

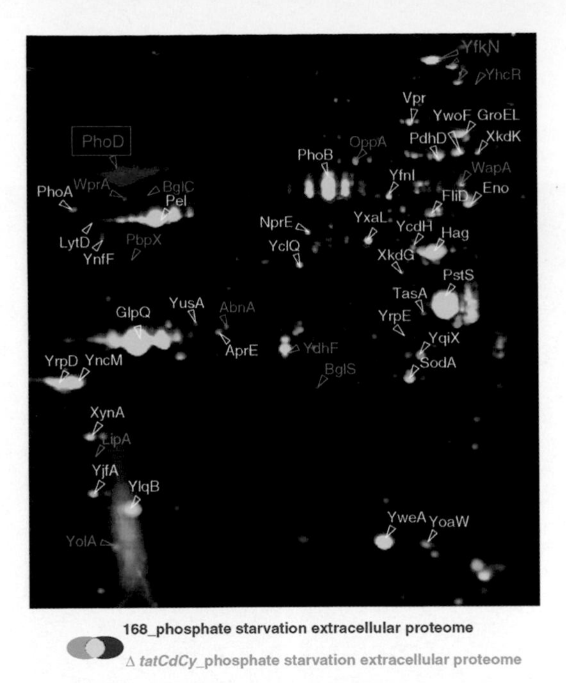


**Figure 12.7** Phosphate starvation proteome of wild type (red image) in comparison to ΔtatCd/ ΔtatCy mutant (green image). Cells of *B. subtilis* 168 and the *ΔtatCd/ΔtatCy* mutant were grown in minimal medium under the conditions of phosphate starvation and proteins in the growth medium were harvested 1 h after entry into the stationary phase. After precipitation with TCA, the extracellular proteins were separated by 2D PAGE and stained with Sypro ruby. Image analysis was performed using the Decodon Delta 2D software that is based on the dual-channel imaging (Bernhardt et al., 1999). Proteins with predicted RR/KR signal petides are labeled in blue and the Tat-dependent PhoD protein is boxed.

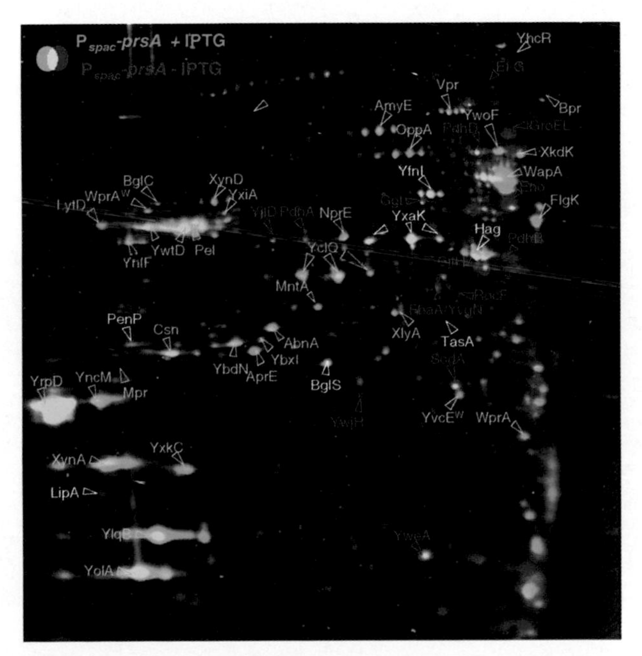


**Figure 12.9** Extracellular proteome of *B. subtilis* IH7211 ($P_{spac}$-*prsA*) grown in absence (red image) or presence (green image) of 1 mM IPTG. Cells were harvested 1 h after the entry into the stationary phase and extracellular proteins in the culture medium were TCA precipitated, separated by 2D PAGE, and stained with Sypro ruby. Image analysis was performed using the Decodon Delta 2D software that is based on the dual-channel imaging (Bernhardt et al., 1999). Proteins that appeared as green spots (higher level in the presence of PrsA) are marked in blue and those that appeared as red spots (higher level in the absence of PrsA) are marked in red. Proteins that appeared as yellow spots are labeled in white (unchanged according to the quantitation data).

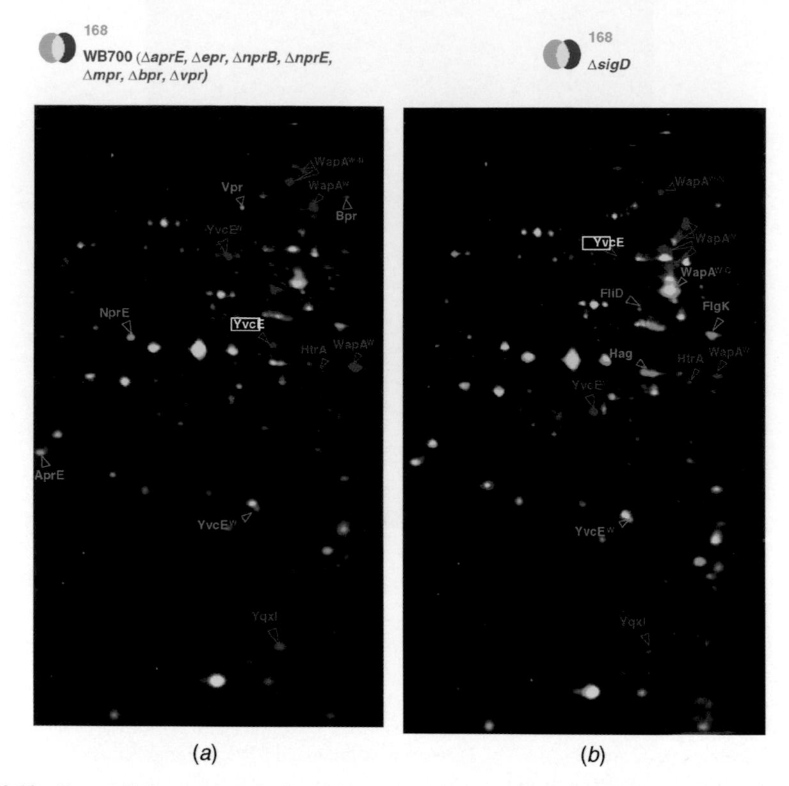


**Figure 12.10** Extracellular proteome of multiple extracellular protease-deficient *B. subtilis* strain WB700 (*a*, red image) and *sigD* mutant (*b*, red image) in comparison to that of wild type (green image). Extracellular proteins were separated by 2D PAGE and stained with Sypro ruby. Image analysis was performed using the Decodon Delta 2D software that is based on dual-channel imaging (Bernhardt et al., 1999). Secreted cell wall proteins, including the large WapA-processing products, unprocessed YvcE, HtrA, and YqxI that are stabilized in the exoprotease, and *sigD* mutants during the stationary-phase mutant, are edited in red and those that are missing in the mutants are labeled in blue.

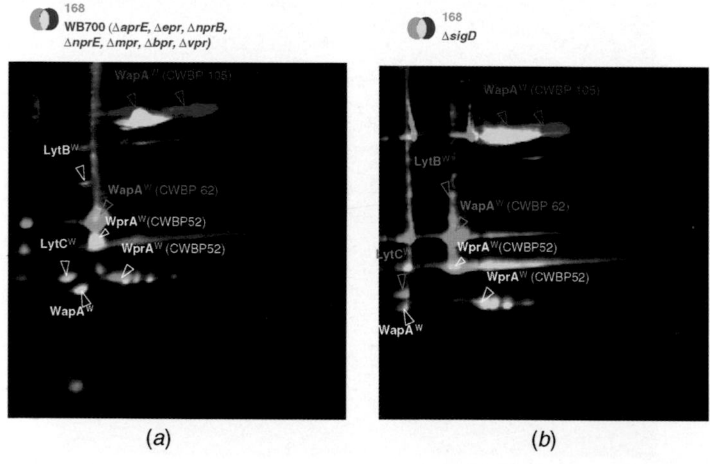


**Figure 12.11** Cell wall proteome of multiple extracellular protease-deficient *B. subtilis* strain WB700 (red image, *a*) and *sigD* mutant (red image, *b*) in comparison to that of wild type (green image). Cell wall proteins were extracted using a LiCl extraction procedure, precipitated with TCA, separated by 2D PAGE, and stained with Sypro ruby. Image analysis was performed using the Decodon Delta 2D software that is based on dual-channel imaging (Bernhardt et al., 1999). The WapA-processing product CWBP105 is increased (labeled red) and WapA-processing product CWBP62 is decreased (labeled blue) in the protease mutant as well as the *sigD* mutant.

**Figure 12.12** Extracellular proteome of *B. subtilis degU32(hy)* mutant (red image) in comparison to that of wild type (green image). Extracellular proteins were separated by 2D PAGE and stained with Sypro ruby. Image analysis was performed using the Decodon Delta 2D software that is based on dual-channel imaging (Bernhardt et al., 1999). Extracellular proteins secreted at higher amounts in the *degU32(hy)* mutant are labeled in red, and those which are repressed in the mutant are labeled in blue.

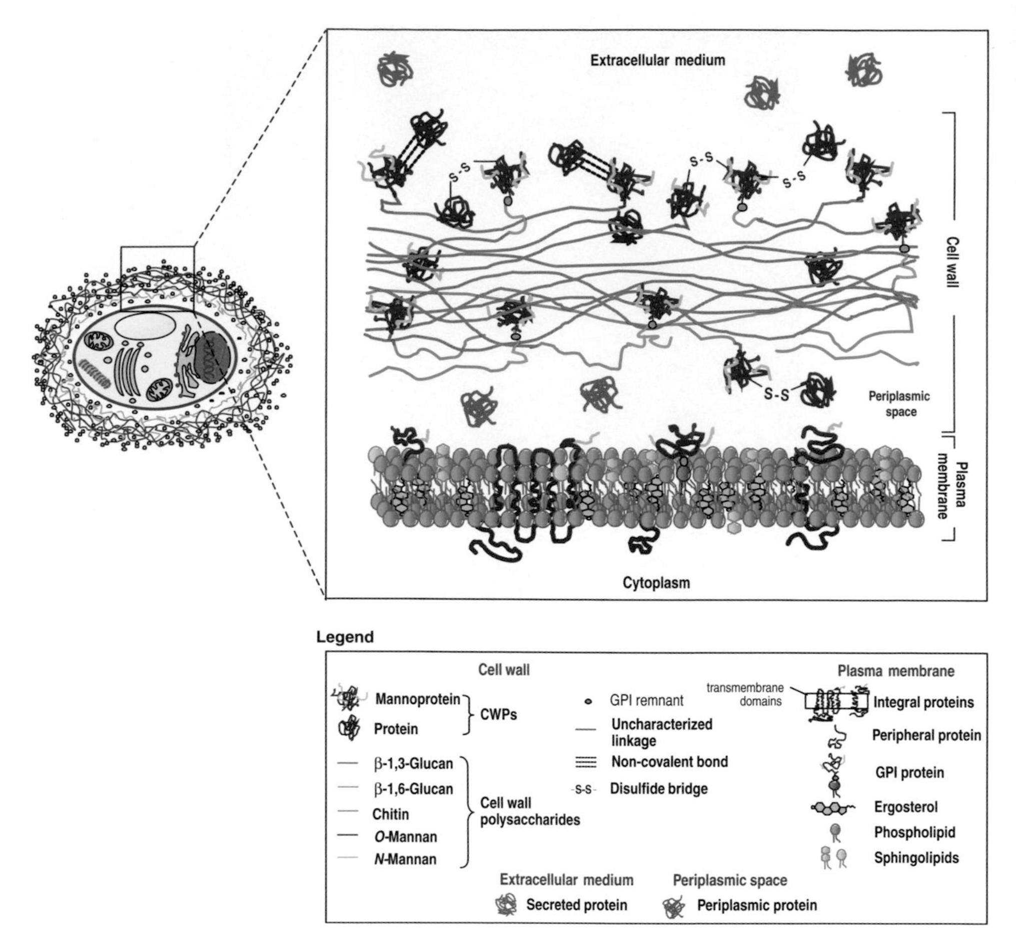


**Figure 17.6** Schematic representation of *C. albicans* cell envelope. A cross section through a yeast cell is illustrated in the left panel. The callout depicts details for the cell envelope of *C. albicans*, composed of an inner plasma membrane and an outer cell wall (right panel). The cell wall—the outermost cellular structure—basically consists of microfibrillar polysaccharides (fuchsia, green, and blue lines) and CWPs (dark blue tortuous lines). β-1,3-Glucan (fuchsia lines) and chitin (green lines) fibrils are glycosidically linked and form the structural framework of the wall. β-1,6-Glucan molecules (blue lines) interconnect certain CWPs (i.e., the GPI-CWPs) with β-1,3-glucan or chitin chains via a GPI remnant (orange circle). Other covalent linkages either between CWPs and β-1,3-glucan or between CWPs and chitin also occur (brown or red lines). Crosslinking of CWPs through disulfide bridges (–S–S–) or noncovalent bonds (black broken lines) is shown. Many CWPs (the so-called mannnoproteins) can carry *O*- and/or *N*-linked glycans (red or light bottle-green lines, respectively). Periplasmic proteins (grey tortuous lines) are trapped between the plasma membrane and the β-1,3-glucan-chitin network. Some proteins (blue tortuous lines) are secreted to extracellular medium. Some mannoproteins (dark brown tortuous lines) are anchored to the plasma membrane through GPI. The plasma membrane is a phospholipid bilayer interspersed with ergosterol that serves as a matrix for integral and peripheral proteins. This model is based on [33, 71, 79, 83]. CWPs denote cell wall proteins and GPI glycosylphosphatidylinositol.

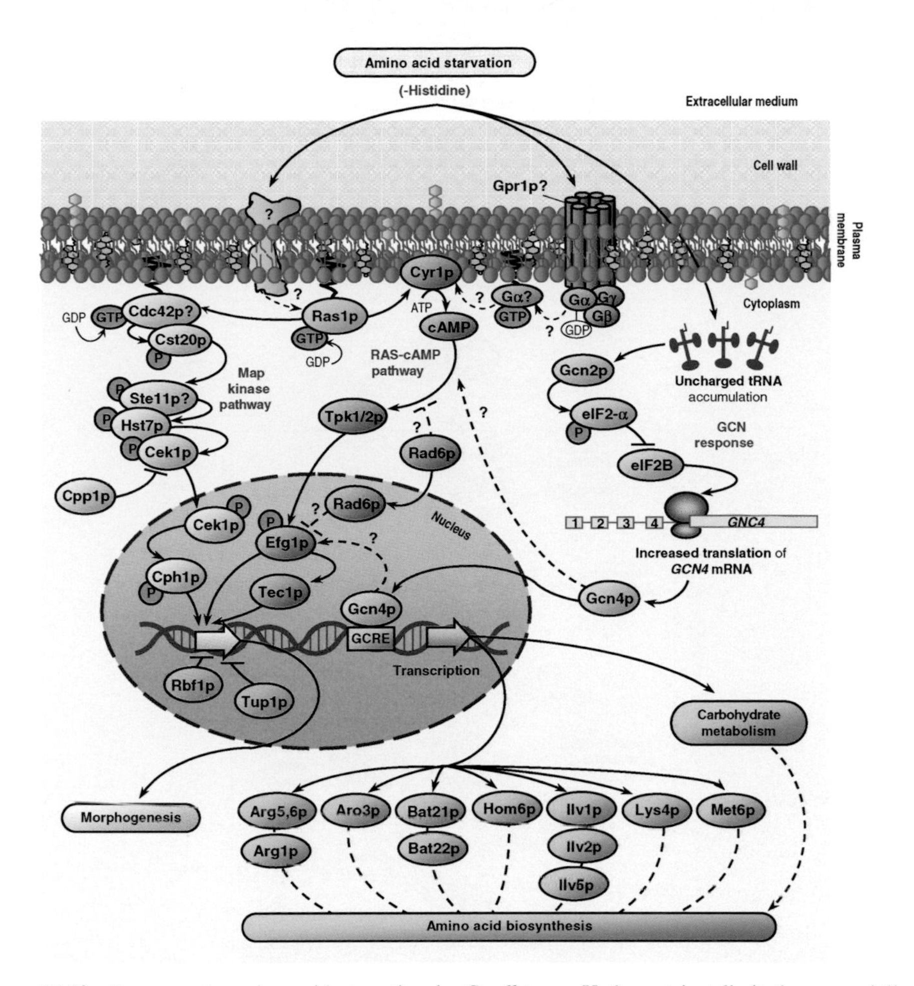


**Figure 17.10** Response to amino acid starvation in *C. albicans*. Under nutrient limitation, especially in nitrogen and/or carbon starvation, two major signaling cascades—MAP (mitogen-activated protein) kinase pathway (in amber) and Ras-cAMP pathway (in pink), defined by the transcription factors Cph1p and Efg1p, respectively, and dependent upon the GTP-binding protein Ras1p—stimulate filamentous growth (pseudohyphal or hyphal growth) [112, 143–147]. When *C. albicans* cells are deprived of any amino acid (e.g., histidine), uncharged tRNAs are accumulated. This leads to activation of the protein kinase Gcn2p, which phosphorylates the α subunit of the translation initiation factor 2 (eIF2). This phosphorylated form in turn inhibits the guanine–nucleotide exchange factor eIF2B. This reduction of eIF2 activity promotes the translation of a specific mRNA encoding the transcription factor Gcn4p, with a long 5'-leader region containing four small open reading frames (ORFs) located upstream of the *GCN4* ORF. Subsequently, this transcription activator interacts with the general control response elements (GCRE)–like conserved sequences in the promoter regions of its target genes. As a result, Gcn4p activates the transcription of genes that encode enzymes required for amino acid biosynthesis (seven different pathways) and carbon metabolism (in green), among others, but not for adenine biosynthesis, gathered from proteomic analysis [40]. This coordinate induction of the expression of genes on diverse amino acid biosynthetic pathways to bypass the external deficiency of a single amino acid is termed general amino acid control (GCN) response (in blue and green), which is independent of morphogenetic signaling (MAP kinase and Ras-cAMP) pathways. In addition, Gcn4p also activates morphogenesis (pseudohyphal growth) by interacting either with a downstream component (Efg1p) or maybe with an upstream one of the Ras-cAMP pathway in response to amino acid starvation, but not under morphogenetic signals [106]. Blunt arrows depict negative effects. Question mark after protein name indicates that the functional homologue of the corresponding *S. cerevisiae* gene placed in the same position on the pathway has been identified in the *C. albicans* genome but it has not yet been located experimentally on this pathway. For clarity, Gcn4p is illustrated inducing the transcription of only one gene (green arrow). Names of amino acid biosynthesis-involved enzymes refer to those in the Appendix in Chapter 18). Others: Cdc42p, GTPase; Cst20p, protein kinase; Ste11p, MAP kinase kinase kinase; Hst7p, MAP kinase kinase; Cek1p, MAP kinase; Cpp1p, protein tyrosine phosphatase; Gpr1p, heterotrimeric G protein receptor; Gα or Gpa2p, heterotrimeric G protein α subunit; Cyr1p, adenylate cyclase; Tpk1/2p, protein kinase A; Rad6p, ubiquitin-conjugate enzyme; Tec1p, transcriptional activator; Tup1p and Rbf1p, transcriptional repressors.

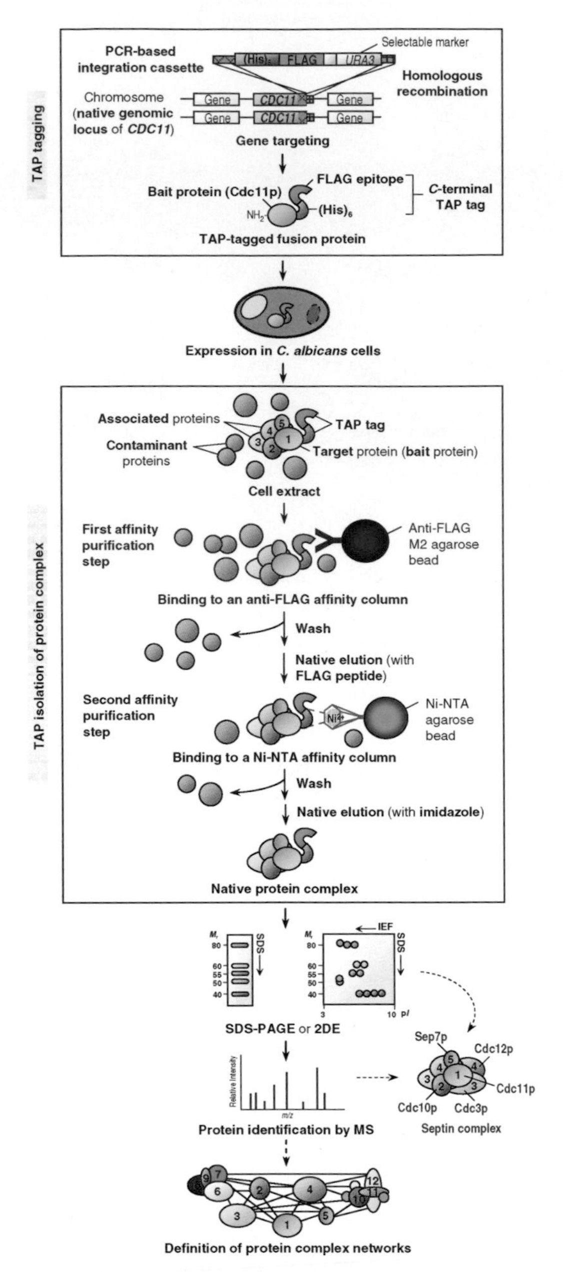


**Figure 17.11** Tandem affinity purification (TAP) strategy for characterization of *C. albicans* septin complex [73]. A TAP tag, consisting of a hexahistidine peptide $(His)_6$ and a FLAG epitope (DYKDDDDK), is fused to the protein of interest (in the example, Cdc11p) at its C terminus. The TAP-tagged fusion protein is expressed in *C. albicans* cells at its physiological level. Cell extracts are then prepared under mild conditions. The fusion protein and its associated endogenous components are recovered from cell lysates by two successive affinity purification steps (TAP) under gentle native conditions. In the first one, the FLAG tag is bound to a column of anti-FLAG M2 agarose beads, and contaminant material is removed by washing. The immobilized protein complex is then eluted with FLAG peptide. The eluate is further incubated with Ni-NTA (nickel nitrilotriacetic acid) agarose beads on a second column. The hexahistidine tag interacts with two of the six ligand-binding sites (not occupied by NTA) in the coordination sphere of the nickel ion. After washing to remove residual contaminants, the highly purified protein complex is eluted with imidazole, which contains part of the histidine structure. The protein complex is then resolved on an SDS–polyacrylamide gel electrophoresis (PAGE) and/or 2DE gel, and proteins interacting with the target protein are identified by MS. The septin complex purified from *C. albicans* cell extracts using this strategy was composed of Cdc3p, Cdc10p, Cdc11p, Cdc12p, and Sep7p in a potential stoichiometric relation 2 : 1–2 : 1 : 2 : $\leq$ 1 [73]. This procedure at high-throughput scale leads to the definition of protein complex networks.

In the study by Guillot et al. [24], all predicted proteins in the acid proteome with corresponding CAIs above 0.8 (14) were identified from the gels. Of the 32 predicted proteins with CAI values between 0.7 and 0.8, 75% were identified, and the coverage of the 107 proteins with CAI values between 0.6 and 0.7 was around 54%. Although the highest fraction of identified proteins had CAI values between 0.5 and 0.6, the coverage in this region was around 27%. In the region from 0.4 to 0.5, less than 6% were detected. Only one protein with a corresponding CAI value below 0.4 (CAI = 0.377, probably HsdM) was detected. The conclusions were twofold: (i) The CAI reflects in vivo expression levels, and (ii) only moderately to highly expressed genes can be detected in Coomassie- and silver-stained gels with normal protein loads.

Until recently, the alkaline region of proteomes had been difficult to access. It was therefore an important achievement when the alkaline IL1403 proteome was resolved by a leading proteome laboratory using fast isoelectric focusing at high voltage and anodic application of the protein mixture [12]. By this protocol the authors were able to identify 185 spots from pH 6 to 12 of the Coomassie-stained IL1403 cytoplasmic proteins (500 μg) after exponential growth in chemically defined SA medium. Even though more than 1000 spots are present in the in silico proteome in this region, only approximately 200 spots could be seen in the 2D gels [12]. This low number, however, is close to the number of proteins with corresponding CAI values above 0.45 (see Fig. 11.1). Nearly 70% of all proteins with CAI > 0.6 in the alkaline proteome are ribosomal proteins [12]. The most abundant protein in the alkaline region, the histonelike protein HslA, has the second highest associated CAI value. All proteins with CAI values above 0.8 (6 proteins) were identified in the study, while only 51% of the proteins with CAI values between 0.7 and 0.8 (18 out of 35) were found. Of the identified proteins, 46 were ribosomal with theoretical pI values ranging from 9.2 to 10.4. As for the identification of the acid IL1403 proteome, many proteins were found as multiple isoforms. Especially RpsC, RplA, and RplF showed the typical pattern of a central spot flanked by two smaller satellites.

The *L. lactis* subsp. *lactis* IL1403 proteome map showed major differences from the map of NCDO 763 (subsp. *cremoris*) grown under similar conditions [24]. It was noted that only 61 of the nearly 400 protein spots that could be detected from either strain in the pH range 4–7 were migrating to identical positions in the gels. Although less than 20% of the proteins showed identical migration, it was found that all of the glycolytic enzymes belonged to this invariant group [24]. In a survey of 100 proteins that did not have positional equivalents in the IL1403 proteome map, 63 proteins from NCDO763 could be identified in the IL1403 map at a slightly different position [24]. This suggested that the genes encoding the proteins had been subjected to genetic drift after the evolutionary separation of the two subspecies and have acquired amino acid substitutions. If these rough numbers may be extrapolated, they signify that 20% of the proteins from the two subspecies have identical amino acid sequences and around 60% have only slightly deviating sequences. This would leave around 20% differences in genome content, which however is greatly overestimated because genes for most of the proteins identified in NCDO763 are found in the IL1403 genome.

### 11.2.4 *Streptococcus* Proteome Reference Maps

Many *Streptococcus* species have been subjected to proteome analysis and reference maps have been reported for a dairy strain of *S. thermophilus* [49] and two oral species, *Streptococcus oralis* [50] and *Streptococcus mutans* [51, 52], involved in dental plaque formation. A comparison between the proteomes of *L. lactis* MG1363 and *S. thermophilus* in the pI range 4–7 [30] shows a similar overall pattern with respect to the highest expressed proteins. Both species have a characteristic proteome pattern where most protein spots are crowded between pIs of 4.3 and 5.5 and molecular masses between 20 and 100 kD. The acidic proteomes of *S. mutans* [51] and *S. oralis* [50] appear to follow a similar pattern, except some of the major bands appear to be shifted toward acidic pIs. The two species *Lactobacillus delbrueckii* and *Lactobacillus sakei* grown under similar conditions showed a totally different proteome pattern [30]. Evolutionary relatedness may thus be manifested also in the overall proteome 2D pattern.

The *S. thermophilus* genome sequence has not yet been published but is more than 95% completed by the Life Science Institute at the Université Catholique de Louvain in Belgium, according to their homepage (http://www.biol.ucl.ac.be/gene/genome/). The *S. thermophilus* reference gel contained 16 μg of whole-cell protein extract of a dairy strain, PD18, grown in rich M17 broth supplemented by lactose [49]. 12 highly expressed protein spots were identified by N-terminal amino acid sequencing. The glycolytic enzymes Pgk, Fba, and GAPDH as well as the chaperone GroEL and superoxide dismutase were among the identified proteins. Fba and GroEL were extracted as double spots and GAPDH as a triple spot.

A Coomassie-stained reference gel of *S. oralis* [50] was reported with whole-cell protein extract from a dental plaque isolate grown in Brain Heart Infusion (BHI) medium at pH 5 until the midexponential growth phase. Thirty-nine different proteins were identified from 65 extracted spots. The work was part of an analysis of acid adaptation, and the spots included both high-end and moderately expressed proteins. The large difference between the number of spots and individual proteins was again due to the presence of many isoforms of specific enzymes: Gap (5 spots), Fba (6 spots), Ldh (5 spots), and GroEL (4 spots) as well as pyruvate oxidase (5 spots) and glutamate dehydrogenase (4 spots).

The proteome of *S. mutans* has been analyzed extensively. In a study of planktonic versus biofilm growth, an autoradiogram of $^{14}$C-labeled proteins was presented as a reference gel with 41 annotated protein spots [51]. Both highly and moderately expressed proteins were identified. As usual, multiple isoforms of Gap (two spots identified) and GroEL (three spots) were identified. Finally, an extensive proteome analysis of cellular and extracellular/exported proteins of *S. mutans* was recently reported [52]. The strain was grown in chemically defined medium in a glucose-limited chemostat at a dilution rate of 0.1 $h^{-1}$ as a model of healthy dental cavities between meals. Gels of four different pI ranges were analyzed to span pIs from 4 to 11. From a wide variety of functional classes and unknown functions, 421 spots were identified by matrix-assisted laser desorption ionization time-of-flight (MALDI-TOF) MS or MS/MS. The most curious findings in this study, as it has been in all other reference maps from the *Streptococcus–Lactococcus* branch, was the presence of multiple protein isoforms. Even larger numbers of isoforms of abundant enzymes were identified in this study: Eno (14 spots), PgmA (3 spots), Gpi (2 spots), Pgk (11 spots), Fba (5 spots), Ldh (2 spots), Gap (15 spots), PfkA (2 spots), TpiA (3 spots), GpmA (2 spots), and Pyk (4 spots). The chaperones GroEL and DnaK were identified as 5 and 6 spots, respectively, but the translation elongation factor TU took the lead with 19 separate spots. The 14 enolase spots were of particular interest because they were found with molecular weights corresponding to monomers, dimers, and trimers. An explanation for the presence of these large stable variants could not be found, except that perhaps subpopulations of isoforms are easier to observe among high-abundant proteins than among lesser abundant proteins. This begs the question whether or not most cellular proteins will exist as subpopulations of molecular isoforms. A reference map for the extracellular proteome was constructed from a TCA precipitate of the chemostat culture fluid. The general pattern in the extracellular proteome was reminiscent of the intracellular proteome, suggesting that many proteins were actually intracellular and released by cell autolysis. All isoforms of Gap were present in large quantities. In addition, however, there were a small number of abundant proteins in the extracellular proteome that were not detected intracellularly. One of these was a Gap homologue (Gpd), previously identified as a plasminogen-binding protein in *Streptococcus equisimilis* [52]. It is interesting that two different Gap proteins were identified for *S. mutans*, which only contains one *gap* homologue, *gapC* (http://www.genome.ad.jp/kegg/pathway.html).

### 11.2.5 Are Multiple Isoforms of Abundant Proteins Technical Artifacts?

All glycolytic enzymes have been observed as multiple migrating species in one or another study. There are studies that detect a multitude of isoforms of many proteins [52], while other studies find fewer proteins with isoforms and fewer isoforms per enzyme [24]. This could point to a dependence of isoforms on the culture conditions. It could also be argued that the prevalence of multiple isoforms was the result of streaking in the isoelectric focusing and these isoforms are thus technical artifacts. It is, however, notable that many other abundant proteins show up as single bands (see [52]), so overloading is not enough for creating the isoforms. Also, the separation of the isoforms as pearls on a string is not common in streaking from overloaded gels. Translational errors, and possibly senescence, are known to produce pI variants of proteins under shortage of amino acids. Proteins expressed from genes with suboptimal codon bias may be exceptionally prone to translation errors. This often happens upon heterologous expression of proteins. Very different changes in pI may result depending on whether the erroneous amino acid substitutions proceed from ionic to nonpolar or between acid and basic residues. If errors in translation are the cause of the differences in migration, it might be possible to find codons for minor tRNA species in the genes encoding the variable proteins. Likewise an analysis of the correlation between the corresponding CAI values and the actual protein expression levels might detect a difference between proteins with and without satellites. Clearly the most interesting explanation for the multiple isoforms would be if the variants have different enzymatic abilities and are the products of posttranslational modifications.

By scrutinizing all the 2D gels from *Lactococcus* [22, 24, 45–53], it is tempting to make subjective judgments about the presence of individual protein forms. An exception is PgiA, which is overlapping the more abundant Tuf protein. Pfk usually appears to be single, while FbaA always has two satellite spots, which looks like mistranslated variants. TpiA appears to be single but GapB is always present in three, sometimes four, widely spaced isoforms. Pgk usually has two small satellite spots while Pmg is single. Pgm is special, however, because it has been found in combination with a larger variant [31]. EnoA and Pyk always have two distinct satellite spots beside them, while Ldh appears to be single or to have a minor satellite at the basic side. There have been reports on induced changes in isoforms for two enzymes, GapB and Pmg, which point at posttranslational modification [31], and the Pyk isoforms are spaced in a manner which appears inconsistent with translational artifacts. So it appears that some isoforms could be related to the translational machinery while others are due to modifications.

## 11.3 EXPRESSION PROTEOMICS OF *LACTOCOCCUS*

The field of expression proteomics in lactic acid bacteria has recently been reviewed [30]. Within the scope of stress physiology, the review resulted in insights into the members of many important stress stimulons. An additional review along these lines would not bring any new insights, so instead of focusing upon the stimulons, the present review will focus upon the few proteome analyses of known regulons and on individual proteins when their regulation appears interesting. Many proteins reoccur in the proteome analysis. Whether this is because the proteins are so distinct that they are noticed right away, such as the GapB isoforms, or because they are involved in the adaptation to many different stress forms is not always clear. Very few studies on proteomics of *Lactococcus* contain data on expression levels, so in most cases it is possible only to make qualitative statements about the distributions between protein isoforms of these proteins or to comment on dramatic changes in intensities. The following summaries are based upon both reported expression data and the subjective inspection of the 2D gels in the cited references.

The study of the *L. lactis* proteome is so far restricted to a few research groups, and the reports therefore reflect the main interests of these groups. This is also reflected in the limited number of topics described in the following review. However, as many different approaches to proteomics have been followed in lactococcal research, including temporal expression studies [22], analysis of mutants [25, 27, 54], analysis of strains over-expressing specific genes [26, 55], and analysis of chemical modifications in isoforms, this review supplies an insight in the potential use of proteomics in physiological research.

### 11.3.1 HrcA Regulon, Including GroEL/GroES and DnaK/DnaJ/GrpE Chaperones

Heat-shock-induced chaperones were among the first *L. lactis* proteins to be visualized by 2DGE [56]. The heat-shock response results in an abrupt decrease in the synthesis rate of most proteins to less than 50% accompanied by an increased synthesis of a small subset of proteins [22]. If proteins are pulse labeled with [$^{35}$S]-methionine before and after a heat shock, chaperones show up in the 2D gels as the most radioactive spots [22, 57]. The GroEL/GroES and DnaK/DnaJ/GrpE chaperones are known to be important under all conditions for folding of nascent polypeptides [58]. The *groESL* and *hrcA-grpE-dnaK* operons and the *dnaJ* gene in *L. lactis* carry CIRCE (controling inverted repeat for chaperone expression) elements [59], required for HrcA repression [60], in the promoter regions [61, 62] and are under the control of the HrcA heat-shock repressor (A. Nielsen and M. Kilstrup, unpublished). HrcA from *L. lactis* requires assistance from either the GroEL/GroES or the DnaK/DnaJ/GrpE chaperone for folding and DNA binding activity ([27]; M. Kilstrup, unpublished information). In *B. subtilis*, only the GroEL/GroES chaperone can aid in HrcA folding [63]. As an effect, HrcA can monitor the available chaperone activity in the cell and represses the expression of the chaperone genes when the activity is sufficient for HrcA folding [64]. The regulatory system is thus a feedback mechanism which controls the surplus chaperone activity. Any stress situation which results in profound protein denaturation is likely to induce elevated chaperone expression.

GroEL and GroES proteins have been found to be massively up regulated under heat stress, with more than 30-fold increase in synthesis rates after 10–15 min [22]. Heat stress at 43°C was chosen as a metabolically active state with continued growth, instead of the standard heat shock at 50–52°C, where all growth ceases. At 50°C, however, the increase in synthesis of the two proteins is still high [65]. Salt stress in GSA medium at a stress level which reduced the growth rate of MG1363 to approximately 50% resulted in a similar temporal expression pattern as for heat stress but at a fivefold lower level of induction [22]. The chaperones were likewise induced during growth at pH 5.5 in GM17 medium adjusted with lactic acid [65] but not in GSA medium adjusted with HCl [23]. In the GSA medium, induction of GroEl and GroES required pH values as low as 4.5 [23]. The reason for the pH difference in the two studies is not known. Ultraviolet irradiation induced GroES expression but curiously not GroEL [66].

DnaK and GrpE/Hsp26 (identified in [23]) showed a similar induction as GroEL and GroES during the first 10 min after heat stress application. After 10 min, the synthesis rate for DnaK and GrpE decreased over the next 15 min, while GroEL and GroES continued at the high rate for at least 30 min [22]. The reason for this difference is that the *groESL* operon belongs to two heat-shock regulons, HrcA and CtsR (see below). The salt stress induction pattern for DnaK and GrpE appeared to be very similar to that of GroEL and GroES under salt stress [22], and so did the induction at pH 5.5 and 4.5 in GSA medium [23].

The HrcA regulon is thus induced under heat, osmotic, and acid stress. The most likely explanation for this induction is that the stress conditions result in denaturation of cellular proteins and that the GroEL/GroES or the DnaK/DnaJ/GrpE chaperone complexes were sequestered in the folding of the proteins. Accordingly, there is no induction of heat-shock proteins under purine limitation stress, a situation that is not likely to provoke denaturation [53]. To test for the importance of the DnaK chaperone complex, a proteome analysis of the heat-shock response of a *dnaKΔ1* deletion mutant was reported [27]. The mutant had elevated levels of GroEL and GroES at normal temperatures but had the capacity to induce the synthesis of the proteins at much higher levels during heat shock than the wild type. This could be explained by a reduced chaperone availability in the *dnaK* mutant, which is too low for folding of the HrcA repressor, even under normal conditions. In the study it could not be concluded whether HrcA can be folded with the aid of only the GroEL/GroES chaperone, as in *Bacillus subtilis* [63], or whether both chaperone complexes can fold the HrcA repressor [27]. The doubt arose because a highly labeled spot corresponding to the predicted location of the truncated DnaK protein was present in the mutant at normal temperatures. The nonnative protein could potentially sequester the GroEL/GroES chaperone and prevent it from folding HrcA. That a high concentration of truncated proteins can sequester enough chaperone activity to induce the HrcA regulon in *L. lactis* was previously shown in a proteome study focused upon the degradation of misfolded protein [67]. It was found that addition of puromycin, which induces premature termination of translation, elicited a weak heat-shock response, with overexpression of GroEL, GroES, and DnaK among other heat-shock proteins.

### 11.3.2 CtsR Regulon, Including ClpB, ClpC, ClpE, ClpX, GroEL, and GroES Chaperones and ClpP Protease

Three large proteins above 80 kD were induced during heat stress with similar kinetics as DnaK and GrpE, with an expression peak at approximately 10 min after the temperature change [22]. Yet, their kinetics were very different from the members of the HrcA regulon during osmotic stress. While GroEL, GroES, DnaK, and GrpE had a peak expression after

10 min, the large proteins showed increases in synthesis during the first 15–25 min, after which they reached a plateau. By analyzing the proteomes of three mutants carrying deletions in the *clpB, clpC,* and *clpE* genes, 2D gels clearly showed that the three proteins Hsp85, Hsp84, and Hsp100 represented ClpE and two isoforms of ClpB, respectively [54]. The smaller isoform of ClpB represented a variant, where translation most likely had started from a secondary start site. The ClpC protein was not identified in the 2D gels, but its position should be at pI 6.29 and 90.6 kD. No highly labeled spot is present in the neighborhood of this position after 10 min. A very faint spot, however, shows up at later time points around coordinates {56.5, 104.5}, which might correspond to ClpC (see Fig. 11.1). None of the *clp* mutants showed induction of other proteins or had altered stress phenotypes except for the mutant lacking ClpE, which had an increased sensitivity to pyromycin [54]. The protease ClpP/Hsp23 appeared from its temporal expression pattern to belong to another regulon than the ClpB and ClpE proteins. Its synthesis rate increased only slowly within the first 15–25 min after heat stress, after which it reached a plateau [22], like the putative ClpC protein. Also under osmotic stress it deviated from the other Clp proteins by having a peak expression after 10 min, much like the HrcA regulon members [22].

The Clp proteins are regulated by the CtsR heat-shock repressor [68, 69] encoded by the first gene in the autoregulated *ctsR-clpC-orf555* operon [25]. The pulse-labeled proteome of a CtsR mutant showed increased synthesis for ClpP and ClpE and several putative members of the CtsR regulon [70]. Interestingly, the GroEL and GroES proteins that belongs to the CtsR regulon were not severely overexpressed in the CtsR mutant, due to the presence of the HrcA repressor, but GroEL appears to be overexpressed in comparison to DnaK in the mutant (compare Figs. 3A and 3B in [25]).

The mechanism for induction of CtsR repression is not as readily comprehensible as for the HrcA repressor, and the current model involves two modulators of CtsR repression (Mcs) and targeted degradation. In *B. subtilis* The CtsR repressor is encoded by the *ctsR-mcsA-mcsB-clpC* operon [71]. Under repressive (unstressed) conditions, the McsA protein stabilizes CtsR:DNA complexes and renders CtsR resistant against degradation, resulting in autorepression of the *ctsR* gene [71]. This feedback loop will confine the CtsR and ClpC levels together with the levels of the CtsR modulators between preset boundaries. To ensure a significant basal level of McsA, McsB, and ClpC, the ClpPX complex keeps CtsR at a low steady-state level by proteolysis [72]. Under stressful conditions, the McsB prosphorylates CtsR, presumably at an arginine residue, which disrupts CtsR:DNA as well as McsA:CtsR:DNA interactions and targets CtsR for degradation by the ClpCP protease [71]. It has not been explained how heat stress or the presence of nonnative proteins can activate the McsB-mediated inactivation and degradation of CtsR, thereby inducing the synthesis of the CtsR regulon members. However, a model was proposed to explain the rebound in repression after prolonged stress, in which the ClpCP protease becomes sequestered by degradation of nonnative proteins during stress, whereby CtsR is rendered more stable and capable of repressing again [71]. The biological logic behind a negative feedforward mechanism, which produces less chaperones and proteases when more is needed, is still to be revealed.

In *L. lactis,* the situation appears to be even more complicated than in *B. subtilis*. The two CtsR modulators are absent in *L. lactis* [73], and deletion of the *clpC* gene is without effect upon CtsR-mediated repression. It is not known whether CtsR is degraded under stressful conditions, but a rebound in repression after prolonged stress, which was also seen in *B. subtilis* (see above), was found to be dependent upon a functional ClpE protein

[73]. It was hypothesized that ClpE stabilizes the CtsR:DNA complex in analogy with the function of McsA in *B. subtilis*. Both proteins have a large Zn finger domain, which are known to be implicated in protein–DNA interactions and stabilization of protein–protein interactions [73]. In accordance with its proposed function, the Zn finger domain of ClpE [54, 74] was needed for the repression rebound [73]. Curiously, it was found that a small fraction of ClpE was processed at high temperatures by ClpP-dependent removal of approximately 6 kDa. If the Zn finger domain of ClpE was mutated, all ClpE was processed [73]. Processed ClpE appears to be more labile than the unprocessed form as the former form was always present at low levels. The ClpP-dependent inactivation of the putative ClpE:CtsR complex by ClpE processing could form the basis for a titration mechanism in analogy to the *B. subtilis* model for repression rebound. If the ClpP protein becomes sequestered by folding of nonnative proteins, it can no longer process ClpE, and the ClpE:CtrR complex represses transcription.

### 11.3.3 The *L. lactis* Heat-Shock Stimulon

Beside being the two major regulons in the heat-shock stimulon, HrcA belongs to a chaperone deficiency stimulon and CtsR possibly to a protease deficiency stimulon, and both are triggered by the accumulation of nonnative proteins in the cytoplasm of *L. lactis* [67]. Although the chaperones appear to have analogous functions in protein renaturation, each chaperone has a different mechanism of action, which results in differences in substrate recognition and consequently in different cellular function. The $GroEL_{14}$/$GroES_{12}$ chaperone is shaped as a hollow barrel whose interior walls can alternate between hydrophobic and hydrophilic, energized by adenosine triphosphate (ATP) hydrolysis [58]. In its hydrophobic conformation, it will bind to nonnative proteins that are exposing patches of hydrophobic residues on their surface [58]. The DnaK/DnaJ/GrpE chaperone, on the other hand, binds preferentially to linear polypeptide regions with a high fraction of hydrophobic residues [58]. The sequestering of the hydrophobic stretches for a defined period, determined by the adenosine diphosphate (ADP) release after ATP hydrolysis, presumably prevents aggregation and precipitation of the substrate protein [58]. The Clp proteins also form barrel-shaped multimers. In complex with the barrel-shaped $ClpP_{12}$ protease subunit, the $Clp_{12}$–$ClpP_{14}$–$Clp_{12}$ complex can degrade nonnative proteins [75]. In the absence of ClpP the $Clp_{12}$ barrels act as chaperones. Whether the $Clp_{12}$ barrels act as Anfinsen cages as the GroEL/GroES chaperone [76] is not known, but the proteins have to be partially unfolded to pass the small opening into the chaperone lumen. The specificity for substrates appears to require a free C-terminal end with a high degree of hydrophobicity [75] or in some cases an N-terminal end following the *N*-rule of protein degradation [75]. All Clp proteins share highly similar ATP binding domains, but different Clp families have different domains which determine the substrate specificity [54, 77]. A Zn finger domain in the N-terminal end of the *L. lactis* ClpE was found to be required for ClpP-dependent processing [73]. The ClpB family is peculiar because it has never been observed as a protease subunit and is believed to function only as a chaperone [77].

The differences in regulation between the chaperone complement of *L. lactis* is evident when the results for the two regulons are summarized: The ClpB and ClpE proteins shows a fast peak in expression with repression rebound in response to heat stress, similar to the HrcA regulon response, while ClpP (and possibly ClpC) shows more relaxed induction kinetics. The fast ClpB and ClpE proteins can however be distinguished from the HrcA regulon in other stress responses. The HrcA regulon shows a fast response to osmotic stress with repression rebound, while the ClpB and ClpE proteins have a more relaxed response [22]. Curiously, ClpP/Hsp23 has a fast response with repression rebound under these conditions [22]. The CtsR regulon members are more sensitive to acid shock, and ClpP and ClpE are induced by pH 5.5, while the HrcA regulon members GroEL, GroES, DnaK, and GrpE require pH 4.5 for induction. Members of both regulons are induced in a *clpP* mutant [67], which is deficient in the breakdown of misfolded proteins as well as in the *dnaKΔ1* mutant, which expresses a truncated DnaK protein [27]. It still remains to be shown whether the remaining unidentified heat-shock proteins, Hsp9, Hsp14, Hsp15, Hsp17, Hsp18, Hsp39, Hsp42, and Hsp48 [22], which are all slowly induced during heat stress and of which some are regulated by salt stress and some are not, belong to either of the two regulons.

### 11.3.4 Family of Small 7-kD Cold-Shock Proteins

Growth under low temperatures poses different challenges to bacteria than heat stress. Whereas protein denaturation appears to be the main problem under elevated temperatures, bacteria have to cope with decreased fluidity of the cytoplasmic membrane and increased formation and stability of secondary structures in mRNA and single-stranded DNA at low temperatures. It has been proposed that members of the 7-kDa family of cold-shock proteins (CSPs) bind to RNA in a cooperative manner and function as RNA chaperones by preventing the formation of secondary structures [78]. This facilitates the translation process, which is otherwise hampered at low temperatures. Other members bind to single-stranded DNA, and others again have been found to confer freeze protection [78]. *Lactococcus lactis* MG1363 harbors at least five different *csp* genes, encoding members of the 7-kDa family, CspA through CspE [79]. The predicted amino acid sequences from the genes show a high degree of similarity between all five Csp proteins. The highest amino acid similarity exists between three Cbs proteins in the acid pI region, CspB (7.3 kDa, pI 4.9), CspD (7.2 kDa, pI 4.4), and CspE (7.1 kDa, pI 4.6), ranging from 80 to 85% identity. The two proteins with alkaline pI, CspA (7.6 kDa, pI 9.2) and CspC (7.6 kDa, pI 9.6), only share 75% identity between each other and less than 65% identity with the acidic Cbs proteins [79]. All *csp* genes are transcribed separately but are clustered on the chromosome [79]. An additional member, CspF, was identified by proteomic techniques around pI 4.7 [80].

In a proteomics analysis of the cold-shock response in MG1363, the authors found that all the acidic Csp proteins, including CspF, were produced in high amounts (between 1.5 and 3.5% of the total cellular protein) 4 h after transfer from 30 to 10°C [80]. CspB, CspD, and CspF were induced from nearly undetectable levels, but CspE was already present in high amounts at 30°C (1.5% of total protein) and was only induced twofold. The two alkaline Csp proteins were identified as weak spots in 2D gels of the 10°C proteome but could not be identified at 30°C [80]. The expression profiles matched the mRNA levels from Northern blots of the individual genes [79]. Analysis of the acidic proteome under various stress conditions showed that heat, acid, or stationary phase stress did not induce any of the four acidic Csp proteins, but a small but significant induction of CspD and CspE was observed under osmotic stress [80]. Curiously, stationary-phase cells but not osmotically stressed cells had increased survival against repeated freeze–thaw cycles. Cold-stressed cells showed the best adaptation toward survival under those conditions [80].

Overproduction of the three acidic Csp proteins, by placing an inducible promoter in front of the respective genes, resulted in Csp levels up to 19% of the total protein content [55]. For the alkaline CspA and CspC proteins, it was not possible to obtain higher levels than 0.5% of the total protein content. When an arginine recidue in CspA was mutated to a proline recidue (proline is present in all the highly abundant acid Csp proteins), the resulting mutant CspA protein accumulated to approximately 9% of the total protein content. CspC also has an arginine residue at the same position, which in combination with a low mRNA half-life was likely to cause the lower abundance of the protein [55]. Csp proteins from other bacteria have previously been recognized as transcriptional activators [78], so it was very interesting that mild overexpression of the alkaline CspC (to 0.4% of total protein) induced all the acidic CspB to CspF to 0.1% levels. None of the acidic Csp proteins induced other Csp proteins, and neither did CspA. However, overexpression of each Csp resulted in induction of non-Csp cold-shock proteins. Among others was an analogue of the ribosomal protein L9 [55]. Previously, both the HPr protein and the CcpA transcriptional activator had been found to be members of the cold-shock stimulon [80]. Among the members of the putative Csp regulons were also proteins that were not part of the cold-shock stimulon. The cellubiose specific enzyme II of the PTS uptake system, CelA, was induced by overproduction of CspB, CspD, or CspE [55]. Overproduction of CspD and CspE resulted partly in protection from repeated freeze–thaw cycles, while no other Csp protein offered protection upon overproduction [79]. The investigation points at a very complex cascade of protective proteins with RNA chaperone and transcriptional regulatory functions, possibly with CspC located as a central controller. Apparently the CspA, CspB, CspC, CspD, and CspE regulons are widely overlapping, and it is not clear whether other regulons are part of the cold-shock stimulon.

### 11.3.5 PurR Regulon

*Lactococcus lactis* is able to synthesize nucleotides, the precursors for DNA and RNA synthesis, both de novo and from preformed nucleobases and nucleosides, supplemented to the medium. It has been found, however, that the purine nucleotide synthesis rate is not sufficient to supply nucleotides for fast-growing cells, and addition of bases results in an increased growth rate. Purine starvation has gained much attention, because when stress-resistant mutants are selected, mutants in the purine metabolism constitute a large fraction of the mutants [81–83]. A proteomics study was performed with a mutant in the purine biosynthesis pathway [53]. The *purD* mutant was grown in chemically defined GSA medium containing a limited source of the purine base hypoxanthine. The [$^{35}$S]-methionine pulse-labeled acid proteome was analyzed after depletion of the purine source, which leads to low GTP and ATP pools. In comparison to the pulse-labeled proteome of the same mutant, grown in abundance of hypoxanthine, a large number of proteins were overexpressed. Among the overexpressed proteins were PurS, PurE, PurM, and PurL, all enzymes of the purine biosynthesis pathway. An induced 83-kDa protein can now be tentatively identified as PurH using the proteome reference map of NCDO763 [24] as a guide. In addition to these enzymes, the two enzymes GlyA and Fhs were induced. At first glance, the two enzymes are not related to purine deficiency, but they are both involved in supplying C1 units for the biosynthesis pathway. C1 units are incorporated as parts of the purine bases at two points in the synthesis pathway by the aid of tetrahydropholate (THF). Fhs is involved in THF interconversion and the GlyA enzyme converts the amino acid serine to glycine by delivering the C1 unit to THF [53]. All the genes were found to belong to the PurR regulon in *L. lactis* [84]. PurR is a transcriptional activator which was previously found to activate the transcription of the *purDEK* and *purCSQLF* operons in response to the availability of purine bases [85]. PurBox'es, required for activation were found at the right locations in front of the *glyA* and *fhs* genes on the *L. lactis* chromosome. In addition PurBox'es were found in front of the *purF* and *purH* genes and the *purMN* operon [53].

The PurR regulator is very interesting for its evolutionary history. It was formed by fusion of a DNA binding region to an ancestral adenine phosphoribosyl transferase [84, 86]. The enzyme converts the purine base adenine to AMP using 5-phosphoribosyl-3-pyrophosphate (PRPP) as a substrate. The ancestral PRPP binding site enables PurR to respond positively to the concentration of PRPP, which is the ultimate precursor in the purine biosynthesis pathway. PurR is thus a feedforward regulator of the purine biosynthesis pathway for its substrate. Upon addition of the purine base hypoxanthine, where the members of the PurR regulon are down regulated, the guanine/hypoxanthine phosphoribosyl transferase removes PRPP and thereby abolishes PurR induction [84, 86].

Under the conditions of purine limitation, the general protein synthesis rate was reduced considerably. A small number of proteins had even more reduced synthesis rates. The abundant ribosomal protein RpsB, the catabolite activator protein CcpA, and HPr/Psd13 were not observed in the pulse-labeled proteome [53]. The translation elongation factor Tsf, the septum-forming protein FtsZ, and the ribosome-associated prolyl isomerase/chaperone Tig were also severely down regulated. In addition, the most acidic of the four major GapB isoforms also disappeared during purine deficiency [84, 86]. The cellular logic underlying the down regulation of these proteins is obscure, but it is interesting that HPr and CcpA, which were highly up regulated during cold shock [80], both disappeared under purine deficiency.

Induction of the PurR regulon members by purine deficiency was expected, but a recent observation that PurH, GlyA, and Fhs are induced under heme-dependent respiration is interesting [31]. A possible explanation could be that the synthesis of PRPP was affected by respiration, rather than its consumption. PRPP is synthesized, with ATP as the donor of the pyrophosphate group, from ribose-5-P, which is a metabolite in the pentose phosphate shunt.

Depletion of a *purD* mutant for purine bases results in deficiency for both adenine and guanine nucleotides (AMP, ADP, ATP, GMP, GDP, GTP, dATP, etc.). A specific class of mutants, which had increased acid and thermotolerance, was only affected in the guanine nucleotide synthesis [81, 82]. The thermotolerance could also be induced by addition of decoynine, an inhibitor of the GMP synthetase (GuaA), thereby artificially lowering the pools of GMP, GDP, and GTP [81]. A proteome analysis of the decoynine stimulon showed that lowering the guanine nucleotide pools to a degree where the cells grew at 50–70% reduced speed resulted in a general lowering of the protein synthesis rate. The synthesis of the most abundant proteins, including Tuf, FusA, Pyk, and EnoA, appeared to be unaffected [53]. Due to the altered protein synthesis rates under the different conditions, all levels were normalized to a constant synthesis rate of glycolytic proteins [53], so the constant synthesis rate was a prerequisite for normalization rather than a result of the quantification. The GuaB protein, which supplies XMP to GuaA for GMP synthesis, was found among those proteins that were not down regulated. The up regulation of this protein under guanine nucleotide deficiency would normally be beneficial to the cells, so it would make sense if *L. lactis* harbors a regulatory mechanism which senses the guanine nucleotide pool.

### 11.3.6 PyrR Regulon

PyrR is the second regulator of nucleotide biosynthesis in *L. lactis* [87, 88]. It regulates genes for the pathway leading to the synthesis of UMP. PyrR originates, like PurR, from an ancestral phosphoribosyl transferase, but in the case of PyrR, the ancestral enzyme converted uracil and PRPP into UMP and pyrophosphate [87, 89]. PyrR does not respond to the PRPP pool but has instead retained the binding site for UMP. In the presence of UMP, the PyrR regulator binds to an RNA region in the PyrR operators [90] and prevents the formation of an antiterminator structure. This provokes termination of transcription within the leader. The use of this attenuation mechanism for expression control of the UMP biosynthesis pathway is thus a classical feedback mechanism which keeps a balance between the production and demand of UMP. The PyrR-mediated regulation of the *pyr* genes is known to respond to the availability of pyrimidines [88]. Recently, the PyrR regulon members CarA, PyrDb(PydB), PyrE, PyrF, PyrR, and Upp were found to be severely down regulated upon growth in GM17 with lactose as energy source, compared to growth on glucose [24]. Lactose is taken up by the lactose-specific enzyme II of the PTS uptake system and is phosphorylated to lactose-6-P. Lactose-6-P is cleaved to glucose and galactose-6-P by the 6-P-beta-galactosidase enzyme. The further degradation of galactose-6-P proceeds via the tagatose pathway: galactose-6-P $\rightarrow$ tagatose-6-P $\rightarrow$ tagatose-1,6-bis-P $\rightarrow$ glyceraldehydes-3-P + dihydroxyacetone-P. Three of the plasmid-encoded enzymes in this pathway were identified and found to be induced by lactose [24]. The intracellular glucose is phosphorylated via the glucose kinase enxyme (Glk) to glucose-6-P and oxidized through glycolysis. Upon growth on glucose, glucose is taken up and phosphorylated to glucose-6-P by the PTS system via the mannose-specific enzyme II (PtnC). The only obvious connection between lactose utilization and the PyrR regulon is the coinduction of enzymes involved in exchange of UDP between galactose-1-P and glucose-1-P moieties (GalE) and from UTP to galactose-1-P (GalT) under growth on lactose [24]. It is not clear, however, why induction of these enzymes should result in higher UMP pool and subsequent attenuation of the genes in the *pyr* pathway.

### 11.3.7 HPr (PtsH) and CcpA

An arrest in the synthesis of Hpr and CcpA, or alternatively a fast specific degradation, was detected under purine deficiency [53]. Detection of labeled Hpr protein was likewise reported to stop between 15 and 25 min after exposure to pH 4.5 in GSA medium [23]. No data on CcpA expression were reported under these conditions. From the location of CcpA at coordinates {19.5, 69.5} [53] in the standard MG1363 reference gel [22], it is possible to follow the expression of the protein during acid stress (Fig. 3 in [23]). It is clear that the amount of labeled CcpA is severely reduced under acid shock.

Coomassie-stained gels were not reported in the studies, so the destinction between decreased synthesis or increased degradation could not be made. In a third study, both HPr and CcpA were induced by cold shock [79]. The two proteins thus appear to be coregulated in all three cases where they have been identified. The *ptsHI* operon and the *ccpA* genes are located on opposite sides of the chromosome [4], so their coregulation must involve a transacting factor or directed proteolysis. The HPr protein is an important enzyme in the PTS sugar transport system, where it transfers phosphorus groups from enzyme I (PtsI) to the sugar-specific enzyme II complexes. In addition to this function, it serves as a messenger between a secondary HPr kinase (PtsK) and the CcpA protein [91]. PtsK phosphorylates Hpr at a specific serine residue under conditions where the cell contains high concentrations of the glycolytic metabolite fructose-1,6-bisphosphate. The P-Ser-HPr combines with CcpA into an active transcriptional activator capable of binding to CRE (catabolite responsive element) sites [91]. Only one isoform of HPr has been identified in 2D gels [23]. In a study of the regulatory properties of the *L. lactis* P-ser-HPr [92], three isoforms were detected by Western blotting using anti–*B. subtilis*-HPr antibodies after separation of crude extracts under nondenaturing conditions. It was found that heating of the protein extracts to 65°C for 10 min prior to electrophoresis resulted in the hydrolysis of the His-P but not Ser-P in both double-phosphorylated $P_2$-(His,Ser)-HPr and single-phosphorylated P-His-HPr protein. It was estimated that wild-type *L. lactis* cells grown in GM17 medium contained a considerable amount of P-His-HPr, slightly less P-Ser-HPr, and a small amount of $P_2$-(His,Ser)-HPr [92]. From inspection of the Western blot, it appears that the phosphorylated forms dominate over the unphosphorylated. Since detection of the phosphoproteome has not been an issue in any of the proteome studies published for *L. lactis*, the conditions for protein extraction have not been optimized for detection of P-His-HPr. However, the more stable P-Ser-HPr isoform should be detectable in the 2D gels. Identification of P-Ser-HPr will be very important for the determination of the ratio between the two isoforms to see if they correlate with the internal pool of fructose-1,6-bisphosphate. The ratio will, however, only be valid under optimized protein extraction conditions.

### 11.3.8 Expression of Glycolytic Proteins

The homolactic fermentation pathway in *Lactococcus* consists of the conversion of glucose-6-phosphate to lactate through the following enzymes: PgiA $\rightarrow$ Pfk $\rightarrow$ FbaA + TpiA $\rightarrow$ GapB $\rightarrow$ Pgk $\rightarrow$ Pmg $\rightarrow$ EnoA $\rightarrow$ Pyk $\rightarrow$ Ldh [4]. Glucose is under most conditions phosphorylated during uptake by the PTS system through the mannose-specific enzyme II (PtnC). Based upon a subjective estimation of the size and intensity of the protein spots in the 2D gels, it appears that FbaA, GapB, EnoA, Pyk, and Ldh are expressed among the highest levels in all species, only surpassed by the two translation elongation factors EF-Tu (Tuf) and EF-G (FusA). The rest of the enzymes are expressed at more moderate levels, but still markedly above the general expression level. Because of the high expression level and the conserved location between the bacterial species, it is usually possible to deduce the location of most glycolytic proteins within the published gels when the gels are of sufficient high quality.

FbaA is identical to Pro32 in [22]. The protein was not identified in the study, but the reported N-terminal amino acid sequence AIVSAEKFVQAA fits the predicted sequence MAIVSAEKFVQAARDNGYAI perfectly. TpiA cannot be identified with certainty in [$^{35}$S]-methionine-labeled cells grown exponentially in chemically defined SA medium. This may be of regulatory significance because it is very abundant in *L. lactis* cells grown in GM17 [24]. Pmg has not been identified in the MG1363 reference map [22], but according to its identification in the 2D gels of [31], a highly labeled spot at coordinates {34, 57.5} is a likely candidate. Under aerated conditions in GM17, the Pmg protein appears as two isoforms of different molecular masses [31]. If hemin is added, resulting in heme-dependent respiration, the larger isoform disappears while the lower isoform is unaffected. As heme-dependent respiration has been shown to protect the bacteria from oxidative stress, the large Pmg isoform could be related to oxidative stress. EnoA is identical to Pro49 in [22, 53]. The genes encoding the Pfk and Pyk enzymes are part of the

Las operon, together with the gene for Ldh [93]. All other genes for glycolytic enzymes are scattered over the chromosome [4]. The *pfk-pyk-ldh* operon is regulated by the CcpA activator [93], but the details about the regulation of the other glycolytic genes are not known.

Expression of different glycolytic enzymes does not appear to follow any logical rule except reflecting that of their normal expression levels. Under heat stress, The Las operon enzymes Pfk and Pyk together with Pgk, FbaA, and Pmg were down regulated, while the extremely abundant GapB and EnoA were unchanged [22]. The apparent difference in regulation could be an artifact based upon saturation or quenching of the radioactive determination. However, the measurements were done using a Packard Instant Imager, which is less prone to those errors. The glycolytic proteins did not show signs of regulation during acid stress [23, 65], salt stress [22], or purine limitation [53]. Under heme-dependent respiration, however, the total amount of the GapB isoforms was lowered compared to aerated conditions in GM17 medium [31].

### 11.3.9 GapA and GapB Isoforms

The major glyceraldehyde-3-phosphate dehydrogenase enzyme in *L. lactis* is GapB [26]. GapA was for a long time, after cloning of the *gapA* gene from MG1363 [46] and report of its high codon adaptation index, expected to be the important glycolytic enzyme. The genome sequence of IL1403 showed that *L. lactis* harbors two genes encoding glyceraldehyde-3-phosphate dehydrogenases, *gapA* and *gapB* [4], and later analysis reported that the GapB protein had the highest CAI of the entire proteome [30]. It was anticipated that the two major isoforms of the Gap protein, previously identified [22], corresponded to the gene products of the *gapA* and *gapB* genes. To identify the GapA product, the proteome of a *L. lactis* strain overproducing the *gapA* gene was investigated [26]. Curiously, instead of an increase in the synthesis rate of one of the two major isoforms, three new isoforms appeared at lower molecular masses, showing the same distribution between isoforms as the original (GapB) isoforms. Mass spectrometry identification confirmed that the omnipresent Gap protein spots were encoded by *gapB* and the induced isoforms by *gapA* [26]. Even though the coverage of the GapA peptides identified by MS was high, no chemical modification could be detected in either isoform. Identification of matching peptides was especially low around amino acids 130–170, which has later been identified as the site of covalent modification (P. Gaudu, personal communication). Induction of GapA has never been reported, but we have observed full induction of the protein in a series of experiments where we used a particular batch of SA medium employing IL1403, MG1363, and various mutants. GapA was strongly induced in all 2D gels (see Fig. 11.5). The nature of the changes in medium composition was never identified.

As mentioned in a previous section, the total amount of GapB protein was found to decrease substantially under heme-dependent respirative conditions in GM17 medium, compared to identical aerated conditions without hemin [31]. Interestingly, the most dramatic change in the proteome was not the general lowering of the GapB level, but a shift between its two major isoforms. Under aerated conditions, the levels of the two isoforms were approximately equal. Under heme-dependent respiration, the acidic isoform was drastically reduced while the basic isoform remained unchanged [31].

Heme-dependent respiration has profound consequences for the physiology of the bacteria. In fermenting lactococci, the majority of the glycolytic flux is channeled through

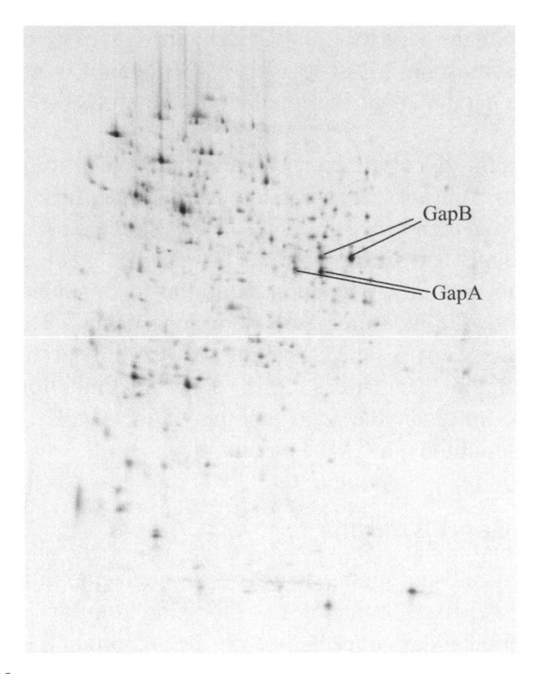


**Figure 11.5** The [$^{35}$S]-methionine-labeled *L. lactis* MG1363 proteome grown in SA medium supplemented with glucose. The culture was part of a series grown in a particular batch of medium which for unknown reasons showed induction of the GapA protein. The location of the GapA and GapB isoforms are indicated.

the lactate dehydrogenase (Ldh) forming almost exclusively lactate as product. Thereby the cell obtains equilibrium between the NADH formed in the glycolytic path and the consumption of NADH by the Ldh enzyme. Under aerated conditions, a fraction of the NADH formed by glycolysis is oxidized to $NAD^+$ by the cytoplasmic NADH oxidase (NoxE) under reduction of $O_2$ to water [94]. If hemin is added under aerated conditions, a new pathway becomes operative, because the heme prosthetic group is inserted into the cytochrome *bd* oxidase, consisting of the CydA and CydB subunits [95]. The cytochrome *bd* complex is able to oxidize reduced dihydro-menaquinone in the membrane with $O_2$ as electron acceptor, forming water in the process. The menaquinone pool is reduced by a membrane-bound NADH dehydrogenase NoxB [5] [NoxA, according to the Kyoto encyclopedia of genes and genomes (KEGG) at http://www.genome.ad.jp/dbget-bin/get_pathway?org_name = lla&mapno = 00190]. A proton gradient is formed over the membrane during this heme-dependent respiration, which can substitute for the $H^+$/ATPase [96]. This protein is normally required for formation of the proton gradient under fermentative conditions. The increased NADH consumption by NoxE and the respiration chain allows a fraction of the glycolytic flux to be directed toward the production of more oxidized products such as acetoin and diacetyl [94, 95]. Under heme-dependent respiration the redirection of the flux away from lactate production resulted in a decreased acidification and thereby in an increase in the biomass production [95]. The growth physiology under respirative conditions in GM17 medium appeared to be divided in two phases [95, 5]. Hemin uptake is repressed in the early growth phase by the CcpA regulator protein [97]. The fermentative pathway is therefore favored under these conditions and the cells produce mainly lactate. Upon entry into the stationary phase, the hemin uptake is induced and the respiration takes over, whereby the cells switch to mixed acid production. There are indications that the lactate produced in the fermentative phase is reutilized in the respirative phase, supplying NADH for the electron transport chain and pyruvate for mixed-acid production [5, 95].

In a proteomics analysis of the MG1363 proteome under aerated and hemin-dependent respirative conditions [31], it was found that subunits of pyruvate dehydrogenase (PdhC and PdhD) were induced together with an enzyme (LipL) for inserting the lipoic acid prosthetic group into the pyruvate dehydrogenase complex. The pyruvate dehydrogenase converts pyruvate into acetyl-CoA under aerobic conditions with the subsequent reduction of $NAD+$ to NADH. Under anaerobic growth, the analogous reaction is performed by the pyruvate formate lyase (Pfl) with formate as the by-product. The acetolactate synthetase (Als), which converts pyruvate to a-acetolactate, was also induced upon respiration. Induction of these enzymes thus directs the pyruvate flux away from the lactate dehydrogenase.

In a *ccpA* mutant, the unrepressed hemin uptake in the first fermentative phase becomes toxic to the cell, and powerful antioxidants like dithiothreitol (DTT) are required for protection [97]. However, a large increase in the long-term survival at low temperature after heme-dependent respiration in wild-type *L. lactis* strains was previously attributed to a change of the bacterial interior to a less oxidative stressful environment [95], so the progression from fermentation to respiration in the late exponential growth phase may be important in coping with oxidative stress. In accordance with an increased oxidative protection condition, a subunit of the thioredoxin reductase (TrxB) was slightly induced under heme-dependent respirative conditions. TrxB is known to be important in the defense against oxidative stress [31].

A proteomics analysis of a *trxB* mutant has recently shed light upon the nature of the two major GapB isoforms (P. Gaudu, personal communication). Surprisingly, the acidic isoform of GapB was undetectable in a mutant lacking TrxB. By MS and the use of the AspN peptidase from *Pseudomonas fragi*, which cleaves at aspertate residues and at oxidized forms of cysteine residues but not at reduced cysteines, it was found that the acidic GapB isoform contained an oxidized cystein residue at the catalytic Cys152. The modification, which changed the $-CH_2-SH$ residue to $-CH_2-SOO^-$, inactivates the enzyme and changes the pI of the enzyme (P. Gaudu, personal communication). The fact that the acidic isoform of GapB is reduced under heme-dependent respiration is in agreement with a more reduced environment under these conditions. However, it is not clear why TrxB is induced under these conditions, when the acidic GapB isoform appears to be correlated to the presence of TrxB.

While the acidic isoform of GapB is explained by the oxidation of Cys152 to a sulfonic acid, there are still important mysteries relating to Gap modifications in related organisms. A proteome analysis of spontaneous erythromycin-resistant isolates of *Streptococcus pneumoniae* showed the appearance of a highly abundant alkaline isoform of Gap, only detectable in minute amounts in susceptible strains [98]. The MS-detected peptides of the new isoform, in comparison with the predicted amino acid sequences from the *Streptococcus equisimilis* and *Streptococcus pyogenes* Gap genes, covered amino acids 1–14, 19–30, 46–60, 116–128, and 309–325. The peptide covering the catalytic cysteines was not identified in the study. The significance of the change in isoform in erythromycin resistance has not been reported, but the strain contained the *mef(A)* gene, often associated with erythomycin resistance [99]. The cellular role of Gap in the

*Lactococcus–Streptococcus* branch is very interesting, and it has been found to be the major outer surface protein in *S. pyogenes* [100] and to act as a plasmin receptor [101] and an ADP-ribosylating enzyme [102]. Clearly, the fascinating story of the GapA and GapB proteins from *L. lactis* has not reached its ending.

## OUTLOOK

Proteomics offers an outstanding expansion of the number of phenotypic traits for microorganisms. Despite the nuisance of the gel-methodology, the analysis of these traits is extremely fast in comparison to the time required to generate an equal amount of genetic and physiological data, in the characterization of a mutant or a cellular response. In accordance with its advantages, proteomics is becoming the natural choice for the initial survey of phenotypes in an increasing number of studies. The obstacles for a general use of proteome analysis used to be the difficulties in 2D-gel handling. However, after the introduction of precast pH gradient gels for first dimension and uncomplicated gel-systems for the second dimension, any laboratory can now perform this part of proteomics.

In *L. lactis*, expression rate proteomics ([$^{35}$S]-methionine incorporation) has been widely used in the characterization of cellular stress responses and mutant phenotypes as well as the responses towards over-expression of genes. This type of proteome analysis has proved to be a very sensitive tool for defining regulatory networks in the transition phase from one physiological state to another, where the synthesis rates change abruptly. Under the same conditions, the absolute protein levels change very slowly and require more than three generation-times to acquire the new proteome structure. But should all proteomic analysis then involve radioactive incorporation, when it is so much more sensitive than protein post-labeling? The answer is no, because expression rates are not directly related to cellular physiology. To the cell, it is the activity of a protein that determines the physiological status, and the protein level, which is currently estimated by staining, is the best estimate of the protein activity. The active protein level may actually be a small subclass of the total protein level, as the protein activity may be modulated by allosteric effectors (not analyzed by proteomics), posttranslational modifications, and protein–protein interactions. As it was pointed above, many of the glycolytic enzymes appear to be modified in *L. lactis*, and the activity of at least one enzyme (GapB) is impaired by the modifications. It is also known that many regulatory proteins convey their functions in the signal networks by covalent modifications. These modifications are identified and quantified by the promising field of phosphoproteomics and other areas of proteomics dealing with posttranscriptional modification. Of equal importance is the development of the cell-map proteomics, where interactions are mapped between all proteins of the proteome. These interactions are as important for the cell physiology as the protein levels, because together they define the levels of the active protein complexes. Thus, the identification and quantification of the active proteome will be the major challenge to proteomics in the future.

Systems biology will in the future develop a common frame for description of the cellular responses, based upon transcriptomic, proteomic, metabolomic, and physiological data from well defined growth experiments. All omics-methodologies are receiving enormous attention, and rightfully so, but it must be kept in mind that the levels of gene expression, cellular mRNAs, cellular proteins, and metabolites in a growing bacterium are absolutely dependent upon the growth rate. So in order to supply meaningful data for systems biology, researchers in both transcriptomics and proteomics are highly dependent upon reproducible growth conditions. However, if proteomics will be able to supply accurate cellular levels of protein isoforms and detailed analysis of their structural modifications, it is likely to become a cornerstone in systems biology.

## REFERENCES


1. Schleifer, K. H., and Ludwig, W., in B. J. B. Wood and W. H. Holzapfel (Eds.), *The Genera of Lactic Acid Bacteria*, Blackie Academic & Professional, Glasgow, 1995, pp. 7–18.
2. Emard, L. O., and Vaughn, R. H., *J. Bacteriol.* 1952, **63**, 487–494.
3. Martinussen, J., and Hammer, K., *Microbiology-UK* 1995, **141**, 1883–1890.
4. Bolotin, A., Wincker, P., Mauger, S., Jaillon, O., Malarme, K., Weissenbach, J., Ehrlich, S. D., and Sorokin, A., *Genome Res.* 2001, **11**, 731–753.
5. Gaudu, P., Vido, K., Cesselin, W., Kulakauskas, S., Tremblay, J., Rezaiki, L., Lamberet, G., Sourice, S., Duwat, P., and Gruss, A., *Ant. Leeuwenhoek Int. J. Gen. Mol. Microbiol.* 2002, **82**, 263–269.
6. Godon, J. J., Delorme, C., Bardowski, J., Chopin, M. C., Ehrlich, S. D., and Renault, P., *J. Bacteriol.* 1993, **175**, 4383–4390.
7. Bauer, S., Tholen, A., Overmann, J., and Brune, A., *Arch. Microbiol.* 2000, **173**, 126–137.
8. Teuber, M., in B. J. B. Wood and W. H. Holzapfel (Eds.), *The Genera of Lactic Acid Bacteria*, Blackie Academic & Professional, Glasgow, 1995, pp. 173–234.
9. Sanders, J. W., Venema, G., and Kok, J., *FEMS Microbiol. Rev.* 1999, **23**, 483–501.
10. Gasson, M. J., *J.Bacteriol.* 1983, **154**, 1–9.
11. Wasinger, V. C., Cordwell, S. J., Cerpapoljak, A., Yan, J. X., Gooley, A. A., Wilkins, M. R., Duncan, M. W., Harris, R., Williams, K. L., and Humphery-Smith, I., *Electrophoresis* 1995, **16**, 1090–1094.
12. Drews, O., Reil, G., Parlar, H., and Gorg, A., *Proteomics* 2004, **4**, 1293–1304.
13. Hermann, T., Pfefferle, W., Baumann, C., Busker, E., Schaffer, S., Bott, M., Sahm, H., Dusch, N., Kalinowski, J., Puhler, A., Bendt, A. K., Kramer, R., and Burkovski, A., *Electrophoresis* 2001, **22**, 1712–1723.
14. Santoni, V., Kieffer, S., Desclaux, D., Masson, F., and Rabilloud, T., *Electrophoresis* 2000, **21**, 3329–3344.
15. Kaufmann, H., Bailey, J. E., and Fussenegger, M., *Proteomics* 2001, **1**, 194–199.
16. Fryksdale, B. G., Jedrzejewski, P. T., Wong, D. L., Gaertner, A. L., and Miller, B. S., *Electrophoresis* 2002, **23**, 2184–2193.
17. Blackstock, W. P., and Weir, M. P., *Trends Biotechnol.* 1999, **17**, 121–127.
18. Patton, W. F., *J. Chromatogr. B Anal. Technol. Biomed. Life Sci.* 2002, **771**, 3–31.
19. Smith, M. W., and Neidhardt, F. C., *J. Bacteriol.* 1983, **154**, 336–343.
20. Bernhardt, J., Buttner, K., Scharf, C., and Hecker, M., *Electrophoresis* 1999, **20**, 2225–2240.
21. Fountoulakis, M., Takacs, M. F., Berndt, P., Langen, H., and Takacs, B., *Electrophoresis* 1999, **20**, 2181–2195.
22. Kilstrup, M., Jacobsen, S., Hammer, K., and Vogensen, F. K., *Appl. Environ. Microbiol.* 1997, **63**, 1826–1837.
23. Frees, D., Vogensen, F. K., and Ingmer, H., *Int. J. Food Microbiol.* 2003, **87**, 293–300.
24. Guillot, A., Gitton, C., Anglade, P., and Mistou, M. Y., *Proteomics* 2003, **3**, 337–354.
25. Varmanen, P., Ingmer, H., and Vogensen, F. K., *Microbiology-Sgm* 2000, **146**, 1447–1455.
26. Willemoes, M., Kilstrup, M., Roepstorff, P., and Hammer, K., *Proteomics* 2002, **2**, 1041–1046.

27. Koch, B., Kilstrup, M., Vogensen, F. K., and Hammer, K., *J. Bacteriol.* 1998, **180**, 3873–3881.
28. Klose, J., *Electrophoresis* 1999, **20**, 643–652.
29. VanBogelen, R. A., Schiller, E. E., Thomas, J. D., and Neidhardt, F. C., *Electrophoresis* 1999, **20**, 2149–2159.
30. Champomier-Verges, M. C., Maguin, E., Mistou, M. Y., Anglade, P., and Chich, J. F., *J. Chromatogr. B Anal. Technol. Biomed. Life Sci.* 2002, **771**, 329–342.
31. Vido, K., le Bars, D., Mistou, M. Y., Anglade, P., Gruss, A., and Gaudu, P., *J. Bacteriol.* 2004, **186**, 1648–1657.
32. Kjeldgaard, N. O., Maaloe, O., and Schaechter, M., *J. Gen. Microbiol.* 1958, **19**, 607–616.
33. Schaechter, M., Maaloe, O., and Kjeldgaard, N. O., *J. Gen. Microbiol.* 1958, **19**, 592–606.
34. Pedersen, S., Bloch, P. L., Reeh, S., and Neidhardt, F. C., *Cell* 1978, **14**, 179–190.
35. O'Farrell, P. H., *J. Biol. Chem.* 1975, **250**, 4007–4021.
36. Maas, W. K., and Clark, A. J., *J. Mol. Biol.* 1964, **8**, 365–.
37. Smith, M. W., and Neidhardt, F. C., *J. Bacteriol.* 1983, **154**, 344–350.
38. Jensen, P. R., and Hammer, K., *Biotechnol. Bioeng.* 1998, **58**, 191–195.
39. Solem, C., and Jensen, P. R., *Appl. Environ. Microbiol.* 2002, **68**, 2397–2403.
40. Koebmann, B. J., Andersen, H. W., Solem, C., and Jensen, P. R., *Ant. Leeuwenhoek Int. J. Gen. Mol. Microbiol.* 2002, **82**, 237–248.
41. Andersen, H. W., Pedersen, M. B., Hammer, K., and Jensen, P. R., *Eur. J. Biochem.* 2001, **268**, 6379–6389.
42. Minton, N. P., *Gene* 1984, **31**, 269–273.
43. Jensen, P. R., and Hammer, K., *Appl. Environ. Microbiol.* 1993, **59**, 4363–4366.
44. Phillips, T. A., Vaughn, R. H., Bloch, P. L., and Neidhardt, F. C., in J. L. Ingraham, K. B. Low, B. Magasanik, M. Schaechter, and H. E. Umbarger (Eds.), *Escherichia coli and Salmonella typhimurium Cellular and Molecular Biology*, Washington, DC, 1987, pp. 919–966.
45. Anglade, P., Demey, E., Labas, V., Le Caer, J. P., and Chich, J. F., *Electrophoresis* 2000, **21**, 2546–2549.
46. Cancilla, M. R., Hillier, A. J., and Davidson, B. E., *Microbiology-UK* 1995, **141**, 1027–1036.
47. Sharp, P. M., and Li, W. H., *Nucleic Acids Res.* 1987, **15**, 1281–1295.
48. Cooper, J. W., and Lee, C. S., *Anal. Chem.* 2004, **76**, 2196–2202.
49. Perrin, C., Gonzalez-Marquez, H., Gaillard, J. L., Bracquart, P., and Guimont, C., *Electrophoresis* 2000, **21**, 949–955.
50. Wilkins, J. C., Homer, K. A., and Beighton, D., *Appl. Environ. Microbiol.* 2001, **67**, 3396–3405.
51. Svensater, G., Welin, J., Wilkins, J. C., Beighton, D., and Hamilton, I. R., *FEMS Microbiol. Lett.* 2001, **205**, 139–146.
52. Len, A. C. L., Cordwell, S. J., Harty, D. W. S., and Jacques, N. A., *Proteomics* 2003, **3**, 627–646.
53. Beyer, N. H., Roepstorff, P., Hammer, K., and Kilstrup, M., *Proteomics* 2003, **3**, 786–797.
54. Ingmer, H., Vogensen, F. K., Hammer, K., and Kilstrup, M., *J. Bacteriol.* 1999, **181**, 2075–2083.
55. Wouters, J. A., Mailhes, M., Rombouts, F. M., De Vos, W. M., Kuipers, O. P., and Abee, T., *Appl. Environ. Microbiol.* 2000, **66**, 3756–3763.
56. Whitaker, R. D., and Batt, C. A., *Appl. Environ. Microbiol.* 1991, **57**, 1408–1412.
57. Hartke, A., Bouche, S., Giard, J. C., Benachour, A., Boutibonnes, P., and Auffray, Y., *Curr. Microbiol.* 1996, **33**, 194–199.
58. Bukau, B., and Horwich, A. L., *Cell* 1998, **92**, 351–366.
59. Zuber, U., and Schumann, W., *J. Bacteriol.* 1994, **176**, 1359–1363.
60. Schulz, A., and Schumann, W., *J. Bacteriol.* 1996, **178**, 1088–1093.

61. Eaton, T. C., Shearman, C., and Gasson, M., *J. Gen. Microbiol.* 1993, **139**, 3253–3264.
62. Kim, S. G., and Batt, C. A., *Gene* 1993, **127**, 121–126.
63. Mogk, A., Homuth, G., Scholz, C., Kim, L., Schmid, F. X., and Schumann, W., *EMBO J.* 1997, **16**, 6518–6527.
64. Mogk, A., Volker, A., Engelmann, S., Hecker, M., Schumann, W., and Volker, U., *J. Bacteriol.* 1998, **180**, 2895–2900.
65. Hartke, A., Frere, J., Boutibonnes, P., and Auffray, Y., *Curr. Microbiol.* 1997, **34**, 23–26.
66. Hartke, A., Bouche, S., Laplace, J. M., Benachour, A., Boutibonnes, P., and Auffray, Y., *Arch. Microbiol.* 1995, **163**, 329–336.
67. Frees, D., and Ingmer, H., *Mol. Microbiol.* 1999, **31**, 79–87.
68. Krüger, E., and Hecker, M., *J. Bacteriol.* 1998, **180**, 6681–6688.
69. Derre, I., Rapoport, G., and Msadek, T., *Mol. Microbiol.* 1999, **31**, 117–131.
70. Frees, D., Varmanen, P., and Ingmer, H., *Mol. Microbiol.* 2001, **41**, 93–103.
71. Kruger, E., Zuhlke, D., Witt, E., Ludwig, H., and Hecker, M., *EMBO J.* 2001, **20**, 852–863.
72. Derre, I., Rapoport, G., and Msadek, T., *Mol. Microbiol.* 2000, **38**, 335–347.
73. Varmanen, P., Vogensen, F. K., Hammer, K., Palva, A., and Ingmer, H., *J. Bacteriol.* 2003, **185**, 5117–5124.
74. Derre, I., Rapoport, G., Devine, K., Rose, M., and Msadek, T., *Mol. Microbiol.* 1999, **32**, 581–593.
75. Horwich, A. L., Weber-Ban, E. U., and Finley, D., *Proc. Nat. Acad. Sci. USA* 1999, **96**, 11033–11040.
76. Csermely, P., *Bioessays* 1999, **21**, 959–965.
77. Schirmer, E. C., Glover, J. R., Singer, M. A., and Lindquist, S., *Trends Biochem. Sci.* 1996, **21**, 289–296.
78. Graumann, P. L., and Marahiel, M. A., *Trends Biochem. Sci.* 1998, **23**, 286–290.
79. Wouters, J. A., Sanders, J. W., Kok, J., De Vos, W. M., Kuipers, O. P., and Abee, T., *Microbiology-UK* 1998, **144**, 2885–2893.
80. Wouters, J. A., Jeynov, B., Rombouts, F. M., De Vos, W. M., Kuipers, O. P., and Abee, T., *Microbiology-UK* 1999, **145**, 3185–3194.
81. Rallu, F., Gruss, A., Ehrlich, S. D., and Maguin, E., *Mol. Microbiol.* 2000, **35**, 517–528.
82. Duwat, P., Ehrlich, S. D., and Gruss, A., *Mol. Microbiol.* 1999, **31**, 845–858.
83. Duwat, P., Cesselin, B., Sourice, S., and Gruss, A., *Int. J. Food Microbiol.* 2000, **55**, 83–86.
84. Kilstrup, M., and Martinussen, J., *J. Bacteriol.* 1998, **180**, 3907–3916.
85. Kilstrup, M., Jessing, S. G., Wichmand-Jorgensen, S. B., Madsen, M., and Nilsson, D., *J. Bacteriol.* 1998, **180**, 3900–3906.
86. Weng, M., Nagy, P. L., and Zalkin, H., *Proc. Natl. Acad. Sci. USA* 1995, **92**, 7455–7459.
87. Turner, R. J., Lu, Y., and Switzer, R. L., *J. Bacteriol.* 1994, **176**, 3708–3722.
88. Martinussen, J., Schallert, J., Andersen, B., and Hammer, K., *J. Bacteriol.* 2001, **183**, 2785–2794.
89. Martinussen, J., Glaser, P., Andersen, P. S., and Saxild, H. H., *J. Bacteriol.* 1995, **177**, 271–274.
90. Bonner, E. R., D'Elia, J. N., Billips, B. K., and Switzer, R. L., *Nucleic Acids Res.* 2001, **29**, 4851–4865.
91. Nessler, S., Fieulaine, S., Poncet, S., Galinier, A., Deutscher, J., and Janin, J., *J. Bacteriol.* 2003, **185**, 4003–4010.
92. Monedero, V., Kuipers, O. P., Jamet, E., and Deutscher, J., *J. Bacteriol.* 2001, **183**, 3391–3398.

93. Luesink, E. J., van Herpen, R. E., M. A., Grossiord, B. P., Kuipers, O. P., and De Vos, W. M., *Mol. Microbiol.* 1998, **30**, 789–798.
94. Hugenholtz, J., Kleerebezem, M., Starrenburg, M., Delcour, J., De Vos, W., and Hols, P., *Appl. Environ. Microbiol.* 2000, **66**, 4112–4114.
95. Duwat, P., Sourice, S., Cesselin, B., Lamberet, G., Vido, K., Gaudu, P., Le Loir, Y., Violet, F., Loubiere, P., and Gruss, A., *J. Bacteriol.* 2001, **183**, 4509–4516.
96. Koebmann, B. J., Nilsson, D., Kuipers, O. P., and Jensen, P. R., *J. Bacteriol.* 2000, **182**, 4738–4743.
97. Gaudu, P., Lamberet, G., Poncet, S., and Gruss, A., *Mol. Microbiol.* 2003, **50**, 183–192.
98. Cash, P., Argo, E., Ford, L., Lawrie, L., and McKenzie, H., *Electrophoresis* 1999, **20**, 2259–2268.
99. Amezaga, M. R., Carter, P. E., Cash, P., and McKenzie, H., *J. Clin. Microbiol.* 2002, **40**, 3313–3318.
100. Pancholi, V., and Fischetti, V. A., *J. Exp. Med.* 1992, **176**, 415–426.
101. Lottenberg, R., Broder, C. C., Boyle, M. D. P., Kain, S. J., Schroeder, B. L., and Curtiss, R., *J. Bacteriol.* 1992, **174**, 5204–5210.
102. Pancholi, V., and Fischetti, V. A., *Proc. Nat. Acad. Sci. USA* 1993, **90**, 8154–8158.

CHAPTER 12

# Proteomic Survey through Secretome of *Bacillus subtilis*

HAIKE ANTELMANN,[1] JAN MAARTEN VAN DIJL,[2] SIERD BRON,[3] and MICHAEL HECKER[1]

[1]Ernst-Moritz-Arndt-Universität Greifswald, Greifswald, Germany
[2]University of Groningen, Groningen, The Netherlands
[3]Groningen Biomolecular Sciences and Biotechnology Institute, Haren, The Netherlands

## 12.1 INTRODUCTION: CONCEPT OF PHYSIOLOGICAL PROTEOMICS

The sequencing of entire genomes opened a new era in biology which provides the opportunity to understand a living cell and thus life in general. This genome sequence, however, represents only the "blueprint of life," not "life itself." Different facets of functional genomics are required to transform the genome sequence into real cellular physiology. Within the different facets of functional genomics, proteomics has and will keep a central position because it deals—as no other discipline does—directly with the proteins, the players of life.

Already in 1975 O'Farrell and Klose published their central papers on the two-dimensional polyacrylamide gel electrophoresis (2D PAGE) technique (O'Farrell, 1975; Klose, 1975). Using this technique a complex mixture of proteins can be separated because each single protein will migrate to its unique position in the 2D gel determined by its charge and molecular mass. Later this technique was introduced into bacterial physiology by Neidhardt and Van Bogelen (2000), who addressed crucial physiological questions by proteomics such as the heat stress response or the phosphate starvation response of *Escherichia coli*. This new approach was extremely attractive for studies on cell physiology because proteomics could visualize cellular events not seen before by more conventional techniques.

Based on the proteomic approach, we started in the mid-1980s to use the panoramic view offered by proteomics for the study of the physiology of *Bacillus subtilis*, the model organism of gram-positive bacteria. The stepwise establishment of physiological proteomics as an essential tool in bacterial physiology can be illustrated in the case of the general stress response of *B. subtilis*. One of the major changes in the 2D protein pattern by stress was the dramatic induction of about 40–50 proteins by a set of different environmental stimuli such as heat, ethanol, or osmotic stress or starvation for glucose,

*Microbial Proteomics: Functional Biology of Whole Organisms*, Edited by Ian Humphery-Smith and Michael Hecker. 

phosphate, or oxygen. In contrast to stress-specific proteins that are induced by one extracellular stimulus only, these proteins were defined as "general stress proteins" (Hecker and Völker, 2001). It was interesting to note that already in the mid-1980s the global view of proteomics led to the discovery of a physiological stress and starvation response of *B. subtilis* that should be, from a physiological point of view, one of the most essential responses of cells during the transition from the growing into a nongrowing state. Up to 20% of the translational capacity or even more was invested for the induction of this group of proteins. Already at that time proteomics proved to be an excellent tool to visualize physiological phenomena not seen before (Richter and Hecker, 1986, Hecker et al., 1988). However, these studies remained on a preliminary, only descriptive level because it was almost impossible or at least extremely difficult to identify the proteins involved in this adaptation process. The progress in N-terminal sequencing in combination with the increasing amount of data stored in the genome databases was the next step in establishing physiological proteomics, allowing the identification of some of the main players in stress/starvation adaptation. The Ctc protein was among the general stress proteins identified by N-terminal sequencing (Völker et al., 1994). The only information available on Ctc was that its gene is under $\sigma^B$ control. Detected as the first bacterial alternative sigma factor almost 25 years ago (Haldenwang and Losick, 1979), the function of $\sigma^B$ remained a matter of speculation until the early 1990s. The interest in $\sigma^B$ declined after knowing that this sigma factor is not involved in the regulation of sporulation (Binnie et al., 1986; Duncan et al., 1987). Our key observation that Ctc is under $\sigma^B$ control led to the suggestion that the entire gene group is controlled by $\sigma^B$. The probabability is that more than 150 general stress proteins provide the nongrowing *B. subtilis* cell with a nonspecific, multiple and preventive stress response in anticipation of "future stress," presumably as an alternative to the sporulation process (Hecker and Völker, 1998; Price et al., 2001). This example should not only underline the power of this new approach in the visualization of new physiological responses even in the early days of proteomics but also show how proteomics became a key technology in the analysis of microbial physiology.

The next event crucial for the establishment of physiological proteomics for *B. subtilis* is the publication of its complete genome sequence in 1997 (Kunst et al., 1997). Since then, it has been possible to identify all proteins present in the 2D gels by large-scale mass spectrometry such as matrix-assissted laser desorption ionization time of flight mass spectrometry (MALDI-TOF) and tandem MS (MS/MS). The performance of high-throughput proteomics was further automated by the usage of spot-handling robots that carry out spot cutting, protein digesting, and peptide spotting. Consequently, the proteomic approach became a routine technique and a major breakthrough in the development of physiological proteomics.

In the future it seems that 2D gel-based proteome analysis will be replaced by non-gel-based procedures, such as high-performace liquid chromatography (HPLC)–MS/MS techniques, which are coming more and more into focus. However, 2D PAGE will keep its attractiveness, particularly for comparative physiological proteomics studies when two or more different physiological conditions or strains (wild type and mutants) have to be compared. This comparative proteomics that should provide quantitative data has to rely on a minimal number of gel fractions covering the majority of proteins, because only a small number of gels can be handled in such studies. The non-gel-based proteomic procedures have to be substantially improved before they can replace the classical gel-based technique in physiological proteomics. It should be added, however, that for many applications in microbial physiology these nongel techniques will more and more compliment the traditional 2D PAGE to visualize proteins not seen before on the 2D gels (e.g., intrinsic membrane proteins).

*B. subtilis* proves to be an excellent model organism for functional genomics and physiological proteomics in gram-positive bacteria. The choice of *B. subtilis* as a model is primarily based on the long-standing interest in this species as a model organism of gram-positive bacteria in general and as a model for understanding cell differentiation in particular. Furthermore, members of the species of *Bacillus,* as producers of many extracellular enzymes and antibiotics, are valuable strains for industrial application. The genome of *B. subtilis* contains about 4100 genes; many more than 1000 of these code for proteins with still unknown function. The high number of unknown global regulators such as alternative sigma factors or two-component systems was surprising. The elucidation of the function of this high number of unknown proteins is a big challenge for future research and a main goal of functional genomics. Joint research programs in Japan and Europe aimed at the construction of a mutant library containing mutations in each single gene of unknown function and in a comprehensive phenotypic screening program should help to get first information on their physiological role (Schumann et al., 2001). Physiological proteomics is one approach whose systematic application should help to predict or even identify the function of many of these still unknown proteins (see below).

From a physiological point of view, only subpopulations of proteins will be synthesized in growing or nongrowing cells because only a part of the genome is expressed at any time. To visualize as many proteins as possible, different stress or starvation stimuli should be imposed to growing cells that will induce quite different stress- and starvation-specific or more general stress/starvation regulons (Hecker and Völker, 2001). From this physiological point of view, two main proteome groups can be defined: proteomes of growing cells containing proteins which in most cases possess housekeeping functions and proteomes of nongrowing cells containing proteins with adaptive functions against the stress or starvation stimuli that triggered the nongrowing state. In growing *B. subtilis* cells about 700 proteins were identified in the main proteome window of isoelectric points (pIs) 4–7 covering almost 50% of the proteins predicted to be expressed in growing cells within this window (Eymann et al., 2004). These vegetative proteins were inserted into metabolic maps showing that many basic metabolic routes are "available at a proteomic scale." This vegetative proteome is now ready for physiological application. This comprehensive proteomic information can be used to analyze the regulation of entire metabolic pathways, which is what we did for the regulation of glycolysis/tricarboxylic acid (TCA) cycle and disposal of a new model for glycolysis regulation in *B. subtilis* (Tobisch et al., 1999; Ludwig et al., 2001, 2002). Many more metabolic pathways are available at a proteomic-scale level and should therefore promote similar studies in the future.

The proteomes of nongrowing cells, in contrast, can be arranged in a comprehensive adaptational network containing many stress- or starvation-inducible stimulons and regulons. Proteomics is an excellent tool to define the single stimulons (set of proteins induced/repressed by one single stimulus) and to dissect the single stimulons into, from a physiological point of view, better defined regulons (see Section 12.3). The dual-channel imaging and gel-warping techniques are the tools used to define the stimulon/regulon network (Bernhardt et al., 1999; Hecker, 2003, for review). All the proteomic data can be assembled into a highly sophisticated adaptational network that is a useful tool for a comprehensive understanding of stress/starvation adaptation. Because stress and starvation are the rule and not the exception in natural ecosystems, the elucidation of stress adaptation describes a crucial chapter of bacterial physiology and microbiology in general. Furthermore, to dissect the entire genome into its basic models of global gene regulation (regulons, modulons) and to assign the still unknown proteins into these functional groups represent a very simple but convincing approach to predict the function of many unknown proteins. A protein strongly induced by oxidative stress, for instance, will have something to do with the adaptation against oxidative stress. This approach provides, however, only the starting point for more detailed studies aimed at the elucidation of the function of each protein and yet have its function clarified.

The proteomic information of growing and nongrowing cells can be used to predict the physiological state of cells grown in a bioreactor or in a biofilm. Using the "proteomic signature" introduced into bacterial physiology by Neidhardt and Van Bogelen (2000), it is possible to predict whether the nongrowing cell suffered from heat or osmotic or oxidative stress. This approach was used to predict the physiological state of growing and nongrowing *B. subtilis* strains during an industrial fermentation process (Antelmann et al., 2004) or in *Bacillus licheniformis* cell populations (Voigt et al., 2004). In the case of *B. licheniformis* proteome signatures were detected suggesting that a severe oxidative stress occurs in stationary-phase cells. This was indicated by (i) a strong induction of the PerR regulon, (ii) induction of protein stress, and finally (iii) by a real protein stress indicated by protein fragmentation.

This brief introduction into physiological proteomics of *B. subtilis* should demonstrate that proteomics is an excellent approach to bring the *virtual life of genes to the real life of the proteins*, thereby aiming at a new understanding of the physiology and life of bacterial cells.

To unleash the full potential of proteomics, the visualization of almost all proteins synthesized in the cell is the final goal. Unfortunately, only a subfraction of cellular proteins can be visualized so far on one single gel. In many proteomic studies, only cytosolic proteins in the main window of pi 4–7 were considered. However, if one adds alkaline proteins or cell-wall-associated or even extracellular proteins to the weakly acid/neutral cytosolic proteins, the majority of proteins synthesized in a bacterial cell can be visualized by this most powerful technique.

In this chapter, we aim to demonstrate how the panoramic view of proteomics can be used to get new information on the protein secretion mechanisms in *B. subtilis*. Collaboration between a group with a long-standing experience in the genetics of protein secretion and a proteomic laboratory offered a great opportunity to gain new and comprehensive information on protein secretion at a proteomewide scale. This proteomic view of protein secretion is a good model example showing how to bring the genome sequence to life.

## 12.2 PROTEOMICS OF PROTEIN SECRETION IN *B. SUBTILIS*

### 12.2.1 Secretome and Exoproteome of *B. subtilis*

Because of the lack of an outer membrane, the gram-positive bacterium *B. subtilis* is able to secrete large amounts of extracellular proteins directly into the growth medium. Consequently, all proteins of the surrounding growth medium represent the extracellular complement of the secretome which can be easily purified using a simple TCA precipitation method. Based on predictions for signal peptides, 300 proteins have the potential to be exported in *B. subtilis* (Tjalsma et al., 2000; van Dijl et al., 2002). During or shortly after translocation of the (pre-) proteins across the membrane, the N-terminal

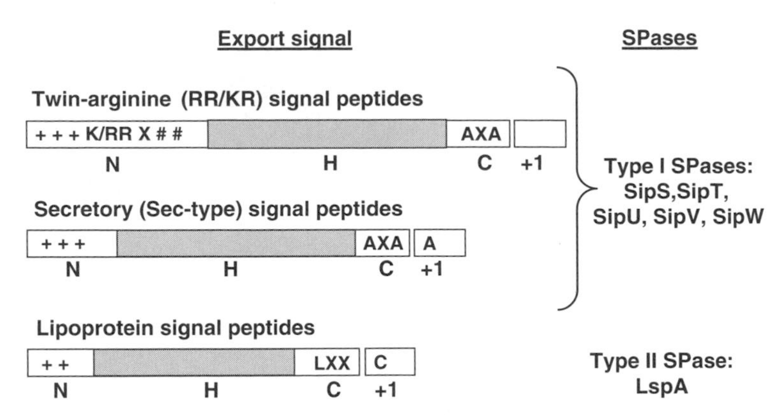


**Figure 12.1** Classification of N-terminal signal peptides. The predicted signal peptides are classified in three major groups: (1) twin-arginine (RR/KR) signal peptides, (2) secretory (Sec-type) signal peptides, and (3) lipoprotein signal peptides. The SPases responsible for the specific cleavage are indicated. The signal peptides are composed of a positively charged N domain, a hydrophobic H domain, and a C domain with the SPase cleavage site. The first amino acid of the mature protein is indicated by +1.

signal peptide is cleaved by the signal peptidases (SPases). Signal peptides can be classified according to the SPase recognition sequence and the predicted export pathways (Fig. 12.1) (Tjalsma et al., 2000, 2004; van Dijl et al., 2001). The first class consists of twin-arginine (RR/KR) signal peptides which direct the proteins into the Tat pathway (Jongbloed et al., 2002). The second and major class is comprised of the secretory (Sec-type) signal peptides that direct the proteins into the major "Sec" pathway. Both the twin-arginine and secretory signal peptides are cleaved by one of the various type I SPases of *B. subtilis* (Tjalsma et al., 2000). The third class of signal peptides is present in the N terminus of prelipoproteins, which are cleaved by the lipoprotein-specific type II SPase (LspA) of *B. subtilis* (Tjalsma et al., 2000). Signal peptides of prelipoproteins are characterized by the "lipobox" containing an invariable cysteine residue that is lipid modified by the diacylglyceryl transferase (Lgt) prior to precursor cleavage by LspA. The exported and modified lipoproteins remain attached to the cytoplasmic membrane by their N-terminal lipid anchor.

The first proteome analyses of secreted proteins of *B. subtilis* were performed in minimal medium containing glucose or alternative carbon sources (Hirose et al., 2000) as well as under the conditions of phosphate starvation (Antelmann et al., 2000). Because of the lower growth rate of *B. subtilis* in minimal medium, there is also a low secretion capacity reflected by the identification of some more than 20 low-expressed extracellular proteins in minimal medium containing glucose (Hirose et al., 2000). In contrast, the very strong induction and secretion of the extracellular phosphatases/phosphodiesterases PhoA, PhoB, and PhoD, the glycerophosphoryl diester phosphodiesterase GlpQ, the 2′,3′-cyclic-nucleotide 2′-phosphodiesterase YfkN, and the lipoproteins YdhF and PstS are reflected in the extracellular proteome characteristic for phosphate starvation (Fig. 12.2*a*) (Antelmann et al., 2000). All these phosphate starvation-specific secreted proteins, which account for 30% of the total phosphate starvation extracellular proteome, are members of the Pho regulon, because these failed to be induced and secreted in a *phoR* mutant (Table 12.1) (Antelmann et al., 2000). As shown by the quantification of the protein abundance using the

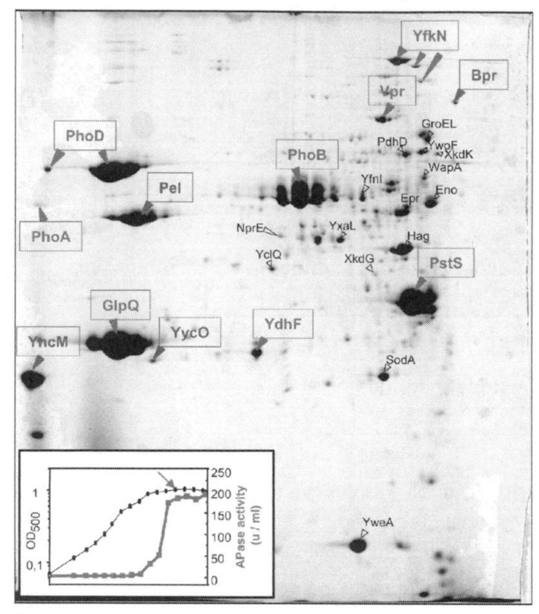


PhoPR-dependent phosphate starvation-induced proteins
PhoPR-independent phosphate starvation- induced proteins

(*a*)

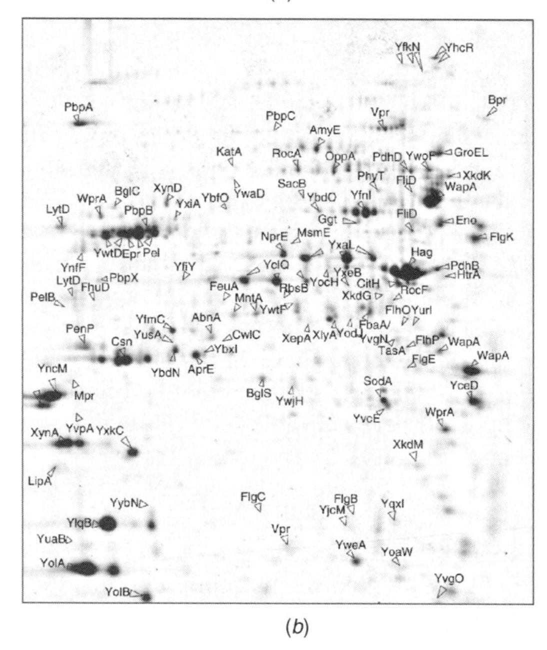


(*b*)

**Figure 12.2** Extracellular proteome of *B. subtilis* 168 (*a*) under conditions of phosphate starvation and (*b*) in complete medium. Cells of *B. subtilis* 168 were grown in minimal medium (*a*) under conditions of phosphate starvation and (*b*) in LB broth. Proteins in the growth medium were harvested 1 h after entry into the stationary phase. After precipitation with TCA, the extracellular proteins were separated by 2D PAGE and stained with Sypro Ruby (Antelmann et al., 2001). The inset in (*a*) shows the growth curve in the phosphate starvation minimal medium (black line) and the measurement of the alkaline phosphate (APase) activity along the growth (red line). The arrow indicates the time point of maximum APase activity when cells were harvested for preparation of the extracellular fraction.

**TABLE 12.1 Quantification of Extracellular Proteins of *B. subtilis* 168 with Export Signals[a]**

| Protein | Function/Similarity | Retention Signal[b] | Percent Abundance[c] | |
|---|---|---|---|---|
| | | | LB-ex | Phosphate-ex |
| **AbnA** | Arabinan-endo 1,5-α-L-arabinase | — | 0.304 | |
| **AmyE** | α-Amylase | — | 0.386 | 0.139 |
| **AprE** | Serine alkaline protease (subtilisin E) | — | 0.441 | 0.226 |
| **BglC** | *endo*-1,4-β-Glucanase, cellulase | — | 0.078 | |
| **BglS** | *endo*-β-1,3-1,4 Glucanase | — | 0.209 | 0.201 |
| **Bpr** | Bacillopeptidase F | — | 0.133 | 0.221 |
| **Csn** | Chitosanase | — | 1.886 | 0.87 |
| **Epr** | Minor extracellular serine protease | — | 0.883 | |
| **FeuA** | Iron-binding protein | Lipid | 0.04 | |
| **FhuD** | Ferrichrome-binding protein | Lipid | 0.042 | |
| **Ggt** | γ-Glutamyltranspeptidase | — | 0.055 | |
| **GlpQ[pho]** | Glycerophosphoryl diester phosphodiesterase | — | 0.036 | 5.787 |
| **HtrA** | Serine protease | — | 0.183 | |
| **LipA** | Lipase | — | 0.246 | 0.56 |
| **LytD** | *N*-Acetylglucosaminidase (major autolysin) | CWB | 0.409 | |
| **MntA** | Manganese-binding protein | Lipid | 0.297 | |
| **Mpr** | Extracellular metalloprotease | — | 0.056 | |
| **MsmE** | Multiple sugar-binding protein | — | 0.006 | |
| **NprE** | Extracellular neutral metalloprotease | — | 0.466 | 0.141 |
| **OppA** | Oligopeptide-binding protein | Lipid | 0.301 | |
| **OpuAC** | Glycine-betaine-binding protein | Lipid | 0.002 | |
| **PbpA** | Penicillin-binding protein 2A | — | 0.87 | |
| **PbpB** | Penicillin-binding protein 2B | — | 0.95 | |
| **PbpC** | Penicillin-binding protein 3 | Lipid | 0.032 | |
| **PbpX** | Penicillin-binding protein | — | 0.112 | |
| **Pel** | Pectate lyase | — | 1.23 | 4.58 |
| **PelB** | Pectate lyase | — | 0.23 | 0.637 |
| **PenP** | β-Lactamase precursor | — | 0.71 | |
| **PhoA[pho]** | Alkaline phosphatase A | — | 0.001 | 0.54 |
| **PhoB[pho]** | Alkaline phosphatase III | — | 0.001 | 5.532 |
| **PhoD[pho]** | Phosphodiesterase/alkaline phosphatase D | — | 0.002 | 9.432 |
| **Phy[SNB]** | Phytase | — | 0.035 | |
| **PstS[pho]** | Phosphate-binding protein | Lipid | 0.001 | 7.871 |
| **RbsB** | Ribose-binding protein | Lipid | 0.006 | |
| **SacB[SNB]** | Levansucrase | — | 0.002 | |
| **TasA** | Antimicrobial spore component | — | 0.107 | 0.224 |
| **Vpr** | Extracellular serine protease | — | 0.405 | 0.398 |
| **WapA** | Cell-wall-associated protein precursor | CWB | 4.45 | 0.536 |
| **WprA** | Cell-wall-associated protein precursor | CWB | 0.606 | 0.112 |
| **XynA** | *endo*-1,4-β-Xylanase | — | 1.601 | 0.724 |
| **XynD** | *endo*-1,4-β-Xylanase | — | 0.194 | 0.123 |
| **YbdN** | Unknown | — | 0.337 | 0.134 |
| **YbdO** | Unknown | — | 0.223 | |
| **YbfO[SNB]** | Similar to erythromycin esterase | — | 0.098 | |
| **YbxI** | Similar to β-lactamase | — | 0.128 | |
| **YcdH** | Zinc-binding protein | Lipid | 0.002 | 0.257 |
| **YclQ** | Ferrichrome-binding protein | Lipid | 1.809 | 0.501 |
| **Y YdhF[pho]** | Similar to unknown proteins from *B. subtilis* | Lipid | 0.001 | 0.747 |

**TABLE 12.1** (*Continued*)

| | | | Percent Abundance[c] | |
|---|---|---|---|---|
| Protein | Function/Similarity | Retention Signal[b] | LB-ex | Phosphate-ex |
| **YdhT** | Mannan *endo*-1,4-β-mannosidase | — | 0.001 | |
| **YfiY** | *iron*(III)-binding protein | Lipid | 0.001 | |
| **YfkN**[pho] | 2′,3′-Cyclic-nucleotide 2′-phosphodiesterase | TM | 0.041 | 1.582 |
| **YflE** | Similar to anion-binding protein | — | 0.001 | |
| **YfmC** | Ferrichrome-binding protein | Lipid | 0.318 | |
| **YfnI** | Probable transmembrane glycoprotein | — | 1.426 | 0.613 |
| **YhcR** | 5′-Nucleotidase | TM | 0.142 | |
| **YjcM** | Unknown | — | 0.005 | |
| **YjfA** | Unknown | — | 0.001 | |
| **YlqB** | Unknown | — | 2.086 | 2.018 |
| **YncM** | Similar to unknown proteins from *B. subtilis* | — | 2.33 | 1.535 |
| **YnfF** | *endo*-Xylanase | — | 0.214 | |
| **YoaW** | Unknown | — | 0.023 | |
| **YocH** | Cell-wall-binding protein | CWB[10] | 0.012 | |
| **YodJ** | D-Alanyl-D-alanine carboxypeptidase | Lipid | 0.06 | |
| **YolA** | Unknown | — | 3.427 | 0.226 |
| **YolB** | Unknown | — | 0.908 | |
| **YqiX** | Amino-acid-binding protein | Lipid | 0.001 | 0.456 |
| **YqxI** | Unknown | — | 0.115 | |
| **YrpD** | Similar to unknown proteins from *B. subtilis* | — | 0.45 | |
| **YrpE** | Similar to unknown proteins | Lipid | 0.001 | 0.154 |
| **YuaB** | Unknown | — | 0.002 | |
| **YurI** | Ribonuclease | — | 0.054 | |
| **YusA** | Putative part of the S box regulon | Lipid | 0.076 | |
| **YvcE** | Cell-wall-binding protein | CWB | 0.354 | |
| **YvgO** | Unknown | — | 0.231 | |
| **YvpA** | Pectate lyase | — | 0.003 | |
| **YwaD** | Aminopeptidase | — | 0.112 | |
| **YweA** | Similar to unknown proteins from *B. subtilis* | — | 0.495 | 3.98 |
| **YwoF** | Unknown | — | 0.223 | 0.633 |
| **YwtD** | DL-Glutamyl hydrolase | CWB | 0.821 | |
| **YwtF** | Transcriptional regulator | — | 0.15 | |
| **YxaL** | Similar to serine/threonine protein kinase | — | 2.27 | 0.645 |
| **YxeB** | Putative binding protein | Lipid | 0.112 | |
| **YxiA** | Arabinan-*endo* 1,5-α-L-arabinase | — | 0.057 | |
| **YxkC** | Unknown | — | 0.864 | |

[a]All listed proteins were identified by MALDI-TOF MS in the extracellular proteome of *B. subtilis* wild-type cells grown in complete medium, 2x SNB medium, or minimal medium under the conditions of phosphate starvation (Antelmann et al., 2001, 2004; Hirose et al., 2000; H. Antelmann, unpublished data).

[b]Identified retention signals present in the mature part of the proteins after processing by specific SPases are lipid modifications (lipid), transmembrane domains (TM), and cell-wall-binding domains (CWB). Absence of known retention signals (Tjalsma et al., 2004).

[c]The protein quantification is related to the extracellular proteome in LB medium (Fig. 12.2*a*) and in phosphate starvation medium (Fig. 12.2*b*). The percentage of protein abundance corresponds to the percentage of the specific spot volume related to the total spot volume in the extracellular proteome of cells grown in LB medium (LB-ex) or in minimal medium under phosphate starvation conditions (Phosphate-ex). The percentage of cytoplasmic protein abundance was determined from the cytoplasmic proteome of cells grown in LB medium (data not shown). The percentage of spot volume was calculated using the quantification tool of the Decodon Delta 2D software after background subtraction.

Decodon Delta 2D software, the PhoD protein was the most strongly secreted protein and alone makes up 10% of the secretome.

The highest level of protein secretion was observed when cells of *B. subtilis* were grown in rich medium [luria broth (LB) medium], in particular during the stationary phase. Thus, we have previously defined the master gel for the extracellular proteome in complete medium (LB) during the stationary phase (Fig. 12.2*b*). In total, 113 different proteins could be identified in the extracellular proteome of *B. subtilis* wild-type and mutant cells grown in complete medium or under the conditions of phosphate starvation. These 113 proteins, which were labeled in the master gel for the extracellular proteome (Fig. 12.2*b*), included 90 proteins identified in the extracellular proteome of the wild type grown in LB: 10 additional lipoproteins detected in the extracellular proteome of the *lgt* mutant (see Lipid Modification by Diacylglyceryl Transferase in Section 12.2.6), 10 mostly PhoPR-dependent proteins present in the extracellular proteome exclusively after phosphate starvation (Figs. 12.2*a,b*) (Antelmann et al., 2000), and 3 new proteins identified in a very rich 2x salt supplemented nutrient broth (SNB) medium (Antelmann et al., 2004).

In addition, these proteins are classified according to the presence or absence of N-terminal signal peptides in Tables 12.1 and 12.2. The relative protein abundance of these extracellular proteins in LB medium as well as in phosphate starvation medium was determined using the Decodon Delta 2D software. The quantification of all 113 proteins in the extracellular proteome (complete medium, Tables 12.1 and 12.2) revealed the highest abundance, between 4 and 0.7% of the total spot volume, for Hag, YolA, WapA, YncM, YlqB, Pel, Csn, XynA, YxaL, YolB, YclQ, PbpA, YxkC, YwtD, FlgK, YfnI, and PenP. Besides Hag and FlgK most of these are proteins with a predicted extracellular localization. In general, most of these 113 exported proteins are mainly involved in the degradation of carbohydrates, proteins, nucleotides, lipids, and phosphate as well as in cell wall metabolism. Other extracellular proteins are lipoproteins that are substrate-binding components of various transport systems, proteins involved in detoxification, flagella-related functions, and phage-related functions (Antelmann et al., 2001; Tjalsma et al., 2004).

### 12.2.2 Mechanisms for "Secretion" of Extracellular Proteins

Of the 113 identified extracellular proteins, 54 were predicted to be secreted because of the presence of an SPase I cleavage site and the lack of retention signals (Tjalsma et al., 2000, 2004). Of these proteins, 40 contain a Sec-type signal peptide and are most likely transported via the Sec pathway. The remaining 14 proteins have a potential RR/KR motif in the N domains, suggesting their potential transport via the Tat pathway (Jongbloed et al., 2002). Strikingly, 59 extracellular proteins are not predicted to be secreted because of the presence of specific retention signals in addition to a signal peptide (18 lipoproteins, 6 cell-wall-binding proteins, and 6 membrane proteins) or the absence of signal peptides (17 cytoplasmic proteins, 6 phage-related proteins, and 7 flagella-related proteins) (Fig. 12.3) (Antelmann et al., 2001, Tjalsma et al., 2004). The mechanisms discussed below may be responsible for the secretion of these unpredicted extracellular proteins.

***Lipoproteins*** The diacylglyceryl transferase Lgt of *B. subtilis* is responsible for the diacylglyceryl modification of the N-terminal cysteine residue of prelipoproteins (Leskelä et al., 1999). Subsequently, these lipid-modified lipoproteins should be retained in the cytoplasmic membrane via the lipid anchor. Of the identified lipoproteins that are secreted by the wild-type *B. subtilis*, four were secreted in complete medium

**TABLE 12.2 Quantification of Extracellular Proteins of *B. subtilis* 168 Without Typical Export Signals**[a]

| Protein | Function/Similarity | percent Extracellular Abundance (LB)[b] | percent Cytoplasmic Abundance (LB)[b] |
|---|---|---|---|
| | *Cytoplasmic Proteins* | | |
| **CitH** | Malate dehydrogenase | 0.475 | 0.633 |
| **Ef-G** | Elongation factor G | 0.105 | 0.904 |
| **Eno** | Enolase | 0.432 | 0.630 |
| **FbaA** | Fructose-1,6-bisphosphate aldolase | 0.432 | 0.645 |
| **GapA** | Glyceraldehyde-3-phosphate | 0.320 | 0.84 |
| **GroEL** | Class I heat-shock protein (chaperonin) | 0.689 | 0.876 |
| **Hag** | Flagellin protein | 3.921 | 1.166 |
| **KatA** | Vegetative catalase 1 | 0.215 | 0.569 |
| **PdhA** | Pyruvate dehydrogenase (E1 $\alpha$ subunit) | 0.229 | 0.594 |
| **PdhB** | Pyruvate dehydrogenase (E1 $\beta$ subunit) | 0.256 | 0.444 |
| **PdhD** | Pyruvate dehydrogenase (E3 subunit) | 0.225 | 0.512 |
| **RocA** | Pyrroline-5 carboxylate dehydrogenase | 0.223 | 0.772 |
| **RocF** | Arginase | 0.167 | 0.916 |
| **SodA** | Superoxide dismutase | 0.367 | 0.735 |
| **YceD** | Similar to tellurium resistance protein | 1.28 | 0.452 |
| **YvgN** | Similar to plant metabolite | 0.406 | 0.645 |
| **YwjH** | Similar to transaldolase (pentose) | 0.17 | 0.363 |
| | *Flagella-Related Proteins* | | |
| **FlgB** | Flagellar basal-body rod protein | 0.067 | |
| **FlgC** | Flagellar basal-body rod protein | 0.006 | |
| **FlgE** | Flagellar hook protein | 0.108 | 0.075 |
| **FlgK** | Flagellar hook-associated protein 1 | 0.778 | |
| **FlhO** | Flagellar basal-body rod protein | 0.057 | |
| **FlhP** | Flagellar hook basal-body protein | 0.048 | |
| **FliD** | Flagellar hook-associated protein 2 | 0.224 | |
| | *Phage-Related Proteins* | | |
| **XepA** | PBSX prophage lytic exoenzyme | 0.065 | |
| **XkdG** | PBSX prophage gene | 0.127 | |
| **XkdK** | PBSX prophage gene | 0.161 | |
| **XkdM** | PBSX prophage gene | 0.157 | |
| **XlyA** | *N*-Acetylmuramoyl-L-alanine amidase | 0.205 | |

[a]All listed proteins were identified by MALDI-TOF MS in the extracellular proteome of *B. subtilis* wild-type cells grown in complete medium (Antelmann et al., 2001; H. Antelmann unpublished data).

[b]The protein quantification is related to the extracellular proteome in LB medium (Fig. 12.2*a*). The percentage of protein abundance corresponds to the percentage of specific spot volume related to the total spot volume in the extracellular proteome of cells grown in LB medium. The percentage of cytoplasmic protein abundance was determined from the cytoplasmic proteome of cells grown in LB medium (data not shown). The percentage of spot volume was calculated using the quantification tool of the Decodon Delta 2D software after background subtraction.

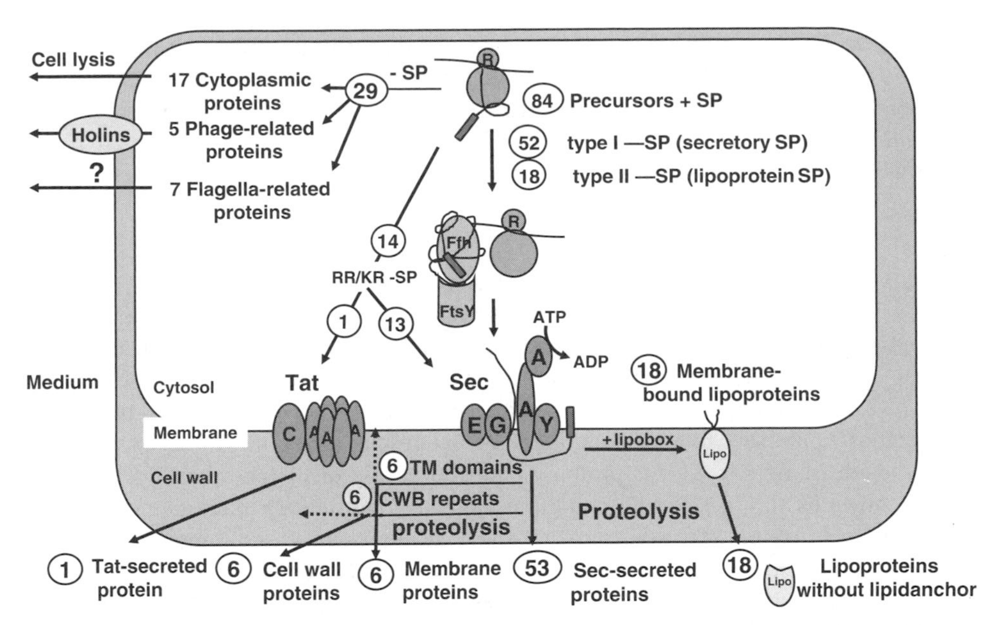


**Figure 12.3** Mechanisms for "secretion" of 113 extracellular proteins. A total 113 different extracellular proteins were identified in the extracellular proteomes of cells grown in complete or phosphate starvation media. Extracellular proteins were classified as "predicted" to be secreted or "unpredicted" based on the presence or absence of signal peptides according to Tjalsma et al. (2000). Accordingly, 54 extracellular proteins were predicted to be secreted because these possess N-terminal signal peptides (SPs) with cleavage sites for type I SPases and lack retention signals (Tjalsma et al., 2000). The remaining 59 are unpredicted because these either lack an SP (17 cytoplasmic proteins, 5 phage-related proteins, and 7 flagella-related proteins) or have retention signals in addition to the SP (18 lipoproteins, 6 transmembrane proteins, and 6 cell wall proteins) (Antelmann et al., 2001).

(OppA, MntA, YfmC, and YclQ) and five were secreted in minimal medium under the conditions of phosphate starvation (YcdH, YdhF, PstS, YqiX, and YrpE) (Antelmann et al., 2001). N-terminal sequence analysis revealed that these lipoproteins lack the N-terminal, lipid-modified, cysteine residue. Thus, lipoproteins are most likely liberated from the cell wall by proteolytic shaving (Antelmann et al., 2001). In addition, some of these secreted lipoproteins can also be extracted in high amounts from the cell membrane fraction of *B. subtilis* that was insoluble in the nondetergent sulfobetaine after extraction with β-D-maltoside (see Section 12.2.4 and Fig. 12.5 below) (Bunai et al., 2004). This indicates that the liberated lipoproteins might belong to highly expressed lipoproteins.

***Cell Wall Proteins*** The fact that cell wall proteins, including the large processing products of WapA, YvcE, and YocH, are stabilized in the extracellular as well as cell wall proteome of a multiple exoprotease-deficient strain (see Role of Extracellular Proteceses in Section 12.2.8) suggests that the release of cell wall proteins into the medium is mediated by proteolytic shaving (Antelmann et al., 2002). Because the same effect on stabilization of cell wall proteins was observed in a *sigD* mutant strain that is impaired in cell wall turnover, cell wall proteins might also be released due to the cell wall turnover (see Role of SigmaD in section 12.2.8).

***Membrane Proteins*** In the case of the membrane proteins YfnI and YflE, an SPase cleavage site is located C terminal of the five membrane-spanning domains. Similarily, the membrane proteins YfkN and YhcR are processed N-terminally as well as C-terminally. Also, the N-terminal processing of HtrA and PbpA is mediated by an unknown protease because these are lacking an SPase cleavage site (Tjalsma et al., 2004).

***Cytoplasmic Proteins*** The 17 cytoplasmic proteins included, for example, components of the pyruvate dehydrogenase complex PdhA,B,D, the enolase Eno, or the arginine catabolic proteins RocA and RocF, which all belong to the most abundant proteins of the cytoplasmic proteome of *B. subtilis*. These relative abundances are presented in Table 12.2. In addition, most of these proteins also represent highly expressed proteins of the previously published cytoplasmic proteome in minimal medium (Büttner et al., 2001). Thus, the release of cytoplasmic proteins is most probably mediated by cell lysis. This is further shown by the fact that all conditional mutants that harbor mutations in essential secretion genes (e.g., *ffh, secA, prsA*) show strongly increased amounts of these cytoplasmic proteins in the extracellular proteome after the depletion of the respective secretion factors (see, e.g., Section 12.2.7 for PrsA), which indicates that the depleted cells are sensitive to lysis (Hirose et al., 2000; Vitikainen et al., 2004, Kobayashi et al., 2003). In addition, typical cytoplasmic proteins that represent highly abundant proteins in the cytoplasmic proteomes were detected in the extracellular proteomes of several pathogenic gram-positive bacteria (Tjalsma et al., 2004).

***Prophage-Related Proteins*** The five prophage-related proteins XkdG, XkdK, XkdM, XepA, and XlyA have the potential to be secreted via prophage-encoded holins (Krogh et al., 1998; Longchamp et al., 1994; Tjalsma et al., 2004).

***Flagella-Related Proteins*** The seven flagella-related proteins FliD, FlgB, FlgC, FlgE, FlgK, FlhO, and FlhP could be exported via a machinery for the assembly of flagella and related to the type III secretion machinery of gram-negative bacteria or released from flagella (Blocker et al., 2003; Hueck 1998; Namba et al., 1989).

### 12.2.3 Cell Wall Proteome of *B. subtilis*

Several *B. subtilis* enzymes involved in cell wall turnover contain a variable number of repeated domains which have affinity for components of the cell wall (Ghuysen et al., 1994; Tjalsma et al., 2000). Some of these cell-wall-binding proteins, including LytD, YwtD, YocH, and several processing products of WapA and YvcE, are also components of the extracellular proteome. Whereas the majority of extracellular proteins is increased during the stationary phase, the large WapA-processing products YvcE and YocH are exclusively present in the extracellular proteome during the exponential growth phase due to the high cell wall turnover. These proteins are subject to proteolytic cleavage during the stationary phase (Antelmann et al., 2002).

Non–covalently linked cell-wall-binding proteins (CWBPs) can be extracted from whole cells according to the LiCl extraction procedure previously described (Rashid et al., 1995). The LiCl-extracted cell wall proteome from stationary-phase cells of the wild-type strain *B. subtilis* 168 grown in complete medium includes most abundantly the

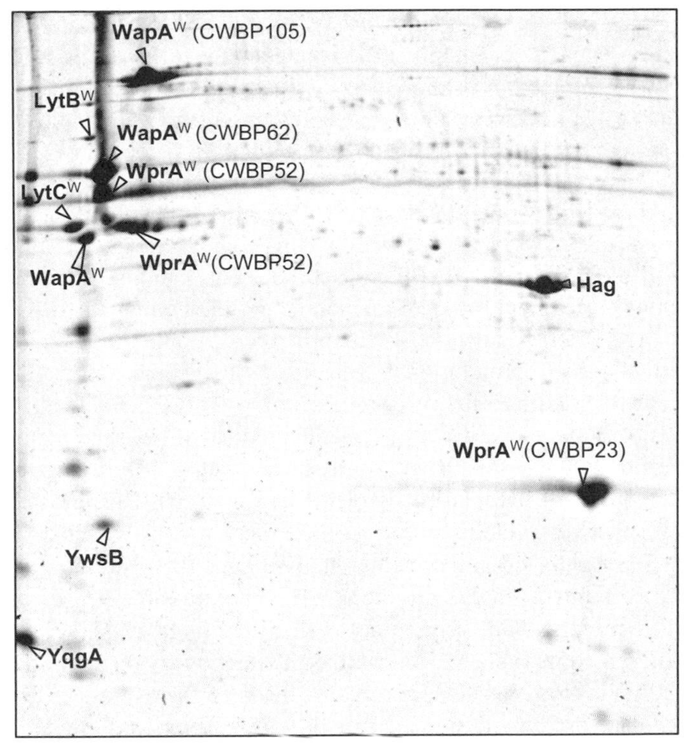


**Figure 12.4** Cell wall proteome of *B. subtilis* 168. Cells of *B. subtilis* 168 grown in LB broth were harvested 1 h after entry into the stationary phase and the cell wall proteins were extracted using a LiCl extraction procedure, precipitated with TCA, separated by 2D PAGE, and stained with Sypro Ruby (Antelmann et al., 2002). Cell wall proteins with wall-binding repeats are indicated with W. Processing products of WapA and WprA are also indicated as CWBPs according to their molecular masses.

WapA processing products (CWBP105 and CWBP62) and the processing products of the major wall-associated protease WprA (CWBP52 and CWBP23) (Fig. 12.4) (Antelmann et al., 2002). Additionally, the major amidase LytC as well as its modifier protein LytB could be extracted from the cell wall. Besides these known CWBPs also flagellin (Hag), YwsB, and YqgA were identified in the cell wall fraction which all have no wall-binding domains. It was remarkable that we were unable to detect the cell wall proteins with the typical cell-wall-binding repeats present in the extracellular proteome (LytD, YocH, YvcE, and YwtD) in the cell wall fraction. This might indicate that the presence of cell-wall-binding repeats is not a guarantee for retention in the cell wall (Tjalsma et al., 2004).

### 12.2.4 Membrane Proteomes and Lipoproteomes of *B. subtilis*

Membrane proteins with predicted putative transmembrane domains are regarded as nonsecretory proteins in previous secretome predictions because the transmembrane domains serve as membrane retention signals (Tjalsma et al., 2000). Due to their strong hydrophobicity, membrane proteins are relatively insoluble in nonionic or zwitterionic detergents and thus cannot be resolved by standard immobilized pH gradients. To describe the membrane proteome of *B. subtilis*, Bunai et al., (2004) prepared a washed cell membrane fraction that was insoluble in the nondetergent sulfobetaine (NDSB) and then extracted membrane proteins using mixtures of detergents, including β-D-maltoside in a stepwise manner. After 1D PAGE, they were able to identify 637 proteins, including 256 membrane proteins, 101 lipoproteins, and 280 cytoplasmic proteins. To optimize the conditions for the proteome analysis, the membrane-protein-containing fraction was separated using three types of 2D PAGE [immobilized pH gradient (IPG), 16-BAC-PAGE, and blue native PAGE] by means of which they could separate 30 lipoproteins that represent 80% of the predicted 38 ABC transporter solute-binding proteins (SBPs) of *B. subtilis*. Using this enrichment protocol membrane protein combined with different types of proteomics, a membrane-anchored "lipoproteome" could be defined that was in accordance with those lipoproteins that were identified in the extracellular proteome, because these are released into the medium due to proteolysis. To verify the membrane proteome results of Bunai et al., we have separated the NDSB-washed membrane fraction that was extracted with β-D-maltoside using our standard 2D PAGE. This was then compared to the standard cytoplasmic proteome (Fig. 12.5). Despite the fact that the majority of cytoplasmic proteins were still present in that membrane fraction, 10 lipoproteins could be detected in the membrane proteome that were not detectable in the cytoplasmic fraction. Interestingly, 9 of these 10 lipoproteins are released into the growth medium of wild-type *B. subtilis* by proteolytic shaving (Table 12.3). As judged from the quantification presented in Table 12.3 as well as the microarray data reported by Bunai et al., (2004), there seems to be no clear correlation between the expression levels of certain lipoproteins and their proteolytic shaving in wild-type *B. subtilis*. For example, PrsA and RbsB were present at relatively high levels in the membrane fraction, but PrsA appears to be absent from the growth medium and only relatively low amounts of extracellular RbsB are detectable. Consistent with the observation that the level of lipoprotein shaving is highly increased in an *lgt* mutant strain, the relative abundance of all 10 lipoproteins in the membrane fraction of this strain was severely reduced (see the discussion of lipid modification in Section 12.2.6). Remarkably, the secretion-specific folding factor PrsA represents the most abundant lipoprotein in the membrane fraction of the wild type (Table 12.3; Fig. 12.5).

### 12.2.5 Protein Export Routes

***SRP- and SecA-Dependent Secretion*** Of the identified extracellular proteins, 84 are synthesized with N-terminal signal peptides, most of which should be translocated via the general secretion (Sec) pathway in an unfolded conformation. Cytoplasmic chaperones and targeting factors like the Ffh protein were found to be homologous to the 54-kDa subunit of the mammalian signal recognition particle (SRP) and the FtsY protein homologous to the mammalian SRP receptor α subunit facilitates targeting of the preproteins to the Sec translocase in the membrane (Honda et al., 1993; Ogura et al., 1995). Both Ffh and FtsY are essential for protein secretion and cell viability (Kobayashi et al., 2003). The Sec translocation machinery consists of the SecA translocation motor and the SecYEG translocation complex. SecA is able to bind adenosine triphosphate (ATP) after the preprotein is bound to SecA, leading to the insertion of SecA into the pore of the SecYEG translocase and the translocation of a short stretch of peptides. In the next step, ATP is hydrolyzed by SecA, leading to the release of the preprotein and the de-insertion of SecA, which can be specifically inhibited by sodium azide.

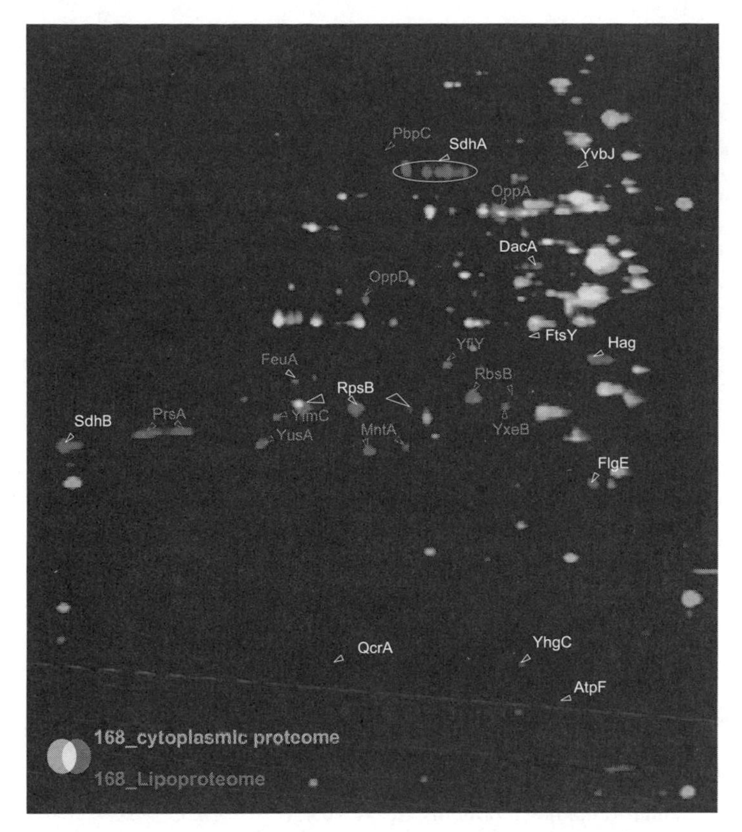


**Figure 12.5** Comparison of membrane-attached lipoproteome (red image) and cytoplasmic proteome (green image) of *B. subtilis*. Cells of *B. subtilis* 168 were grown in LB broth and harvested 1 h after entry into the stationary phase. Cytoplasmic protein extracts were compared to the nondetergent sulfobetaine- (NDSB-) washed membrane fraction obtained after stepwise extraction with β-D-maltoside (corresponding to fraction 4 according to Bunai et al., 2004). Image analysis was performed using the Decodon Delta 2D software that is based on the dual-channel imaging (Bernhardt et al., 1999). Proteins enriched in the membrane fraction that are present at lower amounts in the cytoplasmic fraction were identified. Known or predicted lipoproteins which could be extracted from the membrane fraction are indicated by red labels. (See color insert.)

First attempts to verify the number of proteins exported via the SRP-Sec pathways were performed by Hirose et al. (2000). In this study, a temperature-sensitive *secA* mutant was grown at 30°C (permissive temperature) and 42°C (nonpermissive temperature) and the extracellular proteomes were compared. In addition, the depletion for the targeting factor Ffh was studied using an IPTG-dependent *ffh* mutant grown in the presence or absence of isopropyl thiogalactoside (IPTG). Hirose and co-workers found that all extracellular proteins that were synthesized with secretory signal peptides disappeared from the extracellular proteome of the temperature-sensitive *secA* strain after the upshift to the higher temperature as well as after depletion for Ffh. This indicates that secretion of predicted secretory proteins is directed via the SRP-Sec pathway. In addition, the secretion of cytoplasmic proteins, flagella-related proteins, and phage-related proteins was clearly shown to be independent of the SRP-Sec pathway.

**TABLE 12.3 Lipoprotein Abundance in Extracellular Proteome (ex) and Lipoproteome (mem)**[a]

| | | percent Abundance[b] | | | |
|---|---|---|---|---|---|
| Protein | Function/Similarity | 168-ex | *lgt*-ex | 168-mem | *lgt*-mem |
| **FeuA** | Iron-binding protein | 0.041 | 0.333 | 0.387 | 0.001 |
| **FhuD** | Ferrichrome-binding protein | 0.005 | 0.376 | | |
| **MntA** | Manganese-binding protein | 0.297 | 2.505 | 1.723 | 0.023 |
| **MsmE** | Manganese-binding protein | 0.001 | 0.185 | | |
| **OppA** | Oligopeptide-binding protein | 0.301 | 2.302 | 1.884 | 0.632 |
| **OpuAC**[pst] | Glycine-betaine-binding protein | 0.001 | 0.214 | | |
| **PbpC** | Penicillin-binding protein 3 | 0.032 | 0.315 | 0.105 | 0.001 |
| **PrsA** | Protein secretion (post translocation molecular chaperone) | — | — | 2.865 | 0.015 |
| **RbsB** | Ribose-binding protein | 0.006 | 0.456 | 1.523 | 0.018 |
| **YclQ** | Ferrichrome-binding protein | 1.805 | 1.345 | | |
| **YcdH**[pst] | Zinc-binding protein | 0.257 | 1.373 | | |
| **YdhF**[pst] | Similar to unknown proteins of *B. subtilis* | 0.747 | 1.353 | | |
| **YfiY** | Fe(III)-binding protein | 0.001 | 0.602 | 0.562 | 0.021 |
| **YfmC** | Ferrichrome-binding protein | 0.318 | 1.96 | 0.452 | 0.001 |
| **YodJ** | D-Alanyl-D-alanine carboxypeptidase | 0.065 | 0.453 | | |
| **YrpE**[pst] | Unknown | 0.154 | 0.345 | | |
| **YusA** | Putative part of S box regulon | 0.076 | 1.054 | 0.912 | 0.023 |
| **YqiX**[pst] | Similar to unknown proteins | 0.456 | 1.235 | | |
| **YxeB** | Putative binding protein | 0.112 | 2.53 | 0.394 | 0.011 |

[a]All listed proteins were identified by MALDI-TOF MS in the extracellular proteome of *lgt* mutant cells grown in complete medium or minimal medium under the conditions of phosphate starvation (Antelmann et al., 2001).
[b]The protein quantification is related to the extracellular proteome (ex) and lipoproteome (mem) as shown in Figs. 12.8*a* and *b*. The percentage of lipoprotein abundance corresponds to the percentage of the specific spot volume related to the total spot volume in the extracellular proteome (ex) and lipoproteome (mem) of wild-type (168) and *lgt* mutant cells grown in LB medium. The percentage of spot volume was calculated using the quantification tool of the Decodon Delta 2D software after background subtraction.

To study the SecA-dependent secretion of extracellular proteins, we used a different approach based on the inhibition of the ATPase activity of SecA with sodium azide. To study the extracellular proteome of the wild type after the addition of sodium azide, it was necessary to analyze the secretion of de novo synthesized proteins within 20 min after the exchange of the medium (Fig. 12.6) (Jongbloed et al., 2002). Of the 26 identified de novo synthesized proteins that are secreted in untreated cells, a subset was secreted at strongly reduced levels in the presence of azide. These azide-sensitive proteins included proteins which possess typical SPase cleavage sites. In contrast, no effect of SecA inhibition was observed in the case of the flagella-related proteins and cytoplasmic proteins, which confirms the data obtained upon depletion of SecA and *ffh* (Hirose et al, 2000). However, there was also a set of 13 extracellular proteins with typical Sec-type signal peptides, including 3 lipoproteins that were not reduced upon SecA inhibition. The latter is in contrast to the data obtained in the *secA*-depleted strain by Hirose et al. This suggests that

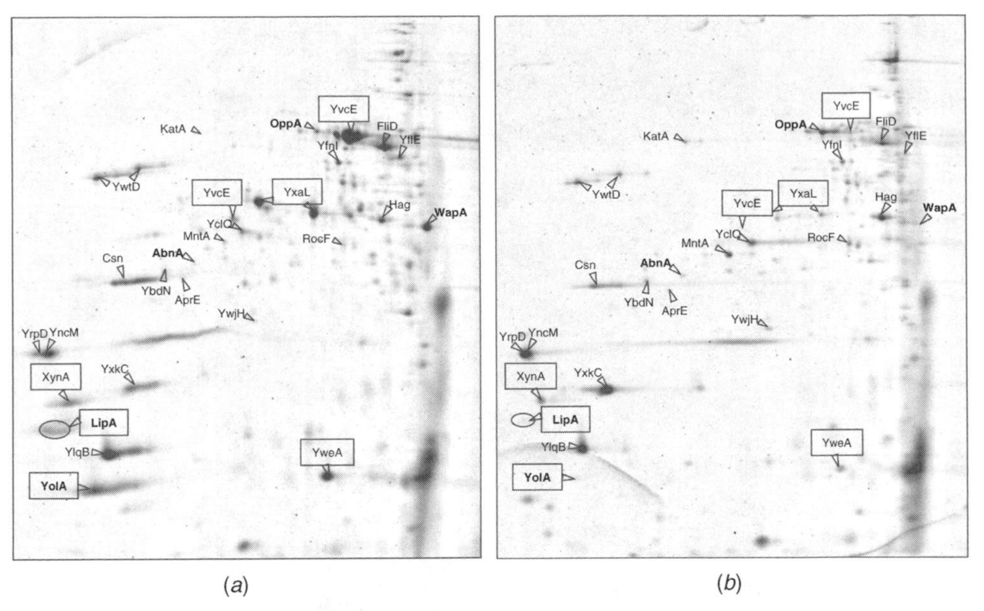


**Figure 12.6** Effects of SecA inhibition by sodium azide on the secretion of de novo synthesized extracellular proteins of *B. subtilis* 168. Cells of *B. subtilis* grown into the stationary phase were washed twice in prewarmed medium and then resuspended in LB broth supplemented (*a*) without and (*b*) with 15 mM sodium azide. After 20 min of growth, the extracellular proteins were separated by 2D PAGE and stained with Sypro ruby. Proteins with predicted RR/KR signal petides are printed in bold and proteins present at decreased amounts after sodium azide addition are boxed.

the depletion for SecA in a temperature-sensitive mutant strain is more effective in reducing SecA activity than sodium azide (Tjalsma et al., 2004).

***Tat-Dependent Secretion*** The presence of twin arginine (RR/KR) signal peptides is an indication that these proteins could be directed into the alternative Tat pathway for protein export, possibly in a Sec-independent manner. The Tat pathway was first discovered in chloroplasts in which it is required for the ΔpH-dependent protein import into the thylakoid lymen (van Dijl et al., 2001). In gram-negative bacteria, such as *E. coli*, the Tat pathway is used mainly for exporting tightly folded proteins, such as metalloproteins that acquire their cofactors and final confirmation in the cytoplasm. *Bacillus subtilis* contains two *tatC* genes denoted *tatCd* and *tatCy*, each of which is preceded by one *tatA* gene and denoted *tatAd* and *tatAy* (Jongbloed et al., 2000). To identify proteins that are secreted via the Tat pathway of *B. subtilis*, Jongbloed and co-workers analyzed the extracellular proteomes of a *tatCd tatCy* double mutant strain and a *total–tat* mutant strain that lacks all known *tat* genes (Jongbloed et al., 2000, 2002). For this purpose cells were grown in complete medium as well under the conditions of phosphate starvation (Fig. 12.7). Extracellular proteome comparison revealed that only the alkaline phosphatase/phosphodiesterase PhoD, which is strongly induced in a PhoPR-dependent manner after phosphate starvation, was secreted in a strictly Tat-dependent manner (Jongbloed et al., 2000, 2002). Interestingly, the genes *tatAd–tatCd* are preceeded by the *phoD* gene and cotranscribed with the *phoD* gene in response to phosphate

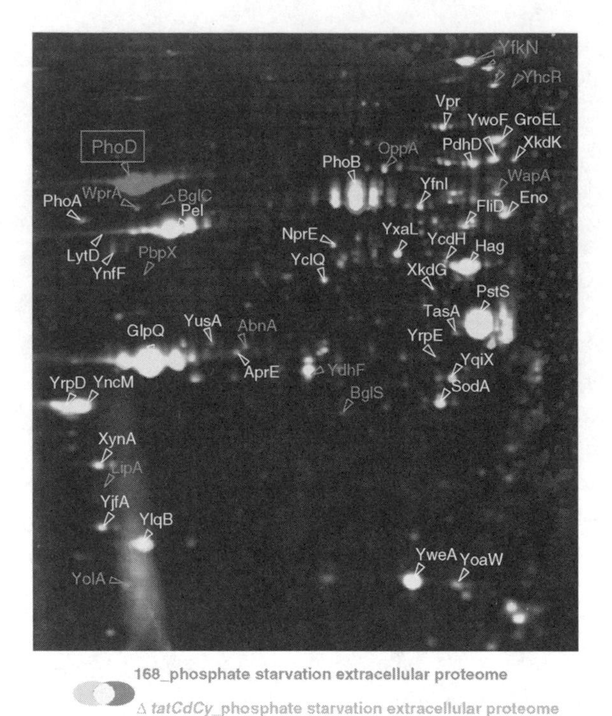


**Figure 12.7** Phosphate starvation proteome of wild type (red image) in comparison to ΔtatCd/ΔtatCy mutant (green image). Cells of *B. subtilis* 168 and the *ΔtatCd/ΔtatCy* mutant were grown in minimal medium under the conditions of phosphate starvation and proteins in the growth medium were harvested 1 h after entry into the stationary phase. After precipitation with TCA, the extracellular proteins were separated by 2D PAGE and stained with Sypro ruby. Image analysis was performed using the Decodon Delta 2D software that is based on the dual-channel imaging (Bernhardt et al., 1999). Proteins with predicted RR/KR signal petides are labeled in blue and the Tat-dependent PhoD protein is boxed.

starvation. Since only TatAdCd, but not TatAyCy, is specifically required for secretion of PhoD after phosphate starvation, it seems that PhoD has a specific coregulated Tat-dependent transport system (van Dijl et al., 2001).

In contrast to PhoD, the secretion of other proteins with predicted RR/KR signal peptides, including LipA, PbpX, WprA, WapA, YdhF, YfkN, YhcR AbnA, BglC, BglS, LytD, OppA, and YolA, was not affected by the *tat* mutations. Thus, the Tat pathway makes a highly selective contribution to the extracellular proteome of *B. subtilis* (Jongbloed et al., 2002).

### 12.2.6 Processing and Modification of Secreted Proteins

***Processing by Type I SPases*** In *B. subtilis,* five chromosomally located *sip* genes code for the type I SPases: SipS, SipT, SipU, SipV, and SipW, of which only SipS and SipT are of major importance for preprotein processing and cell viability (Tjalsma et al., 2000). The extracellular proteome analysis is a powerful tool to study the different substrate specificities of signal peptidases in *B. subtilis*. However, the extracellular proteome comparison between the wild-type and single, double, triple, or quadruple SPase I mutants lacking *sipS*, *sipT*, *sipU*, *sipV*, or *sipW* showed no major differences. The only notable observation was that SipT and SipV were required for cleavage of the membrane protein YfnI (Antelmann et al., 2001; Tjalsma et al., 2004). This confirms the view that the presence of either SipS or SipT is sufficient for efficient precursor processing and that the type I SPases have overlapping substrate specificities in *B. subtilis,* as has been suggested previously (Tjalsma et al., 1998). Recently, different roles have been reported for the three type I SPases SipX, SipY, and SipZ in *Listeria monocytogenes* (Bonnemain et al., 2004). It has been found that SipX and SipZ are specifically involved in the virulence of *L. monocytogenes,* suggesting a possible post-translational control mechanism in extracellular virulence factor expression.

***Processing by SPase II*** The unique type II SPase LspA is required for the processing of lipid-modified lipoproteins. Thus, it was surprising that in the extracellular proteome of the *lspA* mutant, the majority of proteins with SPase I cleavage sites were present in reduced amounts (Antelmann et al., 2001; Tjalsma et al., 2004). Unexpected, increased levels were detected for both WprA processing products (CWBP23 and CWBP52). In addition, two lipoproteins (MntA and YxeB) were also increased in the extracellular proteome of the *lspA* mutant. In contrast, the lipoproteins OppA and YclQ were not affected by the *lspA* mutation. This suggests that the absence of LspA causes pleiotropic effects on the extracellular proteome in the case of lipoprotein processing.

***Lipid Modification by Diacylglyceryl Transferase*** The diacylglyceryl transferase Lgt of *B. subtilis* is responsible for the diacylglyceryl modification of the N-terminal cysteine residue of mature lipoproteins prior to the processing of lipoprotein precursors by SPase II (Leskelä et al., 1999). To further explore the factors required for lipoprotein processing and retention in the cell, the composition of the extracellular proteome of a *lgt* mutant, which is defective in the lipid modification of lipoproteins, was analyzed (Antelmann et al., 2001; Tjalsma et al., 2004). As shown in Figure 12.8*a* and quantified in Table 12.3, the amounts of 13 lipoproteins were significantly increased in the extracellular proteome of the *lgt* mutant in complete medium. Of these the lipoproteins MntA, OppA, YxeB, YclQ, and YusA were among the most strongly released lipoproteins (Table 12.3). In addition, the extracellular levels of the autolytic enzymes YvcE, LytD, XepA, and XlyA were increased by the *lgt* mutation. These observations show that cells lacking the diacylglyceryl transferase shed lipoproteins and autolysins into their growth medium. Because all of these identified lipoproteins are lacking the N-terminal cysteine residue, it seems that (pre-)lipoproteins are released into the medium after alternative proteolytic processing (Antelmann et al., 2001; Tjalsma et al., 2004). If we compare the "lipoproteome" of the wild type with the extracellular proteome of the *lgt* mutant, it is interesting to note that 9 of 10 membrane-extracted lipoproteins of the wild-type are released in high amounts by the *lgt* mutant. This is further shown by the quantification of the most abundant lipoproteins released in complete medium from wild-type cells (MntA, OppA, YusA, or YxeB) that are present in similar amounts in the membrane fraction of the wild type (Table 12.3). To verify if nonmodified lipoproteins can also be detected in the membrane fraction of the *lgt* mutant, the lipoproteome of the wild type was compared with that of the *lgt* mutant (Fig. 12.8*b*). These analyses clearly demonstrate the absence of nine lipoproteins from the membrane fraction of the *lgt* mutant, which further shows that a

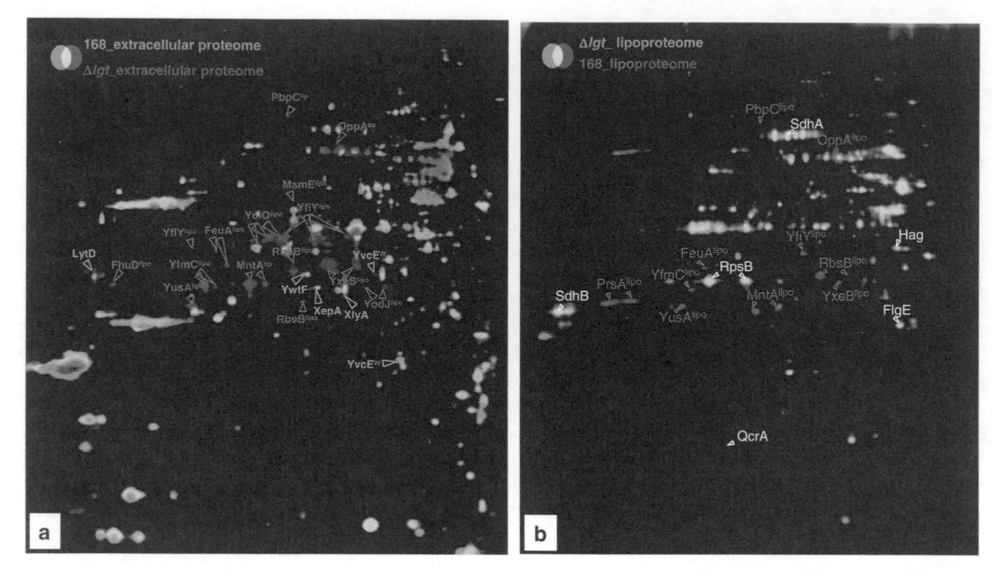


**Figure 12.8** (*a*) Extracellular proteome and (*b*) membrane-attached "lipoproteome" of diacylglyceryl transferase mutant (red image, *a*; green image, *b*) in comparison to that of wild type (red image, *b*; green image, *a*). Cells of *B. subtilis* Δ*lgt*, which lack the diacylglyceryl transferase, were grown in LB broth and harvested 1 h after entry into the stationary phase. After precipitation with TCA, the extracellular proteins were separated by 2D PAGE and stained with Sypro ruby (*a*). The NDSB-washed membrane fraction was obtained after stepwise extraction with β-D-maltoside (*b*; corresponding to fraction 4 according to Bunai et al., 2004). Image analysis was performed using the Decodon Delta 2D software that is based on dual-channel imaging (Bernhardt et al., 1999). Nonlipoproteins that are present at elevated levels in the medium of the Δ*lgt* strain are indicated by blue labels; lipoproteins that are released into the medium of the Δ*lgt* strain (*a*) and absent from the membrane fraction (*b*) are indicated by red labels and marked with lipo.

redistribution of lipoproteins from the membrane to the extracellular medium occurs in the *lgt* mutant.

### 12.2.7 PrsA-Mediated Folding of Extracellular Proteins

The most important known folding catalyst involved in protein secretion is the lipoprotein PrsA, which shows similarity to peptidyl-prolyl cis/trans isomerases and is essential for protein secretion and cell viability of *B. subtilis* (Kontinen et al., 1991; Kontinen and Sarvas, 1993; Jacobs et al., 1993). Therefore, the effects of PrsA depletion on the composition of the extracellular proteome was studied using a conditional IPTG-dependent *prsA* mutant strain grown in the presence or absence of IPTG (Fig. 12.9) (Vitikainen et al., 2004). The first effect that could be observed was that the mutant was very sensitive to cell lysis in the absence of IPTG, as concluded from the finding that the amounts of cytoplasmic proteins were strongly increased in the extracellular proteome. In addition, the amounts of 29 predicted secreted proteins that are synthesized with SPaseI-cleavable signal peptides and three unpredicted phage- and flagella-related proteins were strongly reduced in the absence of PrsA in the extracellular proteome. This

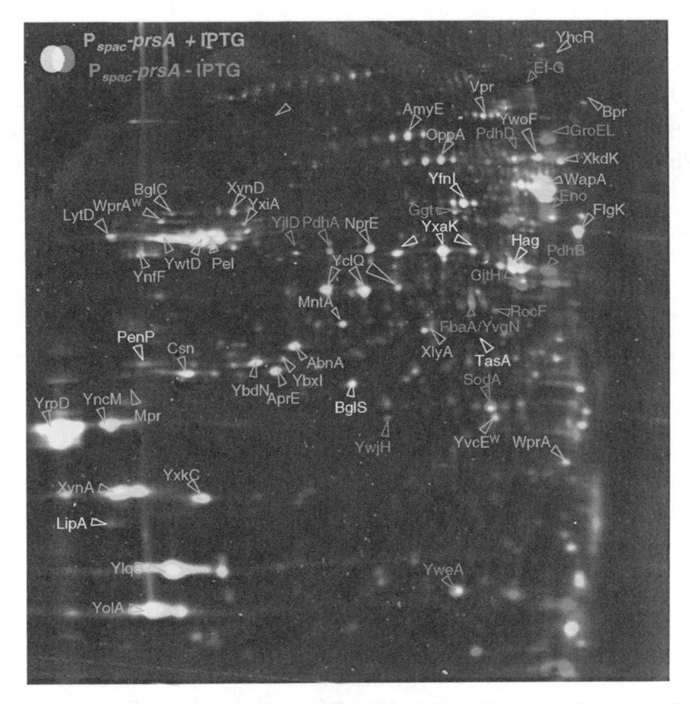


**Figure 12.9** Extracellular proteome of *B. subtilis* IH7211 ($P_{spac}$-*prsA*) grown in absence (red image) or presence (green image) of 1 mM IPTG. Cells were harvested 1 h after the entry into the stationary phase and extracellular proteins in the culture medium were TCA precipitated, separated by 2D PAGE, and stained with Sypro ruby. Image analysis was performed using the Decodon Delta 2D software that is based on the dual-channel imaging (Bernhardt et al., 1999). Proteins that appeared as green spots (higher level in the presence of PrsA) are marked in blue and those that appeared as red spots (higher level in the absence of PrsA) are marked in red. Proteins that appeared as yellow spots are labeled in white (unchanged according to the quantitation data). (See color insert.)

indicated that PrsA is important for the general folding of secretory proteins. Interestingly, there was a specific subset of eight predicted secreted proteins that were not affected after depletion for PrsA. This might indicate that some secreted proteins require catalysts other than PrsA for their folding (Tjalsma et al., 2004).

### 12.2.8 Quality Control Factors

Secretory proteins that are not correctly folded by folding catalysts, such as PrsA, are substrates for one of the 27 proteases that are predicted to be present in the membrane, cell wall, or extracellular medium (Tjalsma et al., 2000). Quality control factors like the HtrA-like proteases/chaperones can either assist in proper folding or degrade the malfolded secretory proteins. The various extracellular proteases were also shown to be responsible for the degradation of heterologous secreted proteins even if such proteins are properly folded (Wu et al., 1991). To identify the natural substrates of HtrA-like and extracellular proteases, we have investigated the extracellular proteome in response to secretion stress that induces the HtrA proteases as well as the extracellular proteome of a multiple exoprotease-deficient strain.

***Secretion Stress Control System*** The HtrA-like proteases HtrA and HtrB are induced in response to secretion stress conditions, which are sensed by the CssRS two-component system with CssS as a membrane-localized sensor kinase and CssR as a response regulator (Hyyrylainen et al., 2001; Darmon et al., 2002). The mechanism that activates CssR in response to secretion stress is unknown. The extracellular proteome was studied after the exposure of cells to secretion stress which was provoked by over-production of the heterologous secretory protein AmyQ as well as in a *htrB* mutant which over-produces HtrA due to reciprocal cross-regulation (Noone et al., 2001; Antelmann et al., 2003). These analyses showed that, in response to secretion stress, only two proteins were induced in the extracellular proteome in a parallel manner, namely HtrA and YqxI. Whereas the induction of HtrA was mediated by the increased CssRS-dependent transcription of the *htrA* gene, the transcription of *yqxI* was not increased by secretion stress. Thus, the induction of YqxI is controlled at the posttranscriptional level. Notably, HtrA is specifically required for the stabilization of YqxI. Since the protease active site of HtrA is dispensable for the stabilization of YqxI, a chaperone-like activity of HtrA could be involved in the appearance of YqxI in the extracellular proteome.

***Role of Extracellular Proteases*** Because some cell wall proteins, such as the large WapA-processing products and YvcE, seem to be degraded during the stationary phase, it might be possible that these are substrates of the extracellular proteases of *B. subtilis*. Therefore, we have analyzed the extracellular proteome of a mutant lacking seven extracellular proteases (AprE, Bpr, Epr, Mpr, NprB, NprE, Vpr) (Wu et al., 1991; Ye et al., 1996; Antelmann et al., 2002). As shown in Figure 12.10*a*, the large WapA-processing products and the unprocessed YvcE protein were increased in the extracellular proteome of the protease mutant during the stationary phase. These results indicate that the released forms of the cell-wall-associated proteins WapA and YvcE are indeed substrates for the extracellular proteases. This mechanism might enable *B. subtilis* to recycle spoiled cell wall proteins that would otherwise be lost as "proteinacious waste" upon cell wall turnover. To verify whether extracellular proteases are also involved in the processing of cell-wall-bound WapA, the cell wall proteome of the protease mutant was analyzed (Fig. 12.11*a*). Interestingly, the level of the large WapA-processing product corresponding to CWBP105 was significantly increased and the amount of the smaller WapA-processing product corresponding to CWBP62 was decreased in the cell wall proteome of the protease mutant. This indicates that extracellular proteases are indeed involved in the specific processing of wall-bound WapA. Specifically, the extracellular and cell wall proteome analysis of an *epr* mutant suggests that the serine protease Epr might be the candidate protease responsible for the WapA processing in the cell wall and the degradation of the secreted large WapA-processing products. Interestingly, besides the unprocessed YvcE and the large WapA precursor, the level of HtrA and YqxI were also increased in the extracellular proteome of the protease mutant (Fig. 12.10*a*). Consequently, both of these proteins are subject to proteolysis upon their export from the cytoplasm.

***Role of SigmaD*** In addition to the extracellular proteases, the alternative motility sigma factor *sigD* that also regulates many cell wall proteins seems to be involved in the quality control of cell wall proteins. This can be concluded from the observation that an

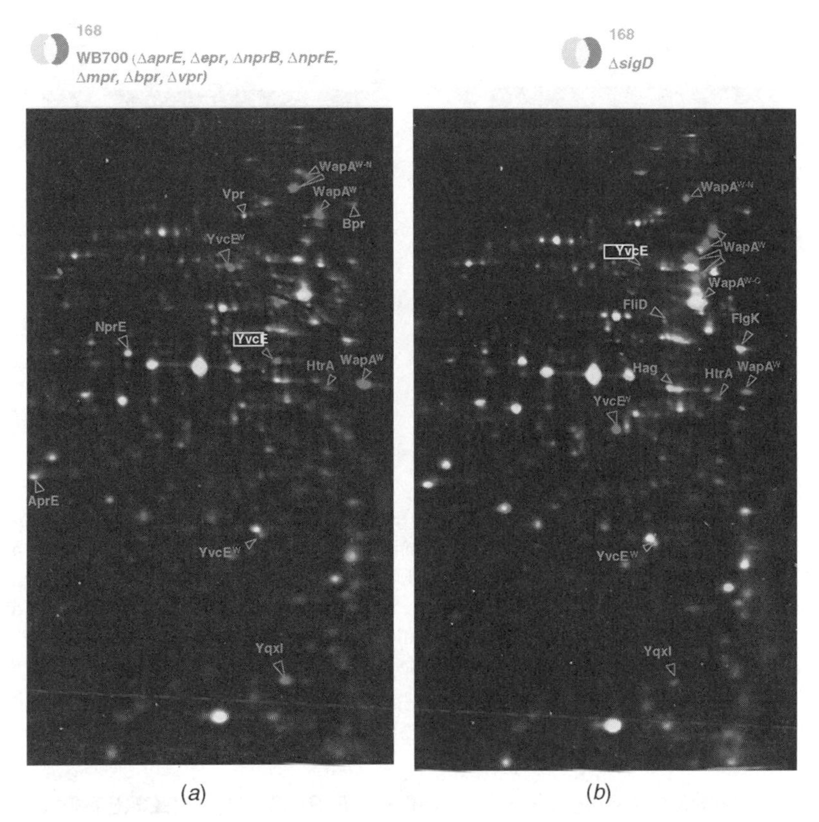


**Figure 12.10** Extracellular proteome of multiple extracellular protease-deficient *B. subtilis* strain WB700 (*a*, red image) and *sigD* mutant (*b*, red image) in comparison to that of wild type (green image). Extracellular proteins were separated by 2D PAGE and stained with Sypro ruby. Image analysis was performed using the Decodon Delta 2D software that is based on dual-channel imaging (Bernhardt et al., 1999). Secreted cell wall proteins, including the large WapA-processing products, unprocessed YvcE, HtrA, and YqxI that are stabilized in the exoprotease, and *sigD* mutants during the stationary-phase mutant, are edited in red and those that are missing in the mutants are labeled in blue. (See color insert.)

artificial cell-wall-binding lipase was stabilized in a *sigD* mutant (Kobayashi et al., 2000). Therefore, the extracellular and cell wall proteomes of a *sigD* mutant were compared to that of the wild type (Figs. 12.10*b* and 12.11*b*; Antelmann et al., 2002). Interestingly, these proteome analyses showed similar results obtained after depletion for extracellular proteases. The amounts of the secreted large WapA-processing products, the unprocessed YvcE, HtrA, and YqxI, were increased in the extracellular proteome. In addition, the cell-wall-bound large form of WapA (CWBP105) was increased and the small form of WapA (CWBP62) was decreased in the cell wall proteome of the *sigD* mutant, which indicates that SigmaD regulates the expression of proteins that are involved directly or indirectly in the processing of WapA and some secreted cell wall proteins (Fig. 12.11*b*). It might be possible that the cell wall turnover is impaired in the absence of the *sigD*-dependent autolysins, which might have an influence on the stability and processing of cell wall proteins.

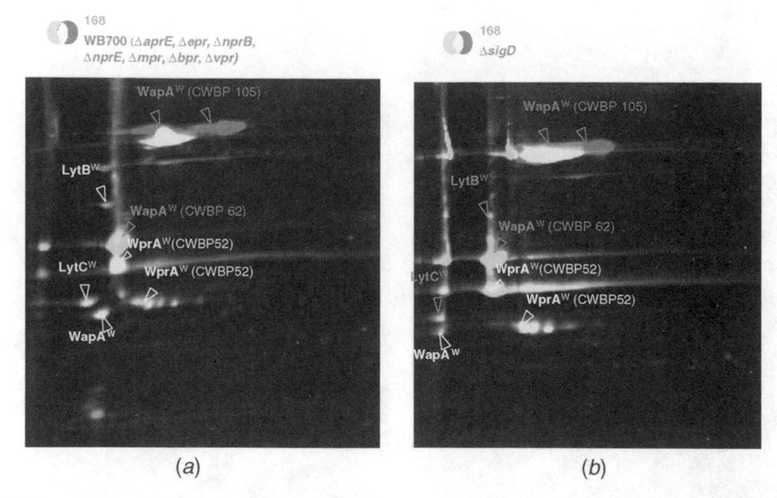


**Figure 12.11** Cell wall proteome of multiple extracellular protease-deficient *B. subtilis* strain WB700 (red image, *a*) and *sigD* mutant (red image, *b*) in comparison to that of wild type (green image). Cell wall proteins were extracted using a LiCl extraction procedure, precipitated with TCA, separated by 2D PAGE, and stained with Sypro ruby. Image analysis was performed using the Decodon Delta 2D software that is based on dual-channel imaging (Bernhardt et al., 1999). The WapA-processing product CWBP105 is increased (labeled red) and WapA-processing product CWBP62 is decreased (labeled blue) in the protease mutant as well as the *sigD* mutant. (See color insert.)

## 12.3 PROTEOMICS AND PHYSIOLOGY OF PROTEIN SECRETION: ROLE OF GLOBAL REGULATORS (PhoPR, DegSU, ScoC, SigD)

Extracellular proteins are mainly involved in the provision of nutrients such as carbohydrates, amino acids, nucleotides, or phosphate. Thus, the main function is to provide alternative growth substrates when the preferred substrate is no longer available in sufficient amounts. This is the physiological role of many extracellular hydrolases such as α-amylase, β-glucanases, xylanases, various proteases, nucleotidases, lipases, or phosphatases. Therefore, it is not surprising that the majority of extracellular proteins are secreted at higher amounts during the stationary phase of growth (Antelmann et al., 2001, 2002).

The kind of nutrient limitation determines the synthesis and secretion of specific degradative extracellular enzymes under nongrowing conditions. This is reflected by the differences in the extracellular proteomes resulting from stationary-phase cells grown under the conditions of phosphate starvation or in LB medium (Figs. 12.2*a,b*; Section 12.2.1). For example, the extracellular phosphatases that are secreted specifically under the conditions of phosphate starvation provide alternative phosphate sources when inorganic phosphate has been exhausted. Another protein that is strongly induced is the lipid-anchored phosphate-binding protein PstS that is part of a high-affinity phosphate uptake system.

Many genes encoding extracellular proteins belong to global regulons. A regulon is a group of genes distributed on the chromosome but controlled by a unique regulator (repressor, activator, alternative sigma factor, etc.). The mechanisms that keep the genes encoding extracellular enzymes silent in growing cells are at least in part known. In glucose-starved cells, genes encoding α-amylases or β-glucanases are derepressed because the global repressor CcpA is no longer active (Henkin et al., 1991; Moreno et al., 2001;

Yoshida et al., 2001). In phosphate-starved cells, the response regulator PhoP is activated by phosphorylation, which causes the induction of transcription of the entire *pho* regulon (Hulett, 1996).

Proteomics is also useful in arranging the extracellular proteins into different regulons. For that reason, the extracellular proteome profile of wild-type cells should be compared with that of mutants in global regulatory genes. Proteins that displayed a change in their abundance in the absence of the regulator might be members of this regulon. These differences in protein amounts between the wild type and mutant can be quantified using the Decodon Delta 2D software, which is based on dual-channel image analysis (Bernhardt et al., 1999). It has to be emphasized, however, that DNA array technologies will step-by-step replace proteomics when the size and structure of regulons have to be defined because only a subfraction of the regulon can be detected using the proteomic approach (Hecker, 2003). In contrast, the transcriptome technique can never replace the proteomic approach in the analysis of protein secretion mechanisms for extracellular proteins, as has been demonstrated in this chapter.

Besides the extracellular PhoPR regulon, the extracellular DegSU (Fig. 12.12) and ScoC regulons were also analyzed using proteomics. Both systems are involved in the regulation of pleiotropic processes occurring during the transition into the stationary phase that include the synthesis of extracellular degradative enzymes, competence, and motility. DegSU is a two-component system with DegS acting as a sensor kinase and DegU as a response regulator. More than 20 years ago, several point mutants in *degU*

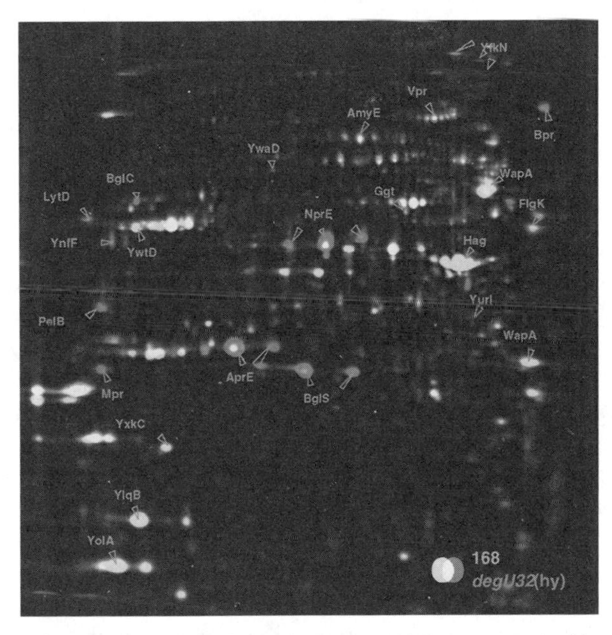


**Figure 12.12** Extracellular proteome of *B. subtilis degU32(hy)* mutant (red image) in comparison to that of wild type (green image). Extracellular proteins were separated by 2D PAGE and stained with Sypro ruby. Image analysis was performed using the Decodon Delta 2D software that is based on dual-channel imaging (Bernhardt et al., 1999). Extracellular proteins secreted at higher amounts in the *degU32(hy)* mutant are labeled in red, and those which are repressed in the mutant are labeled in blue. (See color insert.)

and *scoC* were isolated that caused highly pleiotropic phenotypes, namely, while secreted enzymes are overproduced, competence development and motility are repressed (Henner et al., 1988; Kunst et al., 1988; Msadek et al., 1995). To analyze the extracellular DegSU and ScoC regulons two of these pleiotropic mutants [*degU32(hy)* and *scoC4*] were subjected to proteome analysis. The *degU32(hy)* mutation is characterized by a hyperphosphorylated and thereby hyperactive response regulator DegU. This stable Asp-56 phosphorylation of DegU in the *degU32(hy)* strain could be monitored in the cytoplasmic proteome as an acid-shifted DegU~P spot (Antelmann et al., 2004). The extracellular proteins overproduced in this *degu32(hy)* mutant are probably under positive DegU control. As shown in Figure 12.12, several degradative enzymes were up regulated in the mutant, including seven proteases, the α-amylase AmyE, the β-glucanase BglS, the xylanase YnfF, the pectate lyase PelB, and the nucleotidases YfkN and YurI (Fig. 12.12) (Antelmann et al., 2001; Mäder et al., 2003). As expected, the alkaline protease AprE represented the most abundant protease secreted into the medium. Most of these degradative enzymes were also up regulated in the pleiotropic *scoC4* mutant (Antelmann et al., 2004).

In addition, the levels of eight extracellular proteins involved in motility and chemotaxis were found to be strongly reduced in the *degU32(hy)* mutant, including the cell wall proteins LytD, YwtD, and WapA; the pectate lyase Pel; the flagella-related proteins Hag, FlgK, and FliD; and the unknown proteins YolA, YlqB, and YxkC (Antelmann et al., 2001; Mäder et al., 2003). (Fig. 12.12). These results confirm the previous finding that motility is repressed by DegU. Most of these down-regulated proteins are controlled by the alternative motility sigma factor SigmaD, which was confirmed by the extracellular proteome analysis of the *sigD* mutant. Notably, SigmaD was listed as a quality control factor under Role of SigmaD in Section 12.2.8., because cell wall proteins are stabilized in a *sigD* mutant (Figs. 12.10*b* and 12.11*b*). In addition, the *sigD* mutant showed that all proteins that are repressed in the *degU32(hy)* strain were also absent in the extracellular proteome of the *sigD* mutant. These include the above-mentioned flagella-related proteins, the cell wall proteins, and the proteins of unknown function YolA, YlqB, and YxkC (Fig. 12.10*b*), all of which are members of the SigmaD regulon.

Whereas the up-regulation of degradative enzymes was comparable between both the *degU32(hy)* and the *scoC4* mutant, differences could be observed for the down-regulated proteins. Specifically, repression of Hag, Pel, YlqB, and YolA was detected only in the *degU32(hy)* strain but not in the *scoC4* mutant (Antelmann et al., 2004). This is in contrast to the microarray data for the *scoC4* mutant, which show, for instance, repression of motility genes, including *hag* (Caldwell et al., 2001). These differences in the extracellular proteome profiles between both pleiotropic mutants can be used as proteome signatures for *degU* and *scoC* mutations.

Because extracellular proteins are mainly synthesized during the transition into the stationary phase, it would be interesting to know whether the main components of the secretion machinery are regulated in a similar way. At least one component of the Tat pathway (*tatAd–tatCd*) that is responsible for PhoD secretion is only expressed in response to phosphate starvation and cotranscribed together with the *phoD* gene. To our surprise, the expression of most genes encoding components of the Sec pathway is differently regulated. In contrast to the SecYEG translocase, which seems to be constitutively expressed during the growth in complete medium, the transcription of *secA* is induced upon transition into the stationary phase as revealed by DNA array analyses (T. Koburger, unpublished data). In addition, the transcription of the genes for the signal peptidases SipS and SipT is strongly increased in the postexponential growth phase and both are also controlled by the DegSU system (van Dijl et al., 2001). Finally, the transcription of *sipW* depends on the transition-phase regulators Spo0A and Spo0H (van Dijl et al., 2001).

Despite the fact that many proteins were identified in the extracellular proteome, there are many more proteins predicted to be secreted that remain to be identified. Even if there are probably a few low-abundance proteins secreted but not yet identified, this will not solve the problem. The question is whether environmental conditions exist in the natural ecosystems of *B. subtilis* that require the expression of specific extracellular proteins that are not needed by the cell under the conditions of artificial laboratory studies. Such genes might be silent under laboratory conditions because specific environmental stimuli are necessary for their expression. For example, growth in the neighborhood of plant roots, growth in the form of microcolonies, or even biofilms could provide specific stimuli for gene expression. Further studies are necessary to address this crucial physiological problem.

## 12.4 OUTLOOK

Proteomics offers new opportunities to observe events in the cell never seen before by looking at cells in a new and wider context. In starved cells, the secretome visualized by the proteome approach represents a considerable proportion of the entire proteome. This is an important finding that has to be considered on the way toward the depiction of the entire proteome. The proteomic view of genome-based predictions of extracellular proteins is an excellent example of how to bring the genome sequence to real cell physiology (Antelmann et al., 2001).

Notably, the combination of the panoramic view of proteomics with the molecular genetics of protein secretion led to many new insights as to the protein secretion mechanisms as summarized here. This "advanced proteomic approach" shows that proteomics not only should provide information on the proteins expressed in a cell under defined conditions but also can be used as a starting point to go beyond the descriptive level and to come to a fuller understanding of specific pathways and mechanisms. Furthermore, by analyzing mutants in global regulators or by following the protein secretion pattern in growing and nongrowing cells, new results on the physiology of protein secretion can be obtained.

Last, but not least, the results of proteomic studies on protein secretion are also of key importance from a biotechnological point of view. Species of the genus *Bacillus* are important producers of industrial enzymes. The extracellular protein profile not only provides an overview of the level of proteins secreted into the medium but also tells us if there is a limitation in the growth process or if there is an exhaustion of the secretion machinery. For example, the signature for secretion stress may serve as a reliable indicator for the production of secretory proteins at high levels. Thus, the extracellular protein profile is likely to become a very useful tool for studies aimed at the optimization of protein secretion at an industrial scale.

In conclusion, the proteomics approach has provided completely new fundamental scientific and applied views on the exoproteome of *B. subtilis* by expanding our understanding of the flow of proteins from the cytoplasm to the cell wall membrane or growth medium from the single-protein level to that of the combined secretome level.

## ACKNOWLEDGMENTS

We thank Karin Binder, Sebastian Grund, and Doreen Kliewe for expert technical assistance; Dirk Albrecht for the MS/MS analysis of some extracellular proteins; and the members of the European *Bacillus* Secretion Groups (see: http://www.ncl.ac.uk/ebsg) for stimulating discussions. This work was supported in part by Quality of Life and Management of Living resources Grants QLK3-CT-1999-00413, QLK3-CT-1999-00917, LSHC-CT-2004-503468, and LSHC-CT-2004-5257 from the European Union (to H. A., J. M. v. D., S. B., and M. H.), grants from the Deutsche Forschungsgemeinschaft (DFG), the Bundesministerium für Bildung, Wissenschaft, Forschung und Technologie (BMFT), and the Fonds der Chemischen Industrie (to M. H.), and Genencor International (Palo Alto, California).

## REFERENCES

Antelmann, H., Darmon, E., Noone, D., Veening, J. W., Westers, H., Bron, S., Kuipers, O. P., Devine, K. M., Hecker, M., and van Dijl, J. M. (2003). *Mol. Microbiol.* **49**:143–156.

Antelmann, H., Sapolsky, R., Miller, B., Ferrari, E., Chotani, G., Weyler, W., Gaertner A., and Hecker M. (2004). *Proteomics* **4**:2408–2424.

Antelmann, H., Scharf, C., and Hecker, M. (2000). *J. Bacteriol.* **182**:4478–4490.

Antelmann, H., Tjalsma, H., Voigt, B., Ohlmeier, S., Bron, S., van Dijl, J. M., and Hecker, M. (2001). *Genome Res.* **11**:14984–14502.

Antelmann, H., Yamamoto, H., Sekiguchi, J., and Hecker, M. (2002). *Proteomics* **2**:591–602.

Bernhardt, J., Büttner, K., Scharf, C., and Hecker, M. (1999). *Electrophoresis* **20**:2225–2240.

Binnie, C., Lampe, M., and Losick, R. (1986). *Proc. Natl. Acad. Sci. USA* **83**:5943–5947.

Blocker, A., Komoriya, K., and Aizawa, S. I. (2003). *Proc. Natl. Acad. Sci. USA* **100**:3027–3030.

Bonnemain, C., Raynaud, C., Reglier-Poupet, H., Dubail, I., Frehel, C., Lety, M.-A., Berche, P., and Charbit, A. (2004). *Mol. Microbiol.* **51**:1251–1266.

Bunai, K., Ariga, M., Inoue, T., Nozaki, M., Ogane, S., Kakeshita, H., Nemoto, T., Nakanishi, H., and Yamane, K. (2004). *Electrophoresis* **25**:141–155.

Büttner, K., Bernhardt, J., Scharf, C., Schmid, R., Mäder, U., Eymann, C., Antelmann, H., Völker, A., Völker, U., and Hecker, M. (2001). *Electrophoresis* **22**:2908–2935.

Caldwell, R., Sapolsky, R., Weyler, W., Maile, R. R., Causey, S. C., and Ferrari, E. (2001). *J. Bacteriol.* **183**:7329–7340.

Darmon, E., Noone, D., Masson, A., Bron, S., Kuipers, O. P., Devine, K. M., and van Dijl, J. M. (2002). *J. Bacteriol.* **184**:5661–5671.

Duncan, M. L., Kalman, S. S., Thomas, S. M., and Price, C. W. (1987). *J. Bateriol.* **169**:771–778.

Eymann, C., Dreisbach, A., Albrecht, D., Bernhardt, J., Becher, D., Gentner, S., Tam, L. T., Büttner, K., Buurmann, G., Scharf, C., Venz, S., Völker, U., and Hecker, M. (2004). *Proteomics* **4**:2849–2876.

Ghuysen, J. M., Lamotte-Brasseur, J., Joris, B., and Shockman, G. D. (1994). *FEBS Lett.* **342**:23–28.

Haldenwang, W. G., and Losick, R. (1979). *Nature* **282**:256–260.

Hecker, M., Heim, C., Völker, U., and Wölfel, L. (1988). *Arch. Microbiol.* **150**:564–566.

Hecker, M., and Völker, U. (1998). *Mol. Microbiol.* **29**:1129–1136.

Hecker, M., and Völker, U. (2001). *Adv. Microb. Physiol.* **44**:35–91.

Hecker, M. (2003). *Adv. Biochem. Eng./Biotechnol.* **83**:57–92.

Henner, D. J., Yang, M., and Ferrari, E. (1988). *J. Bacteriol.* **170**:5102–5109.

Henkin, T. M., Grundy, F. J. Nicholson, W. L. and Chambliss, G. H. (1991). *Mol. Microbiol.* **5**:575–584.

Hirose, I., Sano, K., Shiosa, I., Kumano, M., Nakamura, K., and Yamane K. (2000). *Microbiology* **146**:65–75.

Honda, K., Nakamura, K., Nishiguchi, M., and Yamane K. (1993). *J. Bacteriol.* **175**:4885–4894.

Hueck, C. J. (1998). *Microbiol. Mol. Biol. Rev.* **62**:379–433.

Hulett, F. M. (1996). *Mol. Microbiol.* **19**:933–939.

Hyyrylainen, H. L., Bolhuis, A., Darmon, E., Muukkonen, L., Koski, P., Vitikainen, M., Sarvas, M., Pragai, Z., Bron, S., van Dijl, J. M., and Kontinen, V. P. (2001). *Mol. Microbiol.* **41**:1159–1172.

Igo, M., Lampe, M., Ray, C., Schafer, W., Moran Jr., C. P., and Losick, R. (1987). *J. Bacteriol.* **169**:3464–3469.

Jacobs, M., Andersen, J. B., Kontinen, V. P., and Sarvas, M. (1993). *Mol. Microbiol.* **8**:957–966.

Jongbloed, J. D. H., Antelmann, H., Hecker, M., Nijland, R., Pries, F., Koski, P., Quax, W. J., Bron, S., van Dijl, J. M., and Braun P. G. (2002). *J. Biol. Chem.* **277**:44068–44078.

Jongbloed, J. D.H., Martin, U., Antelmann, H., Hecker, M., Tjalsma, H., Venema, G., Bron, S., van Dijl, J. M., and Müller J. (2000). *J. Biol. Chem.* **275**:41350–41357.

Klose, J. (1975). *Humangenetik* **26**:231–243.

Kobayashi, K., Ehrlich, S. D., Albertini, A., Amati, G., Andersen, K. K., Arnaud, M., Asai, K., Ashikaga, S., Aymerich, S., Bessieres, P., et al. (2003). *Proc. Natl. Acad. Sci. USA* **100**:4678–4683.

Kobayashi, G., Toida, J., Akamatsu, T., Yamamoto, H., Shida, T., and Sekiguchi, J. (2000). *FEMS Microbiol. Lett.* **188**:165–169.

Kontinen, V. P., Saris, P., and Sarvas, M. (1991). *Mol. Microbiol.* **5**:1273–1283.

Kontinen, V. P., and Sarvas, M. (1993). *Mol. Microbiol.* **8**:727–737.

Krogh, S., Jorgensen, S. T., and Devine, K. M. (1998). *J. Bacteriol.* **180**:2110–2117.

Kunst, F., Debarbouille, M., Msadek, T., Young, M., Mauel, C., Karamata, D., Klier, A., Rapoport, G., and Dedonder R. (1988). *J. Bacteriol.* **170**:5093–5101.

Kunst, F., Ogasawara, N., Moszer, I., Albertini, A. M., Alloni, G., Azevedo, V., Bertero, M. G., Bessieres, P., Bolotin, A., Borchert, S., et al. (1997). *Nature* **390**:249–256.

Leskelä, S., Wahlström, E., Kontinen, V. P., and Sarvas, M. (1999). *Mol. Microbiol.* **31**:1075–1085.

Longchamp, P. F., Mauel, C., and Karamata, D. (1994). *Microbiology* **140**:1855–1867

Ludwig, H., Homuth, G., Schmalisch, M., Dyka, F. M., Hecker, M., and Stülke, J. (2001). *Mol Microbiol.* **41**:409.

Ludwig, H., Meinken, C., Matin, A., and Stülke, J. (2002). *J Bacteriol.* **184**:5174–5178.

Mäder, U., Antelmann, H., Buder, T., Dahl, M. K., Hecker, M., and Homuth, G. (2003). *Mol. Genet. Genomics* **268**:455–467.

Moreno, M. S., Schneider, B. L., Maile, R. R., Weyler, W., and Saier, M. H. Jr. (2001). *Mol Microbiol.* **39**:1366–1381.

Msadek, T., Kunst, F., and Rapoport, G. (1995). In J. A. Hoch, and T. J. Silhavy (Eds.), *Two-Component Signal Transduction.* Washington, DC:American Society for Microbiology, pp. 447–471.

Namba, K., Yamashita, I., and Vonderviszt, F. (1989). *Nature* **342**:648–654.

Neidhardt, F. C., and Van Bogelen, R. A. (2000). In G. Storz, and R. Hengge-Aronis, (Eds.), *Bacterial Stress Responses.* Washington DC:ASM Press, pp. 445–.

Noone, D., Howell, A., Collery, R., and Devine, K. M. (2001). *J. Bacteriol.* **183**:654–663.

O'Farrell, P. H. (1975). *J. Biol. Chem.* **250**:4007–4021.

Ogura, A., Kakeshita, H., Honda, K., Takamatsu, H., Nakamura, K., and Yamane, K. (1995). *DNA Res.* **2**:95–100.

Price, C. W., Fawcett, P., Ceremonie, H., Su, N., Murphy, C. K., and Youngman, P. (2001). *Mol. Microbiol.* **41**:757–774.

Rashid, M. H., Sato, N., and Sekiguchi, J. (1995). *FEMS Microbiol. Lett.* **132**:131–137.

Richter, A., and Hecker, M. (1986). *FEMS Microbiol. Lett* **36**:69–71.

Schumann, W., Ehrlich, S. D., Ogasawara, N. (Eds.) (2001). *Functional Analysis of Bacterial Genes: A Practical Manual.* Weinheim:Wiley.

Tjalsma, H., Bolhuis, A., van Roosmalen, M. L., Wiegert, T., Schumann, W., Broekhuizen, C. P., Quax, W., Venema, G., Bron, S., and van Dijl, J. M. (1998). *Genes Dev.* **12**:2318–2331.

Tobisch, S., Zühlke, D., Bernhardt, J., Stülke, J., and Hecker, M. (1999). *J. Bacteriol.* **181**: 6996–7004.

Tjalsma, H., Bolhuis, A., Jongbloed, J. D. H., Bron, S., and van Dijl J. M. (2000). *Microbiol. Mol. Biol. Rev.* **64**:515–547.

Tjalsma, H., Antelmann, H., Jongbloed, J. D., Braun, P. G., Darmon, E., Dorenbos, R., Dubois, J. Y., Westers, H., Zanen, G., Quax, W. J., Kuipers, O. P., Bron, S., Hecker, M., and van Dijl, J. M. (2004). *Microbiol. Mol. Biol. Rev.* **68**:207–233.

van Dijl, J. M., Bolhuis, A., Tjalsma, H., Jongbloed, J. D. H., de Jong, A., and Bron, S. (2001). In A. L. Sonenshein, J. A. Hoch, and R. Losick (Eds.), *Bacillus subtilis and Its Closest Relatives.* Washington DC: ASM Press, pp. 337–355.

Vitikainen, M., Lappalainen, I., Seppala, R., Antelmann, H., Boer, H., Taira, S., Savilahti, H., Hecker, M., Vihinen, M., Sarvas, M., and Kontinen, V. P. (2004). *J Biol Chem.* **279**:19302–19314.

Voigt, B., Schweder, T., Becher, D., Ehrenreich, A., Gottschalk, G., Feesche, J., Maurer, K. H., and Hecker, M. (2004). *Proteomics* **4**:1465–1490.

Völker, U., Engelmann, S., Maul, B., Riethdorf, S., Völker, A., Schmid, R., Mach, H., and Hecker, M. (1994). *Microbiology* **140**:741–752.

Wu, X. C., Lee, W., Tran, L., and Wong, S. L. (1991). *J. Bacteriol.* **173**:4952–4958.

Ye, R., Yang, L. P., and Wong, S. L. (1996). In *Proceedings of the International Symposium on Recent Advances in Bioindustry,* pp. 160–169.

Yoshida, K., Kobayashi, K., Miwa, Y., Kang, C.-M., Matsunaga, M., Yamaguchi, H., Tojo, S., Yamamoto, H., Nishi, R., Ogasawara, N., Nakayama, T., and Fujita, Y. (2001). *Nucleic Acids Res.* **29**:683–692.

PART IV

# PROTEOMICS OF PATHOGENIC MICROORGANISMS

CHAPTER 13

# Analyzing Bacterial Pathogenesis at Level of Proteome

PHILLIP CASH

University of Aberdeen, Foresterhill, Aberdeen, Scotland

## 13.1 INTRODUCTION

Understanding the mechanisms of microbial pathogenesis is fundamental to the investigation of infectious diseases, whether this is in humans or plants or animals, and represents a key step toward developing potential protective and therapeutic strategies with which to combat microbial pathogens. A key research question that has to be asked relates to the identity of the microbial components that contribute toward determining pathogenesis and the outcome of infection. However, this is by no means a trivial question due to the multifactorial nature of microbial pathogenesis in which multiple bacterial genetic activities may act as a complex network. Overlaying this is the differential expression of the bacterial genes during the course of the infection. The majority of microbial pathogens that have been studied in detail show differential patterns of gene expression when growing in association with their hosts compared to their growth outside of the host. This switch in the pattern of gene expression as the microbe adapts to the host environment often defines the key events in bacterial pathogenesis. As reviewed by Moxon and Tang [**1**], during their replication in the host the microbes have to survive many distinct environmental insults, for example extremes in nutrient levels and oxygen tension as well as, perhaps most importantly, the diverse defence mechanisms of the hosts immune system. The pathogen actually interacts with its host at many different levels, specifically the pathogen's tropism for specific tissue, the invasion of the host tissues, nutrient acquisition, and survival/avoidance of host clearance mechanisms [**1**]. These complex levels of interaction mean that it is not reasonable to analyze each individual microbial gene activity in isolation but rather one must consider a more holistic approach to define global changes in gene expression. This must then be followed by the careful dissection of the microbial activities to determine those responses due to the external environment and those that are specific to the process of pathogenesis. This chapter will consider the analysis of bacterial pathogenesis from the viewpoint of the activities of the pathogen, but one must also take into account the host response to provide a complete picture of the pathogenic process. One example of the complexity of the microbe–host interaction

*Microbial Proteomics: Functional Biology of Whole Organisms*, Edited by Ian Humphery-Smith and Michael Hecker. 

involves quorum sensing, which will be discussed further below. Quorum sensing is crucial in regulating gene expression for many bacterial pathogens and occurs through the production by the bacterium of small soluble signaling molecules. However, these signaling molecules not only regulate bacterial gene expression but can also trigger an inflammatory response by the host against the invading pathogen [**2**, **3**].

In recent years a range of analytical molecular methods have been used to investigate the basis of bacterial pathogenesis. Rapid progress in whole-genome sequencing of bacteria has provided a plethora of data on gene content and sequence as well as genome organization for many different bacteria, including some human pathogens. As of May 2004 the National Center for Biotechnology Information (NCBI) website (http://www.ncbi.nlm.nih.gov/genomes/MICROBES/Complete.html) lists 162 completed whole microbial genome sequences, a list which contains information on multiple isolates for the same bacterial species. Comparative analyses of these genome sequences provide valuable information on likely genetic differences that might account for distinctive patterns of pathogenesis. In addition, whole-genome comparisons indicate evolutionary relationships between bacteria and allow the tracking of genetic exchange of gene blocks [**4**]. These data illustrate the varied extent to which bacterial isolates are related to each other. Comparisons of the avirulent *Escherichia coli* K12 and virulent *E. coli* O157:H7 genomes showed that the latter possessed a larger genome with a 1.4-Mb sequence that was largely derived from horizontally transferred foreign DNA [**5**]. Moreover, the *E. coli* O157:H7 strain encoded over 1600 additional proteins of which approximately 130 might be related to bacterial virulence. This extensive variation contrasts with comparative data for *Mycobacterium tuberculosis*, which has a more conserved genome. Comparative analysis of the genomes of *M. tuberculosis* and *Mycobacterium bovis*, the progenitor of the BCG vaccine strain, shows them to be 99.95% identical at the DNA level [**6**]. There are, however, extensive regions of polymorphism in the *M. tuberculosis* genome, particularly in the repetitive families of proteins (PE-PGRS/PPE protein family) that may account for the observed differences in pathogenesis [**6**]. High-throughput microarrays covering the complete genome of *Helicobacter pylori* have been used to compare large numbers of clinical isolates at the genome sequence level. These data demonstrate extensive variations in gene content and organization that correlate with bacterial virulence [**7**, **8**]. These few examples show that comparative studies of bacterial genome sequences and organization can reveal potential determinants of pathogenesis and virulence. However, Holden et al. [**9**] highlighted data from comparative genomic studies of nonpathogenic bacteria that demonstrate that some of these nonpathogenic strains carry genes that have previously been shown to correlate with virulence. It was proposed that these genes aid the interaction of the bacterium with its host or environment [**9**]. This suggestion highlights the complexity of the pathogenic processes alluded to by Moxon and Tang [**1**]. The direct comparison of genome sequences provides a valuable contribution to understanding bacterial pathogenesis, but it does not necessarily provide a complete answer.

The close similarity of the *M. tuberculosis* and *M. bovis* genomes led Garnier et al. [**6**] to propose that differences in their pathogenesis could in part be accounted for by differential gene expression. Thus, studies focusing on the bacterial transcriptome and proteome can make significant contributions to this field of research. Genomewide analysis of gene expression for pathogenic variants at the level of the transcriptome can now be achieved for selected bacteria using high-throughput microarrays [**10**, **11**]. This chapter will consider the complementary approach to the analysis of global gene expression at the level of the proteome.

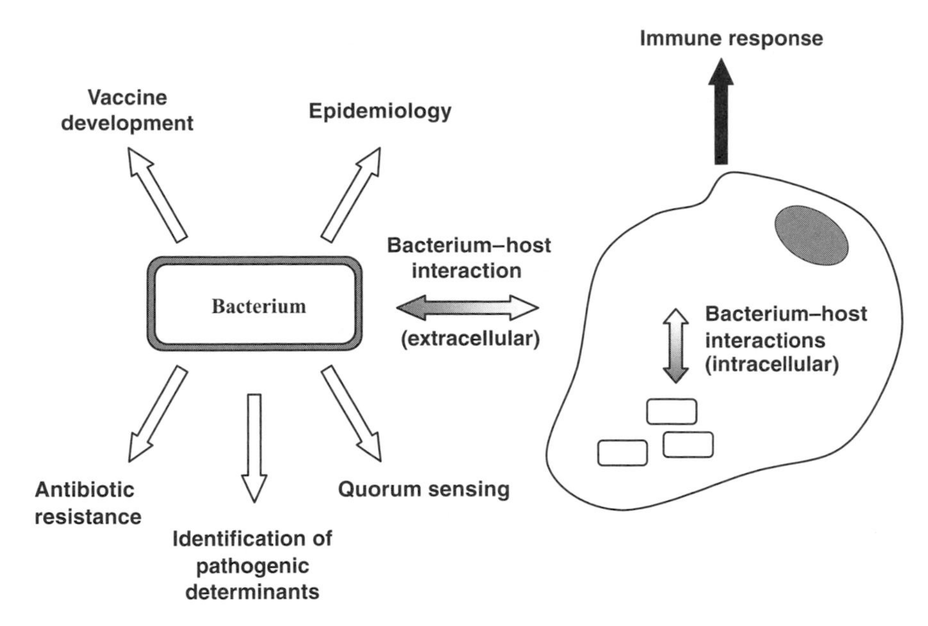


**Figure 13.1** Proteomic applications for analysis of pathogenic bacteria. The diagram illustrates the diverse applications of proteomic technologies that have been used to characterize bacterial pathogens and their interaction with their eukaryotic hosts.

The analysis of cellular gene expression and metabolism at the level of the proteome has many advantages over a purely genomic-based investigation and has been adopted for the study of many biological systems [**12–16**]. These benefits are equally relevant in the field of microbiology [**17**, **18**]. The advantages of proteomics lie in the fact that many cellular processes occur at the level of the proteins, and so analyses at this level are likely to provide significant insights into these activities. However, proteomics should not be considered in isolation from similar studies carried out at the genomic level, and this chapter will highlight the added benefit of looking at the proteome over nucleic-acid-based technologies. Figure 13.1. summarizes the areas in which proteomics has been influential in the study of bacterial pathogenesis; many of the areas detailed in this figure will be expanded upon in the following discussion.

## 13.2 DEVELOPMENT OF COMPREHENSIVE PROTEOME DATABASES FOR BACTERIAL PATHOGENS

A basic assumption for many of the investigations discussed below is that a virulent bacterium differs in one or more ways from a related avirulent bacterium. Specifically, this difference is responsible for the ability of the virulent strain to cause disease. However, pinpointing these key differences is not a straightforward process. Bacteria exhibit extensive variation at the level of their genomic nucleic acid, and a significant amount of this variation can be observed in the expressed proteins. Although the differential synthesis of the bacterial proteins can serve as markers for both taxonomic and epidemiological studies of bacterial pathogens [**19–23**], they may not relate specifically to bacterial pathogenesis and care must be taken to distinguish key protein differences against the "background" of nonspecific variation. A reasonable starting point to approach this problem has been to develop a proteome database that provides a catalog of the proteins expressed by the organism. Depending on the available resources, this database may cover multiple isolates and different growth conditions for the bacteria. Proteome databases can be derived in silico by analyzing the genome sequence [**24–26**], but these are generally limited since they do not provide an indication of the dynamic nature of the proteome and there are also discrepancies with experimental data.

In parallel with the determination of complete bacterial genome sequences there are long-term research programs aimed at deriving comprehensive proteomic databases for representative bacterial groups [**27–32**], including human pathogens, for example *H. pylori* [**33–35**], *M. tuberculosis* [**36–38**], and *Haemophilus influenzae* [**39**, **40**]. The capacity of the proteome to change in response to external stimuli means that a comprehensive proteome database, detailing both the encoded proteins and their expression profiles, is unlikely to be fully achieved. Even though many of the techniques currently being used to determine bacterial proteomes were established over 20 years ago, there are still no complete bacterial proteome databases. The complexity of the data increases as one takes into account various analytical protocols that differ in their coverage of the proteome as well as the differential protein expression as the bacteria respond to the extracellular environment and change their growth phase. A database template has recently been proposed which draws together the complexity of handling and interrogating microbial proteome data [**41**].

Current bacterial proteome databases are commonly centered upon just one or two type isolates of the organism under study. In the context of studying bacterial pathogenesis, it is essential to carefully select the most representative isolate from which to derive a proteomic database. In many instances, a laboratory-adapted strain is chosen, since it generally grows well under laboratory conditions. For similar reasons, this may also be the isolate used for the determination of the genome sequence, thereby making the type isolate even more attractive for investigating the proteome. However, this might not be ideal since there can be many differences between this "type" strain and the clinical isolates that are of interest to the medical microbiologist. For example, *H. pylori* strain 26695 was the first isolate for this bacterium to be sequenced [**42**], and it is commonly used for many proteomic investigations. However, this isolate is defective in the expression of a number of virulence factors, which are likely to play a role in the establishment of in vivo infections [**33**, **43**, **44**]. In addition, multiple qualitative and quantitative differences are apparent in the global protein synthesise when compared by two-dimentional gel electrophoresis (2DE) for strain 26695 and recent clinical isolates of *H. pylori*, some of which affect proteins involved in bacterial pathogenesis (C. Uwins and P. Cash, unpublished data). Similar observations have been reported for the sequenced *H. influenzae* Rd strain and clinical isolates of *H. influenzae* [**45**]. The enzyme tryptophanase is absent from the sequenced Rd strain of *H. influenzae* but detected, by 2DE, for *H. influenzae* NCTC8143, for which fairly extensive proteomic data are available [**39**]. At present, the labor-intensive nature of 2DE and peptide mass mapping makes it prohibitively time consuming to prepare detailed proteomic databases across multiple isolates of related bacteria. Thus, in many instances, databases for bacterial proteomes are generally targeted toward specific biological problems rather than any attempt to be fully comprehensive. The development of high-throughput procedures for analyzing bacterial proteomes will certainly extend these data in the future to give improved proteomic coverage.

Two human bacterial pathogens that have been looked at extensively from the point of view of generating complete proteome databases are *H. pylori* and *M. tuberculosis*. *Helicobacter pylori* is a major cause of gastrointestinal infections and is of significant clinical interest since it has been implicated as a predisposing cause of gastric cancer. The bacterial genome is 1,667,867 bp, and in silico analysis of the genome sequence predicts 1590 open reading frames (ORFs) [**42**]. A number of research groups have published 2D protein maps of *H. pylori* on which protein identities have been annotated [**35**, **46**]. Jungblut et al. [**33**] compared the proteomes of two *H. pylori* strains (26695 and J99) for which complete genome sequences have been determined as well as a type strain used for animal studies. Up to 1800 protein species were resolved by 2DE when silver staining was used to detect the proteins following electrophoresis. One hundred and fifty-two of the detected proteins were identified by peptide mass fingerprinting for the *H. pylori* strain 26695, which covered 126 unique bacterial genes. Among the proteins identified were previously documented virulence determinants and antigens.

The proteome of *M. tuberculosis* has been extensively characterized to identify potential virulence determinants. Two detailed studies of the *M. tuberculosis* proteome are underway [**36**, **37**, **47**], both of which can be accessed by external users via the Internet. Jungblut and his colleagues [**36**, **47**] compared two virulent strains of *M. tuberculosis* and two avirulent vaccine *M. bovis* strains using 2DE and peptide mass mapping. Up to 1800 and 800 proteins were resolved from either cell lysates or culture media, respectively. Two hundred and sixty-three protein spots were identified by peptide mass fingerprinting. In the second *M. tuberculosis* proteome database currently available Rosenkrands et al. [**37**] have looked at the H37Rv strain of *M. tuberculosis* and reported on the identification of 288 proteins and recorded their distribution between culture filtrates and cell lysates. Urquhart et al. [**48**, **49**] compared the predicted proteome map of *M. tuberculosis* with that obtained experimentally from a series of six overlapping zoom 2D gels; the latter resolved 493 unique proteins, equivalent to approximately 12% of the predicted protein coding. Comparisons were made between the predicted and experimentally determined proteome maps for *M. tuberculosis*. When the predicted and observed maps were superimposed, outlier protein spots were found in the experimental 2D protein map that were not present in the predicted proteome; these were typically at the extremes of isoelectric point (pI) and molecular mass. Jungblut et al. [**50**] have also reported discrepancies, between predicted ORFs and observed protein species, among low-molecular-weight proteins. Six genes encoding proteins of $<11.5$ kDa (pI 4.5–5.9) were identified by 2DE and peptide mass fingerprinting that were not predicted from the genome sequence of the H37Rv strain of *M. tuberculosis*. Although the predicted coding sequences were present in the bacterial genome they had not been assigned as ORFs. Rosenkrands et al. [**38**] reported similar observations and identified a 9-kDa (pI 4.9) protein that was not annotated in the genome sequence. In a systematic analysis of low-molecular-mass proteins, 76 proteins with molecular masses between 6 and 15kDa (pI 4–6) were excised and processed for peptide mass fingerprinting [**51**]. Seventy-two of these proteins were identified and found to comprise approximately 50 structural proteins.

When comparing the global proteomes of virulent and avirulent bacterial strains, attention must be paid to physiological differences between the strains that may be unrelated to pathogenic variation. Betts et al. [**52**] have addressed this point to some extent. Comparisons were made of the proteomes of two virulent strains of *M. tuberculosis*, the laboratory-adapted H37Rv strain and CDC1551, a recent clinical isolate [**53**]. The analysis demonstrated that the classic virulent H37Rv strain used in many proteomic analyses retains many of the features present in more recent virulent isolates. A total of 1750 intracellular proteins were resolved by 2DE for each of the strains. The proteomes observed for both isolates were similar over the 12-day growth curve despite the fact that the CDC1551 strain entered the stationary phase in advance of H37Rv. Comparative studies of the proteomes of the bacteria during their growth cycle revealed just 13 consistent spot differences between the two isolates. Seven protein species were specific for CDC1551 and three were specific for H37Rv. Two further protein species were increased in abundance for H37Rv compared to CDC1551. Nine of the proteins exhibiting differences between the isolates were identified by peptide mass mapping. Four of the proteins corresponded to mobility variants of MoxR. The difference in the mobility of the MoxR protein was consistent with a nucleotide change observed in the gene for this protein. One of the *M. tuberculosis* CDC1551 specific proteins identified was a probable alcohol dehydrogenase and one H37Rv specific protein was identified as HisA. The two H37Rv induced spots were electrophoretic variants of alkyl hydro peroxide reductase chain C.

## 13.3 SEARCH FOR PROTEOMIC MARKERS OF BACTERIAL PATHOGENESIS

Proteomics is an excellent, nonspecific method to look for features in pathogenic bacteria that are associated with the ability of the bacteria to cause overt disease. For any pathogenic microbe the outcome of infection is determined by a number of determinants that act together or in close coordination. These pathogenic determinants affect the ability of the organism to colonize the host, to interact and survive within the host, and to direct the synthesis of proteins or metabolites that may be harmful to the host. Modern molecular technologies now allow researchers to probe these events in fine detail and, as discussed below, proteomics plays an important role in these developments. The investigation of complex phenomena such as pathogenesis on a gene-by-gene basis where the role of each gene is investigated in isolation is unreasonable, since it does not allow the analysis of multiple genetic interactions. The extensive genome sequence data now available for many bacterial pathogens open up strategies for analyzing global gene expression that are more appropriate for investigating polygenic phenomena like pathogenesis. Moreover, the capacity of proteomics to analyze global protein synthesis covering both gene expression and posttranslational modifications makes this approach a powerful tool for identifying and characterizing the determinants of bacterial pathogenesis. Proteomic technologies have been used in two broad applications: the comparative analysis of pathogenic and nonpathogenic bacteria grown in vitro in the laboratory and for bacteria grown in association with the eukaryotic host or under conditions that mimic the in vivo host environment.

### 13.3.1 Comparison of In Vitro Grown Bacteria

***Analysis of Total Cell Protein Content*** A simple approach widely used to investigate pathogenesis is the comparison of the proteomes of pathogenic and nonpathogenic bacterial strains grown under standard conditions. Differences in the proteomes of these bacterial strains are determined by comparative 2DE, and variations in protein synthesis that correlate with pathogenesis are then subjected to further analysis. However, this approach has some limitations. First, when comparing naturally occurring bacterial variants, protein differences occur that are unrelated to virulence. Second, bacteria grown in vitro on defined culture media are unlikely to express all of the proteins encoded by the genome at levels characteristic of in vivo growth in the bacteria's natural host. This is apparent when comparing the proteins synthesized by facultative intracellular bacterial pathogens grown either in defined culture or in association with eukaryotic cells. Many proteins are only synthesized in the latter. The analysis of bacteria–host cell interactions through proteomics will be discussed further below.

High-resolution 2DE is a frequently used method to identify pathogenic determinants in the proteome, since it can be used with a wide variety of different bacteria with relatively limited genome sequence data. Many of the initial studies using this approach were largely descriptive, and due to technical limitations in protein identification at the time, the protein spot identities were not determined. A comparison of virulent and avirulent *Mycoplasma pneumoniae* isolates identified three novel proteins expressed by the virulent isolates that were absent from avirulent strains [**54**]. Comparisons of a virulent parental strain of *M. pneumoniae* with two derived avirulent mutant strains revealed both quantitative and qualitative differences in the protein profiles when analyzed by 2DE [**55**]. Pathogenic determinants in *Brucella abortus* have been investigated using a similar approach to that described for *M. pneumoniae* [**56**]. Up to 935 proteins were resolved by 2DE for virulent and vaccine strains of *B. abortus*, equivalent to approximately 43% of the expected 2129 proteins predicted for a genome of the size of *B. abortus* [**57**]. Ninety-two qualitative and quantitative protein differences were found between the 2D protein profiles of the virulent and avirulent strains. It was proposed that the observed differences in protein synthesis in the proteomes of virulent and vaccine isolates may occur due to a variety of factors, including point mutations in coding and regulatory sequences as well genome rearrangements or deletions that indirectly affected the expression of bacterial genes required to maintain cellular homeostasis [**56**].

Clinical isolates of *H. pylori* exhibit a high degree of variation when compared by 2DE [**33**, **58**]. Enroth et al. [**59**] used 2DE to discriminate *H. pylori* isolates on the basis of the nonfractionated cellular proteins resolved by 2DE. *Helicobacter pylori* isolates were collected from patients with different disease symptoms and their proteins compared by 2DE. In agreement with the extensive heterogeneity observed at the level of the genome DNA [**60**], a high level of protein variation was revealed using 2DE. Classification of the data demonstrated some clustering of the isolates according to the disease with which the specific bacteria were associated (i.e., duodenal ulcer, gastric cancer, and gastritis). The authors suggest that there may be some as-yet-unidentified disease-specific proteins contributing to the clustering of the bacterial isolates. However, the significance of this clustering remains unproven since Hazell et al. [**61**] failed to observe any disease-specific clustering among *H. pylori* isolates compared by restriction enzyme digestion analysis of the genomic DNA. Govorun et al. [**62**] used 2DE to compare four clinical isolates of *H. pylori* collected from patients with duodenal ulcer and chronic gastritis. The variable proteins found among these isolates were identified using peptide mass mapping. Variation was observed that affected the mobility in the first dimension of 2DE (GroEL, Tig, SodF, TagD, and Omp18) or in abundance (Cag26, transcription regulator, HP1043, hypothetical protein HP1588, FrxA, TagD, and Pfr). Based on the protein profiles, the four isolates formed two discrete groups, which were unrelated to the origin of isolation in the Russian Federation. These early data indicate that this approach could identify specific patterns of *H. pylori* protein synthesis that influence the outcome of bacterial infection. However, more data still need to be collated from well-characterized clinical isolates.

The *M. tuberculosis* proteome has been extensively studied in order to identify pathogenic determinants for this bacterium through the comparison of virulent *M. tuberculosis* and vaccine *M. bovis* strains [**36**, **47–49**, **63**, **64**]. Relatively few differences in protein synthesis were detected for these bacteria [**36**, **52**], which was consistent with the limited genetic variability observed at the time through the sequence comparison of 26 *M. tuberculosis* genes [**65**]. However, this minimal genomic variability might actually have been underestimated due to the limited genome sequence data available for comparison. More recent studies based on complete bacterial genome sequences indicate a higher degree of genetic variability [**66**]. Urquhart et al. [**48**] used a high-resolution multigel system to look at the proteomes of the virulent *M. tuberculosis* H37Rv and *M. bovis* BCG [Pasteur American Type Culture Collection (ATCC) 35734] strains. Up to 772 protein spots were identified for these two bacterial strains over a broad range of isoelectric points and molecular masses. A bimodal distribution for the proteins in the first dimension was found for both bacterial isolates which is similar to that observed for other bacterial proteomes [**39**, **67**]. No protein identifications were made from these comparisons of *M. tuberculosis* and *M. bovis* so their significance in the context of bacterial pathogenesis is unknown, although they do contribute toward the development of methods in displaying the bacterial proteome. A detailed comparison of virulent *M. tuberculosis* (H37Rv and Erdman) and vaccine (BCG Chicago and BCG Copenhagen) strains has been reported that includes the analysis of proteins extracted from both bacterial cell lysates and infected culture media [**36**]. The majority of the bacterial proteins either detected in the bacterial cells or released into the culture media were common for all four isolates. Up to 31 variant spots were found in pairwise comparisons with the most extensive variation found in the comparison of virulent and vaccine strains. However, differences were also observed between the two virulent strains as well as between the two vaccine strains which were unlikely to be unrelated to pathogenesis. From these comparative data, 96 spots were specific for either the virulent or avirulent strains (56 and 40 protein spots, respectively). Thirty-two of the *M. tuberculosis* specific protein species were identified and corresponded to 27 different proteins. Twelve of the *M. tuberculosis* unique proteins were assigned to genes previously shown to be deleted from avirulent strains. The remaining 20 protein spots were assigned to genes not previously predicted to be deleted from avirulent *M. bovis* strains. The *M. bovis* specific proteins identified in this study were mapped to known *M. tuberculosis* genes. Those proteins specifically synthesized by the virulent *M. tuberculosis* are potential candidates as both pathogenicity determinants and vaccine targets. No novel virulence determinants were found for *M. tuberculosis*, although previously identified virulence determinants were assigned to the proteome.

One problem in interpreting the data obtained from comparative studies of potentially genetically distinct bacterial isolates, such as those described above, is that some of the genomic and proteomic variation is unrelated to pathogenesis. Mahairas et al. [**63**] used an alternative approach to look for pathogenic determinants in *M. tuberculosis* at the level of the proteome. These researchers located genomic differences between *M. bovis* BCG vaccine and virulent isolates of both *M. bovis* and *M. tuberculosis* through the use of subtractive genomic hybridization. Three regions were identified that were deleted in BCG vaccine compared to the virulent strains. One 9.5-kb segment (designated RD1), predicted to encode at least eight ORFs, was absent from six BCG substrains but present in virulent *M. tuberculosis* and *M. bovis* strains as well as in 62 clinical *M. tuberculosis* isolates analyzed. The virulent *M. bovis* and *M. tuberculosis* strains showed indistinguishable protein profiles to each other when compared by 2DE. In contrast, the *M. bovis* vaccine strains expressed at least 10 additional proteins and raised expression levels for a number of other proteins. The RD1 genome region was inserted into the BCG genome to generate BCG::RD1. The protein profile, determined by 2DE, of BCG::RD1 was now indistinguishable from virulent *M. bovis*, suggesting that parts of RD1 specifically suppressed protein synthesis in virulent mycobacteria. Some low-molecular-weight proteins were identified for BCG::RD1 that were consistent in size with the short ORFs encoded in the RD1 sequence.

***Analysis of Bacterial Subproteomes*** A frequently cited limitation to the use of 2DE to characterize cellular proteomes is the limited dynamic range of the technique, which cannot cope with the wide distribution in abundance found for a typical cell. This prevents the display of the entire proteome in a single analytical step. Minor proteins may be enriched based on their physical properties by selective chromatographic methods [**68–70**]. Alternatively, fractionation of the cell into subcellular compartments permits the analysis of "subproteomes." This approach not only enriches for low-abundance proteins but also provides functional information (i.e., cellular location) for the proteins. The majority of investigations employing this approach have concentrated on those proteins either located at the cell surface or secreted into the extracellular environment. Proteins present in either of these fractions are likely to play key roles in bacterial pathogenesis as the bacteria interact with the eukaryotic host and initiate damage via their exposed surface and secreted proteins [**71**]. In addition, the proteins present on the surface of the bacterial cells are exposed to the host immune system, and many highly immunogenic proteins are found in this location in the bacterial cells [**72**]. The latter proteins are thus potential targets for vaccine development [**73**].

Depending on their interaction with the host cell, *Pseudomonas aeruginosa* isolates are classed as either invasive or cytotoxic. The pathogenesis of invasive strains depends on the surface proteins that allow the bacteria to become internalized by the host cell. In contrast, the cytotoxic strains are not internalized but cause symptoms through the release, by the bacterium, of extracellular toxins and proteases into the host tissue. The membrane and secreted proteomes of these two bacterial types have been examined. Approximately 350 proteins were detected, by 2DE, in enriched membrane preparations from *P. aeruginosa* PAO1, an invasive strain. One hundred and eighty-nine protein species, equivalent to 104 genes, were identified by peptide mass mapping; 63% of the identified proteins were either known *Pseudomonas* membrane proteins or homologous to membrane proteins from other bacteria. The remaining proteins had not been previously characterized at the functional level and were either unique to *P. aeruginosa* or matched hypothetical proteins from other bacteria [**74**]. The enriched membrane proteins from a representative cytotoxic strain (*P. aeruginosa* strain 6206) were compared to the PAO1 strain. There was minimal variation in the membrane proteins for these two isolates apart from minor differences consistent with amino acid substitutions. The authors thus concluded that the principal determinants differentiating invasive and cytotoxic strains were not associated with the outer membrane fraction for bacteria grown in vitro. More extensive differences were observed between these two bacterial strains when the secreted proteins were compared [**75**]. Wehmhoner et al. [**76**] have also reported that the secreted proteins showed the most extensive degree of variation between clones of *P. aeruginosa*. Under the experimental conditions employed by Nouwens et al. [**75**], the protein profiles for the two strains were very distinctive, and it was not possible to simply identify the protein by cross matching the 2D protein profiles. In the extracellular fractions from PAO1 a number of isoforms for specific ORFs were identified as well as many mass variants of secreted and membrane proteins, suggesting that there may be high concentrations of bacterial proteases released into the extracellular medium. Consistent with this proposal was the identification of several bacterial proteases (elastase, LasA protease, and endoproteinase) in the extracellular proteins for strain PAO1. When the extracellular protein preparations from strain 6206 (the cytotoxic strain) were analyzed, there were high levels of cytosolic proteins, including GroEL, GroES, trigger factor, and DnaK, as well as some outer membrane flagella proteins. Nouwens et al. [**75**] suggested that the apparent low abundance of known extracellular proteins for the latter strain was consistent either with this strain being defective in its secretory mechanism or with the fact that it relies heavily on cytotoxic proteins for its pathogenesis [**75**]. Although some known virulence determinants were identified as part of this study, there were others that were absent from the extracellular protein preparations, since they required specific growth conditions for their production. These include the type III secreted proteins that require host cell contacts and alkaline phosphatase that requires phosphate-depleted media for expression [**75**]. These latter data illustrate some of the problems of identifying determinants of pathogenesis under the artificial growth conditions used in the laboratory.

Differential patterns of *P. aeruginosa* protein synthesis have also been described between two isolates (TB10839 and TB121838) belonging to the TB clonal lineage that differ markedly in their interaction with host cells [**77**]. The TB10839 isolate proliferated in polymorphonuclear granulocytes and was cytotoxic for macrophages and adhered to epithelial cells. Approximately 4% of the detected proteins showed different levels of expression between the two isolates with the majority of the variation occurring in the secreted proteins. All of theses comparisons were carried out on bacteria grown in vitro. Among the proteins differentially expressed between the two bacterial isolates were several quorum-sensing regulated proteins, including LasB elastase and chitinase, which were down regulated in the TB121838 isolate. Thus, a possible alteration in the quorum-sensing cascade may be responsible for the different patterns of cellular interactions with these two bacteria.

Extensive data have been presented on the membrane and secreted proteomes of *H. pylori*. Membrane proteins of *H. pylori* strain 26695 were identified by biotinylating the surface bacterial proteins of intact bacteria and purifying the labeled proteins from a membrane subfraction by affinity chromatography [**72**]. From these preparations, 82 biotinylated proteins were detected by 2DE and 18 of these were identified by peptide mass mapping. Seven of the identified proteins had been previously associated with *H. pylori* pathogenesis, including urease and one protein encoded in the *H. pylori* Cag pathogenicity island. Hynes et al. [**78**] investigated the effect of the extracellular environment on membrane protein expression by growing *Helicobacter* spp. in the presence of bile, an environmental stimulus to which *Helicobacter* is exposed in vivo. Under these conditions low-molecular-weight (<8-kDa) proteins on the cell surface were differentially expressed. The various species of *Helicobacter* that colonize different niches in the gut exhibited varying responses to bile, and these might reflect the specific pattern of pathogenesis for these bacteria. Baik et al. [**79**] have carried out a systematic review of the sarcosine-insoluble outer membrane fraction recovered from the type strain (26695) of *H. pylori*. Sixty-two protein spots were recovered from the 2D gels and identified by peptide mass mapping. The proteins identified corresponded to 35 genes. Sixteen outer membrane proteins were present in the membrane fraction. There were also a number of cytoplasmic proteins (e.g., phosphoglycerate dehydrogenase, glutamine synthetase, and alkylhydroperoxide reductase) presented in the sarcosine-insoluble membrane fraction.

The authors suggested that the surface properties of *H. pylori* promoted attachment of cytoplasmic proteins [**79**]. Whether this was a natural processing step for *H. pylori* or a technical artifact was not discussed. Bumann et al. [**80**] have also examined secreted proteins released by *H. pylori* 26695 grown in vitro. Thirty-three reproducible proteins were detected of which 26 were identified. As might be expected, the VacA and flagella proteins were present among the identified proteins together with 8 proteins of unknown function. No CagA proteins were detected, but this was believed due to the absence of the type IV secretion system, responsible for CagA secretion, which requires the presence of the host eukaryotic cell to induce its expression [**80**].

The response of the expression of extracellular virulence factors to the environment has been examined for group A *Streptococcus* [**81**]. Various environmental conditions ($O_2$ and $CO_2$ concentrations, temperature, and varying concentrations of NaCl, glucose, and iron) were examined for their effect on extracellular proteins. Differential patterns of extracellular abundance were observed between the varying environmental conditions investigated for a number of the bacterial extracellular virulence factors, including streptococcal inhibitor of complement (Sic), streptococcal pyrogenic exotoxin B (SpeB) and streptococcal pyrogenic exotoxin F (SpeF), endo-β-*N*-acetylglucosaminidase (EndoS), and α-amylase. Many of the environmental conditions examined in this study are analogous to the types of environment that the bacteria might experience in vivo. The authors suggest that these different patterns in the abundance of the extracellular virulence factors could account for the diverse disease outcomes associated with group A streptococcal infection.

***Identification of Antigenic Proteins*** Analyses of bacterial subproteomes, predominantly those covering the surface and secreted proteins, have been a particularly fruitful means of identifying antigenic proteins expressed by bacteria. These data not only provide fundamental information on host–bacteria interactions through the host immune response but also have practical implications for future vaccine development. Membrane-enriched cell fractions have been screened for antigenic proteins in a number of bacterial pathogens, for example *Staphylococcus aureus* [**73**] and *H. pylori* [**79**, **82**, **83**]. Using pooled human sera containing staphylococcal antibodies, Vytvytska et al. [**73**] identified a large number of immunoreactive protein spots associated with surface protein extracts of *S. aureus*. Eighteen of the reactive protein spots were successfully identified and shown to match 15 proteins. The identified proteins included both known (SdrD and IsaA) and novel (Aux1, LP309) immunogenic vaccine candidates. The in vitro growth conditions employed in this study influenced the expression of these surface proteins with LP309, an iron-regulated ATP-binding cassette (ABC) transporter found only when bacteria were grown in iron-depleted growth media. Utt et al. [**83**] used a membrane-enriched fraction of *H. pylori* cells as the antigen against which to screen sera containing *H. pylori* antibodies. One hundred and eighty-six protein species were found to react against the *H. pylori* antibody positive sera. Peptide sequencing identified six highly immunogenic protein spots equivalent to four *H. pylori* proteins in the molecular mass region of 25–30 kDa. These proteins were cell-binding factor 2, urease A, an outer membrane protein (HP1564), and a hypothetical protein (HP0231). One of these proteins, cell-binding protein 2, has previously been proposed as a potential vaccine candidate [**84**].

The identification of secreted proteins that interact with the host immune system has been examined for *M. tuberculosis*. The T-cell response is a major line of host immune defenses against *M. tuberculosis*, and the identification of those bacterial proteins eliciting the T-cell responses will be important for future vaccine development. During in vitro growth, *M. tuberculosis* releases proteins into the media capable of inducing a protective cellular immune response [**85–87**]. Up to 205 protein spots were detected in *M. tuberculosis* culture media by 2DE [**64**], and 34 unique proteins were identified based on their reactions with specific monoclonal antibodies and by amino acid sequencing. One member of a protein cluster at 83–85 kDa, previously recognized as a dominant humoral antigen [**88**], was identified as catalase, KatG [**64**]. The bacterial proteins were harvested from the culture media during the late logarithmic phase of bacterial growth when some bacterial cytoplasmic proteins might also be present in the media and so complicate the specificity of the analyses [**64**, **89**]. Weldingh et al. [**90**] attempted to minimize this problem by analyzing the secreted proteins after only seven days of growth. Among the proteins identified under these conditions were an immunogenic protein (CFP21) that maps to the RD2 genome segment, which is missing from some *M. bovis* BCG strains [**63**]. A 29-kDa immunogenic protein was also detected; this protein is associated with the membranes of several *M. tuberculosis* strains [**91**]. The KatG protein previously identified in the media after prolonged growth of *M. tuberculosis* [**64**] was not found in the media from these short-term growth conditions. The ability of the secreted proteins to stimulate T cells has also been investigated [**92**, **93**]. Although there is some inevitable patient variation, T cells from tuberculosis patients and tuberculin-positive individuals react with a 30-kDa cell-associated antigen [**92**]. When culture media filtrates were screened in a similar assay, antigenic components of 30–100 kDa were detected. T cells from tuberculin-negative individuals showed only a weak or negative reaction [**93**]. A number of bacterial proteins were identified among the antigenic components. Of the 34 identified proteins from cell lysates and culture filtrates, there were both previously documented bacterial antigens as well as 17 proteins considered to be novel T-cell antigens.

## 13.4. CHANGES IN BACTERIAL PROTEOMES DURING IN VIVO INFECTION

### 13.4.1 Bacterial–Host Interactions

Although in vitro analyses of pathogenic bacteria can provide valuable information on likely determinants of pathogenesis, they only offer a limited view of this complex phenomenon. The only reliable way of defining the genetic activities of a bacterium that lead to the development of overt disease is to look at the microbe in association with its natural host. Pathogenic bacteria typically exhibit differential gene expression when grown on artificial media compared to growth in association with their eukaryotic host cells. Specifically, many virulence determinants are significantly induced when bacteria are grown in association with their host(s) [**74**, **80**]. The characterization of these differentially expressed genetic networks is a key area of research, since they may aid the identification of potential targets for new therapeutic drugs. Recombinant DNA technologies play a key role in identifying those bacterial genes specifically expressed when the bacteria are in association with the host cell [**94**]. For example, in vivo expression technology (IVET) can be used for high-throughput screening of bacterial promoters activated when facultative bacterial pathogens are grown with their eukaryotic host either in an intact animal or using in vitro cell culture [**95–97**].

Proteomics has the potential to complement these recombinant DNA approaches, but the technologies widely used to characterize bacterial proteomes suffer from a number of technical limitations for analyzing in vivo gene expression in whole animals.

Consequently, many studies are generally restricted to analyzing bacterial infections of in vitro cell systems. Unfortunately, in vitro grown cell lines are rather poor models for studies on microbial pathogenesis, since they do not reliably cover the interactive networks that occur between differentiated cell types within the tissues of the intact animal. One advantage of using in vitro cell lines, however, is that they can be infected under highly reproducible and controlled conditions and, moreover, radioactive amino acids can be used to increase the sensitivity of detecting bacterial proteins against the background of the host cell protein synthesis. Typically, a combination of antibiotics and radioactive amino acids are used to selectively radiolabel the proteins synthesized by intracellular bacteria [**98–101**]. The data obtained by this experimental approach must be considered in the light of two factors. First, the antibiotics used during the radiolabeling might interfere with the normal interaction of the intracellular bacteria and eukaryotic cell [**102**]. Second, the intracellular and in vitro grown bacteria may exhibit different growth kinetics leading to nonspecific variation in protein synthesis. For *M. bovis* infection of macrophage cells, it has been possible to overcome these limitations and to eliminate the need for one antibiotic, cycloheximide, which is used to inhibit cellular protein synthesis [**102**]. Radiolabeling was carried out in the absence of cycloheximide and the infected macrophage cells were lysed with sodium dodecyl sulfate (SDS) before collecting the bacteria by centrifugation and then washed them with Tween-80. Appropriate controls were included to show that there was minimal carryover of cellular proteins into the bacterial pellet. The recovery of the bacterial proteins under these conditions was sufficient to identify the intracellular bacterial proteins by peptide mass mapping.

Intracellular bacterial pathogens generally gain entry to the eukaryotic host cell by phagocytosis. Once the bacteria have entered the cell's phagosome, they are exposed to various stress conditions, including extremes of acidity, oxygen, and nutrients [**103**]. However, bacterial pathogens that migrate to the cytoplasm from the intracellular vacuoles are exposed to a reduced level of stress. During infection of macrophage cells, *B. abortus*, *Salmonella typhimurium, Yersinia enterocolitica, and Legionella pneumophila* remain within the cellular phagosome. A consistent observation for these bacteria is that specific bacterial proteins are either induced or repressed in intracellular bacteria compared to bacteria growing in artificial culture media. Moreover, some of the bacterial proteins induced during intracellular growth also show increased levels of synthesis when the bacteria are stressed in vitro. For example, the synthesis of the bacterial heat-shock proteins GroEL and DnaK are induced during intracellular growth of *B. abortus* [**98**, **104**] and *S. typhimurium* [**105**]. Two bacterial proteins, homologous to the stress proteins DnaK and CRPA, are induced during *Y. enterocolitica* infection of J774 cells and also in vitro by heat shock and oxidative stress [**99**]. During intracellular infection of U937 cells by *S. typhimurium*, 100 bacterial proteins are repressed and 40 induced compared to in vitro grown bacteria. Many of the proteins induced during intracellular growth are also observed during stress in vitro. However, the changes observed for the former are not simply a summation of the in vitro stress responses [**100**], suggesting that specific response patterns are induced during the intracellular replication phase. In contrast to the above data, *Listeria monocytogenes* infection of J774 cells shows a different pattern of bacterial protein synthesis. None of the 32 proteins induced during intracellular growth are induced by typical in vitro stress conditions, including heat shock and oxidative stress [**106**]. These data are consistent with the intracellular bacteria rapidly migrating out of the phagosome to the cytoplasm of the host cell.

The changes induced in bacterial protein synthesis during intracellular infection have been described for several mycobacteria. During infection of THP-1 macrophage cells by *M. bovis* (BCG strain), proteins are synthesized that are not expressed under in vitro growth conditions [**102**]. These proteins were demonstrated using a combination of radiolabeling and immunoblotting. At least 20 bacterial proteins were either specific or at significantly induced levels for intracellular bacteria. Proteins induced at 24 h postinfection were identified as GroEL homologues, GroEL-1/GroEL-2, InhA, 16-kDa antigen ($\alpha$-crystallin Hsp-X), EF-Tu, and a 31-kDa hypothetical protein. The type and origin of the host cell influence both the initial interaction between bacteria and eukaryotic cell and the bacterial growth kinetics [**107, 108**]. The response of the bacterial proteome to growth in host cells of differing origin has been described for *Mycobacterium avium* [**109**]. *Mycobacterium avium* infection of J774 cells results in the induced synthesis of specific bacterial proteins that was initiated by 6 h postinfection (p.i.) and continued until 4 days p.i. None of the proteins induced during intracellular bacterial growth were induced by in vitro stress conditions. In contrast, when primary bone macrophages were infected with *M. avium*, the bacteria followed different kinetics of replication and protein synthesis. In the latter cells, intracellular bacteria radiolabeled at 5 and 12 days p.i. showed significant differences in their protein synthesis. These data were consistent with the bacteria entering a static growth phase early in infection before commencing a normal replication cycle between 5 and 12 days p.i. At 12 days p.i. the bacteria synthesised a range of proteins similar to that synthesised by *M. avium* growing in J774 cells. The variations in the replication cycle and protein synthesis of the bacteria using macrophage cells of different sources illustrate the care required in selecting the host cell used to investigate intracellular bacterial replication.

The foregoing discussion considered the response of bacterial protein synthesis to the intracellular environment of the host. Similar approaches can, in principle, be used to investigate the host cell response to the bacterial infection, although these have been more limited in number. The effect of the infecting bacteria on host cell phagosome proteins has been examined for *M. tuberculosis*. In *M. tuberculosis*–infected macrophage cells, the phagosomal compartment fails to fuse with the lysosomes, the normal fate of intracellular phagosomes, due to the bacteria arresting the development and processing of the phagosome. Comparisons of the proteins associated with the phagosomal compartments from mycobacterial infected cells with the same structures from uninfected cells showed a number of differences in the cellular proteins present. Fratti et al. [**110**] proposed that one of the key features of the *M. tuberculosis* phagosome was the exclusion of EEA-1 (early endosomal autoantigen), which plays a role in vesicle tethering and also in endosomal fusion. At present, the understanding remains as to the identity of the protein, or proteins, expressed by *M. tuberculosis*, which leads to the exclusion of EEA-1.

### 13.4.2 Quorum Sensing and Biofilm Formation

The entry of pathogenic bacteria into a susceptible host leads to their adaptation to the new environment of the host, that is, changes in nutrient levels and host protective mechanisms. A number of bacterial pathogens respond to this changed environment through a process known as quorum sensing [**111**], which acts through cell–cell communication as the bacterial population density increases. Low-molecular-weight signaling molecules released by the bacteria lead to the transcriptional activation of specific bacterial genes once the signaling molecules reach a critical concentration. The process of quorum sensing and the identification of the activated bacterial genes have been investigated at the level of the proteome for various pathogenic bacteria.

*Pseudomonas aeruginosa* is an important human pathogen and a common cause of respiratory infection in individuals with cystic fibrosis. During infection in these persons, *P. aeruginosa* undergoes a number of structural changes that enable the bacteria to colonize the host [**112**]. Once the bacteria gain entry to the host, their pattern of gene expression is modified in response to the bacterial population density through quorum sensing. Various bacterial genes are switched on through quorum sensing, including those encoding known virulence determinants as well as genes regulating biofilm formation. The latter helps to protect the bacteria from the host defences, thus improving colonization of the host by the bacteria. This will be discussed in more detail below. From the point of view of pathogenesis, quorum sensing allows the bacterial population to reach a sufficient density to establish an infection before there is significant expression of the bacterial pathogenic traits as well as enables the bacteria to protect itself from the host defenses. Support for this model is the demonstration for a number of animal models that bacteria carrying mutations in either of the two quorum-sensing systems present in *P. aeruginosa* are significantly attenuated [**113**, **114**]. In addition, the two *Pseudomonas* signaling molecules have been found in the sputa of patients with cystic fibrosis who are infected with *P. aeruginosa*, suggesting that they may also play a role in human infections by *P. aeruginosa* [**115–117**].

The *P. aeruginosa* quorum-sensing system depends on the synthesis of small diffusible molecules called acyl homoserine lactones (AHLs) that bind to and activate transcriptional regulators to induce specific gene expression at the transcriptional level. Two quorum-sensing systems have been identified for *P. aeruginosa* that are known as Las and Rhl [**118**, **119**]. The Las system includes LasI, which is the synthase for *N*-(3-oxododecanoyl) homoserine lactone (3O-$C_{12}$-HSL) and LasR, which is the transcriptional regulator. The second *P. aeruginosa* quorum-sensing system, Rhl, includes RhlI, the synthase for *N*-butyryl-homoserine lactone ($C_4$-HSL), and Rh1R, a transcriptional regulator. These two systems are not independent of each other since the Las system positively regulates rhlR and rhlI. As the bacterial population density increases, there is a parallel increase in the concentration of the AHL signaling molecules and, once these have reached an intracellular threshold concentration, the signal molecules bind to their respective transcription factors. The activated transcription regulator complexes then induce the expression of a range of bacterial genes.

In addition to the analysis of individual genetic activities, global changes in gene expression in response to quorum sensing have also been investigated globally at the level of both the transcriptome and proteome. These latter studies take advantage of the complete genome sequence of the PAO1 strain of *P. aeruginosa* in order to map those genes that change in expression in response to the extracellular AHL. Whiteley et al. [**120**] employed random mutagenesis of *P. aeruginosa* to identify 35 quorum-sensing-regulated genes. Based on these identifications, the data were extrapolated to predict that over 200 genes were likely to be regulated through quorum sensing in *P. aeruginosa* strain PAO1. Analyses of gene transcription during quorum sensing at the level of the transcriptome have yielded varying numbers for the regulated bacterial genes, with between 163 and 394 genes being positively regulated and 22–38 negatively regulated by AHL [**121–123**].

Arevalo-Ferro et al. [**124**] presented a detailed analysis of the *P. aeruginosa* quorum-sensing system at the level of the proteome. Protein synthesis for wild-type *P. aeruginosa* PAO1 was compared to an isogenic *lasI rhlI* double mutant. Three subcellular fractions of the parental and mutant bacteria grown in minimal medium were analyzed using 2DE; these subcellular fractions comprised the intracellular, extracellular, and cell surface proteins. Combining the data for the three subproteomes demonstrated that 723 of the 1971 detected protein spots showed a significant change in synthesis (>2.5-fold up or down regulated) between the wild-type and mutant bacteria [**124**]. Growing the mutant in the presence of the AHL signal molecules (3-oxo-C12-homoserine lactone and C4-homoserine lactone) further refined these data. Proteins that were rescued under these conditions for the mutant bacteria were consistent with being regulated by quorum sensing. From this analysis, 23.7% of the detected protein spots were regulated through quorum sensing. Although these stricter criteria for identifying quorum-sensing-regulated genes reduced the number of regulated spots, there were still significantly more quorum-sensing-regulated proteins identified by proteomic technologies than by transcriptome analysis. The latter showed only up to 11% of the bacterial genes to be regulated through quorum sensing [**121–123**]. As commented upon by Arevalo-Ferro et al. [**124**], the improved detection of regulated genes by proteomics occurred despite the well-documented limitations of 2DE, that is, the method used to resolve the bacterial proteins. The authors proposed that a significant amount of the regulation through quorum sensing occurred at the posttranscriptional level. A proteomic analysis of quorum-sensing mutants of *P. aeruginosa* strain PAO1 has also been described by Nouwens et al. [**125**] with a particular focus on identifying the differential expression of extracellular bacterial proteins. In addition to the previously documented quorum-sensing-regulated proteins, the synthesis of a number of other bacterial extracellular proteins (an aminopeptidase, an endoproteinase, and a hypothetical protein) were also consistent with being regulated through quorum sensing. Comparison of *las* and *rhl* mutants showed a difference in the specific extracellular proteins that they regulate [**125**].

One limitation of studying quorum sensing at the level of the proteome compared to the transcriptome is the labor-intensive nature of the proteomics methods, which restricts the range of conditions that can be realistically investigated to map the regulated gene activities [**124**]. This factor might account for some of the apparent discrepancies reported for the numbers and range of regulated genes in published studies [**124**, **125**] for which slightly different growth conditions were employed. Nevertheless, the global proteomics approach has significantly expanded our understanding of this important process in the regulation of bacterial gene expression and its likely role in pathogenesis. These data have also demonstrated that quorum sensing regulates a wide variety of bacterial activities beyond just the expression of known virulence determinants. For example, Arevalo-Ferro et al. [**124**] demonstrated that quorum sensing is involved in iron utilization by *P. aeruginosa*. As more functional studies are applied through the use of proteomics, the interplay of different regulatory networks in bacterial gene expression will become increasingly interlinked with each other. A further twist to the story of the *P. aeruginosa* quorum-sensing regulatory system is the occurrence of a further regulatory system. This is the involvement of a non–homoserine lactone (HSL) signaling molecule, 2-heptyl 3-hydroxy 4-quinolone (PQS) [**126**]. It is believed that PQS acts as a connecting signal between the Las and Rhl quorum-sensing systems [**126**]. Guina et al. [**127**] recently described a proteomic approach to look at the role of magnesium limitation in regulating PQS. *Pseudomonas aeruginosa* cultures grown at low magnesium concentrations exhibited quantitative and qualitative changes in approximately 59% of the expressed bacterial proteome [**127**]. The relative abundance of 546 proteins was determined at the high and low magnesium concentrations, and the abundance of 145 proteins changed in response to the magnesium concentration. Many of the proteins showing differences in synthesis were consistent with the previously documented membrane reorganization reported during *P. aeruginosa* infections of person(s) with cystic fibrosis [**127**]. Other proteins showing induced levels of synthesis identified in these experiments were associated with quorum sensing.

Riedel et al. [**128**] presented a proteomic analysis of quorum sensing in *Burkholderia cepacia*, which has a similar quorum-sensing system to that described for *P. aeruginosa* but uses *N*-octanoylhomoserine lactose (C8-HSL) as the primary signaling molecule and *N*-hexanoylhomoserine lactone (C6-HSL) as a minor product. These signaling molecules are synthesised by the AHL synthase, CepI. CepR acts as the transcriptional regulator for the system and, after binding the signaling molecules, modifies the transcription of specific target genes. This *cep* system regulates biofilm formation, motility, and extracellular enzymatic activities [**129**]. Riedel et al. [**128**] examined protein synthesis in the wild-type bacteria compared to an AHL-deficient *cepI* mutant. As for the *Pseudomonas* study described above, proteins expressed in the subcellular fractions (cellular, membrane, and secreted proteins) were analyzed. Of the 985 protein spots detected by 2DE, 55 proteins were differentially expressed between the parent and mutant strains. The pattern of protein expression in the mutant strain could be restored to that of the wild type by the addition of the signal molecules to the extracellular medium, thus supporting their role in the quorum-sensing system. It was estimated that 6% of the bacterial genes were in fact regulated by the *cep* system. Due to the absence of a complete genome sequence for the *B. cepacia* strain used in these analyses, only 19 proteins were reliably identified by homology with other bacterial genes. These identified proteins showed homologies with various hypothetical bacterial proteins, superoxide dismutase and AidA. Despite the limited number of protein identifications for the bacterial proteins, this work illustrates the usefulness of proteomics to study complex changes in global gene expression at the level of the proteome.

One phenotypic change observed in bacteria that is regulated by quorum sensing is the formation of a biofilm [**130**]. A biofilm forms when a predominantly planktonic bacterial population attaches to a surface and becomes encased in an extracellular polysaccharide matrix along with substantial changes in the bacterial phenotype and physiology [**131**]. With respect to microbial pathogenesis, once formed, the bacterial population associated with the biofilm exhibits an increased resistance to antimicrobial agents—a potential benefit for an invading bacterial pathogen. As discussed below, biofilm formation leads to a significant change in the pattern of bacterial gene expression. Proteomic technologies, predominantly high-resolution 2DE, have played a role in defining the differential pattern of gene expression during the formation of biofilms [**124**, **128**, **132–136**].

Sauer et al. [**131**] carried out a detailed analysis of in vitro biofilm formation in *P. aeruginosa* by monitoring protein synthesis as the bacterial population moved from a planktonic state through to the formation of a biofilm on the internal surface of a silicone tube. Based on microscopic inspection of the biofilm during its development, five stages were identified: (1) reversible attachment, (2) irreversible attachment, (3) maturation 1, (4) maturation 2, and (5) dispersion. Unlike many other studies of bacterial biofilms, Sauer et al. [**131**] followed changes in bacterial protein synthesis as the population moved through these different phenotypic stages. Significant differences in protein synthesis were observed at the five stages of biofilm development. During the late stages in biofilm formation, over 53% of the detectable proteome (equivalent to 800 of the 1500 spots resolved by 2DE) differed by more than sixfold between planktonic and maturation 2 cells.

A subset of the proteins exhibiting significant changes between planktonic and biofilm associated bacteria were identified by peptide mass mapping and represented various functional protein categories, including amino acid biosynthesis, membrane and transport proteins, and proteins involved in protection (alkyl hydroperoxide reductase and superoxide dismutase).

Biofilm formation is regulated, in part, by the quorum-sensing system, and *P. aeruginosa* carrying mutations in the Las system fails to produce biofilms despite the exopolysaccharide production being similar to the wild-type bacteria [**113**]. Suaer et al. [**131**] proposed that this observation could be explained by the mutant having a defect in the structure of the exopolysaccharide leading to a defect in biofilm architecture. This proposal was subsequently confirmed by microscopy. Analysis of a *las* mutant and wild-type bacteria at the level of the proteome during biofilm formation demonstrated few differences between the wild-type and Las 1 mutant bacteria during the early attachment stages. Thus, the quorum-sensing system had relatively little effect on the bacterial proteome during the early stage of biofilm development. However, after one day of biofilm development, at least 50 protein spots differed between wild-type and mutant bacteria after the Las system had been activated during the irreversible stage of biofilm development [**131**]. The second *P. aeruginosa* quorum-sensing system (Rhl) plays a role in the maturation 1 stage of biofilm development, at which stage this regulon becomes activated. At this stage of development approximately 39% of the bacterial proteins showed differential levels of synthesis, a number that is larger than might be expected from being regulated by quorum sensing alone. It was suggested that these data might indicate that quorum sensing alone is insufficient to account for the protein changes observed during biofilm development and that there may in fact be another biofilm-specific regulon present in *P. aeruginosa* [**131**]. Sauer and Camper [**134**] drew broadly similar conclusions in a study of in vitro biofilm formation by the plant saprophyte *Pseudomonas putida*. Forty-five bacterial proteins were either up regulated (15) or down regulated (30) during the first 6 h of attachment in the development of the biofilm. Treatment of *P. putida* with the HSL-signaling molecule revealed differential expression of 16 proteins consistent with quorum-sensing regulation. When overlaps between the two types of regulation were examined, only one protein, PotF, an ABC transporter, was common to the two conditions. As with *P. aeruginosa*, the authors concluded that there were additional sensing systems for these two processes [**134**].

## 13.5 CONCLUDING REMARKS

Proteomics has made a significant impact on many areas of the biosciences, providing a complementary approach to nucleic-acid-based technologies in the area of functional genomics. These benefits are also relevant to the analysis of microbial pathogens. Technical limitations of the original but still widely used proteomic technology of 2DE concerning sensitivity and protein classes amenable to analysis can be reduced by modifying the 2DE protocol itself or by adopting completely new analytical approaches. The latter are frequently based on mass spectrometry. Protein identification is still largely dependent on peptide mass mapping using matrix-assisted laser desorption ionization time-of-flight (MALDI-TOF)mass spectrometry. Continued progress in the number of complete bacterial genome sequences that are now available has allowed more reliable protein identifications to be made. In addition, algorithms have been developed for cross-species identification of proteins to allow the identification of proteins from poorly characterized microbes to be identified by homology with proteins from sequenced bacterial genomes [**137**]. Thus, the proteomes of poorly characterized bacteria can be investigated by these methods utilizing the data derived from other bacterial groups. The majority of the data discussed in this chapter are derived from the analysis of protein mixtures resolved under denaturing conditions. There has been no attempt to define the role of protein–protein interactions in the various biological systems investigated. Protein–protein interactions can be determined experimentally using methods based on the yeast two-hybrid system (reviewed by Legarin and Selig [**138**]). An experimental protein interaction map has been described for *H. pylori* which has identified connections between 46.6% of the proteins encoded by the proteome [**139**]. Inspection of the protein interactions revealed by this approach is bringing to light specific biological pathways which contributes toward the prediction of protein function. A bioinformatics approach to looking at protein interactions based on the *H. pylori* interaction map has been described for *Campylobacter jejuni*, a close relation to *H. pylori*. As data from these complementary techniques become more widely available, we will be able to extend our understanding of microbial proteins in general and bacterial pathogenesis in particular.

The technological developments commented upon in this chapter have broadened our view of bacterial proteomes. The various methods provide basic catalogs of proteins synthesised under defined conditions as well as providing information on the abundance of the proteins. Much of the data discussed above complement genomic-based studies on the characterization of pathogenic determinants as well as identify new targets for further study. Newer analytical methods will extend the coverage of the proteome even further as well provide information on potential protein interactions.

The next major development in this area of research rests on the biology of the system itself. The identification and characterization of in vitro expressed virulence determinants is relatively straightforward providing care is taken in interpreting the significance of the data. However, difficulties arise in the characterization of the in vivo expressed bacterial genes and their protein products. Progress can be made using model systems of bacterial infection using in vitro grown cell lines, but these can never reliably mimic the intact animal. Strategies must be developed to either enrich for the bacteria present in the infected tissue, without an intermediate growth step in vitro, or develop methods that are sufficiently sensitive to directly analyze the bacterial proteome in vivo against the background eukaryotic protein synthesis. Once such methods can be applied to a broad range of bacterial pathogens and are widely available to the medical microbiology research community, then there is certain to be an increase in the progress of analyzing bacterial pathogenesis at the level of the proteome.

## REFERENCES

1. Moxon, R., and Tang, C., *Philos. Trans. Roy. Soc. London Ser. B: Biol. Sci.* 2000, **355**, 643–656.
2. Smith, R. S., Harris, S. G., Phipps, R., and Iglewski, B., *J. Bacteriol.* 2002, **184**, 1132–1139.
3. Smith, R. S., Fedyk, E. R., Springer, T. A., Mukaida, N., Iglewski, B. H., and Phipps, R. P., *J. Immunol.* 2001, **167**, 366–374.
4. Sassetti, C., and Rubin, E. J., *Curr. Opin. Microbiol.* 2002, **5**, 27–32.

5. Hayashi, T., Makino, K., Ohnishi, M., Kurokawa, K., Ishii, K., Yokoyamam, K., Han, C. G., Ohtsubo, E., Nakayama, K., Murata, T., Tanaka, M., Tobe, T., Iida, T., Takami, H., Honda, Y., Sasakawa, C., Ogasawara, N., Yasunaga, T., Kuhara, S., Shiba, T., Hattori, M., and Shinagawa, H., *DNA Res.* 2001, **8**, 11–22.
6. Garnier, T., Eiglmeier, K., Camus, J. C., Medina, N., Mansoor, H., Pryor, M., Duthoy, S., Grondin, S., Lacroix, C., Monsempe, C., Simon, S., Harris, B., Atkin, R., Doggett, J., Mayes, R., Keating, L., Wheeler, P. R., Parkhill, J., Barrell, B. G., Cole, S. T., Gordon, S. V., and Hewinson, R. G., *Proc. Natl. Acad. Sci. USA* 2003, **100**, 7877–7882.
7. Salama, N., Guillemin, K., McDaniel, T. K., Sherlock, G., Tompkins, L., and Falkow, S., *Proc. Natl. Acad. Sci. USA* 2000, **97**, 14668–14673.
8. Israel, D. A., Salama, N., Arnold, C. N., Moss, S. F., Ando, T., Wirth, H. P., Tham, K. T., Camorlinga, M., Blaser, M. J., Falkow, S., and Peek, R. M. Jr., *J. Clin. Invest.* 2001, **107**, 611–620.
9. Holden, M., Crossman, L., Cerdeno-Tarraga, A., and Parkhill, J., *Nat. Rev. Microbiol.* 2004, **2**, 91–92.
10. Shimizu, T., Shima, K., and Yoshino, K., Yonezawa, K., Hayashi, H., *J. Bacteriol.* 2002, **184**, 2587–2594.
11. Conway, T., and Schoolnik, G. K., *Mol. Microbiol.* 2003, **47**, 879–889.
12. Canovas, F. M., Dumas-Gaudot, E., Recorbet, G., and Jorrin, J. M., *Proteomics* 2004, **4**, 285–298.
13. Kim, H., Page, G. P., and Barnes, S., *Nutrition* 2004, **20**, 155–165.
14. Onyango, P., *Curr. Cancer Drug Targets* 2004, **4**, 111–124.
15. Park, O. K., *J. Biochem. Mol. Biol.* 2004, **37**, 133–138.
16. Wilson, K. E., Ryan, M. M., Prime, J. E., Pashby, D. P., Orange, P. R., O'Beirne, G., Whateley, J. G., Bahn, S., and Morris, C. M., *J. Neurol. Neurosurg. Psychiat.* 2004, **75**, 529–538.
17. VanBogelen, R. A., in M. Hecker, and S. Mullner (Eds.), *Proteomics of Microorganisms*, Springer, Berlin, 2004, pp. 27–55.
18. Cash, P., in A. T. Bull (Ed.), *Microbial Diversity and Bioprospecting*, ASM Press, Washington DC, 2004, pp. 260–279.
19. Costas, M., *Adv. Electrophoresis* 1992, **5**, 351–408.
20. Morgan, M. G., McKenzie, H., Enright, M. C., Bain, M., and Emmanuel, F. X., *Eur. J. Clin. Microbiol. Infect. Dis.* 1992, **11**, 305–312.
21. Klimpel, K. W., and Clark, V. L., *Infect. Immun.* 1988, **56**, 808–814.
22. Cash, P., Argo, E., and Bruce, K. D., *Electrophoresis* 1995, **16**, 135–148.
23. Gormon, T., and Phan-Thanh, L., *Res. Microbiol.* 1995, **146**, 143–154.
24. Fleischmann, R. D., Adams, M. D., White, O., Clayton, R. A., Kirkness, E. F., Kerlavage, A. R., Bult, C. J., Tomb, J. F., Dougherty, B. A., Merrick, J. M., McKenney, K., Sutton, G., FitzHugh, W., Fields, C., Gocayne, J. D., Scott, J., Shirley, R., Liu, L. I., Glodek, A., Kelley, J. M., Weidman, J. F., Phillips, C. A., Spriggs, T., Hedblom, E., Cotton, M. D., Utterback, T. R., Hanna, M. C., Nguyen, D. T., Saudek, D. M., Brandon, R. C., Fine, L. D., Fritchman, J. L., Fuhrmann, J. L., Geoghagen, N. S. M., Gnehm, C. L., McDonald, L. A., Small, K. V., Fraser, C. M., Smith, H. O., and Venter, J. C., *Science* 1995, **269**, 496–511.
25. Bult, C. J., White, O., Olsen, G. J., Zhou, L., Fleischmann, R. D., Sutton, G. G., Blake, J. A., FitzGerald, L. M., Clayton, R. A., Gocayne, J. D., Kerlavage, A. R., Dougherty, B. A., Tomb, J. F., Adams, M. D., Reich, C. I., Overbeek, R., Kirkness, E. F., Weinstock, K. G., Merrick, J. M., Glodek, A., Scott, J. L., Geoghagen, N. S. M., and Venter, J. C., *Science* 1996, **273**, 1058–1073.

26. Fraser, C. M., Casjens, S., Huang, W. M., Sutton, G. G., Clayton, R., Lathigra, R., White, O., Ketchum, K. A., Dodson, R., Hickey, E. K., Gwinn, M., Dougherty, B., Tomb, J. F., Fleischmann, R. D., Richardson, D., Peterson, J., Kerlavage, A. R., Quackenbush, J., Salzberg, S., Hanson, M., van Vugt, R., Palmer, N., Adams, M. D., Gocayne, J., and Venter, J. C., *Nature* 1997, **390**, 580–586.

27. VanBogelen, R. A., Abshire, K. Z., Pertsemlidis, A., Clark, R. L, and Neidhardt, F. C., in F. C. Neidhardt, R. Curtiss, J. L. Ingraham, et al. (Eds.), *Escherichia coli. and Salmonella:. Cellular and Molecular Biology*, 2nd ed., ASM Press, Washington, DC, 1996, pp. 2067–2117.

28. Tonella, L., Walsh, B. J., Sanchez, J. C., Ou, K., Wilkins, M. R., Tyler, M., Frutiger, S., Gooley, A. A., Pescaru, I., Appel, R. D., Yan, J. X., Bairoch, A., Hoogland, C., Morch, F. S, Hughes, G. J., Williams, K. L., and Hochstrasser, D. F., *Electrophoresis* 1998, **19**, 1960–1971.

29. Ohlmeier, S., Scharf, C., and Hecker, M., *Electrophoresis* 2000, **21**, 3701–3709.

30. Hecker, M., and Engelmann, S., *Int. J. Med. Microbiol.* 2000, **290**, 123–134.

31. Sazuka, T., and Ohara, O., *Electrophoresis* 1997, **18**, 1252–1258.

32. Sazuka, T., Yamaguchi, M., and Ohara, O., *Electrophoresis* 1999, **20**, 2160–2171.

33. Jungblut, P. R., Bumann, D., Haas, G., Zimny-Arndt, U., Holland, P., Lamer, S., Siejak, F., Aebischer, A., and Meyer, T. F., *Mol. Microbiol.* 2000, **36**, 710–725.

34. Bumann, D., Meyer, T. F., and Jungblut, P. R., *Proteomics* 2001, **1**, 473–479.

35. Cho, M. J., Jeon, B. S., Park, J. W., Jung, T. S., Song, J. Y., Lee, W. K., Choi, Y. J., Choi, S. H., Park, S. G., Park, J. U., Choe, M. Y., Jung, S. A., Byun, E. Y., Baik, S. C., Youn, H. S., Ko, G. H., Lim, D., and Rhee, K. H., *Electrophoresis* 2002, **23**, 1161–1173.

36. Jungblut, P. R., Schaible, U. E., Mollenkopf, H. J., Zimny-Arndt, U., Raupach, B., Mattow, J., Halada, P., Lamer, S., Hagens, K., and Kaufmann, S. H., *Mol. Microbiol.* 1999, **33**, 1103–1117.

37. Rosenkrands, I., King, A., Weldingh, K., Moniatte, M., Moertz, E., and Andersen, P., *Electrophoresis* 2000, **21**, 3740–3756.

38. Rosenkrands, I., Weldingh, K., Jacobsen, S., Hansen, C. V., Florio, W., Gianetri, I., and Andersen, P., *Electrophoresis* 2000, **21**, 935–948.

39. Link, A. J., Hays, L. G., Carmack, E. B., and Yates, J. R., *Electrophoresis* 1997, **18**, 1314–1334.

40. Langen, H., Takacs, B., Evers, S., Berndt, P., Lahm, H. W., Wipf, B., Gray, C., and Fountoulakis, M., *Electrophoresis* 2000, **21**, 411–429.

41. Pleissner, K. P., Eifert, T., Buettner, S., Schmidt, F., Boehme, M., Meyer, T. F., Kaufmann, S. H. E., and Jungblut, P. R., *Proteomics* 2004, **4**, 1305–1313.

42. Tomb, J. F., White, O., Kerlavage, A. R., Clayton, R. A., Sutton, G. G., Fleischmann, R. D., Ketchum, K. A., Klenk, H. P., Gill, S., Dougherty, B. A., Nelson, K., Quackenbush, J., Zhou, L., Kirkness, E. F., Peterson, S., Loftus, B., Richardson, D., Dodson, R., Khalak, H. G., Glodek, A., McKenney, K., Fitzegerald, L. M., Lee, N., Adams, M. D., Hickey, E. K., Berg, D. E., Gocayne, J. D., Utterback, T. R., Peterson, J. D., Kelley, J. M., Cotton, M. D., Weidman, J. M., Fujii, C., Bowman, C., Watthey, L., Wallin, E., Hayes, W. S., Borodovsky, M., Karp, P. D., Smith, H. O., Fraser, C. M., and Venter, J. C., *Nature* 1997, **388**, 539–547.

43. Ilver, D., Arnqvist, A., Ogren, J., Frick, I. M., Kersulyte, D., Incecik, E. T., Berg, D. E., Covacci, A., Engstrand, L., and Boren, T., *Science* 1998, **279**, 373–377.

44. McAtee, C. P., Lim, M. Y., Fung, K., Velligan, M., Fry, K., Chow, T., and Berg, D. E., *Clin. Diag. Lab. Immunol.* 1998, **5**, 537–542.

45. Cash, P., Argo, E., Langford, P. R., and Kroll, J. S., *Electrophoresis* 1997, **18**, 1472–1482.

46. Lock, R. A., Cordwell, S. J., Coombs, G. W., Walsh, B. J., and Forbes, G. M., *Pathology* 2001, **33**, 365–374.

47. Mollenkopf, H. J., Jungblut, P. R., Raupach, B., Mattow, J., Lamer, S., Zimny-Arndt, U., Schaible, U. E., and Kaufmann, S. H., *Electrophoresis* 1999, **20**, 2172–2180.
48. Urquhart, B. L., Atsalos, T. E., Roach, D., Basseal, D. J., Bjellqvist, B., Britton, W. L., and Humphery-Smith, I., *Electrophoresis* 1997, **18**, 1384–1392.
49. Urquhart, B. L., Cordwell, S. J., and Humphery-Smith, I., *Biochem. Biophys. Res. Commun.* 1998, **253**, 70–79.
50. Jungblut, P. R., Muller, E. C., Mattow, J., and Kaufmann, S. H., *Infect. Immun.* 2001, **69**, 5905–5907.
51. Mattow, J., Jungblut, P. R., Muller, E. C., and Kaufmann, S. H., *Proteomics* 2001, **1**, 494–507.
52. Betts, J. C., Dodson, P., Quan, S., Lewis, A. P., Thomas, P. J., Duncan, K., and McAdam, R. A., *Microbiology* 2000, **146**, 3205–3216.
53. Valway, S. E., Sanchez, M. P., Shinnick, T. F., Orme, I., Agerton, T., Hoy, D., Jones, J. S., Westmoreland, H., and Onorato, I. M., *N. Engl. J. Med.* 1998, **338**, 633–639.
54. Hansen, E. J., Wilson, R. M., and Baseman, J. B., *Infect. Immun.* 1979, **24**, 468–475.
55. Hansen, E. J., Wilson, R. M., Clyde, W. A., Jr., and Baseman, J. B., *Infect. Immun.* 1981, **32**, 127–136.
56. Sowa, B. A., Kelly, K. A., Ficht, T. A., and Adams, L. G., *Appl. Theor. Electrophor.* 1992, **3**, 33–40.
57. Allardet-Servent, A., Carles-Nurit, M. J., Bourg, G., Michaux, S., and Ramuz, M., *J. Bacteriol.* 1991, **173**, 2219–2224.
58. Dunn, B. E., Perez-Perez, G. I., and Blaser, M. J., *Infect. Immun.* 1989, **57**, 1825–1833.
59. Enroth, H., Akerlund, T., Sille, A., and Engstrand, L., *Clin. Diag. Lab. Immunol.* 2000, **7**, 301–306.
60. Taylor, D. E., Eaton, M., Chang, N., and Salama, S. M., *J. Bacteriol.* 1992, **174**, 6800–6806.
61. Hazell, S. L., Andrews, R. H., Mitchell, H. M., Daskalopoulous, G., Jiang, Q., Hiratsuka, K., and Taylor, D. E., *FEMS Microbiol. Lett.* 1996, **20**, 833–842.
62. Govorun, V. M., Moshkovskii, S. A., Tikhonova, O. V., Goufman, E. I., Serebryakova, M. V., Momynaliev, K. T., Lokhov, P. G., Khryapova, E. V., Kudryavtseva, L. V., Smirnova, O. V., Toropyguine, I. Y., Maksimov, B. I., and Archakov, A. I., *Biochem (Moscow)* 2003, **68**, 42–49.
63. Mahairas, G. G., Sabo, P. J., Hickey, M. J., Singh, D. C., and Stover, C. K., *J. Bacteriol.* 1996, **178**, 1274–1282.
64. Sonnenberg, M. G., and Belisle, J. T., *Infect. Immun.* 1997, **65**, 4515–4524.
65. Sreevatsan, S., Pan, X., Stockbauer, K. E., Connell, N. D., Kreiswirth, B. N., Whittam, T. S., and Musser, J. M., *Proc. Natl. Acad. Sci. USA* 1997, **94**, 9869–9874.
66. Fleischmann, R. D., Alland, D., Eisen, J. A., Carpenter, L., White, O., Peterson, J., DeBoy, R., Dodson, R., Gwinn, M., Haft, D., Hickey, E., Kolonay, J. F., Nelson, W. C., Umayam, L. A., Ermolaeva, M., Salzberg, S. L., Delcher, A., Utterback, T., Weidman, J., Khouri, H., Gill, J., Mikula, A., Bishai, W., Jacobs, W. R. Jr., Venter, J. C., and Fraser, C. M., *J. Bacteriol.* 2002, **184**, 5479–5490.
67. VanBogelen, R. A., Abshire, K. Z., Moldover, B., Olson, E. R., and Neidhardt, F. C., *Electrophoresis* 1997, **18**, 1243–1251.
68. Fountoulakis, M., Langen, H., Evers, S., Gray, C., and Takacs, B., *Electrophoresis* 1997, **18**, 1193–1202.
69. Fountoulakis, M., Langen, H., Gray, C., and Takacs, B., *J. Chromatogr. A* 1998, **806**, 279–291.
70. Birch, R. M., O'Byrne, C., Booth, I. R., and Cash, P., *Proteomics* 2003, **3**, 764–776.
71. McGee, D. J., and Mobley, H. L., *Curr. Top. Microbiol. Immunol.* 1999, **241**, 155–180.

72. Sabarth, N., Lamer, S., Zimmy-Arndt, U., Jungblut, P. R., Meyer, T. F., and Bumann, D., *J. Biol. Chem.* 2002, **31**, 27896–27902.

73. Vytvytska, O., Nagy, E., Bluggel, M., Meyer, H. E., Kurzbauer, R., Huber, L. A., and Klade, C. S., *Proteomics* 2002, **2**, 580–590.

74. Nouwens, A. S., Cordwell, S. J., Larsen, M. R., Molloy, M. P., Gillings, M., Willcox, M. D., and Walsh, B. J., *Electrophoresis* 2000, **21**, 3797–3809.

75. Nouwens, A. S., Willcox, M. D., Walsh, B. J., and Cordwell, S. J., *Proteomics* 2002, **2**, 1325–1346.

76. Wang, S. B., Hu, Q., Sommerfield, M., and Chen, F., *Proteomics* 2004, **4**, 692–708.

77. Arevalo-Ferro, C., Bushmann, J., Reil, G., Gorg, A., Wiehlmann, L., Tummler, B., Eber, L., and Riede, K., *Proteomics* 2004, **4**, 1241–1246.

78. Hynes, S. O., McGuire, J., and Wadstrom, T., FEMS *Immunol. Med. Microbiol.* 2003, **36**, 151–158.

79. Baik, S. C., Kim, K. M., Song, S. M., Kim, D. S., Jun, J. S., Lee, S. G., Song, J. Y., Park, J. U., Kang, H. L., Lee, W. K., Cho, M. J., Youn, H. S., Ko, G. H., and Rhee, K. H., *J. Bacteriol.* 2004, **186**, 949–955.

80. Bumann, D., Aksu, S., Wendland, M., Janek, K., Zimny-Arndt, U., Sabarth, N., Meyer, T. F., and Jungblut, P. R., *Infect. Immun.* 2002, **70**, 3396–3403.

81. Nakamura, T., Hasegawa, T., Torii, K., Hasegawa, Y., Shimokata, K., and Ohta, M., *Arch. Microbiol.* 2004, **181**, 74–81.

82. Nilsson, C. L., Larsson, T., Gustafsson, E., Karlsson, K. A., and Davidsson, P., *Anal. Chem.* 2000, **72**, 2148–2153.

83. Utt, M., Nilsson, I., Ljungh, A., and Wadstrom, T., *J. Immunol. Methods* 2002, **259**, 1–10.

84. McAtee, C. P., Lim, M. Y., Fung, K., Velligan, M., Fry, K., Chow, T. P., and Berg, D. E., *J. Chromatogr. B Biomed. Sci. Appl.* 1998, **714**, 325–333.

85. Pal, P. G., and Horwitz, M. A., *Infect. Immun.* 1992, **60**, 4781–4792.

86. Roberts, A. D., Sonnenberg, M. G., Ordway, D. J., Furney, S. K., Brennan, P J, Belisle, J. T., and Orme, I. M., *Immunology* 1995, **85**, 502–508.

87. Andersen, P., *Infect. Immun.* 1994, **62**, 2536–2544.

88. Laal, S., Samanich, K. M., Sonnenberg, M. G., Zolla-Pazner, S., Phadtare, J. M., and Belisle, J. T., *Clin. Diag. Lab. Immunol.* 1997, **4**, 49–56.

89. Andersen, P., Askgaard, D., Ljungqvist, L., Bennedsen, J., and Heron, I., *Infect. Immun.* 1991, **59**, 1905–1910.

90. Weldingh, K., Rosenkrands, I., Jacobsen, S., Rasmussen, P. B., Elhay, M. J., and Andersen, P., *Infect. Immun.* 1998, **66**, 3492–3500.

91. Rosenkrands, I., Rasmussen, P. B., Carnio, M., Jacobsen, S., Theisen, M., and Andersen, P., *Infect. Immun.* 1998, **66**, 2728–2735.

92. Schoel, B., Gulle, H., and Kaufmann, S. H., *Infect. Immun.* 1992, **60**, 1717–1720.

93. Daugelat, S., Gulle, H., Schoel, B., and Kaufmann, S. H., *J. Infect. Dis.* 1992, **166**, 186–190.

94. Hautefort, I., and Hinton, J. C., *Philos. Trans. Roy. Soc. London Ser. B: Biol. Sci.* 2001, **355**, 601–611.

95. Slauch, J. M., Mahan, M. J., and Mekalanos, J. J., *Methods Enzymol* 1994, **235**, 481–492.

96. Mahan, M. J., Tobias, J. W., Slauch, J. M., Hanna, P. C., Collier, R. J., and Mekalanos, J. J., *Proc. Natl. Acad. Sci. USA* 1995, **92**, 669–673.

97. Wang, J., Mushegian, A., Lory, S., and Jin, S., *Proc. Natl. Acad. Sci. USA* 1996, **93**, 10434–10439.

98. Rafie-Kolpin, M., Essenberg, R. C., and Wyckoff, J. H. 3rd., *Infec. Immun.* 1996, **64**, 5274–5283.
99. Yamamoto, T., Hanawa, T., and Ogata, S., *Microbiol. Immunol.* 1994, **38**, 295–300.
100. Abshire, K. Z., and Neidhardt, F. C., *J. Bacteriol.* 1993, **175**, 3734–3743.
101. Kovarova, H., Stulik, J., Macela, A., Lefkovits, I., and Skrabkova, Z., *Electrophoresis* 1992, **13**, 741–742.
102. Monahan, I. M., Betts, J., Banerjee, D. K., and Butcher, P. D., *Microbiology* 2001, **147**, 459–471.
103. Kwaik, Y. A., and Harb, O. S., *Electrophoresis* 1999, **20**, 2248–2258.
104. Lin, J., and Ficht, T. A., *Infect. Immun.* 1995, **63**, 1409–1414.
105. Buchmeier, N. A., and Heffron, F., *Science* 1990, **248**, 730–732.
106. Hanawa, T., Yammamoto, T., and Kamiya, S., *Infect. Immun.* 1995, **63**, 4595–4599.
107. Mehta, P. K., King, C. H., White, E. H., Murtagh, J. J., and Quinn, F. D., *Infect. Immun.* 1996, **64**, 2673–2679.
108. Barker, K., Fan, H., Carroll, C., Kaplan, G., Barker, J., Hellmann, W., and Cohn, Z. A., *Infect. Immun.* 1996, **64**, 428–433.
109. Sturgill-Koszycki, S., Haddix, P. L., and Russell, D. G., *Electrophoresis* 1997, **18**, 2558–2565.
110. Fratti, R. A., Vergne, I., Chua, J., Skidmore, J., and Deretic, V., *Electrophoresis* 2000, **21**, 3378–3385.
111. De Kievit, T. R., and Iglewski, B. H., *Infect. Immun.* 2000, **68**, 4839–4849.
112. Boucher, J. C., Yu, H., Mudd, M. H., and Deretic, V., *Infect. Immun.* 1997, **65**, 3838–3846.
113. Davies, D. G., Parsek, M. R., Pearson, J. P., Iglewski, B. H., Costerton, J. W., and Greenberg, E. P., *Science* 1998, **280**, 295–298.
114. Wu, H., Song, Z., Givskov, M., Doring, G., Worlitzsch, D., Mathee, K., Rygaard, J., and Hoiby, N., *Microbiology* 2001, **147**, 1105–1113.
115. Erickson, D. L., Endersby, R., Kirkham, A., Stuber, K., Vollman, D. D., Rabin, H. R., Mitchell, I., and Storey, D. G., *Infect. Immun.* 2002, **70**, 1783–1790.
116. Middleton, B., Rodgers, H. C., Camara, M., Knox, A. J., Williams, P., and Hardman, A., *FEMS Microbiol. Lett.* 2002, **207**, 1–7.
117. Singh, P. K., Schaefer, A. L., Parsek, M. R., Moninger, T. O., Welsh, M. J., and Greenberg, E. P., *Nature* 2000, **407**, 762–764.
118. Gambello, M. J., and Iglewski, B. H., *J. Bacteriol.* 1991, **173**, 3000–3009.
119. Ochsner, U. A., and Reiser, J., *Proc. Natl. Acad. Sci. USA* 1995, **92**, 6424–6428.
120. Whiteley, M., Lee, K. M., and Greenberg, E. P., *Proc. Natl. Acad. Sci. USA* 1999, **96**, 13904–13909.
121. Hentzer, M., Wu, H., Andersen, J. B., Riedel, K., Rasmussen, T. B., Bagge, N., Kumar, N., Schembri, M. A., Song, Z., Kristoffersen, P., Manefield, M., Costerton, J. W., Molin, S., Eberl, L., Steinberg, P., Kjelleberg, S., Hoiby, N., and Givskov, M., *EMBO J.* 2003, **22**, 3803–3815.
122. Schuster, M., Lostroh, C. P., Ogi, T., and Greenberg, E. P., *J. Bacteriol.* 2003, **185**, 2066–2079.
123. Wagner, V. E., Bushnell, D., Passador, L., Brooks, A. I., and Iglewski, B. H., *J. Bacteriol.* 2003, **185**, 2080–2095.
124. Arevalo-Ferro, C., Hentzer, M., Reil, G., Gorg, A., Kjelleberg, S., Givskov, M., Riedel, K., and Eberl, L., *Environ. Microbiol.* 2003, **5**, 1350–1369.
125. Nouwens, A. S., Beatson, S. A., Whitchurch, C. B., Walsh, B. J., Schweizer, H. P., Mattick, J. S., and Cordwell, S. J., *Microbiology* 2003, **149**, 1311–1322.
126. McKnight, S. L., Iglewski, B. H., and Pesci, E. C., *J. Bacteriol.* 2000, **182**, 2702–2708.

127. Guina, T., Purvine, S. O., Yi, E. C., Eng, J., Goodlett, D. R., Aebersold, R., and Miller, S. I., *Proc. Natl. Acad. Sci. USA* 2003, **100**, 2771–2776.

128. Riedel, K., Arevalo-Ferro, C., Reil, G., Gorg, A., Lottspeich, F., and Eberl, L., *Electrophoresis* 2003, **24**, 740–750.

129. Huber, B., Riedel, K., Hentzer, M., Heydorn, A., Gotschlich, A., Givskov, M., Molin, S., and Eberl, L., *Microbiology* 2001, **147**, 2517–2528.

130. De Kievit, T. R., Gillis, R., Marx, S., Brown, C., and Iglewski, B. H., *Appl. Environ. Microbiol.* 2001, **67**, 1865–1873.

131. Sauer, K., Camper, A. K., Ehrlich, G. D., Costerton, J. W., and Davies, D. G., *J. Bacteriol.* 2002, **184**, 1140–1154.

132. Miller, B. S., and Diaz-Torres, M. R., *Methods Enzymol.* 1999, **310**, 433–441.

133. Oosthuizen, M. C., Steyn, B., Lindsay, D., Brozel, V. S., and von Holy, A., *FEMS Microbiol. Lett.* 2001, **194**, 47–51.

134. Sauer, K., and Camper, A. K., *J. Bacteriol.* 2001, **183**, 6579–6589.

135. Tremoulet, F., Duche, O., Namane, A., Martinie, B., and Labadie, J. C., *FEMS Microbiol. Lett.* 2002, **215**, 7–14.

136. Welin, J., Wilkins, J. C., Beighton, D., Wrzesinski, K., Fey, S. J., Mose-Larsen, P., Hamilton, I. R., and Svensater, G., *FEMS Microbiol. Lett.* 2003, **227**, 287–293.

137. Lester, P. J., and Hubbard, S. J., *Proteomics* 2002, **2**, 1392–1405.

138. Legrain, P., and Selig, L., *FEBS Lett.* 2000, **480**, 32–36.

139. Rain, J. C., Selig, L., De Reuse, H., Battaglia, V., Reverdy, C., Simon, S., Lenzen, G., Petel, F., Wojcik, J., Schachter, V., Chemama, Y., Labigne, A., and Legrain, P., *Nature* 2001, **409**, 211–215.

CHAPTER 14

# Unraveling *Edwardsiella tarda* Pathogenesis Using the Proteomics Approach

P. S. SRINIVASA RAO, YUEN PENG TAN, JUN ZHENG, and KA YIN LEUNG
National University of Singapore, Singapore

## 14.1 INTRODUCTION

Despite the rapid improvement in sanitation, hygiene, and living conditions, public health is still a major concern to mankind. Infectious disease is one of the major causes of illness and mortality in humans. The emergence and reemergence of new and old pathogens, widespread antibiotic resistance, and threats of bioterrorism are making the study of bacterial pathogenesis very crucial. Hence, it is important to understand the biology of a pathogen, to identify and characterize the key virulence determinants, and to examine the host–pathogen interactions. This study will help in designing suitable preventive measures as well as novel therapeutics and antimicrobials to combat bacterial infections.

For any pathogen to cause infection, it has to enter the host, evade or overcome all the defense barriers, and establish and replicate inside the host. Many of the pathogenic bacteria acquire the ability to adhere to, survive, and replicate within nonphagocytic (e.g., epithelial cells) and phagocytic cells. Some pathogens produce endotoxins or exotoxins to necrotize the tissues; others produce cascades of proteins that will scavenge nutrients and evade attacks against host defense mechanisms. The use of appropriate animal models is essential in the study of these interactions, but the majority of human pathogens (such as enterohemorrhagic *Escherichia coli* and *Pseudomonas aeruginosa*) do not have such models, which often hampers the study of host–pathogen interaction.

Most infectious bacteria produce many multifactorial virulence determinants. It is therefore vital to identify the most important clusters of genes/proteins that are essential for pathogenesis. Using the conventional approach, only one phenotype/factor is studied at a time; this is generally time consuming, and no genetic information will be unveiled. It also fails to give an integrated view of the infection process. Hence, it is necessary to use a genomewide analysis such as functional genomics to get a holistic and systematic view on the infection process.

As the number of completely sequenced bacterial genomes is increasing, the demand for functional analysis of genes is also rising. Proteomics is a very useful tool to study the

*Microbial Proteomics: Functional Biology of Whole Organisms*, Edited by Ian Humphery-Smith and Michael Hecker. 

global changes in the protein profiles of a whole pathogen and also helps in understanding the combined function of many genes and proteins. Proteomics is a powerful platform technology to study bacteria that have complete genome sequences available, while bacterial proteomes have been constructed for comparative analyses [1]. In *Edwardsiella tarda*, the inadequate sequence information did not prevent us from using proteomics to dissect its pathogenesis. Here, we have summarized our past efforts to uncover the key virulence determinants of a less studied organism whose make-up of virulence genes is virtually unknown.

## 14.2 PATHOGEN AND INFECTIONS

*Edwardsiella tarda* is a gram-negative bacillus belonging to the family Enterobacteriaceae, which includes human pathogens such as *E. coli*, *Salmonella*, *Shigella*, and *Yersinia* species. *Edwardsiella tarda* is a relatively new genus, having been described only in 1965 [2]. It is phenotypically tight, has little variability, and is biochemically similar to *Salmonella* species [3]. *Edwardiella tarda* is a ubiquitous organism with a broad host range and wide geographical distribution. It is found in freshwater and marine environments and is known to colonize in animals residing in these ecosystems. It is associated with septicemia and fatal infections in a wide variety of animals, including humans. *Edwardsiella tarda* causes gastro- and extraintestinal infections in humans, including gastroenteritis, diarrhea, bacteremia, septicemia, cellulites, abscesses, neonatal sepsis, and myonecrosis. It has also been isolated from patients of age extremes and patients with hepatobiliary diseases and immunoincompetence [3–5]. *Edwardsiella tarda* causes hemorrhagic septicemia in many farmed and feral fish [6]. It often causes emphysematous putrefactive disease and mortalities in aquaculture industries leading to great economic losses.

## 14.3 PATHOGENESIS AND VIRULENCE FACTORS

The pathogenesis of *E. tarda* is unclear and a small number of potential virulence factors have been reported. *Edwardsiella tarda* can adhere to, invade, and replicate within epithelial cells [7, 8] and fish tissues [9]. The virulent *E. tarda* strains are resistant to serum- and phagocyte-mediated killings, which are the two major defense mechanisms of innate immunity in fish and humans [7, 10, 11]. *Edwardsiella tarda* also produces toxins and exoenzymes such as dermatotoxins, catalases, and hemolysins for disseminating infection [12–14].

## 14.4 *EDWARDSIELLA TARDA* AS MODEL ORGANISM FOR STUDY OF PATHOGENS

Although the above studies help in understanding some of the roles of virulence factors in *E. tarda*, they fail to identify major or key virulence genes/proteins that are essential in pathogenesis. We established a good infection host: the blue gourami (*Trichogaster trichopterus* pallas) to dissect the pathogenicity of *E. tarda*. The median lethal dose ($LD_{50}$) of *E. tarda* PPD130/91 and its attenuated mutants in blue gourami had a difference of at least 3 logs in value ($10^{5.0}$ for the wild type and $>10^{8.0}$ for the highly attenuated mutants). The development of blue gourami as an experimental infection host is an achievement of considerable merit, as to date, only a few experimental hosts are available for systematic studies on common human pathogens including enteric pathogens. At the same time, we used a genomewide approach such as transposon tagging to systematically screen and rank a large number of potential virulence-related genes [15, 16]. The proteomics approach was also used to examine the expression of virulence factors between pathogenic and nonpathogenic strains and their production under two regulators [17, 18]. As a result, we discovered two secretion systems [a type III secretion system (TTSS) and a novel secretion system (*E. tarda* virulence protein, EVP)] that define much *E. tarda* pathogenesis. Surprisingly, these secretion systems are the most common gene clusters present in many plant and animal pathogens. Thus, *E. tarda* and our infection host have allowed us to fill the gap and to generate the much needed information to understand the biology of pathogen *E. tarda* and other pathogens.

## 14.5 GENOMEWIDE ANALYSIS OF VIRULENCE DETERMINANTS

Despite the fact that a thorough understanding of the biology of a pathogen can be achieved by examining the entire genome, this task is expensive and labor intensive. The alternative approach is to identify the virulence genes via a genomewide analysis, a cost-effective approach. Virulence genes involved in bacterial pathogenesis can be identified and characterized using various alternative genomics approaches. Some of them are (i) transposon mutagenesis [19], (ii) signature-tagged mutagenesis [20], (iii) in vivo expression technology [21], (iv) suppression subtractive hybridization [22, 23], (v) differential fluorescence induction [24], and (vi) representational differential analysis [25].

## 14.6 USING TRANSPOSON TAGGING TO IDENTIFY KEY VIRULENCE DETERMINANTS

Strauss and co-workers [26] employed transposon mutagenesis and isolated *E. tarda* mutants that were noncytotoxigenic to HEp-2 cells and defective in hemolysin production. These mutants were unable to enter into HEp-2 cells and had transposon insertions in genes having sequence similarities to the transport and activation of hemolysin genes. This study used a narrow phenotype to screen mutants in an in vitro study.

We used a similar systematic approach (Tn*phoA* mutagenesis) but utilized a broader phenotype for screening virulence-related genes by using an in vivo infection host in blue gourami fish. Secreted and outer membrane proteins are known to play an important role in causing infection and that was why we employed Tn*phoA* transposon mutagenesis to target these proteins. We generated 450,000 mutants of which 490 formed alkaline phosphatase fusions ($\text{PhoA}^{+}$) [15, 16]. These 490 mutants having mutations in secreted or surface proteins were individually injected intramuscularly into the blue gourami at a dose equivalent to 10 times the $LD_{50}$ of the wild type. We identified 14 genes (derived from 15 Tn*phoA* mutants) that are important in *E. tarda* pathogenesis [16]. They encode for virulence determinants that are homologous to regulators for arming the bacteria (PstSCAB-PhoU and EsrB), enzymes for the survival in the host (catalase KatB and glutamate decarboxylase GadB), and proteins for adhesion and motility (FimA). Mutations in *pstSCAB-phoU* (*pst* operon) or *esrB* are highly attenuated with $LD_{50}$ of about 3 logs higher when compared to the wild type. They are believed to represent two important switches or global regulators that control many virulence proteins. Proteomics was then used to unravel the mystery and lead us to the discovery of the TTSS and EVP gene clusters [17]. Transposon tagging was thus very useful in selecting key virulence genes and in providing the clues for proteomics to follow up.

## 14.7 PROTEOMICS APPROACH TO STUDYING *E. TARDA* VIRULENCE

As more bacterial genomes are being sequenced, global analysis of genes and proteins are urgently needed. New platform technologies in functional genomics such as microarray and proteomics are making studies of complex interactions of multiple genes and proteins much easier. Many research groups are actively constructing proteome databases for various important bacterial pathogens, such as pathogenic *E. coli*, *Helicobacter pylori*, *Mycobacterium tuberculosis*, and *Haemophilus influenzae* [1, 27–30]. Proteomics not only is used to identify proteins but also is capable of defining pathogenic determinants and locating immunogenic proteins [31].

In the following sections, we report on how comparative proteomics propelled our *E. tarda* research to higher ground. A comparison of extracellular proteins of virulent and avirulent *E. tarda* wild types revealed the involvement of TTSS in *E. tarda* pathogenesis. Our proteomics analysis therefore confirmed the result of transposon tagging that TTSS plays a key role in the virulence of this bacterium. Furthermore, an additional gene cluster encoded for a novel secretion system was discovered when we compared the extracellular and total proteomes of highly attenuated mutants (mutations in the *pst* operon) and the wild type. It is hard to identify *E. tarda* proteins when the genome sequence information is not available. However, the combination of gene sequencing, tandem mass spectrometry (MS/MS), and N-terminal Edman sequencing helped us to identify most of the virulence proteins in *E. tarda*.

### 14.7.1 Comparison of Extracellular Proteins of Virulent and Avirulent Strains

Comparative proteomics of virulent and avirulent bacteria is a very efficient approach to identifying proteins that are differentially expressed. It is based on the hypothesis that virulent strains produce additional proteins (virulence determinants) that avirulent strains lack. These virulence proteins are potential candidates for further studies of *E. tarda* pathogenesis. The extracellular proteins (ECPs) from various virulent and avirulent *E. tarda* strains were isolated [18]. These samples were separated using one-dimensional polyacrylamide gel electrophoresis (1D PAGE) and showed a distinct pattern for all virulent strains which had protein bands around 55, 24, and 18 kD but which were absent in all avirulent strains.

One representative virulent and one representative avirulent strain of *E. tarda* were chosen for 2D PAGE. The ECP profile showed that nine of the protein spots that were present in the virulent strain were absent in the avirulent strain and one was present in both. These spots were cut from the gel, digested using trypsin, and further analyzed using matrix-assisted laser desorption ionization time-of-flight (MALDI-TOF) MS. The masses obtained were searched in the peptide database and the only common spot was found to be a flagellin protein (Fig. 14.1a). This result validated our proteomics study as both the strains had flagella and were motile when observed under light microscopy. A protein spot

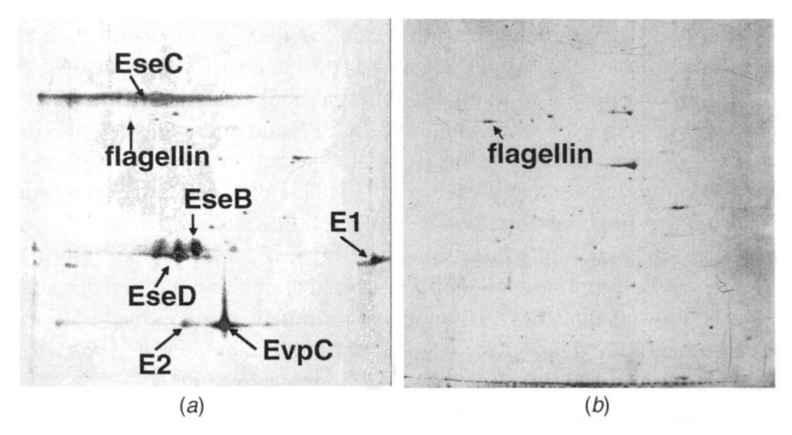


**Figure 14.1** Proteomics analysis of extracellular proteins of (*a*) *E. tarda* PPD130/91 and (*b*) highly attenuated mutant 135 in Dulbecco's modified Eagle's medium (DMEM).

that was present only in the virulent *E. tarda* strain was identified as the *E. tarda* secretion system effector protein B (EseB; Fig. 14.1*a*), which has high homology to the SseB protein of *Salmonella typhimurium*. SseB forms a translocon apparatus in TTSS and belongs to *Salmonella* pathogenicity island 2 (SPI2). The complete gene sequence of the EseB protein and the whole TTSS gene cluster was later performed by genome walking [32]. Two more Ese proteins (EseC, EseD) were eventually identified (Fig. 14.1*a*) when their peptide mass fingerprints were compared with the open reading frames (ORFs) of the TTSS gene cluster. Southern blot results revealed that the *eseD* gene was present in the virulent but not in the avirulent strains. Since SPI2 in *Salmonella* species is required for survival and replication in macrophages, virulent *E. tarda* may also have a similar system to resist phagocyte-mediated killing.

### 14.7.2 Comparative Proteomics of Highly Attenuated Mutants and Wild Type

In a concurrent study, we compared the bacterial proteomes between the wild type and three highly attenuated mutants (135, *pstC*::Tn*phoA*; 227, *pstB*::Tn*phoA* and 280, *pstS*::Tn*phoA*) to identify virulence determinants [17]. All the highly attenuated mutants had insertions in the *pst* operon and had $LD_{50}$s more than 3 log units higher in the blue gourami than that of the wild type [16]. Since the $LD_{50}$s from the highly attenuated mutants of *E. tarda* were similar to those from the avirulent strains [7, 18], we speculated that the disruption in the *pst* system might affect most of the important virulence factors.

The comparative analysis of extracellular proteomes from the wild type and the highly attenuated mutants helped in identifying several protein spots that were differentially expressed or missing in the protein profile of the highly attenuated mutants (Figs. 14.1*a* and 14.1*b* [17]). Based on a quadruple TOF MS/MS, some of these proteins were identified as (i) EseB, (ii) EseC, (iii) EseD, (iv) EvpC, and (v) flagellin. Two of the proteins could not be identified (spots E1 and E2 in Fig. 14.1*a*). Of these proteins, EseBCD are homologs of SseBCD of *S. typhimurium*, which function as translocon components and are required for the translocation of effector proteins into infected host cells [33]. The identification of EseB, EseC, and EseD proteins in the ECP fraction indicates that they may be associated with the *E. tarda* surface, as in the case of SseBCD in *S. typhimurium*, where they are loosely associated with the bacterial membrane [34]. The TTSS is known to play an important role in the virulence of many plant and animal pathogens [35]. In *E. tarda,* the TTSS is also vital for pathogenesis as mutations in *eseB* and *esrB* led to attenuation [17, 18].

At the same time, a novel temperature-regulated secretion system (encoded by *evpA-C* gene cluster) was found to secrete protein EvpC [17]. EvpA and EvpC were identified as the homologs of the putative Eip18 and Eip20 of *Edwardsiella ictaluri*, respectively. Genes *eip18* and *eip20* belong to the same operon in *E. ictaluri* [36]. Both Eip18 and Eip20 are known to be immunogenic antigens. Sequencing and a bioinformatics analysis of *evpAC* genes showed that they are widely distributed in many bacterial strains [17]. They include human pathogens such as *Salmonella enterica*, *Yersinia pestis*, *E. coli*, and *P. aeruginosa*, plant pathogens such as *Agrobacterium tumefaciens*, *Pseudomonas syringae*, and other symbionts such as *Rhizobium leguminosarum*. But the functions of these proteins are still unknown.

Further analyses were carried out on the TTSS and EVP genes using the genomics approach such as genome walking to obtain the up- and down-stream sequences. The sequencing revealed the presence of two genomic islands in *E. tarda*; hence the TTSS and EVP (*evpA-H*) gene clusters were successfully obtained. Several of these genes were further mutated using insertional and deletion mutation strategies. Analysis of proteomic profiles of these mutants showed that the EVP cluster is an independent secretion system and does not utilize TTSS apparatus to secrete EvpC. Mutations in *eseB* and *evpABC* genes resulted in attenuation of mutants, compared to the wild type, in blue gourami. The mutants were also deficient in replicating within the phagocytes [17, 18]. Further complementation of these mutants led to recovery of phenotypes, indicating that these genes are required for pathogenesis.

From comparative proteomics analyses we found two major protein secretion clusters, TTSS and EVP, and some other proteins which could not be identified. These composed most of the major secreted proteins in ECP of *E. tarda* (Fig. 14.1). These groups of proteins were shown to be required for pathogenesis in *E. tarda*.

## 14.8 FUTURE EXPERIMENTS AND CONCLUSIONS

Currently, we are attempting to completely sequence the TTSS and EVP clusters. We are also trying to determine the regulation and relationship of TTSS and EVP as well as the functions of their effector molecules that are required to communicate with the host. We would like to unveil the function of every gene present in these two secretion systems. Experiments are in progress to study the structures and mechanism of the secretion systems, such as the assembly of the secretion apparatus (translocon complex), the role of chaperones in secretion, and the function of effectors. These will help in the discovery of antimicrobials, vaccines, and diagnostics not only for *E. tarda* but also for other related pathogens.

Our studies show that the functional genomics approach (proteomics together with transposon mutagenesis) help in unraveling the pathogenesis of *E. tarda* (Fig. 14.2). This is a fast, reliable, holistic, and integrated approach which can be used to study other pathogens. A comparative proteomics analysis of virulent and avirulent *E. tarda* combined with mutational analysis help in successfully identifying two different gene clusters, TTSS and EVP. The EVP homologs are present in most pathogenic bacteria, but their precise

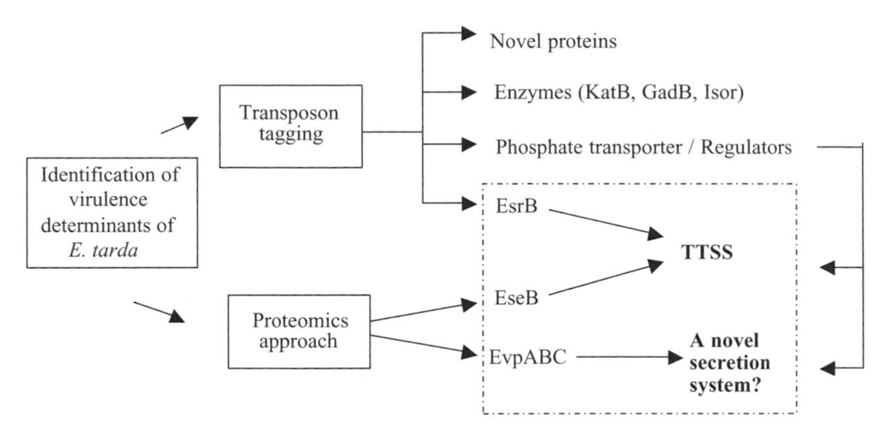


**Figure 14.2** Use of functional genomics to unravel *E. tarda* pathogenesis.

functions are still unknown. In the *E. tarda* model, the EVP gene cluster played roles in pathogenesis, replication inside phagocytes, and as a novel secretion system. This study not only assists us in understanding the pathogenesis of *E. tarda* to a greater extent but also enables this knowledge to be applied to other pathogens as EVP and TTSS clusters are present in many pathogenic bacteria.

## ACKNOWLEDGMENTS

The authors are grateful to the National University of Singapore and the Lee Hiok Kwee Donation for providing the research grant for this work. They also acknowledge the assistance of Dr. Yoshiyuki Yamada for providing the 2D gels.

## REFERENCES

1. Cash, P., *Adv. Biochem. Eng. Biotechnol.* 2003, **83**, 93–115.
2. Ewing, W. H., McWhorter, A. C., Escobar, M. R., and Lubin, A. H., *Int. Bull. Bacteriol. Nomencl. Taxon.* 1965, **15**, 33–38.
3. Janda, J. M., and Abbott, S. L., *Clin. Infect. Dis.* 1993, **17**, 742–748.
4. Plumb, J. A., in V. Inglis, R. J. Roberts, and N. R. Bromage (Eds.), *Bacterial Diseases of Fish*, University Press, Cambridge, 1993, pp. 61–79.
5. Slaven, E. M., Lopez, F. A., Hart, S. M., and Sanders, C. V., *Clin. Infect. Dis.* 2001, **32**, 1430–1433.
6. Thune, R. L., Stanley, L. A., and Cooper, R. K., *Annu. Rev. Fish Dis.* 1993, **3**, 37–68.
7. Ling, S. H., Wang, X. H., Xie, L., Lim, T. M., and Leung, K. Y., *Microbiology* 2000, **146**, 7–19.
8. Phillips, A. D., Trabulsi, L. R., Dougan, G., and Frankel, G., *FEMS Microbiol. Lett.* 1998, **161**, 317–323.
9. Ling, S. H., Wang, X. H., Lim, T. M., and Leung, K. Y., *FEMS Microbiol. Lett.* 2001, **194**, 239–243.

10. Janda, J. M., and Abbott, S. L., Kroske-Bystrom, S., Cheung, W. K., Powers, C., Kokka, R. P., and Tamura, K., *J. Clin. Microbiol.* 1991, **29**, 1997–2001.
11. Srinivasa Rao, P. S., Lim, T. M., and Leung, K. Y., *Infect. Immun.* 2001, **69**, 5689–5697.
12. Hirono, I., Tange, N., and Aoki, T., *Mol. Microbiol.* 1997, **24**, 851–856.
13. Srinivasa Rao, P. S., Yamada, Y., and Leung, K.Y., *Microbiology* 2003, **149**, 2635–2644.
14. Ullah, M. A., and Arai, T., *Fish Pathol.* 1983, **18**, 65–70.
15. Mathew, J. A., Tan, Y. P., Srinivasa Rao, P. S., Lim, T. M., and Leung, K. Y., *Microbiology* 2001, **147**, 449–457.
16. Srinivasa Rao, P. S., Lim, T. M., and Leung, K. Y., *Infect. Immun.* 2003, **71**, 1343–1351.
17. Srinivasa Rao, P. S., Yamada, Y., Tan, Y. P., and Leung, K. Y., *Mol. Microbiol.*, 2004, **53**, 573–586.
18. Tan, Y. P., Lin, Q., Wang, X. H., Joshi, S., Hew, C. L., and Leung, K. Y., *Infect. Immun.* 2002, **70**, 6475–6480.
19. Berg, C. M., Berg, D. E., and Groisman, E. A., in E. D. Berg, and M. M. Howe (Eds.), *Mobile Elements*, American Society for Microbiology Press, Washington, DC, 1994, pp. 880–925.
20. Hensel, M., Shea, J. E., Gleeson, C., Jones, M. D., Dalton, E., and Holden, D. W., *Science* 1995, **269**, 400–403.
21. Mahan, M. J., Slauch, J. M., and Mekalanos, J. J., *Science* 1993, **259**, 686–688.
22. Bogush, M. L., Velikodvorskaya, T. V., Lebedev, Y. B., Nikolaev, L. G., Lukyanov, S. A., Fradkov, A. F., Pliyev, B. K., Boichenko, M. N., Usatova, G. N., Vorobiev, A. A., Andersen, G. L., and Sverdlov, E. D., *Mol. Gen. Genet.* 1999, **262**, 721–729.
23. Zhang, Y. L., Ong, C. T., and Leung, K. Y., *Microbiology* 2000, **146**, 999–1009.
24. Valdivia, R. H., and Falkow, S., *Science* 1997, **277**, 2007–2011.
25. Claus, H., Friedrich, A., Frosch, M., and Vogel, U., *J. Bacteriol.* 2000, **182**, 1296–1303.
26. Strauss, E. J., Ghori, N., and Falkow, S., *Infect. Immun.* 1997, **65**, 3924–3932.
27. Govorun, V. M., Moshkovskii, S. A., Tikhonova, O. V., Goufman, E. I., Serebryakova, M. V., Momynaliev, K. T., Lokhov, P. G., Khryapova, E. V., Kudryavtseva, L. V., Smirnova, O. V., Toropyguine, I. Y., Maksimov, B. I., and Archakov, A. I., *Biochem. (Moscow)* 2003, **68**, 42–49.
28. Jungblut, P. R., Schaible, U. E., Mollenkopf, H. J., Zimny-Arndt, U., Raupach, B., Mattow, J., Halada, P., Lamer, S., Hagens, K., and Kaufmann, S. H., *Mol. Microbiol.* 1999, **33**, 1103–1117.
29. Mo, L., Rosenshine, I., Tung, S. L., Wang, X. H., Friedberg, D., Hew, C. L., and Leung, K. Y., *Appl. Environ. Microbiol.*, 2004, **70**, 5274–5282.
30. Thoren, K., Gustafsson, E., Clevnert, A., Larsson, T., Bergstrom, J., and Nilsson, C. L., *J. Chromatogr. B. Anal. Technol. Biomed. Life Sci.* 2002, **782**, 219–226.
31. Krah, A., and Jungblut, P. R., *Methods Mol. Med.* 2004, **94**, 19–32.
32. Tan, Y. P., Zheng, J., Tung, S. L., Rosenshine, I., and Leung, K. Y., *Microbiology* 2005, **151**, 2301–2313.
33. Nikolaus, T., Deiwick, J., Rappl, C., Freeman, J. A., Schroder, W., Miller, S. I., and Hensel, M., *J. Bacteriol.* 2001, **183**, 6036–6045.
34. Klein, J. R., and Jones, B. D., *Infect. Immun.* 2001, **69**, 737–743.
35. Hueck, C. J., *Microbiol. Mol. Biol. Rev.* 1998, **62**, 379–433.
36. Moore, M. M., Fernandez, D. L., and Thune, R. L., *Dis. Aquat. Org.* 2002, **52**, 93–107.

CHAPTER 15

# Structural Proteomics and Computational Analysis of a Deadly Pathogen: Combating *Mycobacterium tuberculosis* from Multiple Fronts

MICHAEL STRONG and CELIA W. GOULDING
UCLA-DOE Institute of Genomics and Proteomics, Los Angeles, California

## 15.1 INTRODUCTION

Tuberculosis (TB) is a devastating respiratory disease caused by the bacterium *Mycobacterium tuberculosis*. This disease of the lungs has been estimated to result in two million to three million human deaths per year. While this number itself is astounding, approximately one-third of the world's population is currently infected with the *M. tuberculosis* bacterium (World Health Organization, 2002). Although the majority of those infected will never develop symptoms of the active disease, a number of new developments, including the emergence of several multidrug-resistant *M. tuberculosis* strains, have caused the World Health Organization to declare tuberculosis a global health emergency (World Health Organization, 2002).

Biomedical research has been dramatically transformed with the introduction of high-throughput DNA-sequencing technologies. These technologies have enabled scientists to identify the complete set of genes that defines a particular organism. This set of genes, otherwise known as the genome, can be thought of as a molecular blueprint of an organism since the genome contains the complete set of instructions necessary to "build" a particular organism. In the simplest case of bacterial genomes, each gene encodes a particular protein, which in turn performs a particular molecular function. Proteins can be thought of as the molecular workhorses of an organism, since proteins carry out most of the activities within biological cells. As a result, proteins are frequently the targets of antimicrobial drugs.

Currently, the complete genome sequences of more than 200 organisms are known, providing researchers with the genetic blueprint of organisms ranging from the smallest viral organisms to higher eukaryotic organisms such as humans. Although genome-sequencing efforts have enabled researchers to identify the complete set of genes and corresponding proteins for a particular organism, a more relevant biological question is, "how do these proteins work together within their cellular environment?" Typically,

*Microbial Proteomics: Functional Biology of Whole Organisms*, Edited by Ian Humphery-Smith and Michael Hecker. 

proteins do not operate in isolation but, instead, work synergistically with a variety of other proteins, often as members of common protein pathways or protein complexes.

In conjunction with the rapid increase of completely sequenced genomes, a profound shift toward the study of complete biological systems has emerged. As a complement to traditional research projects involving individual genes and individual proteins, the fields of genomics and proteomics have, by definition, focused on the investigation of the complete set of organismal genes and proteins, respectively. This focus has facilitated the investigation of complete cellular systems and has led to exciting new directions in biological research. Two disciplines that have embraced such a holistic approach are the fields of structural genomics and computational biology.

Structural genomics involves the large-scale determination and analysis of protein structures on a genomewide scale. There are three main objectives of structural genomics: (1) the determination of three-dimensional (3D) structures for new protein families, (2) the assignment of functions to proteins of previously unknown function based on the analysis of their atomic structures, and (3) the construction of genomewide protein–protein interaction networks. The determination of structures for new protein families facilitates the construction of 3D models for homologous proteins without structural data. Significant technological developments are currently underway to ensure that structural genomics is a high-throughput endeavor. Robots, for example, are being developed to automate all the steps in the experimental pathway from gene cloning to 3D protein structure determination.

To better understand the molecular mechanisms of *M. tuberculosis*, a structural genomics consortium has been set up to investigate the proteome of the deadly *M. tuberculosis* pathogen. The goal of the *M. tuberculosis* structural genomics project is to elucidate the biology and pathogenicity of *M. tuberculosis* through structural biology and functional genomics and to provide a foundation for drug discovery (Goulding et al., 2002a; Terwilliger et al., 2003). In this chapter, we will describe how computational biology in combination with structural genomics may reveal potential drug targets against TB and be used to build a map of interconnected proteins within *M. tuberculosis*.

## 15.2 COMPUTATIONAL PROTEOMICS OF *M. TUBERCULOSIS*

The field of computational biology, much like that of structural genomics, holds great promise for the study of microorganisms at the systems level. Since most proteins do not operate in isolation, a more holistic approach is necessary to truly understand the complexity of protein relationships within their cellular environment. Most proteins, in fact, function within the context of other proteins. These relationships range from proteins that interact physically, forming common complexes, to proteins that participate in shared biochemical pathways. Proteins that participate in common complexes or shared pathways are often referred to as functionally linked, since they have a functional dependence on one another.

A number of bioinformatic and computational methods have been developed to identify functionally linked proteins on a genomewide basis. These methods utilize genomic information available from hundreds of genome-sequencing efforts. Complementing traditional experimental methods such as yeast two-hybrid experiments (Uetz et al., 2000), microarray experiments (Brown and Botstein, 1999), and co-immunoprecipitation experiments, these computational methods facilitate the genome wide identification of functionally linked proteins based on sequence information alone.

Four of the most useful computational methods for identifying functionally linked genes and proteins are the Rosetta Stone (Marcotte et al., 1999a), Phylogenetic Profile, (Pellegrini et al., 1999) conserved Gene Neighbor (Dandekar et al., 1999; Overbeek et al., 1999), and Operon (Strong et al., 2003b) methods. These methods enable the identification of functionally linked genes and proteins on a genomewide basis. Functionally linked proteins may represent members of a common protein complex or a shared biochemical pathway or may indicate proteins that serve related functions within a cell. In combination, these methods provide a very powerful approach for identifying genomewide functional linkages among proteins. For example, Strong et al. (2003a,b) applied the four described methods to identify functionally linked genes and proteins throughout the *M. tuberculosis* genome.

The first of these computational methods, termed the Rosetta Stone method, identifies individual genes that occur as a single fusion gene in another organism (Marcotte and Marcotte, 2002; Marcotte et al., 1999a). Named after the historical stone tablet that was used to decipher the ancient Egyptian hieroglyphs in the early 1800s, this computational method enables the identification of previously cryptic functional linkages among proteins on a genomewide scale. An example of two genes linked by the Rosetta Stone method are the *M. tuberculosis leuC* and *leuD* genes. Although *leuC* and *leuD* exist as two separate genes in the *M. tuberculosis* genome, these genes occur as a single fusion gene in the yeast *Saccharomyces cerevisiae* (Fig. 15.1). The *S. cerevisiae* fusion gene, encoding α-isopropylmalate isomerase, contains one region that is similar, or homologous, to the *M. tuberculosis leuC* gene and another region that is homologous to the *M. tuberculosis leuD* gene. Both *leuC* and *leuD* encode proteins involved in leucine biosynthesis. Studies in other bacterial organisms have also shown that these two proteins interact physically to form a common heterodimeric protein complex (Fultz and Kemper, 1981).

The Phylogenetic Profile method identifies genes that occur in a correlated manner across many genomes (Pellegrini et al., 1999). A profile of each gene consists of a record of the presence or absence of identifiable homologues in all other sequenced genomes. Again using the *leuC* and *leuD* genes for demonstration purposes, we see that *leuC* and *leuD* have similar phylogenetic profiles, indicated by the presence or absence of these genes in each genome. In Figure 15.1 we see that the *leuC* and *leuD* genes are both present in the *Streptomyces coelicolor*, *M. tuberculosis*, and *Salmonella typhi* genomes, but both genes are absent in the *Streptococcus pyogenes* and *Treponema pallidum* genomes. Genes that occur in

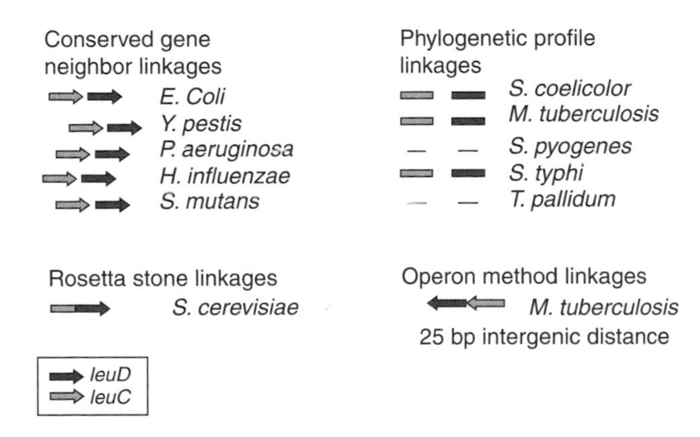


**Figure 15.1** Overview of four computational methods. The *M. tuberculosis leuC* and *leuD* genes are linked by the conserved Gene Neighbor, Rosetta Stone, Phylogenetic Profile, and Operon methods.

a correlated manner are likely to have an evolutionarily selected dependence on each other and are thus linked by the Phylogenetic Profile method (Pellegrini et al., 1999).

Genes linked by the conserved Gene Neighbor method occur in close chromosomal proximity in multiple genomes (Dandekar et al., 1999; Overbeek et al., 1999). Continuing with the example of the *leuC* and *leuD* genes, we see that *leuC* and *leuD* occur next to each other on the chromosome in the genomes of *Escherichia coli*, *Yersinia pestis*, *Pseudomonas aeruginosa*, *Haemophilus influenzae*, and *Streptococcus mutans*. The clustering of genes of related function is a common characteristic in bacteria (Moreno-Hagelsieb et al., 2001) but is also observed to a lesser extent in some eukaryotic organisms. Conservation of gene neighbors has also been used to study the evolution of bacterial genomes (Wolf et al., 2001).

The fourth method links genes that are likely to belong to common operons in bacteria, based on the distance between genes in the same orientation (Salgado et al., 2000; Strong et al., 2003b). This method, designated the Operon method, links genes that are separated by a minimal intergenic distance. In order for genes to be linked by the Operon method, they must occur in the same orientation and in close proximity on the chromosome, both characteristics of microbial operons (Moreno-Hagelsieb and Collado-Vides, 2002b; Salgado et al., 2000; Strong et al., 2003b). Microorganisms such as *M. tuberculosis* often have their genes organized into multigenic operons (Madigan et al., 2000). Genes that are organized into common operons often share related cellular roles, either as members of a shared biochemical pathway or protein complex, or may have related cellular functions (Lodish et al., 1995). Databases of known and predicted operons are available for other organisms such as *E. coli* (Salgado et al., 2001), and a number of groups have utilized predicted operon structure for the inference of gene function (Overbeek et al., 1999; Ermolaeva et al., 2001; Pellegrini et al., 2001; Strong et al., 2003b).

The Operon method is distinct from that of the conserved Gene Neighbor method since the Operon method relies only on information contained in a single genome, whereas the conserved Gene Neighbor method relies on information contained in multiple genomes. The Operon method, therefore, is useful in cases of species-specific genes since linkages established by the Operon method are not dependent on the presence of identifiable homologues in other organisms. The Phylogenetic Profile method and the conserved Gene Neighbor method have also been used to confirm operon predictions based on intergenic distances (Moreno-Hagelsieb and Collado-Vides, 2002a, b) as well as to examine the coevolution of gene clusters (Zheng et al., 2002). Genes that constitute a fusion gene in one organism have also been found to occur as members of a common operon in another organism (Yanai et al., 2002), much like the *leuC* and *leuD* example. Homologues of the *leuC* and *leuD* genes have also been shown to occur in common operons in other organisms (Tamakoshi et al., 1998).

Although for the sake of simplicity the characterized *leuC* and *leuD* genes have been used for demonstration purposes, many functional linkages occur among uncharacterized proteins. These linkages can be used to infer the function of previously uncharacterized proteins on a genomewide basis and can be used to infer and identify networks of functionally linked proteins. An example of the inference of protein function based on computationally inferred functional linkages involves the uncharacterized *M. tuberculosis* gene Rv1879. Rv1879 is linked to the glutamine synthetase paralog *glnA3* by both the Rosetta Stone and Operon method (Strong et al., 2003b), indicating that a fusion gene homologous to both Rv1879 and *glnA3* exists in another organism and that the genome organization of these two genes suggests that they are part of a common operon in *M. tuberculosis*. Based on this evidence, we infer that Rv1879 has a function related to that of the glutamine synthetase homologue *glnA3*. The reliability of individual computational methods are assessed by comparing the annotations of computationally linked proteins (Marcotte et al., 1999b). Proteins linked by the four methods are much more likely to share common annotations, and therefore functions, than randomly linked proteins (Strong et al., 2003b). These methods provide a non-homology-based approach for identifying protein function and complement traditional homology-based methods such as Basic Local Alignment Search Tool (BLAST) (Altschul et al., 1990). These computational methods also allow the function of a protein to be defined within the context of its cellular linkages (Eisenberg et al., 2000b).

Many functional linkages connect previously uncharacterized genes to important metabolic pathways, complexes, and drug targets (Strong et al., 2003b). For example, the uncharacterized gene Rv1130 is linked to the isocitrate lyase gene *icl*, an important persistence factor and proposed drug target (McKinney et al., 2000; Sharma et al., 2000). The uncharacterized genes Rv2164c, Rv2165c, and Rv2166c are linked *pbpB*, which encodes for a penicillin-binding protein, suggesting a role in cell wall metabolism. The uncharacterized genes Rv2926c and Rv2927c are linked to a gene involved in RNA degradation (*rnc*). And the uncharacterized genes Rv1503c and Rv1504c are linked to components of the arabinogalactan biosynthesis pathway (Strong et al., 2003b), the major target of the anti- tubercular drug ethambutol (Belanger et al., 1996).

## 15.3 FROM FUNCTIONAL LINKAGES TO GENOME MAPS

To better understand the molecular mechanisms leading to the pathogenesis of *M. tuberculosis*, Strong et al. (2003a) developed a novel method to visualize computationally inferred linkages across a particular genome. Typically, computational methods such as the Rosetta Stone, Phylogenetic Profile, conserved Gene Neighbor, and Operon methods yield thousands of high-confidence functional linkages for a single genome. A major challenge, therefore, has been the visualization and interpretation of these linkages on a genomewide basis.

The traditional method for investigating genomewide functional linkages has utilized node and edge graphs, as shown in Figure 15.2*a*. These traditional graphs represent each protein as a single circle or node in the graph and each pair of functionally linked proteins as a line or edge connecting two protein nodes (Marcotte et al., 1999b; Uetz et al., 2000). The resulting collection of connected nodes is often referred to as a network.

While this method of network representation has its merits when dealing with dozens or even hundreds of functional linkages, it is often more difficult to investigate graphs representing thousands of linkages due to the overlap of graphical properties. To address this difficulty, an alternative method for network representation has been developed by Strong et al. (2003a) that utilizes a two-dimensional matrix to represent functional linkages inferred throughout a particular genome (Fig. 15.2b). This matrix is analogous to one developed to indicate experimentally identified protein interactions in the T7 bacteriophage (Grigoriev, 2001).

In the case of *M. tuberculosis*, Strong et al. (2003a) identified and mapped 9766 high-confidence functional linkages on a two-dimensional scatter plot, better known as a genomewide functional linkage map. The organization of the genome map corresponds directly to the organization of genes along the *M. tuberculosis* chromosome, so that the order of the genes along each axis corresponds to the natural gene order. For example, in Figure 15.2, both the $x$ and $y$ axes represent a monotonically ordered list of genes

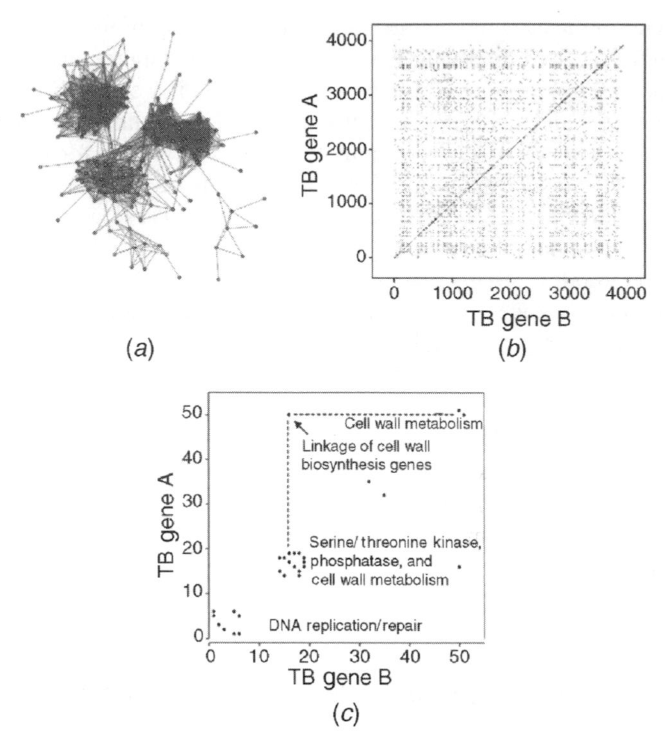


**Figure 15.2** Comparison of two methods for illustrating protein networks: (*a*) traditional protein network composed of nodes and edges, (*b*) genomewide functional linkage map representing functional linkages throughout the *M. tuberculosis* genome. The *M. tuberculosis* genome consists of a single, circular chromosome consisting of approximately 4000 genes. Each axis of the map represents a monotonically ordered list of genes corresponding to the order of genes along the *M. tuberculosis* chromosome, starting at the origin of replication. Functional linkages are indicated by a single point on the graph; for example the point at coordinate 1, 5 indicates a functional linkage between the first gene (*dnaA*) and the fifth gene (*gyrB*) from the origin of replication. (*c*) Expanded view of the genomewide functional linkage map representing only the first 50 genes. (Adapted from Strong et al., 2003a.)

corresponding to the order of genes along the *M. tuberculosis* chromosome. Genes are numbered in sequential order, starting at the origin of replication; in the case of *M. tuberculosis*, the genome contains about 4000 genes (Cole et al., 1998). Genome-sequencing centers such as the Pasteur Institue and Sanger Center contain genome depositories where information on chromosomal gene order can be obtained.

Each functional linkage represented in Figure 15.2*b* is inferred by two or more independent computational methods. These functional linkages are indicated by a single point or dot at the coordinate corresponding to the functional link. For example, the functional linkage between the first gene (*dnaA*) and the fifth gene (*gyrB*) is indicated by a single dot or point at coordinate 1, 5. Points on the functional linkage map are symmetric since a link from gene A to gene B is also indicated as a link from gene B to gene A.

A number of interesting features of chromosomal organization are revealed in the genomewide functional linkage map, including features that would not typically be apparent using traditional node and edge graphs. For instance, there is a dense clustering of functionally linked genes near the diagonal, indicating functionally linked genes that are in close chromosomal proximity. Many of these linkages are likely to correspond to potential *M. tuberculosis* operons.

Figure 15.2*c* represents a zoomed-in region of the genomewide functional linkage map representing only the first 50 genes. Again, the clustering of genes along the diagonal is readily apparent. Many of these clusters contain genes of related function, for example, the cluster of genes Rv0001 through Rv0006 are involved in DNA replication and repair. The cluster of genes Rv0014c through Rv0019c contains two serine/threonine kinase genes, one phosphatase gene, two cell wall metabolism genes, and an uncharacterized gene. While at first glance the relationship among these six genes is not obvious, both computational and structural analyses of these proteins reveals that these proteins may be involved in a signaling pathway involved in cell wall metabolism. The serine/threonine kinase gene Rv0014c *pknB* contains a unique protein domain termed the PASTA domain (derived from pencillin-binding protein and serine/threonine kinase associated domain) that, based on structural analysis, is hypothesized to sense unlinked peptidoglycan near the cell surface (Yeats et al., 2002). This structural evidence links *pknB* to a function related to that of the cell wall metabolism genes. The phosphatase protein Rv0018c may also play a role in this pathway, and the uncharacterized gene Rv0019c, which contains an FHA (forkhead associated) domain, may also be involved, since FHA domains are known to mediate phosphorylation-dependent protein–protein interactions (Pallen et al., 2002).

To further facilitate the identification of functionally related genes on a genomewide basis, subsequent hierarchical clustering is applied to cluster the genomewide functional linkage maps (Strong et al., 2003a). The hierarchical clustering algorithm clusters genes that have similar functional linkage profiles. A functional linkage profile represents the presence or absence of a functional linkage in the form of a binary vector. In the example shown in Figure 15.3*a*, the hypothetical gene, gene A, is functionally linked to gene B, gene C, and gene D, as indicated by a 1 in the profile. The absence of a functional linkage is indicated by a 0. Genes that have similar function linkage profiles cluster by this method. This method is analogous to methods traditionally used to cluster microarray data (Eisen et al., 1998), but in this case, data are clustered based on the similarity of functional linkage profiles instead of the similarity of expression profiles.

The resulting hierarchical clustered map reveals distinct clusters of genes. These clusters, also referred to as functional modules, often contain genes of related cellular function and are analogous to functional modules identified in traditional protein networks (Snel et al., 2002). In the case of *M. tuberculosis*, clusters are observed for a wide variety of pathways and complexes ranging from degradation pathways to two-component systems (Fig. 15.3*b*). The largest clusters, or functional modules, correspond to proteins involved in the degradation of fatty acids, energy metabolism, and polyketide synthesis. Genes are clustered using the programs Cluster and Treeview (Eisen et al., 1998) originally developed to cluster microarray data.

A few of the clustered groups of genes are shown in Figure 15.3*c*. As shown, both components of known protein pathways and known protein complexes cluster by this method. Here we see that seven of the eight members of the adenosine triphosphate (ATP) synthase complex cluster by this method, as do all seven members of the arginine biosynthesis pathway. Clusters of genes are also observed that contain a mix of annotated and nonannotated genes. These clusters aid in the inference of protein function for uncharacterized *M. tuberculosis* genes. For example, Figure 15.3*c* shows a cluster of six genes involved in chaperone or heat-shock activity, along with an uncharacterized gene

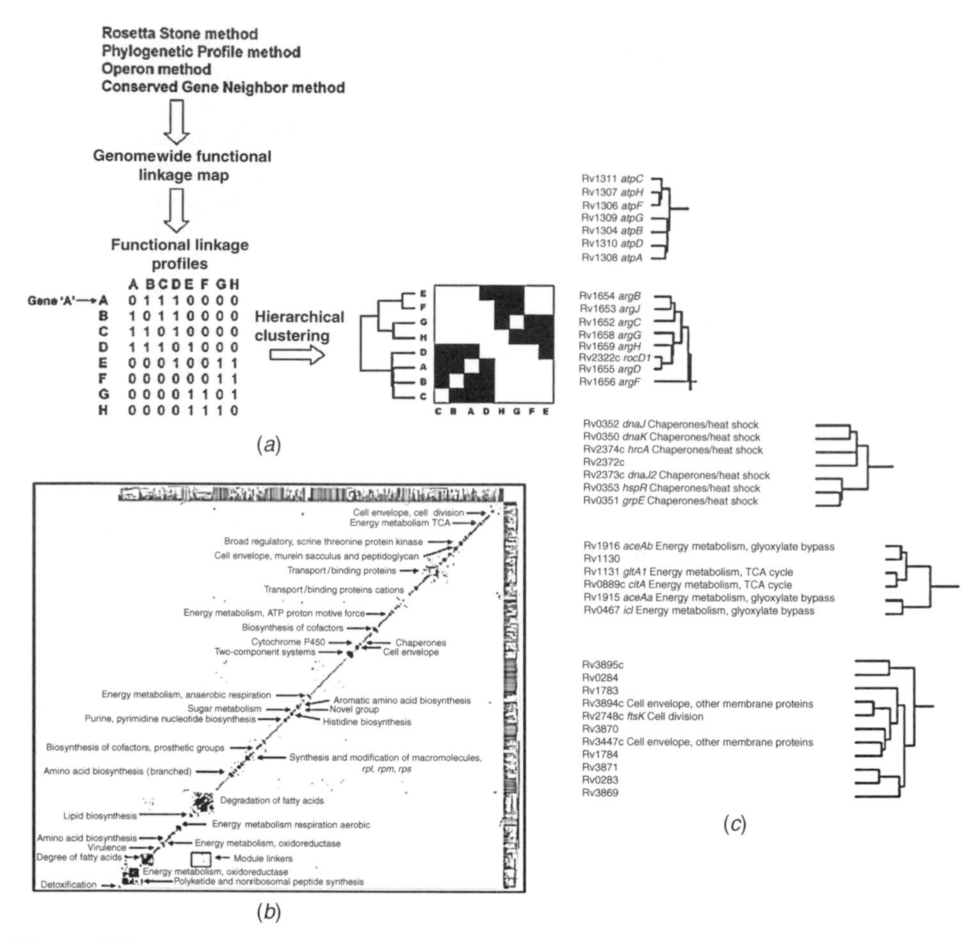


**Figure 15.3** Overview of the hierarchical clustering method. (*a*) Functional linkage profile is created for each gene, representing genomewide functional linkages in the form of a bit vector. Genes are clustered according to the similarity of their functional linkage profiles. (*b*) Clustered map of the *M. tuberculosis* genome. Genes of related function often have similar functional linkage profiles and therefore cluster together. (*c*) Five examples of *M. tuberculosis* gene clusters. (Adapted from Strong et al., 2003a.)

Rv2372c. We hypothesize that this gene has a function related to that of the chaperone proteins. In another example we see five genes involved in the tricarboxylic acid (TCA) cycle or closely related glyoxylate bypass cluster with the uncharacterized gene Rv1130. Although Rv1130 is uncharacterized, it does have some homology to a putative *PrpD* domain (Strong et al., 2003a). The *PrpD* domain is thought to be involved in the conversion of propionate to pyruvate (Horswill and Escalante-Semerena, 2001), a substrate of the TCA cycle, further supporting this functional inference. Clusters containing mostly nonannotated genes are also observed, such as the large cluster shown in Figure 15.3*c*. These clusters may suggest novel protein pathways or protein complexes in *M. tuberculosis* and provide direction for the design of future biochemical experiments. Interaction and metabolic networks have also been clustered in a similar manner, yielding additional insights into the modular nature of biological networks (Ravasz et al., 2002; Rives and Galitski, 2003).

As we venture into the postgenomic era, it is likely that methods for identifying functionally linked genes and proteins on a genomewide basis will become increasingly useful. In addition to databases of experimentally identified protein interactions (Bader et al., 2003; Mewes et al., 2002; Xenarios et al., 2002), a number of databases have been created to catalog inferred protein linkages for many organisms (Bowers et al., 2004; Mellor et al., 2002; von Mering et al., 2003).

## 15.4 STRUCTURAL PROTEOMICS OF *M. TUBERCULOSIS*

This section will summarize some of the *M. tuberculosis* structures from the TB structural genomics consortium and other independent laboratories. To illustrate proteins involved with the cell envelope of *M. tuberculosis*, the structures described concentrate on secreted proteins, transport proteins, signaling proteins, and proteins that are involved in the synthesis of cell wall components. There are examples of potential protein anti-TB drug targets, and some of the structures have been combined with computational methods in an attempt to reconstruct biochemical pathways.

### 15.4.1 Secretion Systems

Approximately 50% of the predicted secreted proteins in *M. tuberculosis* have signal peptides (Gomez et al., 2000). These proteins are secreted by a type II secretion mechanism and one of the key proteins, SecA1, is described below. When proteins are secreted into the extracellular environment, they must fold correctly, which sometimes involves the formation of intermolecular disulfide bonds. Hence, the oxidoreductase (Dsb) proteins are important structural targets due to their involvement in oxidatively refolding secreted proteins and reduction of thiols for heme binding. Two of these proteins, DsbE and DsbF, are probably secreted by the *sec*-dependent type II secretion system.

***SecA1 (Rv3240c)*** In mycobacteria, one of the major secretion mechanisms is the type II, *sec*-dependent pathway. This system is catalyzed by a multiprotein translocase, which recognizes the signal sequence of a preprotein and uses ATP binding and hydrolysis as the driving force for transport (Mori and Ito, 2001). In *M. tuberculosis*, there are two SecA homologues, which is unusual among bacterial species. *secA*1 is an essential gene and is equivalenit to the *E. coli* SecA (Sharma et al., 2003). *secA*2, in contrast, is not essential, though it is required for full virulence (Braunstein et al., 2001). The crystal structure of SecA1 reveals two domains, a motor domain and a translocation domain (Sharma et al., 2003). SecA1 forms a dimer, which is predicted to interact with the SecYEG pore and function as a "molecular ratchet" that utilizes ATP hydrolysis for physical movement of the preprotein. In Figure 15.4, one can imagine movement of the monomers so that the central pore of the dimer "breathes", enabling proteins to be transported into the extracellular envirnoment.

***DsbE, DsbD, and DsbF (Rv2878, Rv2874, and Rv1677)*** DsbE (MPT53) is a secreted oxidoreductase protein which has a Cys–XX–Cys active site. The structure revealed a thioredoxin-like fold with its two active-site cysteines in their reduced form (Fig. 15.4) (Goulding et al., 2004). Structural and functional analysis of DsbE suggests that it has a similar function to *E. coli* DsbA, which catalyzes the oxidation of reduced,

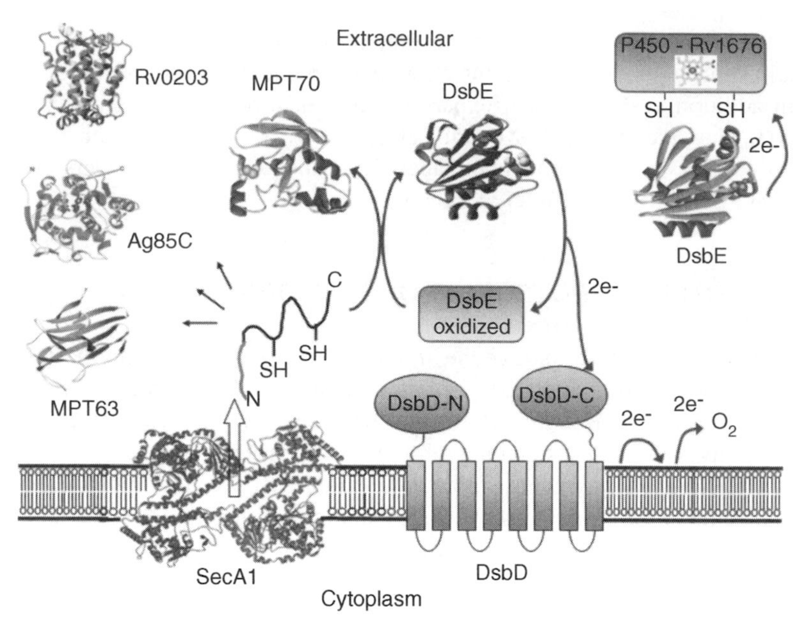


**Figure 15.4** Structures and mechanism of secretion system and some secreted proteins. SecA1 is part of the *sec*-dependent secretion system which secretes proteins with signal peptides (red section of unfolded secreted protein) form the cytoplasm into the extracellular environment. The *M. tuberculosis* membrane is depicted in the figure. If the secreted proteins contain disulfide bonds, oxidized DsbE ensures that the disulfide bonds are correctly formed. The reduced DsbE is reoxidized by one of the extracellular domains of DsbD, so that DsbE completes its redox cycle. The electron transfer in the membrane sometimes is taken up molecular oxygen. MPT70 is one protein which is predicted to interact with DsbE and the structure of MPT70 contains a disulfide bond and hence may be a substrate of DsbE. DsbF is a close homologue to DsbE. DsbF is believed to be a reducing protein which is predicted to interact with P450 (Rv1676). DsbF may function in the maturation pathway of cytochrome P450, as DsbF may reduce cysteines in P450 for heme attachment. There are also structures of three secreted proteins that may be secreted by the *sec*-dependent mechanism, Rv0203, Ag85C, and MPT63. The electron transfer is shown with blue arrows.

unfolded secreted proteins to form folded proteins with correctly formed disulfide bonds (Goulding et al., 2004). As 60% of the 161 predicted secreted proteins of *M. tuberculosis* may contain disulfide bonds, DsbE may be involved in the oxidative refolding of these secreted proteins. The transmembrane interacting partner which completes the redox cycle of DsbE is predicted to be Rv2874 (DsbD). DsbD has N- and C-terminal extracellular domains and an eight-transmembrane helical domain (Fig. 15.4). The C-terminal extracellular domain has been solved and consists of a thioredoxin and jelly-roll domain (Goldstone et al., 2005). Efforts are being made to determine which of the DsbD domains interacts with DsbE.

Secreted DsbF is an ortholog of DsbE. The structure of DsbF reveals that its active-site cysteines are mostly in their oxidized form (Fig. 15.4) (C. W. Goulding, unpublished data). DsbF is predicted to be cotranscribed with a heme-binding cytochrome P450 (Rv1676). Due to the active site of DsbF being in its disulfide form, one may infer that DsbF functions in a similar fashion to *E. coli* DsbE, in which DsbF would reduce the active-site cysteines of apo-Rv1676 in preparation for heme binding. One could postulate that the function of DsbF may extend to heme attachment in all cytochrome P450's in *M. tuberculosis*.

### 15.4.2 Secreted Proteins

Proteins are secreted by *M. tuberculosis* in response to environmental changes to protect against oxidative damage, to colonize a host successfully, or to scavenge for essential nutrients from the host. Some of these proteins play biological roles in maintaining the cell envelope, whereas others act as virulence factors. Additionally, proteins secreted by intracellular pathogens such as *M. tuberculosis* play a central role in determining pathways of antigen presentation and recognition by T cells involved in protective immunity. Most of the secreted proteins presented in this section contain signal peptides and are therefore secreted by the SecA1-dependent pathway. The only proteins described below which may not be secreted by the *sec*-dependent pathway is glutamate synthetase and superoxide dismutase. Superoxide dismutase has been shown to be secreted by the SecA2-dependent mechanism (Braunstein et al., 2003).

***Antigen 85 A, B, and C and Related FbpA (Rv1886, Rv0129, Rv3803c, and Rv3804c)*** The antigen 85 complex comprises of three closely related enzymes, Antigen (Ag) 85A (31 kDa), Ag85B (30 kDa), and Ag85C (31.5 kDa). These secreted proteins are antigenic and involved in cell wall maintenance (Wiker and Harboe, 1992). These proteins allow for rapid invasion of macrophages via direct interaction between the host immune systems and the invading bacillus (Armitige et al., 2000; Mariani et al., 2000). Ag85 proteins have mycolyl-transferase activity and catalyze the transfer of the fatty acid mycolate from one trehalose monomycolate (TMM) to another, resulting in the formation of trehalose dimycolate and free trehalose, thus helping to build the mycobacterial cell wall (Daffe, 2000).

The structure of Ag85C was the first protein of the Ag85 complex to be solved (Ronning et al., 2000) revealing a single-domain, monomeric protein with a $\alpha/\beta$-hydrolase fold (Fig. 15.4). The elongated active-site pocket is a negatively charged cavity that binds trehalose with a carbohydrate binding motif. Also in the vicinity of the active site is a catalytic triad contributing the nucleophile for the mycolyl transfer reaction. The Ag85 complex stimulates the uptake of mycobacteria by human macrophages and has been implicated in the interaction with human fibronectin. Each Ag85 protein contains a fibronectin-binding sequence located on an exposed surface.

The structure of Ag85B (Anderson et al., 2001) is similar to that of Ag85C, as was expected from the 73% sequence similarity, but the bound ligands of Ag85B gave a clue about potential drug binding. The trehalose-bound structure of Ag85B revealed that two molecules of trehalose are bound per monomer (Anderson et al., 2001). The two molecules were located at opposite ends of the active site, one representing the liberated trehalose and the other the incoming TMM. This knowledge enables rational drug design to inhibit the Ag85 complex proteins. More recently, the structure of Ag85A has been solved. The active sites among all three Ag85 proteins are identical, though surface residue disparity is quite variable among the three proteins (Ronning et al., 2004). Each Ag85 is expressed under different conditions (Mariani et al., 2000), and the surface residue disparity between the Ag85 proteins suggests that all three Ag85 proteins may be required for mycobacteria to evade the host immune system.

FbpA is related to the trimeric antigen 85C complex as it has 40% sequence identity to Ag85 components, and antibodies to Ag85 recognize FbpA. It has been shown to have a similar structure to Ag85B and C except that FbpA does not contain the catalytic elements required for mycolyl-transferase activity, since the binding site for trehalose derivatives is absent (Wilson et al., 2003). This leads one to speculate that FbpA has a nonenzymatic role, which may involve host tissue attachment to fibronectin or carbohydrates.

***MPT63 (Rv1926c)*** Rv1926c is a major secreted protein of unknown function which is specific to mycobacteria (Wiker et al., 1991; Young et al., 1991) and stimulates humoral immune responses in guinea pigs infected with *M. tuberculosis* (Manca et al., 1997). The structure of Rv1926c consists of an antiparallel β sandwich (Fig. 15.4) with structural similarity to cell surface binding proteins (i.e., arrestin, adaptin, invasin), some of which are involved in endocytosis (Goulding et al., 2002b). Structural similarity suggests that Rv1926c (MPT63) may play a role in host–bacteria interactions. Recently this hypothesis was experimentally supported as it has been shown that Rv1926 activates host mast cells. This is mediated by the activation of the GPI-anchored protein CD48 by Rv1926c though a direct interaction between this protein and Rv1926c remains undetermined (Munoz et al., 2003).

***MPT70 (Rv2875)*** MPT70 (Rv2875) and its homologue MPT83 (Rv2873, 63% sequence identical) are highly immunogenic during the infection of mice (Hewinson et al., 1996). The structure of MPB70, the *Mycobacterium bovis* ortholog of MPT70, has been solved by nuclear magnetic resonance (NMR) and reveals a complex and novel bacterial fold (Carr et al., 2003). The structure of MPB70 (MPT70) consists of an antiparallel seven-stranded β barrel with eight α helices that pack together on one side (Fig. 15.4). One side of the barrel is solvent exposed whereas the other is decorated with four α helices. The N-terminal helix is anchored to the β barrel by a disulfide bond. MPB70 has similar topology to FAS1 domains (Clout et al., 2003), and FAS1-containing extracellular matrix proteins (e.g., fasciclin I) appear to bridge interactions between the cell surface and extracellular matrix (Billings et al., 2002), thus implicating MPT70 and MPT83 in bacteria–host cell interactions. One may also infer that DsbE (Rv2878c, see above) may play a role in oxidatively refolding MPT70 as it is secreted, ensuring correct disulfide bond formation (Goulding et al., 2004).

***Unknown Function (Rv0203)*** Rv0203 is a mycobacteria-specific, secreted protein of unknown function. The structure is mainly α helical and forms a cagelike tetramer (Fig. 15.4). Further biochemical studies indicate that Rv0203 is a heme-binding protein which suggests that Rv0203 is a hemophore involved in a heme uptake pathway in mycobacteria (C. W. Goulding, unpublished data). If this is so, there must be a cytoplasmic enzyme which breaks down heme to $Fe^{2+}$; this protein has been identified as Rv3592, a heme-degrading protein with high sequence homology to the *Staphylococcus aureus* HemO protein (Skaar et al., 2004). The presence of these two proteins suggests that there is an unidentified heme uptake pathway in mycobacteria.

***Glutamine Synthetase GlnA (Rv2220)*** Glutamine synthetase (GS) is a secreted protein which is released during the early stages of infection (Harth et al., 1994) and is thought to be necessary for the synthesis of poly(L-glutamine-L-glutamate) chains (Hirschfield et al., 1990), which are a large component of the mycobacterial cell wall. GS is essential for mycobacterial growth in both human macrophages and animal models (Tullius et al., 2003) and therefore has been proposed a potential drug target (Harth and

Horwitz, 1999, 2003). Recently, a serine/threonine protein kinase (PknG) has been linked to cellular glutamate/glutamine levels and has been shown to promote survival in macrophages; hence GS is an important virulence factor in *M. tuberculosis* (Cowley et al., 2004; Walburger et al., 2004). The structure of GS consists of a dodecamer, with each subunit containing a bifunnel in which ATP and glutamate bind at opposite ends. A metal ion is also bound to one end of the bifunnel (Gill et al., 2002). *Mycobacterium tuberculosis* GS is an example of a prokaryotic type I GS (dodecamer), whereas human GS is a type II GS (seven- or eight-subunit oligomer) (Eisenberg et al., 2000a). Because of the difference in subunit architecture, a comparison of types I and II GS structures may one day lead to the structure-based drug design of *M. tuberculosis* GS inhibitors.

***Superoxide Dismutase SodC (Rv0432)*** Finally, superoxide dismutases (SODs) are metalloenzymes that form part of the *M. tuberculosis* defense system against toxic oxygen species. *Mycobacterium tuberculosis* SOD binds both $Mn^{2+}$ and $Fe^{2+}$ (depending on a single mutation in the sequence). SOD is also found in both the cytoplasm and the extracellular environment (Zhang et al., 1991). The structure of Fe-SOD has been solved and reveals a dimer of dimers (Cooper et al., 1995). As *M. tuberculosis* SOD does not have a signal peptide, it was not originally clear how this protein is secreted. Recently SecA2, the homologue of the housekeeping secretion system SecA1, was shown to be involved in the secretion of SOD. SOD is thought to protect *M. tuberculosis* against oxidative attack by macrophages (Braunstein et al., 2003).

### 15.4.3 Membrane Protein Channels

In this section, two membrane protein channels, MscL and MspA, are described. These proteins transport ions and nutrients from the extracellular environment across the *M. tuberculosis* membrane and into the cytoplasm. The overall shapes of these channels appear similar at first sight, though the top half of the membrane-spanning region of MscL corresponds to the bottom half of MspA, as shown in Figure 15.5*a*. The most striking difference between the two structures is that MscL is comprised entirely from α helices while MspA consists solely of β sheets (Fig. 15.5*a*).

***MscL (Rv0985c)*** MscL is a mechanosensitive ion channel which plays a critical role in transducing physical stresses at the cell membrane to electrochemical responses. MscL is organized as a homopentamer. The extracellular side of the water-filled channel leads to an 18-Å diameter pore which is lined with hydrophilic residues (Fig. 15.5*a*). This pore narrows at the cytoplasmic site, funneling to a hydrophobic pore eyelet that may act as the channel gate (Chang et al., 1998).

***Mycobacterial outer Membrane Channel: MspA*** Porins are channels that cross the outer membrane and are involved in the uptake of small hydrophilic nutrients. The porin in *Mycobacterium smegmatis*, MspA consists of a homooctomeric globetlike conformation with a single central channel (Faller et al., 2004). As in the MscL structure, the extracellular domain has a larger diameter pore of 40 Å which constricts to a pore eyelet (diameter 28 Å) at the cytoplasm (Fig. 15.5*a*). A homologue to MspA in *M. tuberculosis* has not been identified as porins have notoriously different amino acid sequences. It is likely that *M. tuberculosis* possesses an analogous porin since the conductance of 1 M KCl has been observed across the *M. tuberculosis* membrane.

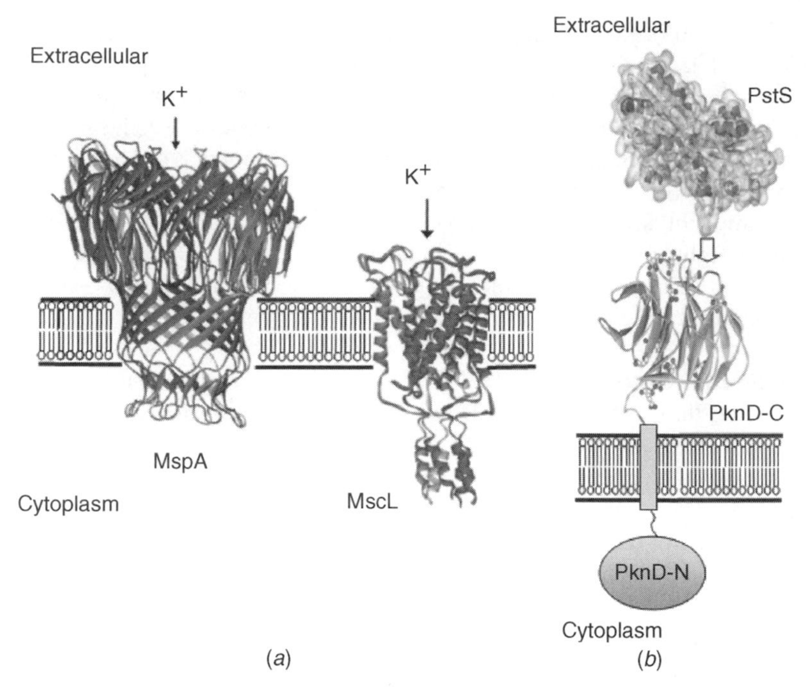


**Figure 15.5** Structures of transport and signaling proteins. (*a*) The two transport membrane proteins are depicted. Both MspA and MscL transport ions and nutrient from the extracellular environment across the membrane and into the cytoplasm. *Mycobacterium tuberculosis* membrane is depicted in the figure. The overall shapes of these channels appear similar at first sight, though the bottom half of the membrane-spanning region of MspA corresponds to the top half of MscL. The most striking difference between the two structures is that MscL is comprised entirely from $\alpha$ helices while MspA consists solely of $\beta$ sheets. (*b*) Structures of serine–threonine protein kinase, PknD, and a secreted ABC phosphate transport receptor, PstS, are depicted. PknD has an N-terminal cytoplasmic domain and a membrane-spanning region. The structure of the C-terminal extracellular domain of PknD consists of a beta-propeller structure with a negatively charged pore. The structure of PstS shows a ribbon diagram with a transparent electrostatic molecular surface. It has been predicted that PknD and PstS may interact and a proposed interaction is shown by the arrow. A positively charged node on PstS may insert itself into the negatively charged pore of the extracellular domain of PknD.

### 15.4.4 Extracellular-to-Intracellular Signaling Mechanism

There are 11 serine–threonine protein kinases (Ser/Thr kinases) in *M. tuberculosis* (Cole et al., 1998) belonging to the eukaryotic kinase superfamily (Hanks and Hunter, 1995). Most Ser/Thr kinases have both an extracellular and an intracellular domain which are connected by a transmembrane region. These Ser/Thr kinases respond to extracellular signals which are translated into cellular responses. These reactions are normally mediated by dephosphorylation events, which are carried out by protein phosphatases (Av-Gay and Everett, 2000). Ser/Thr kinases in bacteria have been shown to be involved in the regulation of development, stress responses, and pathogenicity. Two Ser/Thr kinases are highlighted in this section along with their possible interaction partners.

***Serine–Threonine Protein Kinases Rv0014c (PknB) and Rv0931c (PknD)*** As discussed earlier, PknB is thought to regulate cell division and growth (Strong et al.,

2003a; Yeats et al., 2002), and the structure of its N-terminal intracellular kinase domain has been solved (Ortiz-Lombardia et al., 2003; Young et al., 2003). The extracellular domain contains repeats which have similarity to the targeting domain of penicillin-binding proteins (PASTA domain) (Yeats et al., 2002). The intracellular kinase domain complexed with ATP-$\gamma$-S adopts the characteristic two-lobed structure of eukaryotic Ser/Thr kinases (Huse and Kuriyan, 2002). This transmembrane Ser/Thr kinase probably plays a role in signal transduction pathways within *M. tuberculosis* and, in response to environmental changes, may lead to oligomerization of the extracellular domain. The intracellular domain of a Ser/Thr protein phosphatase (Rv0018c) has been recently solved (Pullen et al., 2004). This protein is predicted to interact with PknB, and therefore the 3D structure is of great interest to study protein–protein interactions.

PknD is contained within the phosphate uptake operon and has been associated with phosphate transport (Av-Gay and Everett, 2000). The extracellular C-terminal domain of PknD consists of a sensor domain which forms a funnel-shaped $\beta$-propeller structure containing six blades arranged cyclically around a central pore (Good et al., 2004). The central pore is 8 Å in diameter and is lined with one aspartate residue from each of the six blades (Fig. 15.5*b*). This C-terminal domain has a flexible linker connecting the trans-membrane domain, and it is postulated that signaling is mediated through a direct or indirect change in structure or localization that alters the activity of the attached, intercellular kinase domain.

***Secreted Interacting Partner: PstS-1 (Rv0934c)*** PknD (Rv0931c) appears to be cotranscribed with PstS (Rv0932c). PstS is also part of the phosphate uptake system, along with several other components located within its chromosomal vicinity (Av-Gay and Everett, 2000). PstS-1 (Rv0934c) is one of these components and is also a secreted ATP-binding cassette (ABC) phosphate transport receptor similar to PstS. PstS-1 has 30% sequence identity to PstS, suggesting that their structures will be similar. The structure of PstS-1 reveals two similar $\beta/\alpha$ folding domains with a phosphate ion bound within the cleft between the two domains (Vyas et al., 2003). The surface area of PstS-1 reveals a nodelike protrusion in its structure, which would be positively charged if residues for PstS were modeled in the structure. One may postulate that the nodelike protrusion might fit into the negatively charged central pore of PknD to initiate intracellular signaling of PknD as shown in Figure 15.5*b*.

### 15.4.5 Proteins Involved in Synthesis of Cell Wall Components

Many proteins are involved in maintaining the cell envelope components, including a mixture of peptidoglycans, mycolic acids, lipids, and other metabolites. This section describes some of the enzymes which are involved in the biosynthesis of these components.

***Cyclopropane Synthases CcmA1 (Rv3392c), CcmA2 (Rv0503c), and PcaA (Rv0470c)*** Cyclopropane synthases introduce a cyclopropane ring at either the distal or proximal position to mycolic acids. This addition is critical to the structure and function of the cell envelope (George et al., 1995; Yuan et al., 1995). The cyclopropane synthase, encoded by *pcaA*, has been identified as a persistence factor (Glickman et al., 2000; Stewart et al., 2003), and the structures of three cyclopropane synthases, including PcaA, have been solved (the other two are CcmA1 and CcmA2). All three cylopropane synthases show a seven-stranded $\alpha/\beta$ fold, similar to other methyl transferases. The fold contains a conserved interaction site for *S*-adenosyl-L methionine (SAM) and a hydrophobic, lipid-binding pocket. CcmA2 and PcaA both act at the proximal position producing cis and trans cyclopropane rings, respectively (George et al., 1995; Glickman et al., 2000). The structures of these two proteins are extremely similar, whereas CcmA1 acts at the distal position (Yuan et al., 1995). Structural comparison of CcmA1 with CcmA2 and PcaA shows that there is one differing region which is a conserved basic/hydrophobic region that has been shifted in CcmA1 compared to CcmA2 and PcaA. It is hypothesized that this is the region where the cyclopropane synthase protein partner (AcpM) binds and presents the acyl chain of acyl-AcpM. Hence, acyl-AcpM may bind closer to the active site of CcmA2 and PcaA, favoring the reaction at the proximal position, whereas acyl-AcpM, in contrast, would sit further away from the active site to favor the reaction at the distal position. The structures of this diverse family of cyclopropane synthases are surprisingly similar, which may lead to the design of an inhibitor that could block all cyclopropanation reactions.

***Rv0046c Ino1*** Ino1 (inositol 1-phosphate synthase) is a key enzyme in phosphatidylinositol (PI) synthesis. PI is a key precursor of many *M. tuberculosis* glycolipid cell wall components. The structure of Ino1 comprises of two domains connected by two hinge regions. Domain I has a Rossmann fold (NADH cofactor) while domain II contains residues that form the interface of the tetramer (Norman et al., 2002). Interestingly, a zinc ion bridges the nicotinamide ring of $NAD^+$ to the proteins. This placement raises the question whether the zinc is structural or functional?

***Rv3465 (RmlC)*** RmlC (dTDP-4-keto-6-deoxyglucose epimerase) is the third enzyme in the biosynthetic pathway of dTDP-L-rhamnose which is essential for cell wall synthesis. This protein may also serve as a good TB drug target since L-rhamnose is a sugar that is not present in the human host and other anti-TB drugs are known to target cell wall biosynthetic genes. The structure of *M. tuberculosis* RmlC reveals that the size of the active-site pocket varies slightly between other bacterial structures of RmlC and provides a good starting point for virtual drug screening (Kantardjieff et al., 2004).

***Rv1484 and Rv1595 (InhA and NadC)*** NADH-dependent enoyl-ACP reductase (InhA) participates in FAS-II fatty acid biosynthesis pathway, which produces mycolic acids. Inactivation of InhA alone is sufficient to inhibit mycolic acid biosynthesis and induce cell lysis (Vilcheze et al., 2000), and the antitubercular drug isoniazid (INH) specifically targets InhA (Banerjee et al., 1994). The structure of InhA was solved in complex with NADH (Dessen et al., 1995), leading to the belief that INH interacts with NADH. When the structure of the complex of InhA-NADH-INH was solved, it was observed that INH was covalently bound to NADH in the presence of a divalent metal, hence inhibiting InhA (Rozwarski et al., 1998, 1999). The enzyme that modifies INH is NadC [quinolinic acid phosphoribosyltransferase (QAPRTase)]. This is an important drug target as it indirectly inhibits mycolic acid synthesis and hence kills *M. tuberculosis*. The structure of apo-QAPRTase with its substrate, products, and inhibitors bound shows conformational changes which elucidate the catalytic mechanism (Sharma et al., 1998).

### 15.4.6 Cytochrome P450 and Partner Proteins

In the above sections, two secreted heme-binding proteins were discussed. One was a predicted cytochrome P450 enzyme (Rv1676) which is connected to the predicted reducing oxidoreductase, DsbF. The other was a proposed hemophore protein, Rv0203, which may scavenge heme from the host. In this section, we expand on the heme-binding proteins in

*M. tuberculosis* by discussing cytochrome P450s, and hemoglobins as well as their protein partners.

***Cytochrome P450 Proteins CYP51 (Rv0764c) and CYP121 (Rv2276)*** Cytochrome P450s are *b*-heme-containing enzymes that bind and reduce molecular oxygen, leading to mono-oxygenation of substrate and production of water. These proteins usually have oxidative roles in lipid metabolism, making cytochrome P450s potential drug targets against TB (Munro et al., 2003). Two cytochrome P450 structures have been determined, CYP51 (Podust et al., 2001) and CYP121 (Leys et al., 2003). CYP121 is a novel cytochrome P450 structure which contains mixed heme conformations and putative proton relay pathways from the protein surface to the heme. Further investigation into the mechanism of cytochrome P450 with a potential redox partner, FprA (see below) is ongoing.

***Rv3106 (FprA)*** FprA is an oxidoreductase that catalyzes the transfer of reducing equivalents from NADPH to a protein acceptor. The structures of both the oxidized and reduced forms of FprA were solved (Bossi et al., 2002) and revealed a homodimer with a FAD domain and an NADPH domain exhibiting dinucleotide-binding fold topology. Interestingly, the FprA-NADP$^+$ complex shows that the NADP$^+$ nicotinamide ring exhibits an unusual cis conformation and that NADP$^+$ is covalently modified. FprA is a paralog of adrenodoxin reductase and therefore may be involved with either iron metabolism or cytochrome P450 reductase activity.

***Hemoglobin: Rv1542c (trHbN) and Rv2470 (trHbO)*** Truncated hemoglobins are a class of small oxygen-binding hemoproteins forming a distinct group within the hemoglobin superfamily. The *M. tuberculosis* genome contains two such proteins: hemoglobin N (trHbN) from group I and hemoglobin O (trHbO) from group II. TrHbN has been shown to play a protective role against nitric oxide (NO), and a precise function for trHbO is not known. The sequence identity between the two *M. tuberculosis* trHb's is only 18%, but they both have similar hemoglobin-type folds, though trHbN is a dimer and trHbO is a dodecamer (Milani et al., 2001, 2003). The main structural difference between the two truncated hemoglobins are the bulky residues in trHbO that transform the long tunnel cavity in trHbN (leading from solvent to distal heme site) into two distinct small cavities in trHbO. This indicates that these two truncated hemoglobins may have different functional roles in *M. tuberculosis*. Interestingly, trHbO is predicted to interact with the heme-degrading protein Rv3592, which was discussed in the previous section on Rv0203.

### 15.4.7 Protection against Oxidative Damage

This last section describes proteins that protect *M. tuberculosis* from oxidative damage due to interactions with human macrophages.

***Rv0137c (MsrA)*** MsrA (methionine sulfoxide reductase A) repairs oxidative damage to methionine residues arising from reactive oxygen and nitrogen species. As a result, this protein acts a primary defense against oxidative damage (Moskovitz et al., 2001). Structural analysis of *M. tuberculosis* MsrA shows that there only two reactive cysteines (Taylor et al., 2003) and that a methionine residue is bound at the active site of a neighboring molecule. This structure provides insight into the mechanism of protein-bound methionine sulfoxide recognition and repair.

***Biosynthesis of Mycothiol: Rv1170 and Rv0819 (MshB and MshD)*** Mycobacteria do not have glutathione as a reducing agent controlling the levels of cellular oxygen species and maintaining redox cycles. Instead they have mycothiol. Loss of mycothiol in mycobacteria is associated with slow growth and increased sensitivity to both reactive oxygen species and antibiotics. Hence, enzymes in the mycothiol biosynthetic pathway may serve as anti-TB drug targets since there are no human homologues. Two of the enzymes in this pathway have been solved, MshB and MshD.

MshB is a two-domain metal-dependent deacetylase containing a large putative active site that is enclosed by large loops. The proposed reaction mechanism is similar to that of carboxypeptidase and thermolysin (Maynes et al., 2003; McCarthy et al., 2004). MshD (mycothiol synthase) is the last enzyme in the mycothiol biosynthetic pathway. The structure of MshD consists of two GNAT (Gen5-related *N*-acetyltransferase) domains. The cofactor of MshD, acetyl-CoA, wraps around a "canyon" created by the twisted two-stranded β sheet separating the two GNAT domains. Modeling of the substrate binding revealed that the substrate may have rotational freedom within the proposed active site (Vetting et al., 2003). Hence, a more complete picture will only emerge after the further structural analysis of MshC and MshC/MshD with substrate.

## 15.5 CONCLUSION

Together the fields of structural biology and computational biology provide a powerful combination for investigating microorganisms at the genomic and proteomic level. Complementing one another, these fields are likely to aid in our quest to understand the underlying mechanisms of the cell. In the case of *M. tuberculosis,* this strategy has enabled us to identify functionally linked proteins on a genomewide basis, which in turn has enabled us to build a map of functionally linked proteins. As the field of structural genomics and computational biology continues to mature, it is likely that approaches, as outlined above, will become increasingly important.

## REFERENCES

Altschul, S. F., Gish, W., Miller, W., Myers, E. W., and Lipman, D. J. (1990). Basic local alignment search tool. *J. Mol. Biol.* **215**:403–410.

Anderson, D. H., Harth, G., Horwitz, M. A., and Eisenberg, D. (2001). An interfacial mechanism and a class of inhibitors inferred from two crystal structures of the *Mycobacterium tuberculosis* 30 kDa major secretory protein (antigen 85B), a mycolyl transferase. *J. Mol. Biol.* **307**:671–681.

Armitige, L. Y., Jagannath, C., Wanger, A. R., and Norris, S. J. (2000). Disruption of the genes encoding antigen 85A and antigen 85B of *Mycobacterium tuberculosis* H37Rv: Effect on growth in culture and in macrophages. *Infect. Immun.* **68**:767–778.

Av-Gay, Y., and Everett, M. (2000). The eukaryotic-like Ser/Thr protein kinases of *Mycobacterium tuberculosis. Trends Microbiol* **8**:238–244.

Bader, G. D., Betel, D., and Hogue, C. W. (2003). BIND: The Biomolecular Interaction Network Database. *Nucleic Acids Res.* **31**:248–250.

Banerjee, A., Dubnau, E., Quemard, A., Balasubramanian, V., Um, K. S., Wilson, T., Collins, D., de Lisle, G., and Jacobs, W. R. Jr. (1994). *InhA*, a gene encoding a target for isoniazid and ethionamide in *Mycobacterium tuberculosis. Science* **263**:227–230.

Belanger, A. E., Besra, G. S., Ford, M. E., Mikusova, K., Belisle, J. T., Brennan, P. J., and Inamine, J. M. (1996). The *embAB* genes of *Mycobacterium avium* encode an arabinosyl transferase involved in cell wall arabinan biosynthesis that is the target for the antimycobacterial drug ethambutol. *Proc. Natl. Acad. Sci. USA* **93**:11919–11924.

Billings, P. C., Whitbeck, J. C., Adams, C. S., Abrams, W. R., Cohen, A. J., Engelsberg, B. N., Howard, P. S., and Rosenbloom, J. (2002). The transforming growth factor-beta-inducible matrix protein (beta)ig-h3 interacts with fibronectin. *J. Biol. Chem.* **277**:28003–28009.

Bossi, R. T., Aliverti, A., Raimondi, D., Fischer, F., Zanetti, G., Ferrari, D., Tahallah, N., Maier, C. S., Heck, A. J., Rizzi, M., and Mattevi, A. (2002). A covalent modification of NADP+ revealed by the atomic resolution structure of FprA, a *Mycobacterium tuberculosis* oxidoreductase. *Biochemistry* **41**:8807–8818.

Bowers, P. M., Pellegrini, M., Thompson, M. J., Fierro, J., Yeates, T. O., and Eisenberg, D. (2004). Prolinks: A database of protein functional linkages derived from coevolution. *Genome Biol.* **5**:R35.

Braunstein, M., Brown, A. M., Kurtz, S., and Jacobs, W. R. Jr. (2001). Two nonredundant SecA homologues function in mycobacteria. *J. Bacteriol.* **183**:6979–6790.

Braunstein, M., Espinosa, B. J., Chan, J., Belisle, J. T., and Jacobs, W. R. Jr. (2003). SecA2 functions in the secretion of superoxide dismutase A and in the virulence of *Mycobacterium tuberculosis*. *Mol Microbiol.* **48**:453–464.

Brown, P. O., and Botstein, D. (1999). Exploring the new world of the genome with DNA microarrays. *Nat. Genet.* **21**:33–37.

Carr, M. D., Bloemink, M. J., Dentten, E., Whelan, A. O., Gordon, S. V., Kelly, G., Frenkiel, T. A., Hewinson, R. G., and Williamson, R. A. (2003). Solution Structure of the *Mycobacterium tuberculosis* complex protein MPB70: From Tuberculosis Pathogenesis to Inherited Human Corneal Disease. *J. Biol. Chem.* **278**:43736–43743.

Chang, G., Spencer, R. H., Lee, A. T., Barclay, M. T., and Rees, D. C. (1998). Structure of the MscL homolog from *Mycobacterium tuberculosis*: A gated mechanosensitive ion channel. *Science* **282**:2220–2226.

Clout, N. J., Tisi, D., and Hohenester, E. (2003). Novel fold revealed by the structure of a FAS1 domain pair from the insect cell adhesion molecule fasciclin I. *Structure (Camb.)* **11**:197–203.

Cole, S. T., Brosch, R., Parkhill, J., Garnier, T., Churcher, C., Harris, D., Gordon, S. V., Eiglmeier, K., Gas, S., Barry, C. E. 3rd, Tekaia, F., Badcock, K., Basham, D., Brown, D., Chillingworth, T., Connor, R., Davies, R., Devlin, K., Feltwell, T., Gentles, S., Hamlin, N., Holroyd, S., Hornsby, T., Jagels, K., Krogh, A., McLean, J., Moule, S., Murphy, L., Oliver, K., Osborne, J., Quail, M. A., Rajandream, M. A., Rogers, J., Rutter, S., Seeger, K., Skelton, J., Squares, R., Squares, S., Sulston, J. E., Taylor, K., Whitehead, S., and Barrell, B. G. (1998). Deciphering the biology of *Mycobacterium tuberculosis* from the complete genome sequence. *Nature* **393**:537–544.

Cooper, J. B., McIntyre, K., Badasso, M. O., Wood, S. P., Zhang, Y., Garbe, T. R., and Young, D. (1995). X-ray structure analysis of the iron-dependent superoxide dismutase from *Mycobacterium tuberculosis* at 2.0 Angstroms resolution reveals novel dimer-dimer interactions. *J. Mol. Biol.* **246**:531–544.

Cowley, S., Ko, M., Pick, N., Chow, R., Downing, K. J., Gordhan, B. G., Betts, J. C., Mizrahi, V., Smith, D. A., Stokes, R. W., and Av-Gay, Y. (2004). The *Mycobacterium tuberculosis* protein serine/threonine kinase PknG is linked to cellular glutamate/glutamine levels and is important for growth in vivo. *Mol. Microbiol.* **52**:1691–1702.

Daffe, M. (2000). The mycobacterial antigens 85 complex—from structure to function and beyond. *Trends Microbiol.* **8**:438–440.

Dandekar, T., Schuster, S., Snel, B., Huynen, M., and Bork, P. (1999). Pathway alignment: Application to the comparative analysis of glycolytic enzymes. *Biochem. J.* **343** (Pt 1):115–124.

Dessen, A., Quemard, A., Blanchard, J. S., Jacobs, W. R. Jr., and Sacchettini, J. C. (1995). Crystal structure and function of the isoniazid target of *Mycobacterium tuberculosis*. *Science* **267**: 1638–1641.

Eisen, M. B., Spellman, P. T., Brown, P. O., and Botstein, D. (1998). Cluster analysis and display of genome-wide expression patterns. *Proc. Natl. Acad. Sci. USA* **95**:14863–14868.

Eisenberg, D., Gill, H. S., Pfluegl, G. M., and Rotstein, S. H. (2000a). Structure-function relationships of glutamine synthetases. *Biochim. Biophys. Acta* **1477**:122–145.

Eisenberg, D., Marcotte, E. M., Xenarios, I., and Yeates, T. O. (2000b). Protein function in the post-genomic era. *Nature* **405**:823–826.

Ermolarva, M. D., White, O., and Salzbaerg, S. L. (2001). Prediction of operons in microbial genomes. *Nucleic Acids Res.* **29**:1216–1221.

Faller, M., Niederweis, M., and Schulz, G. E. (2004). The structure of a mycobacterial outer-membrane channel. *Science* **303**:1189–1192.

Fultz, P. N., and Kemper, J. (1981). Wild-type isopropylmalate isomerase in *Salmonella typhimurium* is composed of two different subunits. *J. Bacteriol.* **148**:210–219.

George, K. M., Yuan, Y., Sherman, D. R., and Barry, C. E., 3rd (1995). The biosynthesis of cyclopropanated mycolic acids in *Mycobacterium tuberculosis*. Identification and functional analysis of CMAS-2. *J. Biol. Chem.* **270**:27292–27298.

Gill, H. S., Pfluegl, G. M., and Eisenberg, D. (2002). Multicopy crystallographic refinement of a relaxed glutamine synthetase from *Mycobacterium tuberculosis* highlights flexible loops in the enzymatic mechanism and its regulation. *Biochemistry* **41**:9863–9872.

Glickman, M. S., Cox, J. S., and Jacobs, W. R. Jr. (2000). A novel mycolic acid cyclopropane synthetase is required for cording, persistence, and virulence of *Mycobacterium tuberculosis. Mol. Cell.* **5**:717–727.

Goldstone, D., Baker, E. N., and Metcalf, P. (2005). Crystallization and preliminary diffraction studies of the C-terminal domain of the DipZ homologue from *Mycobacterium tuberculosis. Acta Crystallograph Sect F Struct Biol Cryst Commun.*, **61**:243–245.

Gomez, M., Johnson, S., and Gennaro, M. L. (2000). Identification of secreted proteins of *Mycobacterium tuberculosis* by a bioinformatic approach. *Infect. Immun.* **68**:2323–2327.

Good, M. C., Greenstein, A. E., Young, T. A., Ng, H. L., and Alber, T. (2004). Sensor domain of the *Mycobacterium tuberculosis* receptor Ser/Thr protein kinase, PknD, forms a highly symmetric beta propeller. *J. Mol. Biol.* **339**:459–469.

Goulding, C. W., Apostol, M., Anderson, D. H., Gill, H. S., Smith, C. V., Kuo, M. R., Yang, J. K., Waldo, G. S., Suh, S. W., Chauhan, R., Kale, A., Bachhawat, N., Mande, S. C., Johnston, J. M., Lott, J. S., Baker, E. N., Arcus, V. L., Leys, D., McLean, K. J., Munro, A. W., Berendzen, J., Sharma, V., Park, M. S., Eisenberg, D., Sacchettini, J., Alber, T., Rupp, B., Jacobs, W. Jr., and Terwilliger, T. C. (2002a). The, T. B. structural genomics consortium: Providing a structural foundation for drug discovery. *Curr. Drug Targets Infect. Disord.* **2**:121–141.

Goulding, C. W., Gleiter, S., Apostol, M. I., Parseghian, A., Bardwell, J., Gennaro, M. L., and Eisenberg, D. (2004). Gram-positive DsbE proteins function differently from gram-negative DsbE homologs: A structure to function analysis of DsbE from *Mycobacterium tuberculosis. J. Biol. Chem.* **279**:3516–3524.

Goulding, C. W., Parseghian, A., Sawaya, M. R., Cascio, D., Apostol, M. I., Gennaro, M. L., and Eisenberg, D. (2002b). Crystal structure of a major secreted protein of *Mycobacterium tuberculosis*-MPT63 at 1.5-A resolution. *Protein Sci.* **11**:2887–2893.

Grigoriev, A. (2001). A relationship between gene expression and protein interactions on the proteome scale: Analysis of the bacteriophage T7 and the yeast *Saccharomyces cerevisiae. Nucleic Acids Res.* **29**:3513–3519.

Hanks, S. K., and Hunter, T. (1995). Protein kinases 6. The eukaryotic protein kinase superfamily: Kinase (catalytic) domain structure and classification. *FASEB J.* **9**:576–596.

Harth, G., Clemens, D. L., and Horwitz, M. A. (1994). Glutamine synthetase of *Mycobacterium tuberculosis*: Extracellular release and characterization of its enzymatic activity. *Proc. Natl. Acad. Sci. USA* **91**:9342–9346.

Harth, G., and Horwitz, M. A. (1999). An inhibitor of exported *Mycobacterium tuberculosis* glutamine synthetase selectively blocks the growth of pathogenic mycobacteria in axenic culture and in human monocytes: Extracellular proteins as potential novel drug targets. *J. Exp. Med.* **189**:1425–1436.

Harth, G., and Horwitz, M. A. (2003). Inhibition of *Mycobacterium tuberculosis* glutamine synthetase as a novel antibiotic strategy against tuberculosis: Demonstration of efficacy in vivo. *Infect. Immun.* **71**:456–464.

Hewinson, R. G., Michell, S. L., Russell, W. P., McAdam, R. A., and Jacobs, W. R. Jr. (1996). Molecular characterization of MPT83: A seroreactive antigen of *Mycobacterium tuberculosis* with homology to MPT70. *Scand. J. Immunol.* **43**:490–499.

Hirschfield, G. R., McNeil, M., and Brennan, P. J. (1990). Peptidoglycan-associated polypeptides of *Mycobacterium tuberculosis*. *J. Bacteriol.* **172**:1005–1013.

Horswill, A. R., and Escalante-Semerena, J. C. (2001). In vitro conversion of propionate to pyruvate by *Salmonella enterica* enzymes: 2-Methylcitrate dehydratase (PrpD) and aconitase enzymes catalyze the conversion of 2-methylcitrate to 2-methylisocitrate. *Biochemistry* **40**:4703–4713.

Huse, M., and Kuriyan, J. (2002). The conformational plasticity of protein kinases. *Cell* **109**:275–282.

Kantardjieff, K. A., Kim, C. Y., Naranjo, C., Waldo, G. S., Lekin, T., Segelke, B. W., Zemla, A., Park, M. S., Terwilliger, T. C., and Rupp, B. (2004). *Mycobacterium tuberculosis* RmlC epimerase (Rv3465): A promising drug-target structure in the rhamnose pathway. *Acta Crystallogr. D Biol. Crystallogr.* **60**:895–902.

Leys, D., Mowat, C. G., McLean, K. J., Richmond, A., Chapman, S. K., Walkinshaw, M. D., and Munro, A. W. (2003). Atomic structure of *Mycobacterium tuberculosis* CYP121 to 1.06 A reveals novel features of cytochrome P450. *J. Biol. Chem.* **278**:5141–5147.

Lodish, H., Baltimore, D., MBerk, A., Szipursky, S., Matsudaira, P., and Ndarnell, J. (1995). *Molecular Cell Biology*. New York: Scientific American Books.

Madigan, M., Martinko, J., and Parker, J. (2000). *Brock Biology of Microorganisms*. Englewood cliffs, NJ: Prentice-Hall.

Manca, C., Lyashchenko, K., Wiker, H. G., Usai, D., Colangeli, R., and Gennaro, M. L. (1997). Molecular cloning, purification, and serological characterization of MPT63, a novel antigen secreted by *Mycobacterium tuberculosis*. *Infect. Immun.* **65**:16–23.

Marcotte, C. J., and Marcotte, E. M. (2002). Predicting functional linkages from gene fusions with confidence. *Appl Bioinformatics* **1**:93–100.

Marcotte, E. M., Pellegrini, M., Ng, H. L., Rice, D. W., Yeates, T. O., and Eisenberg, D. (1999a). Detecting protein function and protein-protein interactions from genome sequences. *Science* **285**:751–753.

Marcotte, E. M., Pellegrini, M., Thompson, M. J., Yeates, T. O., and Eisenberg, D. (1999b). A combined algorithm for genome-wide prediction of protein function. *Nature* **402**:83–86.

Mariani, F., Cappelli, G., Riccardi, G., and Colizzi, V. (2000). *Mycobacterium tuberculosis* H37Rv comparative gene-expression analysis in synthetic medium and human macrophage. *Gene* **253**:281–291.

Maynes, J. T., Garen, C., Cherney, M. M., Newton, G., Arad, D., Av-Gay, Y., Fahey, R. C., and James, M. N. (2003). The crystal structure of 1-D-myo-inositol 2-acetamido-2-deoxy-{alpha}-D-glucopyranoside deacetylase (MshB) from *Mycobacterium tuberculosis* reveals a zinc hydrolase with a lactate dehydrogenase fold. *J. Biol. Chem.* **278**:47166–47170.

McCarthy, A. A., Peterson, N. A., Knijff, R., and Baker, E. N. (2004). Crystal structure of MshB from *Mycobacterium tuberculosis*, a deacetylase involved in mycothiol biosynthesis. *J. Mol. Biol.* **335**:1131–1141.

McKinney, J. D., Honer zu Bentrup, K., Munoz-Elias, E. J., Miczak, A., Chen, B., Chan, W. T., Swenson, D., Sacchettini, J. C., Jacobs, W. R., Jr., and Russell, D. G. (2000). Persistence of *Mycobacterium tuberculosis* in macrophages and mice requires the glyoxylate shunt enzyme isocitrate lyase. *Nature* **406**:735–738.

Mellor, J. C., Yanai, I., Clodfelter, K. H., Mintseris, J., and DeLisi, C. (2002). Predictome: A database of putative functional links between proteins. *Nucleic Acids Res.* **30**:306–309.

Mewes, H. W., Frishman, D., Guldener, U., Mannhaupt, G., Mayer, K., Mokrejs, M., Morgenstern, B., Munsterkotter, M., Rudd, S., and Weil, B. (2002). MIPS: A database for genomes and protein sequences. *Nucleic Acids Res.* **30**:31–34.

Milani, M., Pesce, A., Ouellet, Y., Ascenzi, P., Guertin, M., and Bolognesi, M. (2001). *Mycobacterium tuberculosis* hemoglobin N displays a protein tunnel suited for O2 diffusion to the heme. *EMBO J.* **20**:3902–3909.

Milani, M., Savard, P. Y., Ouellet, H., Ascenzi, P., Guertin, M., and Bolognesi, M. (2003). A TyrCD1/TrpG8 hydrogen bond network and a TyrB10TyrCD1 covalent link shape the heme distal site of *Mycobacterium tuberculosis* hemoglobin O. *Proc. Natl. Acad. Sci. USA* **100**:5766–5771.

Moreno-Hagelsieb, G., and Collado-Vides, J. (2002a): Operon conservation from the point of view of *Escherichia coli*, and inference of functional interdependence of gene products from genome context. *In Silico Biol.* **2**:87–95.

Moreno-Hagelsieb, G., and Collado-Vides, J. (2002b). A powerful non-homology method for the prediction of operons in prokaryotes. *Bioinformatics* **18** (Suppl. 1):S329–S336.

Moreno-Hagelsieb, G., Trevino, V., Perez-Rueda, E., Smith, T. F., and Collado-Vides, J. (2001). Transcription unit conservation in the three domains of life: A perspective from *Escherichia coli*. *Trends Genet.* **17**:175–177.

Mori, H., and Ito, K. (2001). The Sec protein-translocation pathway. *Trends Microbiol.* **9**:494–500.

Moskovitz, J., Bar-Noy, S., Williams, W. M., Requena, J., Berlett, B. S., and Stadtman, E. R. (2001). Methionine sulfoxide reductase (MsrA) is a regulator of antioxidant defense and lifespan in mammals. *Proc. Natl. Acad. Sci.* USA **98**:12920–12925.

Munoz, S., Hernandez-Pando, R., Abraham, S. N., and Enciso, J. A. (2003). Mast cell activation by *Mycobacterium tuberculosis*: Mediator release and role of CD48. *J. Immunol.* **170**:5590–5596.

Munro, A. W., McLean, K. J., Marshall, K. R., Warman, A. J., Lewis, G., Roitel, O., Sutcliffe, M. J., Kemp, C. A., Modi, S., Scrutton, N. S., and Leys, D. (2003). Cytochromes P450: Novel drug targets in the war against multidrug-resistant *Mycobacterium tuberculosis*. *Biochem. Soc. Trans.* **31**:625–630.

Norman, R. A., McAlister, M. S., Murray-Rust, J., Movahedzadeh, F., Stoker, N. G., and McDonald, N. Q. (2002). Crystal structure of inositol 1-phosphate synthase from *Mycobacterium tuberculosis*, a key enzyme in phosphatidylinositol synthesis. *Structure (Camb.)* **10**:393–402.

Ortiz-Lombardia, M., Pompeo, F., Boitel, B., and Alzari, P. M. (2003). Crystal structure of the catalytic domain of the PknB serine/threonine kinase from *Mycobacterium tuberculosis*. *J. Biol. Chem.* **278**:13094–13100.

Overbeek, R., Fonstein, M., D'Souza, M., Pusch, G. D., and Maltsev, N. (1999). Use of contiguity on the chromosome to predict functional coupling. *In Silico Biol.* **1**:93–108.

Pallen, M., Chaudhuri, R., and Khan, A. (2002): Bacterial, FHA domains: Neglected players in the phospho-threonine signalling game? *Trends Microbiol.* **10**:556–563.

Pellegrini, M., Marcotte, E. M., Thompson, M. J., Eisenberg, D., and Yeates, T. O. (1999). Assigning protein functions by comparative genome analysis: Protein phylogenetic profiles. *Proc. Natl. Acad. Sci. USA* **96**:4285–4288.

Pellegrini, M., Thompson, M., Fierro, J., and Bowers, P. (2001). Computational method to assign microbial genes to pathways. *J. Cell. Biochem. Suppl.* **37**:106–109.

Podust, L. M., Poulos, T. L., and Waterman, M. R. (2001). Crystal structure of cytochrome P450 14alpha -sterol demethylase (CYP51) from *Mycobacterium tuberculosis* in complex with azole inhibitors. *Proc. Natl. Acad. Sci. USA* **98**:3068–3073.

Pullen, K. E., Ng, H. L., Sung, P. Y., Good, M. C., Smith, S. M., and Alber, T. (2004). An alternate conformation and a third metal in PstP/Ppp, the *M. tuberculosis* PP2C-Family Ser/Thr protein phosphatase. *Structure* **12**:1947–1954.

Ravasz, E., Somera, A. L., Mongru, D. A., Oltvai, Z. N., and Barabasi, A. L. (2002). Hierarchical organization of modularity in metabolic networks. *Science* **297**:1551–1555.

Rives, A. W., and Galitski, T. (2003). Modular organization of cellular networks. *Proc. Natl. Acad. Sci. USA* **100**:1128–1133.

Ronning, D. R., Klabunde, T., Besra, G. S., Vissa, V. D., Belisle, J. T., and Sacchettini, J. C. (2000). Crystal structure of the secreted form of antigen 85C reveals potential targets for mycobacterial drugs and vaccines. *Nat. Struct. Biol.* **7**:141–146.

Ronning, D. R., Vissa, V., Besra, G. S., Belisle, J. T., and Sacchettini, J. C. (2004). *Mycobacterium tuberculosis* antigen 85A and 85C structures confirm binding orientation and conserved substrate specificity. *J. Biol. Chem.* **23**:2972–2981.

Rozwarski, D. A., Grant, G. A., Barton, D. H., Jacobs, W. R., Jr., and Sacchettini, J. C. (1998). Modification of the NADH of the isoniazid target (InhA) from *Mycobacterium tuberculosis*. *Science* **279**:98–102.

Rozwarski, D. A., Vilcheze, C., Sugantino, M., Bittman, R., and Sacchettini, J. C. (1999). Crystal structure of the *Mycobacterium tuberculosis* enoyl-ACP reductase, InhA, in complex with NAD + and a C16 fatty acyl substrate. *J. Biol. Chem.* **274**:15582–15589.

Salgado, H., Moreno-Hagelsieb, G., Smith, T. F., and Collado-Vides, J. (2000). Operons in *Escherichia coli*: Genomic analyses and predictions. *Proc. Natl. Acad. Sci. USA* **97**:6652–6657.

Salgado, H., Santos-Zavaleta, A., Gama-Castro, S., Millan-Zarate, D., Diaz-Peredo, E., Sanchez-Solano, F., Perez-Rueda, E., Bonavides-Martinez, C., and Collado-Vides, J. (2001). RegulonDB (version 3.2): Transcriptional regulation and operon organization in *Escherichia coli* K-12. *Nucleic Acids Res.* **29**:72–74.

Sharma, V., Arockiasamy, A., Ronning, D. R., Savva, C. G., Holzenburg, A., Braunstein, M., Jacobs, W. R., Jr., and Sacchettini, J. C. (2003). Crystal structure of *Mycobacterium tuberculosis* SecA, a preprotein translocating ATPase. *Proc. Natl. Acad. Sci. USA* **100**:2243–2248.

Sharma, V., Grubmeyer, C., and Sacchettini, J. C. (1998). Crystal structure of quinolinic acid phosphoribosyltransferase from *Mycobacterium tuberculosis*: A potential TB drug target. *Structure* **6**:1587–1599.

Sharma, V., Sharma, S., Hoener zu Bentrup, K., McKinney, J. D., Russell, D. G., Jacobs, W. R., Jr., and Sacchettini, J. C. (2000). Structure of isocitrate lyase, a persistence factor of *Mycobacterium tuberculosis*. *Nat. Struct. Biol.* **7**:663–668.

Skaar, E. P., Gaspar, A. H., and Schneewind, O. (2004). IsdG and IsdI, heme-degrading enzymes in the cytoplasm of *Staphylococcus aureus*. *J. Biol. Chem.* **279**:436–443.

Snel, B., Bork, P., and Huynen, M. A. (2002). The identification of functional modules from the genomic association of genes. *Proc. Natl. Acad. Sci. USA* **99**:5890–5895.

Stewart, G., Robertson, B., and Young, D. (2003): Tuberculosis: A problem with persistence. *Nat. Rev. Microbiol.* **1**:97–105.

Strong, M., Graeber, T. G., Beeby, M., Pellegrini, M., Thompson, M. J., Yeates, T. O., and Eisenberg, D. (2003a). Visualization and interpretation of protein networks in *Mycobacterium tuberculosis* based on hierarchical clustering of genome-wide functional linkage maps. *Nucleic Acids Res.* **31**:7099–7109.

Strong, M., Mallick, P., Pellegrini, M., Thompson, M. J., and Eisenberg, D. (2003b). Inference of protein function and protein linkages in *Mycobacterium tuberculosis* based on prokaryotic genome organization: A combined computational approach. *Genome Biol.* **4**:R59.

Tamakoshi, M., Yamagishi, A., and Oshima, T. (1998). The organization of the leuC, *leuD* and *leuB* genes of the extreme thermophile *Thermus thermophilus*. *Gene* **222**:125–132.

Taylor, A. B., Benglis, D. M. Jr., Dhandayuthapani, S., and Hart, P. J. (2003). Structure of *Mycobacterium tuberculosis* methionine sulfoxide reductase A in complex with protein-bound methionine. *J. Bacteriol.* **185**:4119–4126.

Terwilliger, T. C., Park, M. S., Waldo, G. S., Berendzen, J., Hung, L. W., Kim, C. Y., Smith, C. V., Sacchettini, J. C., Bellinzoni, M., Bossi, R., De Rossi, E., Mattevi, A., Milano, A., Riccardi, G., Rizzi, M., Roberts, M. M., Coker, A. R., Fossati, G., Mascagni, P., Coates, A. R., Wood, S. P., Goulding, C. W., Apostol, M. I., Anderson, D. H., Gill, H. S., Eisenberg, D. S., Taneja, B., Mande, S., Pohl, E., Lamzin, V., Tucker, P., Wilmanns, M., Colovos, C., Meyer-Klaucke, W., Munro, A. W., McLean, K. J., Marshall, K. R., Leys, D., Yang, J. K., Yoon, H. J., Lee, B. I., Lee, M. G., Kwak, J. E., Han, B. W., Lee, J. Y., Baek, S. H., Suh, S. W., Komen, M. M., Arcus, V. L., Baker, E. N., Lott, J. S., Jacobs, W. Jr., Alber, T., and Rupp, B. (2003). The, T. B. structural genomics consortium: A resource for *Mycobacterium tuberculosis* biology. *Tuberculosis (Edinb.)* **83**:223–249.

Tullius, M. V., Harth, G., and Horwitz, M. A. (2003). Glutamine synthetase GlnA1 is essential for growth of *Mycobacterium tuberculosis* in human THP-1 macrophages and guinea pigs. *Infect. Immun.* **71**:3927–3936.

Uetz, P., Giot, L., Cagney, G., Mansfield, T. A., Judson, R. S., Knight, J. R., Lockshon, D., Narayan, V., Srinivasan, M., Pochart, P., Qureshi-Emili, A., Li, Y., Godwin, B., Conover, D., Kalbfleisch, T., Vijayadamodar, G., Yang, M., Johnston, M., Fields, S., and Rothberg, J. M. (2000). A comprehensive analysis of protein-protein interactions in *Saccharomyces cerevisiae*. *Nature* **403**:623–627.

Vetting, M. W., Roderick, S. L., Yu, M., and Blanchard, J. S. (2003). Crystal structure of mycothiol synthase (Rv0819) from *Mycobacterium tuberculosis* shows structural homology to the GNAT family of *N*-acetyltransferases. *Protein Sci.* **12**:1954–1959.

Vilcheze, C., Morbidoni, H. R., Weisbrod, T. R., Iwamoto, H., Kuo, M., Sacchettini, J. C., and Jacobs, W. R., Jr. (2000). Inactivation of the inhA-encoded fatty acid synthase II (FASII) enoyl-acyl carrier protein reductase induces accumulation of the FASI end products and cell lysis of *Mycobacterium smegmatis*. *J. Bacteriol.* **182**:4059–4067.

von Mering, C., Huynen, M., Jaeggi, D., Schmidt, S., Bork, P., and Snel, B. (2003). STRING: A database of predicted functional associations between proteins. *Nucleic Acids Res.* **31**:258–261.

Vyas, N. K., Vyas, M. N., and Quiocho, F. A. (2003). Crystal structure of *M tuberculosis* ABC phosphate transport receptor: Specificity and charge compensation dominated by ion-dipole interactions. *Structure (Camb.)* **11**:765–774.

Walburger, A., Koul, A., Ferrari, G., Nguyen, L., Prescianotto-Baschong, C., Huygen, K., Klebl, B., Thompson, C., Bacher, G., and Pieters, J. (2004). Protein kinase G from pathogenic mycobacteria promotes survival within macrophages. *Science* **304**:1800–1804.

Wiker, H. G., and Harboe, M. (1992). The antigen 85 complex: A major secretion product of *Mycobacterium tuberculosis*. *Microbiol. Rev.* **56**:648–661.

Wiker, H. G., Harboe, M., and Nagai, S. (1991). A localization index for distinction between extracellular and intracellular antigens of *Mycobacterium tuberculosis*. *J. Gen. Microbiol.* **137**( Pt. 4): 875–884.

Wilson, R. A., Rai, S., Maughan, W. N., Kremer, L., Kariuki, B. M., Harris, K. D., Wagner, T., Besra, G. S., and Futterer, K. (2003). Crystallization and preliminary X-ray diffraction data of *Mycobacterium tuberculosis* FbpC1 (Rv3803c). *Acta Crystallogr. D Biol. Crystallogr.* **59**:2303–2305.

Wolf, Y. I., Rogozin, I. B., Kondrashov, A. S., and Koonin, E. V. (2001). Genome alignment, evolution of prokaryotic genome organization, and prediction of gene function using genomic context. *Genome Res.* **11**:356–372.

World Health Organization (2002). *Tuberculosis Fact Sheet*. Geneva, Switzerland: WHO.

Xenarios, I., Salwinski, L., Duan, X. J., Higney, P., Kim, S. M., and Eisenberg, D. (2002). DIP, the Database of Interacting Proteins: A research tool for studying cellular networks of protein interactions. *Nucleic Acids Res.* **30**:303–305.

Yanai, I., Wolf, Y. I., and Koonin, E. V. (2002). Evolution of gene fusions: Horizontal transfer versus independent events. *Genome Biol.* **3**:1–13.

Yeats, C., Finn, R. D., and Bateman, A. (2002). The, PASTA domain: A beta-lactam-binding domain. *Trends Biochem. Sci.* **27**:438.

Young, D., Garbe, T., Lathigra, R., Abou-Zeid, C., and Zhang, Y. (1991). Characterization of prominent protein antigens from mycobacteria. *Bull Int. Union Tuberc Lung Dis.* **66**:47–51.

Young, T. A., Delagoutte, B., Endrizzi, J. A., Falick, A. M., Alber, T. (2003). Structure of *Mycobacterium tuberculosis* PknB supports a universal activation mechanism for Ser/Thr protein kinases. *Nat. Struct. Biol.* **10**:168–174.

Yuan, Y., Lee, R. E., Besra, G. S., Belisle, J. T., and Barry, C. E., 3rd (1995). Identification of a gene involved in the biosynthesis of cyclopropanated mycolic acids in *Mycobacterium tuberculosis*. *Proc. Natl. Acad. Sci. USA* **92**:6630–6634.

Zhang, Y., Lathigra, R., Garbe, T., Catty, D., and Young, D. (1991). Genetic analysis of superoxide dismutase, the 23 kilodalton antigen of *Mycobacterium tuberculosis*. *Mol. Microbiol.* **5**:381–391.

Zheng, Y., Roberts, R. J., and Kasif, S. (2002). Genomic functional annotation using co-evolution profiles of gene clusters. *Genome Biol.* **3**:1–9.

CHAPTER 16

# Proteomic Studies of Plant-Pathogenic Oomycetes and Fungi

CATHERINE R. BRUCE, PIETER VAN WEST, and LAURA J. GRENVILLE-BRIGGS
University of Aberdeen, Aberdeen, Scotland, United Kingdom

## 16.1 INTRODUCTION

There are at least 10,000 species of fungi and oomycetes known to be associated with plants [1]. Their interactions can be either symbiotic, that is, beneficial to both host and fungus, or pathogenic, whereby the microbe is able to cause some form of disease on the host plant. Studying their interactions is of paramount importance. Economic costs implicated in revenue losses and controlling plant diseases are overwhelming. The social and economic impacts of plant diseases affect both the most affluent and impoverished world nations. Therefore, plant pathogens play an important role in shaping the history and management of modern agricultural practices. In the mid-1840s the deaths of a quarter of a million Irish people, and as a result mass immigration to the United States, was due to the outbreak of late blight on potato. The causal agent, *Phytophthora infestans*, decimated the Irish potato crop in the worst late blight epidemic ever recorded [2]. *Phytophthora* species represent pernicious plant pathogens of major economic importance; however, to date they have been intractable to functional genetics, a situation which is now being offset through the application of proteomic technologies in combination with genomic approaches. Here, we review recent results obtained through proteomic investigations in the area of plant-pathogenic oomycetes and fungi.

### 16.1.1 Genetics of Fungal and Oomycete Plant–Pathogen Interactions

The interactions that occur between pathogens and their host plants can be broadly divided into two categories. The first are nonspecific interactions, whereby the pathogen is opportunistic and invades through wound sites or natural openings, such as stomata. These are interactions in which a particular pathogen often has a broad host range [1]. The second type of interaction is a specific one governed by the gene-for-gene hypothesis. Here, we can distinguish between an incompatible interaction, whereby the plant is resistant, or a compatible interaction, whereby the plant is susceptible. In an incompatible interaction,

*Microbial Proteomics: Functional Biology of Whole Organisms*, Edited by Ian Humphery-Smith and Michael Hecker. 

the product of a host resistance gene recognizes, either directly or indirectly, the product of an avirulence gene from the pathogen, leading to a defense response in the plant. In a compatible interaction the necessary alleles of either the resistance gene or the avirulence gene are not expressed or are mutated; therefore recognition by the host plant does not occur, and as a result the pathogen is able to cause disease. Although genetic analysis has identified a large number of plant-pathogenic systems in which this premise holds true and several major resistance (*R*) genes have been cloned in plants, many of the corresponding avirulence genes have not yet been characterized and the mechanism of recognition is largely unknown (for reviews see [3, 4]). Aside from the resistance response that has been well studied in many host plants, the avirulence side of the incompatible interaction has been studied in both the oomycetes and the true fungi through the use of genetic crosses and map-based cloning. Within the oomycetes, many avirulence loci have been identified in this way, such as the *atr1Nd* locus in *Hyaloperonospora parasitica* [5], *avr1b* in *Phytophthora sojae* [6], and several avirulence loci from *P. infestans* [7, 8]. However, cloning and proof of function studies have proved difficult in these organisms [9]. Genetic approaches have so far been more successful in the cloning of avirulence genes from fungal plant pathogens such as *Magnaporthe grisea* [10], *Melampsora lini* [11], *Ustilago hordeii* [12], *Blumeria graminis* [13], and *Leptosphareia maculans* [14]. Fungal avirulence genes have also been identified using reverse-genetics approaches. For example, the first fungal avirulence gene *avr9* from *Cladosporium fulvum* was cloned after extracellular proteins from *C. fulvum* were identified in the apoplastic fluid from infected tomato leaves [15]. The *nip1* gene from *Rhynchosporium secalis* was also cloned in a similar manner [16]. This clearly demonstrates the power of reverse-genetics approaches in these systems.

Approaches to investigate the secreted proteome of plant-pathogenic oomycetes and fungi to identify secreted avirulence compounds are now underway (see below) (Li, Bruce, Gow, and van West, unpublished data).

The ingression of major resistance (*R*) genes into cultivars of many plant species has allowed some level of disease control to be achieved in the field. However, experience has shown that the ingression of a single major *R* gene does not always offer durable resistance. Often these *R* genes are overcome by the pathogen in the field [17]. In recent years, work has therefore switched to the elucidation of essential pathogenicity factors and requirements for establishing disease, in terms of pathogen biology. It is anticipated that this will provide a better understanding of the essential processes required for disease in important crop systems and hence provide new targets for the development of durable disease control measures.

### 16.1.2 Pathogenicity

Pathogenicity is defined as the capability of a pathogen to cause disease. Virulence is the degree of pathogenicity exhibited by a given pathogen [1]. To be capable of causing disease on a plant, a pathogen must fulfill several important criteria. First, it needs to locate an appropriate host and then gain entry, while avoiding or overcoming the host defense mechanism(s). Once inside, it must gain enough nutrients to grow and reproduce and disseminate its progeny to start the whole cycle again [1].

Oomycetes from the genus *Phytophthora* and *Pythium* produce motile zoospores that are able to swim in films of water in the soil to locate plant roots. Interestingly, they are able to sense electrical signals surrounding root tips and exploit electric gradients to target specific sites for initiating infection [18]. Several fungal pathogens, including the grey mold fungus *Botrytis cinerea* and *Aspergillus flavus,* utilize natural openings, including stomata or wound sites, to gain entry into the plant [1], whereas other pathogens, such as the rice blast fungus *M. grisea* and *Cochliobolus* species, make use of specialized structures called appressoria. These penetration structures differentiate from a germinated spore in response to plant surface cues [19, 20]. The appressorium allows sufficient turgor pressure to be built up to pierce the plant cell wall [19]. Pathogens such as *Phytophthora* produce appressoria and probably also use enzymes such as cutinases, glucanases, and other cell-wall-degrading enzymes to soften the plant cuticle and allow easier entry [21]. Once inside the host plant, necrotrophic pathogens such as *Cochliobolus heterostrophus* secrete enzymes or toxins that result in the death of host tissue, the disintegration of cells, and the release of nutrients. These pathogens are then able to continue to feed necrotrophically. Fungi such as *C. fulvum* carry out their entire disease cycle within the intercellular spaces of the plant tissues. Others like *Verticillium* species, which cause vascular wilts of a range of plants, spread through the vascular bundles and surrounding parenchymal cells [1]. Some pathogens, including *Venturia inaequalis*, live an entirely subcuticular lifestyle, whereas *Claviceps purpurea*, the causal agent of ergot in cereals and grasses, lives both intercellularly and intracellularly [1].

Biotrophic plant pathogens have developed specialized feeding structures called haustoria, which allow invagination of the host cell membrane. A close membrane interface is formed between the host and pathogen cell without damaging the host cell membrane. Subsequently, the pathogen is able to feed by drawing nutrients across the double membrane. Pathogens that produce haustoria include the powdery mildews such as *Blumeria graminis*, which is the causal agent of wheat and barley mildew. These mildews produce haustoria that invaginate only the membranes of the epidermal cells. Hyphae do not penetrate further cell layers within the leaf [1]. The downy mildews, including the oomycete *Hyaloperonospora parasitica*, which causes downy mildew of *Brassica* crops, grow intercellularly with haustoria produced predominantly in the parenchymal cells. Hemibiotrophic pathogens, such as *M. grisea* and *P. infestans*, exhibit a transient form of biotrophy before switching to more necrotrophic forms of growth in the later stages of disease. They produce haustoria during the early biotrophic stage of the disease cycle [19, 22].

Finally, pathogens must be able to reproduce and disseminate their progeny in search of new hosts. Many plant-pathogenic fungi produce a wealth of different spore types, both asexual and sexual, that are dispersed by the wind or by rain splashes. Several plant pathogens also produce thick-walled spores that can survive undisturbed in the soil for years until environmental signals trigger germination [1].

Most of the above processes are likely to be controlled by complex sets of factors that contribute to the virulence of each pathogen. Global approaches to understand the biology of these processes, such as proteomics, have the advantage of being able to identify multiple factors governing processes and dissect complex biological pathways.

### 16.1.3 Global Approaches to Studying Plant–Pathogen Interactions

Global approaches such as the use of emerging technologies such as genomics, transcript profiling, suppression subtraction hybridization, and complementary deoxyribonucleic acid–(cDNA–AFLP) allow the researcher to investigate gene expression on a much larger scale than has been previously possible. Proteomic studies can aid tremendously in the functional analysis of gene products and cellular pathways and can also be used to discover proteins involved in particular disease stages. Even with a limited knowledge of DNA sequence information proteomic studies can aid in these objectives. Nonetheless, good DNA sequence databases are invaluable for downstream analysis. This may be one reason why much of the proteomic work so far in the area of plant–fungus interactions has focused on the plant side. Besides, in most plant–pathogen systems, it is easier to obtain sufficient biological material from the host plant rather than from the microbe. Meanwhile, several pathogen genomes are in the process of being or have now been sequenced. These databases should provide the resources required for effective global approaches to answering key questions in plant pathology. Several initiatives to use proteomic techniques to study the biology of microbes interacting with plants are underway and here we discuss some of these results. We anticipate that many more initiatives will follow soon.

## 16.2 PLANT-PATHOGENIC OOMYCETES

Within the group of oomycetes there are several genera that contain economically important plant pathogens. These include *Aphanomyces*, *Bremia*, *Hyaloperonospora*, *Peronospora*, *Plasmopara*, *Phytophthora*, and *Pythium*. Most molecular research has been performed with *Phytophthora* species. More than 60 *Phytophthora* species have been described [2], all of which cause plant disease of crops, shrubs, and trees on a global scale [23]. *Phytophthora infestans* was responsible for the Irish potato famine in the 1840s. Analysis of the epidemic was pivotal in the development of the science of plant pathology, predating even the germ theory of disease. More than 150 years after its discovery, *P. infestans* costs the global potato industry $3 billion per annum [24, 25]. Some *Phytophthora* species are able to infect a wide range of hosts, for example *Phytophthora nicotianae* and *Phytophthora cinnamomi* have host ranges of over 900 species of plants [2]. *Phytophthora* diseases continue to endanger food production with warnings issued in 2002 of a potential catastrophic potato famine in Russia as recently as 2001 [25]. Also, newly emerging species such as *Phytophthora ramorum* are also predicted as likely to cause widespread epidemics. This pathogen causes sudden death of oak trees and is spreading to redwood and other tree and shrub species [26].

Analyses of proteomes of several *Phytophthora* spp. have recently been employed to investigate a variety of aspects of developmental and (pre-) infection processes. Investigations are centering on metabolic processes of the pathogen prior to infection of the host plant. Moreover, proteome profiles of specific cell types and specialized structures of *Phytophthora* species that are crucial for infection are being analyzed. The asexual life cycle of *Phytophthora* species is well documented [27–29]. Typical infection of the host plant takes place when sporangia release biflagellate, motile zoospores. On finding a host, the zoospores shed their flagella, encyst, and produce a germ tube. An appressorium differentiates at the tip and produces a penetration peg, which pierces the cuticle of the leaf, allowing development of the infection vesicle in the epidermal cell. Branching hyphae expand to neighboring cells through the intracellular space, forming specialized feeding structures called haustoria. Mycelia eventually produce sporangiophores, which release asexual sporangia [29].

### 16.2.1 Asexual-Stage-Specific Proteins in *P. infestans* Identified by Proteomics

Kramer et al. [30] analyzed the proteome profiles of four distinct stages of the asexual life cycle of *P. infestans*: hyphae, cysts, germinating cysts, and appressoria. Several proteins were identified on two-dimensional (2D) gels that showed stage-specific differences in their relative amounts. In a similar study of *Phytophthora palmivora*, the causal agent of blackpod on cocoa, it was observed that approximately 1% of proteins are specific for each developmental stage of the life cycle [31]. This confirmed the observation by Kramer et al. [30] that a number of proteins are specific for distinct developmental stages. Furthermore, these results suggest that de novo protein synthesis is important for the transition between the asexual life-cycle stages of *Phytophthora* species.

Recently, Grenville-Briggs et al. (submitted) investigated the nutritional requirements of appressoria from *P. infestans* by using a combination of proteomics and suppression subtraction hybridization (SSH). Comparing protein abundance and transcript levels in mycelia, sporangia, zoospores, germinating cysts, and germinating cysts with appressoria revealed that a number of proteins and transcripts were specific or more abundant in germinated cysts with appressoria. Peptide mass fingerprinting and mass spectroscopy identified five proteins involved in biosynthesis of methionine, tryptophan, arginine, and branched-chain amino acids. Real-time polymerase chain reaction (RT-PCR) expression analysis confirmed that the corresponding genes were up regulated during appressoria. At present, it is speculated that these genes are most likely to be starvation rather than appressoria specific. In addition, it was noted that expression of these genes is down regulated during the biotrophic stages of the interaction with the host. However, expression is up regulated by 48 h postinoculation, which is about the time at which *P. infestans* switches to a more necrotrophic mode of growth, suggesting that additional amino acid synthesis is required by the pathogen itself in the later stages of infection.

### 16.2.2 Dissecting Membrane Proteomes in Zoospores and Cysts

The main mode of dispersal of *Phytophthora* and many other plant-pathogenic oomycetes is via motile zoospores that are released in large numbers from sporangia into soil water. In this environment, they use a combination of chemical and electrical cues to target root sites for infection [18, 23]. Upon reaching their site of infection, they undergo a process of encystment, shedding their flagella and secreting adhesives to allow attachment to plant tissue. Within a few minutes a thin cellulosic wall is formed and cysts begin to germinate [18, 32]. It is anticipated that molecules on the cell surface play an important role in signaling as well as transport of nutrients. Therefore, Mitchell et al. [33] used a proteomics approach to identify differences in the protein composition of plasma membranes and endomembranes of zoospores and cysts of *P. nicotianae*. Solubilized proteins from microsomal fractions of zoospores and cysts were analyzed by two-dimensional electrophoresis (2DE). A number of protein spots were found to be specific membrane proteins from either zoospores (5 spots) or cysts (23 spots). Quantitative differences were also observed. The zoospore membrane proteins were further analyzed using immunocytochemistry and immunoblotting with 10 monoclonal antibodies raised against *P. nicotianae* spores that reacted with the zoospore surface. Reactivity of the antibodies with *P. nicotianae* cysts, zoospores, sporulating hyphae, and vegetative hyphae showed that the antibodies bound different sets of membrane proteins and were accordingly grouped into four classes. Group 1 antibodies bound proteins in the membrane of the bladder of the water expulsion vacuole or a region of the plasma membrane closely associated with it. Group 2 antibodies adhered to a high-molecular-weight (>200-kDa) protein localized in the plasma membrane of zoospores and cysts and the cleavage membranes of sporangia. Group 3 antibodies bound a diverse set of proteins, including a set found in the plasma membrane of zoospores and cysts, the peripheral cisternae, the spongiome membranes of the water expulsion vacuole, the cleavage membranes of sporangia, and the plasma membrane and apical vesicle membranes in hyphae and germinating cysts. Group 4 antibodies reacted with a set of proteins found in the plasma membrane of zoospores and cysts, the spongiome of the water expulsion vacuole, and the cleavage membranes of sporangia but which were absent in the peripheral cisternae, hyphae, and germinating cysts. The surface proteins found in these four distinct groups are excellent candidates for involvement in pathogenicity and can now be investigated further to elucidate their molecular roles.

### 16.2.3 Extracellular Proteins from *P. infestans*

The extracellular proteome of *P. infestans* mycelia has also been analyzed by 2DE to identify secreted proteins, which are anticipated to be important molecules in the initiation of disease [34]. An algorithm, PexFinder, was developed to identify extracellular proteins from *P. infestans* expressed sequence tag (EST) data sets by identifying N-terminal signal peptides. Using this algorithm, 142 nonredundant Pex (*Phytophthora* extracellular protein) cDNAs were identified of which 55% were novel with no significant matches in the public databases. The extracellular proteome of *P. infestans* mycelia was analyzed by 2DE to validate PexFinder. Thirty protein spots were excised from a 2D gel of mycelial culture filtrate, digested with trypsin, and analyzed using matrix-assisted linear desorption ionization time-of-flight mass spectrometry (MALDI-TOF MS) and peptide mass fingerprinting. Of these, nine proteins unambiguously matched Pex cDNAs. These included an acidic chitinase, glutathionine-*S*-transferase, enoyl-coenzyme A hydratase, arabinofuranosidase/β-xylosidase, peptidylprolyl isomerase, and two unknown proteins. Interestingly, an ortholog of Avr1b from *P. sojae* was also identified in the secreted protein fraction from *P. infestans* and is currently under investigation to determine its role during the interaction (Armstrong, Whisson, and Birch, personal communications). The Avr1b protein from *P. sojae* triggers a hypersensitive response in host plants carrying the *rps1b* resistance gene [6]. The presence of the *P. infestans* orthologue in culture filtrates indicates that a more complete secreted proteome of *P. infestans* could lead to the identification of potential effector molecules. This work is currently underway (Bruce, Li, Gow, and van West, unpublished data).

### 16.2.4 Interspecies Proteome Comparisons

Studies have also been carried out to compare proteome profiles between different oomycete species [31]. It was found that two species *P. palmivora* and *P. infestans*, exhibit similar protein profiles. A comparison of sporangial proteomes from both species showed relatively high conservation in the protein profile. Equivalent numbers of protein spots were detected on 2D gels of sporangia of both species, with 30% of spots of both species having similar or identical positions. In theory, this observation could therefore facilitate identification of proteins from *Phytophthora* species, for which little genomic information is available, by initially characterizing the *P. infestans* homologous protein spot.

## 16.3 PLANT-PATHOGENIC FUNGI

Many fungal pathogens have complex life cycles and proteomics therefore offers a powerful tool in dissecting particular aspects of their biology. The genomes of several important fungal pathogens have been, and some are currently being, sequenced and annotated. Furthermore, nonphytopathogenic fungi are also being sequenced, allowing for valuable comparisons. This section will highlight how the use of proteomics has enabled us to gain valuable understanding and insight into areas of fungal phytopathology that would otherwise be difficult to elucidate using other methods while also highlighting its effectiveness at answering broad questions within this field.

### 16.3.1 *Magnaporthe grisea* and Rice Interactions

*Magnaporthe grisea* is a filamentous ascomycete that is able to cause disease on over 50 species of grasses, including economically important crops such as rice, millet, barley, and wheat. Rice blast is the most serious disease of rice and is a major problem in most rice-growing areas of the world [19]. Interactions between rice and *M. grisea* occur in a gene-for-gene specific manner, and this pathosystem has been developed as a genetically tractable model of fungal plant disease and resistance [35]. A wealth of genomic information about the host plant and its pathogen exists. There is also a 2D rice database at the National Institute of Agrobiological Sciences in Japan, at http://gene64.dna.affrc.go.jp/RPD/database_en.html, which allows easy identification of host proteins from 2D gels containing complex mixtures of proteins. Proteomic analysis of the rice blast pathogen and of infected material is underway at the University of Exeter (United Kingdom) and the Consortium for the Functional Genomics of Microbial Eukaryotes (COGEME) (Talbot, personal communication). This consortium provides facilities for genome analysis for eukaryotic microbes, including important model organisms and pathogens (for further details see www.cogeme.man.ac.uk). COGEME has a proteomics facility at the University of Aberdeen. It provides proteomic analysis of a variety of eukaryotic microbes and has so far performed proteome analysis of *Saccharomyces cerevisiae, Candida albicans, Candida glabrata, M. grisea*, and *Schizosaccharomyces pombe* (for more information see www.cogeme.abdn.ac.uk).

Recent studies of the rice–*M. grisea* interaction have identified predominantly plant proteins that are up regulated during the rice blast infection [36]. The first reported study of *M. grisea* proteins induced by host cells was published by Kachroo et al. [37] and not only showed the usefulness of 2D analysis of proteins in understanding plant–fungal pathogen interactions but also highlighted novel proteins that appear to be triggered in *M. grisea* by exposure to susceptible rice cells only. Proteins from cellular and extracellular material were obtained after exposure to resistant or susceptible rice cells and compared to control samples that had not been exposed to host extracts. They observed at least 17 different proteins, in the intracellular fraction, that were induced in response to susceptible host extracts. These proteins were not present in the untreated control or in the sample that had been treated with the resistant host extract. While this study did not identify the nature of the protein spots observed, it did, however, highlight global changes in the pathogen protein profile in response to host cells. It was interesting that the proteins observed in this study were induced only by the susceptible host and in fact that no change in the protein profile was observed upon incubation with resistant host cells. This suggests that the proteins present after incubation with susceptible host cells may be involved in establishment of a successful infection.

### 16.3.2 Differentiation-Related Proteins from Broad Bean Rust Fungus *Uromyces vicae-fabae*

While many plant-pathogenic fungi infect their host plant cells by direct penetration of the cuticle and cell wall, *Uromyces* species differentiate a series of infection structures in order to invade the leaf through the stomata. The signal for infection structure differentiation is induced by surface topography [38]. Differentiation of appressoria, substomatal vesicles, infection hyphae, and haustorial mother cells can be triggered in vitro, making *U. vicae-fabae* amenable to laboratory studies. Proteins that are differentiation related and occur immediately prior to the establishment of biotrophy may be essential for pathogenesis. In an early study, Deising et al. [39] studied protein patterns of sequentially formed infection structures using 2DE. Seven hundred and thirty-three protein spots were identified, representing the whole developmental sequence. During infection structure differentiation 55 proteins were newly formed, altered in quantity, or disappeared. Major differences in protein profiles were seen during uredospore germination and during the formation of infection hyphae. Uredospore germination was characterized by a decrease of acidic proteins. The authors found that 9 proteins were newly formed during differentiation of infection hyphae, with 15 spots significantly increased and 12 protein spots down regulated. Haustorial mother cell formation was accompanied by an increase of 6 spots, a decrease of 6 spots, and the disappearance of 1 spot. This early study did not characterize the discovered proteins in detail. However, it does provide a foundation for a proteomic approach to studying differential protein expression in the rust fungi. It is clear that several proteins are produced at the time point just prior to establishment of infection, and these may therefore represent potential pathogenicity factors. Identification of these proteins may lead to a greater understanding of the infection process and pathogenicity of this group of plant pathogens.

### 16.3.3 Secreted Fungal Proteins

It is anticipated that secreted proteins from fungal and oomycete pathogens will be important in the establishment of disease. Several secreted proteins from plant pathogens have been shown to be important in the establishment of infection or activation of the plant defense response. For example, the avirulence genes *avr4* and *avr9* from the tomato pathogen *C. fulvum* encode small protein elicitors that are secreted by the fungus and induce a resistance response in plants expressing the corresponding resistance genes [40]. *Cladosporium fulvum* also secretes small extracellular proteins (ECPs), five of which have been characterized that are also potential avirulence factors. Production of these proteins is induced during in-planta growth and all stages of infection [41, 42]. In addition several other classes of secreted fungal proteins aid the establishment of infection, for example, the hydrophobins, such as MPG1 from *M. grisea*, and the HCf genes from *C. fulvum* play a role in the initiation of infection [43, 44]. These proteins are small-secreted hydrophobic polypeptides, characterized by the conserved spacing of eight cysteine residues. Hydrophobins play distinct roles in fruiting body development, conidiogenesis, aerial hyphae formation, and infection structure elaboration in diverse fungal species. The MPG1 hydrophobin of *M. grisea* is highly expressed during appressorium formation, disease development, and conidiation. The MPG1 protein interacts with hydrophobic surfaces and may act as a developmental sensor for appressorium formation. It is also necessary for full pathogenicity [43].

Many fungal pathogens also secrete cell-wall-degrading enzymes (CWDEs), such as polysaccharide hydrolases that aid the infection process by allowing easier penetration of the host tissue. However, some fungi that are able to produce appressoria also produce CWDEs to aid in the establishment of infection. Therefore, CWDEs play an important part in the infection process of many fungi, such as *Fusarium* species, from which pectate lyases have been identified [45] and *Botrytus cinerea* from which several polygalacturonases have been cloned [46]. Rust fungi penetrate their host through the stomata. However, the bean rust fungus *U. vicae-fabae* also produces appressoria prior to penetration, and a variety of CWDEs have been detected within the first 24 h of infection of bean plants and during growth of the rust in vitro. The identified proteins include cellulases [47], pectin methylesterases [48], and polygalacturonate lyases [49]. Schmidt and Wolf [47] detected cellulase and xylanase activity within the intercellular wash fluids of infected bean plants from two days after infection with the highest levels of enzyme being detected seven days after infection. The authors used 2DE to detect the presence of 13 cellulase isoforms of the *U. vicae-fabae* seven days after infection. They speculate that while CWDEs are required during the initial stages of infection to facilitate penetration of the host cell, they may also function in the later stages of infection by breaking down the host cell wall, thus providing an extra source of carbon for the pathogen.

***Understanding Specific Events Using Proteomics: Extracellular Matrix and Spore Adhesion*** For many fungal pathogens, secreted proteins are important for the infection process, from the earliest stages, where spores need to adhere to the plant surface, right through to the point of obtaining nutrients from living cells, as in the case of obligate biotrophes.

Many fungal pathogens secrete an extracellular matrix (ECM) which may function in several ways. The ECM provides adhesion of spores to the host surface as well as a medium that allows the buildup, concentration, and localization of enzymes needed to establish infection. The ECM is also likely to be important in preventing spore desiccation and may also provide protection from toxic metabolites secreted by the host defense machinery [50]. Many of the proteins present in the ECM are glycoproteins that have been traditionally difficult to separate. Apoga et al. [51] have developed a method for isolating and sequencing proteins from the ECM of filamentous fungi. Biotinylation was used to label the proteins present in the ECM of conidia, germlings, and hyphae of *Bipolaris sorokiniana*, and a glycine–HCl buffer was used to extract proteins from these cells. Interestingly, no protein bands were detected on 1D gels of conidial samples; however, several proteins were identified from germlings and hyphae. Some of the bands initially identified on 1D gels were present only in the germling sample and were absent from hyphae. Two-dimensional electrophoresis confirmed that some of the bands present on 1D gels contained proteins and were resolved into several spots. Four protein spots were digested with trypsin and sequenced by tandem mass spectrometry (MS/MS). No significant homologies were found to public databases. One of the isolated proteins contained stretches of acidic amino acid residues. It is interesting to note that the cyst surface proteins identified by Gornhardt et al. [52] from the oomycete pathogen *P. infestans* also have stretches of acidic amino acid residues, as do surface proteins of the human-pathogenic fungus *C. albicans* [53]. In a further study of the protein composition of the ECM of *B. sorokiniana,* Apoga et al. [54] labeled the surface proteins of germlings with $^{125}I$ and analyzed them using 2DE. At least 40 labeled proteins were detected on the 2D gels, indicating a large level of complexity at the cell surface. These proteins have yet to be identified. However, the authors postulate that they are likely to be involved in a wide range of processes, such as adhesion, tolerance to toxins or desiccation, and enzymatic degradation of host tissue to allow the pathogen to gain entry into the plant cell.

Glycoproteins in the ECM of *Colletotrichum* have recently been identified and localized using Western blotting of protein gels and enzyme-linked immunosorbent assay (ELISA) by Hutchison et al. [55]. Their results indicate that cell walls of *Colletotrichum lindemuthianum* are heterogeneous in nature, with some glycoproteins spreading further from the cell surface than others. They also showed that the ECM surrounding germ tubes and appressoria is similar but very different from the ECM surrounding the conidia and intracellular hyphae formed within host cells. Differences such as these may be exploited to design novel antifungal agents that prevent adhesion and initiation of disease. Due to the apparent complex nature of the ECM in many fungal species, proteomics is the ideal approach to separate out individual proteins and study differences between cell types and species.

***Extracellular Proteome of* A. flavus*: Dissecting Nutritional Pathways*** The biotechnology industry has been mining a broad spectrum of structural proteins and enzymes that are secreted by filamentous fungi for a number of years, and studies focusing on the production of individual enzymes are widespread. Recently, Medina et al. [56] have exploited the fact that proteomics allows the identification of proteins from complex mixtures and that it can be used to compare differential expression patterns. They investigated the extracellular proteome of *A. flavus* to identify proteins involved in degrading a defense compound from plants. *Aspergillus flavus* is a filamentous fungus that is associated with postharvest disease. It commonly infects corn kernels and groundnuts while still in the field. Being a largely opportunistic pathogen, the incidence of infection is increased in damaged kernels or in plants that are environmentally stressed [1]. *Aspergillus flavus* can degrade the flavenoid rutin as its only carbon source using an extracellular enzyme system. Flavenoids are secondary metabolites that occur widely throughout the plant kingdom. Some flavenoids have been shown to have antimicrobial properties, particularly against *A. flavus* and *Aspergillus parasiticus*. Some species of *Aspergillus*, including *A. flavus*, are capable of degrading the flavenoid rutin. Several of the enzymes involved in the degradation pathway have been identified or proposed, although sequence information is only available for one of these, a quercetinase from *Aspergillus japonicus*. A proteomic analysis of the extracellular proteins of *A. flavus* grown either in potato dextrose or in a medium containing mineral salts and rutin was performed. Sodium dodecyl sulfate–polyacrylamide gel electrophoresis (SDS-PAGE) revealed a large number of protein bands present in the secreted fraction of rutin grown cultures compared to the control in potato dextrose. Two-dimensional gel electrophoresis revealed there to be approximately 70 protein spots present in the rutin-induced samples, compared with only about 20 in the control sample. All visible protein spots were cut out and analyzed by MALDI-TOF MS. Fifteen rutin-induced and 7 non-rutin-induced proteins were identified; however, over 90 proteins remain unidentified, highlighting the importance of genomic or EST databases to support proteomic approaches. In addition to the proteins required to degrade rutin, Medina et al. [56] also found a number of other enzymes and proteins, including an alanyl dipeptidyl peptidase, alkaline proteases, α-amylase precursors, catalases, a β-galactosidase, a gutaminase, an α-mannosidase, a neutral trehalase, and a xanthine dehydrogenase. Four of these enzymes were also present in gels from cultures grown in potato dextrose agar. Since their proteomic studies were accompanied by enzyme assays based on the proposed degradation pathway of rutin, the authors were able to build up a picture of *A. flavus* metabolism in the presence of rutin compared to potato dextrose. This study highlighted the fact that proteomics is a powerful tool for the identification of global changes under different environmental conditions.

## 16.4 FUTURE PERSPECTIVES

Proteomics is a valuable tool for the study of plant–fungal and plant–oomycete interactions. With this high-throughput technology, large numbers of expressed proteins can be studied. Proteomic studies are particularly informative when there is good sequence data available to allow accurate identification of proteins via MALDI-TOF MS. Studies that involve comparisons between complex states such as differential nutritional requirements or cell differentiation are particularly amenable to proteomic analysis. In addition, specific questions relating to the mode of growth of phytopathogens have been successfully addressed using proteomics approaches. With the integration of genomic and proteomic knowledge, characterization of proteins identified through 2DE should be much easier in the future. We will then be able to address interesting biological questions using functional analysis of identified proteins through emerging methods such as RNA interference and gene knockouts, which are already possible in some of the organisms discussed in this chapter. Targeted control methods may then also emerge, allowing significant alleviation of disease symptoms, with the use of minimal chemical treatments, thus preserving our environment.

## ACKNOWLEDGMENTS

The authors would like to thank The Royal Society for providing funding and a personal fellowship to P. van West (Royal Society University Research Fellowship) and the Biotechnology and Biological Sciences Research Council (BBSRC) for financial support to Catherine Bruce and Laura Grenville-Briggs (02/B1/P/08009 and 1/P17160). We are also grateful to Alison Horsburgh for critically reading the manuscript.

## REFERENCES

1. Agrios, G. N., *Plant Pathology*, 4th ed. Academic, San Diego, 1997.
2. Erwin, D. C., and Ribeiro, O. K., in *Phytophthora Diseases Worldwide*, American Phytopathological Society, St. Paul, MN, 1996, pp. 1–7.
3. Lauge, R., and de Wit, P. J. G. M., *Fungal Genet. Biol.* 1998, **24**, 285–297.
4. Greenberg, J. T., and Yao, N., *Cell. Microbiol.* 2004, **6**, 201–211.
5. Rehmany, A. P., Grenville, L. J., Gunn, N. D., Allen, R. L., et al., *Fungal Genet. Biol.* 2003, **38**, 33–42.
6. Shan, W., Cao, M., Leung, D., and Tyler, B. M., *Mole. Plant-Microbe Interact.* 2004, **17**, 394–403.
7. van der Lee, T., Robold, A., Testa, A., van 't Klooster, J. W., et al., *Genetics* 2001, **157**, 949–956.
8. Whisson, S. C., van der Lee, T., Bryan, G. J., Waugh, R., et al., *Mol. Genet. Genomics* 2001, **266**, 289–295.

9. Tyler, B. M., *Annu. Rev. Phytopathol.* 2002, **40**, 137–167.
10. Sweigard, J. A., Carroll, A. M., Farrall, L., Chumley, F. G., et al., *Mol. Plant-Microbe Interact.* 1998, **11**, 404–412.
11. Dodds, P. N., Lawrence, G. J., Cantanzariti, A. M., Ayliffe, M. A., et al., *Plant Cell* 2004, **16**, 755–768.
12. Linning, R., Lin, D., Lee, N., Abdennadher, M., et al., *Genetics* 2004, **166**, 99–111.
13. Pedersen, C., Rasmussen, S. W., and Giese, H., *Fungal Genet. Biol.* 2002, **35**, 235–246.
14. Balesdent, M. H., Attard, A., Kuhn, A. L., and Rouxel, T., *Phytopathology* 2002, **92**, 1122–1133.
15. van Kan, J. A. L., van den Ackerveken, G. F. J. M., and de Wit, P. J. G. M., *Mol. Plant-Microbe Interact.* 1991, **4**, 52–59.
16. Rohe, M., Gierlich, A., Hermann, H., Hahn, M., et al., *EMBO J.* 1995, **14**, 4168–4177.
17. Crute, I. R., and Pink, D., *Plant Cell* 1996, **8**, 1747–1755.
18. van West, P., Morris, B. M., Reid, B., Appiah, A. A., et al., *Mol. Plant-Microbe Interact.* 2002, **15**, 790–798.
19. Talbot, N. J., *Trends Microbiol.* 1995, **9**, 9–16.
20. Mendgen, K., Hahn, M., and Deising, H., *Annu. Rev. Phytopathol.* 1996, **34**, 367–386.
21. van West, P., and Vleeshouwers, V. G. A. A., in N. J. Talbot (Ed.), *Annual Plant Reviews*, Vol 11: *Plant-Pathogen Interactions*. Blackwell Scientific, 2004, pp. 219–242.
22. Vleeshouwers, V. G. A. A., van Dooijeweert, W., Govers, F., Kamoun, S., et al., *Planta* 2000, **210**, 853–864.
23. van West, P., Appiah, A. A., and Gow, N. A. R., *Physiol. Mol. Plant Pathol.* 2003, **62**, 99–113.
24. Duncan, J., *Microbiol. Today* 1999, **26**, 114–116.
25. Schiermeier, Q., *Nature* 2001, **410**, 1011.
26. Knight, J., *Nature* 2002, **415**, 251.
27. van West, P., de Jong, A. J., Judelson, H. S., Emons, A. M. C., et al., *Fungal Genet. Biol.* 1998, **23**, 126–138.
28. Hardham, A. R., *Australasian Plant Pathol.* 2001, **30**, 91–98.
29. Judelson, H. S., *Fungal Genet. Biol.* 1997, **22**, 65–76.
30. Kramer, R., Freytag, S., and Schmelzer, E., *Eur. J. Plant Pathol.* 1997, **103**, 43–53.
31. Shepherd, S. J., van West, P., and Gow, N. A. R., *Mycol. Res.* 2003, **107**, 395–400.
32. Warburton, A. J., and Deacon, J. W., *Fungal Genet. Biol.* 1998, **25**, 54–62.
33. Mitchell, H. J., Kovac, K. A., and Hardham, A. R., *Mycol. Res.* 2002, **106**, 1211–1232.
34. Torto, T. A., Li, S., Styer, A., Huitema, E., et al., *Genome Res.* 2003, **13**, 1675–1685.
35. Valent, B., *Phytopathology* 1990, **80**, 33–36.
36. Konishi, H., Ishiguro, K., and Komatsu, S., *Proteomics* 2001, **1**, 1162–1171.
37. Kachroo, P., Lee, K. H., Schwerdel, C., Bailey, J. E., et al., *Electrophoresis* 1997, **18**, 163–169.
38. Hoch, H. C., Staples, R. C., Whitehead, B., Comeau, J., et al., *Science* 1987, **235**, 1659–1662.
39. Deising, H., Jungblut, P. R., and Mendgen, K., *Arch. Microbiol.* 1991, **155**, 191–198.
40. Joosten, M. H. A. J., and de Wit, P. J. G. M., *Annu. Rev. Phytopathol.* 1999, **37**, 335–367.
41. Lauge, R., Joosten, M. H. A. J., van den Ackervecken, G. F. J. M., van den Broek, H. W. J., et al., *Mol. Plant-Microbe Interact.* 1997, **10**, 725–734.
42. Lauge, R., Goodwin, P. H., de Wit, P. J. G. M., and Joosten, M. H. A. J., *Plant J.* 2000, **23**, 735–745.
43. Talbot, N. J., Kershaw, M. J., Wakely, G. E., de Vries, O. M. H., et al., *Plant Cell* 1996, **8**, 985–999.
44. Whiteford, J. R., and Spanu, P. D., *Fungal Genet. Biol.* 2001, **32**, 159–168.

45. Gonzalez-Candelas, L., and Kolattukudy, P. E., *J. Bacteriol.* 1992, **174**, 6343–6349.
46. Johnston, D. J., and Williamson, B., *Mycol. Res.* 1992, **96**, 343–349.
47. Schmidt, C. S., and Wolf, G. A., *Eur. J. Plant Pathol.* 1999, **105**, 285–295.
48. Frittrang, A. K., Deising, H., and Mendgen, K., *J. Gen. Microbiol.* 1992, **138**, 2213–2218.
49. Deising, H., Rauscher, M., Haug, M., and Helier, S., *Can. J. Bot.* 1995, **73**, 624–631.
50. Epstein, L., and Nicholson, R. L., in G. C. Carrol and P. Tudzynski (Eds.), *Plant Relationships, The Mycota V, Part A,* Springer-Verlag, Berlin, 1997, pp. 11–25.
51. Apoga, D., Ek, B., and Tunlid, A., *FEMS Microbiol. Lett.* 2001, **197**, 145–150.
52. Gornhardt, B., Rouhara, I., and Schmelzer, E., *Mol. Plant-Microbe Interact.* 2000, **13**, 32–42.
53. Staab, J. F., Ferrer, C. A., and Sundstrom, P., *J. Biol. Chem.* 1996, **271**, 6298–6305.
54. Apoga, D., Jansson, H-B., and Tunlid, A., *Mycol. Res.* 2001, **105**, 1251–1260.
55. Hutchison, K. A., Green, J. R., Wharton, P. S., and O'Connell, R. J., *Mycol. Res.* 2002, **106**, 729–736.
56. Medina, M. L., Kiernan, U. A., and Francisco, W. A., *Fungal Genet. Biol.* 2004, **41**, 327–335.

# CHAPTER 17

# *Candida albicans* Biology and Pathogenicity: Insights from Proteomics

AIDA PITARCH, CÉSAR NOMBELA, and CONCHA GIL
Complutense University, Madrid, Spain

## 17.1 ANTECEDENTS

### 17.1.1 Awakening of *Candida albicans* as Organism of Clinical and Biological Relevance

For many years, *C. albicans* remained in relative obscurity and was somewhat banished from the scientific community's mind. This reluctance to undertake research into diverse key issues about this fungus was possibly fueled by the fact that this normal commensal inhabitant of the human microbial flora was not considered to be a life-threatening infectious agent. However, over the past few decades, candidiasis has given up being a trivial infection to become a major clinical problem for modern biomedicine, mainly because of its disturbingly high morbidity and mortality rates, especially in association with systemic candidiasis [1–4].

What were the factors that predisposed this change of patterns at that time? Perhaps, the alarming increase in its incidence could have been partly associated with improved methods for its detection. Be that as it may, there is no doubt that both an expanding immunocompromised population (i.e., AIDS or cancer patients) and recent advances in medical care directed toward prolonging patient survival (e.g., use of broad-spectrum antibiotics, aggressive chemotherapy, cytotoxic drugs, organ or bone marrow transplants, other surgical procedures or indwelling catheters, as well as longer hospitalization [5, 6]) decisively contributed—and go on contributing—to it (Fig. 17.1). At present, the frequency of this fatal opportunistic mycosis still continues to be a course for concern [7–14] and, unfortunately, its resolution is hindered by the reduced effectiveness and serious side effects of the few currently available drugs, the appearance of antifungal-drug resistance, and the lack of accurate and prompt diagnostic procedures [15–18].

Beyond its clinical relevance and basically attributable to its ability to switch among different cellular forms—mostly between budding and filamentous forms—during its proliferation in human fluids and/or tissues [19], this polymorphic fungus was paradoxically regarded in the late twentieth century as an attractive eukaryotic model organism for

*Microbial Proteomics: Functional Biology of Whole Organisms*, Edited by Ian Humphery-Smith and Michael Hecker. 

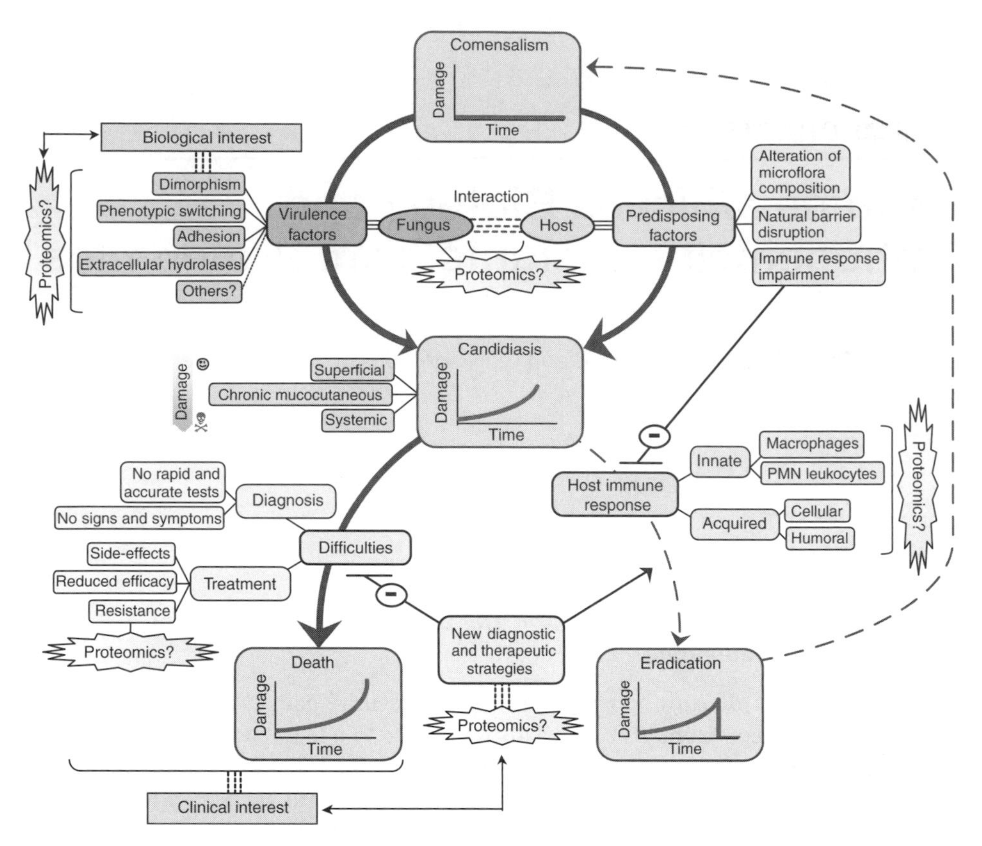


**Figure 17.1** Overview of the current setting for human candidiasis. *Candida albicans*, an endogenous commensal resident on the mucosal surfaces, can cause host damage (*candidiasis*) by mechanisms mediated by both host (*predisposing factors*) and fungus (*virulence factors*). Owing to its nature as an opportunistic pathogen, a host dysfunction/s (resulting, e.g., from antibiotic use, surgical procedures, and/or immunosuppressive therapy) is an essential prerequisite (thick arrow) for the establishment of disease. Depending on the underlying host defect, *C. albicans* can cause infections ranging from benign superficial (oral or vaginal) candidiasis to life-threatening systemic candidiasis. Due to the host's debilitated immune response, often displayed among patients with systemic candidiasis, unable to eradicate the infection (dashed arrows), together with difficulties in the diagnosis and treatment of these systemic infections, the outcome of these patients is often fatal (thick arrow). Arrowheads in the continuous lines indicate the current direction of the host–fungus interaction, which frequently ends in growing host damage (e.g., host's tissue and organ invasion) or even in host death, particularly among patients with systemic candidiasis. Thick arrows represent leading ways. Blunt arrows depict negative effects. At present, proteomics appears to be a promising technology to tackle the search for new diagnostic and therapeutic strategies for these infections as well as to study diverse key issues about the host–fungus interaction. The different goals of proteomics research into *C. albicans* that have already been reported in the literature or might be achieved in the near future are highlighted with blue stars.

studying morphogenesis and cellular differentiation in basic research. Hence, all these reasons have stimulated today's scientists to take heed of *C. albicans* in an attempt to discover new diagnostic and therapeutic strategies for this infectious agent as well as to understand its biology and mechanisms involved in its dimorphism and virulence (Fig. 17.1).

### 17.1.2 Why Proteomics?

Most of the knowledge on the biology, pathogenicity, virulence, and antifungal-drug resistance mechanisms of *C. albicans* was gained at the gene level from the 1980s onward by applying molecular biology techniques to this fungus, such as transformation systems, gene disruption strategies, gene reporter assays, expression-controlled systems, strain typing, and chromosome analysis (karyotyping), to name but a few (reviewed in [20, 21]). In spite of this, as the desired results have not yet been reached with these more traditional methodologies, other alternative approaches, such as proteomics, have emerged with the aim of finding a solution to the problems raised.

Intriguingly, given that proteins (i.e., end gene products) are the engines of most of the physiological and pathological processes of living cells and considering the fact that proteomics (a dynamic multidimensional platform) in turn offers an overview of the complex world of proteins within a cell under different expression conditions at a given time point [17, 18], this technology could, therefore, substantially complement the existing molecular understanding of this pathogen and result in new insight both into its biological complexity and into the host–fungus interaction at a higher level.

### 17.1.3 History of Proteomics in *C. albicans* Research

***Historical Landmarks in Advancement of C. albicans Proteomics*** The origin of proteomics in *C. albicans* research dates back to the early 1980s. The first study that displayed two-dimensional *C. albicans* protein maps, corresponding to differential expression of yeast and hyphal soluble proteins, was published in 1980 by Manning and Mitchell [24] or, to be more precise, 5 years later than the development of this innovative technique [two-dimensional gel electrophoresis (2DE)] by O'Farrell [25] and Klose [26]. No protein identification could be accomplished in that pioneering proteomic analysis. Surprisingly, 16 years had to pass prior to the characterization of the first 2DE-separated *C. albicans* proteins by classical procedures, such as immunostaining [27] or N-terminal sequencing [28] (Fig. 17.2). Nevertheless, the revolutionary technology of mass spectrometry (MS) as applied to proteins from 2DE gels was not introduced to investigations on the *C. albicans* proteome until 2000 [29–31], and even then, only 14 different protein identities were attained. Soon afterward, more comprehensive reference 2DE *C. albicans* maps were gradually developed [32–41], which certainly now represent the keystone for upcoming *C. albicans* proteomic discoveries.

Some of these maps began to be compiled in a 2DE database named COMPLUYEAST-2DPAGE, which was created at the Complutense University of Madrid (Spain) by the present authors at the end of 2000 [42]. Currently, this includes a dozen cartographic maps of different *C. albicans* subproteomes publicly available at http://babbage.csc.ucm.es/2d/2d.html or http://www.expasy.ch/ch2d/2d-index.html. This library is a 2DE federated database that provides direct links to the SWISS-PROT protein knowledgebase, which is in turn cross-referenced with more than 60 different data resources [43]. This attribute enables users to retrieve broad information about each individual protein spot identified in these 2DE maps through the so-called active hypertext cross-references. In 2003, members of Aberdeen University (United Kingdom) and the Consortium for the Functional Genomics of Microbial Eukaryotes (COGEME) constructed a further *C. albicans* 2DE data bank. This contains one extensive reference *C. albicans* 2DE map with a total of 316 different proteins identified [40] and can be accessed at http://www.abdn.ac.uk/cogeme/.

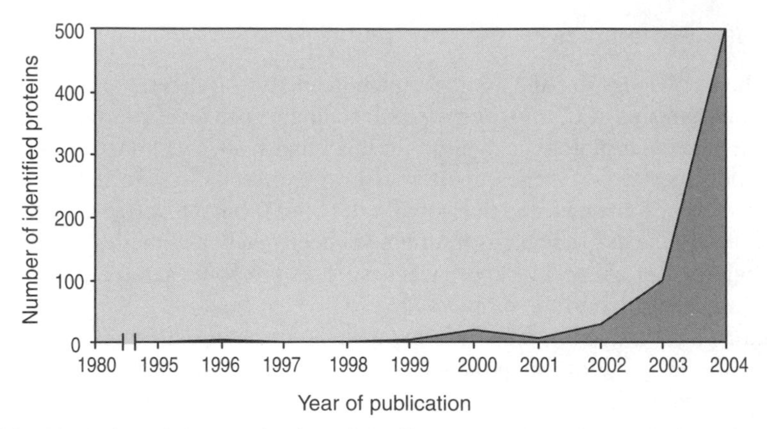


**Figure 17.2** Evolution of characterization of *C. albicans* proteome from 1980 to 2004. The graph represents the number of *C. albicans* proteins identified over the past 25 years by proteomic strategies that have been reported in the scientific literature, obtained from a Medline search (PubMed, reachable using Entrez retrieval system, http://www.ncbi.nlm.nih.gov/entrez) with keywords *Candida albicans* or *candidiasis* and *proteomic(s)/proteome* or *2DE*. See the text and the Appendix in Chapter 18 for further details.

All the proteins reported in the literature to be characterized by proteomic approaches focused on diverse aspects of *C. albicans* (discussed both in the following sections and in the Chapter 18) are listed in the Appendix in Chapter 18.

Why was proteomics only applied to *C. albicans* research a short time ago? What was the bottleneck that delayed the successful progress of this discipline in these initial inquiries? The dearth of DNA sequence resources appears to be arguably the main motive since the high-throughput establishment of entire proteomes requires annotated DNA and/or protein sequence databases [44]. For this reason, several public as well as private initiatives were conceived over the past few years to sequence the *C. albicans* genome (Fig. 17.3; reviewed in [42]). In March 2004, the diploid genome of *C. albicans* was reported as fully sequenced [45] (http://www-sequence.stanford.edu/group/candida). This accomplishment was the result of joint physical mapping and genome sequencing efforts.

Physical mapping of *C. albicans* chromosomes, achieved at Minnesota University (United States) and publicly accessible at http://alces.med.umn.edu/Candida.html, made possible the location of all the genes of *C. albicans* and the assignment of sequence contigs to chromosomes [46, 47]. Sequencing of *C. albicans* genome began at the Stanford Genome Technology Center (Stanford University, United States) in 1996 and was carried out as a whole-genome shotgun (WGS) of the heterozygous diploid genome of *C. albicans*, given that a haploid or homozygous form for *C. albicans* is not known to exist. However, certain difficulties were triggered during assembly of the WGS sequence, that is, as compared with conventional genome assemblies. In the standard assembly (assembly 6), the option of diploidy and heterozygosity was not contemplated, whereby sequences from the two alleles in heterozygous regions were improperly put into different contigs, clearly exceeding the estimated "haploid" genome size. To solve this obstacle, this was converted into a diploid sequence assembly (assembly 19), which led to the combination of separated contigs from assembly 6 into diploid contigs. Notwithstanding these problems, there is no doubt that the public release of the *C. albicans* genome sequence at different stages of

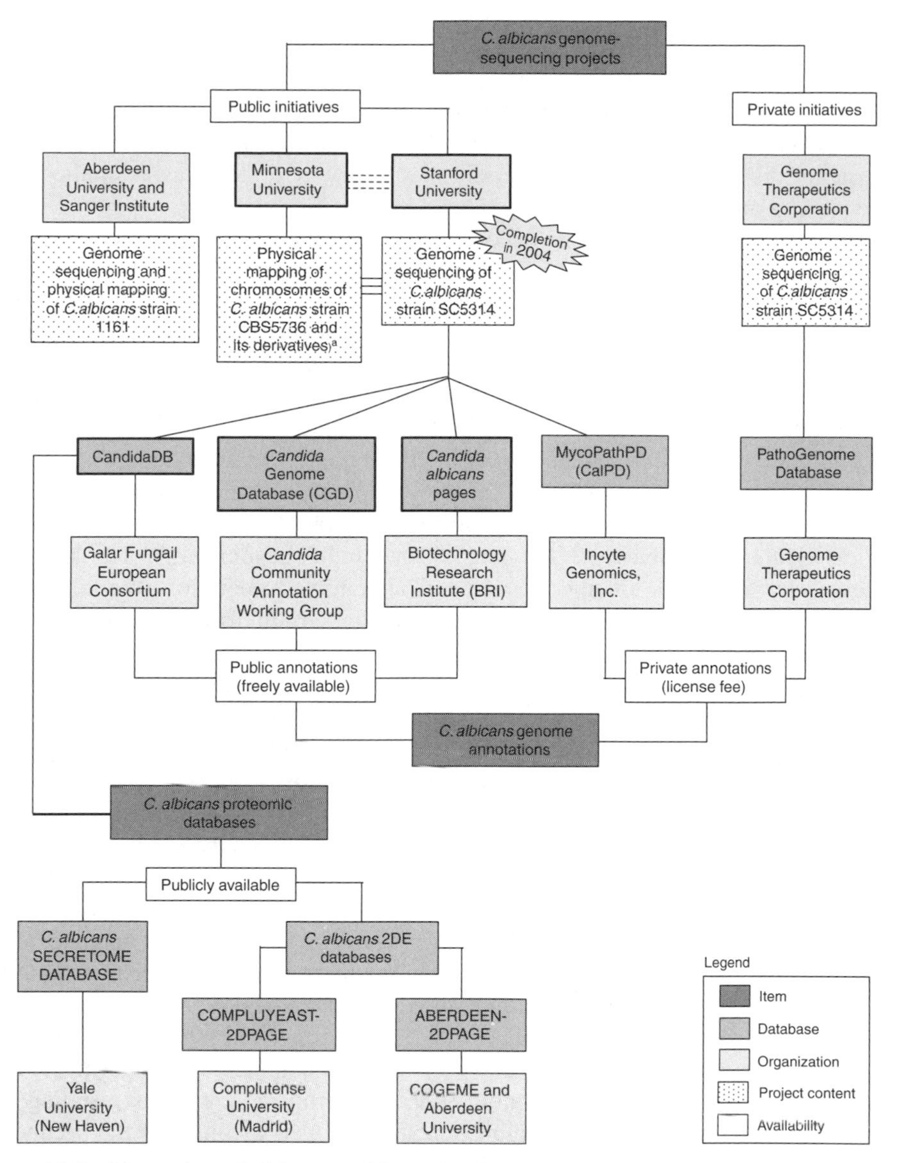


**Figure 17.3** Integration of different public and private *C. albicans* genome-sequencing projects and annotated genomic and proteome databases derived from them [42]. The thick line connects the keystones that have been and/or will be presumably the main impetus for the high-throughput expansion currently under way of proteomics into *C. albicans* research. See the text for further details. URL addresses. (i) *C. albicans* genome-sequencing project from Minnesota University, http://alces.med.umn.edu/Candida.html; Sanger Institute and Aberdeen University, http://www.sanger.ac.uk/Projects/C_albicans/; Stanford University, http://www-sequence.stanford.edu/group/candida/; Genome Therapeutics Corporation, http://www.genomecorp.com/programs/pathogenome.shtml; (ii) specialized *C. albicans* genomic or proteome databases: CandidaDB, http://genolist.pasteur.fr/CandidaDB; *Candida* Genome Database (CGD), http://www.candidagenome.org; *Candida albicans* pages, http://candida.bri.nrc.ca; MycoPathPD™ (CalPD™), http://www.incyte.com/sequence/proteome/databases/MycoPathPD.shtml; PathoGenome™ Database, http://www.genomecorp.com/programs/pathogenome.shtml; *C. albicans* secretome database, http://info.med.yale.edu/intmed/infdis/candida/; COMPLUYEAST-2DPAGE database, http://www.expasy.ch/ch2d/2d-index.html or http://babbage.csc.ucm.es/2d/2d.html; Aberdeen-2DPAGE database, http://www.abdn.ac.uk/cogeme/. The legend "a" refers to [47]. COGEME denotes Consortium for the Functional Genomics of Microbial Eukaryotes.

completion—for example, assembly 19 was made public in 2002—has quickened the pace of *C. albicans* research. It is of note that publication of its whole sequence has been one of the most significant events in the history of proteomics as applied to the understanding of this fungus.

Initial annotation of *C. albicans* sequence was carried out on data from Stanford's sequencing project or from private initiatives (Fig. 17.3). Interestingly, a high percentage of its sequence—approximately 95% of its genes (assembly 6)—was found to be in the public domain at an annotated genomic database named CandidaDB (http://genolist.pasteur.fr/CandidaDB) since 2002. The latter was created by members of the Pasteur Institute (France) and the Galar Fungail European Consortium [42]. This public release, albeit as a draft, represented a milestone in the evolution of *C. albicans* proteomics that was strengthened when the sequencing of its full genome was concluded (Fig. 17.2). In August 2004, an annotation of final *C. albicans* diploid genomic sequence, released as assembly 19, was also freely available online at a curated *C. albicans* database called *Candida* Genome Database (CGD; http://www.candidagenome.org). The latter was developed by members of the *Candida* Community Annotation Working Group (CAWG), an international, collaborative annotation group formed during the 2002 *Candida* and candidiasis conference [48].

In 1996 the first eukaryotic genome was announced as completely sequenced, that is, that of *Saccharomyces cerevisiae* [49]. Between 1996 and 2002, the *C. albicans* proteome analysis could be assisted partially by information available for the *S. cerevisiae* genome, namely exploiting the high degree of homology among most sequences from both closely related yeast organisms [28, 30, 32, 36].

Two types of useful data collections for future investigations on the *C. albicans* proteome have benefited from Stanford's sequencing project and the availability of its public annotation in CandidaDB: 2DE (see above) and secretome databases (Fig. 17.3). The potential *C. albicans* secretome—soluble secreted proteins—was defined computationally by using *C. albicans* genomic database (CandidaDB) information and prediction algorithms [50] and is currently compiled at http://info.med.yale.edu/intmed/infdis/candida.

Taking into consideration that this vital shortcoming (the shortage of sequence resources) was in attendance until recently, imagine the complex challenges to study the *C. albicans* proteome and to be familiar with the pertinent inferences resulting from its investigations that the first proteomic scientists as well as those of the second and third generations (see below) confronted without the current public annotated databases derived from the *C. albicans* DNA sequencing efforts. Today's scientists are witnessing a significant change in progression of proteomics into the *C. albicans* realm, due both to completion of the *C. albicans* genome sequence and to expansion of MS into protein chemistry and increasing technological advances in proteomic discipline.

***Chronological Outlook on C. albicans Proteomics*** The main contributions to the field of the *C. albicans* proteome over the past 25 years are depicted in chronological order, with reference to pivotal *C. albicans* issues, in Figure 17.4. According to ups and downs that *C. albicans* research has had in the ambit of proteomics, we propose a schematic division of its history into four periods, or "ages," as follows:

- *"Early Ages" (1980–1981)* This period lasted scarcely two years and comprised the pioneering differential expression studies. The only topics tackled at that time were dimorphism [24, 51] and host antibody response [52].

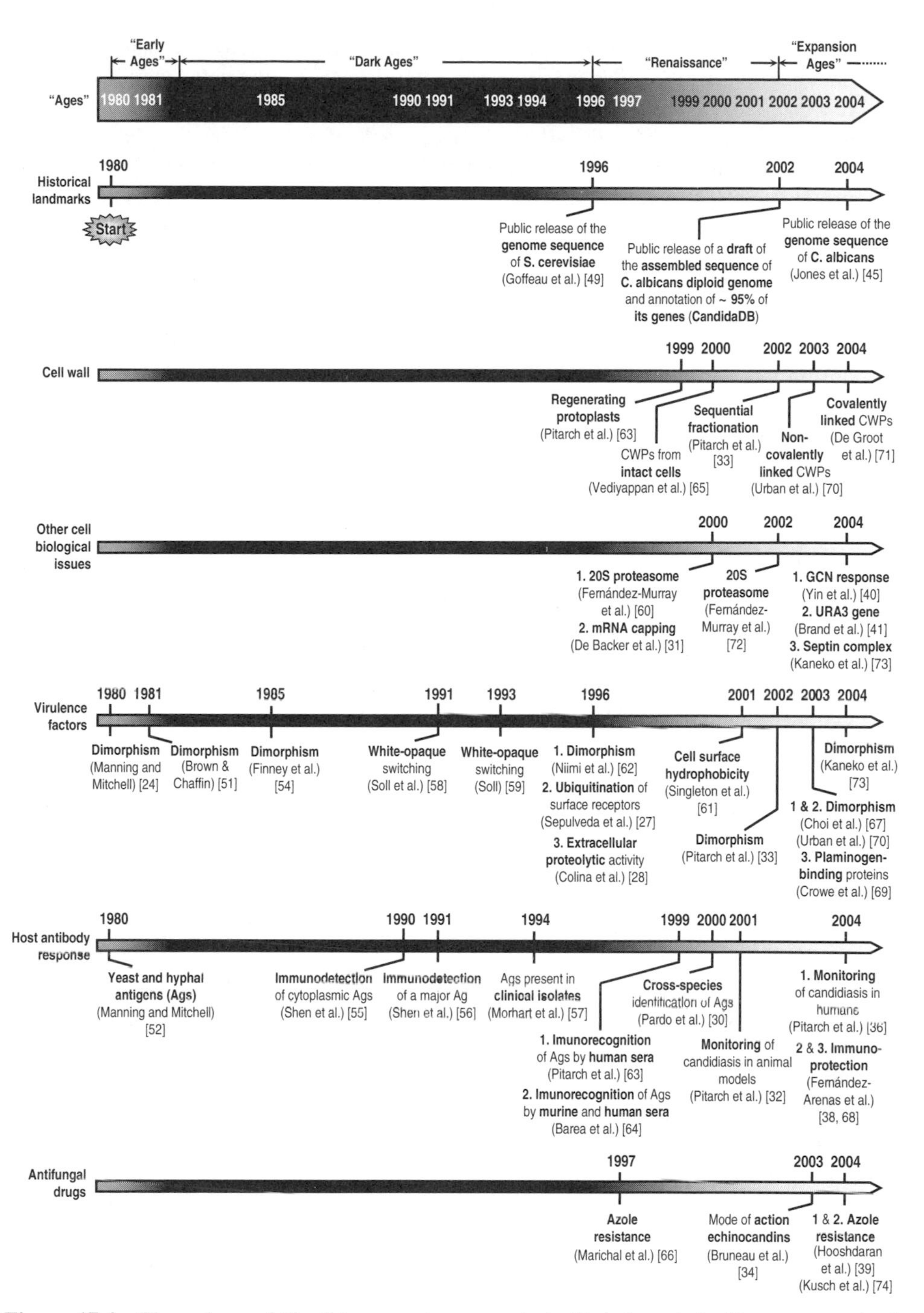


**Figure 17.4** Chronology of *C. albicans* proteome analysis. Periods and significant events in the history of *C. albicans* proteomics as well as the major proteomic contributions to the study of diverse pivotal issues about *C. albicans* are illustrated. See the text for further details. Proteomic studies focused on host antibody response and antifungal drugs are covered in Chapter 18. Contributors are shown in parentheses and references in square brackets. Abbreviations: CWPs, cell wall proteins; GCN, general amino acid control; Ag, antigen.

- *"Dark Ages" (1982–1995)* This lengthy stage was marked by a paucity of new data pertaining to *C. albicans* proteome analysis. Maybe, the deficiency in sequence resources as stated above along with some drawbacks associated with 2DE technology [53] could in part explain the lack of enthusiasm for this discipline among the early investigators. This virtual abandonment of potential *C. albicans* proteomic projects at that time can call to mind the Early Middle Ages in Europe—often portrayed as the Dark Ages. Despite a few isolated contributions pertaining to yeast-to-hypha transition [54], the host immune response [55–57], and phenotypic switching [58, 59], the modest knowledge acquired previously on the *C. albicans* proteome remained almost unaltered for these 14 years. This resulted in alternative technologies, such as molecular biology gaining ground in *C. albicans* research during this period.
- *"Renaissance" (1996–2001)* The *C. albicans* proteome study began to undergo a gradual rebirth from 1996 onward. Intriguingly, 1996, as indicated above, marks the key time both for the public release of the *S. cerevisiae* sequenced genome and for the beginning of *C. albicans* genome sequencing at Stanford University. Although some, albeit few, proteins detected in these analyses could be confidently identified using the *C. albicans* sequence databases existing at that time [29, 31, 60, 61] many proteins were unable to be characterized because of the low number of *C. albicans* proteins annotated in the databases [28, 29]. Nonetheless, the *S. cerevisiae* genome sequence data provisionally supported the cross-species identification—both by Edman degradation [28, 31, 32] and by MS [30]—of certain *C. albicans* proteins found in the proteomic approaches achieved during that time. This new setting favored a considerable increase in the number of publications as well as topics investigated in this arena. In fact, not only the major issues from the previous stages (dimorphism [62] and host response [30, 32, 63, 64]) but also novel pivotal *C. albicans* affairs—such as cell wall [63, 65], 20S proteasome [60], messenger ribonucleic acid (mRNA) capping [31], further virulence factors (cell surface hydrophobicity [61] and extracellular hydrolases [28]), and antifungal-drug action [31] or resistance [66] mechanisms—were considered at that juncture. Furthermore, some pioneering reference 2DE *C. albicans* maps containing several protein identities [29, 32] as well as the first 2DE database for *C. albicans*, the so-called COMPLUYEAST-2DPAGE database, were also created.
- *"Age of Expansion" (2002–)* From 2002—that is, the key time when a draft of the assembled sequence of *C. albicans* diploid genome (assembly 19) was publicly released at Standford University and the sequence of about 95% of *C. albicans* genes became freely available at the genomic database CandidaDB—to nowadays, analyses on the *C. albicans* proteome have been gaining weight in diverse areas, such as morphogenesis [33, 67], diagnosis [36, 37], immune protection [38, 68], other virulence factors [69], cell wall [33, 70, 71], 20S proteasome [72], general amino acid control (GCN) response [40], ectopic expression of *URA3* gene [41], septin complex [73], and antifungal-drug action [34] or resistance mechanisms [39, 74]. As a result, and perhaps also because of recent technological advances in proteomics, reference 2DE *C. albicans* maps were increasing both in number and in quality [33–41, 74], the COMPLUYEAST-2DPAGE database was in turn growing steadily, and other 2DE *C. albicans* data bank (ABERDEEN-2DPAGE) and a secretome database were also established. However, this progression in

the systematic *C. albicans* proteome characterization has only just begun and remains a long road before this burgeoning impetus reaches its zenith. Therefore, today's proteomic scientists working on the *C. albicans* field are now living in the "Early Ages of Expansion" of *C. albicans* proteomic research. See the next chapter for our own personal view into the "Middle and Later Ages of Expansion."

The goal of the following sections is to integrate knowledge about *C. albicans* biology (cell wall, 20S proteasome, mRNA capping, GCN response, ectopic expression of *URA3*, and septin cytoskeleton) and virulence (dimorphism, phenotypic switching, adherence, and extracellular hydrolytic activity) that has been attained at the proteomic level over the past 25 years, with special emphasis on 2DE and MS-associated contributions, since they have become both the hallmark of proteomic research and the main source of publications on the *C. albicans* proteome. However, proteomic studies concentrated on host antibody response, immune protection, diagnosis, and treatment (antifungal-drug action and resistance mechanisms) of *C. albicans* infections are covered more fully in the next chapter. For earlier reviews on proteomic analyses as applied to *C. albicans*, the reader is directed to references [42, 75–77].

## 17.2 CELL WALL: KEY COMPLEX STRUCTURE WITH NO MAMMALIAN COUNTERPART

### 17.2.1 Challenging Target for Wide Range of Proteomic Projects

The cell wall of *C. albicans* is a cellular compartment of substantial interest. Its peculiar attributes include its privileged location within the cell, plasticity, essential condition, and distinctiveness compared to mammalian cells. These features have moved proteomic researchers to pay it special attention, especially to its protein components [cell wall proteins (CWPs)] and, as a result, to bypass its complexity to explore it (Fig. 17.5).

Since the fungal cell wall is the outermost cellular structure, it constitutes the initial point of contact between the fungus and its host. This host–fungus interaction is directly related to pathogenicity, virulence, adhesion to host substrates, immunomodulation of the immune response, and antigenicity [78–82]. In particular, *C. albicans* CWPs represent the major antigens and host recognition biomolecules. Consequently, several proteomic approaches have recently focused on their analysis to investigate their implication in virulence—adhesion, hydrophobicity, or host recognition [27, 61, 69]—and in antigenicity [27, 37, 63]. These studies are argued in Section 17.4 and the next chapter, respectively.

The *C. albicans* cell wall is unanimously thought to be a plastic and dynamic structure that determines the cell shape of the fungus and changes during cellular and transitional growth. Hence, this cellular compartment plays a pivotal role in morphogenesis and virulence [79, 83–85]. Knowledge of the alterations in CWP composition occurring during the yeast-to-hypha conversion via a proteomics approach [33, 70] is also referred to in Section 17.4.

In addition to its biological interest, the cell wall appears to be a promising location to search for new specific antifungal drugs, because this structure, essential for *C. albicans*, is

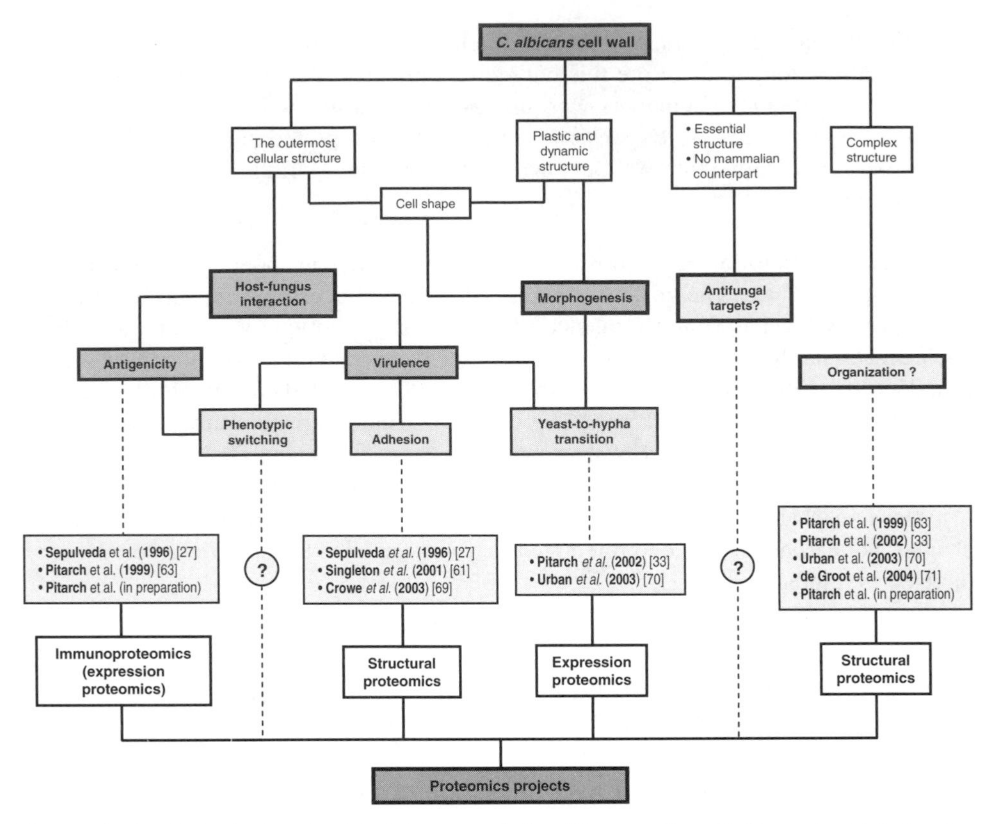


**Figure 17.5** Biological and clinical interest of *C. albicans* cell wall. Its main attributes (top boxes), host–fungus interactions mediated by this structure (middle left boxes), its clinical relevance (middle right box), and proteomic projects focused on its study for basic and applied research (bottom boxes) are depicted. Question marks indicate that proteomic research into the corresponding topic has not yet been reported in the literature. See the text for further details. References are in square brackets.

absent from mammalian cells [86–88]. Accordingly, proteomics could offer insight into this quest.

This section highlights how proteomics is helping to tackle the study of this complex structure of *C. albicans* at the protein level and to unravel new hypotheses in its cell biology that may be important for both basic and applied research [33, 37, 63, 70, 71].

### 17.2.2 How Complex is Cell Envelope of *C. albicans*?

The *C. albicans* cell wall is an intricate multilayered structure located outside the plasma membrane and composed of polysaccharides [i.e., (i) β-glucans (47–60%) formed by β-1,3 and β-1,6 linkages, (ii) mannans (32–36%) found solely in covalent association with proteins, and (iii) chitin (0.6–9%)], proteins (6–25%), and small amounts of lipids (1–7%) [79]. Its main constituents, β-1,3 and β-1,6-glucans, CWPs (i.e., mannoproteins and proteins), and chitin, are interconnected through covalent linkages and also, to a lesser extent, by noncovalent associations in various ways, resulting in a high complexity.

Therefore, how are these components distributed within this complex structure? β-1,3-Glucan molecules form an elastic three-dimensional microfibrillar network encircling the whole cell, to which several chitin chains are covalently anchored mostly on its inner side (Fig. 17.6). Both polymers supply rigidity to the cell wall, thereby affording to maintain cell integrity and shape. In turn, CWPs are found predominantly at the outside of this framework and, in smaller amounts, throughout the cell wall, thereby determining its porosity [33, 79, 89, 90]. These CWPs can be loosely associated with other cell wall components or covalently linked to (i) β-1,3-glucan either directly by an alkali-labile linkage (such as the so-called Pir-CWPs for their internal repeats) or indirectly via β-1,6-glucan through a phosphodiester bridge [a remnant of their glycosyl phosphatidylinositol (GPI) anchor, referred to as GPI CWPs] or (ii) chitin indirectly by a β-1,6-glucan moiety (i.e., GPI CWPs) [33, 91–96]. However, other types of linkages, hitherto uncharacterized, among CWPs and structural wall components could also be present in this cell structure.

As discussed below, recent proteomic researches [33, 37, 63, 70, 71] (as well as numerous pivotal classical biochemical experiments) (reviewed in [79, 83]) have also contributed to this view. For further coverage, the reader is directed to recent reviews on this topic [79, 80, 83, 84, 97, 98].

### 17.2.3 *Candida albicans* Cell Wall Proteome Comes True!

***Encouraging Strategies for Tackling Its Study*** Until recently, there have been relatively few data on the analysis of the *C. albicans* cell wall proteome, which has been an Achilles heel over the years, due in part to a dearth of effective strategies to profile this complex cellular compartment. As drawn from its molecular organization (Fig. 17.6) and bearing in mind that CWPs commonly show low abundance, low solubility, high heterogeneity, hydrophobicity, and interconnections with wall structural polysaccharides, it is unsurprising that CWPs are notoriously tricky to resolve by 2DE [75, 76].

However, several current proteomic studies [27, 33, 37, 61, 63, 65, 69–71] have demonstrated the usefulness of three different methods of CWP extraction designed to reduce the sample complexity, increase the enrichment and solubilization of these CWPs, and facilitate the subsequent characterization of this subproteome (reviewed in [76, 99]). These procedures are based on (i) the isolation of CWPs from intact cells under mild alkaline-reducing conditions [27, 65]; (ii) the extraction or sequential fractionation of CWPs from isolated cell walls, after cell breakage, by chemical and/or enzymatic treatments [33, 61, 69–71]; or (iii) the secretion of CWPs into the medium when protoplasts (cells whose walls have been eliminated previously by enzymatic digestion in an isotonic medium) are in the early stages of regenerating their cell walls. The latter exploits the fact that the covalent incorporation of many CWPs into the nascent cell wall is delayed during the early stages, and they are thus secreted into the medium [37, 63]. The advantage of this method is that it does not destroy covalent linkages, thus circumventing the underlying problems with chemical and enzymatic extractions (such as potential protein modifications or the presence of proteins bearing glucan and chitin side-chain residues that hinder their resolution by 2DE [100]). All these approaches are illustrated in Figure 17.7.

Interestingly, the sequential fractionation of CWPs according to the type of interactions that they establish with other structural wall components provides a comprehensive view of the high degree of sophistication of this subproteome (see below) and to define the first

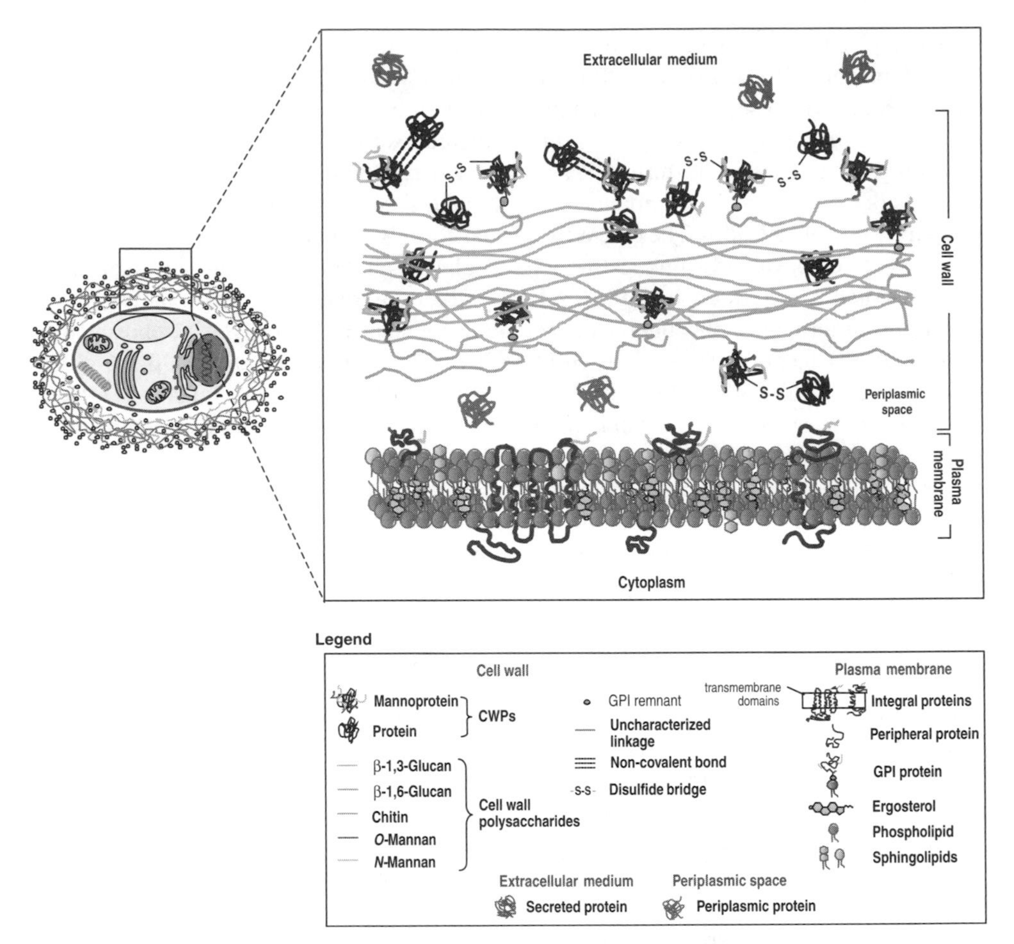


**Figure 17.6** Schematic representation of *C. albicans* cell envelope. A cross section through a yeast cell is illustrated in the left panel. The callout depicts details for the cell envelope of *C. albicans*, composed of an inner plasma membrane and an outer cell wall (right panel). The cell wall—the outermost cellular structure—basically consists of microfibrillar polysaccharides (fuchsia, green, and blue lines) and CWPs (dark blue tortuous lines). β-1,3-Glucan (fuchsia lines) and chitin (green lines) fibrils are glycosidically linked and form the structural framework of the wall. β-1,6-Glucan molecules (blue lines) interconnect certain CWPs (i.e., the GPI-CWPs) with β-1,3-glucan or chitin chains via a GPI remnant (orange circle). Other covalent linkages either between CWPs and β-1,3-glucan or between CWPs and chitin also occur (brown or red lines). Crosslinking of CWPs through disulfide bridges (–S–S–) or noncovalent bonds (black broken lines) is shown. Many CWPs (the so-called mannnoproteins) can carry *O*- and/or *N*-linked glycans (red or light bottle-green lines, respectively). Periplasmic proteins (grey tortuous lines) are trapped between the plasma membrane and the β-1,3-glucan-chitin network. Some proteins (blue tortuous lines) are secreted to extracellular medium. Some mannoproteins (dark brown tortuous lines) are anchored to the plasma membrane through GPI. The plasma membrane is a phospholipid bilayer interspersed with ergosterol that serves as a matrix for integral and peripheral proteins. This model is based on [33, 71, 79, 83]. CWPs denote cell wall proteins and GPI glycosylphosphatidylinositol. (See color insert.)

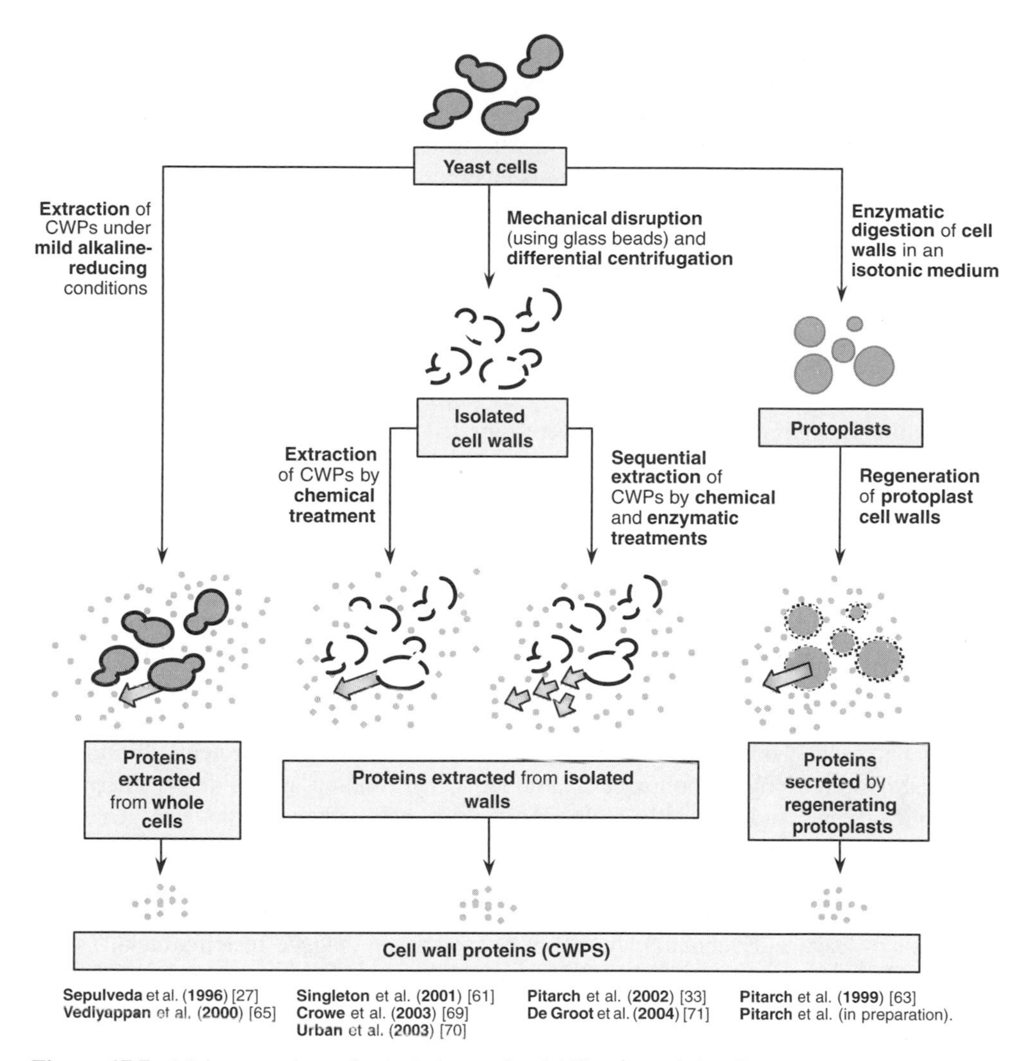


**Figure 17.7** Main procedures for isolation and solubilization of *C. albicans* cell wall proteome [76, 99]. The extraction of CWPs (small blue circles) either from intact cells or from isolated cell walls (black thick broken lines) and the secretion of CWPs from regenerating protoplasts are indicated with a green arrow. The outermost cellular structure of a yeast cell (the cell wall) is depicted with a black thick line, and that of a protoplast (the plasma membrane) with a blue thin line. The regeneration of protoplast cell wall (the nascent cell wall) is illustrated with a black streaked line. See the text for further details. Proteomic projects that have exploited some of these strategies are reported at the bottom and references are in square brackets.

reference 2DE maps of several selectively enriched *C. albicans* CWP fractions (available at http://babbage.csc.ucm.es/2d/d.html) [33]. Using this strategy, CWPs loosely associated, either by noncovalent bonds or disulfide bridges, with other wall constituents were the first elements to be released from isolated *C. albicans* walls. This was achieved by using detergents and reducing agents [sodium dodecyl sulfate (SDS) and dithiothreitol (DTT)]. Thereafter, CWPs covalently anchored to or closely enmeshed in the internal β-1,3-glucan-chitin microfibrillar network were extracted either by mild alkali (30 mM NaOH) or by enzymatic treatment first with β-1,3-glucanases (Quantazyme) and then with chitinases. Whereas a large number of *C. albicans* cell-surface-associated proteins were successfully identified by MS analysis (see below), the characterization of covalently linked CWPs—highly glycosylated proteins—was a challenging task since their glycosylation seemed, in most cases, to hamper their in-gel digestion with proteases or, more exactly, with trypsin. Amazingly, further enzymatic steps that involved in-gel deglycosylation of these CWPs previous to their tryptic proteolysis enabled the removal of this obstacle and the identification of a putative *C. albicans* β-1,3-glucanase covalently attached to β-1,3-glucan [33].

To bypass the difficulties associated with in-gel digestion of such highly glycosylated CWPs, a non-gel-based proteomic approach was exploited [71]. *Candida albicans* CWPs released from SDS-resistant cell walls by chemical (HF-pyridine or 30 mM NaOH) or enzymatic (Quantazyme) procedures were digested with trypsin, and after that, tryptic peptides were separated by liquid chromatography (LC) and identified by tandem MS (LC-MS/MS). This stratagem made it possible to characterize 14 different CWPs covalently linked to the β-1,3-glucan framework (see below).

Diverse strategies, such as biotinylation of intact cells, binding experiments (incubation of intact cells with cytosolic extracts), and in situ localization using tagged fusion proteins [70], as well as immunodetection of known cytosolic markers [33], have been pursued to assess whether proteins extracted from isolated walls were bona fide CWPs. In an attractive proteomic analysis, cell-surface-associated proteins were specifically labeled with a membrane-impermeable biotin derivative, extracted from isolated walls using detergents, and then separated from nonlabeled intracellular proteins by affinity chromatography using immobilized neutravidin [70]. Binding assays demonstrated that cytosolic proteins, and especially Tsalp (thiol-specific antioxidant-like protein), generated, for instance, from potential lysed cells did not bind unspecifically to the *C. albicans* cell surface previous to its biotinylation. These findings suggest that CWPs identified in this study appeared to be located at the cell surface in vivo, whereby they could not come from cell lysis or leakage. Eventually, the use of a reporter gene-tagged fusion protein, Tsa1p-GFP (green fluorescent protein), allowed confirmation of the location of one protein classically related to the intracellular compartment (Tsa1p) at the hyphal cell surface by using confocal microscopy.

***Unraveling Plot of Its Complexity and Biology*** Isolation of the *C. albicans* cell wall proteome by successive extractions of its CWPs by virtue of their linkages with other constituents supported the hypothesis that a heterogeneous population of proteins could be implicated in cell envelope construction and be located throughout the cell wall, that is, not only at its surface but also in its inner layers [33]. Detailed studies using Western blot assays shed some light on the potential mechanisms of incorporation, assembly, and retention of some *C. albicans* CWPs into the cell wall as well as on their interactions with wall polysaccharides. Certain CWPs could intriguingly become anchored to the cell wall by different mechanisms simultaneously at the same stage of the fungal cell cycle. For example, a protein related to a member of the Pir-CWP family (Pir2p/Hsp150p) appeared to be double anchored both to β-1,3-glucan through an alkali-labile linkage and to chitin by a β-1,3-glucanase-resistant bond.

All these exciting findings helped to outline how CWPs might be organized into the *C. albicans* cell wall. The model proposed for their retention throughout this structure is presented in Figure 17.8. Whereas a high percentage of CWPs could be loosely associated

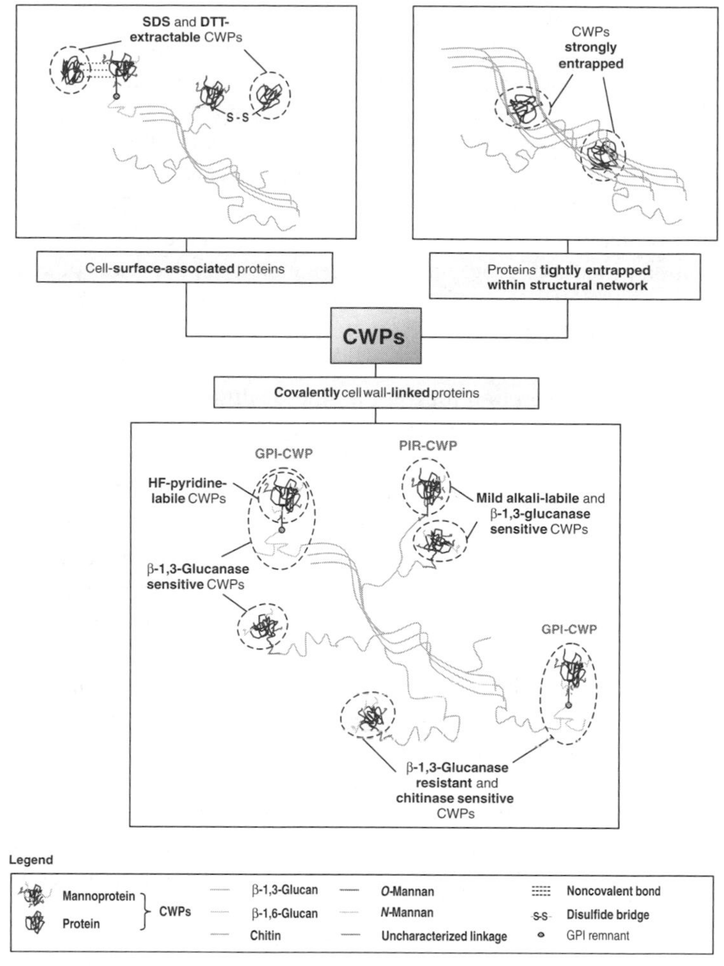


**Figure 17.8** Molecular model for potential mechanisms of retention of *C. albicans* CWPs into cell wall. Dashed circles show *C. albicans* CWPs retained into the cell wall and extracted chemically or enzymatically as follows. SDS and/or DTT-extractable CWPs are loosely associated with the cell surface through noncovalent interactions (black broken lines) or disulfide bridges (–S–S–). Alternatively, proteins tightly trapped inside the glucan–chitin network (fuchsia and green lines) can only be released after treatment with glucanases and/or chitinases. Mild alkali-labile CWPs are directly linked to β-1,3-glucan (fuchsia line) through their *O*-glycosyl side chains (red line), such as Pir-CWPs, or by other uncharacterized linkages (brown line). HF-pyridine-sensitive CWPs (i.e., GPI-CWPs) are indirectly attached to β-1,3-glucan via β-1,6-glucan (blue line) by their GPI remnant (orange circle). β-1,3-Glucanase-sensitive CWPs include mild-alkali- and HF-pyridine-labile CWPs as well as other CWPs attached to β-1,3-glucan through other linkages (brown line). β-1,3-Glucanase-resistant and chitinase-sensitive CWPs are linked to chitin (green line) either indirectly through a β-1,6-glucan moiety, such as some GPI-CWPs, or directly via some hitherto uncharacterized linkage (*brown line*). This model is based on [33, 70, 71, 79, 83, 94–96]. HF denotes hydrogen fluoro acid, CWP cell wall protein, GPI glycosylphosphatidylinositol, and PIR protein with internal repeats.

with cell surface by noncovalent and/or disulfide bridges, some of them and many other CWPs, however, might be tightly entrapped within the internal structural β-1,3-glucan-chitin network and/or covalently attached to (i) β-1,3-glucan through mild alkali-sensitive and/or β-1,3-glucanase-labile bonds and/or (ii) chitin by β-1,3-glucanase-resistant and chitinase-extractable linkages [33]. Further proteomic analysis of covalently linked CWPs revealed the presence of at least 2 mild alkali-labile CWPs and 12 GPI CWPs, of which 10 were attached to β-1,6-glucan via HF-pyridine-sensitive linkage [71].

Furthermore, proteomic studies of *C. albicans* proteins secreted during the early stages of the regeneration process of protoplast walls resulted in an overview of some of the cell wall precursors prior to their retention in the nascent cell wall, that is, those gene expression products involved in de novo wall biosynthesis, as well as of their putative interactions [37, 63].

Matrix-assisted laser desorption ionization time-of-flight (MALDI-TOF) MS and MS/MS analyses of *C. albicans* CWPs isolated by sequential fractionation led to the characterization of many new proteins on the surface of *C. albicans* cells that were noncovalently bound to the β-1,3-glucan-chitin framework [33]. More than 30 bona fide cell-surface-associated proteins, representing around 90 protein spots, were identified using this strategy, including (i) several classical CWPs with a predictable signal sequence which were involved in cell wall construction and maintenance and (ii) many other noncanonical proteins without conventional secretory signal sequences. The latter group included heat-shock, chaperone, and folding proteins, elongation factors, glycolytic and fermentative enzymes, phosphorylases, and unknown function proteins, among others, which might play a key role in adhesion to host substrates and eliciting host immune responses, that is, in host–pathogen interactions, as well as in the assembly of other cell wall components (see the Appendix in Chapter 18). Similar results were achieved by peptide mass fingerprinting (PMF) in further proteomic studies both of cell-surface-associated proteins biotinylated and purified from *C. albicans* cell walls by affinity chromatography [70] and of proteins secreted from protoplasts in active cell wall regeneration [37] (see the Appendix in Chapter 18).

The presence of these noncanonical CWPs, classically considered to be confined to the intracellular compartment, unraveled novel concepts in the cell wall biology of *C. albicans*, which prompted the stimulating and revolutionary theory that the targeting of these proteins to the *C. albicans* cell surface might imply the existence of alternative secretory pathways or signal peptides hitherto undiscovered, gene splice variants, moonlighting proteins with different functions according to their subcellular location, and/or proteins with no catalytic role recruited as structural components into the cell wall during the evolution of this fungal cellular compartment [33].

Liquid Chromatography MS/MS analysis of tryptic digests of CWPs covalently incorporated into the β-1,3-glucan network revealed the presence of glycosyl hydrolases related to cell wall polymer crosslinking, biosynthesis, and remodeling as well as of adhesion and defense proteins (Chapter 18 Appendix) [71]. Accordingly, these findings suggested that these covalently linked CWPs of *C. albicans* could interestingly be implicated both in cell wall maturation and expansion and in host–pathogen interactions.

A more complete view of *C. albicans* cell wall constituents should allow the discovery of new specific antifungal targets, an understanding of the diverse host–*C. albicans* interaction mechanisms (discussed in Section 17.4), and development of further serodiagnostic tests for candidiasis (argued in Chapter 18).

## 17.3 OTHER OUTSTANDING CELL BIOLOGY ISSUES: AN INFINITE WORLD

There seem to be no boundaries to *C. albicans* proteome research. Other remarkable topics of *C. albicans* biology, such as 20S proteasome, mRNA capping, general amino acid control (GCN) response, expression of *URA3* gene in ectopic gene loci, and septin complex, have also been tackled by proteomic approaches (Fig. 17.4) [31, 40, 41, 60, 72, 73]. This section will survey the relevance of these contributions to the understanding of *C. albicans*.

### 17.3.1 The 20S Proteasome: Central Proteolytic Component of Ubiquitin–Proteasome Pathway

The 20S proteasome is considered to be the central proteolytic enzyme of the ubiquitin–proteasome pathway, which in turn is the main nonlysosomal proteolytic system involved in many cellular processes, such as stress response, signal transduction, transcription, cell differentiation, and enzymatic regulation, among others [101–103]. The *C. albicans* 20S proteasome related to ubiquitination and proteosomal degradation is another issue that has recently motivated the attention of proteomic investigators [60, 72].

By successive chromatographic steps, the *C. albicans* 20S proteasome was isolated and appeared, under electron microscopy, as symmetrical barrel-shaped particles composed of four stacked rings (Fig. 17.9) [60]. This enzyme exhibited a multicatalytic nature. The 2DE analysis of purified 20S proteasome showed that it was made up of at least 14 polypeptides, 4 of which were found to be α-type subunits, as evidenced by internal sequencing of their tryptic fragments (α3, α5, and α6 subunits) or by immunorecognition with specific antibodies (α7 subunit; Chapter 18 Appendix). Two-dimensional immunoblot assays of acid-phosphatase-treated 20S proteasome using a specific anti-α7 antibody demonstrated that one of them (i.e., α7/Prs1p/C8) was a multiple-phosphorylated subunit. Further bioinformatic studies suggested that this α7/Prs1p sub-unit possessed eight potential target sites for protein kinases.

Phosphoproteomic analyses of the *C. albicans* 20S proteasome (see Fig. 17.9 for details) revealed that α3, α5, and α6 proteasome subunits were both the main in vitro substrates of the protein kinase CK2, an essential serine/threonine kinase, and the principal in vivo phosphorylated components [72]. However, the α7 subunit seemed not to be phosphorylated either in vitro or in vivo, suggesting that the previous observations might be due to contaminating activities existing in the phosphatase preparation. The α3/Pre9p/C9 subunit proved to be the major in vivo phosphate acceptor. Phosphoamino acid assays indicated that α3/Pre9p was phosphorylated in serine, both in vivo and in vitro. Two-dimensional phosphopeptide maps of endoproteinase Glu-C or tryptic digests from in vivo $^{32}$P-labeled α3 subunit, phosphorylated α3/Pre9p by *C. albicans* CK2, and phosphorylated recombinant α3 subunit by recombinant human CK2-α displayed identical patterns containing only one phosphoserine peptide. This finding supported the hypothesis that CK2 could be the enzyme accountable for in vivo phosphorylation of this subunit. Further directed mutagenesis and phosphopeptide mapping analysis uncovered the α3 sub-unit phosphorylation site in vivo and in vitro, that is, serine 248 residue in its C terminus. All these results support the suggestion that proteasome activity could be regulated by phosphorylation of certain α subunits. In turn, proteasome phosphorylation might play a role in peptidasic activity, modulation of binding to regulatory particles, or translocation through the nucleus.

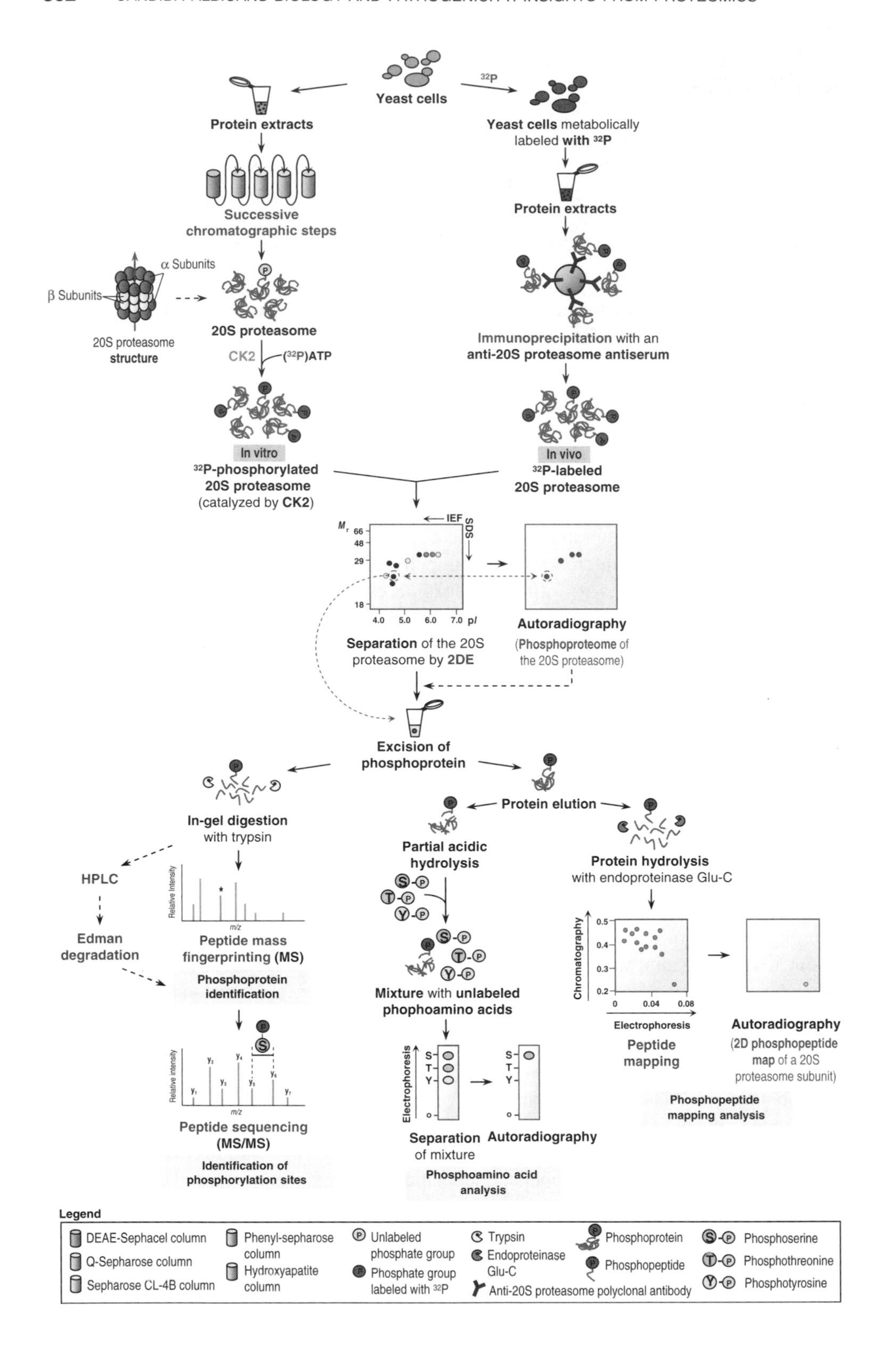

Yeast cells
32P
Yeast cells metabolically labeled with 32P
Protein extracts
Successive chromatographic steps
α Subunits
β Subunits
20S proteasome structure
20S proteasome
CK2
(32P)ATP
Protein extracts
Immunoprecipitation with an anti-20S proteasome antiserum
In vitro
32P-phosphorylated 20S proteasome (catalyzed by CK2)
In vivo
32P-labeled 20S proteasome
IEF
SDS
Separation of the 20S proteasome by 2DE
Autoradiography
(Phosphoproteome of the 20S proteasome)
Excision of phosphoprotein
In-gel digestion with trypsin
HPLC
Edman degradation
Peptide mass fingerprinting (MS)
Phosphoprotein identification
Peptide sequencing (MS/MS)
Identification of phosphorylation sites
Protein elution
Partial acidic hydrolysis
Mixture with unlabeled phophoamino acids
Electrophoresis
Separation of mixture
Autoradiography
Phosphoamino acid analysis
Protein hydrolysis with endoproteinase Glu-C
Chromatography
Electrophoresis
Peptide mapping
Autoradiography
(2D phosphopeptide map of a 20S proteasome subunit)
Phosphopeptide mapping analysis
Legend
DEAE-Sephacel column
Q-Sepharose column
Sepharose CL-4B column
Phenyl-sepharose column
Hydroxyapatite column
Unlabeled phosphate group
Phosphate group labeled with 32P
Trypsin
Endoproteinase Glu-C
Anti-20S proteasome polyclonal antibody
Phosphoprotein
Phosphopeptide
Phosphoserine
Phosphothreonine
Phosphotyrosine

### 17.3.2 mRNA Capping: Crucial Process for *C. albicans* with Potential Molecular Targets for Novel Antifungal Drugs

Acquisition of the 5′ cap during mRNA synthesis is essential for stability, processing, nuclear export, and efficient translation of mRNA, in short for the growth and survival of *C. albicans*. Consequently, its study has attracted considerable interest from scientists who continue to search for novel molecular targets for antifungal drugs. One of the key enzymes implicated in this process is the mRNA 5'-guanylyltransferase or GTase (Cgt1p) [104, 105].

The impact of inactivation of the *CGT1* gene on the *C. albicans* proteome was evaluated by comparative 2DE analysis ("expression proteomics") between a heterozygous *cgt1/CGT1* strain, that is, a mutant depleted of one *CGT1* allele, and its parental strain [31]. This approach, clearly promoted both by the vast influence of *CGT1* on the synthesis of all mRNAs and by its potential as an antifungal molecular target, revealed the existence of a large number of differentially expressed proteins. This in turn reflected the predictable pleiotropic effect of *CGT1* depletion on the overall protein expression. Edman degradation and MS analyses led to the identification of three up-regulated proteins (Chapter 18 Appendix 1): a ribosomal protein (Rps5p) that might be associated with a putative rise in translation efficiency after *CGT1* disruption, a translation elongation factor (Tef1p/Ef-1αp), and a heat-shock protein (Ssa2p). The up-regulation of Tef1p/Ef-1αp and Ssa2p intriguingly supported the biochemical results of an increase in resistance to hygromycin B (an antibiotic that affects translational elongation) and to heat stress, respectively, observed in the *cgt1/CGT1* mutant.

### 17.3.3 General Amino Acid Control: Response to Amino Acid Starvation

A subset of genes encoding enzymes in diverse amino acid biosynthetic pathways is subject to a common control in *C. albicans* that leads to their coordinate activation in

**Figure 17.9** Phosphoproteomic approach for studying in vitro and in vivo phosphorylation of *C. albicans* 20S proteasome [72]. 20S proteasome is a hollow barrel-shaped particle (Top left of the figure) composed of (i) two outer rings each containing seven different α subunits, which control the entry of the substrates and (ii) two inner rings each containing seven distinct β subunits, which delineate the catalytic chamber. Its phosphoproteome can be analyzed either (i) in vitro by the protein kinase CK2-mediated phosphorylation of the 20S proteasome isolated from yeast cells by successive chromatography steps or (ii) in vivo by growing yeast cells in inorganic $^{32}$P and immunoprecipitating the corresponding cell lysates with an anti–*C. albicans* 20S proteasome polyclonal antiserum. The phosphorylation state of the 20S proteasome components, phosphorylated both in vitro and in vivo, is then assessed by 2DE followed by autoradiography. Phosphoproteins of interest are excised from a 2DE gel, in-gel digested and identified by MS or HPLC Edman degradation. Their phosphorylation sites can subsequently be characterized by peptide sequencing using MS/MS. Alternatively, phosphoproteins can be eluted from the 2DE gel, partially hydrolyzed with HCl, mixed with unlabeled phosphoamino acids, and electrophoresed on a cellulose sheet followed by autoradiography, with the purpose of detecting which phosphoamino acids (in the example, only phosphoserine) are present in the acid hydrolysate by comparing with the standard unlabeled phosphoamino acids [phosphoserine (S), phosphothreonine (T), and phosphotyrosine (Y)]. Eluted phosphoproteins can also be hydrolyzed with endoproteinase Glu-C separated first by electrophoresis on a cellulose TCL (thin-layer chromatography) plate and then by ascending chromatography and examined by autoradiography to map phosphopeptides present in such phosphoprotein (in the example, only one phosphopeptide). CK2 denotes protein kinase CK2 and o origin.

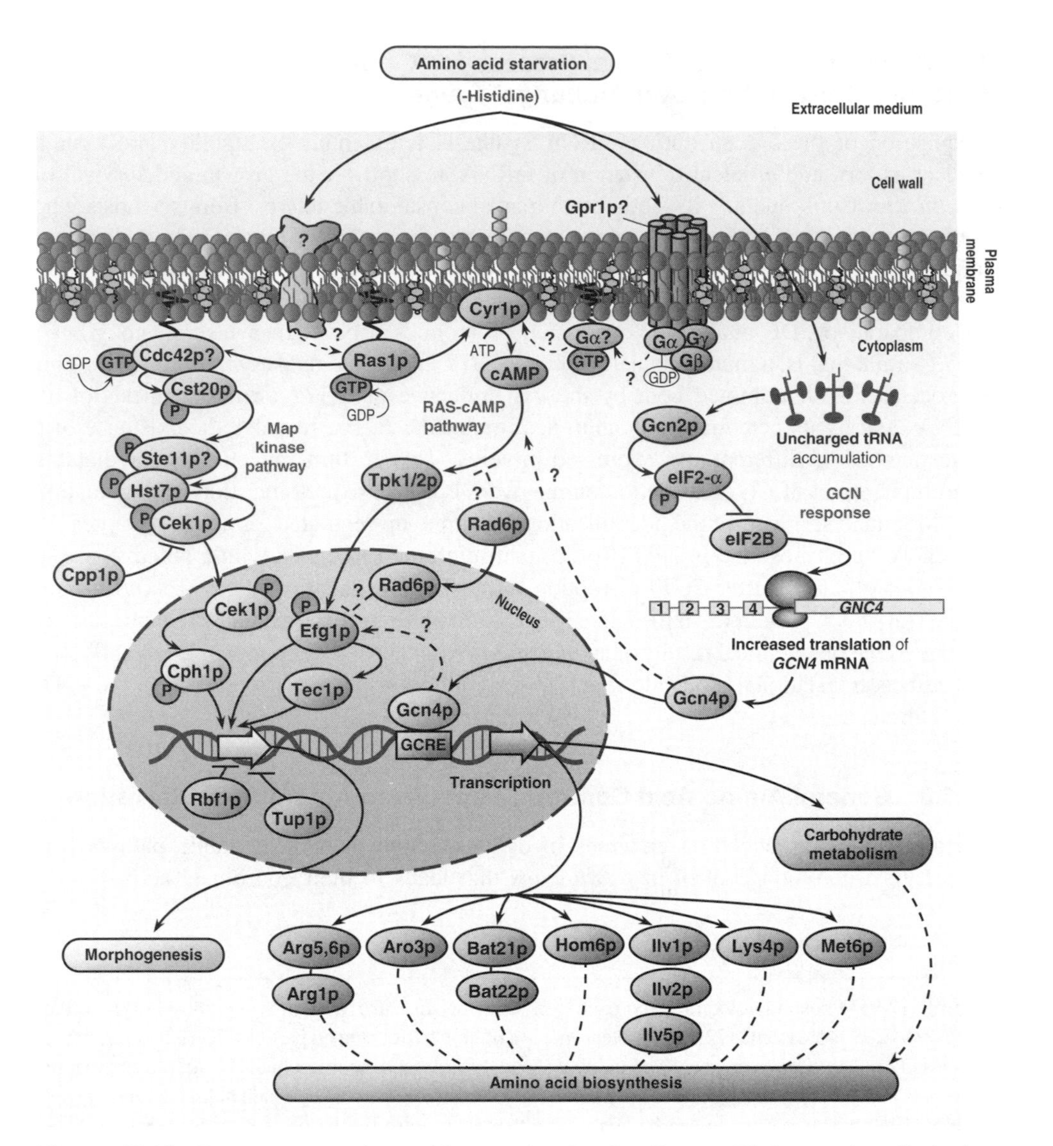


**Figure 17.10** Response to amino acid starvation in *C. albicans.* Under nutrient limitation, especially in nitrogen and/or carbon starvation, two major signaling cascades—MAP (mitogen-activated protein) kinase pathway (in amber) and Ras-cAMP pathway (in pink), defined by the transcription factors Cph1p and Efg1p, respectively, and dependent upon the GTP-binding protein Ras1p—stimulate filamentous growth (pseudohyphal or hyphal growth) [112, 143–147]. When *C. albicans* cells are deprived of any amino acid (e.g., histidine), uncharged tRNAs are accumulated. This leads to activation of the protein kinase Gcn2p, which phosphorylates the α subunit of the translation initiation factor 2 (eIF2). This phosphorylated form in turn inhibits the guanine–nucleotide exchange factor eIF2B. This reduction of eIF2 activity promotes the translation of a specific mRNA encoding the transcription factor Gcn4p, with a long 5'-leader region containing four small open reading frames (ORFs) located upstream of the *GCN4* ORF. Subsequently, this transcription activator interacts with the general control response elements (GCRE)–like conserved sequences in the promoter regions of its target genes. As a result, Gcn4p activates the transcription of genes that encode enzymes required for amino acid biosynthesis (seven different pathways)

response to a deficiency of any single amino acid [40, 106, 107]. This cross-pathway control is often referred to as general amino acid control, or GCN, response [108, 109]. Amino acid starvation in *C. albicans* stimulates the bZIP (basic region leucine zipper) transcription factor, Gcn4p, which in turn promotes both filamentous growth (inducing the Ras-cAMP pathway) and the transcription of amino acid biosynthetic genes (in an morphogenetic signaling-independent manner) through its interaction with general control response elements (GCRE)–like sequences in their promoters (Fig. 17.10) [106].

Gcn4p-dependent changes in the *C. albicans* proteome that take place in response to histidine starvation—imposed by treatment of cells with 3-aminotriazole (a histidine analogue)—were defined as those displayed in a wild-type strain but not in a homozygous *gcn4/gcn4* strain, that is, a deletion mutant in which the two *GCN4* alleles were disrupted sequentially [40]. This approach demonstrated that starvation for a single amino acid (i.e., histidine) in *C. albicans* repressed 24 proteins and stimulated 31 proteins dependent upon Gcn4p (Chapter 18 Appendix). It brought to light the idea that genes encoding enzymes of at least seven different amino acid biosynthetic pathways were induced as part of the GCN response in *C. albicans* (Fig. 17.10). Furthermore, some enzymes of carbohydrate metabolic pathways were also (up or down) regulated to presumably increase the source of precursors for amino acid biosynthesis, reflecting the breadth of the physiological response to starvation for a single amino acid. However, neither hypha-specific proteins nor morphogenetic signaling components dependent upon Gcn4p were identified under conditions of amino acid starvation. This may be due to their potential location at the cell wall or low abundance, respectively, indicating the need for further studies of key *C. albicans* subproteomes.

A comparison of Gcn4p-dependent proteomic response to histidine starvation between *C. albicans* and *S. cerevisiae* cells revealed that adenine biosynthetic pathways were activated by the GCN response in *S. cerevisiae* but not in *C. albicans*. Accordingly, purine biosynthesis appears to be regulated in a different way in both yeast organisms, suggesting that this divergence in their metabolism might be associated with their distinct ecological niches.

and carbon metabolism (in green), among others, but not for adenine biosynthesis, gathered from proteomic analysis [40]. This coordinate induction of the expression of genes on diverse amino acid biosynthetic pathways to bypass the external deficiency of a single amino acid is termed general amino acid control (GCN) response (in blue and green), which is independent of morphogenetic signaling (MAP kinase and Ras-cAMP) pathways. In addition, Gcn4p also activates morphogenesis (pseudohyphal growth) by interacting either with a downstream component (Efg1p) or maybe with an upstream one of the Ras-cAMP pathway in response to amino acid starvation, but not under morphogenetic signals [106]. Blunt arrows depict negative effects. Question mark after protein name indicates that the functional homologue of the corresponding *S. cerevisiae* gene placed in the same position on the pathway has been identified in the *C. albicans* genome but it has not yet been located experimentally on this pathway. For clarity, Gcn4p is illustrated inducing the transcription of only one gene (green arrow). Names of amino acid biosynthesis-involved enzymes refer to those in the Appendix in Chapter 18). Others: Cdc42p, GTPase; Cst20p, protein kinase; Ste11p, MAP kinase kinase kinase; Hst7p, MAP kinase kinase; Cek1p, MAP kinase; Cpp1p, protein tyrosine phosphatase; Gpr1p, heterotrimeric G protein receptor; Gα or Gpa2p, heterotrimeric G protein α subunit; Cyr1p, adenylate cyclase; Tpk1/2p, protein kinase A; Rad6p, ubiquitin-conjugate enzyme; Tec1p, transcriptional activator; Tup1p and Rbf1p, transcriptional repressors. (See color insert.)

### 17.3.4 *URA3* Gene: Selectable Marker in Gene Disruption Studies

The *URA3* gene, which codes for an enzyme in the uridine biosynthetic pathway, that is, orotidine 5'-monophosphate (OMP) decarboxylase, has extensively been exploited as a positive selectable marker for (i) disruption of *C. albicans* genes (using the "Ura-blaster" method, among others, which takes advantage of a *C. albicans ura3* auxotroph [110] and (ii) evaluation of their virulence [111, 112]. In the Ura-blaster approach (Fig. 17.11), the *URA3* gene is transferred from its native locus to an ectopic chromosomal location within the deleted gene. However, ectopic expression of the *URA3* from within the disrupted gene can alter *C. albicans* OMP decarboxylase activity, hyphal formation, adherence, and virulence [113]. All these misleading *URA3* positional effects can intriguingly be circumvented by placing *URA3* at its native locus or some other common genomic location, for example, at a highly expressed locus, in the mutant strain(s) generated [41, 111, 113, 114].

The effects of *URA3* deletion on overall changes in protein expression were assessed by comparison of the cytosolic proteome from a homozygous *ura3-iro1/ura3-iro1* strain—that is, a mutant depleted of the two copies of both *URA3* and the 3' half of *IRO1* (encoding a putative transcription factor)—with the counterpart from its wild-type strain [41]. This study revealed the presence of 14 differentially regulated proteins (aside from Ura3p) after inactivating the *URA3* gene. Whereas only three of them were clearly implicated in purine and pyrimidine metabolism, the rest were related to heme biosynthesis, aromatic amino acid turnover, translation, or transcription (Chapter 18 Appendix). To elucidate whether these changes were dependent either upon *URA3* (the selectable marker) or upon *IRO1* (the target gene), the proteome of a strain generated after reintegrating one copy of *URA3* at the high-expression *RPS10* locus in the homozygous strain mentioned previously was also analyzed. This approach provided evidence for partial or full restoration of protein levels of all except one (i.e., Toa2p, a transcription factor) to those of the wild-type strain. These findings supported the notion that *URA3*, rather than *IRO1*, is the leading regulator of the phenotype of this deletion mutant and its chromosomal location appears to have a significant influence on the cellular proteome in unforeseen ways. Hence, confusing phenotypes resulting from the use of auxotrophic selectable markers may stem from unexpected collateral effects on *C. albicans* cell metabolism and physiology.

**Figure 17.11** Tandem affinity purification (TAP) strategy for characterization of *C. albicans* septin complex [73]. A TAP tag, consisting of a hexahistidine peptide $(His)_6$ and a FLAG epitope (DYKDDDDK), is fused to the protein of interest (in the example, Cdc11p) at its C terminus. The TAP-tagged fusion protein is expressed in *C. albicans* cells at its physiological level. Cell extracts are then prepared under mild conditions. The fusion protein and its associated endogenous components are recovered from cell lysates by two successive affinity purification steps (TAP) under gentle native conditions. In the first one, the FLAG tag is bound to a column of anti-FLAG M2 agarose beads, and contaminant material is removed by washing. The immobilized protein complex is then eluted with FLAG peptide. The eluate is further incubated with Ni-NTA (nickel nitrilotriacetic acid) agarose beads on a second column. The hexahistidine tag interacts with two of the six ligand-binding sites (not occupied by NTA) in the coordination sphere of the nickel ion. After washing to remove residual contaminants, the highly purified protein complex is eluted with imidazole, which contains part of the histidine structure. The protein complex is then resolved on an SDS–polyacrylamide gel electrophoresis (PAGE) and/or 2DE gel, and proteins interacting with the target protein are identified by MS. The septin complex purified from *C. albicans* cell extracts using this strategy was composed of Cdc3p, Cdc10p, Cdc11p, Cdc12p, and Sep7p in a potential stoichiometric relation $2 : 1\text{–}2 : 1 : 2 : \leq 1$ [73]. This procedure at high-throughput scale leads to the definition of protein complex networks. (See color insert.)

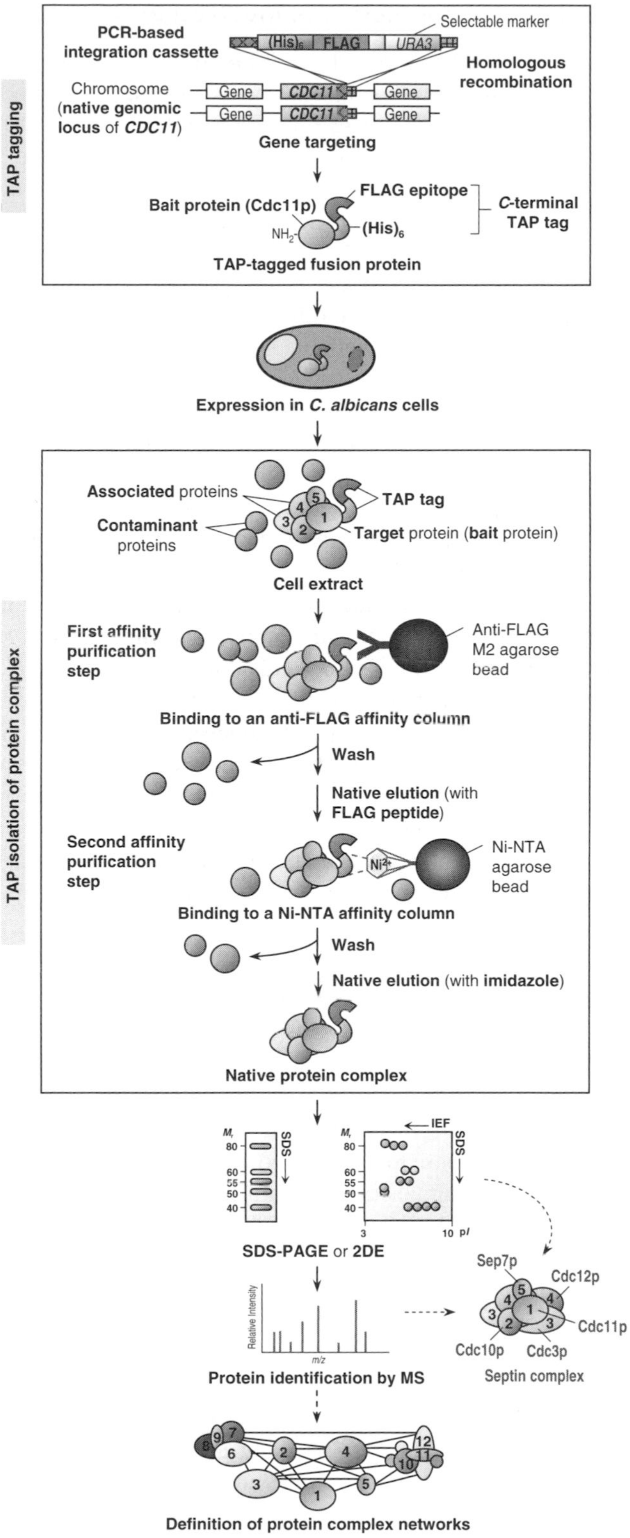

TAP tagging

PCR-based integration cassette
(His)6 FLAG URA3
Selectable marker
Homologous recombination
Chromosome (native genomic locus of *CDC11*)
Gene *CDC11* Gene
Gene *CDC11* Gene
Gene targeting
Bait protein (Cdc11p)
FLAG epitope
$NH_2$-
(His)6
*C*-terminal TAP tag
TAP-tagged fusion protein

Expression in *C. albicans* cells

TAP isolation of protein complex

Associated proteins
TAP tag
Contaminant proteins
Target protein (bait protein)
Cell extract
First affinity purification step
Anti-FLAG M2 agarose bead
Binding to an anti-FLAG affinity column
Wash
Native elution (with FLAG peptide)
Second affinity purification step
$Ni^{2+}$
Ni-NTA agarose bead
Binding to a Ni-NTA affinity column
Wash
Native elution (with imidazole)
Native protein complex

$M_r$ 80 60 55 50 40 SDS
IEF
$M_r$ 80 60 55 50 40 SDS
3 10 p*I*
SDS-PAGE or 2DE

Relative Intensity
*m/z*
Protein identification by MS

Sep7p Cdc12p Cdc11p Cdc10p Cdc3p
Septin complex

Definition of protein complex networks

### 17.3.5 Septins: Cytoskeletal Elements Involved in Cytokinesis

The septins comprise a family of highly conserved cytoskeletal proteins that form heteromeric complexes, can assemble into filaments, and seem to bind and hydrolyze guanidine triphosphate (GTP) (for reviews on septins, refer to [115–120]). These play a key role both in diverse cellular processes of *C. albicans*, including cytokinesis, cell septation, bud site selection, morphogenesis, and chitin deposition, and in pathogenesis and invasiveness of this fungus [121–123]. Septins encoded by cell division cycle (CDC) genes (*CDC3*, *CDC10*, *CDC11*, and *CDC12*) assemble into a ring of neck filaments located subjacent to the plasma membrane at the presumptive site of septation during budding and filamentous development of *C. albicans* (see Evolution of Understanding Dimorphic Transition Delivered by Proteomic Approaches in Section 17.4.2, for further details) [112, 121, 122].

A tandem affinity purification (TAP) tag consisting of six consecutive histidine residues $(His)_6$ and FLAG epitope and fused to the C terminus of the septin Cdc11p enabled the successful purification both of Cdc11p and of its associated proteins expressed at their physiological level and derived from *C. albicans* cell lysates (Fig. 17.11) [73]. The 2DE and MS analyses of this affinity-purified *C. albicans* septin complex indicated that four proteins (Cdc3p, Cdc10p, Cdc12p, and Sep7p) interacted with Cdc11p (Chapter 18 Appendix). Further insertion of the $(His)_6$-FLAG tag into each of these four septins identified and subsequent purification of their related protein complexes from *C. albicans* cellular extracts shed some light on the possible composition and stoichiometry of the *C. albicans* septin complex. This appeared to contain two subunits of Cdc3p and Cdc12p, one or two subunits of Cdc10p, one subunit of Cdc11p, and less than one subunit of Sep7p. Strikingly, Sep7p might be multiple phosphorylated (inferred from its overlapping spots on 2DE protein patterns) and, as a result, exhibits a regulatory role in septin complex formation and stability. This appealing procedure is a useful tool not only for protein complex or protein purification but also for proteome exploration, since it offers an efficient way to characterize functional protein assemblies and to uncover novel protein functions.

## 17.4 VIRULENCE FACTORS: UNMASKING MR. HYDE

*Candida albicans* is apparently a benign member of the human microflora on the mucosal surfaces of the oral cavity, gastrointestinal tract, and vaginal canal. This harmless organism can evolve from being a commensal within the human host into a deadly opportunistic pathogen, in response to a change(s) in host physiology, such as alteration of the endogenous microflora composition, natural barrier disruption, and/or immune response impairment (Fig. 17.1). Nevertheless, all these host factors are crucial but not sufficient to account for *C. albicans* invasion. This double-edged organism (i.e., commensal and opportunistic pathogen) is in need of certain flexible attributes or strategies in order to access host tissues, evade the immune system and drug therapy, and cause disease under favorable conditions. It is thus possible that a set of factors, rather than a single property, contributes to *C. albicans* pathogenicity and virulence. In this sense, several *C. albicans* virulence traits have been postulated, including responses to environmental changes (i.e., phenotypic switching and dimorphism), adherence to host cells and tissues, and secretion of hydrolytic enzymes (reviewed in [85, 124–134]) (Fig. 17.12).

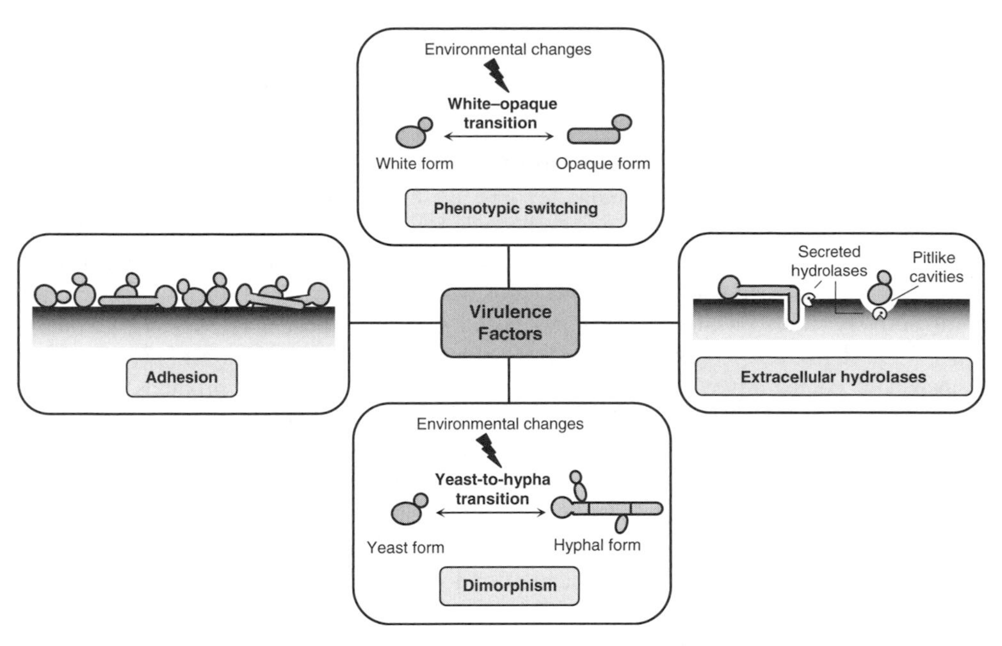


**Figure 17.12** Putative stratagems developed by *C. albicans* for invading host tissues and evading host immune system under favorable conditions. These include (i) the ability to switch among different cell phenotypes, (ii) morphogenetic adaptation to the host environment (yeast-to-hypha transition), (iii) adherence to host structures, and (iv) secretion of hydrolytic enzymes (small symbols) that act on host substrates. See the text for further details.

Given their vital significance, the study of these virulence factors has undoubtedly become a major research aim for proteomics (Fig. 17.4). This endeavors to unmask the underlying mechanisms that allow *C. albicans* to change from commensal to pathogenic behavior (from "Dr. Jekyll" to "Mr. Hyde") to control *C. albicans* infection, and to find novel targets for vaccine and antifungal drug design that may be useful for the development of better therapies. This section will integrate the pivotal proteomic contributions focused on this topic.

### 17.4.1 Phenotypic Switching: Spontaneous and High-Frequency Generation of Variety of General Phenotypes

***White–Opaque Switching System*** Of all switch phenotypes reported, the white–opaque system observed in strain WO-1 [135] (a clinical isolate) has been specially useful as a model for studying the molecular basis of switching in *C. albicans*. The white–opaque transition, in which smooth, white, round-ovoid colonies (white phase) switch reversibly to flat, gray, elongated, or bean-shaped colonies (opaque phase), takes place spontaneously at high frequency (Fig. 17.12). In addition to this alteration in colony morphology, switching has an effect on other phenotypic traits, such as antigen expression, adhesion to human epithelial cells, sensitivity to neutrophils and oxidants, secretion of proteinase activity, environmental constraints on the yeast-to-hypha transition, and drug susceptibility (reviewed in [130, 134, 136]. Consequently, this attribute might intriguingly endow *C. albicans* with a singular phenotypic flexibility to be capable of living both as a commensal and pathogen, adapting quickly to environmental cues or specific sites in the body, invading tissues, avoiding immune surveillance, and/or developing drug resistance.

***Correlation between White–Opaque Switching and Its Differential Proteome Display*** Approaches that include "expression proteomics" have been exploited to compare the entire *C. albicans* proteome from white- and opaque-phase cells in an attempt to detect proteins unique to either phase that could serve as a regulatory function in switching as well as to elucidate the mechanisms involved [58, 59]. Proteome studies with radioactively labeled white- and opaque-phase cytoplasmic proteins revealed both phase-specific proteins, that is, one white- and two opaque-specific protein spots [58]. Likewise, subsequent comparative 2DE analysis between in vitro translation products of mRNA isolated from white- and opaque-phase cells also demonstrated the existence of translated proteins that were specific to opaque-phase mRNA, that is, three protein spots [59].

As a result of the changes undergone both in cell surface and in antigen expression during the white–opaque transition, it is possibly safe to speculate that further research into the cell wall proteome and maybe into other putative relevant subproteomes of both phase cells could result in the identification of novel switching-related proteins and more precise data on this issue (Fig. 17.5).

### 17.4.2 Yeast-to-Hyphal Transition: Morphogenetic Adaptation to Host Environment

***Polymorphic Trait of C. albicans*** Although this section focuses on the yeast-to-hyphal transition, it should be mentioned that *C. albicans* is a polymorphic—or pleomorphic—organism that can assume an assortment of morphologies, including (i) budding yeastlike cells (unicellular forms), (ii) pseudohyphae (elongated buds that appear as filamentous cell chains with constrictions at their septa), (iii) hyphae or mycelia (chains of filamentous cells with parallel walls at their septa), and (iv) clamydospores (thick-walled, sporelike cells developed on pseudohyphal or hyphal support cells) (Fig. 17.13) [19]. The initial stage in the yeast-to-hypha conversion is referred to as germ tube formation.

The ability to switch between yeast and hyphal growth under specific host environmental stimuli, a process designated as dimorphism, is considered to be important for the invasiveness of *C. albicans*. Surprisingly, both types of morphology are usually found in clinical lesions, where both possibly play a role in pathogenesis [137]. Nonetheless, hyphal formation seems to be more critical as hyphae adhere more strongly to host cells, promote tissue penetration, and offer a mechanism to escape from macrophages [19, 124, 138]. Owing to their thigmotropic or chemotropic nature, hyphae could be capable of detecting breaks in the host's epithelial or endothelial surfaces and penetrating them to reach host tissues [139, 140]. This dimorphic transition can be regulated in vitro by a wide range of environmental conditions, mimicking those found inside the host (for example, temperature, pH, and nutrients such as serum, *N*-acetylglucosamine, or proline) [19, 85].

Recently, comprehensive reference 2DE maps from *C. albicans* yeast and hyphal proteins have been established [33–36, 40] which are a potentially useful tool for further proteomic studies directed at monitoring expression changes occurring during yeast-to-hypha transition, for instance, those undertaken at different induction times. Some of them are

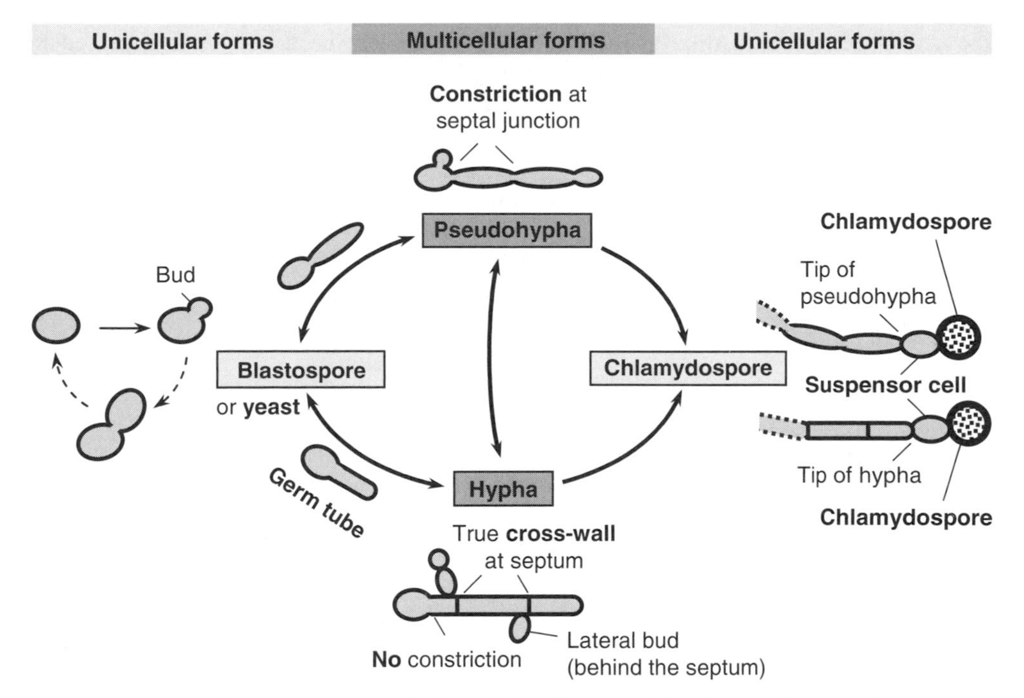


**Figure 17.13** Different cell morphologies of *C. albicans*. This pleomorphic organism has the capacity to adopt (i) unicellular forms, that is, yeast cells (ovoid, budding cells, also referred to as blastospores) and chlamydospores (large, thick-walled, spherical cells located at the tips of pseudohyphae or hyphae, to which they are attached by specialized blastospore-like cells, i.e., the so-called suspensor cells), and (ii) multicellular forms, that is, pseudohyphae (chains of elongated ovoid cells with constrictions at their cell–cell junctions) and hyphae (chains of tubelike cells with true cross walls at their septa). Germ tubes (hypha progenitors) do not have constrictions at their mother cell junctions. Yeast cells, pseudohyphae, and hyphae are often present in infected tissues. The formation of chlamydospores only takes place in vitro under certain conditions. The morphological form commonly found in the laboratory is the yeast. The potential development relationships among these morphological forms of *C. albicans* are shown with arrows. The double-headed arrows indicate reversible morphological conversion. The initial stage in the transition from yeast to filamentous form is pictured over the arrows.

available at http://babbage.csc.ucm.es/2d/2d.html (diverse 2DE maps of CWPs and cytoplasmic proteins from *C. albicans* yeast and hyphal forms) or http://www.abdn.ac.uk/cogeme/ (one 2DE map of *C. albicans* yeast cytoplasmic proteins). The fact that a good number of elements present in these proteomes turned out to be involved in metabolism, energy generation, translation apparatus, and heat-shock response implies that both *C. albicans* budding and filamentous forms are equipped with a great arsenal of housekeeping gene products in high copy number.

***Evolution of Understanding Dimorphic Transition Delivered by Proteomic Approaches*** From the first steps in *C. albicans* proteome analysis in 1980 to the present time, several research groups [24, 33, 51, 54, 62, 67, 70, 73] have taken advantage of "expression-proteomics"-based approaches to uncover proteins related to the yeast-to-hypha transition as well as to shed light on the mechanisms involved in morphogenesis and indirectly in virulence (Fig. 17.4). In so doing, protein profiles of the whole proteome or specific subproteomes of *C. albicans* yeast cells and hyphae can be compared both qualitatively and quantitatively by 2DE and, in turn, yield data both on differential protein expression and on protein expression levels, respectively. Here the purpose is to characterize dimorphic switching-associated proteins by MS (Fig. 17.14) [53]. Nevertheless, the identification of relevant proteins (Chapter 18 Appendix) had to be deferred inexorably until the advent of DNA sequence resources (see Section 17.1.3) [33, 67]. As we shall now see, the conclusions gathered from these investigations have been gradually refined.

*Looking for Needle in a Haystack* No proteins synthesized de novo in *C. albicans* mycelium cells were found in most 2DE analyses achieved with samples from total cellular extracts [24, 51, 62, 67]. Although 33 yeast- and 10 hypha-specific cytoplasmic protein spots were apparently detected during temperature-regulated dimorphism, the information drawn from further immunoabsorption and filamentation-defective strain experiments, however, supported the hypothesis that "mycelium-specific" proteins were actually modifications of preexisting yeast proteins [24]. Furthermore, the extra screening also contended that certain yeast proteins could play a regulatory role in hyphal

**Figure 17.14** Flow chart of general strategy for protein expression profiling during yeast-to-hypha transition by 2DE. The dimorphic transition from yeast to hyphal cells can be induced in vitro by growth at 37°C and physiological pH and/or exposure to serum or *N*-acetylglucosamine (see Table 17.1). After induction, *C. albicans* yeast and hyphal proteins are extracted from whole cells [24, 51, 54, 62, 67], subcellular compartments (e.g., cell wall, as illustrated in the example [33, 70]), or multiprotein complexes (e.g., DNA-binding proteins [62] or septin complex [73]) and then separated in the same conditions and in triplicate by 2DE both from a single preparation and from different preparations obtained on different occasions. A computer-assisted image analysis of the two series of 2DE gels is first performed to itemize the yeast and hyphal protein spots and to establish their expression level (normalized as a relative optical density or volume). The set of 2DE gels from yeast forms is then overlapped and matched to that from hyphae using one gel of the yeast series as master gel and 30–40 common spots present in all gels involved as landmarks in an attempt to detect protein spots uniquely or differentially expressed in yeast or hyphal forms. Subsequently, these variations in protein expression are statistically analyzed and then confirmed in the same way by matching between the synthetic gel obtained for yeast forms and the counterpart for hyphae. A synthetic gel of yeast proteins, for instance, is created by merging all replicate gels from yeast forms generated and contains most of the important features from them. Specific or differentially expressed protein spots are excised from preparative 2DE gels and in-gel digested with proteases. The resulting peptides are mass fingerprinted by MALDI-TOF MS and/or sequenced by MS/MS. The information gathered from mass spectra is then used to identify the protein by database searching. Further direct functional assays can be carried out. The spectra illustrated in the example correspond to an up-regulated hyphal protein, Pdc11p (pyruvate decarboxylase) [33]. The fragmentation of the tryptic peptide with a molecular mass of 1352.6 Da (red peak in the blue spectrum) resulted in the amino acid sequence NIVEFHSDYTK deduced from MS/MS data (red spectrum) and the unambiguous identification of Pdc11p. This scheme is based on [33] and can clearly be scaled up to other approaches targeted at *C. albicans* that include "expression proteomics" (see Sections 17.3 and 17.4.1 and Chapter 18). The double-headed arrows indicate gel matching. GlcNAc denotes *N*-acetylglucosamine.

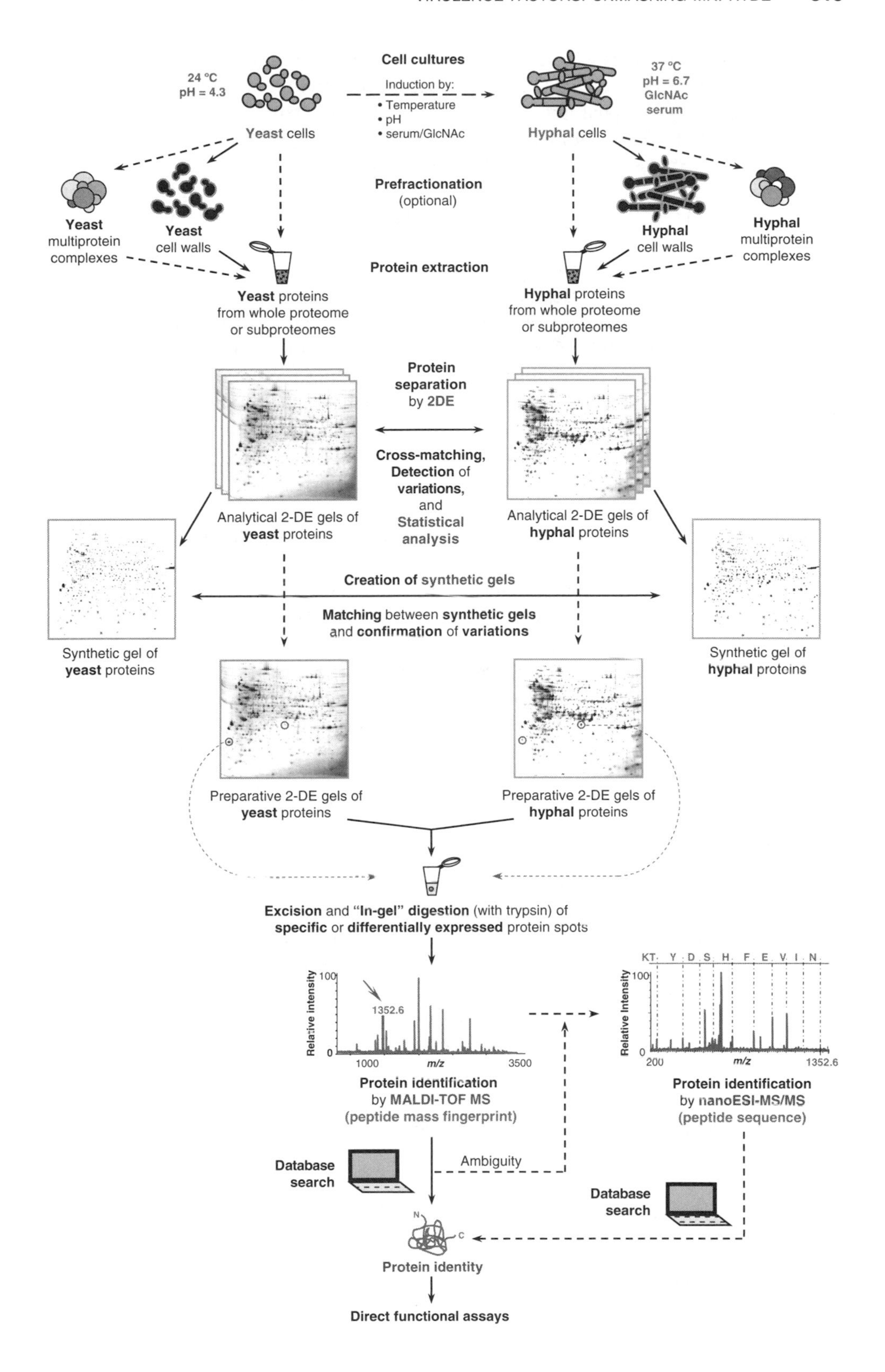
Cell cultures
24 °C
pH = 4.3
Induction by:
• Temperature
• pH
• serum/GlcNAc
37 °C
pH = 6.7
GlcNAc
serum
Yeast cells
Hyphal cells
Prefractionation
(optional)
Yeast multiprotein complexes
Yeast cell walls
Hyphal cell walls
Hyphal multiprotein complexes
Protein extraction
Yeast proteins from whole proteome or subproteomes
Hyphal proteins from whole proteome or subproteomes
Protein separation by 2DE
Cross-matching, Detection of variations, and Statistical analysis
Analytical 2-DE gels of yeast proteins
Analytical 2-DE gels of hyphal proteins
Creation of synthetic gels
Matching between synthetic gels and confirmation of variations
Synthetic gel of yeast proteins
Synthetic gel of hyphal proteins
Preparative 2-DE gels of yeast proteins
Preparative 2-DE gels of hyphal proteins
Excision and "In-gel" digestion (with trypsin) of specific or differentially expressed protein spots
Relative intensity
100
0
1352.6
1000
m/z
3500
KT Y D S H F E V I N
200
1352.6
Protein identification by MALDI-TOF MS (peptide mass fingerprint)
Protein identification by nanoESI-MS/MS (peptide sequence)
Database search
Ambiguity
N
C
Protein identity
Direct functional assays

conversion. Similar results were contemplated in another 2DE study focused on the search for yeast and mycelial antigens [52]. The subsequent detection of five yeast-specific and two down-regulated hyphal protein spots in a detailed temporal assay enabled the clarification that the dearth of mycelium-specific proteins appeared to be attributable to a change in gene expression (below the detection limit of that procedure) occurring during *C. albicans* morphogenesis, and not owing to modified forms of proteins already present in budding cells [51]. This notion was due to the fact that the 2DE hyphal protein patterns displayed neither new spots nor spots with different migration in relation to their yeast counterparts.

Taken as a whole, few differences in the establishment of alternative morphologies were visualized in 2DE profiles from crude *C. albicans* cellular extracts, that is, at a global scale [24, 51, 54, 62, 67]. For example, only one yeast- and one hyphal-specific cytoplasmic protein were discriminated during pH-regulated dimorphism after considering the upshot of additional mycelium-defective mutant analyses [54]. This insignificant variation could be the result of posttranslational modifications, turnover rates, or technical shortcomings of 2DE associated with the detection of low-abundance regulatory proteins. Likewise, comparison of pH-dependent germ tube induction by glucose and galactose with pH-independent induction by *N*-acetylglucosamine also revealed minor expression changes in the respective 2DE protein patterns [62]. Surprisingly, differentially expressed proteins proved to be growth or pH specific and not morphogenesis-related effects, thus failing to characterize any proteins uniquely associated with either morphology. All these proteomic findings suggested that a large number of dimorphism-involved proteins seemed to be present at low expression levels and/or of transient (temporal and spatial) appearance in the cell [54]. Why then does someone go on looking for a needle in a haystack? This challenging remark rapidly drew attention to the dissection of key *C. albicans* subproteomes, as opposed to the entire proteome, in order to study accurate clues as to the establishment of both morphologies [33, 62, 70, 73] (see below).

The MALDI-TOF MS analysis of six of the 11 up-regulated hyphal cytoplasmic protein spots visualized during serum and temperature-regulated dimorphism has recently led to the identification of three proteins (Chapter 18 Appendix) [67], two of which, Pra1p (pH regulated antigen), and Phr1p (pH response protein), were previously characterized as morphogenesis-implicated molecules [141, 142]. Both proteins were reported to be highly glycosylated in mycelial cells, which was inferred from their vertical streak patterns. Most strikingly, comparative proteome studies using deletion mutants indicated that Pra1p expression relied on the presence of Phr1p, but not vice versa, suggesting that the *PRA1* gene appeared to be downstream of the *PHR1* gene [67]. However, the effects of its regulation mechanism on hyphal conversion will require more investigation.

Further details of these expression proteomic works are shown in Table 17.1. Taking into account the differences in strains, dimorphic transition inducers, induction times, and detection methods, among others, introduced to each of them, it is not unexpected that slight deviations in the number of specific and/or differentially regulated proteins were found among these inquiries (Table 17.1).

*Shortening Complexity to Unearth Potential Key Pieces* As gathered above, the study of key subcellular compartments, macromolecular structures, or multiprotein complexes in which morphological switching-related gene products could be located seemed to be perhaps the only way to access and identify these low-copy proteins. To address this inference, previous subcellular fractionation or affinity techniques were

**TABLE 17.1 A Schematic Comparison among Different Documented Proteomic Studies That Have Focused on Analysis of Differential Protein Expression During *C. albicans* Yeast-to-Hyphal Transition**[a]

| | Manning and Mitchell, 1980 [24] | Brown and Chaffin, 1981 [51] | Finney et al., 1985 [54] | Niimi et al., 1996 [62] | Pitarch et al., 2002 [33] | Choi et al., 2003 [67] | Urban et al., 2003 [70] | Kaneko et al., 2004 [73] |
|---|---|---|---|---|---|---|---|---|
| *C. albicans* strain(s) | 4918, 2252[b] | B311 | 3153A, MD20[c] | ATCC 10261 | SC5314, | SC5314, CAMB43[d] CAS10[d] | SC5314, | iCDC11-HF[e] |
| **Hyphal induction** | | | | | | | | |
| Induction system | Temperature | Temperature | pH | pH, GlcNAc | pH | Temperature, serum | Temperature | Temperature, pH |
| Induction time | 45 min, 3 h[f] | 1 h and 10 min, 2 h and 10 min, 3 h and 10 min, 4 h and 10 min | 1 h and 40 min, 2 h and 30 min, 3 h and 20 min[g] | 1 h, 2 h | 6 h | 24 h | 24 h | 1 h and 30 min |
| **Cell fraction** | Cytoplasm | Cytoplasm | Cytoplasm | Cytosolic (S100) fraction DNA-binding protein fraction[h] | Cell wall | Cytoplasm | Cell wall | Septin complex |
| **Assay type** | | | | | | | | |
| Temporal | No | Yes | Yes | Yes | No | No | No | No |
| Spatial | No | No | No | No[i], Yes[j] | Yes | No | Yes | Yes |
| **2DE protein pattern** | | | | | | | | |
| pH gradient (first dimension) | Carrier ampholytes | Carrier ampholytes | Carrier ampholytes | Carrier ampholytes[i] | Immobilines (IPGs) | Immobilines (IPGs) | n.a.[h] | Immobilines (IPGs) |
| Detection method (second dimension) | Radioactivity ([$^{35}$S]-sulfate labeling) | Radioactivity ([$^{3}$H]-Leu and [$^{35}$S]-Met labeling at 4 times) | Radioactivity ([$^{35}$S]-Met labeling at 3 times) | Silver stain | Silver stain | Silver stain | | Silver stain |

(*continued*)

TABLE 17.1 (*Continued*)

| | Manning and Mitchell, 1980 [24] | Brown and Chaffin, 1981 [51] | Finney et al., 1985 [54] | Niimi et al., 1996 [62] | Pitarch et al., 2002 [33] | Choi et al., 2003 [67] | Urban et al., 2003 [70] | Kaneko et al., 2004 [73] |
|---|---|---|---|---|---|---|---|---|
| No. of visible protein spots | ~200 | ~230 | 374 | >400[i] | 1560–1660[k] | >900 | ~18–21 bands[h,l] | ~15 |
| No. of specific protein spots in: | | | | | | | | |
| Yeast forms | 33 | 5[m] | 1 | No | Variable[n] | No | 13 bands[h,o] | No |
| Hyphal forms | No | No | 1 | No,[i] 2 bands[h,j] | Variable[n] | No | 13 bands[h,o] | No |
| No. of hyphal protein spots: | | | | | | | | |
| Up regulated | No | No | No | No, [i]4 bands[h,j] | 94/13/ > 12/24[q] | 11 | n.r. | No |
| Down regulated | No | 2 | 2[p] | No | 38/28/ > 25/2[q] | No | n.r. | No |
| **Identification** | | | | | | | | |
| Method | n.r. | n.r. | n.r. | n.r. | MS, MS/MS | MS, MS/MS | MS, micro sequencing | MS |
| Hyphal proteins: | | | | | | | | |
| Up regulated | n.a. | n.a. | n.a. | n.a. | Different groups[r] | Phr1p, Pra1p, Tsa1p | n.a. | n.a. |
| Down regulated | n.a. | n.a. | n.a. | n.a. | Different groups[s] | n.a. | n.a. | n.a. |
| Specific | n.a. | n.a. | n.a. | n.a. | n.r. | n.a. | Different groups[o,t] | n.a. |
| Yeast proteins, specific | n.a. | n.a. | n.a. | n.a. | n.r. | n.a. | Different groups[o,u] | n.a. |

[a]GlcNAc denotes *N*-acetylglucosamine, IPGs immobilized pH gradients, Leu leucine, Met methionine, n.a. not applicable, and n.r. not reported.

[b]Nonfilamentous and non-temperature-dependent strain, which forms buds at both 24 and 37°C. The use of this isogenic dimorphism-deficient mutant led to the discrimination between proteins induced by temperature and those by morphological change.

[c]Nonfilamentous and non-pH-dependent strain, which produces budding cells at both low and high pH (pH 4.5 and 6.7, respectively). The use of this isogenic dimorphism-deficient mutant led to the discrimination between pH-specific proteins and morphogenesis-related proteins.

[d]CAMB43 and CAS10 are *PRA1*- and *PHR1*-delected mutant strains, respectively. Their use enabled the confirmation of the identity of two up-regulated hyphal proteins (Pra1p and Phr1p).

[e]Strain expressing Cdc11p tagged with a tandem polyhistidine and FLAG epitope, which was used for further purification of yeast and hyphal septin complex.

[f]Pulse-chase experiments at 45 min to minimize reutilization of isotope released by yeast protein catabolism and continuous labeling at 3 h.

[g]Corresponding to preevagination (100 min), evagination (150 min), and postevagination (200 min) periods.

[h]Protein separation by sodium dodecyl sulfate polyacrylamide gel electrophoresis (SDS-PAGE).

[i]In experiments with *C. albicans* S100 fraction.

[j]In experiments with *C. albicans* DNA-binding protein fraction.

[k]About 1560 and 1660 protein spots were displayed on 2DE gels from the yeast and hyphal walls, respectively, of which approximately 700 spots were SDS and reducing-agent extractable, about 290 and 210 spots were solubilized under mild alkali conditions from SDS-resistant yeast and hyphal walls, respectively, near 450 spots were released with β-1,3-glucanases, and about 65 and 250 protein spots were β-1,3-glucanase resistant and exochitinase extractable in yeast and mycelial walls, respectively. For further details, see Section 17.2.3.

[l]Twenty-one protein bands were identified in yeast walls and 18 in hyphal walls.

[m]Two of which were detected in hyphal cells at the first 50-min interval, but not subsequently.

[n]No SDS- and reducing-agent-extractable proteins were specific from yeast or hyphal walls. At least five mild alkali-released CWPs were exclusively detected in hyphal forms. Yeast- and hyphal-specific proteins were found among β-1,3-glucanase-extractable CWPs. About 185 β-1,3-glucanase-resistant and exochitinase-extractable CWP spots were only displayed in mycelial walls.

[o]Their specific expression during yeast-to-hypha transition was only confirmed for Tsa1p (a hyphal-specific CWP).

[p]They appear to represent different charged species of the same polypeptide and to be pH rather than phenotype specific.

[q]Relation corresponding to number of SDS- and reducing agent-extractable/mild alkali-released/β-1,3-glucanase-extractable/β-1,3-glucanase-resistant and exochitinase-extractable CWP spots.

[r]Include two heat-shock proteins (Hsp90p and Ssa4p), one folding protein (the 73-kDa form of Pdi1p), five glycolytic enzymes (Fba1p, Gap1p, Pgk1p, Eno1p, and Cdc19p), two fermentative enzymes (Pdc11p and Adh1p), and three proteins with miscellaneous functions (Psa1p, Ino1p, and the pH 4.7 form of Ipf8762p).

[s]Include one putative β-1,3-glucanase, one heat-shock protein (Ssz1p/Pdr13p), one folding protein (the 70-kDa form of Pdi1p), three glycolytic enzymes (Fba1p, Gap1p, and Cdc19p released under mild alkali conditions), two fermentative enzymes (Pdc11p and Adh1p released under mild alkali conditions), and one protein with unknown function (the pH 4.8 form of Ipf8762p).

[t]Include two mannoproteins (Pra1p and Phr1p), two heat-shock proteins (Ssa2p and Hsp90p), one glycolytic enzyme (Pgk1p), one fermentative enzyme (Pdc11p), one pyrophosphorylase (Gph1p), and six proteins with miscellaneous functions (Ubp1p, Atp1p, Sam2p, Ebp1p, Csh1p, and Tsa1p).

[u]Include two heat-shock proteins (Ssb1p and Ssc1p), one glycolytic enzyme (Cdc19p), one fermentative enzyme (Adh1p), two pyrophosphorylases (Ugp1p and Psa1p/Srb1p), two elongation factors (Eft2p and Eft3p), and five proteins with miscellaneous functions (Tkl1p, Hem13p, Bmh1p, Rps6p, and Sdh2p).

brought into play to reduce sample complexity. This was achieved by removing the major components, thereby enriching samples for proteins of interest, which would otherwise be masked on 2DE gels by the more abundant soluble proteins [33, 62, 70, 73].

It is known that several transcription factors and proteins mediating signal transduction to these factors can regulate the expression of yeast and hypha-specific genes (see Fig. 17.10; reviewed in [112, 143–147]). Amazingly, proteins with DNA-binding activity can be isolated by heparin–agarose affinity chromatography, taking advantage of the polyanionic nature of heparin. Using this attractive approach, a small group of DNA-binding proteins, or even proteins from a DNA-binding complex, proved to be differentially synthesized during dimorphic transition (Table 17.1). This suggested that these low-abundance proteins, undetected on S100 cell-free extracts, might be pivotal regulators of cellular response to physiological, cell-cycle-related, and morphogenetic changes associated with the early phases of *C. albicans* germ tube formation [62].

On the other hand, it is generally accepted that the yeast-to-hypha conversion involves alterations in composition and organization of the *C. albicans* cell wall (reviewed in [79, 84, 148]). Most of the proteins present in this compartment (CWPs) can be extracted by cell wall fractionation according to their interactions with other structural wall components [33, 83, 94, 149, 150] (further details are dealt with in the Section 17.2.3). The combination of this type of CWP enrichment along with comparative 2DE analyses revealed that the cell wall remodeling that takes place during yeast-to-hypha transition was correlated with both qualitative and quantitative changes in the composition of CWPs (Table 17.1) [33]. While cell-surface-associated proteins turned out to be similar but differentially regulated in budding and filamentous forms (Fig. 17.15), the assembly of CWPs covalently linked to wall polysaccharides (β-glucans and chitin) diverged in both morphologies. These outstanding observations clarified that the CWP–chitin linkage was probably the major retention mechanism of most CWPs covalently attached to mycelial cell walls. The MS and MS/MS analyses led to the identification of 15 CWPs differentially regulated in yeast and hyphal cells (Chapter 18 Appendix and Table 17.1). Remarkably, the finding that some glycolytic enzymes were strongly expressed in mycelial walls raised the question as to whether both morphologies exhibited a different metabolism or whether an extra input of energy was needed for the wall remodeling, both for the incorporation of new building blocks into the cell wall and for local biosynthesis of covalent linkages between cell wall components that occur during germ tube formation.

Differences in expression patterns of noncovalently linked *C. albicans* CWPs from yeast and hyphal cells were also observed in a further proteomic approach using a membrane-impermeable biotin derivative in conjunction with affinity purification (see Section 17.2.3 for details) [70]. Although 26 different CWPs were found in either one of the two growth forms and 4 CWPs in both morphologies (Chapter 18 Appendix and Table 17.1), their conclusive differential expression during dimorphic transition was only confirmed for one of them, Tsa1p (thiol-specific antioxidant-like protein). This was achieved by protein localization studies using a reporter gene, that is, *GFP* (green fluorescent protein). These additional experiments demonstrated that Tsa1p was localized in the nucleus and cytosol in yeast forms, and intriguingly, this appeared to be moved partially at the cell surface in hyphae, indicating that its expression at the cell wall could be regulated by differential translocation.

Last, but not least, it is also thought that septins—cytoskeletal elements (see Section 17.3.5)—play an essential part in *C. albicans* morphogenesis [112, 121, 122]. *Candida*

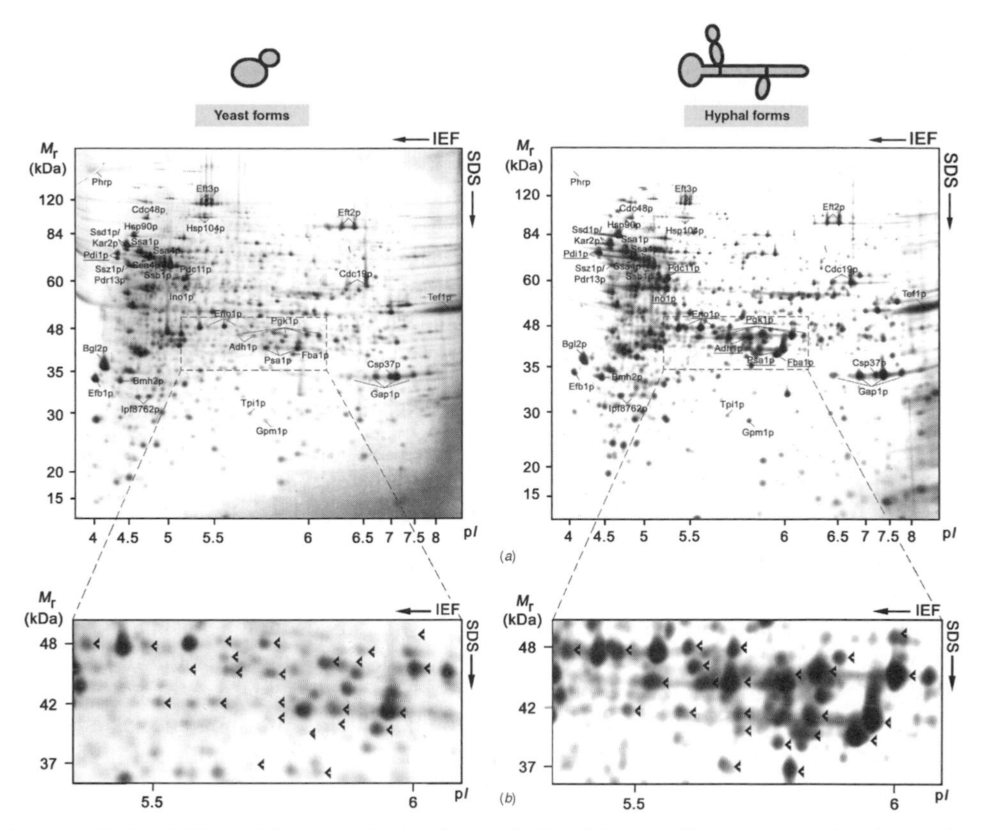


**Figure 17.15** Differential expression analysis of *C. albicans* cell-surface-associated proteins during dimorphic transition [33]. (*a*). Silver-stained 2DE gels of SDS- and reducing-agent-extractable proteins from *C. albicans* yeast (left panel) and hyphal (right panel) forms. No major qualitative changes (appearance and disappearance of spots) between both morphologies were found. However, 94 up- and 38 down-regulated hyphal cell-surface-associated protein spots were consistently observed during the yeast-to-hypha transition. One of the most significant differences in protein expression levels between both morphological forms is indicated with a dashed rectangle. Labeled spots were identified by MALDI-TOF MS and/or MALDI-TOF/TOF MS. Up-regulated proteins identified are underlined. Pdi1p (protein disulfide isomerase): 73-kDa and 70-kDa protein spots were up and down regulated, respectively, during hyphal formation. Spot names refer to those in the Appendix (in Chapter 18). These gels are available at the COMPLUYEAST-2DPAGE database (http://babbage.csc.ucm.es/2d/2d.html). (*b*). Enlargements of the sections highlighted in (a). Protein spots differentially expressed during germ tube formation are depicted with arrowheads.

*albicans* septins seem to (i) promote hyphal morphogenesis by recruiting certain septin-associated proteins required for hyphal development (e.g., the integrinlike protein Int1p [151]), (ii) determine the hyphal shape by serving as a boundary that confines actin to the growing tip, and (iii) guide incipient tip morphogenesis by differentially regulating the secretion machinery during budding and filamentous growth [122]. Interestingly, the pattern and location of the septin ring differ in the diverse cell morphologies of *C. albicans* [121, 122]. Whereas in yeast and pseudohyphal cells septins are located in a tight ring at the mother cell-bud neck, in hyphae a septin ring is also initially formed at the point of evagination but then disorganizes and disappears to be replaced by a second ring. The latter is formed in the first cell cycle within the germ tube at the future site of cytokinesis. Despite this molecular distinction between these morphological forms, similar 2DE protein profiles of the *C. albicans* septin complexes purified from yeast cells and hyphae by TAP (see Section 17.3.5 and Fig. 17.11) were, however, displayed [73]. The fact that the septin complex was not affected by cell morphology at the proteome level could be related to preservation of septin structure and function during *C. albicans* morphogenesis.

Overall, all these results demonstrate how fractionation can increase the sensitivity of proteomic assays and permit the detection of protein expression changes associated per se with dimorphic transition as well as the unearthing of its key pieces. What repercussions will it have on future studies? There is no doubt that other strategic *C. albicans* subproteomes will also be investigated in years to come in an effort to gain a finer understanding of this virulence trait at the protein level.

### 17.4.3 Adhesion to Host Structures: First Stage of Infection Process

***Plethora of Host Recognition Biomolecules*** Adhesion is presumably a crucial step in the pathogenic process and a requirement for colonization, which is a vital prelude to invasion. Many fungal cell surface constituents or moieties, such as specific receptors, hydrophobic proteins, or mannoproteins, can mediate the binding of *C. albicans* to epithelial or endothelial cells, extracellular matrix (ECM) proteins (e.g., laminin, fibronectin, collagen), serum proteins (such as fibrinogen, plasminogen, or iC3b and C3d complement fragments), and/or medical devices implanted in the host (developing the so-called biofilms), to name but a few. These host recognition biomolecules have been referred to as adhesins. Their wide variety may be indicative of the diversity of sites that *C. albicans* is able to colonize and invade. Some of them are described below, and the reader is directed to excellent reviews on this topic [78, 79, 131, 152–158].

***Quest for Novel Host Recognition Biomolecules Using Proteomics*** There is an incipient interest in "structural proteomics" to simultaneously screen for *C. albicans* cell surface proteins that can be associated with host recognition and mediate attachment to host ligands by maintaining protein–protein interactions, that is, adhesin–ligand interactions. As contended next, encouraging proteomic analyses that combine 2DE with Western blotting using specific antibodies have contributed to the characterization of novel *C. albicans* proteins or moieties related to adherence [27, 61, 69].

*Plasminogen-Binding Proteins* Like other microbial pathogens, *C. albicans* may adhere to host plasminogen and induce its activation, which in turn would trigger a host-derived proteolytic system capable of potentially increasing the alacrity of this fungus for tissue invasion and degradation (Fig. 17.16) [131, 159, 160]. For instance, the ability of *C. albicans* cells to penetrate the in vitro blood–brain barrier is enhanced in the presence of bound plasmin, the active form of plasminogen [160].

Eight recently identified proteins (four glycolytic and one fermentative enzymes, two redox proteins, and one elongation factor) accounted for most of the plasminogen-binding capacity observed in *C. albicans* CWP extracts as detected using 2DE followed by ligand blotting with plasminogen and MALDI-TOF MS analysis (Chapter 18

Appendix) [69]. The bulk of binding was attributed to five *C. albicans* CWPs that contained carboxy-terminal lysine residues. This consistency was supported by biochemical assays that demonstrated that binding of plasminogen to CWPs relied on these residues and not on carbohydrate moieties of mannoproteins, since binding was (i) inhibited with the lysine analogue ε-amino-caproic acid (εACA), (ii) reduced by treatment with basic carboxypeptidase, and (iii) preserved in *C. albicans* mutant strains defective in protein glycosylation. Additional functional analysis suggested that plasminogen bound to the *C. albicans* cell surface seemed to be activated, to plasmin, by host plasminogen activators as a consequence of the apparent absence of an endogenous one (Fig. 17.16). Although plasmin-bound *C. albicans* cells were capable of degrading fibrin in vitro, they did not exhibit an increased ability either to penetrate and damage endothelial cells or to invade endothelial matrix, highlighting the need for further research to elucidate the role(s) of plasminogen activation in *C. albicans* invasive process in vivo.

*Ubiquitinated Cell Surface Receptors* Among the specific surface receptors displayed by *C. albicans*, the integrinlike receptors can bind host proteins with RGD sequences, such as laminin, fibrinogen, fibronectin, or iC3b and C3b complement fragments, through protein–protein interactions [79, 131, 151, 161].

Two-dimensional electrophoresis and immunoblot experiments with polyclonal antibodies against ubiquitin—a small polypeptide [79, 162]—and three purified integrin-like surface receptors (i.e., the 37-kDa laminin receptor, the 58-kDa fibrinogen-binding mannoprotein, and the candidal C3d receptor) indicated that these *C. albicans* surface receptors were ubiquitinated [27]. This attractive observation prompted the proposal of a potential key role for ubiquitination in modulating the activity of these receptors and, therefore, in the adhesion of *C. albicans* to host structures.

*Hydrophobic Proteins* Cell surface hydrophobicity (CSH) appears to play a significant function in adhesion to host cells, ECM proteins, and catheters as well as in circumventing the surveillance of phagocytes [163–165].

The combination of hydrophobic interaction chromatography (HIC)–high-performance liquid chromatography (HPLC) along with 2DE, Western blotting, and LC-MS/MS analysis led to the enrichment of hydrophobic cell surface proteins and the subsequent unambiguous identification of Csh1p (38-kDa cell surface hydrophobic protein), a novel *C. albicans* gene product of unknown function (with a high sequence similarity to aryl alcohol dehydrogenases) that could mediate binding to host protein ligands (such as fibronectin) [61].

### 17.4.4 Production of Extracellular Hydrolytic Enzymes: Weapons That Assist Invasiveness

***Extracellular Hydrolases with Broad Substrate Specificities*** It has long been recognized that *C. albicans* is able to produce extracellular hydrolytic enzymes related to virulence with broad substrate specificities. The secreted aspartyl proteases (Saps), phospholipases, and lipases are conceivably the most significant hydrolases secreted by *C. albicans*. These are found in both budding and filamentous forms and appear to promote the erosion of pitlike cavities in host surfaces and penetrative growth observed in yeast and hyphal cells, respectively (Fig. 17.12) [166]. These hydrolytic enzymes may contribute to

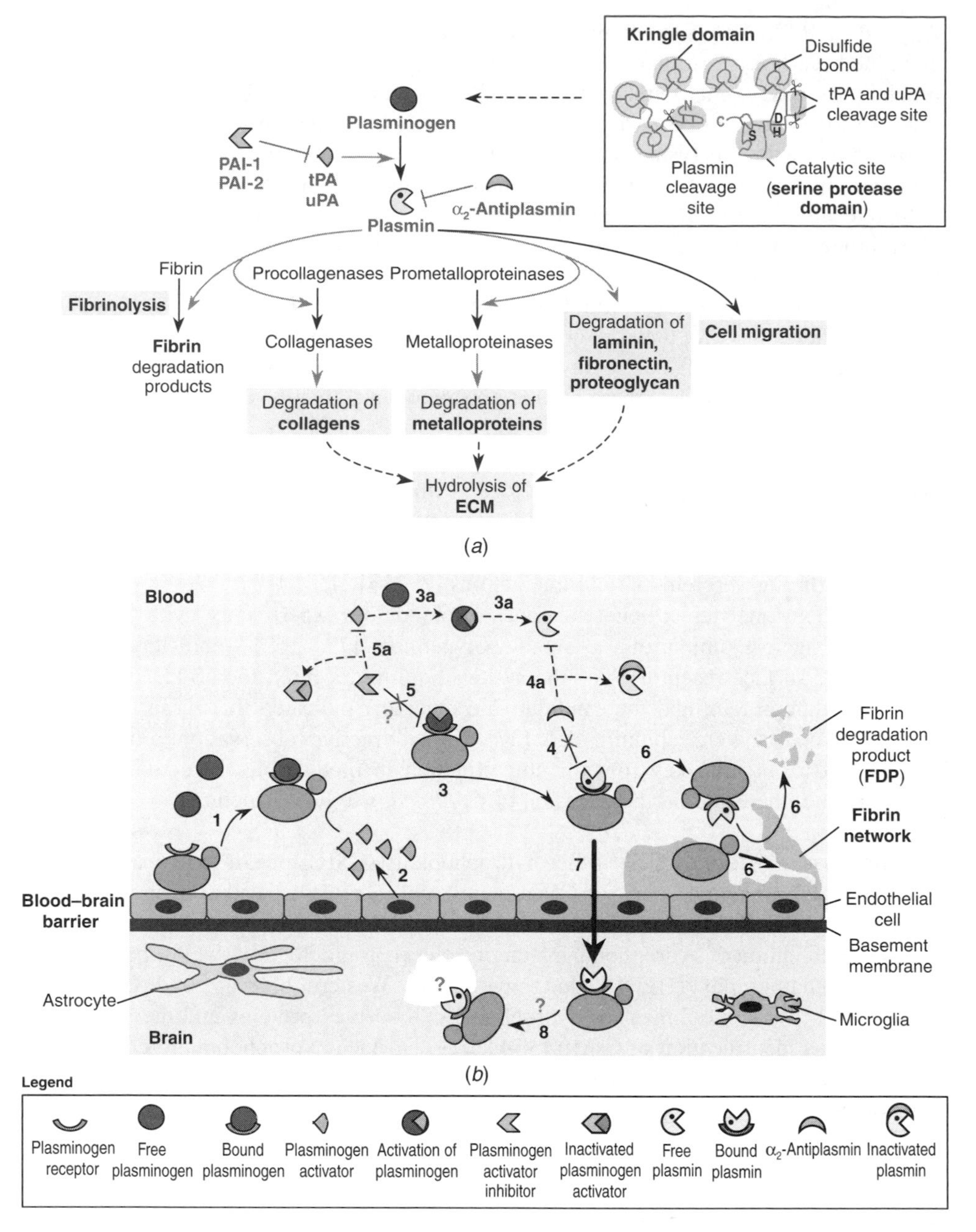


**Figure 17.16** Interaction of *C. albicans* with human plasminogen system. (*a*) Schematic view of human plasminogen system. Plasminogen is an abundant zymogen protein in human plasma and extracellular fluids that contains five kringle domains (triple-looped structures) with lysine binding sites (inset). It is converted to the active broad-spectrum serine protease plasmin by tissue- or urokinase-type plasminogen activators (tPA or uPA, respectively), which are also serine proteases. Plasmin can dissolve intravascular fibrin clots (fibrinolysis), break down extracellular matrix (ECM) proteins (e.g., laminin, fibronectin, and proteoglycan), activate latent matrix prometalloproteinases and procollagenases (proenzymes) to hydrolyze ECM and basement membranes, and promote

the penetration and invasion of host barriers, adhesion to host mucosal surfaces, inactivation of host defense molecules, and exploitation of polymers as a nutrient source (reviewed in [79, 132, 167–171]).

The extracellular proteolytic activity of *C. albicans* is attributed to Saps, which are encoded by a family of 10 *SAP* genes. These Saps, especially Sap2p, can hydrolyze various host substrates, including structural proteins, proteinase inhibitors, proteins of cascade systems, humoral defense (like secretory IgA), cell surface, and ECM, to name just a few [132].

***Assessment of Extracellular Proteolytic Activity of C. albicans by Using Proteomic Approaches*** Proteomic analysis of a concentrated culture filtrate obtained from *C. albicans* cells grown in a medium with mucin as a nitrogen or carbon source led to the identification of Sap2p as the major mucin-degrading enzyme and the detection of a less prominent protein that might be a potential breakdown product of mucin [28]. This latter notion was gathered from the observations that its N-terminal amino acid sequence showed no homology with mature Saps, and its zymogram and metabolic labeling were negative. The extracellular mucinolytic activity of *C. albicans* Sap2p was successfully evaluated by using a zymogram, in which the proteins from culture supernatant separated by a native isoelectric focusing (IEF) gel were overlaid on mucin-coated paper (Fig. 17.17). This appealing proteomic approach enabled the detection of Sap2p as the only component responsible for proteolysis of mucin by *C. albicans* in vitro. As a consequence, this secreted enzyme could be involved in the virulence of *C. albicans* as a result, in part, of facilitating the hydrolysis of mucus, penetration of the mucosal barriers, and subsequent adherence to

cellular migration and/or tissue remodeling. Both plasminogen activator inhibitors (PAIs) and $\alpha_2$-antiplasmin, the main inhibitor of plasmin, avoid indiscriminate proteolytic activity and subsequent tissue damage. Red blunt arrows indicate negative effects and blue arrows proteolytic activation. (*b*) Hypothetical mechanism of human plasminogen–*C. albicans* interaction in pathogenesis of invasive candidiasis validated by proteomic and biochemical studies [69, 160]. *Candida albicans* cells bind plasminogen to their surface through certain receptors that contain C-terminal lysine residues which interact with plasminogen's lysine binding sites located in the kringle structures (step 1). This induces the secretion of plasminogen activators from human cells (step 2), which trigger the conversion of plasminogen to plasmin (steps 3). This enables *C. albicans* cells to acquire a host protease on their surface that, unlike unbound plasmin (step 4a), is protected from its physiological inactivation by human $\alpha_2$-antiplasmin (blunt arrow with a cross; step 4). In addition, the regulators of plasminogen activators (PAIs) appear not to effectively inhibit the plasminogen activation on the *C. albicans* cell surface, mediated by plasminogen activators (step 5). This acquisition of extracellular proteolytic activity could thus allow *C. albicans* to evade without difficulty the fibrin network deposited by the host to confine the infection focus (step 6) and assist the migration of *C. albicans* through normal tissue barriers, such as the blood–brain barrier (step 7), and its dissemination into human tissues (step 8). For clarity, *C. albicans* cells are depicted with only one plasminogen receptor on their surfaces. Blunt arrows indicate negative effects. Dashed (blunt) arrows show the way in physiological conditions, and numbers followed by the letter "a" correspond to the counterparts of those mentioned previously in the plasminogen–*C. albicans* interaction.

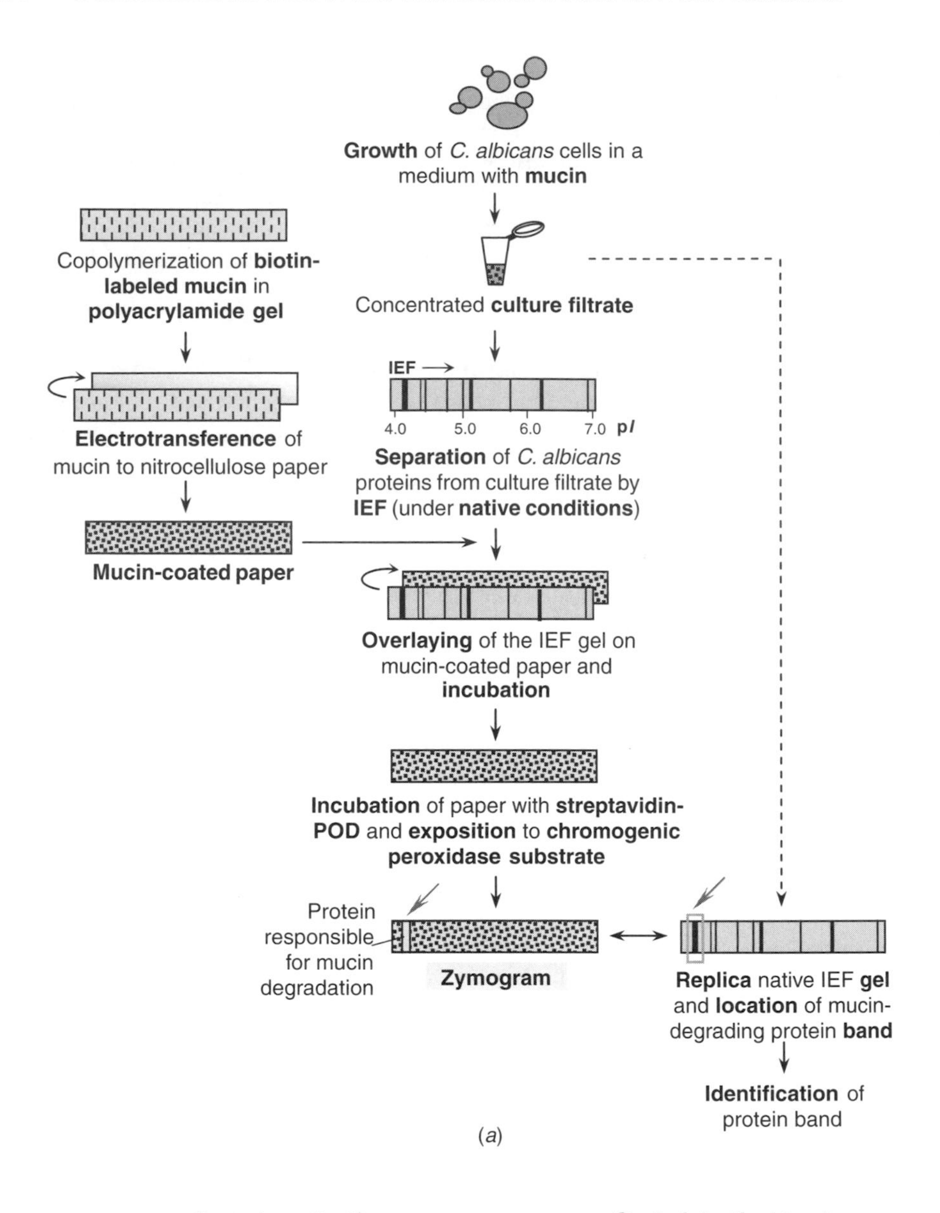
Growth of C. albicans cells in a medium with mucin
Copolymerization of biotin-labeled mucin in polyacrylamide gel
Concentrated culture filtrate
IEF
4.0 5.0 6.0 7.0 pI
Electrotransference of mucin to nitrocellulose paper
Separation of C. albicans proteins from culture filtrate by IEF (under native conditions)
Mucin-coated paper
Overlaying of the IEF gel on mucin-coated paper and incubation
Incubation of paper with streptavidin-POD and exposition to chromogenic peroxidase substrate
Protein responsible for mucin degradation
Zymogram
Replica native IEF gel and location of mucin-degrading protein band
Identification of protein band

(*a*)

Gastrointestinal tract
Secretion of Sap2p
1
Mucus
Epithelium layer
Basement membrane
ECM
Tissue
Gastrointestinal tract
Breakdown of mucus
2
Penetration of mucin barrier
3
4
Adherence to epithelial cells
5
Invasion of epithelial cells
Tissue

(*b*)

and invasion of epithelial cells. Furthermore, this in vitro study promotes the suggestion that Sap2p could be the enzyme responsible for in vivo progressive extracellular digestion of the mucin layer reported previously in the jejunum of infant mice inoculated intragastrically with *C. albicans* [172].

## 17.5 CONCLUDING REMARKS

In recent years, there has been a refinement of our understanding of *C. albicans* biology and pathogenicity which has, in part, been attained by using high-throughput proteomic approaches. In fact, the potential of these studies to provide a more global and detailed view into how *C. albicans* functions at the molecular level in ways that are not predictable from genomic investigations, becoming more and more tangible. Furthermore, one can expect that the combination of proteomic analyses with other complementary technologies, such as molecular biology, will soon give us answers to key questions about the biological complexity and invasive process of *C. albicans*. In short, more accurate clues about the basic networks that drive the existence of this opportunistic fungus are likely to be forthcoming as we delve deeper into the enigmatic organism's "functional" biology at a molecular level.

## ACKNOWLEDGMENTS

We thank the Merck, Sharp & Dohme Special Chair in Genomics and Proteomics, Comunidad Autónoma de Madrid (Strategic Groups; CPGE 1010/2000), Comisión Interministerial de Ciencia y Tecnología (BIO-2003-00030), and Fundación Ramón Areces for financial support of our laboratory.

**Figure 17.17** Extracellular mucinolytic activity of *C. albicans*. (*a*) Strategy for detecting secretory mucinolytic activity of *C. albicans* in vitro. Proteins secreted into the medium in the presence of mucin are separated by IEF (or 2DE) under native conditions. A zymogram from the resulting IEF gel, in which biotin-labeled mucin is used as the substrate, leads to the assessment of the mucinolytic activity from the separated proteins. The mucin-degrading proteins are detected as clear bands (green line) located on their corresponding isoelectric point (pI) values against a dark blue-violet background (pink mottled rectangle) as a result of the removal of mucin molecules previously labeled with biotin on their protein moeities, that is, by their hydrolysis [28]. Proteins vouched for mucin proteolysis can then be identified by Edman degradation or MS. (*b*) Degradation of gastrointestinal mucin by *C. albicans*-secreted aspartyl protease 2 (Sap2p). Mucus (pink mottled surface) is mainly composed of mucins (black dots), which are glycoproteins involved in protection against invasion by pathogens, owing to their ability to form a gel. Sap2p (small green symbols) is the major mucinolytic enzyme secreted by *C. albicans* that may be implicated as a virulence factor in the progressive extracellular digestion of the intestinal mucous barrier and in promoting penetration of the mucous barrier by *C. albicans* and its subsequent adhesion to epithelial cells (shaded cells), invasion of epithelial cells (green arrow), and spread within the host. The events depicted are based both on in vitro proteomic studies [28] and on in vivo assays [172]. ECM denotes extracellular matrix and POD peroxidase.

## REFERENCES


1. Gudlaugsson, O., Gillespie, S., Lee, K., Vande, B. J., et al., *Clin. Infect. Dis.* 2003, **37**, 1172–1177.
2. Pfaller, M. A., Jones, R. N., Doern, G. V., Sader, H. S., et al., *J. Clin. Microbiol.* 1998, **36**, 1886–1889.
3. al Soub, H., and Estinoso, W., *J. Hosp. Infect.* 1997, **35**, 141–147.
4. Fraser, V. J., Jones, M., Dunkel, J., Storfer, S., et al., *Clin. Infect. Dis.* 1992, **15**, 414–421.
5. Maertens, J., Vrebos, M., and Boogaerts, M., *Eur. J. Cancer Care (Engl.)* 2001, **10**, 56–62.
6. Kullberg, B. J., and Filler, S. G., in R. A. Calderone (Ed.), *Candida and Candidiasis*, ASM Press, Washington, DC, 2002, pp. 327–340.
7. Vincent, J. L., Bihari, D. J., Suter, P. M., Bruining, H. A., et al., *JAMA* 1995, **274**, 639–644.
8. Pfaller, M. A., Diekema, D., Jones, R. N., Sader, H. S., et al., *J. Clin. Microbiol.* 2001, **39**, 3254–3259.
9. Rolston, K., *Oncol. (Huntingt.)* 2001, **15**, 11–14.
10. Koch, S., Hohne, F. M., and Tietz, H. J., *Mycoses* 2004, **47**, 40–46.
11. Blot, S. I., Hoste, E. A., Vandewoude, K. H., and Colardyn, F. A., *J. Crit. Care* 2003, **18**, 130–131.
12. Leleu, G., Aegerter, P., and Guidet, B., *J. Crit. Care* 2002, **17**, 168–175.
13. Sallah, S., Wan, J. Y., Nguyen, N. P., Vos, P., and Sigounas, G., *Cancer* 2001, **92**, 1349–1353.
14. Patterson, J. E., *Transpl. Infect. Dis.* 1999, **1**, 229–236.
15. Sanglard, D., and Bille, J., in R. A. Calderone (Ed.), *Candida and Candidiasis*, ASM Press, Washington, DC, 2002, pp. 349–386.
16. Kontoyiannis, D. P., and Lewis, R. E., *Lancet* 2002, **359**, 1135–1144.
17. Jones, J. M., *Clin. Microbiol. Rev.* 1990, **3**, 32–45.
18. Pfaller, M. A., *Mycopathologia* 1992, **120**, 65–72.
19. Odds, F. C., *Candida and Candidiasis*, Bailliere Tindall, London, 1988, pp. 42–59.
20. Pla, J., Gil, C., Monteoliva, L., Navarro-Garcia, F., et al., *Yeast* 1996, **12**, 1677–1702.
21. De Backer, M. D., Magee, P. T., and Pla, J., *Annu. Rev. Microbiol.* 2000, **54**, 463–498.
22. Wilkins, M. R., Pasquali, C., Appel, R. D., Ou, K., et al., *Bio/Technology* 1996, **14**, 61–65.
23. Pennington, S. R., Wilkins, M. R., Hochstrasser, D. F., and Dunn, M. J., *Trends Cell. Biol.* 1997, **7**, 168–173.
24. Manning, M., and Mitchell, T. G., *J. Bacteriol.* 1980, **144**, 258–273.
25. O'Farrell, P. H., *J. Biol. Chem.* 1975, **250**, 4007–4021.
26. Klose, J., *Humangenetik* 1975, **26**, 231–243.
27. Sepulveda, P., Lopez-Ribot, J. L., Gozalbo, D., Cervera, A., et al., *Infect. Immun.* 1996, **64**, 4406–4408.
28. Colina, A. R., Aumont, F., Deslauriers, N., Belhumeur, P., and de Repentigny, L., *Infect. Immun.* 1996, **64**, 4514–4519.
29. Valdes, I., Pitarch, A., Gil, C., Bermudez, A., et al., *J. Mass Spectrom.* 2000, **35**, 672–682.
30. Pardo, M., Ward, M., Pitarch, A., Sanchez, M., et al., *Electrophoresis* 2000, **21**, 2651–2659.
31. De Backer, M. D., de Hoogt, R. A., Froyen, G., Odds, F. C., et al., *Microbiology* 2000, **146**(2), 353–365.
32. Pitarch, A., Diez-Orejas, R., Molero, G., Pardo, M., et al., *Proteomics* 2001, **1**, 550–559.
33. Pitarch, A., Sanchez, M., Nombela, C., and Gil, C., *Mol. Cell Proteomics* 2002, **1**, 967–982.
34. Bruneau, J. M., Maillet, I., Tagat, E., Legrand, R., et al., *Proteomics* 2003, **3**, 325–336.

35. Hernandez, R., Nombela, C., Diez-Orejas, R., and Gil, C., *Proteomics* 2004, **4**, 374–382.
36. Pitarch, A., Abian, J., Carrascal, M., Sanchez, M., et al., *Proteomics* 2004, **4**, 3084–3106.
37. Pitarch, A., Jimenez, A., Nombela, C., and Gil, C., *Mol. Cell Proteomics* 2006, **5**, 79–96.
38. Fernandez-Arenas, E., Molero, G., Nombela, C., Diez-Orejas, R., and Gil, C., *Proteomics* 2004, **4**, 3007–3020.
39. Hooshdaran, M. Z., Barker, K. S., Hilliard, G. M., Kusch, H., et al., *Antimicrob. Agents Chemother.* 2004, **48**, 2733–2735.
40. Yin, Z., Stead, D., Selway, L., Walker, J., et al., *Proteomics* 2004, **4**, 2425–2436.
41. Brand, A., MacCallum, D. M., Brown, A. J., Gow, N. A., and Odds, F. C., *Eukaryot. Cell* 2004, **3**, 900–909.
42. Pitarch, A., Sanchez, M., Nombela, C., and Gil, C., *J. Chromatogr. B Analyt. Technol. Biomed. Life Sci.* 2003, **787**, 129–148.
43. Boeckmann, B., Bairoch, A., Apweiler, R., Blatter, M. C., et al., *Nucleic Acids Res.* 2003, **31**, 365–370.
44. Liska, A. J., and Shevchenko, A., *Proteomics* 2003, **3**, 19–28.
45. Jones, T., Federspiel, N. A., Chibana, H., Dungan, J., et al., *Proc. Natl. Acad. Sci. USA* 2004, **101**, 7329–7334.
46. Chibana, H., Magee, B. B., Grindle, S., Ran, Y., et al., *Genetics* 1998, 149, 1739–1752.
47. Chibana, H., Beckerman, J. L., and Magee, P. T., *Genome Res.* 2000, **10**, 1865–1877.
48. Sturtevant, J., *Mycopathologia* 2004, **158**, 141–146.
49. Goffeau, A., Barrell, B. G., Bussey, H., Davis, R. W., et al., *Science* 1996, **274**, 546–567.
50. Lee, S. A., Wormsley, S., Kamoun, S., Lee, A. F., et al., *Yeast* 2003, **20**, 595–610.
51. Brown, L. A., and Chaffin, W. L., *Can. J. Microbiol.* 1981, **27**, 580–585.
52. Manning, M., and Mitchell, T. G., *Infect. Immun.* 1980, **30**, 484–495.
53. Graves, P. R., and Haystead, T. A., *Microbiol. Mol. Biol. Rev.* 2002, **66**, 39–63.
54. Finney, R., Langtimm, C. J., and Soll, D. R., *Mycopathologia* 1985, **91**, 3–15.
55. Shen, H. D., Choo, K. B., Lin, W. L., Lin, R. Y., and Han, S. H., *Electrophoresis* 1990, **11**, 878–882.
56. Shen, H. D., Choo, K. B., Yu, K. W., Ling, W. L., et al., *Int. Arch. Allergy Appl. Immunol.* 1991, **96**, 142–148.
57. Morhart, M., Rennie, R., Ziola, B., Bow, E., and Louie, T. J., *J. Clin. Microbiol.* 1994, **32**, 766–776.
58. Soll, D. R., Anderson, J., and Bergen, M., in R. Prasad (Ed.), *Candida albicans, Cellular and Molecular Biology*, Springer-Verlag, Berlin, 1991, pp. 20–45.
59. Soll, D. R., in H. V. Bossche, F. C. Odds, and D. Kerridge (Eds.), *Dimorphic Fungi in Biology and Medicine*, Plenum, New York, 1993, pp. 73–82.
60. Fernandez-Murray, P., Biscoglio, M. J., and Passeron, S., *Arch. Biochem. Biophys.* 2000, **375**, 211–219.
61. Singleton, D. R., Masuoka, J., and Hazen, K. C., *J. Bacteriol.* 2001, **183**, 3582–3588.
62. Niimi, M., Shepherd, M. G., and Monk, B. C., *Arch. Microbiol.* 1996, **166**, 260–268.
63. Pitarch, A., Pardo, M., Jimenez, A., Pla, J., et al., *Electrophoresis* 1999, **20**, 1001–1010.
64. Barea, P. L., Calvo, E., Rodriguez, J. A., Rementeria, A., et al., *FEMS Immunol. Med. Microbiol.* 1999, **23**, 343–354.
65. Vediyappan, G., Bikandi, J., Braley, R., and Chaffin, W. L., *Electrophoresis* 2000, **21**, 956–961.
66. Marichal, P., Vanden Bossche, H., Odds, F. C., Nobels, G., et al., *Antimicrob. Agents Chemother.* 1997, **41**, 2229–2237.

67. Choi, W., Yoo, Y. J., Kim, M., Shin, D., et al., *Yeast* 2003, **20**, 1053–1060.
68. Fernandez-Arenas, E., Molero, G., Nombela, C., Diez-Orejas, R., and Gil, C., *Proteomics* 2004, **4**, 1204–1215.
69. Crowe, J. D., Sievwright, I. K., Auld, G. C., Moore, N. R., et al., *Mol. Microbiol.* 2003, **47**, 1637–1651.
70. Urban, C., Sohn, K., Lottspeich, F., Brunner, H., and Rupp, S., *FEBS Lett.* 2003, **544**, 228–235.
71. de Groot, P. W., de Boer, A. D., Cunningham, J., Dekker, H. L., et al., *Eukaryot. Cell* 2004, **3**, 955–965.
72. Fernandez-Murray, P., Pardo, P. S., Zelada, A. M., and Passeron, S., *Arch. Biochem. Biophys.* 2002, **404**, 116–125.
73. Kaneko, A., Umeyama, T., Hanaoka, N., Monk, B. C., et al., *Yeast* 2004, **21**, 1025–1033.
74. Kusch, H., Biswas, K., Schwanfelder, S., Engelmann, S., et al., *Mol. Genet. Genomics* 2004, **271**, 554–565.
75. Niimi, M., Cannon, R. D., and Monk, B. C., *Electrophoresis* 1999, **20**, 2299–2308.
76. Pitarch, A., Sanchez, M., Nombela, C., and Gil, C., *J. Chromatogr. B Analyt. Technol. Biomed. Life Sci.* 2003, **787**, 101–128.
77. Rupp, S., *Curr. Opin. Microbiol.* 2004, **7**, 330–335.
78. Calderone, R., Diamond, R., Senet, J. M., Warmington, J., et al., *J. Med. Vet. Mycol.* 1994, **32**(Suppl. 1), 151–168.
79. Chaffin, W. L., Lopez-Ribot, J. L., Casanova, M., Gozalbo, D., and Martinez, J. P., *Microbiol. Mol. Biol Rev.* 1998, **62**, 130–180.
80. Martinez, J. P., Gil, M. L., Lopez-Ribot, J. L., and Chaffin, W. L., *Clin. Microbiol. Rev.* 1998, **11**, 121–141.
81. Lopez-Ribot, J. L., Casanova, M., Murgui, A., and Martinez, J. P., *FEMS Immunol. Med. Microbiol.* 2004, **41**, 187–196.
82. Calderone, R. A., *Trends Microbiol.* 1993, **1**, 55–58.
83. Klis, F. M., de Groot, P., Hellingwerf, K., *Med. Mycol.* 2001, **39**(Suppl. 1), 1–8.
84. Valentin, E., Mormeneo, S., and Sentandreu, R., *Contrib. Microbiol.* 2000, **5**, 138–150.
85. Brown, A. J., in R. A. Calderone (Ed.), *Candida and Candidiasis*, ASM Press, Washington, DC, 2002, pp. 87–94.
86. Georgopapadakou, N. H., and Tkacz, J. S., *Trends Microbiol.* 1995, **3**, 98–104.
87. Groll, A. H., De Lucca, A. J., and Walsh, T. J., *Trends Microbiol.* 1998, **6**, 117–124.
88. Odds, F. C., Brown, A. J., and Gow, N. A., *Trends Microbiol.* 2003, **11**, 272–279.
89. Alloush, H. M., Lopez-Ribot, J. L., Masten, B. J., and Chaffin, W. L., *Microbiology* 1997, **143**(2), 321–330.
90. Gozalbo, D., Gil-Navarro, I., Azorin, I., Renau-Piqueras, J., et al., *Infect. Immun.* 1998, **66**, 2052–2059.
91. Surarit, R., Gopal, P. K., and Shepherd, M. G., *J. Gen. Microbiol.* 1988, **134**(6), 1723–1730.
92. Kapteyn, J. C., Montijn, R. C., Dijkgraaf, G. J., Van den, E. H., and Klis, F. M., *J. Bacteriol.* 1995, **177**, 3788–3792.
93. Sarthy, A. V., McGonigal, T., Coen, M., Frost, D. J., et al., *Microbiology* 1997, **143**(2), 367–376.
94. Kapteyn, J. C., Hoyer, L. L., Hecht, J. E., Muller, W. H., et al., *Mol. Microbiol.* 2000, **35**, 601–611.
95. Kandasamy, R., Vediyappan, G., and Chaffin, W. L., *FEMS Microbiol. Lett.* 2000, **186**, 239–243.
96. Angiolella, L., Micocci, M. M., D'Alessio, S., Girolamo, A., et al., *Antimicrob. Agents Chemother.* 2002, **46**, 1688–1694.

97. Marcilla, A., Valentin, E., and Sentandreu, R., *Int. Microbiol.* 1998, **1**, 107–116.
98. Chauhan, N., Li, D., Singh, P., Calderone, R., and Kruppa, M., in R. A. Calderone (Ed.), *Candida and Candidiasis*, ASM Press, Washington, DC, 2002, pp. 159–178.
99. Molina, M., Gil, C., Pla, J., Arroyo, J., and Nombela, C., *Microsc. Res. Tech.* 2000, **51**, 601–612.
100. Pardo, M., Monteoliva, L., Pla, J., Sanchez, M., et al., *Yeast* 1999, **15** 459–472.
101. Coux, O., Tanaka, K., and Goldberg, A. L., *Annu. Rev. Biochem.* 1996, **65**, 801–847.
102. Hilt, W., and Wolf, D. H., *Mol. Biol. Rep.* 1995, **21**, 3–10.
103. Hershko, A., and Ciechanover, A., *Annu. Rev. Biochem.* 1998, **67**, 425–479.
104. Yamada-Okabe, T., Mio, T., Matsui, M., Kashima, Y., et al., *FEBS Lett.* 1998, **435**, 49–54.
105. Schwer, B., Lehman, K., Saha, N., and Shuman, S., *J. Biol. Chem.* 2001, **276**, 1857–1864.
106. Tripathi, G., Wiltshire, C., Macaskill, S., Tournu, H., et al., *EMBO J.* 2002, **21**, 5448–5456.
107. Pereira, S. A., and Livi, G. P., *Cell. Biol. Int.* 1995, **19**, 65–69.
108. Hinnebusch, A. G., *Microbiol. Rev.* 1988, **52**, 248–273.
109. Irniger, S., and Braus, G. H., *Curr. Genet.* 2003, **44**, 8–18.
110. Fonzi, W. A., and Irwin, M. Y., *Genetics* 1993, **134**, 717–728.
111. Staab, J. F., and Sundstrom, P., *Trends Microbiol.* 2003, **11**, 69–73.
112. Berman, J., and Sudbery, P. E., *Nature* 2002, **3**, 918–930.
113. Cheng, S., Nguyen, M. H., Zhang, Z., Jia, H., et al., *Infect. Immun.* 2003, **71**, 6101–6103.
114. Lay, J., Henry, L. K., Clifford, J., Koltin, Y., et al., *Infect. Immun.* 1998, **66**, 5301–5306.
115. Cooper, J. A., and Kiehart, D. P., *J. Cell. Biol.* 1996, **134**, 1345–1348.
116. Longtine, M. S., DeMarini, D. J., Valencik, M. L., Al Awar, O. S., et al., *Curr. Opin. Cell. Biol.* 1996, **8**, 106–119.
117. Trimble, W. S., *J. Membr. Biol* 1999, **169**, 75–81.
118. Mitchison, T. J., and Field, C. M., *Curr. Biol.* 2002, **12**, R788–R790.
119. Lew, D. J., *Curr. Opin. Cell. Biol.* 2003, **15**, 648–653.
120. Longtine, M. S., and Bi, E., *Trends Cell. Biol.* 2003, **13**, 403–409.
121. Sudbery, P. E., *Mol. Microbiol.* 2001, **41**, 19–31.
122. Warenda, A. J., and Konopka, J. B., *Mol. Biol. Cell.* 2002, **13**, 2732–2746.
123. Warenda, A. J., Kauffman, S., Sherrill, T. P., Becker, J. M., et al., *Infect. Immun.* 2003, **71**, 4045–4051.
124. Cutler, J. E., *Annu. Rev. Microbiol.* 1991, **45**, 187–218.
125. Hogan, L. H., Klein, B. S., and Levitz, S. M., *Clin. Microbiol. Rev.* 1996, **9**, 469–488.
126. Calderone, R. A., and Fonzi, W. A., *Trends Microbiol.* 2001, **9**, 327–335.
127. Haynes, K., *Trends Microbiol.* 2001, **9**, 591–596.
128. van Burik, J. A., and Magee, P. T., *Annu. Rev. Microbiol.* 2001, **55**, 743–772.
129. Navarro-Garcia, F., Sanchez, M., Nombela, C., and Pla, J., *FEMS Microbiol. Rev.* 2001, **25**, 245–268.
130. Odds, E. C., *Mycoses* 1997, **40**(Suppl. 2), 9–12.
131. Calderone, R., and Gow, N. A., in R. A. Calderone (Ed.), *Candida and Candidiasis*, ASM Press, Washington, DC, 2002, pp. 67–86.
132. Hube, B., and Naglik, J., in R. A. Calderone (Ed.), *Candida and Candidiasis*, ASM Press, Washington, DC, 2002, pp. 107–122.
133. Yang, Y. L., *J. Microbiol. Immunol. Infect.* 2003, **36**, 223–228.
134. Soll, D. R., in R. A. Calderone (Ed.), *Candida and Candidiasis*, ASM Press, Washington, DC, 2002, pp. 123–144.

135. Slutsky, B., Staebell, M., Anderson, J., Risen, L., et al., *J. Bacteriol.* 1987, **169**, 189–197.
136. Soll, D. R., *Acta Trop.* 2002, **81**, 101–110.
137. Mitchell, A. P., *Curr. Opin. Microbiol.* 1998, **1**, 687–692.
138. Lo, H. J., Kohler, J. R., DiDomenico, B., Loebenberg, D., et al., *Cell* 1997, **90**, 939–949.
139. Sherwood, J., Gow, N. A., Gooday, G. W., Gregory, D. W., and Marshall, D., *J. Med. Vet. Mycol.* 1992, **30**, 461–469.
140. Gow, N. A., *Curr. Top. Med. Mycol.* 1997, **8**, 43–55.
141. Sentandreu, M., Elorza, M. V., Sentandreu, R., and Fonzi, W. A., *J. Bacteriol.* 1998, **180**, 282–289.
142. Saporito-Irwin, S. M., Birse, C. E., Sypherd, P. S., and Fonzi, W. A., *Mol. Cell Biol* 1995, **15**, 601–613.
143. Ernst, J. F., *Microbiology* 2000, **146**(8), 1763–1774.
144. Whiteway, M., *Curr. Opin. Microbiol.* 2000, **3**, 582–588.
145. Brown, A. J., in R. A. Calderone (Ed.), *Candida and Candidiasis*, ASM Press, Washington, DC, 2002, pp. 95–106.
146. Liu, H., *Curr. Opin. Microbiol.* 2001, **4**, 728–735.
147. Navarro-Garcia, F., Eisman, B., Román, E., and Pla, J., *Med. Mycol.* 2001, **39**, 87–100.
148. Lipke, P. N., and Ovalle, R., *J. Bacteriol.* 1998, **180**, 3735–3740.
149. Mrsa, V., Seidl, T., Gentzsch, M., and Tanner, W., *Yeast* 1997, **13**, 1145–1154.
150. Kapteyn, J. C., Montijn, R. C., Vink, E., de la Cruz, J., et al., *Glycobiology* 1996, **6**, 337–345.
151. Gale, C. A., Bendel, C. M., McClellan, M., Hauser, M., et al., *Science* 1998, **279**, 1355–1358.
152. Calderone, R. A., and Braun, P. C., *Microbiol. Rev.* 1991, **55**, 1–20.
153. Hostetter, M. K., *Clin. Microbiol. Rev.* 1994, **7**, 29–42.
154. Fukazawa, Y., and Kagaya, K., *J. Med. Vet. Mycol.* 1997, **35**, 87–99.
155. Sundstrom, P., *Curr. Opin. Microbiol.* 1999, **2**, 353–357.
156. Pendrak, M. L., Yan, S. S., and Roberts, D. D., *Arch. Biochem. Biophys.* 2004, **426**, 148–156.
157. Sundstrom, P., *Cell Microbiol.* 2002, **4**, 461–469.
158. Cotter, G., and Kavanagh, K., *Br. J. Biomed. Sci.* 2000, **57**, 241–249.
159. Lottenberg, R., Minning-Wenz, D., and Boyle, M. D., *Trends Microbiol.* 1994, **2**, 20–24.
160. Jong, A. Y., Chen, S. H., Stins, M. F., Kim, K. S., et al., *J. Med. Microbiol.* 2003, **52**, 615–622.
161. Hostetter, M. K., *Curr. Opin. Microbiol.* 2000, **3**, 344–348.
162. Finley, D., and Chau, V., *Annu. Rev. Cell Biol* 1991, **7**, 25–69.
163. Hobden, C., Teevan, C., Jones, L., and O'Shea, P., *Microbiology* 1995, **141**(8), 1875–1881.
164. Hazen, K. C., and Glee, P. M., *Curr. Top. Med. Mycol.* 1995, **6**, 1–31.
165. Masuoka, J., Wu, G., Glee, P. M., and Hazen, K. C., *FEMS Immunol. Med. Microbiol.* 1999, **24**, 421–429.
166. Cole, G. T., Seshan, K. R., Lynn, K. T., and Franco, M., *Mycol. Res.* 1993, **97**, 385–408.
167. Naglik, J. R., Challacombe, S. J., and Hube, B., *Microbiol. Mol. Biol Rev.* 2003, **67**, 400–428.
168. Ghannoum, M. A., *Clin. Microbiol. Rev.* 2000, **13**, 122–143.
169. Hube, B., *Curr. Top. Med. Mycol.* 1996, **7**, 55–69.
170. Hube, B., and Naglik, J., *Microbiology* 2001, **147**, 1997–2005.
171. Bein, M., Schaller, M., and Korting, H. C., *Curr. Drug Targets.* 2002, **3**, 351–357.
172. Cole, G. T., Seshan, K. R., Pope, L. M., and Yancey, R. J., *J. Med. Vet. Mycol.* 1988, **26**, 173–185.

## CHAPTER 18

# Contributions of Proteomics to Diagnosis, Treatment, and Prevention of Candidiasis

AIDA PITARCH, CÉSAR NOMBELA, and CONCHA GIL

Complutense University, Madrid, Spain

The previous chapter outlined an overview of the main contributions of proteomics to the study of *Candida albicans* and candidiasis over the past 25 years and highlighted how proteomic approaches may provide critical information on *C. albicans* cell biology and pathogenicity. In this chapter, the powerful potential of this discipline in the diagnosis, prevention, and treatment of candidiasis, particularly *C. albicans* infections, is discussed. Our own personal view of upcoming insights into this promising realm of the *C. albicans* proteome focusing on its later stages, that is, the "Middle and Later Expansion Ages," is also presented. For earlier reviews on proteomic investigations as applied to *C. albicans*, the reader is referred to [1–4].

## 18.1 HOW PROBLEMATIC IS CANDIDIASIS TO MODERN MEDICINE?

Candidiasis is an opportunistic fungal infection caused by *Candida* spp. (especially *C. albicans*, commonly represented in normal flora on human mucosal surfaces) that currently ranks among the most frequent of nosocomial diseases [5–7]. It comprises a wide spectrum of diseases that range from superficial mucosal lesions (e.g., oropharyngeal, esophageal, and vulvovaginal candidiasis) to invasive or systemic forms of infection, including candidemia (bloodstream infection), localized deep-seated candidiasis (primary infection of a single deep organ, such as kidney, brain, spleen, biliary tree, lung, heart, and eye, among others, acquired either through hematogenous spread or by direct inoculation), and disseminated candidiasis (infection of multiple deep organs as the result of bloodstream invasion and hematogenous seeding) (reviewed in [8–12]). Whereas superficial mucosal candidiasis often occurs both in normal individuals and in human immunodeficiency virus (HIV)–infected patients, systemic candidiasis is mainly found in immunocompromised patients. Furthermore, systemic forms of candidiasis are associated with dreadfully high morbidity and mortality in severely ill patients, reaching mortality rates of 30–80% depending on underlying patient conditions [13–18]. Lamentably, this

*Microbial Proteomics: Functional Biology of Whole Organisms*, Edited by Ian Humphery-Smith and Michael Hecker. 

alarming clinical setting is directly related to limitations in the diagnosis and treatment of these fungal infections.

It is of note that systemic candidiasis rarely produces specific clinical signs and symptoms, whereby its diagnosis is frequently delayed until the infection is in advanced stages or unfortunately following autopsy [19–22]. In addition, routine blood cultures, which are still regarded as the gold standard in the diagnosis of these opportunistic infections, can require several days to become positive or even remain negative in patients with disseminated candidiasis. For this reason, in recent years, various innovative non-culture-based approaches which involve the detection and/or measurement of *Candida* metabolites (D-arabinitol), polysaccharides (e.g., mannans, β-1,3-glucan), antigens and/or their related antibodies, or DNA [by polymerase chain reaction (PCR) amplification] in body fluids of infected patients are gaining a noteworthy interest, but none of them has yet achieved widespread clinical use (reviewed in [23–27]). However, there are further alternatives. For instance, given that systemic candidiasis is difficult to diagnose, immunological strategies to prevent it in high-risk patients could clearly circumvent this stumbling block.

The therapy of candidiasis also constitutes another challenging task for physicians, because of clinical failures, reduced efficacy and toxicity of the limited therapeutic arsenal available, and/or emergence of antifungal resistance, to name just a few (see Section 18.3 for further details) [19, 28, 29].

In a nutshell, candidiasis represents a major clinical problem that will undoubtedly remain until some significant advance in its diagnosis, treatment, and/or prophylaxis is attained. Hence, novel strategies to detect, combat, and/or prevent these infections are urgently necessary. At present, there is an increasing interest in the application of proteomics to inquiries into candidiasis with the goal of (i) identifying novel biomarkers for screening these opportunistic fungal infections and predicting patients' outcome, through assays of serum specimens, (ii) discovering new or alternative therapies to their management, and (iii) understanding the mechanisms involved in antifungal resistance.

## 18.2 HOST IMMUNE RESPONSE: ADAPTING IT FOR IMMUNODIAGNOSTIC AND IMMUNOTHERAPEUTIC PURPOSES

### 18.2.1 From a Host Point of View: Array of Intricate and Interconnected Responses to Cope with *C. albicans*

The strategies developed by the human host to counteract the invasive nature of *C. albicans* are numerous, disparate, and highly complex. Certain clinical observations, especially an elevated incidence of mucocutaneous and systemic candidiasis in AIDS (T-cell deficiency) and neutropenic patients, respectively, have supported the notion that cellular immunity could clearly be a decisive host defense mechanism against mucosal candidiasis, whereas innate immunity, mediated by macrophages and polymorphonuclear leukocytes, might analogously be imperative in host resistance to systemic candidiasis. However, it is not such a simple affair. It appears that the diverse anti-*Candida* host responses could be interconnected, cross-regulated, and overlapped, whereby innate, humoral, and cell-mediated immunity may plausibly play a critical role in the resolution of candidiasis and in anti-*Candida* protection [30–34].

What role then do antibodies play in host defense against *C. albicans* infections? Certain antibodies may be involved in the inhibition of yeast-to-hypha transition or adherence of *C. albicans* to host structures, the neutralization of *C. albicans* extracellular proteases, or the blockage of *C. albicans* immununomodulating polysaccharides, among others (reviewed in [32, 35, 36]). In addition, it has recently been acknowledged that some antibodies, albeit not all, may confer protection against both superficial and systemic candidiasis. Protective and nonprotective antibodies for *C. albicans* are likely to possess different isotype and epitope specificity [32, 35–38]. In fact, vaccines that induce protective antibodies may protect against experimental *C. albicans* infection [39, 40]. Interestingly, a human recombinant antibody against the *C. albicans* heat-shock protein Hsp90p (Mycofab) is currently in human clinical trials to assess its efficacy for treatment of invasive candidiasis in patients receiving antifungal therapy, that is, amphotericin B [41].

This section will emphasize the pivotal contributions of proteomics to the investigation of the anti–*C. albicans* host response. So far, these studies have focused on analysis of the host antibody response to *C. albicans* infections for both immunodiagnostic and/or immunotherapeutic purposes. It must be borne in mind that in addition to the potential of certain anti-*Candida* antibodies in protecting the host against candidiasis, some antibodies and/or their related antigens could be useful for the development of serological tests for the diagnosis and clinical follow-up of these fungal infections [25, 26].

### 18.2.2 Identification of Antigens That Elicit Antibody Response by Immunoproteomics for Clinical Purposes

The host antibody response to candidiasis has become a subject of great interest to proteomic projects since their first steps into the *C. albicans* field in 1980 to the present day [42–54]. This attraction lies in the desire to identify *C. albicans* immunogenic proteins that induce a specific antibody response and may in turn have utility in the diagnosis, monitoring, and/or immunotherapy of these infections, especially systemic candidiasis.

***Immunoproteomics: Sensible Platform to Implement Serological Screenings of C. albicans Infections*** To tackle this issue, cross-absorption experiments were originally coupled with two-dimensional gel electrophoresis (2DE) [42]. However, over the last few years, several approaches have been based on immunoproteomics. Immunoproteomics is based on the separation of *C. albicans* proteins by 2DE and subsequent detection of those with antigenic properties by immunoblotting using serum samples from systemic candidiasis patients or immune sera. This approach has been implemented as a means for (i) obtaining an overview of different anti-*Candida* antibodies that are being produced in humans [47, 48, 51, 54] or in animal models of experimentally induced systemic candidiasis [48, 50, 52, 53], respectively, or, more exactly, defining the so-called "immune proteome" (or "immunome") of *C. albicans*, and (ii) discovering potential disease-specific markers, prognostic factors, and/or candidates that can confer protective immunity for their inclusion in future vaccines [2]. The general strategy currently used in these studies is illustrated in Figure 18.1.

The combination of 2DE with Western blot assays using specific antibodies raised against *C. albicans* antigens was already documented to be an extremely sensitive platform for the detection of immunoreactive *C. albicans* proteins in 1990 [43]. In that initial study, a supernatant of hybridoma cells containing monoclonal antibodies against *C. albicans* was used to discriminate cytoplasmic antigens.

At present, this association of proteome analysis with serology, referred to as immunoproteomics, is widely being exploited to look for serological biomarkers and

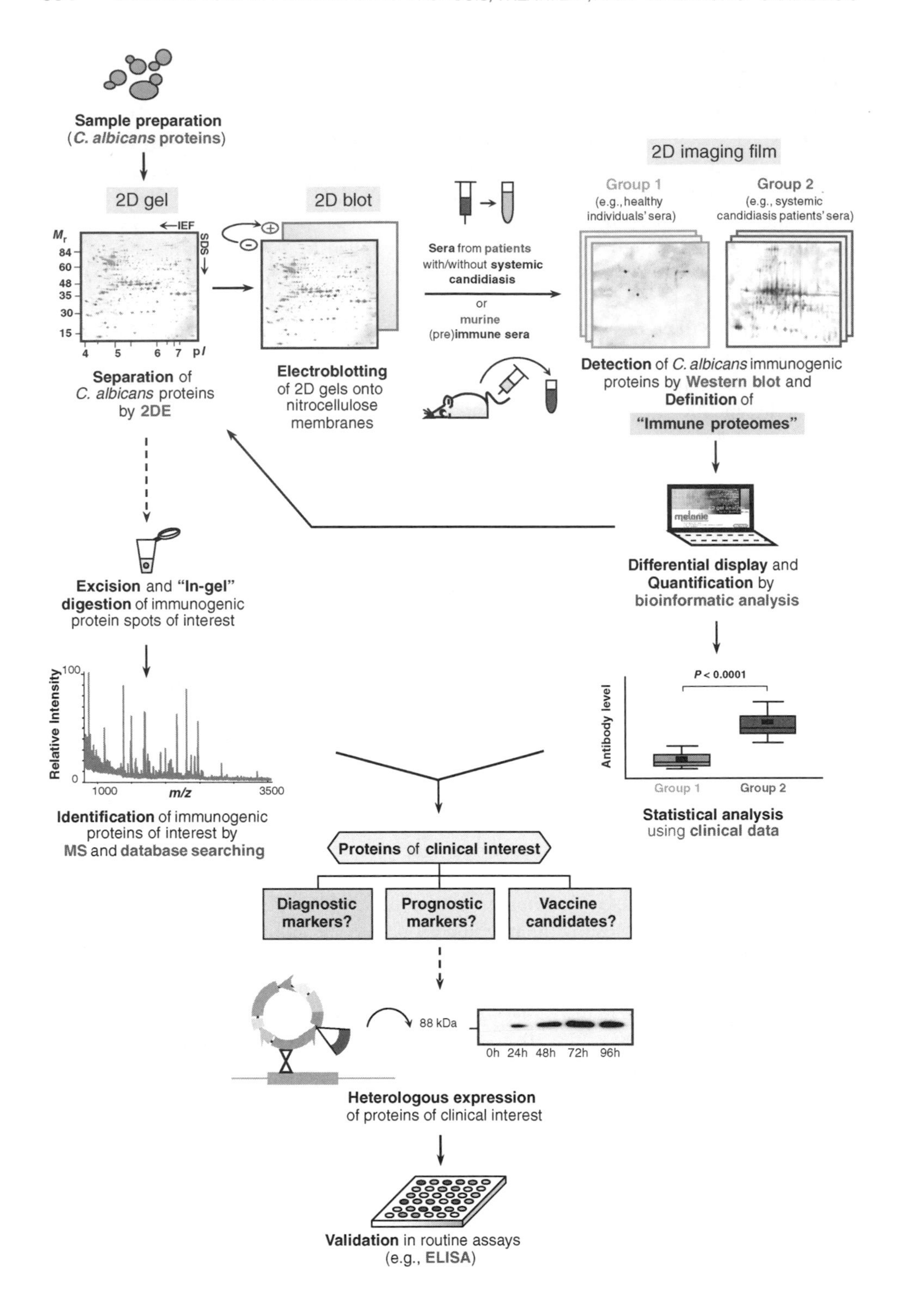

Sample preparation (C. albicans proteins)
2D gel
←IEF
SDS→
Mr
84
60
48
35
30
15
4
5
6
7
pI
Separation of C. albicans proteins by 2DE
2D blot
Electroblotting of 2D gels onto nitrocellulose membranes
Sera from patients with/without systemic candidiasis
or
murine (pre)immune sera
2D imaging film
Group 1 (e.g., healthy individuals' sera)
Group 2 (e.g., systemic candidiasis patients' sera)
Detection of C. albicans immunogenic proteins by Western blot and Definition of
"Immune proteomes"
Differential display and Quantification by bioinformatic analysis
P < 0.0001
Antibody level
Group 1
Group 2
Statistical analysis using clinical data
Excision and "In-gel" digestion of immunogenic protein spots of interest
Relative Intensity
100
0
1000
m/z
3500
Identification of immunogenic proteins of interest by MS and database searching
Proteins of clinical interest
Diagnostic markers?
Prognostic markers?
Vaccine candidates?
88 kDa
0h 24h 48h 72h 96h
Heterologous expression of proteins of clinical interest
Validation in routine assays (e.g., ELISA)

therapeutic targets, not only in systemic candidiasis (see below) but also in other microbial infections, cancer, or autoimmune diseases [55–61].

### *Translating C. albicans Immune Proteome into Promising Clinical Markers and Therapeutic Targets*

*Diagnostic Markers* As mentioned previously, the dearth of accurate and reliable diagnostic tests for systemic candidiasis has highlighted the importance of searching for good disease markers. A useful procedure for their identification has been the analysis of serum for antibodies against *C. albicans* antigens. Highly sensitive techniques such as proteomics have minimized the false-negative rate associated with low antibody sensitivity in immunocompromised patients. These patients are usually characterized by delayed, reduced, or absent antibody response [26, 51, 54, 62].

Several immunoproteomics-based approaches have included comparative analysis of antibody profiles present in sera from subjects with and without systemic candidiasis. These are screened individually to establish the spectrum of anti-*Candida* antibody specificities for these infections, that is, the common occurrence of circulating antibodies to *C. albicans* antigens [47–49, 51, 54]. This stratagem has successfully assisted the quest for serological markers (i.e., disease-specific antibodies and/or antigens), which might contribute to the improvement of the current screening and diagnosis of systemic *C. albicans* infections (A. Pitarch et al., unpublished results and [49, 51, 54, 63, 64]).

The first proteomic investigations using serum specimens from infected and noninfected individuals revealed the presence of certain cytoplasmic proteins that elicited an immune response in systemic candidiasis patients but not in healthy individuals [47, 48]. At least 17 different *C. albicans* immunoreactive cytoplasmic proteins with $M_r$ ranging from 30 to 84 kDa were detected using serum samples from six systemic candidiasis patients suffering from malignancies, one of which [immunostained as enolase

**Figure 18.1** Flow chart of typical approach for analyzing *C. albicans* immune proteome (or immunome) and identifying *C. albicans* immunogenic proteins of clinical interest by immunoproteomics. *Candida albicans* proteins are separated by 2DE and transferred onto nitrocellulose membranes by electroblotting. Serum samples from (i) systemic candidiasis patients and healthy individuals [47, 48, 51, 54]; (ii) patients or animals upon artificial immunization at different stages of systemic *Candida* infection, such as before and after treatment [51], or at 10 and 20 days postinfection with *C. albicans* [50]; or (iii) immunized protected and nonprotected animals [52, 53], among others, are screened individually by Western blotting for antibodies that react against 2DE-separated *C. albicans* proteins. In this method, each membrane is hybridized first with one human or murine immune serum used as the primary antibody and then with horseradish-peroxidase-labeled anti-human IgG or anti-mouse total Ig, IgG, and/or IgG2a, respectively, as the secondary antibody and, after that, immunoreactive protein spots are detected by enhanced chemiluminescence (ECL). The different 2D *C. albicans* antigen recognition patterns ("immune proteomes" or "immunomes") obtained with serum specimens from the paired aforementioned groups are compared by bioinformatic analysis to decode them into potential (i) diagnostic markers, (ii) prognostic markers, or (iii) vaccine candidates, respectively. The immunogenic protein spots of interest can be assessed by statistical analysis using clinical data and/or located on a preparative 2DE gel, extracted from it, digested with a protease (e.g., trypsin), and identified by MS analysis (or alternatively by Edman sequencing) and database searching. Relevant proteins of clinical interest can then be expressed as recombinant proteins to be validated in routine assays and included in a future serological test kit and/or vaccine design [63, 64].

(Eno1p)] was recognized by all sera [47]. A recent immunoproteomic analysis has indicated that about 6.5% of spots visualized on a silver-stained 2DE gel of *C. albicans* cytoplasmic extracts reacted with infected patients' sera, suggesting that at least 45 different antigens could stimulate the human immune system during systemic candidiasis [51].

In pioneering studies, only two glycolytic enzymes, enolase [47, 48] and glyceraldehyde-3-phosphate dehydrogenase (Gap1p) [47], were identified as *C. albicans* immunogens by immunostaining. At that time, the high-throughput protein characterization was considered as a challenging task since many proteins had not yet been annotated in public *C. albicans* databases. (See Chapter 17 for further details). Intriguingly, four *C. albicans* antigens which are candidates for diagnostic markers could be cross-species identified by matching *C. albicans* amino acid sequences deduced from nanoESI-MS/MS (electrospray ionization tandem mass spectrometry) data to proteins from *Saccharomyces cerevisiae,* a yeast organism whose genome was already entirely sequenced (Fig. 18.2) [49]. All of them corresponded to products of open reading frames (ORFs) predicted from the *C. albicans* genome sequence, three of which had not previously been proven to encode immunogenic proteins (see the Appendix). Shortly afterward, 10 immunorelevant proteins (mainly heat-shock proteins and glycolytic enzymes), two of which were also novel antigens, were characterized further by immunodetection, N-terminal sequencing, and/or peptide mass mapping [50, 65].

Most strikingly, the recent availability of an annotated *C. albicans* genomic database (the so-called CandidaDB) in combination with matrix-assisted laser desorption ionization time-of-flight (MALDI-TOF) MS and/or nanoESI ion trap MS analyses afforded the identification of a total of 82 *C. albicans* immunorelevant protein spots. These represented 42 different housekeeping enzymes, of which 26 were novel antigens in *C. albicans* and 35 were novel targets of the human antibody response to systemic candidiasis [51]. Accordingly, the *C. albicans* immune proteome naturally exposed during systemic candidiasis appears to be composed of diverse chaperones, glycolytic and fermentative enzymes, other metabolic enzymes, elongation factors, ribosomal proteins, porins, and redox enzymes, among others (Fig. 18.3*a* and the Appendix). All enzymes involved in the highly conserved and highly expressed glycolytic pathway turned out to be immunogens in patients with systemic *C. albicans* infections.

To search for additional disease-specific serological markers, the antibody response induced by *C. albicans* cell wall proteins (CWPs) during systemic candidiasis was also examined by immunoproteomics [47, 54]. Owing to their privileged location within the cell (i.e., host–fungus interface; see Chapter 17), some of these CWPs could clearly be major elicitors of a specific immune response [66, 67]. Immunoproteomic analysis of CWPs, or more precisely proteins secreted from protoplasts in active cell wall regeneration, resulted in the detection of four immunorelevant proteins, one of which was a 34-kDa acidic glycosylated immunodominant protein [recently identified as a glucan-β-1,3-glucosidase (Bgl2p) [54]] that reacted with all sera from systemic candidiasis patients [47]. In a more detailed study, several CWPs were identified as *C. albicans* antigens, including Bgl2p and several glycolytic enzymes with potential diagnostic value [54]. Their specificities were further assessed against a panel of serum specimens from patients with different underlying conditions using statistical analysis. These results suggested that serum anti-Bgl2p antibodies are a novel independent marker of diagnosis for systemic candidiasis.

Low levels of circulating antibodies directed against some abundant *C. albicans* glycolytic enzymes (i.e., enolase Eno1p [47, 48, 51] and pyruvate kinase Pgk1p [51]) and

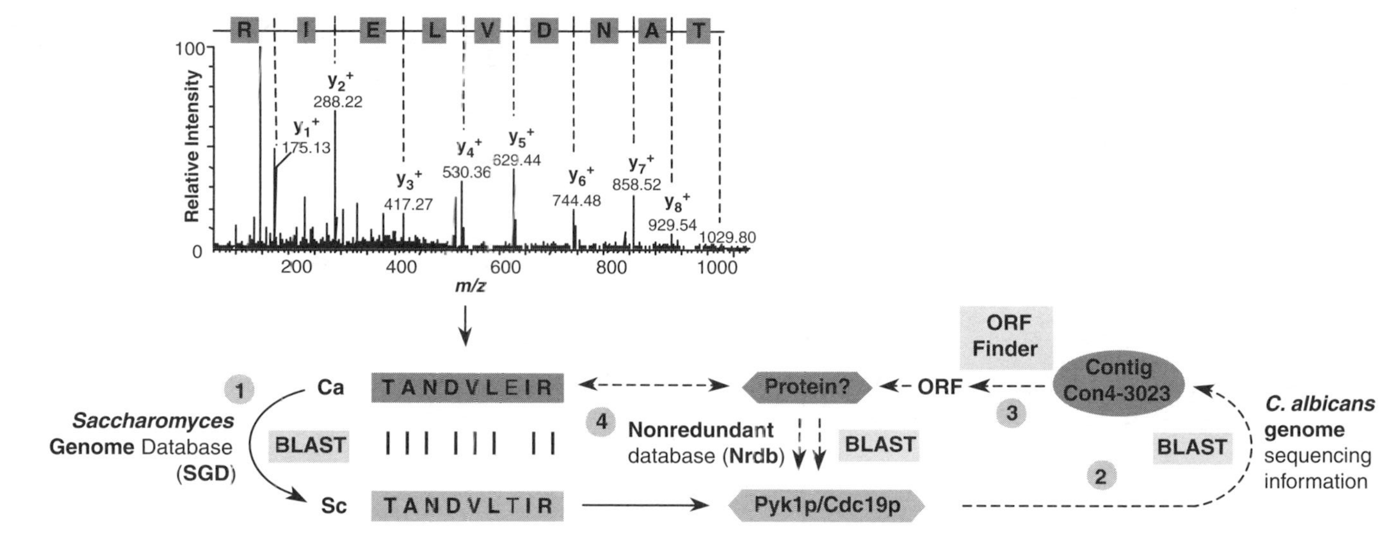


**Figure 18.2** Cross-species protein identification by MS/MS. This proteomic approach is exploited for identifying proteins which are not present in public domain databases by cross-species comparison to proteins from organisms with entirely sequenced genomes. At present, this approach is not clearly used for identifying *C. albicans* proteins (refer to Chapter 17). In the example illustrated, the partial amino acid sequence TANDVLEIR of a *C. albicans* immunogenic protein was deduced de novo from the series of *y*-ions deriving from an MS/MS spectrum obtained for the doubly charged precursor ion corresponding to the tryptic peptide of 1029.8 Da [49]. This sequence was used to search the *S. cerevisiae* Genome Database (SGD; http://genome.www.stanford.edu/Saccharomyces/) for homologue proteins with Basic Local Alignment Search Tool (BLAST; http://www.ncbi.nlm.nih.gov/entrez) and retrieved a homologue amino acid sequence (TANDVLTIR) contained within *S. cerevisiae* pyruvate kinase 1 or Pyk1p/Cdc19p (step 1). The corresponding protein sequence of *S. cerevisiae* Pyk1p/Cdc19p was in turn used to search the *C. albicans* genome-sequencing database (on-going at Stanford University at that time; http://www-sequence.stanford.edu/group/candida) for the *C. albicans* DNA sequence—within the *C. albicans* genome—coding for the counterpart of *S. cerevisiae* Pyk1p/Cdc19p with the BLAST algorithm and returned the contig or DNA fragment Con4-3023 (step 2). This *C. albicans* DNA sequence was analyzed using the ORF Finder program (http://www.ncbi.nlm.nih.gov/gorf) to locate the ORF coding for the counterpart of Pyk1p/Cdc19p (step 3). The amino acid sequence translated from this ORF contained the peptide experimentally determined from MS/MS data. When a search with this translated sequence was conducted with the BLAST algorithm in the nonredundant database (Nrdb; http://www.ncbi.nlm.nih.gov), the same protein from *S. cerevisiae* that was identified previously with the BLAST algorithm in the SGD was found as the top hit (step 4). Therefore, the *C. albicans* immunogenic protein under consideration could unambiguously be identified as Pyk1p/Cdc19p. Vertical bars denote identical amino acids.

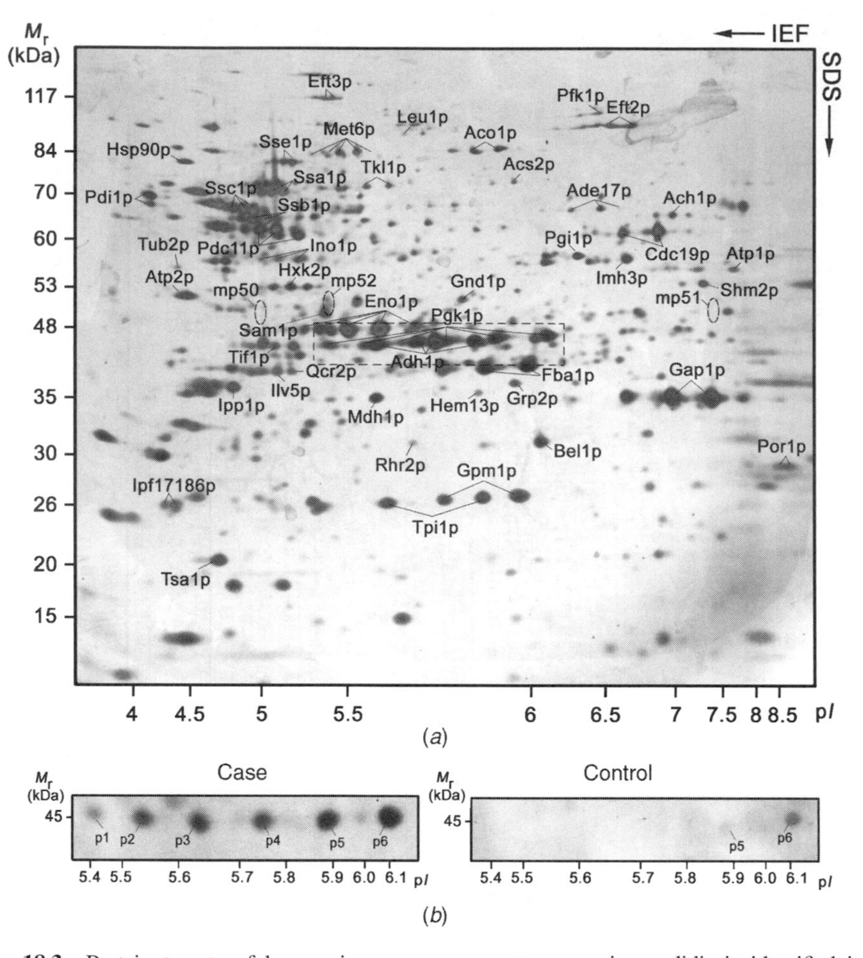


**Figure 18.3** Protein targets of human immune response to systemic candidiasis identified in the *C. albicans* proteome. (*a*) Silver-stained 2DE map of soluble *C. albicans* cytoplasmic proteins showing position and identity of *C. albicans* immunogenic proteins detected using systemic candidiasis patients' sera (A. Pitarch et al., unpublished results; [51]). Immunoreactive *C. albicans* protein spots were identified by MALDI-TOF MS and/or nano ESI ion trap MS. Spot names refer to those in the Appendix. The relative 2D position of three rodlike protein spots (mp50, mp51, and mp52) immunostained on 2D blots but not visualized in silver-stained 2DE gels is indicated with dashed circles. These data are available at the COMPLUYEAST-2DPAGE database (http://babbage.csc.ucm.es/2d/2d.html). The dashed rectangle depicts the close-up section of the 2D immunoblots illustrated in (*b*). (*b*) Expanded sections of representative 2D Western blotting patterns displaying a different antibody response to posttranslationally modified forms of the *C. albicans* Pgk1p (phosphoglycerate kinase 1; p1–p6) in patients with (left panel) and without (right panel) systemic candidiasis [51]. No reactivity was detected against the most acidic isoforms of Pgk1p (labeled as p1–p4) using control sera, suggesting that posttranslational modifications of Pgk1p may mediate variations in the epitopic specificity of anti-Pgk1p antibodies elicited during systemic candidiasis with regard to those mounted during *C. albicans* comensalism or colonization.

fermentative enzymes (i.e., alcohol dehydrogenase Adh1p and pyruvate decarboxylase Pdc11p [51]) were found to be present in subjects with no evidence of candidiasis or healthy individuals. Further studies using individual sera from a sizable group of both systemic candidiasis patients and noncandidiasis controls revealed the occurrence of more nonspecific reactions for other abundant *C. albicans* antigens (A. Pitarch et al., unpublished results and [54]). The presence of these natural anti-*Candida* antibodies was related to the commensal nature of *C. albicans* and/or cross-reactivity of these ubiquitous, abundant, and phylogenetically conserved proteins with antibodies elicited by other human commensal or infectious agents. Remarkably, some acidic isoforms of Pgk1p and Adh1p proved to be specifically recognized by systemic candidiasis patients' serum specimens (Fig. 18.3*b*) [51]. These outstanding findings led to the proposal that specific epitopes of these acidic isoforms might be of diagnostic value for systemic candidiasis. This finding could plausibly be explained by a differentiation of the host immune response having taken place against these posttranslationally-modified protein species.

Consequently, immunoproteomics may easily uncover antibodies directed against posttranslational modifications of specific immunogenic targets (Fig. 18.3*b*). For instance, several isoforms of an immunodominant 41-kDa protein were recognized by a monoclonal antibody generated against the major antigen identified previously by radioimmunoassay using sera from patients with invasive candidiasis [44]. Furthermore, the anti-*Candida* antibody response seems not to be uniformly directed to all protein species of an antigen [47, 48, 50, 51, 54].

Taken together, the large panel of immunorelevant *C. albicans* antigens identified in these proteomic studies (see the Appendix) [51, 54] represent potential diagnostic markers of systemic candidiasis which could be used to develop serological test kits based on recombinant antigens for screening for these infections (Fig. 18.1). Two of these candidates have recently been produced using a yeast expression system and used to assess the circulating levels of their related antibodies in patients with and without systemic candidiasis using immunoassays [63, 64]. These had good diagnostic and analytical performances.

*Prognostic Markers* Some immunoproteomics-based strategies have also been directed toward the detection and assessment of changes occurring in the serum antibody profiles present in systemic candidiasis patients (or in animal models upon artificial immunization) along the disease evolution to search for immunorelevant *C. albicans* proteins specifically recognized by antibodies produced during the progression of systemic *Candida* infection. Here the object is discovering useful serological markers with prognostic value, monitoring the efficacy of antifungal treatment regimens, and/or predicting patients' outcome [50, 51].

Immune sera from two mouse strains with different susceptibilities to systemic candidiasis obtained on different days postinfection with *C. albicans* and assayed against 2DE-separated cytoplasmic proteins resulted in different serological responses to systemic candidiasis along the course of infection [50]. Whereas elevated anti-enolase (Eno1p) antibody titers appeared to help to extend the survival rate of mice with systemic candidiasis, high levels of antibodies against methionine synthase (Met6p), pyruvate kinase (Pgk1p), and a member of the heat shock protein family (Ssb1p) seemed to be related to lethal *C. albicans* infections (Fig. 18.4). In view of this observation the measurement of the serum levels of these specific anti-*Candida* antibodies might be of prognostic value for systemic candidiasis.

A profiling of *C. albicans* immunogenic proteins that elicits an antibody response during the evolution of systemic candidiasis in patients with underlying hematological malignancies was undertaken using serum specimens collected at the time of diagnosis (before treatment) and at one month after the fungal infection had started (when all patients had already received antifungal therapy) [51]. Differences observed in the *C. albicans* antigen recognition patterns correlated with infection progression and efficacy of antifungal treatment. This strategy led to the identification of a large panel of *C. albicans* antigens, including some glycolytic and fermentative proteins, heat-shock proteins (Hsp90p and members of the Hsp70p family),

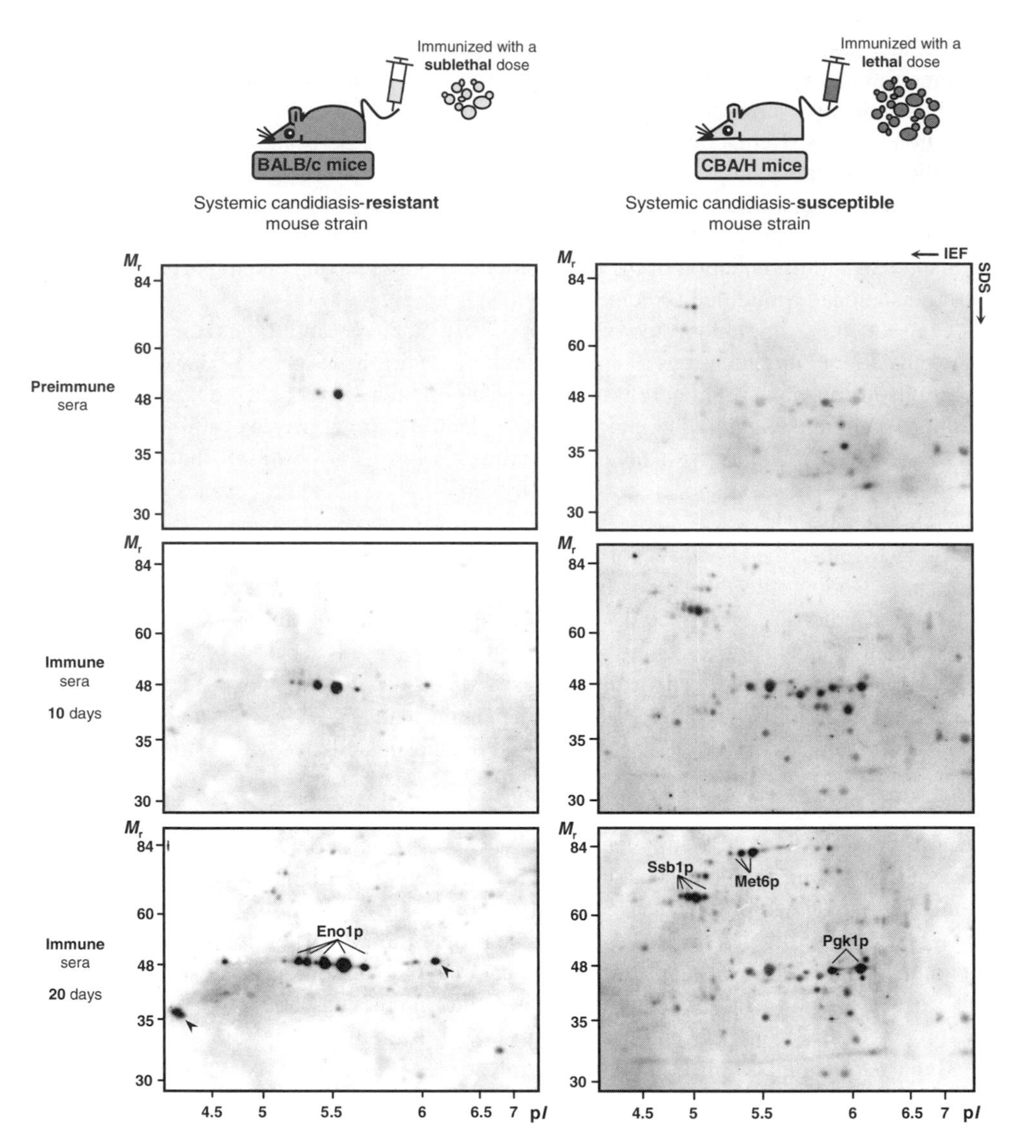


**Figure 18.4** Immunoproteomic analysis of serological response to systemic candidiasis along course of infection in murine model [50]. *Candida albicans* protoplast lysates were separated by 2DE and analyzed by Western blotting with (i) preimmune sera generated from uninfected mice and (ii) immune sera obtained at 10 and 20 days postinfection from two mouse strains with low and moderate susceptibilities to candidiasis (i.e., BALB/c and CBA/H mice) infected with sublethal and lethal doses, respectively. Relying both on the mouse strain and on the course of the systemic *C. albicans* infection, different 2D *C. albicans* antigen recognition profiles were displayed. The main antibody response mounted by BALB/c mice immunized with a sublethal dose was against Eno1p (left panel). Anti-Eno1p antibodies might consequently exhibit a potential immunoprotective effect. In contrast, high levels of antibodies directed against a large number of proteins, such as heat-shock and metabolic proteins, were present in CBA/H mice immunized with a lethal dose (right panel). Proteins showing a significant increase in reactivity associated with infection progression (i.e., potential prognostic markers) are labeled or indicated with arrowheads. Spot names refer to those in the Appendix. See Figure 18.3 for the identity of the remaining protein spots, not labeled for clarity.

methionine synthase, and a 52-kDa mannoprotein, among others, that could be useful for predicting the outcome of patients undergoing systemic candidiasis. The clearance of fungal infection as a result of the antifungal therapy administered, and thus a good prognosis for these patients, might particularly be associated with the rise of high serum anti-Eno1p antibody levels, falling or even the disappearance of anti-Met6p and anti-Pgk1p antibody titers, and/or the maintenance or development of certain antibodies related to putative protective capacity (i.e., anti-Hsp90p, anti-Tpi1p, and antimannan antibodies), to name but a few. A recent immunoproteomic study has revealed that antibodies directed against the *C. albicans* cell wall-associated form of Eno1p are an independent prognostic marker for systemic candidiasis [54].

*Vaccine Candidates* Last, but not least, some immunoproteomics-based approaches have also been undertaken to screen the profile of anti–*C. albicans* antibodies of IgG2a isotype (related to protection) in sera from immunized protected animals. Since some types of antibody-mediated immune responses may play an essential role in anti–*C. albicans* protection (see above), a comprehensive and integrated view of all those *C. albicans* antigens that induce protective antibodies could be useful in identifing potential vaccine candidates. Such an approach would provide a basis for further design of prophylactic or therapeutic strategies for systemic candidiasis.

Immunization of mice with a low virulent *C. albicans* mutant strain using several doses and regimens of inoculation led to the generation of different immune sera that were classified into protective and nonprotective according to median survival time (MST) of the immunized mice [52]. Different antibody profiles were defined using an immunoproteomics-based approach. The results suggested that a protective serum could be composed of IgG2a antibodies against at least seven cytoplasmic proteins, including high levels of antienolase antibodies. Nonprotective serum might have low levels of IgG2a antibodies against enolase and relying on the MST of the immunized mice, high levels of non-IgG2a antibodies against methionine synthase (see the Appendix).

Further studies using a low virulent *C. albicans* mutant strain as live vaccine in a murine model of systemic candidiasis led to the identification of nine antigens that induced protective IgG2a antibodies in vaccinated animals, two of which were novel protective antigens (see the Appendix) [53].

Immunoproteomic and bioinformatic analyses of the serological response to candida cell wall immunome or immune proteome in patients with systemic candidiasis have recently suggested that serum antibodies against *C. albicans* glucan β-1,3-glucosidase (Bgl2p) and the wall-associated form of endose (Eno1p) confer protection against systemic candidiasis [54]. These findings promoted the proposal that these antibodies and/or their related antigens could serve as the basis to design new immunotherapy- and/or vaccine-based strategies, respectively, for prevention and/or treatment of these infections.

## 18.3 ANTIFUNGAL DRUGS: LIMITED THERAPEUTIC ARSENAL ASSOCIATED WITH POSSIBLE RESISTANCE PROBLEMS

Antifungal treatments against candidiasis show certain limitations because of the relatively few active agents available, undesirable side effects, fungistatic character of some agents that hampers therapy in immunocompromised patients, appearance of refractory fungal species, and development of resistance [29, 68, 69]. This clinical setting has heightened the need to design fungicidal, nontoxic, and new antifungal agents with novel mechanisms of action, based in part on the knowledge of *C. albicans* biology and virulence.

It has also stimulated today's investigators to embark on proteomic and genomic approaches with the goal of detecting further protein targets and providing further insight into the modes of action of and resistance mechanisms to antifungal agents in clinical or preclinical use with a view of extending their efficacy [70–73]. This section will concentrate on the key contributions of proteomics to study the mechanisms of action of and resistance to antifungals that are currently in force or passing through the final development stages of clinical trials.

### 18.3.1 Modes of Action of Current and Emerging Antifungal Drugs

***Antifungal Targets and Agents Available for Treating Candidiasis*** Taking into consideration that *C. albicans* is a eukaryotic organism (like human beings), it is comprehensible that this fungus has few and far-between specific targets for antifungal drugs that are not shared by mammalian cells, that is, targets for drugs of selective antifungal action. In fact, most antifungal drugs in common usage (see below) target ergosterol, the major sterol of the fungal plasma membrane, in some way, since this is absent from mammalian cells, whose plasma membranes contain cholesterol instead, and is also essential for fungal growth.

The current antifungal agents (i) interact with ergosterol in fungal plasma membranes and impair membrane integrity (polyenes); (ii) inhibit ergosterol biosynthesis (azoles, morpholines, and allylamines); (iii) interact with microtubules (griseofulvin); (iv) inhibit DNA and RNA biosynthesis (flucytosine or fluorocytosine); or (v) block β-1,3-glucan biosynthesis (echinocandins, such as caspofungin or micafungin) (Fig. 18.5). At present, polyenes (amphotericin B), triazoles (fluconazole, ketoconazole, itraconazole, and voriconazole), and pyrimidine analogues (flucytosine) are used to treat life-threatening candidiasis. Remarkably, caspofungin, licensed in the United States in 2001 for aspergillosis therapy, was approved by the U.S. Food and Drug Administration (FDA) in 2003 for the treatment of candidemia, esophageal candidiasis, and invasive *Candida* infections unresponsive to other standard antifungal drugs [74–76]. Although caspofungin is now licensed for clinical use both in the United States and in nearly all Europe, and micafungin in Japan, other echinocandins, however, are still being evaluated in clinical research.

The antifungal drugs in development with novel modes of action act at the level of (i) the cell wall composition, either inhibiting biosynthesis of chitin (nikkomycin Z and protoberberines) or binding to mannan (pradimicins); (ii) the plasma membrane [efflux pump inhibitors, cationic antimicrobial peptides, aureobasidins, proton adenosine triphosphatase (ATPase) inhibitors, etc.]; or (iii) protein biosynthesis (sordarins) (Fig. 18.5). The reader is directed to outstanding reviews detailing modes of action of antifungals [29, 69, 77–83].

***Evaluation of Action Mechanisms of Antifungal Agents by Proteomics*** Proteomic analyses of *C. albicans* cells treated independently with four antifungal drugs, that is, two echinocandin-class compounds (mulundocandin and one of its derivatives) and two triazole antifungals (fluconazole and itraconazole), followed by clustering of the results supported the hypothesis that antifungal agents with a common mechanism of action resulted in similar effects at the proteome level, whereas unrelated antifungal drugs effected different proteome changes [70]. These findings demonstrated how a simple proteomic approach could become a useful tool for classifying antifungal agents according to their mode of action as well as for studying drugs of unknown mechanisms of action.

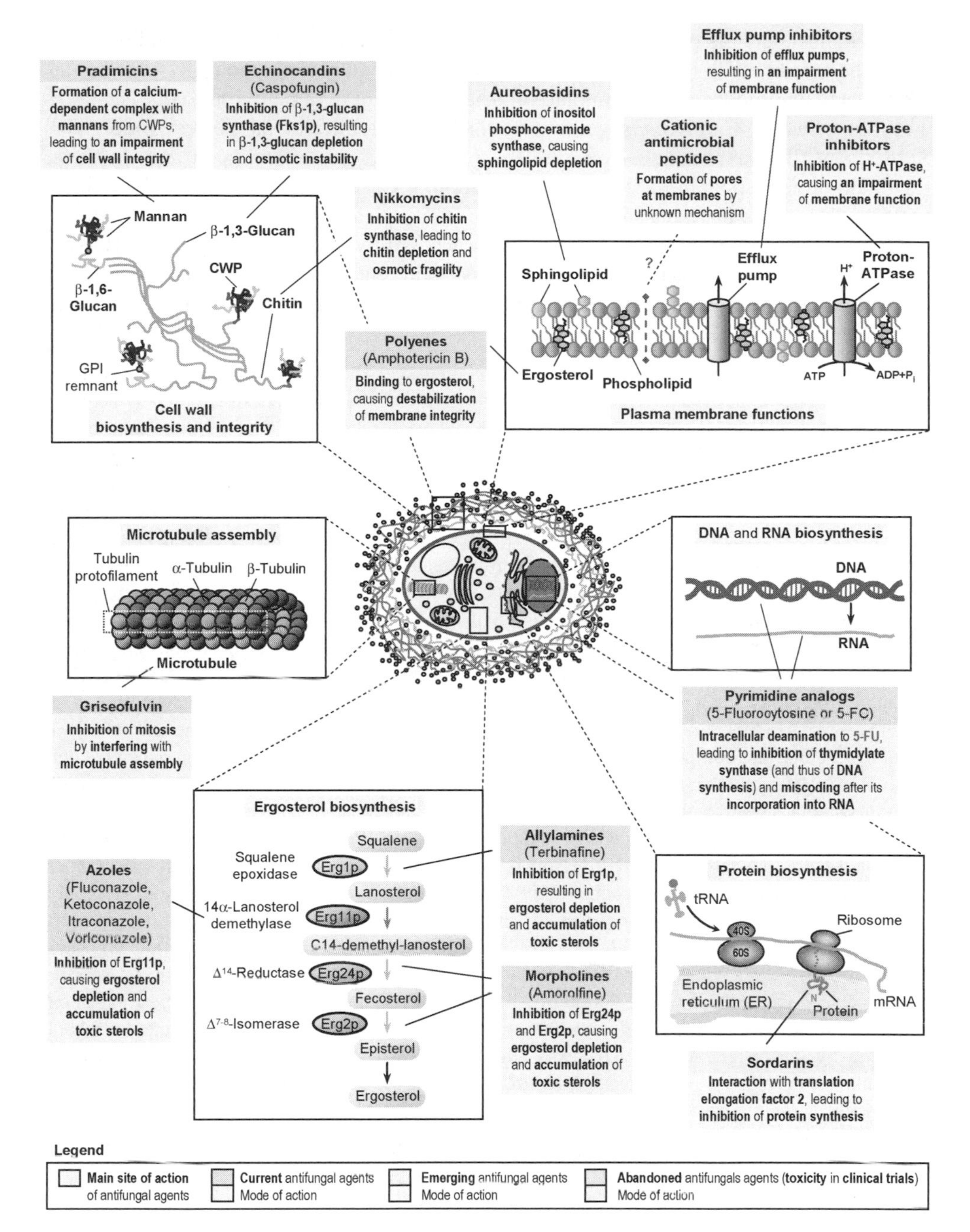


**Figure 18.5** Main targets of antifungal agents for treatment of candidiasis. A schematic overview of the sites of action of current and emerging antifungal agents active against candidiasis is shown. Details for each site are also depicted. The most relevant antifungal compounds and their modes of action are indicated. See the text for further details. CWP denotes cell wall protein and 5-FU 5-fluorouracil (a toxic fluorinated pyrimidine antimetabolite generated from 5-FC by the action of a cytosine deaminase).

Mechanistic details of inhibition of β-1,3-glucan biosynthesis by echinocandins, which target the membrane β-1,3-glucan synthase, still continue to be ambiguous [77]. To evaluate the effect of mulundocandin, membrane-associated proteins of dimethyl sulfoxide (DMSO)– and drug–treated *C. albicans* cells were isolated by differential centrifugation [70]. Subsequent 2DE experiments revealed the presence of 46 up-regulated and 22 down-regulated protein spots in the microsomal fraction of *C. albicans* cells incubated with this antifungal agent. Thirty-one different proteins (27 up- and 4 down-regulated proteins), representing a total of 46 protein spots, were successfully identified by MALDI-TOF MS and/or nanoESI-MS/MS. Many of these participated in central cellular metabolism and stress response, thereby reflecting the possible metabolic pathways involved (see the Appendix).

**Figure 18.6** Effect of echinocandins on *C. albicans* cells. (*a*) The β-1,3-glucan synthase complex of *C. albicans*. β-1,3-Glucan (unique and essential to fungi) is the most abundant structural polymer of the *C. albicans* cell wall and is responsible for the shape and mechanical strength of the wall. This polymer is synthesized at the plasma membrane by the β-1,3-glucan synthase complex, which uses UDP-glucose as a substrate and is composed of (i) a soluble regulatory subunit, the GTPase Rho1p, and (ii) a plasma-membrane-bound catalytic subunit, consisting at least of Fks1p, given that Fks2p, present in the *S. cerevisiae* counterpart, appears not to be expressed in growing *C. albicans* cells [111–114]. The generated β-1,3-glucan chains are extruded from the plasma membrane to the cell wall where they are covalently linked to other cell wall components. Rho1p activates not only Fks1p but also the protein kinase C (Pkc1p)–mediated signal transduction pathway (in blue) that is involved in maintenance of cellular integrity [115]. (*b*) Proposed model of effect of mulundocandin, an echinocandin derivative, on *C. albicans* cells [70]. Echinocandins are fungicidal against most *Candida* species and target Fks1p, an essential protein in *C. albicans*, inhibiting noncompetitively the synthesis of β-1,3-glucan [74, 113]. This gives rise to a loss of mechanical strength of the cell wall and an increase in osmotic instability of *C. albicans* cells, thus compromising the maintenance of cellular integrity and leading to their eventual destruction. The high-osmolarity glycerol (HOG) pathway (in amber) is activated in response to this external osmotic stress, resulting in a rise in the production and intracellular accumulation of glycerol as a compatible osmolyte to avoid cell dehydration [115, 117]. Proteins involved in stress response, associated with both cell surface (e.g., Hsp90p, Eno1p, and Fba1p) and the HOG pathway (Gpp2p) [115, 116], are up regulated as a compensatory reaction in response to this osmotic lysis-susceptible wall [70]. Echinocandins have two secondary effects in *C. albicans*: (i) a decrease in the ergosterol content of *C. albicans* plasma membrane [70,118] by repressing the biosynthesis of certain ergosterol precursors, such as acetylCoA (via down regulation of Acs2p and up regulation of Cit1p) or isopentyl pyrophosphate (through down regulation of Mvd1p) [70]; (ii) an increased chitin deposition in cell walls as a result of a compensatory mechanism for the depletion of β-1,3-glucan content, probably mediated by the cellular integrity pathway (in blue) [118–120], as well as a potential enhanced crosslinking of CWPs via β-1,6-glucan to chitin, mimicking the effect described in *S. cerevisiae* [121]. This model is mainly based on proteomic and biochemical analyses [70, 118]. Question mark after protein name indicates that the functional homologue of the corresponding *S. cerevisiae* gene placed in the same position on the pathway has been identified in the *C. albicans* genome but it has not yet been located experimentally on this pathway. Msn2p and Msn4p-like transcription factors, present in the HOG pathway in *S. cerevisiae*, seem not to participate in the osmotic stress response in *C. albicans* (question mark on amber circle) [122]. Abbreviations: CWP, cell wall protein; HMG, hydroxymethylglutaryl; DHAP, dihydroxyacetone phosphate; DHA, dihydroxyacetone; G3P, glyceraldehyde 3-phosphate; PEP, phosphoenolpyruvate; Gpd1p, glycerol-3-phosphate dehydrogenase; Gpp2p, glycerol phosphate phosphatase; Fba1p, fructose-biphosphate aldolase; Eno1p, enolase; Cit1p, citrate synthase; Acs2p, acetyl-CoA synthetase; Mvd1p, mevalonate pyrophosphate decarboxylasc; Chs3p, chitin synthase; Sho1p, osmosensor; Cdc42p, GTPase; Sln1p, Nik1p, and Chk1p, histidine kinases; Ypd1p, phosphorelay protein; Cst20p and Ssk1p, protein kinases; Bck1p, Ste11p, and Ssk2p, mitogen-activated protein (MAP) kinase kinase kinase (MAPKKK); Mkk2p and Pbs2p, MAP kinase kinase (MAPKK); Mkc1p and Hog1p, MAP kinase (MAPK); Rlm1p, transcription factor.

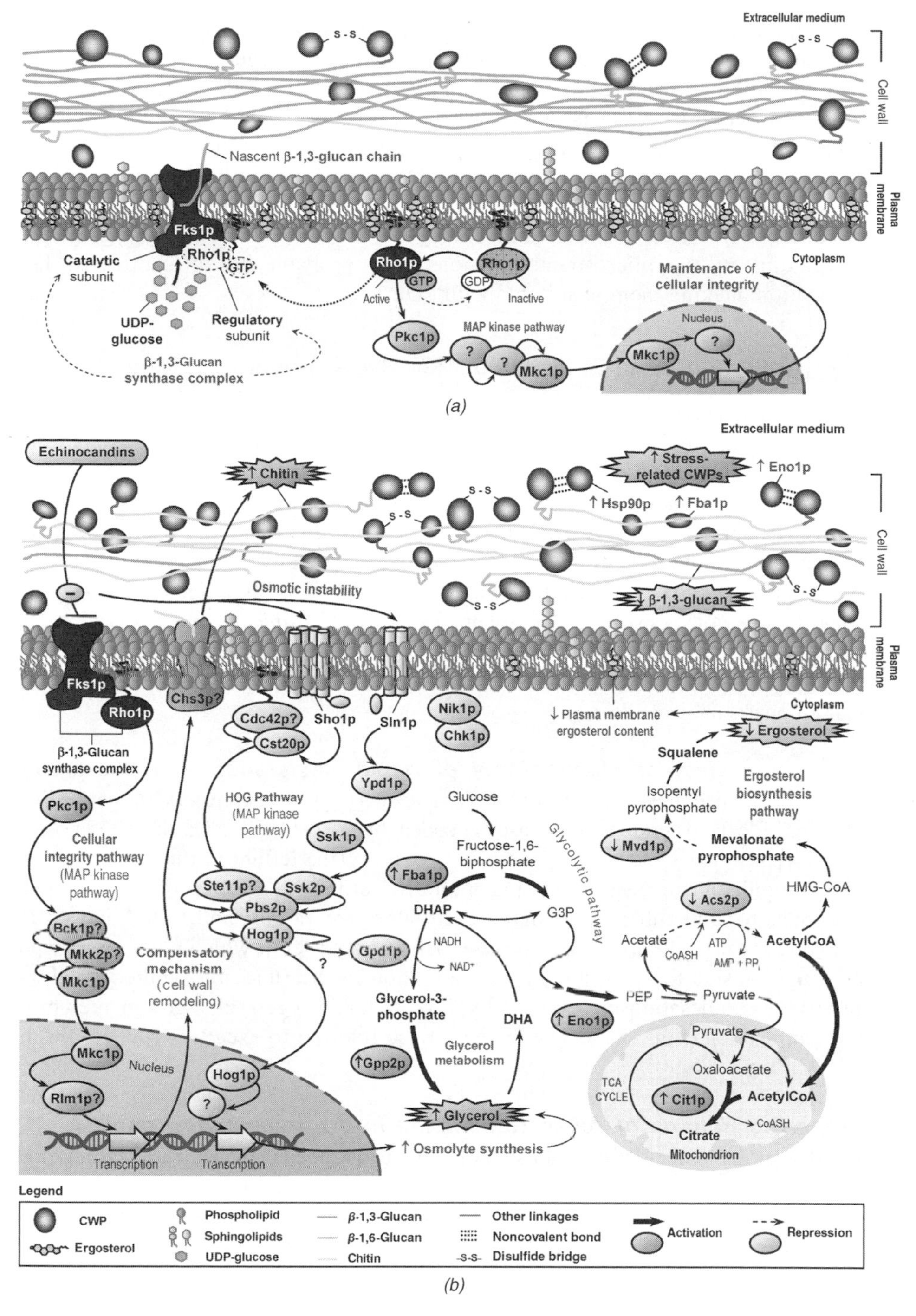


Intriguingly, of the 27 up-regulated proteins identified, 3 were osmotic stress-associated proteins that might be involved in a compensatory mechanism in response to inhibition of β-1,3-glucan biosynthesis. This is supported by the fact that no production of this major constituent of the *C. albicans* cell wall (see Chapter 17) would give rise to an osmotically fragile cell wall (Fig. 18.6). Furthermore, 6 overexpressed proteins, 2 of which were osmotic stress-associated proteins, corresponded to CWPs that could be related to stress at the cell surface (see the Appendix). Two of the proteins downregulated by echinocandins and upregulated by azole agents were implicated in ergosterol biosynthesis or in the generation of acetyl-CoA (an essential metabolite for this biosynthetic pathway), which might account for the decrease in ergosterol content displayed in the *C. albicans* plasma membrane after echinocandin treatment. Despite these attractive findings and identification of 1 up-regulated mitochondrial membrane protein, no integral plasma membrane proteins were characterized among the proteins differentially regulated by echinocandins in the microsomal fractions, due probably to difficulties in their solubilization, making them unsuitable with 2DE.

### 18.3.2 Antifungal Drug Resistance: "Mr. Hyde" Resurfaces Again!

The recent emergence of clinical *C. albicans* isolates resistant to antifungals, due in part to their large-scale empirical prophylactic and therapeutic use in immunosuppressed patients, has brought on a rising frequency of therapy failures as well as changes in the prevalence of *Candida* species causing disease. Antifungal resistance has been reported for polyenes, flucytosine, and particularly azoles [28, 29, 84]. This phenomenon has become an important problem in certain individuals, especially HIV-patients undergoing oropharyngeal candidiasis (OPC), the most frequent opportunistic infection among these patients. Fluconazole-resistant *Candida* spp., especially *C. albicans* and *C. glabrata*, are frequently isolated in this population [69, 85, 86]. Accordingly, active investigation is being directed to understanding molecular mechanisms of resistance to antifungal agents, with a special emphasis on azole drugs.

***Underlying Molecular Mechanisms of Azole Resistance*** It is thought that *Candida* species become resistant to azoles as a result of (i) overexpression of their cellular target (i.e., 14α-lanosterol demethylase, encoded by *ERG11/CYP51*); (ii) alterations in Erg11p/Cyp51p that yield reduced affinity to azoles; (iii) failure in their accumulation due to upregulation of genes encoding multidrug efflux transporters [such as *MDR1*, which codes for a member of the major facilitator superfamily, and *CDR1* and *CDR2*, which code for the ATP-binding cassette (ABC) transporters], which actively transport azole drugs outside the cell, and/or (iv) alteration in the sterol biosynthetic pathway (defective Δ-5,6-desaturase, encoded by *ERG3*) that triggers changes in membrane sterol composition (Fig. 18.7). The reader is also referred to excellent reviews on this area [29, 69, 84, 87–90].

***Deciphering Enigma of Azole Resistance by Proteomic Approaches*** This knowledge of the molecular mechanisms of azole resistance has provided a starting point for conducting "expression proteomics"-based studies. Susceptible and resistant clinical *Candida* isolates to azole drugs were compared with a view to gaining further insight into the alterations that take place at the higher level during the evolution of resistance as well as into the identification of key control and effector proteins [71–73].

*Up Regulation of Gene That Encodes Target Enzyme* The increase in the content of the target enzyme molecules for azoles (i.e., 14α-lanosterol demethylase or Erg11p/Cyp51p) is considered as a possible origin of azole resistance (see above; Fig. 18.7). In a clinical isolate of *C. glabrata*, fluconazole resistance was associated with *ERG11/CYP51* gene

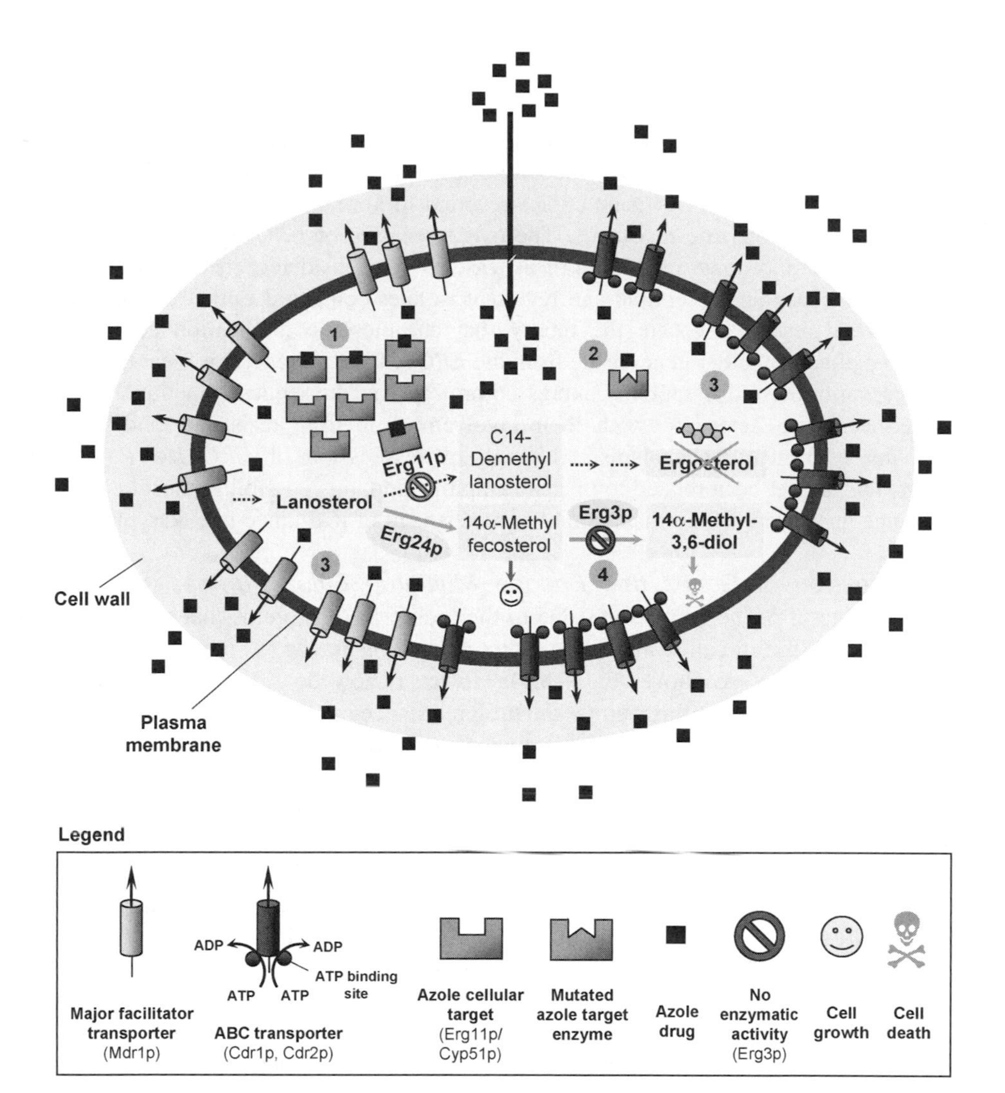


**Figure 18.7** Molecular mechanisms of azole resistance in *C. albicans*. In a susceptible isolate, azole drugs are transported into the cell, possibly by passive diffusion, and then bind to their cellular target (cytochrome P450 14α-lanosterol demethylase or Erg11p/Cyp51p, located in the membrane of the endoplasmic reticulum), inhibiting its activity, that is, the transformation of lanosterol into ergosterol precursors. This gives rise to depletion of ergosterol (a fungal-specific component of plasma membrane) and accumulation of 14α-methyl-3,6-diol (a toxic metabolite for the cell) by action of the sterol $\Delta^{5,6}$-desaturase or Erg3p (green arrows). Nevertheless, *C. albicans* cells, subject to the selective pressure of azoles, have evolved at least four resistance mechanisms. These include (i) an increase in intracellular content of the azole target enzyme by up regulation of *ERG11/CYP51* (point 1); (ii) a reduction in the affinity of azoles to their cellular target by mutation(s) in *ERG11/CYP51* (point 2); (iii) a rise in the active efflux of azoles from the cell, and thus a decrease in their intracellular accumulation, by up regulation of membrane multidrug transporter genes (*MDR1* and/or *CDR1/2*) (point 3); and/or (iv) an alteration of other enzymes in the ergosterol biosynthetic pathway, such as inactivation of $\Delta^{5,6}$-desaturase by a loss-of-function mutation in *ERG3*, which avoids the production of toxic sterols in the presence of azoles and results in changes in sterol composition of the plasma membrane (point 4). Whereas Cdr1/2p, ATP-binding cassette (ABC) transporters (red tubes) with two ABC domains (red circles), are effective against a wide spectrum of azole drugs, Mdr1p, the major facilitator transporter (yellow tubes) appears to specifically reduce the accumulation of fluconazole. For clarity, the plasma membrane is shown as a blue circumference and the cell wall as a pink ring (see Chapter 17 for further details).

amplification linked to chromosome duplication. This correlated with increased mRNA and enzyme levels [71]. After successive subcultures of this isolate in fluconazole-free medium, this increase in the number of gene copies reverted to normal, but the resulting revertant retained partial fluconazole resistance. The overexpression of *ERG11/CYP51* induced up regulation of 25 and down regulation of at least 76 genes as determined by 2DE protein profiles of fluconazole-susceptible and revertant isolates compared to that of the resistant isolate. These results advocate the theory that chromosome duplication could trigger pleiotropic effects on other genes aside from the *ERG11/CYP51* gene. It is clear that further identification of these differentially expressed proteins and subsequent functional analyses will be needed to determine both their involvement in drug resistance and potential underlying resistance mechanisms. Although overexpression of *ERG11/CYP51* in *C. albicans* has not yet been related to its gene amplification, this possibility must be borne in mind since gene amplification is a common mechanism of resistance in eukaryotic cells.

*Overexpression of Genes that Encode Multidrug Efflux Pumps* A decreased intracellular level of the azole drugs due to the induction of proteins that transport them outside the cell (the so-called multidrug efflux pumps), is maybe the most common cause of azole resistance (see above; Fig. 18.7). However, how do the regulatory networks control the expression of efflux pumps and drug resistance in *C. albicans*? What additional alterations occur in the clinical *C. albicans* isolates that become resistant to fluconazole due to the constitutive activation of the major facilitator (*MDR1*) gene?

To find an answer to these questions, an exciting proteomic approach was initiated using three matched pairs of fluconazole-susceptible and resistant clinical *C. albicans* isolates, in which fluconazole resistance was associated with overexpression of *MDR1* or *CDR1/2* (efflux transporter genes) [72]. MALDI-TOF MS analysis led to the identification of six putative aldo-keto reductases of unknown function (see the Appendix) specifically up regulated when *MDR* but not *CDR1/2* was overexpressed. It brought to light the proposal that the expression of the major facilitator (Mdr1p) and ABC transporters (Cdr1/2p) in *C. albicans* and, therefore, that of diverse multidrug efflux pumps could be controlled by different regulatory networks. More direct functional assays of these findings by using forced gene overexpression and gene deletion approaches demonstrated that extra alterations not associated with the fluconazole-resistant phenotype could also take place. It enabled the suggestion that these *C. albicans* regulatory networks appeared to control not only expression of the specific efflux pumps but also that of genes implicated in physiological functions not related to drug resistance. These putative aldo-keto reductases might generate metabolite(s) that are transported outside the cell through the major facilitator transporter Mdr1p (Fig. 18.8).

*Accumulation of Several Alterations* The high rates of resistance to azoles reported in some clinical *C. albicans* isolates seem to be the result not of a single alteration (described above and in Fig. 18.7) but of multiple mechanisms. The accumulation of several alterations on account of continuous selective pressure of these agents, that is, long periods in their presence, somehow supports the gradual, stepwise increase in azole resistance in *C. albicans* [29, 69, 91].

The overexpression both of efflux pump genes (*MDR1* and *CDR1/2*) and of the target enzyme gene (*ERG11/CYP51*), combined with the loss of allelic variation and acquisition of a point mutation in *ERG11/CYP51,* has recently been shown to be responsible for the overall fluconazole resistance phenotype of a clinical *C. albicans* isolate [92, 93].

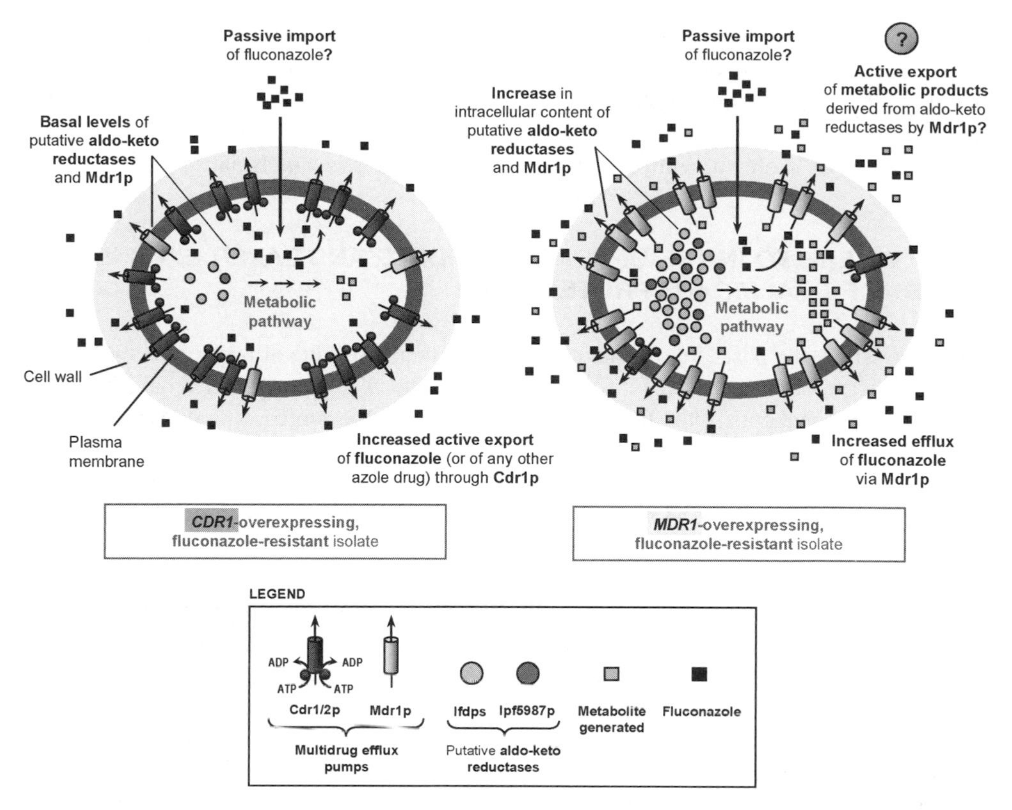


**Figure 18.8** Novel face for multidrug efflux pump-mediated fluconazole resistance in *C. albicans*. Increased levels of certain membrane proteins that actively export azoles out of the cell, such as the major facilitator Mdr1p (specific for fluconazoles) and ATP-binding cassette (ABC) transporters Cdr1/2p (effective against many azoles), lead to a reduction in intracellular content of azoles (probably imported into the cell by passive diffusion) and the appearance of azole resistance. Left panel: Fluconazole-resistant isolate that overexpresses *CDR1/2* but not *MDR1*. Right panel: Isolate in which fluconazole resistance is correlated with constitutive activation of *MDR1* but not *CDR1/2*. In resistant isolates, genes encoding putative aldo-keto reductases (Ifd1p, Ifd2p, Ifd4p, Ifd5p, Ifd6p, and Ipf5987p) not related to fluconazole resistance are coordinately up regulated with *MDR1* (right panel) but not with *CDR1/2* (left panel), indicating that different regulatory networks appear to control *MDR1* and *CDR1/2* expression. These aldo-keto reductases could speculatively provide some metabolic product(s) that, like fluconazole drugs, are removed out of the cell through Mdr1p. The discovery of analogous sequences in the promoters of their corresponding genes might reveal the potential site to which the transcription activator associated with the control of the *MDR1* regulatory network is bound. This hypothetical model is based on proteomic analysis [72]. For clarity, the plasma membrane is depicted as a blue circumference and the cell wall as a pink ring (refer to Chapter 17 for further details).

Comparative proteome analysis between this resistant isolate and a susceptible one revealed the existence of 11 up- and 4 down-regulated proteins in association with fluconazole resistance (see Appendix) [73]. Up-regulated proteins included five reductases (such as some of the aldo-keto reductases of unknown function mentioned above), four carbohydrate metabolism enzymes, and two proteins related directly or indirectly to ergosterol biosynthesis. However, the membrane-associated multidrug transports (Mdr1p and Cdr1/2p) were not identified among proteins that displayed a change in their expression levels, probably because of their low copy numbers or low solubility in whole-cell extracts. This emphasizes the importance of further studies focused on the proteomic analysis not only of entire cells but also of organelle membranes from multiple matched sets of azole-susceptible and azole-resistant *C. albicans* isolates.

## 18.4 WHERE TO IN THE UPCOMING "MIDDLE AND LATER EXPANSION AGES" OF *C. ALBICANS* PROTEOMICS?

With hindsight, the past 25 years have borne witness to the value of proteomic approaches with respect to *C. albicans*. The different proteomic analyses accomplished have helped to unravel novel concepts in the *C. albicans* cell wall biology, proteasome phosphorylation, messenger ribonucleic acid (mRNA) synthesis, response to nitrogen starvation, expression of a gene in an ectopic gene loci, septin cytoskeleton, yeast-to-hypha transition, phenotypic switching, adhesion to host plasminogen, ubiquitination of surface receptors, cell surface hydrophobicity, extracellular proteolytic activity, antigenicity, immune protection, modes of action of certain antifungal agents, and resistance mechanisms to azoles as well as the discovery of promising diagnostic and prognostic markers for screening systemic candidiasis. In addition, potential vaccine candidates and antifungal drug targets for treating these infections have been elucidated.

Much of this fruitful research has undoubtedly been fueled in part by the progressive completion of *C. albicans* genome sequencing and the recent advances in protein separation and identification techniques. Today's investigators, with the whole-genome sequence data of *C. albicans* in their hands, have already entered into the "Expansion Ages" of *C. albicans* proteomics with the key tools for going as far as they can, in an attempt to learn how to better control candidiasis (by its diagnosis, treatment, and/or prevention). However, this expansion is still in its early stages.

What new areas of investigation into *C. albicans* could be targeted by proteomic approaches in the short-term and long-term future or, more accurately, in the next "Middle and Later Expansion Ages"? With what new challenges could proteomic scientists working on the *C. albicans* field be confronted in this fourth era of *C. albicans* proteomics? In short, where to in the coming years? It is plausible that future proteomic studies will focus on the analysis of diverse cell organelles, macromolecular structures, or protein complexes (e.g., cell wall, membrane systems, lipid rafts, mitochondria, nucleus, cytoskeleton, endocytic apparatus, proteasome, phagosome) as opposed to the entire cells in different conditions, with the purpose of gaining a better knowledge of the complex cellular processes that take place in *C. albicans* during infection. On the other hand, one does not have to be an oracle to sustain the conviction that proteomic research in the foreseeable future will be in part aimed at exploring changes in protein expression patterns under different sets of conditions, such as in the presence of oxidizing agents, high temperature, hyperosmolarity, nutrient limitation, antifungal drugs, to name just a few, using quantitative proteomic technologies—for example, fluorescence 2D differential gel electrophoresis (2D DIGE) [94] or peptide liquid chromatography (LC) MS/MS coupled with stable-isotope protein-labeling methods, such as isotope-coded affinity tagging (ICAT) or stable-isotope labeling with amino acids in cell culture (SILAC) [95–98].

There is no doubt that multifaceted proteomic options could be tackled to investigate *C. albicans* and candidiasis. Tandem affinity purification (TAP) or multidimensional LC combined with MS and protein chips, among others [97, 99–105], will serve as a backbone to study protein complexes and protein–protein interactions, respectively, in several items, such as *Candida*–macrophage interaction, host-cell-mediated response, signaling pathways, apoptosis, mechanisms of chlamydospore formation or antifungal drug tolerance, mating, and biofilm development. Posttranslational modifications (PTMs) in *C. albicans* proteins, especially phosphorylation, due to its relevance in the regulation of enzymatic activity, complex formation, and degradation of proteins, will presumably be characterized in next to no time, making use, for example, of immobilized metal affinity chromatography (IMAC) or multidimensional chromatography and MS/MS [106, 107].

It is unsurprising that proteomic analyses, albeit still in their infancy, as applied to other non-*albicans Candida* species of industrial or clinical interest, such as *Candida magnoliae* [108, 109] and *C. glabrata* [71, 110], respectively, also undergo a stimulating progress in the near future.

Regardless of our rather limited view into the crystal ball, the future for research into the *C. albicans* proteome is exciting. Hopefully, all key issues of *C. albicans* will soon feel the impact of the proteomic discipline, which will become an indispensable tool in *C. albicans* research to elucidate new models not predictable from genomic studies.

## ACKNOWLEDGMENTS

We thank the Merck, Sharp & Dohme Special Chair in Genomics and Proteomics, Comunidad Autónoma de Madrid (Strategic Groups; CPGE 1010/2000), Comisión Interministerial de Ciencia y Tecnología (BIO-2003-00030), and Fundación Ramón Areces for financial support of our laboratory.

## REFERENCES

1. Niimi, M., Cannon, R. D., and Monk, B. C., *Electrophoresis* 1999, **20**, 2299–2308.
2. Pitarch, A., Sanchez, M., Nombela, C., and Gil, C., *J. Chromatogr. B Analyt. Technol. Biomed. Life Sci.* 2003, **787**, 101–128.
3. Pitarch, A., Sanchez, M., Nombela, C., and Gil, C., *J. Chromatogr. B Analyt. Technol. Biomed. Life Sci.* 2003, **787**, 129–148.
4. Rupp, S., *Curr. Opin. Microbiol.* 2004, **7**, 330–335.
5. Harbarth, S., Rued, C., Francioli, P., Widmer, A., and Pittet, D., *Schweiz Med. Wochenschr.* 1999, **23**, 1521–1528.
6. Pfaller, M. A., Jones, R. N., Doern, G. V., Sader, H. S., et al., *J. Clin. Microbiol.* 1998, **36**, 1886–1889.
7. Vincent, J. L., Bihari, D. J., Suter, P. M., Bruining, H. A., et al., *JAMA* 1995, **274**, 639–644.
8. Odds, F. C., *Candida and Candidiasis*, Bailliere Tindall, London, 1988, pp. 115–230.
9. Garber, G., *Drugs* 2001, **61**, 1–12.
10. Ruhnke, M., in R. A. Calderone (Ed.), *Candida and Candidiasis*, ASM Press, Washington, DC, 2002, pp. 307–326.
11. Kullberg, B. J., and Filler, S. G., in R. A. Calderone (Ed.), *Candida and Candidiasis*, ASM Press, Washington, DC, 2002, pp. 327–340.
12. Filler, S. G., and Kullberg, B. J., in R. A. Calderone (Ed.), *Candida and Candidiasis*, ASM Press, Washington, DC, 2002, pp. 341–348.
13. Fraser, V. J., Jones, M., Dunkel, J., Storfer, S., et al., *Clin. Infect. Dis.* 1992, **15**, 414–421.

14. al Soub, H., and Estinoso, W., *J. Hosp. Infect.* 1997, **35**, 141–147.
15. Patterson, J. E., *Transpl. Infect. Dis.* 1999, **1**, 229–236.
16. Pfaller, M. A., Diekema, D., Jones, R. N., Sader, H. S., et al., *J. Clin. Microbiol.* 2001, **39**, 3254–3259.
17. Leleu, G., Aegerter, P., and Guidet, B., *J. Crit. Care* 2002, **17**, 168–175.
18. Blot, S. I., Hoste, E. A., Vandewoude, K. H., and Colardyn, F. A., *J. Crit. Care* 2003, **18**, 130–131.
19. Edwards, J. E., *N.Engl. J. Med.* 1991, **324**, 1060–1062.
20. Bodey, G., Bueltmann, B., Duguid, W., Gibbs, D., et al., *Eur. J. Clin. Microbiol. Infect. Dis.* 1992, **11**, 99–109.
21. Iwasaki, H., Misaki, H., Nakamura, T., Ueda, T., and Arisawa, M., *Int. J. Hematol.* 2000, **71**, 266–272.
22. Groll, A. H., Shah, P. M., Mentzel, C., Schneider, M., et al., *J. Infect.* 1996, **33**, 23–32.
23. Jones, J. M., *Clin. Microbiol. Rev.* 1990, **3**, 32–45.
24. Pfaller, M. A., *Mycopathologia* 1992, **120**, 65–72.
25. Reiss, E., and Morrison, C. J., *Clin. Microbiol. Rev.* 1993, **6**, 311–323.
26. Ponton, J., Moragues, M. D., and Quindos, G., in R. A. Calderone (Ed.), *Candida and Candidiasis*, ASM Press, Washington, DC, 2002, pp. 395–427.
27. Richardson, M. D., and Carlson, P., in R. A. Calderone (Ed.), *Candida and Candidiasis*, ASM Press, Washington, DC, 2002, pp. 387–394.
28. Kontoyiannis, D. P., and Lewis, R. E., *Lancet* 2002, **359**, 1135–1144.
29. Sanglard, D., and Bille, J., in R. A. Calderone (Ed.), *Candida and Candidiasis*, ASM Press, Washington, DC, 2002, pp. 349–386.
30. Casadevall, A., Cassone, A., Bistoni, F., Cutler, J. E., et al., *Med. Mycol.* 1998, **36**(Suppl. 1), 95–105.
31. Cassone, A., De Bernardis, F., Ausiello, C. M., Gomez, M. J., et al., *Res. Immunol.* 1998, **149**, 289–299.
32. Polonelli, L., Casadevall, A., Han, Y., Bernardis, F., et al., *Med. Mycol.* 2000, **38**(Suppl. 1), 281–292.
33. Romani, L., in R. A. Calderone (Ed.), *Candida and Candidiasis*, ASM Press, Washington, DC, 2002, pp. 223–242.
34. Cole, G. T., Halawa, A. A., and Anaissie, E. J., *Clin. Infect. Dis.* 1996, **22**, S73–88.
35. Cutler, J. E., Granger, B. L., and Han, Y., in R. A. Calderone (Ed.), *Candida and Candidiasis*, ASM Press, Washington, DC, 2002, pp. 243–258.
36. Casadevall, A., *Infect. Immun.* 1995, **63**, 4211–4218.
37. Casadevall, A., Feldmesser, M., and Pirofski, L. A., *Curr. Opin. Microbiol.* 2002, **5**, 386–391.
38. Matthews, R. and Burnie, J., *Trends Microbiol.* 1996, **4**, 354–358.
39. Han, Y., Morrison, R. P., and Cutler, J. E., *Infect. Immun.* 1998, **66**, 5771–5776.
40. Han, Y., Ulrich, M. A., and Cutler, J. E., *J. Infect. Dis.* 1999, **179**, 1477–1484.
41. Matthews, R. C., Burnie, J. P., *Vaccine* 2004, **22**, 865–871.
42. Manning, M. Mitchell, T. G., *Infect. Immun.* 1980, **30**, 484–495.
43. Shen, H. D., Choo, K. B., Lin, W. L., Lin, R. Y., and Han, S. H., *Electrophoresis* 1990, **11**, 878–882.
44. Shen, H. D., Choo, K. B., Yu, K. W., Ling, W. L., et al., *Int. Arch. Allergy Appl. Immunol.* 1991, **96**, 142–148.
45. Morhart, M., Rennie, R., Ziola, B., Bow, E., and Louie, T. J., *J. Clin. Microbiol.* 1994, **32**, 766–776.

46. Sepulveda, P., Lopez-Ribot, J. L., Gozalbo, D., Cervera, A., et al., *Infect. Immun.* 1996, **64**, 4406–4408.
47. Pitarch, A., Pardo, M., Jimenez, A., Pla, J., et al., *Electrophoresis* 1999, **20**, 1001–1010.
48. Barea, P. L., Calvo, E., Rodriguez, J. A., Rementeria, A., et al., *FEMS Immunol. Med. Microbiol.* 1999, **23**, 343–354.
49. Pardo, M., Ward, M., Pitarch, A., Sanchez, M., et al., *Electrophoresis* 2000, **21**, 2651–2659.
50. Pitarch, A., Diez-Orejas, R., Molero, G., Pardo, M., et al., *Proteomics* 2001, **1**, 550–559.
51. Pitarch, A., Abian, J., Carrascal, M., Sanchez, M., et al., *Proteomics* 2004, **4**, 3084–3106.
52. Fernandez-Arenas, E., Molero, G., Nombela, C., Diez-Orejas, R., and Gil, C., *Proteomics* 2004, **4**, 1204–1215.
53. Fernandez-Arenas, E., Molero, G., Nombela, C., Diez-Orejas, R., and Gil, C., *Proteomics* 2004, **4**, 3007–3020.
54. Pitarch, A., Jimenez, A., Nombela, C., and Gil, C., *Mol. Cell Proteomics* 2006, **5**, 79–96.
55. Hanash, S., *Nature* 2003, **422**, 226–232.
56. Le Naour, F., *Proteomics* 2001, **1**, 1295–1302.
57. Krah, A., and Jungblut, P. R., *Methods Mol. Med.* 2004, **94**, 19–32.
58. Haas, G., Karaali, G., Ebermayer, K., Metzger, W. G., et al., *Proteomics* 2002, **2**, 313–324.
59. Marshall, T., and Williams, K. M., *Br. J. Biomed. Sci.* 2002, **59**, 47–64.
60. Seliger, B., and Kellner, R., *Proteomics* 2002, **2**, 1641–1651.
61. Purcell, A. W., and Gorman, J. J., *Mol. Cell Proteomics* 2004, **3**, 193–208.
62. Buckley, H. R., Richardson, M. D., Evans, E. G., and Wheat, L. J., *J. Med. Vet. Mycol.* 1992, **30**(Suppl. 1), 249–260.
63. Pitarch, A., Nombela, C., and Gil, C., *Mol. Cell Proteomics* (submitted).
64. Pitarch, A., Nombela, C., and Gil, C., in preparation.
65. Valdes, I., Pitarch, A., Gil, C., Bermudez, A., et al., *J. Mass Spectrom.* 2000, **35**, 672–682.
66. Martinez, J. P., Gil, M. L., Lopez-Ribot, J. L., and Chaffin, W. L., *Clin. Microbiol. Rev.* 1998, **11**, 121–141.
67. Lopez-Ribot, J. L., Casanova, M., Murgui, A., and Martinez, J. P., *FEMS Immunol. Med. Microbiol.* 2004, **41**, 187–196.
68. Kauffman, C. A., and Carver, P. L., *Drugs* 1997, **53**, 539–549.
69. White, T. C., Marr, K. A., and Bowden, R. A., *Clin. Microbiol. Rev.* 1998, **11**, 382–402.
70. Bruneau, J. M., Maillet, I., Tagat, E., Legrand, R., et al., *Proteomics* 2003, **3**, 325–336.
71. Marichal, P., Vanden Bossche, H., Odds, F. C., Nobels, G., et al., *Antimicrob. Agents Chemother.* 1997, **41**, 2229–2237.
72. Kusch, H., Biswas, K., Schwanfelder, S., Engelmann, S., et al., *Mol. Genet. Genomics* 2004, **271**, 554–565.
73. Hooshdaran, M. Z., Barker, K. S., Hilliard, G. M., Kusch, H., et al., *Antimicrob. Agents Chemother.* 2004, **48**, 2733–2735.
74. Kartsonis, N. A., Nielsen, J., and Douglas, C. M., *Drug Resist. Updat.* 2003, **6**, 197–218.
75. Letscher-Bru, V., and Herbrecht, R., *J. Antimicrob. Chemother.* 2003, **51**, 513–521.
76. Denning, D. W., *Lancet* 2003, **362**, 1142–1151.
77. Odds, F. C., Brown, A. J., and Gow, N. A., *Trends Microbiol.* 2003, **11**, 272–279.
78. Groll, A. H., Piscitelli, S. C., and Walsh, T. J., *Adv. Pharmacol.* 1998, **44**, 343–500.
79. Dodds, E. S., Drew, R. H., and Perfect, J. R., *Pharmacotherapy* 2000, **20**, 1335–1355.
80. Groll, A. H., De Lucca, A. J., and Walsh, T. J., *Trends Microbiol.* 1998, **6**, 117–124.
81. Theis, T., and Stahl, U., *Cell Mol. Life Sci.* 2004, **61**, 437–455.

82. Gupta, A. K., and Tomas, E., *Dermatol. Clin.* 2003, **21**, 565–576.
83. Wingard, J. R., and Leather, H., *Biol Blood Marrow Transplant.* 2004, **10**, 73–90.
84. Sanglard, D., and Odds, F. C., *Lancet Infect. Dis.* 2002, **2**, 73–85.
85. Law, D., Moore, C. B., Wardle, H. M., Ganguli, L. A., et al., *J. Antimicrob. Chemother.* 1994, **34**, 659–668.
86. Rex, J. H., Rinaldi, M. G., and Pfaller, M. A., *Antimicrob. Agents Chemother.* 1995, **39**, 1–8.
87. Ghannoum, M. A., and Rice, L. B., *Clin. Microbiol. Rev.* 1999, **12**, 501–517.
88. Lupetti, A., Danesi, R., Campa, M., Del Tacca, M., and Kelly, S., *Trends Mol. Med.* 2002, **8**, 76–81.
89. Morschhauser, J., *Biochim. Biophys. Acta* 2002, **1587**, 240–248.
90. Casalinuovo, I. A., Di Francesco, P., and Garaci, E., *Eur. Rev. Med. Pharmacol. Sci.* 2004, **8**, 69–77.
91. Perea, S., Lopez-Ribot, J. L., Kirkpatrick, W. R., McAtee, R. K., et al., *Antimicrob. Agents Chemother.* 2001, **45**, 2676–2684.
92. White, T. C., *Antimicrob. Agents Chemother.* 1997, **41**, 1482–1487.
93. White, T. C., *Antimicrob. Agents Chemother.* 1997, **41,** 1488–1494.
94. Tonge, R., Shaw, J., Middleton, B., Rowlinson, R., et al., *Proteomics* 2001, **1**, 377–396.
95. Gygi, S. P., Rist, B., Gerber, S. A., Turecek, F., et al., *Nat. Biotechnol.* 1999, **17**, 994–999.
96. Ong, S. E., Blagoev, B., Kratchmarova, I., Kristensen, D. B., et al., *Mol. Cell Proteomics* 2002, **1**, 376–386.
97. Aebersold, R., and Mann, M., *Nature* 2003, **422**, 198–207.
98. Goshe, M. B., and Smith, R. D., *Curr. Opin. Biotechnol.* 2003, **14**, 101–109.
99. Gavin, A. C., Bosche, M., Krause, R., Grandi, P., et al., *Nature* 2002, **415**, 141–147.
100. Ho, Y., Gruhler, A., Heilbut, A., Bader, G. D., et al., *Nature* 2002, **415**, 180–183.
101. Link, A. J., Eng, J., Schieltz, D. M., Carmack, E., et al., *Nat. Biotechnol.* 1999, **17**, 676–682.
102. Han, D. K., Eng, J., Zhou, H., and Aebersold, R., *Nat. Biotechnol.* 2001, **19**, 946–951.
103. Zhu, H., Bilgin, M., and Bangham, R., Hall, D., et al., *Science* 2001, **293**, 2101–2105.
104. Zhu, H., Bilgin, M., and Snyder, M., *Annu. Rev. Biochem.* 2003, **72**, 783–812.
105. MacBeath, G., *Nat. Genet.* 2002, **32**, 526–532.
106. Gaberc-Porekar, V., and Menart, V., *J. Biochem. Biophys. Methods* 2001, **49**, 335–360.
107. Kalume, D. E., Molina, H., and Pandey, A., *Curr. Opin. Chem. Biol* 2003, **7**, 64–69.
108. Lee, D. Y., Park, Y. C., Kim, H. J., Ryu, Y. W., and Seo, J. H., *Proteomics* 2003, **3**, 2330–2338.
109. Kim, H. J., Lee, D. Y., Lee, D. H., Park, Y. C., et al., *Proteomics* 2004, **4**, 3588–3599.
110. Weig, M., Jansch, L., Gross, U., De Koster, C. G., et al., *Microbiology* 2004, **150**, 3129–3144.
111. Kondoh, O., Tachibana, Y., Ohya, Y., Arisawa, M., and Watanabe, T., *J. Bacteriol.* 1997, **179**, 7734–7741.
112. Mio, T., Adachi-Shimizu, M., Tachibana, Y., Tabuchi, H., et al., *J. Bacteriol.* 1997, **179**, 4096–4105.
113. Douglas, C. M., D'Ippolito, J. A., Shei, G. J., Meinz, M., et al., *Antimicrob. Agents Chemother.* 1997, **41**, 2471–2479.
114. Douglas, C. M., *Med. Mycol.* 2001, **39**(Suppl. 1), 55–66.
115. Navarro-García, F., Eisman, B., Román, E., Nombela, C., and Pla, J., *Med. Mycol.* 2001, **39**(Suppl. 1), 87–100.
116. Pahlman, A.-K., Granath, K., Ansell, R., Hohmann, S., and Alder, L., *J. Biol Chem.* 2001, **276**, 3555–3563.
117. Xu, J-R., *Fungal Genet. Biol.* 2000, **31**, 137–152.
118. Pfaller, M., Riley, J., and Koerner, T. *Eur. J. Clin. Microbiol. Infect. Dis.* 1989, **8**, 1067–1070.

119. Popolo, L., Gualtieri, T., and Ragni, E., *Med. Mycol.* 2001, **39**(Suppl. 1), 111–121.

120. Navarro-García, F., Alonso-Monge, R., Rico, H., Pla, J., *et al.*, *Microbiology* 1998, **144**, 411–424.

121. Kapteyn, J. C., Ram, A. F., Groos, E. M., and Kollar, R., *J.Bacteriol.* 1997, **179**, 6279–6284.

122. Nicholls, S., Straffon, M., Enjalbert, B., Nantel, A., *et al.*, *Eukaryot. Cell.* 2004, **3**, 1111–1123.

123. Pitarch, A., Sanchez, M., Nombela, C., and Gil, C., *Mol. Cell Proteomics* 2002, **1**, 967–982.

124. Yin, Z., Stead, D., Selway, L., Walker, J., et al., *Proteomics* 2004, **4**, 2425–2436.

125. Crowe, J. D., Sievwright, I. K., Auld, G. C., Moore, N. R., et al., *Mol. Microbiol.* 2003, **47**, 1637–1651.

126. Urban, C., Sohn, K., Lottspeich, F., Brunner, H., and Rupp, S., *FEBS Lett.* 2003, **544**, 228–235.

127. Brand, A., MacCallum, D. M., Brown, A. J., Gow, N. A., and Odds, F. C., *Eukaryot. Cell* 2004, **3**, 900–909.

128. De Backer, M. D., de Hoogt, R. A., Froyen, G., Odds, F. C., et al., *Microbiology* 2000, **146**(2), 353–365.

129. Colina, A. R., Aumont, F., Deslauriers, N., Belhumeur, P., and de Repentigny, L., *Infect. Immun.* 1996, **64**, 4514–4519.

130. Fernandez-Murray, P., Biscoglio, M. J., and Passeron, S., *Arch. Biochem. Biophys.* 2000, **375**, 211–219.

131. Fernandez-Murray, P., Pardo, P. S., Zelada, A. M., and Passeron, S., *Arch. Biochem. Biophys.* 2002, **404**, 116–125.

132. Choi, W., Yoo, Y. J., Kim, M., Shin, D., et al., *Yeast* 2003, **20**, 1053–1060.

133. de Groot, P. W., de Boer, A. D., Cunningham, J., Dekker, H. L., et al., *Eukaryot. Cell* 2004, **3**, 955–965.

134. Sepulveda, P., Lopez-Ribot, J. L., Gozalbo, D., Cervera, A., et al., *Infect. Immun.* 1996, **64**, 4406–4408.

135. Kaneko, A., Umeyama, T., Hanaoka, N., Monk, B. C., et al., *Yeast* 2004, **21**, 1025–1033.

136. Singleton, D. R., Masuoka, J., and Hazen, K. C., *J. Bacteriol.* 2001, **183**, 3582–3588.

## APPENDIX: Synopsis of different *C. albicans* proteins identified by proteomic approaches that have been documented in literature

In the table below a unfilled box (without √) indicates that the *C. albicans* protein under consideration has not yet been found to be associated with the corresponding topic by proteomic approaches. *Candida albicans* protein name and accession number are according to CandidaDB (http://genolist.pasteur.fr/CandidaDB/). MTB denotes metabolism, aa amino acyl, HSP heat-shock protein, and GCN general amino acid control. See Chapter 17 for further details into the cell wall, other cell biology issues, and virulence.

| *C. albicans* Proteins | | | | Main *C. albicans*–Related Topics | | | | | | | | | | | | | | References |
|---|---|---|---|---|---|---|---|---|---|---|---|---|---|---|---|---|---|---|
| | | | | | Other Cell Biology Issues | | | | | Virulence[a] | | | Host Response | | | Anti Fungal Drugs | | |
| | | | | | | | | | | | | | Antibody response | | | | | |
| Functional Category | Short Name | Full Name/Description | *Candida* DB Accession Number | Cell Wall | 20S Proteasome | mRNA Capping | GCN Response | *URA3* | Septin Cytoskeleton | Dimorphism | Adherence | Hydrolases | Humans | Animal Models | Protective Response | Modes of Action | Resistance | References |
| I. Carbohydrate Metabolism – Glycolysis | **Pfk1p** | 6-Phosphofructokinase | CA1834 | | | | | | | | | | √[b] | | | | | [b] |
| | **Hxk2p** | Hexokinase II | CA0127 | | | | | | | | | | √ | | | | | [51] |
| | **Pgi1p** | Glucose-6-phosphate isomerase | CA3559 | | | | | | | | | | √ | | | | | [51] |
| | **Fba1p** | Fructose-biphosphate aldolase | CA5180 | √[c,d] | | | | | | √[e] | √[f] | | √ | √ | √ | √[g] | | [50–54,70,123–125] |
| | **Tpi1p** | Triose phosphate isomerase | CA5950 | √[c,d] | | | | | | | | | √ | √ | √ | | | [50–54,123] |
| | **Gap1p** | Glyceraldehyde-3-phosphate dehydrogenase | CA5892 | √[c,d,h] | | | | | | √[e] | √[f] | | √ | √ | | | √[i] | [47,50,51,53,54,73,123–126] |
| | **Pgk1p** | Phosphoglycerate kinase | CA1691 | √[c,d,h] | | | | | | √[e,,j] | √[f] | | √ | √ | | | | [47,50,51,53,54,123–126] |
| | **Gpm1p** | Phosphoglycerate mutase | CA4671 | √[c,d] | | | | | | | √[k] | | √[l] | | | | | [49,51,53,54,123–125] |
| | **Eno1p** | Enolase | CA3874 | √[c,d] | | | | | | √[e] | | | √ | √ | √ | | | [47,48,50–54,123] |
| | **Cdc19p** | Pyruvate kinase | CA3483 | √[c,h] | | | | | | √[e,m] | | | √[l] | √ | √ | | √[i] | [49–53,73,123,124,126] |
| Fermentation | **Adh1p** | Alcohol dehydrogenase | CA4765 | √[c,h] | | | | | | √[e,m] | √[k] | | √ | √ | | | | [50–53,123,125,126] |
| | **Adh5p** | Probable alcohol dehydrogenase | CA2391 | | | | | | | | | | | √ | | | | [53] |
| | **Ald5p** | Aldehyde dehydrogenase | CA4159 | | | | √[n,o] | | | | | | | | | | | [127] |
| | **Pdc11p** | Pyruvate decarboxylase | CA2474 | √[c,h] | | | | | | √[e,j] | | | √ | √ | √ | | √[i] | [51–53,73,123,124,126] |
| Tricarboxylic Acid (TCA) Pathway | **Cit1p** | Citrate synthase | CA3909 | | | | | | | | | | | | | √[g] | | [70] |
| | **Aco1p** | Aconitate hydratase 1 | CA3546 | | | | | | | | | | √[l] | √ | | | | [49–51,124] |
| | **Aco2p** | Aconitate hydratase 2 | CA4077 | | | | | | | | | | | | | | | [124] |
| | **Idhp** | Isocitrate dehydrogenase | | | | | | | | | | | | | | √[g] | | [70] |
| | **Lpd1p** | Dihydrolipoamide dehydrogenase | CA2998 | | | | | | | | | | | √ | | √[p] | | [53,70] |
| | **Mdh1p** | Mitochondrial malate dehydrogenase | CA5164 | | | | | | | | | | √ | √ | | | | [51,53] |
| | **Sdh2p** | Succinate dehydrogenase | | √[h] | | | | | | √[m] | | | | | | | | [126] |
| Pentose Phosphate Pathway | **Gnd1p** | Phosphogluconate dehydrogenase | CA5239 | | | | | | | | | | √[b] | | | | √[i] | [73][b] |
| | **Tkl1p** | Transketolase | CA3924 | √[h] | | | | | | √[m] | | | √ | √ | | √[g] | | [51–53,70,126] |
| | **Zwf1p** | Glucose-6-phosphate dehydrogenase | CA2634 | | | | | | | | | | | √ | | | | [53] |
| Glycerol MTB | **Hor2p/ Gpp2p** | Glycerol phosphate phosphatase | | | | | | | | | | | | | | √[g] | | [70] |
| | **Rhr2p** | DL-Glycerol phosphatase | CA5788 | | | | | | | | | | √[b] | √ | √ | | | [52,53,124][b] |
| Miscellaneous | **Acs2p** | Acetyl-coenzyme-A synthetase | CA2858 | | | | | | | | | | √ | √ | √ | | | [51,53] |
| | **Gph1p** | Glycogen phosphorylase | CA5206 | √[h] | | | | | | √[j] | | | | | | | | [126] |
| | **Icl1p** | Isocitrate lyase | CA4446 | | | | | | | | | | | | | √[g] | | [70] |
| | **Ino1p** | *myo*-Inositol-1-phosphate synthase | CA5986 | √[c] | | | | | | √[e] | | | √ | | | √[g] | | [51,70,123] |
| | **Pmi40p** | Mannose-6-phosphate isomerase | CA0988 | | | | | | | | | | | | | | | [124] |
| | **Pmm1p** | Phosphomannomutase | CA2198 | | | | √[o,q] | | | | | | | √ | | | √[r] | [53,73,127] |
| | **Psa1p/ Srb1p** | GDP-mannose pyrophosphorylase | CA3208 | √[c,h] | | | | | | √[e,m] | | | | | | | | [123,126] |
| | **Pyc2p** | Pyruvate carboxylase | CA1463 | √[h] | | | | | | | | | | | | | | [126] |
| | **Ugp1p** | UTP-glucose-1-phosphate uridylyl transferase (UGPase) | CA0435 | √[h] | | | | | | √[m] | | | | | | | | [126] |

| | | | | | | | | | | | | | | | | | |
|---|---|---|---|---|---|---|---|---|---|---|---|---|---|---|---|---|---|
| II. Fatty Acid Metabolism | **Ach1p** | Acetyl-coenzyme-A hydrolase | CA0345 | | | | | | | | | | √ | | | | | [51] |
| | **Erg10p/ Pot14p** | Acetyl-coenzyme-A acetyltransferase | CA0290 | | | | | | | | | | | | | √g | √i | [70,73] |
| | **Erg19p/ Mvd1p** | Mevalonate diphosphate decarboxylase | CA3853 | | | | | | | | | | | √ | | | | [70] |
| III. Amino Acid Metabolism | **Aat1p** | Aspartate aminotransferase | CA2661 | | | | | | | | | | | | | | √r | [73] |
| | **Arg1p** | Argininosuccinate synthetase | CA5818 | | | | √s | | | | | | | | | | | [124] |
| | **Arg5,6p** | Acetylglutamate kinase and acetylglutamyl-phosphate reductase | CA2836 | | | | √s | | | | | | | | | | | [124] |
| | **Aro3p** | Dehydro-3-deoxyphospho heptonate aldolase | CA1751 | | | | √s | | | | | | | | | | | [124] |
| | **Aro8p** | Aromatic amino acid aminotransferase I | CA4804 | | | | | √o,q | | | | | | | | | | [127] |
| | **Aro10p** | Phenylpyruvate decarboxylase | | | | | | √q,t | | | | | | | | | | [127] |
| | **Bat21p** | Branched-chain amino acid aminotransferase | CA0330 | | | | √s | | | | | | | | | | | [124] |
| | **Bat22p** | Branched-chain amino acid aminotransferase | CA5040 | | | | √s | | | | | | | | | | | [124] |
| | **Cys4p** | Cystathionine beta-synthase | CA4195 | | | | | | | | | | | | | √g | | [70] |
| | **Gdh1p** | NADP-dependent glutamate dehydrogenase | | | | | | | | | | | | | | √g | | [70] |
| | **Hom6p** | Homoserine dehydrogenase | CA4181 | | | | √s | | | | | | | | | | | [124] |
| | **Ilv1p** | Threonine dehydratase | CA2318 | | | | √s | | | | | | | | | | | [124] |
| | **Ilv2p** | Acetolactate synthase | CA0428 | | | | √s | | | | | | | | | | | [124] |
| | **Ilv5p** | Ketol-acid reducto-isomerase | CA1983 | | | | √s | | | | | | √ | √ | | √g | √r | [51,53,70,73, 124] |
| | **Leu1p** | 3-Isopropylmalate dehydratase | CA5842 | | | | | | | | | | √ | | | | | [51] |
| | **Lys4p** | Homoaconitase | CA4869 | | | | √s | | | | | | | | | | | [124] |
| | **Met3p** | ATP sulfurylase | CA5238 | | | | | | | | | | | | | √g | | [70] |
| | **Met6p** | Methionine synthase | CA0653 | √d | | | √s | | | | | | √ | √l | | √g | | [49–54,70,124] |
| | **Met17p** | *O*-Acetylhomoserine sulfhydrylasc | | | | | | | | | | | | | | √g | | [70] |
| | **Sah1p** | *S*-adenosyl-L-homocysteine hydrolase | CA3018 | | | | √u | | | | | | √ | | | √g | √i | [51,70,73,124] |
| | **Sam2p** | *S*-adenosylmethionine synthase 2 | CA0959 | √h | | | | | | √j | | | | | | | | [126] |
| | **Shm2p** | Serine hydroxymethyltransferase | CA0895 | | | | | | | | | | √ | | | | | [51] |
| IV. Nucleotide Metabolism | **Ade2p** | Phosphoribosylaminoimidazole carboxylase | CA6139 | | | | | √q,t | | | | | | | | | | [127] |
| | **Ade17p** | 5-Aminoimidazole-4-carboxamide ribotide transformylase | CA4513 | | | | | | | | | | √ | | | √g | | [51,70] |
| | **Apa2p** | ATP adenylyltransferase II | CA2765 | | | | √u | | | | | | | | | | | [124] |
| | **Apt1p** | Adenine phosphoribosyltransferase 1 | CA4551 | | | | | | | | | | | | | √g | | [70] |
| | **Imh3p** | Inosine-5′-monophosphate dehydrogenase | CA1245 | | | | | | | | | | √ | √ | √ | | | [50,51,53] |
| | **Hpt1p** | Hypoxanthine guanine phosphoribosy transferase | CA3787 | | | | | √o,v | | | | | | | | | | [127] |
| | **Ura3p** | Orotidine-5-monophosphate decarboxylase | CA2801 | | | | | √t,v | | | | | | | | | | [127] |
| | **Ura5p** | Orotate phosphoribosyltransferase | CA2056 | | | | | √o,q | | | | | | | | | | [127] |
| | **Ynk1p** | Nucleoside diphosphate kinase | CA2645 | | | | | | | | | | | | | | √r | [73] |
| V. Energetic Intermediary Metabolism | **Atp1p** | F1F0-ATPase complex, F1 α subunit | CA4457 | √h | | | | | | √j | | | √ | | | | | [51,126] |
| | **Atp2p** | F1F0-ATPase complex, F1 β subunit | CA4302 | | | | | | | | | | √ | | | | | [51] |
| | **Coq5p** | *C*-Methyltransferase | CA3482 | | | | √u | | | | | | | | | | | [124] |
| | **Ipp1p** | Inorganic pyrophosphatase | CA0870 | | | | | | | | | | √ | | | | | [51] |
| | **Qcr2p** | Ubiquinol-cytochrome *c* reductase 40-kDa chain II | CA2065 | | | | | | | | | | √ | | | √g | | [51,70] |

| | | | | | | | | | | | | | | | | | | | |
|---|---|---|---|---|---|---|---|---|---|---|---|---|---|---|---|---|---|---|---|
| VI. RNA Biosynthesis | | **Rbp8p** | RNA polymerase I, II, III subunit | | | | | √[u] | | | | | | | | | | | [124] |
| | | **Toa2p** | Transcription initiation factor IIA, gamma subunit | CA5187 | | | | | √[w,x] | | | | | | | | | | [127] |
| VII. Protein Bsinthesis | Ribosomal proteins | **Bel1p** | 40S small-subunit ribosomal protein | CA4588 | | | | √[u] | | | | | | √ | √ | √ | | | [51–53,100] |
| | | **Rpl13p** | Ribosomal protein L13A | CA2818 | | | | | | | | | | | √ | | | | [53] |
| | | **Rps5p** | Ribosomal protein S5 | CA0632 | | | √[y] | | | | | | | | | | | | [128] |
| | | **Rps6p** | Ribosomal protein S6 | | √[h] | | | | | | √[m] | | | | | | | | [126] |
| | | **Rps12p** | Acidic ribosomal protein S12 | CA5920 | | | | √[u] | √[o,w] | | | | | | | | | | [124,127] |
| | | **Rpp0p** | Acidic ribosomal protein p0 | | | | | | | | | | | | | | √[p] | | [70] |
| | | **Yst1p** | Ribosomal protein Rps0B | CA5021 | | | | √[u] | | | | | | | | | | | [124] |
| | aa-tRNA synthetases | **Frs2p** | Phenylalanyl tRNA synthetase β subunit | CA4173 | | | | | | | | | | | | | √[g] | | [70] |
| | | **Ala1p** | Alanyl-tRNA synthetase | CA1427 | | | | √[u] | | | | | | | | | | | [124] |
| | Translation factors | **Tef1p** | Translation elongation factor 1α | CA0362 | √[c,h] | | √[y] | | | | | √[k] | | | | | | | [123,125,126,128] |
| | | **Efb1p** | Translation elongation factor 1β | CA4862 | √[c] | | | | | | | | | | | | | | [123] |
| | | **Eft2p** | Translation elongation factor 2 | CA2810 | | | | √[t,u] | √[q] | | √[m] | | | √ | | | | | [51,123,124,126,127] |
| | | **Eft3p** | Translation elongation factor 3 | CA3081 | √[c,d,h] | | | | | | √[m] | | | √ | | | | | [51,54,123,126] |
| | | **Tif1p** | Translation initiation factor 4A | CA2939 | | | | | | | | | | √ | | | | | [51] |
| VIII. Protein Degradation | | **Sap2p** | Secreted aspartyl proteinase 2 | CA3138 | | | | | | | | | √[z] | | | | | | [129] |
| | | **α3/Pre9p/Y13/C9** | 20S proteasome α 3 subunit | CA4643 | | √[aa] | | | | | | | | | | | | | [130,131] |
| | | **α5/Pup2p/ζ** | 20S proteasome α5 subunit | CA0663 | | √[aa] | | | | | | | | | | | | | [130,131] |
| | | **α6/Pre5p/C2** | 20S proteasome α6 subunit | CA5308 | | √[aa] | | √[s] | | | | | | | | | | | [124,130,131] |
| | | **α7/Prs1p/C8** | 20S proteasome α7 subunit | | | √ | | | | | | | | | | | | | [130] |
| | | **Rpt4p/Pcs1p** | 26S proteasome regulatory subunit | CA4522 | | | | | | | | | | | | | √[g] | | [70] |
| | | **Skp1p** | Kinetochore protein complex CBF3, subunit D | CA1285 | | | | √[u] | | | | | | | | | | | [124] |
| | | **Ubp1p** | Ubiquitin carboxyl-terminal hydrolase | CA5665 | √[h] | | | | | | √[j] | | | | | | | | [126] |
| IX. Protein Folding and Stabilization | Heat-Shock Response | **Ssa1p** | Member of HSP70 family | CA2857 | √[c] | | | | | | | | | √ | √ | | | | [50,51,123] |
| | | **Ssa2p** | Member of HSP70 family | | √[h] | | √[y] | | | | √[j] | | | | | | | | [126,128] |
| | | **Ssa4p** | Member of HSP70 family | CA1230 | √[c] | | | | | | √[e] | | | | | | | | [123] |
| | | **Ssb1p** | Member of HSP70 family | CA3534 | √[c,h] | | | √[s] | | | √[m] | | | √ | √ | | | | [50,51,123,124,126] |
| | | **Ssc1p** | Member of HSP70 family | CA4474 | √[c,h] | | | √[s] | | | √[m] | | | √ | | | | | [51,124,126] |
| | | **Ssd1p/Kar2p** | Member of HSP70 family | CA0915 | √[c] | | | | | | | | | | | | | | [123] |
| | | **Sse1p** | Member of HSP70 family | CA1911 | | | | | | | | | | √ | | | | | [51] |
| | | **Ssz1p/Pdr13p** | Member of HSP70 family | CA4844 | √[c] | | | | | | √[bb] | | | | | | | | [123] |
| | | **Hsp70p** | Heat-shock protein, 70 kDa | | √[h] | | | | | | | | | | | | | | [126] |
| | | **Hsp90p** | Heat-shock protein, 90 kDa | CA4959 | √[c,h] | | | | | | √[e,j] | | | √ | | | √[g] | | [51,70,123,126] |
| | | **Hsp140p** | Heat-shock protein, 104 kDa | CA5135 | √[c] | | | | | | | | | | | | | | [123] |
| | Others | **Pdi1p** | Protein disulfide isomerase | CA1755 | √[c] | | | | | | √[cc] | | | √ | | | | | [51,123] |
| | | **Rbp1p** | Rapamycin-binding protein (putative peptidyl-prolyl cis–trans isomerase) | CA3676 | | | | √[u] | | | | | | | | | | | [124] |
| X. Protection – Detoxification | | **Cta1p** | Peroxisomal catalase | CA3011 | √ | | | | | | | √[k] | | | | | | | [125] |
| | | **Grp2p** | Reductase | CA2644 | | | | | | | | | | √ | | | | √[r] | [51,73] |
| | | **Grx3p** | Glutaredoxin-like protein | CA1161 | | | | √[u] | | | | | | | | | | | [124] |
| | | **Hem13p** | Coproporphyrinogen III oxidase | CA0517 | √[h] | | | | √[q,t] | | √[m] | | | √ | | | | | [51,126,127] |
| | | **Pga2p/Sod4p** | Putative Cu/Zn superoxide dismutase | CA4835 | √[dd,ee] | | | | | | | | | | | | | | [133] |
| | | **Tsa1p** | Thiol-specific antioxidant protein | | √[h] | | | | | | √[j,ff] | √[k] | | √[b] | | | √[g] | | [70,125,126,132][b] |

| | | | | | | | | | | | | | | | | | | |
|---|---|---|---|---|---|---|---|---|---|---|---|---|---|---|---|---|---|---|
| XI. Signal Transduction | **Bmh2p** | Homolog of mammalian 14-3-3 protein | CA5050 | √ *c,h* | | | | | | √ *m* | | | | | | | | [123,126] |
| | **Sgt2p** | Small glutamine-rich tetratricopeptide repeat containing protein | CA3796 | | | | | √ *n,o* | | | | | | | | | | [127] |
| XII. Cellular Transport | **Por1p** | Mitochondrial outer membrane porin | CA0919 | | | | | | | | | | √ | | | | | [51] |
| | **Sec13p** | Protein transporter | CA3392 | | | | | | | | | | | √ | | | | [53] |
| | **Vps4p** | Vacuolar sorting protein | CA1340 | | | | √ *s* | | | | | | | | | | | [124] |
| XIII. Binding Function | **Als1p** | Agglutinin-like sequence | CA0316 | √ *dd,ee* | | | | | | | | | | | | | | [133] |
| | **Als4p** | Agglutinin-like sequence | | √ *dd,ee* | | | | | | | | | | | | | | [133] |
| | **Csp37p** | Cell surface protein, 37 kDa | CA1075 | √ *c* | | | | | | | | | | | | | | [123] |
| | **Ebp1p** | Echinocandin-binding protein | | √ *h* | | | | | | √ *j* | | | | | | | | [126] |
| | **Rbt5p** | Repressed by *TUP1* protein 5 | CA2558 | √ *dd,ee* | | | | | | | | | | | | | | [133] |
| | **Zpr1p** | Zinc finger protein | CA1309 | | | | √ *u* | | | | | | | | | | | [124] |
| | - | Candidal C3d receptor | | √ | | | | | | | √ *gg* | | | | | | | [134] |
| | - | Fibrinogen-binding mannoprotein, 58 kDa | | √ | | | | | | | √ *gg* | | | | | | | [134] |
| | - | Laminin receptor, 37 kDa | | √ | | | | | | | √ *gg* | | | | | | | [134] |
| XIV. Cytoskeleton | **Aip2p** | Actin interacting protein 2 | CA2406 | | | | √ *s* | | | | | | | | | | | [124] |
| | **Cdc3p** | Cell division control protein | CA0844 | | | | | | √ | | | | | | | | | [135] |
| | **Cdc10p** | Cell division control protein | CA4259 | | | | | | √ | | | | | | | | | [135] |
| | **Cdc11p** | Septin | CA2610 | | | | | | √ | | | | | | | | | [135] |
| | **Cdc12p** | Septin | CA5049 | | | | | | √ | | | | | | | | | [135] |
| | **End3p** | Protein required for endocytosis and cytoskeletal organization | CA1556 | | | | √ *s* | | | | | | | | | | | [124] |
| | **Sep7p/ Shs1p** | Septin | CA6210 | | | | | | √ | | | | | | | | | [135] |
| | **Tpm2.3p** | Tropomyosin, isoform 2 | CA5116 | | | | √ *u* | | | | | | | | | | | [124] |
| XV. Cell wall Organization and Biogenesis | **Bgl2p** | β-1,3-Glucosyltransferase (glucan-β-1,3-glucosidase) | CA1541 | √ *c,d* | | | | | | | | | √ | | | | | [54,123] |
| | **Cht2p** | Chitinase 2 | CA1051 | √ *dd,ee* | | | | | | | | | | | | | | [133] |
| | **Crh11p** | Probable membrane protein | CA0375 | √ *dd,ee* | | | | | | | | | | | | | | [133] |
| | **Ecm33.3 p** | GPI-anchored protein | | √ *dd,ee* | | | | | | | | | | | | | | [133] |
| | **Pga4p** | β-1,3-Glucanosyl transferase (putative GPI-anchored protein) | CA4800 | √ *dd,co* | | | | | | | | | | | | | | [133] |
| | **Pga24p/ Ycw1p** | Flocculin (putative cell wall protein) | CA1678 | √ *dd,ee* | | | | | | | | | | | | | | [133] |
| | **Phr1p** | pH Response protein | CA4857 | √ *c,h,dd,ee* | | | | | | √ *j,ff* | | | | | | | | [123,126,132, 133] |
| | **Pir2p/Hs p150p** | Member of Pir-CWP family/150-kDa heat-shock glycoprotein | | √ *c* | | | | | | | | | | | | | | [123] |
| | **Pir1p** | Protein with internal repeats | | √ *dd,ee* | | | | | | | | | | | | | | [133] |
| | **Scw1p** | *endo*-β-1,3-Glucanase | | √ *dd,ee* | | | | | | | | | | | | | | [133] |
| | **Ssr1p** | Secretory stress response protein | CA5213 | √ *dd,ee* | | | | | | | | | | | | | | [133] |
| | - | Putative β-1,3-Glucanase | - | √ *c,dd* | | | | | | √ *bb* | | | | | | | | [123] |
| XVI. Unknown Function | **Ifd1p** | Putative alcohol dehydrogenase-oxidoreductase | | | | | | | | | | | | | | | √ *i,hh* | [72,73] |
| | **Ifd2p** | Putative oxidoreductase | CA1959 | | | | | | | | | | | | | | √ *hh* | [72] |
| | **Ifd4p/ Csh1p** | Putative aryl-alcohol dehydrogenase/cell surface hydrophobicity protein 1 | CA2416 | √ *h* | | | | | | √ *i* | √ *ii* | | | | | | √ *i,hh* | [72,73,126,136] |
| | **Ifd5p** | Putative aryl-alcohol dehydrogenase | CA0924 | | | | | | | | | | | | | | √ *i,hh* | [72,73] |
| | **Ifd6p** | Putative aryl-alcohol dehydrogenase | CA2417 | | | | | | | | | | | | | | √ *i,hh* | [72,73] |
| | **Ipf18418p** | Unknown function | CA2756 | | | | √ *s* | | | | | | | | | | | [124] |
| | **Ipf17186p** | Unknown function | CA0828 | | | | | | | | | | √ | | | | | [51] |

| | | | | | | | | | | | | | | | | | | |
|---|---|---|---|---|---|---|---|---|---|---|---|---|---|---|---|---|---|---|
| | **Ipf16470p** | Unknown function | CA2002 | | | | √[u] | | | | | | | | | | | [124] |
| | **Ipf16194p** | Unknown function | CA3352 | | | | | | | | | | | √ | | | | [53] |
| | **Ipf14775p** | Unknown function | CA0866 | | | | √[u] | | | | | | | | | | | [124] |
| | **Ipf14662p** | Unknown function with homology to *d*-xylose reductase | CA1592 | | | | | | | | | | | √ | | | | [53] |
| | **Ipf13867p** | Unknown function | CA4437 | | | | √[s] | | | | | | | | | | | [124] |
| | **Ipf11900p** | Unknown function | CA3730 | | | | √[u] | | | | | | | | | | | [124] |
| | **Ipf10482p** | Unknown function | CA4049 | | | | √[s] | | | | | | | | | | | [124] |
| | **Ipf9484p** | Unknown function | CA2434 | | | | √[s] | | | | | | | | | | | [124] |
| | **Ipf8762p** | Unknown function | CA4220 | √[c] | | | | | | √[jj] | | | | | | | | [123] |
| | **Ipf6878p** | Unknown function | CA4129 | | | | √[u] | | | | | | | | | | | [124] |
| | **Ipf6037p** | Unknown function | CA5171 | | | | | √[o,q] | | | | | | | | | | [127] |
| | **Ipf5987p** | Unknown function | CA5544 | | | | | | | | | | | | | | √[hh] | [72] |
| | **Ipf4328p** | Unknown function | CA0210 | | | | | √[o,w] | | | | | | | | | | [127] |
| | **Ipf3584p** | Unknown function | CA5065 | | | | √[s] | | | | | | | | | | | [124] |
| | **Ipf3144p** | Unknown function | CA1319 | | | | √[u] | | | | | | | | | | | [124] |
| | **Ipf1399p** | Unknown function | CA4928 | | | | √[u] | | | | | | | | | | | [124] |
| | **Ipf864p** | Unknown function | CA5347 | | | | √[u] | | | | | | | | | | | [124] |
| | **Pga29p** | Unknown function | CA5476 | √[dd,ee] | | | | | | | | | | | | | | [133] |
| | - | Kiaa 0551 protein | | | | | | | | | | | | | | √[g] | | [70] |
| **XVII. Miscellaneous** | **Cdc48p** | Member of ATPases associated with diverse cellular activities (AAA) superfamily | CA3333 | √[c] | | | √[u] | | | | | | | | | | | [123, 124] |
| | **Pra1p** | pH-Regulated antigen | CA4399 | √[h] | | | | | | √[j,ff] | | | | | | | | [126, 132] |
| | **Snz1p** | Stationary-phase protein | CA4184 | | | | √[s] | | | | | | | | | | | [124] |
| | - | Rehydrin-like protein | | | | | | | | | | | | | | √[g] | | [70] |
| | - | Homolog to glycosyl transferase AmsE | | | | | | | | | | | | | | √[p] | | [70] |
| **Overall** | **Number of identified proteins (*n*=182)** | | | **66** | **4** | **3** | **47** | **13** | **5** | **20** | **12** | **1** | **47** | **28** | **9** | **27** | **17** | |
| | **Number of publications (*n*=25)** | | | **7** | **2** | **1** | **1** | **1** | **1** | **2** | **3** | **1** | **5** | **4** | **2** | **1** | **2** | |

[a]Phenotypic switching has not been included because no protein identification was accomplished in the proteomic approaches reported in the literature.
[b]Unpublished result (Pitarch et al., manuscript in preparation).
[c]Protein identified in a proteomic analysis of *C. albicans* cell wall proteins isolated by sequential fractionation [123].
[d]Protein identified in a proteomic analysis of *C. albicans* proteins secreted from protoplasts in active cell wall regeneration [54].
[e]Up-regulated hyphal cell wall-associated protein [SDS- (sodium docedyl sulfate) and reducing-agent-extractable cell wall protein] [123].
[f]Plasminogen-binding protein without C-terminal lysine residues [125].
[g]Up-regulated protein in the microsomal fraction of *C. albicans* after 2 h incubation with mulundocandin (an echinocandin-class compound) [70].
[h]Protein identified in a proteomic analysis of cell-surface-associated proteins biotinylated and purified from *C. albicans* cell walls by affinity chromatography [126].
[i]Up-regulated protein in a clinical *C. albicans* isolate that has become resistant to fluconazole due both to the activation of genes that code for the target enzyme (*ERG11/CYP51*) and multidrug efflux transporters (*MDR1* and *CDR1/2*) and to the loss of allelic variation and acquisition of a point mutation in *ERG11/CYP51* [73].
[j]Cell-surface-associated protein identified in hyphae [126].
[k]Plasminogen-binding protein with C-terminal lysine residues [125].
[l]Immunogenic protein cross-species identified by matching of MS/MS data to *S. cerevisiae* proteins [49].
[m]Cell-surface-associated protein identified in yeast cells [126].
[n]One of its two protein species was not present in a mutant depleted of the two alleles both of *URA3* and of the 3′ half of *IRO1* (the homozygous *ura3-iro1/ura3-iro1* strain) [127].
[o]Its expression was fully restored when one copy of *URA3* was reintegrated at the *RPS10* locus [127].
[p]Down-regulated protein in the microsomal fraction of *C. albicans* after 2 h incubation with mulundocandin (an echinocandin-class compound) [70].
[q]Up-regulated protein in a mutant depleted of the two alleles both of *URA3* and of the 3' half of *IRO1* (the homozygous *ura3-iro1/ura3-iro1* strain) [127].
[r]Down-regulated protein in a clinical *C. albicans* isolate that has become resistant to fluconazole due both to the activation of genes that code for the target enzyme (*ERG11/CYP51*) and multidrug efflux transporters (*MDR1* and *CDR1/2*) and to the loss of allelic variation and acquisition of a point mutation in *ERG11/CYP51* [73].
[s]Up-regulated protein in response to histidine starvation in a Gcn4p-dependent manner [124].
[t]Its expression was partially restored when one copy of *URA3* was reintegrated at the *RPS10* locus [127].
[u]Down-regulated protein in response to histidine starvation in a Gcn4p-dependent manner [124].

[v]Protein not detected in a mutant depleted of the two alleles both of *URA3* and of the 3' half of *IRO1* (the homozygous *ura3-iro1/ura3-iro1* strain) [127].
[w]Down-regulated protein in a mutant depleted of the two alleles both of *URA3* and of the 3' half of *IRO1* (the homozygous *ura3-iro1/ura3-iro1* strain) [127].
[x]Its expression was not restored when one copy of *URA3* was reintegrated at the *RPS10* locus [127].
[y]Protein overexpressed in a mutant depleted of one allele of *CGT1* (the *cgt1/CGT1* heterozygote) and identified by MS and/or Edman degradation [128]. *CGT1* encodes a *C. albicans* mRNA capping enzyme.
[z]Enzyme accountable for proteolysis of mucin—a highly glycosylated protein from mucus— by *C. albicans* in vitro [129].
[aa]In vivo phosphorylated protein and in vitro substrate of the protein kinase CK2 [131].
[bb]Down-regulated hyphal cell-wall-associated protein (SDS- and reducing-agent-extractable cell wall protein) [123].
[cc]Its 73- and 70-Da forms were up and down regulated, respectively, in hyphal walls [123].
[dd]Cell wall protein (CWP) covalently attached to β-1,3-glucan network [123,133].
[ee]Protein identified in a proteomic analysis of tryptic digests of CWPs covalently incorporated to β-1,3-glucan network [133].
[ff]Up-regulated protein in hyphal cytoplasmic cell extracts [132].
[gg]Ubiquitinated cell surface receptor [134].
[hh]Up-regulated protein in a fluconazole-resistant clinical *C. albicans* isolate in which drug resistance correlated with constitutive activation of the major facilitator gene (*MDR1*) [72].
[ii]The 38-kDa protein involved in cell surface hydrophobicity (CSH) [136].
[jj]Its pH 4.7 and 4.8 forms were up and down regulated, respectively, in hyphal walls [123].

CHAPTER 19

# Identification of Protein Candidates for Developing Bacterial Ghost Vaccines against *Brucella*

VITO G. DELVECCHIO,[1] TIM ALEFANTIS,[1] RODOLFO A. UGALDE,[2] DIEGO COMERCI,[2] MARIA INES MARCHESINI,[2] AKBAR KHAN,[3] WERNER LUBITZ,[4] and CESAR V. MUJER[1]

[1]Vital Probes, Mayfield, Pennsylvania
[2]Universidad Nacional De General San Martin, Buenos Aires, Argentina
[3]Defense Threat Reduction Agency, Alexandria, Virginia
[4]Biotech Innovation Research Development & Consulting GmbH & Co KEG, Wien, Austria

## 19.1 INTRODUCTION

Members of the genus *Brucella* are the etiologic agents of brucellosis in livestock and humans (Corbel, 1997). The disease is also known as Malta fever in humans and is a chronic febrile disease characterized by undulant fever, chills, night sweats, headache, weakness, musculoskeletal pains, fatigue, prostration, and mental depression (Young, 1983). Among the various species, *Brucella melitensis*, *Brucella abortus*, and *Brucella suis* are the main causes of human brucellosis (Corbel, 1997; Taylor and Perdue, 1989). Brucellosis is a worldwide zoonotic disease responsible for significant economic consequences due to the loss of livestock because of spontaneous abortion and infertility (Boschiroli et al., 2001). In Argentina and Central America, it is estimated that brucellosis of livestock results in an annual economic loss of $60 million and $25 million, respectively (Samartino, 2002; Moreno, 2002). Finally, certain species of *Brucella* are listed as potential weapons of mass destruction because of their highly pathogenic nature, ability to cause harmful and chronic diseases in animals and human, and potential to disrupt national and global food supplies. Given these factors it is important to advance research on both the prevention and treatment of *Brucella* infection through new vaccine strategies and therapeutic interventions.

*Brucella* species are gram-negative coccobaccilli that range in size from 0.5 to 0.7 μm in diameter and 0.5 to 1.5 μm in length (Corbel and Brinley-Morgan, 1984). They are non–spore forming, nonmotile, and not encapsulated. Currently, there are six recognized species. These include *B. melitensis*, which primarily infects sheep and goats; *B. abortus* is

*Microbial Proteomics: Functional Biology of Whole Organisms*, Edited by Ian Humphery-Smith and Michael Hecker. 

pathogenic to cattle; *B. suis* is found in swine; *Brucella ovis* in sheep; *Brucella canis* in dogs; and *Brucella neotomae* in the desert wood rat (Corbel and Brinley-Morgan, 1984; Corbel, 1997). Recent isolation of *Brucella* from marine mammals has indicated a more extensive ecological range than originally thought (Bricker et al., 2000; Jahans et al., 1997; Verger et al., 2000). Although members of the *Brucella* genus are designated as separate species, DNA hybridization and genomic studies have indicated that they are members of the same species which differ mainly in their host preference (Verger et al., 1985).

*Brucella* is comprised of facultative intracellular pathogens that establish an intimate relationship with their eukaryotic host cells. They are thought to initially penetrate the oral mucosa and then invade the lymphatic and circulatory systems of the host (Corbel, 1997). *Brucella* organisms have the potential to be used as formidable biological warfare agents since they can be easily aerosolized and are highly invasive, are extremely infectious requiring only 1–10 colony-forming units (CFU) per person in the case of *B. melitensis*, and can also be transmitted in dairy products. Since the incubation period of the disease varies from weeks to several months, rapid diagnosis is difficult and delayed mitigation of an intentional outbreak could result in thousands of people being infected.

Live, attenuated vaccines of *Brucella* such as *B. melitensis* Rev1 and *B. abortus* S19 and RB51 are currently used to control brucellosis in livestock such as cattle, goats, and swine (WHO, 1997). However, these vaccines are unsafe for use in other animals, including bison, reindeer, elk, and caribou. All current vaccines have the potential to be pathogenic to humans and thus cannot be included in a human vaccination strategy. Some vaccines are not as effective as one would desire. Thus, an effective, preferably nonliving, vaccine that is able to stimulate a broad protective immune response is needed. Herein, the recent advances in the global identification of *Brucella* proteins from virulent and knockout mutants using current proteomics technology and the potential use of bacterial ghosts (Haidinger et al., 2003; Lubitz, 2001; Marchart et al., 2003) are discussed in relation to a rational approach in the design of a potent *Brucella* vaccine and other novel antimicrobial therapies.

## 19.2 HOST IMMUNE EVASION STRATEGIES OF *BRUCELLA*

The design of an effective and safe *Brucella* vaccine depends not only on identification and characterization of immunogenic proteins but also on understanding of how these pathogens invade, multiply, and survive in host cells. The mechanisms by which *Brucella* evades the host antibacterial defenses have been the subject of numerous studies in the last decade (for a recent review see Ko and Splitter, 2003). *Brucella* takes advantage of two host cellular pathways to enter the host cell in order to evade the immune system. One pathway is opsonization, whereby the bacterial cell is coated with antibacterial antibodies and complement. It is then phagocytized as part of the innate host immune response. The other pathway involves direct contact of nonopsonized *Brucella* with host cells allowing for adherence and invasion through such processes as receptor-mediated endocytosis (Detilleux et al., 1991; Gross et al., 2000). Experimental evidence has indicated that *Brucella* cells are able to replicate inside host macrophages and nonprofessional phagocytes while, through some unknown mechanism, avoiding both the host humoral and cell-mediated immune responses (Detilleux et al., 1990; Jones and Winter, 1992; Riley and Roberson, 1984; Roop et al., 2004). This ability to evade killing and reside inside the host cell following phagocytosis is the primary factor leading to chronic infection. Establishing a residence within the intracellular organelles of host cells allows the bacteria to escape the extracellular host defense mechanisms (Roop et al., 2004). *Brucella*-infected macrophages also serve as vehicles for systemic spread of the infection.

Establishment of *Brucella* residence within macrophages initially occurs through an inhibition of phagosome–lysosome fusion via rapid acidification (Gorvel and Moreno, 2002; Kohler et al., 2001). They also control the formation of VirB-dependent endoplasmic reticulum (ER)–derived organelles called *Brucella* replicative organelles in which they replicate and survive intracellularly (Celli et al., 2003; for a recent review, see Celli and Gorvel, 2004). *Brucella* enters the cell by endocytosis forming a *Brucella*-containing vesicle (BCV) that then fuses with the ER. Once inside the ER, *Brucella* forms *Brucella* replicative organelles by budding from the ER membrane. The *Brucella* replicative organelles support *Brucella* replication and escape fusion with the lysosomes. Studies with different mutations have begun to elucidate the specific mechanisms of *Brucella* intracellular survival. One study examined infection with *Brucella* mutants of a two-component regulatory system, *bvrR–bvrS* (Sola-Landa et al., 1998; Gorvel and Moreno, 2002). Infection with these mutants resulted in an inhibition of phagosome–lysosome fusion ultimately leading to the failure of *Brucella* to localize to the ER compartment and to multiply. Likewise, *virB* mutants deficient in a type IV secretion system were unable to reach the ER (Comerci et al., 2001; Delrue et al., 2001). They were killed when the phagosome in which they are contained fused with the lysosome (Celli et al., 2003 ). *Brucella* VirB10 is an inner-membrane-bound protein possessing a C-terminal periplasmic domain with homology to proteins of the type IV secretion system. *Brucella virB10* polar and nonpolar mutants were able to penetrate the host cell but were unable to replicate in phagocytic and nonphagocytic cells in vitro. Interestingly, phagosomes containing *virB10* polar mutants fused with lysosomes and were directed to a degradation pathway. By contrast, nonpolar mutants were able to avoid interactions with the endocytic pathway but were unable to reach the ER to multiply. This suggested that VirB10 proteins are involved in the maturation of BCV (Comerci et al., 2001).

## 19.3 *BRUCELLA MELITENSIS* PROTEOMES

The completion of the *B. melitensis* genome (DelVecchio et al., 2002a) has paved the way for a comprehensive proteomic analysis of *Brucella*. The *Brucella* proteome initiative was aimed at providing a global view of expressed proteins which would serve as a reference map for future studies of metabolic pathways associated with in vivo infection, pathogenicity, virulence, host specificity, and evolutionary relatedness (Mujer et al., 2002; Wagner et al., 2002). The global identification of *B. melitensis* 16M proteins expressed in laboratory-grown culture has been ongoing (Wagner et al., 2002; Eschenbrenner et al., 2002) and is the subject of recent reviews (DelVecchio et al., 2002b; Mujer et al., 2003). Using overlapping and narrow-range immobilized pH gradient strips for 2D gel electrophoresis, 937 gene products ranging in pH from 3.5 to 11 have been identified by peptide mass fingerprinting (DelVecchio et al., unpublished data). These proteins represented 269 discrete open reading frames (ORFs) in the *B. melitensis* genome. The corresponding ORFs of the identified proteins were evenly distributed over each strand of the two circular *B. melitensis* chromosomes (Wagner et al., 2002).

A comparative proteomic analysis of the virulent *B. melitensis* 16M and the vaccine strain Rev1 was conducted using two-dimensional (2D) gel electrophoresis to further understand the mechanisms of virulence in *B. melitensis* (Eschenbrenner et al., 2002). Computer-assisted analyses indicated that the two strains have significant protein expression and metabolic differences. These differences in the vaccine strain may result in a host immune response that is not ideal for long-term protection against future challenges of the wild-type strain. Differentially expressed proteins associated with iron metabolism, sugar binding, lipid degradation, and amino acid metabolism were observed (Eschenbrenner et al., 2002). In Rev1, bacterioferritin accumulation was found to be significantly higher than in the virulent strain 16M. In addition, the iron-regulated outer membrane protein and an Fe(III)-binding periplasmic protein were also overexpressed in Rev1. These two proteins are typically derepressed during low iron availability. This overexpression of proteins needed for iron availability suggested the presence of a misregulated system for iron capture and metabolism that may result in unnecessary expenditure of energy (Eschenbrenner et al., 2002). To compensate, Rev1 up regulated other pathways, that is, proteins involved in the $\beta$ oxidation of fatty acids and protein synthesis, to generate more adenosine triphosphate (ATP). A higher accumulation in Rev1 of enoyl co-enzyme A hydratase and acyl-CoA dehydrogenase proteins, two enzymes that are necessary in the synthesis of acetyl-CoA, supported this observation.

Global proteome analysis of other *Brucella* species was also undertaken. For instance, a comparative analysis of *B. abortus* 2308 and *B. melitensis* 16M proteomes at pH 4.0–6.0 has been conducted using 2D gel electrophoresis (Horn, 2002). A total of 575 and 549 protein spots were noted for *B. melitensis* and *B. abortus*, respectively. However, since the *B. abortus* genome is not completely annotated, matched spots between the two species have been assigned putative identification based on *B. melitensis* (Wagner et al., 2002). The proteomes of other *Brucella* species have been investigated by 2D gel electrophoresis, including those of *B. suis*, *B. ovis*, *B. canis*, and *B. neotomae*. It is clear from the 2D protein patterns that one species of *Brucella* can be distinguished from the other by comparing respective proteomes (DelVecchio and Mujer, 2004).

A comprehensive proteomics analysis of the isogenic mutants of *B. abortus* 2308 lacking functional *virB1*, *virB10*, and *virB11* genes has been conducted (DelVecchio and Mujer, 2004). Ten proteins were detected in the culture filtrate extract or "secretome" (proteins secreted into the growth medium) of the wild type 2308 but were absent in the secretomes of polar and nonpolar *virB10* mutants. The proteins that were differentially expressed included DnaK chaperone, choloyl-glycine hydrolase, aspartate aminotransferase, polyribonucleotide nucleotidyltransferase, phosphoserine aminotransferase, 6-phosphoglucolactonase, alkyl hydroperoxide reductase, dihydrolipoamide dehydrogenase, cytosol aminopeptidase, and leucine–isoleucine–valine–threonine– alanine-binding protein. These data suggest that the secreted proteins pass through the type IV secretion system encoded by the *virB* operon. Some of these proteins are potential candidates for putative virulence factors and may be useful in the development of improved diagnostic assays, antimicrobial drugs, and vaccines.

## 19.4 PROTEOMICS-BASED APPROACHES TO VACCINE DEVELOPMENT

Several strategies have been pursued in the development of next-generation vaccines. These include development of subunit vaccines (Oliveira and Splitter, 1996), vaccination with live bacterial vectors expressing *Brucella* antigens (Oñate et al., 1999), overexpression of protective homologous antigen (Vemulapalli et al., 2000), and immunization with plasmid DNA encoding the protective antigen (Oñate et al., 2003). Critical to any vaccine development strategy is the identification of proteins which have the potential to elicit an immunogenic response. Several *Brucella* immunogenic antigens have been found in the outer membrane. Together with certain secretome proteins, bacterial cell surface antigens are prime candidates for use in future vaccines as they represent the initial point of contact between the pathogen and its host. Thus, the identification of immunogenic membrane proteins is important in vaccine design and development. Proteomics methods using 2D gel electrophoresis and Western immunoblotting with antisera collected from infected animals provide a means of identification of the humoral immunogenic proteins. Narrow-pH-range immobilized pH gradient (IPG) strips are used to zoom in and achieve a greater resolution for selecting protein vaccine candidates. Teixeira-Gomes et al. (1997a) used a similar method to identify several immunogenic proteins from laboratory-grown *B. melitensis* strain B1115 and *B. ovis*. In these and other 2D studies, protein spots were identified by Edman sequencing and Western blotting (Teixeira-Gomes et al., 1997a, b). A more comprehensive listing of proteins that were reported to activate the immune response is presented in Table 19.1.

Currently, our laboratory is conducting a comprehensive identification of membrane proteins of *B. abortus* 2308 as well as selecting a few attenuated knockout mutants using 2D gel electrophoresis (Figs. 19.1 *a–d*). Three knockout mutants were used. *Brucella abortus* Δ*pgm* is an unmarked deletion mutant of the gene coding for phosphoglucomutase that catalyzes the interconvertion of glucose-6-phosphate to glucose-1-phosphate (Fig. 19.1*b*). The mutant does not synthesize the sugar nucleotide uridine diphosphate (UDP)–glucose and is unable to form any polysaccharide containing glucose, galactose, or any other sugar whose synthesis proceeds through a glucose–nucleotide intermediate. The Δ*pgm* mutant has a rough-lipopolysaccharide phenotype, is avirulent in mice, and induces protection levels comparable to those induced by vaccine strain S19 (Ugalde et al., 2003). These characteristics suggest that Δ*pgm* could be a useful vaccine strain to improve the immunological status of livestock, particularly in those countries with a large cattle industry. *Brucella abortus* Δ *pcs* is a knockout mutant of the gene coding for phosphatidylcholine synthase, the enzyme that condenses choline directly with cytidine diphosphate (CDP)–diacylglyceride to form phosphatidylcholine, a rare phospholipid in bacterial membranes but one of the major membrane components in *Brucella* (Fig. 19.1*c*). This mutant displays alteration in the outer membrane protein profiles that is currently being characterized (D. J. Comerci, manuscript in preparation). *Brucella abortus* Δ*cgh* is an unmarked deletion mutant of the gene coding for a choloylglycine hydrolase–like protein (Fig. 19.1*d*). This mutant has a severe defect in outer membrane biogenesis, displays reduced adherence to and invasiveness of macrophage cells, and is resistant to high concentration of the cationic peptide polymyxin B (M. I. Marchesini, unpublished). Analysis of these mutants has indicated that proteins involved in virulence are differentially expressed when compared to the virulent strain. Identification of these proteins and comparison of the survival and physiology of virulent versus mutant *Brucella* strains will aid in understanding survival inside the host macrophages.

Western blots of 2D gels have been probed with antisera from rabbits, goat, and cattle infected with *Brucella*. The use of different antisera for Western blot analysis is necessary since the specific host immune response may differ between species. Indeed, in our laboratory, the patterns of immunogenic proteins differ in 2D blots probed with

**TABLE 19.1** ***Brucella*** **Immunogenic Proteins**

| Immunogenic Protein | Reference |
|---|---|
| Alanyl aminopeptidase | Contreras-Rodriguez et al., 2003 |
| Amino acid ABC-type transporter | Teixeira-Gomes et al., 1997a |
| BA 14K | Chirhart-Gilleland et al., 1998 |
| Bacterioferritin | Denoel et al., 1997a; Al-Mariri et al., 2002 |
| BCSP 31 | Bricker et al., 1988 |
| BP 26 | Debbarh et al.,1996 |
| ClpP (27 kDa) | Teixeira-Gomes et al., 1997a |
| CP24 | Vizcaino et al., 1996 |
| DNA K | Teixeira-Gomes et al., 1997a |
| Dihydrolipoamide succinyltransferase | Teixeira-Gomes et al., 1997a |
| GroEL | Stevens et al., 1997 |
| GroES | Oliveira et al., 1996 |
| HtrA | Roop et al., 1994 |
| Leu/Ile/Val-binding protein precursor | Teixeira-Gomes et al., 1997a |
| L7/L12 ribosomal protein | Oliveira and Splitter, 1996 |
| Lumazine synthase (18 kDa) | Velikovsky et al., 2002, 2003 |
| Malate dehydrogenase | Teixeira-Gomes et al., 1997a |
| NikA (58 kDa) | Teixeira-Gomes et al., 1997a |
| Omp 89 kDa | Limet et al., 1993 |
| Omp 31 | Kittelberger et al., 1995 |
| Omp 28 | Lindler et al., 1996 |
| Omp 25 | Bowden et al., 1998 |
| Omp 19 | Kovach et al., 1997 |
| P39 | Denoel et al., 1997; Letesson et al., 1997 |
| P15 | Letesson et al., 1997 |
| Succinyl-CoA synthetase $\alpha$ subunit | Teixeira-Gomes et al., 1997a |
| Superoxide dismutase (Cu, Zn) | Muñoz-Montesino et al., 2004; Onate et al., 1999; Tabatabai and Pugh, 1994; Vemulapalli et al., 2000 |
| UvrA | Oliveira et al., 1996 |
| YajC | Vemulapalli et al., 1998 |
| 32.2-kDa protein | Cespedes et al., 2000 |
| 31-kDa protein | Teixeira-Gomes et al., 1997a |
| 22.9-kDa protein | Cespedes et al., 2000 |
| 20-kDa protein | Zygmunt et al., 1992 |
| 17-kDa protein | Hemmen et al., 1995 |

rabbit antisera from those probed with bovine antisera (Fig. 19.2). Protein identification is currently underway to reveal which proteins are uniquely immunogenic in each species. It is anticipated that the use of specific immunogenic proteins in vaccines design will result in more effective and safer vaccines for domesticated and wild animals as well as humans.

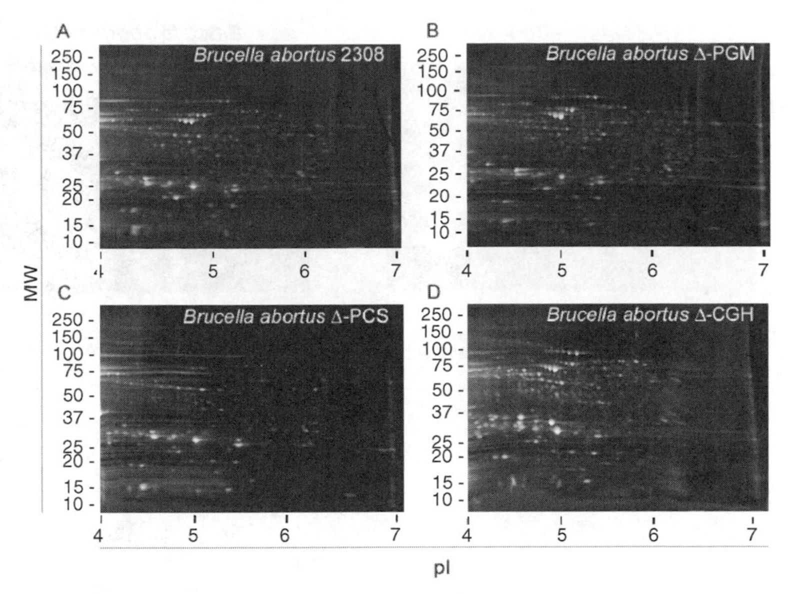


**Figure 19.1** Membrane proteomes of virulent and knockout mutant strains of *B. abortus* visualized using 2D gel electrophoresis. Each strain was grown under identical laboratory conditions and the membrane protein fraction was collected by ultracentrifugation. Purified membrane proteins (25 μg total) were focused on a pH 4–7 IPG strip and then run in the second dimension on a 6–12% SDS PAGE gel. Proteins in each gel were visualized after Sypro ruby staining. Four strains are represented (virulent strain 2308, knockout strain ΔPGM (phosphoglucomutase), knockout strain ΔCGH (choloylglycine hydrolase), and knockout strain ΔPCS (phosphatidylcholine synthase). The differences between the virulent 2308 strain and each knockout strain are easily visualized using proteomics.

## 19.5 BACTERIAL GHOSTS AS NEW BRUCELLOSIS VACCINE STRATEGY

The immune response to *Brucella* infection has been thoroughly characterized (Ko and Splitter, 2003). Both the humoral and cell-mediated immune response of the host are activated during infection; however, they are not totally effective since the bacteria continue to grow and multiply inside host macrophages. Because of this intracellular immune evasion strategy utilized by *Brucella*, currently used vaccines have limited potency. In order to design a more effective vaccine, a new strategy is necessary to control the infection by simultaneously killing the *Brucella* inside the infected macrophages and inducing a strong immune response.

One relatively new vaccine delivery system makes use of bacterial ghosts (BGs) to produce nonliving and nonpathogenic shells of gram-negative bacteria (for a review see Lubitz, 2001; Ebensen et al., 2004; Paukner et al., 2004). Bacterial ghosts are formed by the expression of the cloned *E* gene from bacteriophage ΦX174 placed under the transcriptional control of a chemical or thermosensitive inducible promoter system (Jechlinger et al., 1999; Szostak et al., 1996). The *E* gene codes for a 91-amino-acid protein subunit. Individual subunits combine and fuse with inner and outer membranes of gram-negative bacteria forming an *E*-specific lysis tunnel. The tunnel can vary in diameter from 40 to 200 nm through which the cytoplasmic content is expelled (Witte et al., 1990)

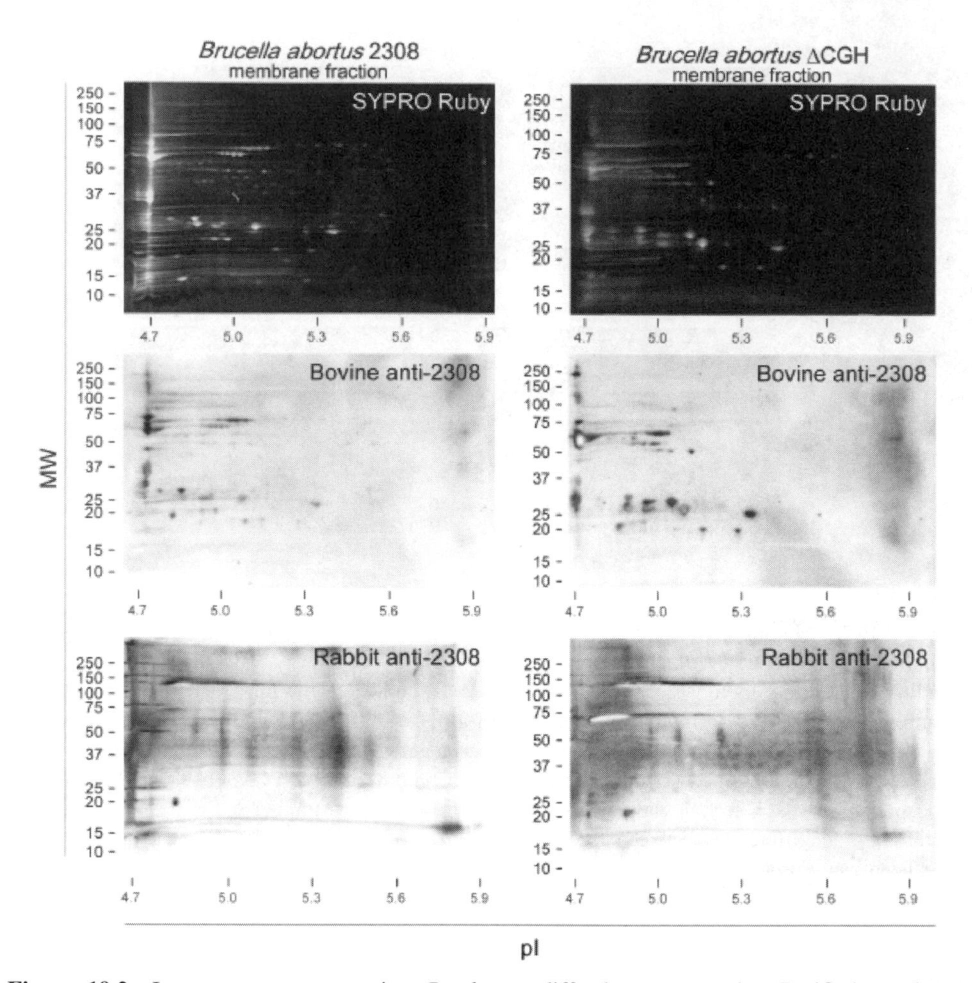


**Figure 19.2** Immune responses against *B. abortus* differ between species. Purified membrane proteins of the virulent (2308) or knockout mutant (ΔCGH) of *B. abortus* were separated using 2D gel electrophoresis [Isoelectric point (pI) 4.7–5.9]. Two-dimensional gels were then used for either visualization of total protein (Sypro ruby staining) or for Western blot analysis. Blots were probed with either whole sera from *B. abortus*–infected rabbit or cattle. The proteins visualized after Western blot analysis differ depending on the species of animal from which the primary sera was generated.

(Fig. 19.3). The subsequent BGs are virtually devoid of nucleic acids, ribosomes, or other cytoplasmic components, yet inner and outer membrane constituents are preserved. After *E*-mediated cell lysis, each BG production batch contains no viable cells and endonucleases can be incorporated into BGs to ensure that no nucleic acids are present (Haidinger et al., 2003). Bacterial ghosts can be produced in large quantities by fermentation and are stable as freeze-dried material for long periods of time without cold storage. They also have adjuvant properties. The simplicity of BG production and the versatility in incorporating and packaging both target antigens and nucleic acids make them extremely suitable for vaccine development.

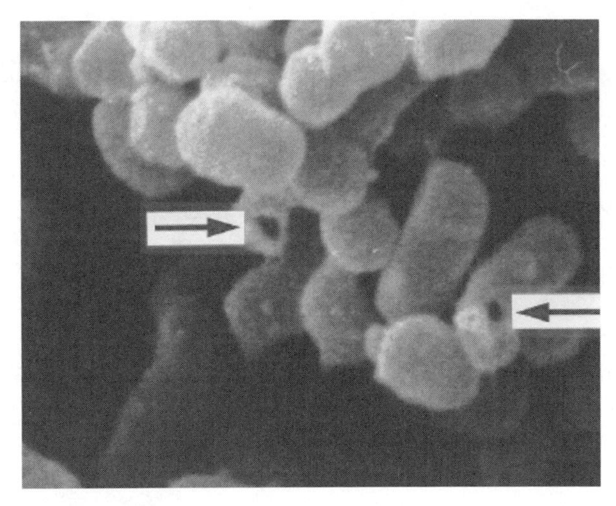

**Figure 19.3** Scanning electron micrograph of *Mannheimia haemolytica* ghosts. The arrows indicate the *E*-specific lysis tunnel.

*Brucella* candidate immunogenic proteins can be placed on the outer and inner membranes, exported into the periplasmic space, or expressed as S-layer fusion proteins which form shelllike self-assembly structures filling either the periplasmic space or cytosol (Lubitz et al., 1999). Thus, synthesis of BGs engineered to express specific immunogenic *Brucella* proteins offers an effective and safe approach for vaccine development (Fig. 19.4). Additionally, BGs can also be used as a delivery system for antimicrobial compounds since it is possible to fill the internal space with a substitute matrix and seal its lysis tunnel (Paukner et al., 2003). Thus, BGs can be used as an antibiotic carrier that can directly target macrophages, including those infected with *Brucella*. Once engulfed, antibiotics against *Brucella* are released, thereby killing and clearing the bacteria from the infected macrophage. In this capacity, BGs represent a modern-day bacterial Trojan horse. Alternatively, since species infects a specific host, antibiotic-loaded BGs can be produced using each of the known *Brucella* species as ghosts ensuring that a more host-specific vaccine is generated.

## 19.6 CONCLUSIONS AND FUTURE DIRECTIONS

The global characterization of the *Brucella* proteome has been initiated following the landmark sequencing of its genome. So far, a baseline reference map of an annotated proteome has been generated that will be useful for future studies of the proteomes of other *Brucella* species and vaccinal strains. Proteomics analysis will continually advance our understanding of the biology of this organism, particularly in areas concerning pathogenicity, attenuation of virulence, infection inside the host, host responses, and evolutionary relatedness. More particularly, proteomics data will help define the systematic classification of the various species, a subject of current debate in the *Brucella* scientific community. While a majority of the proteins so far identified are soluble in nature, progress is being made in the identification of membrane and secretome proteins. Characterization of these proteins is most valuable in our search for a safer and more

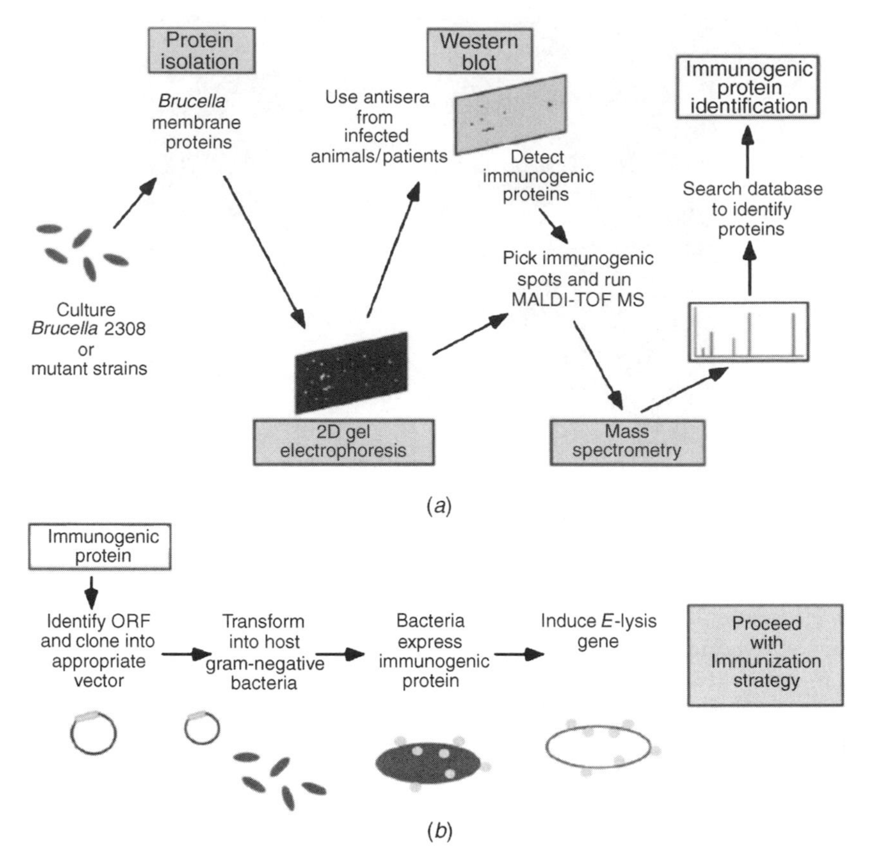


**Figure 19.4** Proposed strategy to combine proteomics and bacterial ghost vaccine technology to develop new vaccines against *Brucella*. (*a*). The strategy for identifying immunogenic proteins from *Brucella* begins with the isolation of membrane proteins from bacterial cell cultures. Isolated membrane proteins are then separated using 2D electrophoresis and used for Western blot analysis. Immunogenic proteins detected by Western blotting are picked from a Sypro ruby-stained gel and identified using mass spectrometry. (*b*). Identified immunogenic proteins are cloned into a specialized vector for creating bacterial ghosts. Host ghost cells are transformed with the vector and begin to produce the candidate immunogenic protein. After an initial time period where the immunogenic protein is made, induction of the *E* lysis gene kills the cell by creating a hole in the bacterial cell membrane, which results in a release of almost all cytoplasmic content. The void and nonliving bacterial ghost can then be used in vaccination.

effective vaccine for humans and for domesticated and wild animals. Further, they are useful for diagnostics and rapid fingerprinting of the different *Brucella* species.

While the design of effective vaccines is essential, the delivery of antimicrobial drugs to infected cells is equally effective as an antimicrobial for these intracellular pathogens. The same strategy presented here for determining immunogenic proteins for vaccination can be utilized to search for drug delivery mechanisms. For example, the membrane protein profile of infected versus noninfected macrophages could be easily examined using ICAT (isotope-coded affinity tagging) and mass spectrometry (Gygi et al., 1999, 2002). Unique surface receptor proteins up regulated by infected macrophages could then be identified and used as target proteins for BG binding. Bacterial ghosts that are engineered to contain a ligand for the unique surface receptor and contain an anti-*Brucella* antibiotic would then specifically target and enter infected macrophages.

The marriage of two relatively new technologies, *Brucella* proteomics and BG production, may yet revolutionize our current approaches to developing more efficient and safer vaccines and drug delivery systems for *Brucella*. *Brucella* ghosts can potentially be used for mitigating initial infection and as part of a clearance system designed to eliminate *Brucella*-infected cells.

## REFERENCES


Al-Mariri, A., Tibor, A., Lestrate, P., Mertens, P., De Bolle, X., and Letesson, J. J. (2002). *Yersinia enterocolitica* as a vehicle for a naked DNA vaccine encoding *Brucella abortus* bacterioferritin or P39 antigen. *Infec. Immun.* **70**:1915–1923.

Bowden, R. A., Cloeckaert, A., Zygmunt, M. S., and Dubray, G. (1998). Evaluation of immunogenicity and protective activity in BALB/c mice of the 25 kDa major outer membrane protein of *Brucella melitensis* (Omp25) expressed in *Escherichia coli*. *J. Med. Microbiol.* **47**: 39–48.

Boschiroli, M. L., Foulongne, V., and O'Callaghan, D. (2001). Brucellosis, a worldwide zoonosis. *Curr. Opin. Microbiol.* **4**:58–64.

Bricker, B. J., Tabatabai, L. B., Deyoe, B. L., and Mayfield, J. E. (1988). Conservation and antigenicity in a 31-kDa *Brucella* protein. *Vet. Microbiol.* **18**:313–325.

Bricker, B. J., Ewalt, D. S., MacMillan, A. P., Foster, G., and Brew, S. (2000). Molecular characterization of *Brucella* strains isolated from marine mammals. *J. Clin. Microbiol.* **38**:1258–1262.

Celli, J., de Chastellier, C., Franchini, D. M., Pizarro-Cerda, J., Moreno, E., and Gorvel, J. P. (2003). *Brucella* evades macrophage killing via *Vir*B-dependent sustained interactions with the endoplasmic reticulum. *J. Exp. Med.* **198**:545–556.

Celli, J., and Gorvel, J. P. (2004). Organelle robbery: *Brucella* interactions with the endoplasmic reticulum. *Curr. Opin. Microbiol.* **7**:93–97.

Cespedes, S., Andrews, E., Folch, H., and Oñate, A. (2000). Identification and partial characterization of a new protective antigen of *Brucella abortus*. *J. Med. Microbiol.* **49**:165–170.

Chirhart-Gilleland, R. L., Kovach, M. E., Elzer, P. H., Jennings, S. R., and Roop II, R. M. (1998). Identification and characterization of a 14-kilodalton *Brucella abortus* protein reactive with antibodies from naturally and experimentally infected hosts and T lymphocytes from experimentally infected BALB/c mice. *Infect. Immun.* **66**:4000–4003.

Comerci, D. J., Martinez-Lorenzo, M. J., Sieira, R., Gorvel, J. P., and Ugalde, R. A. (2001). Essential role of the *Vir*B machinery in the maturation of the *Brucella abortus*-containing vacuole. *Cell. Microbiol.* **3**:159–168.

Contreras-Rodriguez, A., Ramirez-Zavala, B., Contreras, A., Schurig, G. G., Sriranganathan, N., and Lopez-Merino, A. (2003). Purification and characterization of an immunogenic aminopeptidase of *Brucella melitensis*. *Infect. Immun.* **71**:5238–5244.

Corbel, M. J. (1997). Brucellosis:An overview. *Emerg. Infect. Dis.* **3**:213–221.

Corbel, M. J., and Brinley-Morgan, W. J. (1984). Genus *Brucella* Meyer and Shaw 1920, 173AL. In N. R. Krieg and J. G. Holt (Eds.), *Bergey's Manual of Systematic Bacteriology*, Vol. 1. Baltimore, MD: Williams & Wilkins, pp. 377–388.

Debbarh, H. S., Zygmunt, M. S., Dubray, G., and Cloeckaert, A. (1996). Competitive enzyme-linked immunosorbent assay using monoclonal antibodies to the *Brucella melitensis* BP26 protein to evaluate antibody responses in infected and *B. melitensis* Rev1 vaccinated sheep. *Vet. Microbiol.* **53**:325–337.

Delrue, R. M., Martinez-Lorenzo, M., Lestrate, P., Danese, I., Bielarz, V., Mertens, P., De Bolle, X., Tibor, A., Gorvel, J. P., and Letesson, J. J. (2001). Identification of *Brucella* spp. genes involved in intracellular trafficking. *Cell. Microbiol.* **3**:487–497.

DelVecchio, V. G., Kapatral, V., Redkar, R. J., Patra, G., Mujer, C., Los, T., Ivanova, N., Anderson, I., Bhattacharyya, A., Lykidis, A., Reznik, G., Jablonski, L., Larsen, N., D'Souza, M., Bernal, A., Mazur, M., Goltsman, E., Selkov, E., Elzer, P. H., Hagius, S., O'Callaghan, D., Letesson, J. J., Haselkorn, R., Kyrpides, N., and Overbeek, R. (2002a). The genome sequence of the facultative intracellular pathogen *Brucella melitensis*. *Proc. Natl. Acad. Sci. USA.* **99**:443–448.

DelVecchio, V. G., Wagner, M. A., Eschenbrenner, M., Horn, T. A., Kraycer, J. A., Estock, F., Elzer, P., and Mujer, C. V. (2002b). *Brucella* proteomes—a review. *Vet. Microbiol.* **90**:593–603.

DelVecchio, V. G., Kapatral, V., Patra, G., and Mujer, C. V. (2002c). The genome of *Brucella melitensis*. *Vet. Microbiol.* **90**:575–580.

DelVecchio, V. G., and Mujer, C. V. (2004). Comparative proteomics of *Brucella* species. In I. López-Goñi and I. Moriyón (Eds.), *Frontier in the Molecular and Cellular Biology of Brucella*. Horizon Scientific Press Norfolk. pp. 103–115.

Denoel, P. A., Vo, T. K., Weynants, V. E., Tibor, A., Gilson, D., Zygmunt, M. S., Limet, J. N., and Letesson, J. J. (1997a). Identification of the major T-cell antigens present in the *Brucella melitensis* B115 protein preparation, Brucellergene OCB. *J. Med. Microbiol.* **46**:801–806.

Denoel, P. A., Vo, T. K., Tibor, A., Weynants, V. E., Trunde, J. M., Dubray, G., Limet, J. N., and Letesson, J. J. (1997b). Characterization, occurrence, and molecular cloning of a 39-kilodalton *Brucella abortus* cytoplasmic protein immunodominant in cattle. *Infect Immun.* **65**:495–502.

Detilleux, P. G., Deyoe, B. L., and Cheville, N. F. (1990). Penetration and intracellular growth of *Brucella abortus* in nonphagocytic cells *in vitro*. *Infect. Immun.* **58**:2320–2328.

Detilleux, P. G., Deyoe, B. L., and Cheville, N. F. (1991). Effect of endocytic and metabolic inhibitors on the internalization and intracellular growth of *Brucella abortus* in Vero cells. *Am. J. Vet. Res.* **52**:1658–1664.

Ebensen, T., Paukner, S., Link, C., Kudela, P., De Domenico, C., Lubitz, W., and Guzman, C. A. (2004). Bacterial ghosts are an efficient delivery system for DNA vaccines. *J. Immunol.* **172**: 6858–6865.

Eschenbrenner, M., Wagner, M. A., Horn, T. A., Kraycer, J. A., Mujer, C. V., Hagius, S., Elzer, P., and DelVecchio, V. G. (2002). Comparative proteome analysis of *Brucella melitensis* vaccine strain Rev 1 and a virulent strain, 16M. *J. Bacteriol.* **184**:4962–4970.

Gorvel, J. P., and Moreno, E. (2002). *Brucella* intracellular life: From invasion to intracellular replication. *Vet. Microbiol.* **90**:281–297.

Gross, A., Terraza, A., Marchant, J., Bouaboula, M., Ouahrani-Bettache, S., Liautard, J. P., Casellas, P., and Dornand, J. (2000). A beneficial aspect of a CB1 cannabinoid receptor antagonist: SR141716A is a potent inhibitor of macrophage infection by the intracellular pathogen *Brucella suis*. *J. Leukoc. Biol.* **67**:335–344.

Gygi, S. P., Rist, B., Gerber, S. A., Turecek, F., Gelb, M. H., and Aebersold, R. (1999). Quantitative analysis of complex protein mixtures using isotope-coded affinity tags. *Nat. Biotechnol.* **17**:994–999.

Gygi, S. P., Rist, B., Griffin, T. J., Eng, J., and Aebersold, R. (2002). Proteome analysis of low-abundance proteins using multidimensional chromatography and isotope-coded affinity tags. *J. Proteome Res.* **1**:47–54.

Haidinger, W., Mayr, U. B., Szostak, M. P., Resch, S., and Lubitz, W. (2003). *Escherichia coli* ghost production by expression of lysis geneE and *Staphylococcal nuclease*. *Appl. Environ. Microbiol.* **69**:6106–6113.

Horn, T. A. (2002). Comparative proteomic study of *Brucella melitensis* strain 16M and *Brucella abortus* strain 2308. MS Thesis, University of Scranton, Scranton, PA.

Hemmen, F., Weynants, V., Scarcez, T., Letesson, J. J., and Saman, E. (1995). Cloning and sequence analysis of a newly identified *Brucella abortus* gene and serological evaluation of the 17-kilodalton antigen that it encodes. *Clin. Diagn. Lab. Immunol.* **2**:263–267.

Jahans, K. L., Foster, G., and Broughton, E. S. (1997). The characterization of *Brucella* strains isolated from marine mammals. *Vet. Microbiol.* **57**:373–382.

Jechlinger, X. W., Szostak, M. P., Witte, A., and Lubitz, W. (1999). Altered temperature induction sensitivity of lambda pR/cI857 system for controlled gene *E* expression in *Escherichia coli. FEMS Microbiol. Lett.* **173**:347–352.

Jones, S. M., and Winter, A. J. (1992). Survival of virulent and attenuated strains of *Brucella abortus* in normal and gamma interferon-activated murine peritoneal macrophages. *Infect. Immun.* **60**:3011–3014.

Kittelberger, R., Hilbink, F., Hansen, M. F., Ross, G. P., de Lisle, G. W., Cloeckaert, A., and de Bruyn, J. (1995). Identification and characterization of immunodominant antigens during the course of infection with *Brucella ovis. J. Vet. Diagn. Investig.* **7**:210–218.

Ko, J., and Splitter, G. A. (2003). Molecular host-pathogen interaction in brucellosis: Current understanding and future approaches to vaccine development for mice and humans. *Clin. Microbiol. Rev.* **16**:65–78.

Kohler, S., Laysaac, M., Naroeni, A., Gentschev, I., Rittig, M., and Liautard, J. P. (2001). Secretion of listeriolysin by *Brucella suis* inhibits its intramacrophagic replication. *Infect. Immun.* **69**:2753–2756.

Kovach, M. E., Elzer, P. H., Robertson, G. T., Chirhart-Gilleland, R. L., Christensen, M. A., Peterson, K. M., and Roop II, R. M. (1997). Cloning and nucleotide sequence analysis of a *Brucella abortus* gene encoding an 18 kDa immunoreactive protein. *Microbial Pathogenesis.* **22**:241–246.

Letesson, J. J., Tibor, A., van Eynde, G., Wansard, V., Weynants, V., Denoel, P., and Saman, E. (1997). Humoral immune responses of *Brucella*-infected cattle, sheep, and goats to eight purified recombinant *Brucella* proteins in an indirect enzyme-linked immunosorbent assay. *Clin. Diag. Lab. Immun.* **4**:556–564.

Limet, J. N., Cloeckaert, A., Bezard, G., Van Broeck, J., and Dubray, G. (1993). Antibody response to the 89 kDa outer membrane protein of *Brucella* in bovine brucellosis. *J. Med. Microbiol.* **39**:403–407.

Lindler, L. E., Hadfield, T. L., Tall, B. D., Snellings, N. J., Rubin, F. A., Van De Verg, L. L., Hoover, D., and Warren, R. L. (1996). Cloning of a *Brucella melitensis* group 3 antigen gene encoding Omp28, a protein recognized by the humoral immune response during human brucellosis. *Infect. Immun.* **64**:2490–2499.

Lubitz, W. (2001). Bacterial ghosts as carrier and targeting systems. *Exp. Opin. Biol. Ther.* **1**: 765–771.

Lubitz, W., Witte, A., Eko, F. O., Kamal, M., Jechlinger, W., Brand, E., Marchart, J., Haidinger, W., Huter, V., Felnerova, D., Stralis-Alves, N., Lechleitner, S., Melzer, H., Szostak, M. P., Resch, S., Mader, H., Kuen, B., Mayr, B., Mayrhofer, P., Geretschlager, R., Haslberger, A., and Hensel, A. (1999). Extended recombinant bacterial ghost system. *J. Biotechnol.* **73**:261–273.

Marchart, J., Dropmann, G., Lechleitner, S., Schlapp, T., Wanner, G., Szostak, M. P., and Lubitz, W. (2003). *Pasteurella multocida-* and *Pasteurella haemolytica*-ghosts:New vaccine candidates. *Vaccine* **21**:3988–3997.

Moreno, E. (2002). Brucellosis in Central America. *Vet. Microbiol.* **90**:31–38.

Mujer, C. V., Wagner, M. A., Eschenbrenner, M., Horn, T. A., Kraycer, J. A., Redkar, R., and DelVecchio, V. G. (2002). Global analysis of *Brucella melitensis* proteomes: Its potential use in vaccine development, identification of virulence proteins and establishing evolutionary relatedness. *Ann. the N. Y. Acad. Sci.* **969**:97–101.

Mujer, C. V., Wagner, M. A., Eschenbrenner, M., Estock, F., and DelVecchio, V. G. (2003). Principles and applications of proteomics in *Brucella* research. In V. G. DelVecchio and V. Krcmery (Eds.)

*Applications of Genomics and Proteomics for Analysis of Bacterial Biological Warfare Agents*, Vol. 352. Amsterdam: IOS Press, pp. 25–42.

Muñoz-Montesino, C., Andrews, E., Rivers, R., González-Smith, A., Moraga-Cid, G., Folch, H., Céspedes, S., and Oñate, A. A. (2004). Intraspleen delivery of a DNA vaccine coding for superoxide dismutase (SOD) of *Brucella abortus* induces SOD-specific $CD4^+$ and $CD8^+$ T cells. *Infect. Immun.* **72**:2081–2087.

Oliveira, S. C., and Splitter, G. A. (1996). Immunization of mice recombinant L7/L12 ribosomal protein confers protection against *Brucella abortus* infection. *Vaccine* **14**:959–962.

Oliveira, S. C., Harms, J. S., Banai, M., and Splitter, G. A. (1996). Recombinant *Brucella abortus* proteins that induce proliferation and gamma-interferon secretion by CD4+ T cells from *Brucella*-vaccinated mice and delayed-type hypersensitivity in sensitized guinea pigs. *Cell Immunol.* **172**:262–268.

Oñate, A. A., Céspedes, S., Cabrera, A., Rivers, R., González, A., Muñoz, C., Folch, H., and Andrews, E. (2003). A DNA vaccine encoding Cu, Zn superoxide dismutase of *Brucella abortus* induces protective immunity in BALB/c mice. *Infect. Immun.* **71**:4857–4861.

Oñate, A. A., Vemulapalli, R., Andrews, E., Schurig, G. G., Boyle, S., and Folch, H. (1999). Vaccination with live *Escherichia coli* expressing *Brucella abortus* Cu/Zn superoxide dismutase protects mice against virulent *B. abortus*. *Infect. Immun.* **67**:986–988.

Paukner, S., Kohl, G., Jalava, K., and Lubitz, W. (2003). Sealed bacterial ghosts—novel targeting vehicles for advanced drug delivery of water-soluble substances. *J. Drug Target.* **11**:151–161.

Paukner, S., Kohl, G., and Lubitz, W. (2004). Bacterial ghosts as novel advanced drug delivery systems: Antiproliferative activity of loaded doxorubicin in human Caco-2 cells. *J. Control Release* **94**:63–74.

Riley, L. K., and Robertson, D. C. (1984). Ingestion and intracellular survival of *Brucella abortus* in human and bovine polymorphonuclear leukocytes. *Infect. Immun.* **46**:224–230.

Roop II, R. M., Fletcher, T. W., Sriranganathan, N. M., Boyle, S. M., and Schurig, G. G. (1994). Identification of an immunoreactive *Brucella abortus* HtrA stress response protein homolog. *Infect Immun.* **62**:1000–1007.

Roop II, R. M., Bellaire, B. H., Valderas, M. W., and Cardelli, J. A. (2004). Adaptation of the brucellae to their intracellular niche. *Mol. Microbiol.* **52**:621–630.

Samartino, L. E. (2002). Brucellosis in Argentina. *Vet. Microbiol.* **90**:71–80.

Sola-Landa, A., Pizarro-Cerda, J., Grillo, M. J., Moreno, E., Moriyon, I., Blasco, J. M., Gorvel, J. P., and Lopez-Goñi, I. (1998). A two-component regulatory system playing a critical role in plant pathogens and endosymbionts is present in *Brucella abortus* and controls cell invasion and virulence. *Mol. Microbiol.* **29**:125–138.

Stevens, M. G., Olsen, S. C., Pugh, G. W., and Mayfield, J. E. (1997). Role of immune responses to a GroEL heat shock protein in preventing brucellosis in mice vaccinated with *Brucella abortus* strain RB51. *Comp. Immunol. Microbiol. Infect. Dis.* **20**:147–153.

Szostak, M. P., Hensel, A., Eko, F. O., Klein, R., Auer, T., Mader, H., Haslberger, A., Bunka, S., Wanner, G., and Lubitz, W. (1996). Bacterial ghosts: Nonliving candidate vaccines. *J. Biotechnol.* **44**:161–170.

Tabatabai, L. B., and Pugh, G. W. Jr. (1994). Modulation of immune responses in Balb/c mice vaccinated with *Brucella abortus* Cu-Zn superoxide dismutase synthetic peptide vaccine. *Vaccine.* **12**:919–924.

Taylor, J. P., and Perdue, J. N. (1989). The changing epidemiology of human brucellosis in Texas, 1977–1986. *Am. J. Epidemiol.* **130**:160–165.

Teixeira-Gomes, A. P., Cloeckaert, A., Bézard, G., Bowden, R. A., Dubray, G., and Zygmunt, M. S. (1997a). Identification and characterization of *Brucella ovis* immunogenic proteins using two-dimensional electrophoresis and immunoblotting. *Electrophoresis.* **18**:1491–1497.

Teixeira-Gomes, A. P., Cloeckaert, A., Bézard, G., Dubray, G., and Zygmunt, M. S. (1997b). Mapping and identification of *Brucella melitensis* proteins by two-dimensional electrophoresis and microsequencing. *Electrophoresis* **18**:156–162.

Ugalde, J. E., Comerci, D. J., Leguizamon, M. S., and Ugalde, R. A. (2003). Evaluation of *Brucella abortus* phosphoglucomutase (pgm) mutant as a new live rough-phenotype vaccine. *Infect Immun.* **71**:6264–6269.

Velikovsky, C. A., Cassataro, J., Giambartolomei, G. H., Goldbaum, F. A., Estein, S., Bowden, R. A., Bruno, L., Fossati, C. A., and Spitz, M. (2002). A DNA vaccine encoding lumazine synthase from *Brucella abortus* induces protective immunity in BALB/c mice. *Infect Immun.* **70**:2507–2511.

Velikovsky, C. A., Goldbaum, F. A., Cassataro, J., Estein, S., Bowden, R. A., Bruno, L., Fossati, C. A., and Giambartolomei, G. H. (2003). *Brucella* lumazine synthase elicits a mixed Th1-Th2 immune response and reduces infection in mice challenged with *Brucella abortus* 544 independently of the adjuvant formulation used. *Infect. Immun.* **71**:5750–5755.

Vemulapalli, R., Duncan, A. J., Boyle, S. M., Sriranganathan, N., Toth, T. E., and Schurig, G. G. (1998). Cloning and sequencing of *yajC* and *secD* homologs of *Brucella abortus* and demonstration of immune responses to YajC in mice vaccinated with *B. abortus* RB51. *Infect Immun.* **66**:5684–5691.

Vemulapalli, R., He, Y., Cravero, S., Sriranganathan, N., Boyle, S. M., and Schurig, G. G. (2000). Overexpression of protective antigen as a novel approach to enhance vaccine efficacy of *Brucella abortus* strain RB51. *Infect. Immun.* **68**:3286–3289.

Verger, J. M., Grayon, M., Cloeckaert, A., Lefevre, M., Ageron, E., and Grimont, F. (2000). Classification of *Brucella* strains isolated from marine mammals using DNA-DNA hybridization and ribotyping. *Res. Microbiol.* **151**:797–799.

Verger, J. M., Grimont, F., Grimont, P. A. D., and Grayon, M. (1985). *Brucella*, a monospecific genus as shown by deoxyribonucleic acid hybridization. *Int. J. Sys. Bact.* **35**:292–295.

Vizcaino, N., Cloeckaert, A., Dubray, G., and Zygmunt, M. S. (1996). Cloning, nucleotide sequence, and expression of the gene coding for a ribosome releasing factor-homologous protein of *Brucella melitensis*. *Infect. Immun.* **64**:4834–4837.

Wagner, M. A., Eschenbrenner, M., Horn, T. A., Kraycer, J. A., Mujer, C. V., Hagius, S., Elzer, P., and DelVecchio, V. G. (2002). Global analysis of the *Brucella melitensis* proteome: Identification of proteins expressed in laboratory-grown culture. *Proteomics* **2**:1047–1060.

Witte, A., Wanner, G., Blasi, U., Halfmann, G., Szostak, M., and Lubitz, W. (1990). Endogenous transmembrane tunnel formation mediated by phi X174 lysis protein, E. *J. Bacteriol.* **172**: 4109–4114.

World Health Organization (WHO) (1998). *The Development of New/Improved Brucellosis Vaccines*, WHO document EMC/ZDI/98.14, Report of WHO Meeting, Dec. 11–12, 1997. Geneva, Switzerland: WHO, pp.1–44.

Young, E. J. (1983). Human brucellosis. *Rev. Infect. Dis.* **5**:821–842.

Zygmunt, M. S., Gilbert, F. B., and Dubray, G. (1992). Purification, characterization, and seroactivity of a 20-kilodalton *Brucella* protein antigen. *J. Clin. Microbiol.* **30**:2662–2667.

CHAPTER 20

# Genomics and Proteomics in Reverse Vaccines

GUIDO GRANDI

Chiron Vaccines, Siena, Italy

## 20.1 PROBLEM OF ORPHAN VACCINE INFECTIOUS DISEASES

The vaccines currently on the market prevent diseases caused by 25 different pathogens, among them viruses, parasites, and bacteria [1]. The list of vaccines is by far much shorter than the list of pathogens against which vaccines are not available yet. Indeed, pathogens causing devastating diseases such as malaria, tuberculosis, AIDS, hepatitis, meningitis, and sepsis and responsible for the death of several million people each year worldwide are still awaiting the development of efficacious vaccines.

The reasons for vaccine unavailability are economic, technical, or both. A paradigmatic example is *Plasmodium falciparum*. Not only does industry have to face the unfavourable economics, the potential target population being mostly concentrated in developing countries unable to cover the cost of efficacious vaccination programs, but also antimalaria vaccine discovery represents a tremendous challenge because of the complex life cycle of the parasite, which keeps changing its morphology and, when in the human host, hides itself in different intracellular compartments [2]. A second important example is *Mycobacterium tuberculosis*, for which, in addition to the same economic issue pointed out for malaria, one of the major obstacles to vaccine development is represented by the complexity of the efficacy trials that, because of the long period of latency before the outcome of the disease, would require several years to be completed. Finally, the discovery of vaccines against several bacterial pathogens has been hampered by the ability of these bacteria to modify the sequence and/or the chemical nature of their surface antigens. For these pathogens the available technologies for vaccine discovery have proven to be inadequate.

## 20.2 CHANGE OF PARADIGM: REVERSE VACCINOLOGY

The search for vaccine candidates is somewhat of a look-for-a-needle-in-the-haystack undertaking. In fact, it basically consists in the identification of the very few protective antigens among the several hundreds of components of a given pathogen.

*Microbial Proteomics: Functional Biology of Whole Organisms*, Edited by Ian Humphery-Smith and Michael Hecker. 

With their pioneering work published in *Science* in 2000 [3], Pizza and co-workers proposed a revolutionizing approach to vaccine discovery. The approach, named "reverse vaccinology" [4], stems from the simple and straightforward consideration that, if in a given pathogen there exist conserved protein antigens with protective immunological properties, they must be among the proteins encoded by the genome. Therefore, by providing the complete list of proteins of the organisms, the knowledge of genome sequences offers the opportunity to systematically analyze each protein until the ones having the desired properties are unraveled. A key to the success of reverse vaccinology is the availability of (1) robust high-throughput expression and purification systems that make the rapid production of large batteries of recombinant proteins a feasible task and (2) a reliable screening assay that correlates as much as possible with protection in humans and that allows testing several hundred recombinant proteins in a reasonably short time frame. Also important to the success of the approach is the use of selection criteria capable of discriminating between proteins likely to become vaccine candidates (e.g., surface proteins and virulence factors) from proteins of poor vaccinological interest (e.g., cytoplasmic, housekeeping enzymes). Such selection criteria can in fact reduce quite substantially the number of proteins to be tested for protective activity. In its most classical application, reverse vaccinology can be outlined as follows: (1) genome sequencing of the pathogen of interest, (2) selection of potential vaccine candidates by in silico analysis of the genome, (3) high-throughput cloning, expression, and purification of selected candidates, and (4) identification of vaccine candidates by systematic analysis of all purified proteins using appropriate in vitro and/or in vivo assays. Subsequently, it became clear that gene selection could be optimized by applying, in addition to the in silico analysis, other criteria that make the entire process more efficient [1, 5]. These criteria include DNA microarray and proteomics analyses.

In the following sections, I will present a few examples of successful applications in our laboratories of reverse vaccinology on bacterial pathogens and show the usefulness of the combined application of different gene selection strategies in candidate discovery.

## 20.3 EXAMPLES OF REVERSE VACCINES

### 20.3.1 Group B *Neisseria meningitidis*

Meningococcal meningitis and sepsis are caused by *N. meningitidis*, a gram-negative, capsulated bacterium classified into five major pathogenic serogroups (A, B, C, Y, and W135) on the basis of the chemical composition of their capsular polysaccharides [6, 7]. Very effective vaccines based on the capsular polysaccharides against meningococcus C are already on the market and anti–meningococcus A/C/Y/W polyvalent vaccines are expected to be launched in the near future. However, there are no vaccines available for the prevention of MenB disease responsible for a large proportion (from 32 to 80%) of all meningococcal infections [8]. The use of capsular polysaccharide is not feasible since in MenB its chemical composition is identical to a widely distributed human carbohydrate [$\alpha(2 \rightarrow 8)$ *N*-acetyl neuraminic acid or polysialic acid] that, being a self-antigen, is a poor immunogen in humans. Furthermore, the use of this polysaccharide in a vaccine may elicit autoantibodies [9, 10]. An alternative approach to vaccine development is based on the use of surface-exposed proteins, which in some instances have been shown to elicit protective bactericidal antibodies [11, 12]. However, many

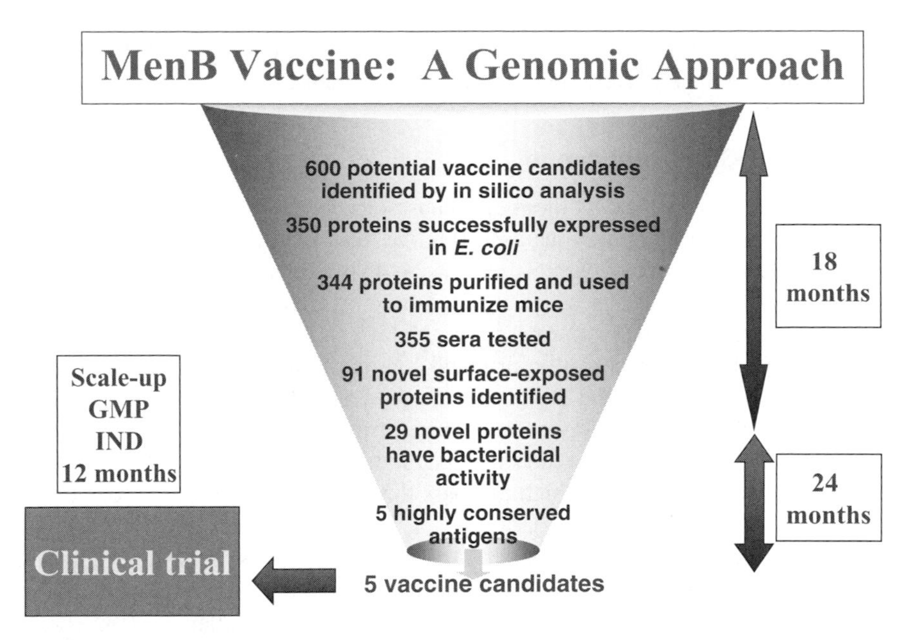


**Figure 20.1** Schematic representation of reverse vaccinology applied to *Neisseria meningitidis* group B. Once genome sequence was deciphered, computer analysis was used to predict genes potentially coding for secreted and membrane-associated proteins. All selected genes were expressed in *E. coli* and the corresponding recombinant proteins were purified. Each protein was then tested in the in vitro bactericidal assay [1], and those proteins that were capable of eliciting bactericidal antibodies were checked for conservation on a large panel of clinical isolates. Five conserved, protective antigens are currently under evaluation in phase II clinical studies. Also indicated is the time (months) necessary to complete the different experimental steps.

of the major surface protein antigens show sequence and antigenic variability, thus failing to confer protection against heterologous strains. Therefore, the challenge for anti-MenB vaccine research is the identification of highly conserved antigens eliciting protective immune responses (bactericidal antibodies) against a broad range of MenB isolates.

To achieve this goal, reverse vaccinology was applied for the first time (Fig. 20.1) [3]. The MenB genome sequence [13] was submitted to computer analysis to identify genes potentially encoding surface-exposed or exported proteins. Of the 650 proteins thus predicted, approximately 50% were successfully expressed in *Escherichia coli* as either His-tag or glutathione *S*-transferase (GST) fusion proteins. The recombinant proteins were purified and used to immunize mice and the immune sera were tested for bactericidal activity, an assay which strongly correlates with protection in humans [14]. Twenty-eight sera turned out to be bactericidal. To test sequence conservation of the protective antigens, 34 different *N. meningitidis* clinical isolates were selected and the nucleotide sequences of the corresponding genes of all protective antigens were compared. This kind of analysis led to the identification of five highly conserved antigens. Most importantly, the sera against these five antigens and, in particular, combinations of these sera were capable of killing all the meningococcal strains so far utilized in the complement-mediated bactericidal assay, thus making the antigens particularly promising for vaccine formulations. Phase II clinical studies are currently in progress to establish the ability of these antigens to induce bactericidal antibodies in humans.

### 20.3.2 Group B Streptococcus

Group B streptococcus (GBS) is the major cause of neonatal sepsis in the industrialized world, accounting for 0.5–3.0 deaths/1000 live births. Eighty percent of the GBS infections in newborns occur within the first 24–48 h after delivery [15]. This group is known as early-onset disease and is generally caused by direct transmission of the bacteria from the mother to the baby during passage through the vagina. A second peak of infections, which begins a week after birth and continues through the first month of life, is known as late-onset disease and is not associated with colonization of the mother. This latter group is often caused by infection acquired from other sources in the clinic.

Protection in humans against invasive GBS disease correlates with high titers of maternal anticapsule antibodies which can pass through the placenta and protect the child in the first months of life.

Experiments in mice have demonstrated that glycoconjugates of capsular polysaccharide with tetanus toxoid carrier protein can induce an immune response in pregnant females which can confer protection in the pups against lethal GBS challenge [16]. These data suggest that immunizing women before they become pregnant could effectively prevent the majority of invasive GBS disease in newborns. Unfortunately, there are at least nine capsular serotypes, and antibodies against any one of these fail to confer protection against the other serotypes [17–21]. Hence, a vaccine based on capsular polysaccharide would need to contain several serotype-specific antigens. An alternative approach would be to identify few conserved protein antigens eliciting protective immunity against most, preferably all, GBS serotypes. We therefore decided to apply reverse vaccinology with the aim of identifying such conserved, protective antigens.

In line with the reverse vaccinology strategy, our approach to anti-GBS vaccine discovery started with the sequencing of the genome of the pathogen. In collaboration with The Institute for Genome Research (TIGR), we determined the complete genome sequence of serotype V strain 2603 v/r of GBS [22]. The genome consists of 1,160,267 bp and contains 2175 genes. Biological roles were assigned to 1333 (61%) of the predicted proteins. A further 623 genes matched with genes of unknown function whose sequence was available in public databases, whereas 219 additional genes had no database matches.

A series of computer programs were used to identify potential signal peptides (PSORT, SignalP), transmembrane spanning regions (TMPRED), lipoproteins and cell-wall-anchored proteins (Motifs), and homology to known surface proteins in other bacteria (FastA). This analysis ultimately led to the selection of 473 genes that were subjected to the high-throughput and expression purification procedure. Each of the 473 genes was amplified by the polymerase chain reaction (PCR) and cloned in parallel into *E. coli* expression vectors containing sequences coding either for a C-terminal 6 $\times$ His tag or for an N-terminal GST domain. Overall, 357 recombinant proteins were successfully purified at a degree of purity compatible with the subsequent in vivo assays.

Among the assays potentially available for candidate screening, the most indicative is the maternal active immunization assay. According to this assay [16], female mice are first immunized with the recombinant antigens, then mated, and the resulting offspring are challenged with a lethal dose of GBS within the first 48 h of life. For the pups to be protected, immunization has to induce in the mothers sufficiently high levels of antibodies

**TABLE 20.1 In vivo Protective Activity of GBS Antigens in Mouse Active Maternal Immunization Model Using Three Different GBS Challenge Strains**

| | GBS Strain Percentage of Protection (Alive/Treated) | | |
|---|---|---|---|
| Antigen | COH1 (Type III) | CJB111 (Type V) | DK21 (Type II) |
| **GBS80** | 75% (52/69) | 70% (27/30) | 10% (3/30) [a] |
| **GBS322** | 53% (33/62) | 83% (25/30) | 67% (27/40) |
| **GBS67** | 20% (6/30)[a] | 57% (84/148) | 47% (68/145) |
| **GBS80 + 322 + 67** | 85% (36/40) | 95% (38/40) | 83% (25/30) |
| **Control** | 5% (10/195) | 27% (70/254) | 24% (57/232) |

[a]Antigen missing in this strain.

which can cross the placenta and reach the newborn mice. Using this model, four new antigens were found to confer a statistically significant protection (survival rate >30% over the background, with $P_{val} < 0.05$) against at least one of the GBS strains used for challenge. Interestingly, antigen combinations were capable of protecting up to 100% of the animals from lethal doses of different GBS strains (Table 20.1), indicating the additive, if not synergistic, effect of antigen coadministration. A particular combination of these antigens is currently in the development phase.

### 20.3.3 Group A Streptococcus

Group A streptococcus (GAS) can colonize human throat and skin, causing, in general, relatively mild diseases. When colonizing the throat, GAS is responsible for the classical strep throat which is the bane of parents of young children. Although in most cases the disease is mild, it is nevertheless very costly in terms of health care visits and days of school lost [23]. Like GBS, GAS can also cause severe invasive disease, including scarlet fever and a frequently lethal toxic shock syndrome. Perhaps of more importance are rheumatic fever (RF), rheumatic heart disease (RHD), and glomerulonephritis, the autoimmune sequelae that can follow, in some countries at high frequencies, throat and skin infection and scarlet fever. Last but not least, GAS is also known as the flesh-eating bacterium since it is one of the major causes of necrotizing fasciitis which can turn a small wound into massive necrosis and necessitates emergency measures, including extensive surgical intervention and tissue reconstruction. Overall, it has been estimated that more than 600 million people are infected annually by GAS worldwide, 500,000 thousands of whom die because of GAS invasive disease and RHD.

There is no vaccine available for GAS, and attempts to identify protective antigens have so far been unsuccessful [24]. Although encapsulated, the capsular polysaccharide is hyaluronic acid, which is found in many human tissues and is thus not immunogenic. A major immunodominant antigen, the M protein, has shown type-specific protection in both humans and animal models. However, there are over 124 known serotypes of this protein; therefore, although M-protein-based vaccines are being attempted [25], their efficacy still awaits confirmation in the clinics and they are unlikely to provide broad coverage against GAS infections.

Supported by the positive results obtained with both MenB and GBS, we decided to investigate whether reverse vaccinology could lead to the identification of protective protein antigens that, differently from M protein, were sufficiently conserved among the different GAS isolates. Taking advantage of the fact that when we started the project the genome sequences of three GAS serotypes were available, we were in the position to immediately start gene selection using both the in silico analysis (see above) and DNA microarray technology (see below). By using these selection criteria, the final candidate list for expression in *E. coli* contained 286 genes. Two hundred eighty-five recombinant proteins were successfully cloned and expressed in *E.coli* as either His-tagged proteins or GST fusions. Forty of the His-tagged proteins were insoluble and had to be solubilized in urea before use.

In order to test the candidates for their capacity to induce protective immunity, we decided to use an adult mouse model of invasive disease based on the intraperitoneal challenge of CD1 mice using a virulent M1 serotype strain [median lethal dose ($LD_{50}$) 10 colony-forming units (CFUs)]. Protection in this model implies the elicitation of circulating opsonic/bactericidal antibodies capable of preventing systemic infection in the animals. Mouse immunization with the homologous M protein conferred 100% protection in this model. Therefore, our goal was to identify antigens the combination of which could perform at least as well as our gold-standard positive control.

As shown in Figure 20.2*a*, from the screening, which is still under way, six antigens have so far been identified showing a statistically significant protective activity.

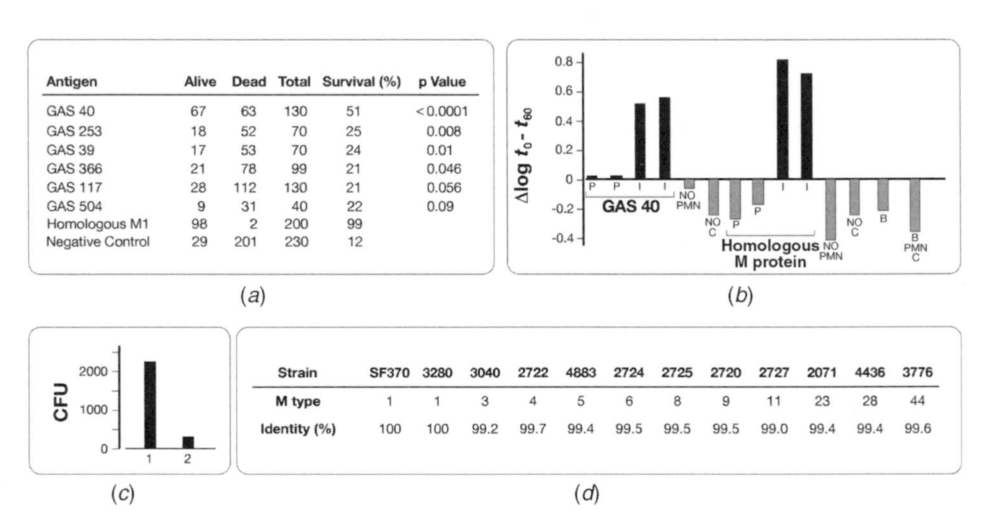

| Antigen | Alive | Dead | Total | Survival (%) | p Value |
|---|---|---|---|---|---|
| GAS 40 | 67 | 63 | 130 | 51 | <0.0001 |
| GAS 253 | 18 | 52 | 70 | 25 | 0.008 |
| GAS 39 | 17 | 53 | 70 | 24 | 0.01 |
| GAS 366 | 21 | 78 | 99 | 21 | 0.046 |
| GAS 117 | 28 | 112 | 130 | 21 | 0.056 |
| GAS 504 | 9 | 31 | 40 | 22 | 0.09 |
| Homologous M1 | 98 | 2 | 200 | 99 | |
| Negative Control | 29 | 201 | 230 | 12 | |

(*a*)



| Strain | SF370 | 3280 | 3040 | 2722 | 4883 | 2724 | 2725 | 2720 | 2727 | 2071 | 4436 | 3776 |
|---|---|---|---|---|---|---|---|---|---|---|---|---|
| M type | 1 | 1 | 3 | 4 | 5 | 6 | 8 | 9 | 11 | 23 | 28 | 44 |
| Identity (%) | 100 | 100 | 99.2 | 99.7 | 99.4 | 99.5 | 99.5 | 99.5 | 99.0 | 99.4 | 99.4 | 99.6 |

(*d*)

**Figure 20.2** Results of reverse vaccinology applied to GAS. (*a*) In vivo protective activity of GAS antigens: CD-1/c mice were first immunized with purified recombinant GAS antigens and subsequently challenged i.p. with 30 CFUs of a highly virulent M1 GAS strain ($10 \times LD_{50}$). Indicated are six protective antigens which have been selected so far. Antigen screening is still in progress. (*b*) Killing activity of anti-GAS40 and antihomologous M-protein sera. Sera from mice immunized with GAS40 and M protein were used in the opsonophagocytosis assay [26] in the presence of human PMNs, rabbit complement, and SF370 GAS strain. P: preimmune serum; I: immune serum; B: bacteria. (*c*) Bactericidal activity of blood from mice immunized with GAS40. BALB/c mice were immunized three times at two-week intervals with 20 μg of purified GAS40, and 15 days after the last immunization mice were bled and SF370 GAS strain ($10^4$ bacteria) was added to 100 μL of blood. Viable bacteria were colony counted after 3 h exposure to blood. Indicated is the total number of viable bacteria found in 10 mice. (1) Negative control. (2) GAS40 group. (*d*) Conservation of GAS 40 among GAS clinical isolates. GAS40 gene was PCR amplified from different GAS clinical isolates and the amplified genes were sequenced. Indicated is the percent of identity of the coded proteins with respect to the reference strain SF370.

Particularly promising is one antigen, GAS40, which conferred a survival rate of 51% in all experiments so far carried out. Furthermore, the antigen turned out to elicit protective antibodies, as judged by the opsonophagocytosis assay (Fig. 20.2*b*) and the bactericidal assay using the blood from the animals immunized with the protein (Fig. 20.2*c*). Importantly, the antigen also appears to be extremely well conserved in all strains so far subjected to sequence analysis (Fig. 20.2*d*). Experiments are in progress to test whether combinations of the six antigens can protect all mice from the lethal challenge of GAS.

### 20.3.4 *Chlamydia trachomatis*

Like all obligate intracellular pathogens, for its survival and propagation the gram-negative bacterium *C. trachomatis* must accomplish several essential tasks, which include adhering to and entering host cells, creating an intracellular niche for replication, exiting host cells for subsequent invasion of neighbouring cells, and avoiding host defense mechanisms [27]. To carry out all these functions, *C. trachomatis* has developed a unique biphasic life cycle involving two developmental forms, a sporelike infectious form (elementary bodies, EBs) and an intracellular replicative form (reticulate bodies, RBs). Adhesion, host cell colonization capabilities, and ability to cope with the host defense mechanisms when outside the cell presumably rely in large part on EB surface organization.

Its unique life-cycle organization renders *C. trachomatis* very successful in avoiding host immune responses and establishing a chronic infection leading often to serious diseases [27]. Chronic infection of the ocular mucosa can result in blindness, whereas in the female infection of the upper genital tract can lead to pelvic inflammatory disease, ectopic pregnancy, and sterility. Indeed, *C. trachomatis* infection is one of the most serious causes of both male and female sterility in industrialized countries. Sexually transmitted disease (STD) induced by *C. trachomatis* has been also implicated as a risk factor for the sexual transmission of other serious pathogens such as the human immunodeficiency virus (HIV) [28].

It has been estimated that approximately 90 million people become infected by *C. trachomatis* every year worldwide. This high infection incidence explains why anti-chlamydia vaccination against STD administered to 12-year-old subjects falls into the "more favorable" category (category II) of recommended vaccines according to a committee convened by the U.S. Institute of Medicine to study priorities for vaccine development (*Vaccines for the 21st Century*, Academy National Press). Antichlamydia vaccines would have further potential value if capable of preventing chlamydial ocular infection in those developing countries where trachoma is still an important cause of blindness.

Despite years of effort by several research groups around the world, a vaccine against human chlamydial infection is still unavailable. This may be attributed to several reasons, among which worth mentioning are the difficulty in culturing large quantities of the pathogen (limiting the purification of antigens to be tested in vaccine studies) and the inability to carry out any kind of genetic analysis. As a consequence of these limitations, vaccine studies have been restricted to very few chlamydial antigens, mostly tested in the mouse, intravaginally challenged with either human or mouse-adapted chlamydial isolates. From these studies, as well as from epidemiological data and vaccine trials in humans, it has been established that protection against chlamydial infection most likely correlates with both the elicitation of a $CD4^+$ T-cell-specific cytotoxic activity and a neutralizing antibody response. The data also indicate that none of the antigens so far tested are capable of conferring a consistent, robust protection in the mouse, the only efficacious vaccination being represented by the intravaginal administration of live chlamydia which protect the mouse against subsequent *C. trachomatis* challenges [29].

With this background, we decided to undertake a reverse vaccinology approach with the aim of identifying new antigens that, given alone or in combination, could confer mouse protection to a level similar to that obtained with live chlamydiae.

*Chlamydia trachomatis* genome was scanned in search of genes encoding putative surface-exposed antigens. Ninety-three genes were selected and cloned in *E. coli* and the corresponding recombinant proteins purified [30]. Each purified protein was injected into mice and the immune sera were used in two types of assays: FACS analysis of EBs to confirm their surface exposure [30, 31] and in vitro neutralization of infection [30, 32]. We found that 48 proteins out of 93 were positive to the FACS assay and 13 proteins elicited antibodies with neutralizing activity in vitro [30]. The 48 FACS-positive (surface-exposed) proteins are currently under screening in the mouse model of infection. At present, a pool of five antigens appears to confer a statistically significant protection. Interestingly, this pool included some antigens whose antiserum was neutralizing in vitro and whose homologues in *C. pneumoniae* were also protective in the mouse and/or hamster model of *C. pneumoniae* infection [32].

To the best of our knowledge, this is the first reported example of systemically injected protein antigens eliciting in the mouse a protective activity comparable to primary infection with live chlamydial EBs.

## 20.4 ADDITIONAL GENE SELECTION CRITERIA IN REVERSE VACCINOLOGY: ROLE OF TRANSCRIPTOMICS AND PROTEOMICS

As already pointed out, the success of reverse vaccinology largely relies on the ability to select, among the list of all genes, those most likely to encode protective antigens. Therefore, the use of the proper selection procedure is critical to shorten the time to vaccine candidate identification (many less recombinant proteins have to be produced and tested in the screening assay) and to reduce the risk of losing potentially good candidates. Historically, the first selection strategy used in reverse vaccinology has been the scanning of the genomes with algorithms designed to predict gene functions likely to correlate with the capacity of the encoded proteins to elicit protective immune responses. For instance, in extracellular pathogens, protective antigens usually belong to the category of surface-exposed and secreted proteins. Up to now, several algorithms are capable of predicting, with sufficient reliability, this group of functional proteins.

In silico selection has the great advantage of being relatively rapid since no experimental work is needed once the genome is fully sequenced. However, it has two drawbacks. First, it is currently unable to predict if, when, and to what extent proteins are expressed. From a reverse vaccinology viewpoint, this limitation translates into an unnecessary extra work load, since proteins that are poorly expressed in the pathogen are expected to be bad vaccine candidates and therefore could be excluded from the high-throughput screening (see below). The second drawback of the in silico approach is that it relies on algorithms not yet fully refined. Therefore, a nonnegligible proportion of potential vaccine candidates could be lost if antigen selection were solely based on computer prediction. For the above reasons, we are currently coupling in silico analysis to two other selection methods, transcriptome analysis and proteome analysis. The potential of these technologies in vaccine candidate selection is discussed below.

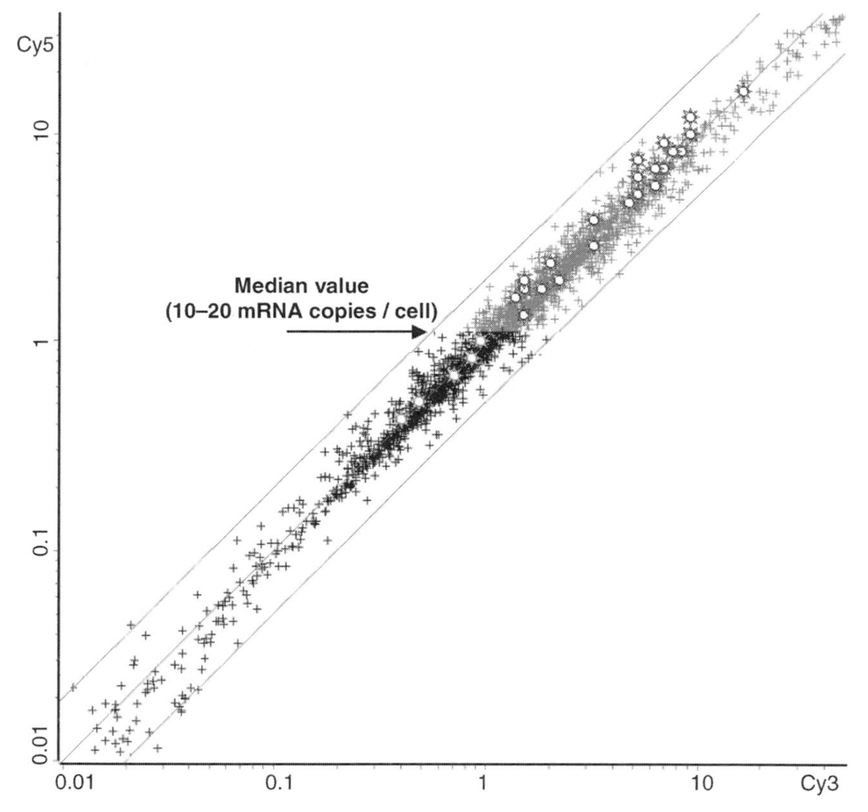


**Figure 20.3** Identification of highly expressed genes in meningococcus B by DNA microarrays. Bacteria were grown in GC medium and when the culture reached an $OD_{600}$ value of 0.3, cells were collected and RNA purified [33]. RNA was labeled with either Cy3 or Cy5 and hybridized with the whole collection of meningococcal genes (2152) spotted on a glass slide [33]. Conventionally, here highly expressed genes are considered those genes (gray spots) giving relative fluorescence signals above the median value of all the signals of the slide. Asterisks correspond to the fluorescence intensities of those genes selected by reverse vaccinology and proven to be vaccine candidates.

### 20.4.1 Transcriptomics

Global transcription profile analysis can help to select vaccine candidates in two ways.

First, since it provides quantitative indications on the transcription activity of each gene and since transcription levels often (even if not always) correlate with the relative abundance of the corresponding protein, it can be combined with in silico analysis to predict surface-exposed, highly expressed proteins. This is clearly illustrated in Figure 20.3, where the transcription activity of all *menB* genes during the exponential growth phase is analyzed. Under these experimental conditions, 50% of the genes gives a fluorescent signal above the median value, corresponding to an expression level higher than 5–10 messenger ribonucleic acid (mRNA) copies per cell. If we now look at the 28 MenB protective antigens (see Section 20.3), we find that, apart from a minor fraction, which included the least promising candidates ultimately excluded from the final vaccine formulation, all of them are expressed at a level above the threshold median value. This indicates that the a priori exclusion of the poorly expressed genes would have avoided the cloning, expression, purification, and screening of approximately 200 (30% of the 650 selected by the computer) useless proteins without losing any good candidate. On the basis of this retrospective analysis, our current strategy is to look at the highly transcribed genes first and then to select, among them, the surface-exposed ones as predicted by the available algorithms.

Second, transcription analysis can provide indications on proteins whose abundance increases during infection. These proteins are of potential vaccinological interest since preexisting immune responses against them might result in the amelioration or prevention of the disease.

An example of the usefulness of DNA microarray in vaccine discovery has been recently published by our group [33]. The first step toward meningococcal infection is the colonization of the nasopharyngeal mucosa after bacterial invasion through the respiratory tract. It is reasonable to assume that when meningococcus comes into contact with epithelial cells, it modulates the expression of specific genes to reorganize part of its surface protein repertoire, thus facilitating the adhesion/invasion process. To elucidate the molecular events leading to bacterial colonization, we used DNA microarray technology to follow the changes in gene expression profiles when *N. meningitidis* interacts with human epithelial cells. Bacteria were incubated with human epithelial cells, cell-adhering bacteria were recovered and total RNA was purified at different times. In a parallel experiment, RNA was prepared from bacteria grown in the absence of epithelial cells. After labeling with different fluorochromes, the two RNA samples were mixed together and used in hybridization experiments on DNA microarrays carrying the entire collection of polymerase chain reaction (PCR)–amplified *menB* genes (2152 genes). Relative emission intensities of the two fluorochromes after laser excitation of each spot (gene) were used to assess transcription activity of each gene under the two experimental conditions analyzed. From these kinds of experiments, we found that adhesion to epithelial cells altered the expression of several genes, including adhesion genes, host–pathogen cross-talk genes (genes encoding transport machineries for ammonium, amino acids, chloride, sulfate, and iron), amino acids and selenocysteine biosynthesis genes, DNA metabolism genes (methylases, nucleases, transposases, helicases, and ligases), and hypothetical genes (107 genes).

Twelve proteins whose transcription was found to be particularly activated during adhesion were expressed in *E. coli*, purified, and used to produce antisera in mice. Sera were finally tested for their capacity to mediate complement-dependent killing of MenB. Five of these sera (against the products of the hypothetical genes NMB0315 and NMB1119, the adhesin MafA, the MIP-related protein, and N-acetylglutamate synthetase) showed substantial bactericidal activity against the different menigococcal strains.

These data show how DNA microarray analysis can be successfully exploited to identify new protective antigens against MenB. Being particularly expressed during the infection process, these antigens are expected to be ideal candidates for vaccine formulations.

### 20.4.2 Proteomics

As already pointed out, the efficacy of in silico antigen selection depends on the reliability of the available algorithms. In general, these algorithms are good predictors. However, systematic analyses supported by extensive experimental data aimed at quantifying the reliability of the algorithms are missing. For instance, the computer programs used for predicting protein localization are being developed by selecting characteristic structural motifs of proteins belonging to specific cellular compartments and by using them to scan proteins of unknown localization. The presence or absence of such motifs within the proteins under examination are taken to assign the probability of finding the proteins in particular cellular compartments. This process assumes that all proteins belonging to a specific cellular compartment share at least one of the structural signatures used for the scanning analysis. This rule is not necessarily always fulfilled.

With the aim of evaluating the reliability of protein localization algorithms when applied to bacterial systems, we experimentally identified the MenB outer membrane and the soluble proteins by analyzing the carbonate-insoluble and carbonate-soluble total proteins by 2D electrophoresis coupled to mass spectrometry. After proteomic analysis, we subjected the proteins experimentally found in the insoluble and soluble preparations to protein location-predicting algorithms, asking the question whether the experimental data coincided with computer prediction. Since protein fractionation of mechanically disrupted bacterial cells in carbonate solutions has been reported as the method of choice to separate outer membrane proteins (carbonate insoluble) from cytoplasmic and inner membrane proteins [34], the presence of computer-predictable outer membrane proteins in the insoluble protein gel and of computer-predictable cytoplasmic and inner membrane proteins in the soluble protein gel would indicate that the algorithms in use are highly reliable.

Before analyzing the results of this exercise, it is necessary to point out that in addition to proteins exclusively found either in the membrane protein gel or in the cytoplasmic protein gel, there was a considerable number of proteins that partitioned in both gels (Fig. 20.4).

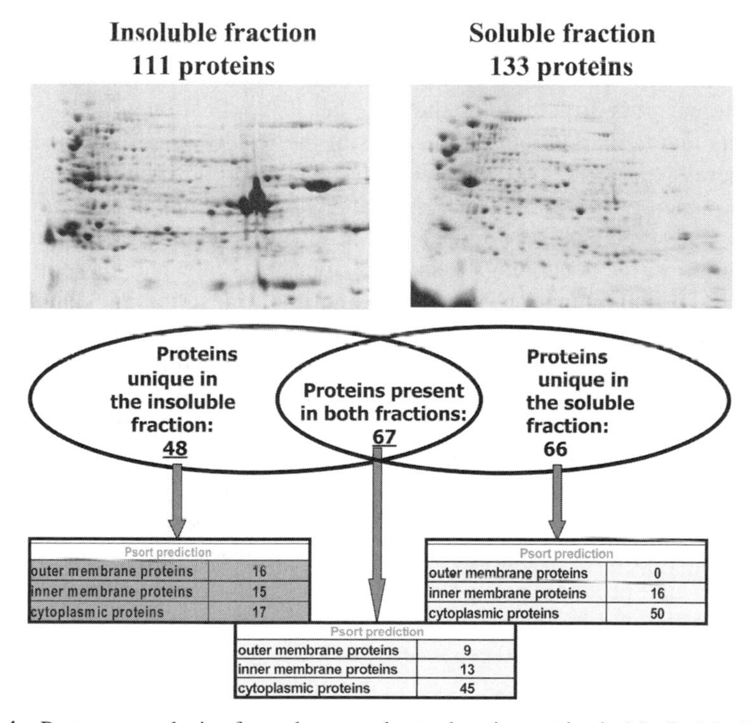


**Figure 20.4** Proteome analysis of membrane and cytoplasmic proteins in MenB. MenB cells were resuspended in sodium carbonate [34] and physically disrupted using a French press apparatus. Soluble and insoluble fractions were separated by centrifugation and resolved by 2D electrophoresis. Protein spots were collected and identified by matrix-assisted laser desorption ionization time-of-flight mass spectrometry.

However, in most cases, the relative abundance of these proteins in the two gels, as judged by computer analysis of the intensities of the corresponding protein spots, varied quite substantially. We therefore classified as "insoluble proteins" those proteins exclusively or predominantly found in the insoluble protein gel and as "soluble proteins" the proteins exclusively or predominantly found in the carbonate-soluble protein gel.

With these premises, the results of protein compartmentalization prediction of the experimentally defined, soluble and insoluble proteins can be summarized as follows. In general, algorithms appear to be reliable. This conclusion is supported by two pieces of evidence: (1) All computer-predictable outer membrane proteins are found in the carbonate-insoluble protein gel and (2) carbonate-soluble proteins are all predicted as either cytoplasmic (>85%) or inner membrane proteins. However, 31% of the carbonate-insoluble proteins are predicted as cytoplasmic or inner membrane. While it cannot be ruled out that these proteins represent experimental artifacts, the data open the possibility that a certain group of membrane proteins still remains hidden to the current computer predictors. In this context, it is worth mentioning that we recently found some of these proteins to be in fact surface associated, as judged by FACS analysis of bacterial cells using specific polyclonal antibodies (Grifantini and Grandi, unpublished).

In conclusion, we believe that the application of proteomics analysis for membrane protein identification represents a useful tool to unravel new potential vaccine candidates.

## 20.5 CONCLUSIONS

Reverse vaccinology represents a very attractive approach to vaccine discovery. Like all new technologies, it is difficult to predict whether its application will ultimately lead to new products on the market, but on the basis of the results so far obtained, and in part presented here, expectations are very high. Originally proposed as the combination of genome sequencing, in silico analysis, and high-throughput protein expression, purification, and screening, it is now clear that reverse vaccinology can greatly benefit from other innovative technologies such as transcriptomic and proteomic analysis. Indeed, the proper combination of in silico analysis, transcriptomics, and proteomics allows the selection, out of the large repertoire of genes in a given pathogen, of a relatively small number of candidates likely to become potential vaccines. Our current strategy for antigen selection starts with the identification of highly transcribed genes by DNA microarray (usually they represent 50–60% of all genes). Once these genes are identified, in silico analysis is applied to select only those coding for putative secreted and membrane-associated proteins. This process reduces the original gene pool down to a more manageable and reliable size of approximately 400 genes (20–30% of the genome). Finally, proteome analysis of membrane preparation is applied to include in the candidate list proteins not predicted to be membrane associated by current algorithms. At the end of this selection process, the pool of best candidates usually includes approximately 550–600 genes ready to move to the high-throughput expression step and to the screening assay.

I would like to point out that this antigen selection flowchart, which is currently the most effective, may become obsolete in a few years. The tremendous progress in the understanding of the physiology and system biology of bacterial pathogens will probably refine the algorithms of bioinformatics to such an extent that extremely reliable predictions of vaccine candidates will simply require a mouse click of our personal computer.

## REFERENCES

1. Grandi, G., Antibacterial vaccine design using genomics and proteomics, *Trends Biotechnol.* 2001, **19**, 181–188.
2. Hoffman, S. L., Subramanian, G. M., Collins, F. H., and Venter, J. C., Plasmodium, human and Anopheles genomics and malaria, *Nature* 2002, **415**, 702–709.
3. Pizza, M., Scarlato, V., Masignani, V., Giuliani, M. M., Aricò, B., Comanducci, M., Jennings, G. T., Baldi, L., Bartolini, E., Capecchi, B., Galeotti, C. L., Luzzi, E., Manetti, R., Marchetti, E., Mora, M., Nuti, S., Ratti, G., Santini, L., Savino, S., Scarselli, M., Storni, E., Zuo, P., Broecker, M., Hundt, E., Knapp, B., Blair, E., Mason, T., Tettelin, H., Hood, D. W., Jeffries, A. C., Saunders, N. J., Granoff, D. M., Venter, J. C., Moxon, E. R., Grandi, G. , and Rappuoli, R., Identification of vaccine candidates against serogroup B meningococcus genome sequencing, *Science* 2000, **287**, 1816–1820.
4. Rappuoli, R., Reverse vaccinology, *Curr.Opin.Microbiol.* 2000, **3**, 445–450.
5. Grandi, G., Bioinformatics, DNA microarrays and proteomics in baccine discovery: Competing or complementary technologies?, in G. Grandi (Ed.), *Genomics, Proteomics and Vaccines*, Wiley, New York, 2004.
6. Gotschlich, E. C., Liu, T. Y., and Artenstein, M. S., Human immunity to the Meningococcus. 3. Preparation and immunochemical properties of the group A, group B and group C meningococcal polysaccharides, *J. Exp. Med.* 1969, **129**, 1349–1365.
7. Gotschlich, E. C., Goldschneider, I., and Artenstein, M. S., Human immunity to the meningococcus, IV. Immunogenicity of group A and group C meningococcal polysaccharides in human volunteers, *J. Exp. Med.* 1969, **129**, 1367–1384.
8. Scholten, R. J., Bijlmer, H. A., Poolman, J. T., Kuipers, B., Caugant, D. A., Van Alphen, L., Dankert, J., and Valkenburg, H. A., Meningococcal disease in The Netherlands, 1958–1990: A steady increase in the incidence since 1982 partially caused by new serotypes and subtypes of *Neisseria meningitidis*. *Clin. Infect. Dis.* 1993, **16**, 237–246.
9. Hayrinen, J., Jennings, H., Raff, H. V., Rougon, G., Hanai, N., Gerardy-Schahn, R., and Finne, J., Antibodies to polysialic acid and its *N*-propyl derivative: Binding properties and interaction with human embryonal brain glycopeptides, *J. Infect. Dis.* 1995, **171**, 1481–1490.
10. Finne, J., Bitter-Suermann, D., Goridis, C., and Finne, U., AnIgG monoclonal antibody to group B meningococci cross reacts with developmentally regulated polysialic acid units of glycoproteins in neural and extraneural tissues, *J. Immunol.* 1987, **138**, 4402–4407.
11. Poolman, J. T., Development of a meningococcal vaccine, *Infect. Agents Dis.* 1995, **4**, 13–28.
12. Martin, D., Cadieux, N., Hamel, J., and Brodeur, B. R., Highly conserved *Neisseria meningitidis* surface protein confers protection against experimental infection, *J. Exp. Med.* 1997, **185**, 1173–1183.
13. Tettelin, H., Saunders, N. J., Nelson, K. E., Jeffries, A. C., Heidelberg, J., Eisen, J. A., Ketchum, K. A., Hood, D. W., Dodson, R. J., Nelson, W. C., Gwinn, M. L., DeBoy, R., Peterson, J. D., Hickey, E. K., Haft, D. H., Salzberg, S. L., White, O., Fleischmann, R. D., Dougherty, B. A., Mason, T., Ciecko, A., Parksey, D. S., Blair, E., Cittone, H., Clark, E. B., Cotton, M. D., Utterback, T. R., Khouri, H., Qin, H., Vamathevan, J., Gill, J., Scarlato, V., Masignani, V., Pizza, M., Grandi, G., Sun, L., Smith, H. O., Fraser, C. M., Moxon, E. R., Rappuoli, R., and Venter, J. C., Complete genome sequence of *Neisseria meningitidis* serotype B strain MC58. *Science* 2000, **287**, 1809–1815.
14. Goldscheider, I., Gotschlich, E. C., and Artenstein, M. S., Human immunity to meningococcus. I. The role of humoral antibodies. *J. Exp. Med.* 1969, **129**, 1307–1326.
15. Schuchat, A., Epidemiology of group B streptococcal disease in the United States: Shifting paradigms, *Clin. Microbiol. Rev.* 1998, **11**, 497–513.

16. Paoletti, L. C., Wessels, M. R., Rodewald, A. K., Shroff, A. A., Jennings, H. J., and Kasper, D. L., Neonatal mouse protection against infection with multiple group B streptococcal (GBS) serotypes by maternal immunization with a tetravalent GBS polysaccharide-tetanus toxoid conjugate vaccine, *Infect. Immun.* 1994, **62**, 3236–3243.
17. Berg, S., Trollfors, B., Lagergard, T., Zackrisson, G., and Claesson, B. A., Serotypes and clinical manifestations of group B streptococcal infections in western Sweden, *Clin. Microbiol. Infect.* 2000, **6**, 9–13.
18. Davies, H. D., Raj, S., Adair, C., Robinson, J., and McGeer, A., Population-based active surveillance for neonatal group B streptococcal infections in Alberta, Canada: Implications for vaccine formulation, *Pediatr. Infect. Dis. J.* 2001, **20**, 879–884.
19. Hickman, M. E., Rench, M. A., Ferrieri, P., and Baker, C. J., Changing epidemiology of group B streptococcal colonization, *Pediatrics* 1999, **104**, 203–209.
20. Lin, F. Y., Clemens, J. D., Azimi, P. H., Regan, J. A., Weisman, L. E., Philips, J. B. 3rd, Rhoads, G. G., Clark, P., Brenner, R. A., and Ferrieri, P., Capsular polysaccharide types of group B streptococcal isolates from neonates with early-onset systemic infection. *J. Infect. Dis.*, 1998, **177**, 790–792.
21. Suara, R. O., Adegbola, R. A., Mulholland, E. K., Greenwood, B. M., and Baker, C. J., Seroprevalence of antibodies to group B streptococcal polysaccharides in Gambian mothers and their newborns, *J. Natl. Med. Assoc.* 1998, **90**, 109–114.
22. Tettelin, H., Masignani, V., Cieslewicz, M. J., Eisen, J. A., Peterson, S., Wessels, M. R., Paulsen, I. T., Nelson, K. E., Margarit, I., Read, T. D., Madoff, L. C., Wolf, A. M., Beanan, M. J., Brinkac, L. M., Daugherty, S. C., DeBoy, R. T., Durkin, A. S., Kolonay, J. F., Madupu, R., Lewis, M. R., Radune, D., Fedorova, N. B., Scanlan, D., Khouri, H., Mulligan, S., Carty, H. A., Cline, R. T., Van Aken, S. E., Gill, J., Scarselli, M., Mora, M., Iacobini, E. T., Brettoni, C., Galli, G., Mariani, M., Vegni, F., Maione, D., Rinaudo, D., Rappuoli, R., Telford, J. L., Kasper, D. L., Grandi, G., and Fraser, C.M., Complete genome sequence and comparative genomic analysis of an emerging human pathogen, serotype V *Streptococcus agalactiae*, *Proc. Natl. Acad. Sci. USA* 2002, **99**, 12391–12396.
23. Cunningham, M. W., Pathogenesis of group A streptococcal infections, *Clin. Microbiol. Rev.* 2000, **13**, 470–511.
24. Dale, J. B., Group A streptococcal vaccines, *Infect. Dis. Clin. North. Am.* 1999, **13**, 227–243, viii.
25. Hu, M. C., Walls, M. A., Stroop, S. D., Reddish, M. A., Beall, B., and Dale, J.B., Immunogenicity of a 26-valent group A streptococcal vaccine, *Infect. Immun.* 2002, **70**, 2171–2177.
26. Baltimore, R. S., Kasper, D. L., Baker, C. J., and Goroff, D. K., Antigenic specificity of opsonophagocytic antibodies in rabbit anti-sera to group B streptococci, *J. Immunol.* 1977, **118**, 673–678.
27. Stephers, R. S. (Ed.), *Chlamydia. Intracellular Biology, Pathogenesis, and Immunity*, ASM Press, Washington, DC, 1999.
28. Ho, J. L., He, S., Hu, A., Geng, J., Basile, F. G., Almeida, M. G., Saito, A. Y., Laurence, J., and Johnson, W. D. Jr., Neutrophils from human immunodeficiency virus (HIV)-seronegative donors induce HIV replication from HIV-infected patients' mononuclear cells and cell lines: An in vitro model of HIV transmission facilitated by *Chlamydia trachomatis*, *J. Exp. Med.* 1995, **181**, 1493–1505.
29. Ramsey, K. H., Cotter, T. W., Salyer, R. D., Miranpuri, G. S., Yanez, M. A., Poulsen, C. E., DeWolfe, J. L., and Byrne, G. I., Prior genital tract infection with a murine or human biovar of *Chlamydia trachomatis* protects mice against heterotypic challenge infection, *Infect. Immun.* 1999, **67**, 3019–3025.
30. Bonci et al., submitted for publication.
31. Montigiani, S., Falugi, F., Scarselli, M., Finco, O., Petracca, R., Galli, G., Mariani, M., Manetti, R., Agnusdei, M., Cevenini, R., Donati, M., Nogarotto, R., Norais, N., Garaguso, I., Nuti, S.,

Saletti, G., Rosa, D., Ratti, G., and Grandi, G., Genomic approach for analysis of surface proteins in *Chlamydia pneumoniae*, *Infect. Immun.* 2002, **70**, 368–379.

32. Finco et al., *Vaccine*, in press.
33. Grifantini, R., Bartolini, E., Muzzi, A., Draghi, M., Frigimelica, E., Berger, J., Ratti, G., Petracca, R., Galli, G., Agnusdei, M., Giuliani, M. M., Santini, L., Brunelli, B., Tettelin, H., Rappuoli, R., Randazzo, F., and Grandi, G., Previously unrecognized vaccine candidates against group B meningococcus identified by DNA microarrays, *Nat. Biotechnol.* 2002, **20**, 914–921.
34. Molloy, M. P., Herbert, B. R., Slade, M. B., Rabilloud, T., Nouwens, A. S., Williams, K. L., and Gooley, A. A., Proteomic analysis of the *Escherichia coli* outer membrane, *Eur. J. Biochem.* 2000, **267**, 2871–2881.
35. Lei, B., Liu, M., Chesney, G. L., and Musser, J. M., Identification of new candidate vaccine antigens made by *Streptococcus pyogenes*: Purification and characterization of 16 putative extracellular lipoproteins, *J. Infect. Dis.* 2004, **189**, 79–89.

PART V

# PROTEOME DATABASES, BIOINFORMATICS, AND BIOCHEMICAL MODELING

CHAPTER 21

# Databases and Resources for in silico Proteome Analysis

MANUELA PRUESS, PAUL KERSEY, TAMARA KULIKOVA, and ROLF APWEILER

European Bioinformatics Institute, Hinxton, Cambridge, United Kingdom

## 21.1 INTRODUCTION

The enormous development of genome-sequencing techniques in recent years has led to a huge increase in the number of organisms whose complete genome sequences are available in public databases, most of them microbial—of approximately 180 organisms sequenced to date, 147 are bacteria, 18 archaea, and 7 eukaryotic microbes. An advancement in refined proteomics techniques has followed quickly, generating huge amounts of data. On the one hand, there are experimentally derived data; on the other, there are protein sequences predicted from the genomes. There is particular interest in the growing number of proteomes (the sets of all proteins derived from each genome) because of the insights into the functionality of the whole organism available through their study. All these data have to be stored in a sensible way to allow for maximum information to be extracted from them. The sequences need to be analyzed to provide a solid basis for further investigations; predicted protein sequences in particular have to be aligned with known sequences in order to derive functional characterization. For these purposes, special resources and tools are necessary. Most tools analyze proteins (with different emphases), some describe protein–protein interactions, and others analyze complete proteomes in silico, mainly in a statistical way, making use of data from many different resources. They constitute important instruments toward a better understanding of the function of the proteome and thus a more complete biology of the organisms.

In this chapter, we will first describe how genomic DNA sequences are stored and how they are presented in the Genome Reviews database and other similar resources. Then we will show how proteomes are presented in the Proteome Analysis database and the Integr8 resource, how they are analyzed, and what associated tools for protein clustering and domain identification are available; finally we will characterize a selection of other databases and resources which represent useful tools for the analysis of proteins, their interactions, and proteome analysis. Web addresses for the databases and tools mentioned in the text are presented in the Appendix at the end of the chapter.

*Microbial Proteomics: Functional Biology of Whole Organisms*, Edited by Ian Humphery-Smith and Michael Hecker. 

## 21.2 GENOME ANALYSIS

Since the first complete genome of a cellular organism (that of *Haemophilus influenzae*) was sequenced in 1995, the sequencing of complete prokaryotic genomes advanced rapidly. Numerous microorganisms of agricultural, environmental, and industrial importance have been or are in the process of having their entire genomes sequenced. The largest published prokaryotic genomes at the time of writing are those of *Bradyrhizobium japonicum* (a nitrogen-fixing symbiotic bacterium, 9,105,828 bp) [1] in bacteria and *Methanosarcina acetivorans* (an acetate-utilizing methanogen prokaryote, 5,751,492 bp) in archaea [2]. By now we can say that the diversity of prokaryotes as we know it is reasonably well covered by completely sequenced species [3]. Availability of sequences from different genomes has made comparisons between genomes possible; and as soon as the first data became available, newly published papers on complete genomes have reported results of such comparisons. Sequences and the annotation are submitted to and stored in the databases of the International Nucleotide Sequence Database (INSD) Collaboration, which serve as repositories of the nucleotide sequence data.

### 21.2.1 Nucleotide Sequence Databases

The INSD Collaboration provides a data repository that contains nucleic acid sequence data submitted by the scientific community with the purpose of making it freely available. Three databases participate in this collaboration: The EMBL Nucleotide Sequence Database at the European Molecular Biology Laboratory in the United Kingdom [4], the DNA Data Bank of Japan (DDBJ) at the National Institute of Genetics (NIG) in Japan [5], and GenBank at the National Center for Biotechnology Information (NCBI) in the United States [6]. These collect, organize, and distribute nucleotide sequence data and automatically update each other every 24 h with the new sequences that they have received. This daily exchange is facilitated by shared rules, a unified taxonomy, and a common set of unique identifiers for protein and DNA sequences. The submitted data are very heterogeneous; they vary with respect to the source of the material (e.g., genomic DNA, cDNA), the intended quality (e.g., finished sequences, single-pass sequences), the extent of sequence annotation, and the intended completeness of the sequence relative to its biological target (e.g., complete versus partial coverage of a gene or a genome). The INSD are mainly archival databases; the curation is limited to syntax checking and some basic checks on the integrity of the data. An inevitable consequence of the fact that the databases are archival is some inconsistency in the annotation.

### 21.2.2 Resources for Analyzing Genomes

Analysis of any newly assembled bacterial genome starts with the identification of protein-coding genes. Despite the increasing number of available complete genome sequences, which provide useful comparisons with closely related species during the annotation process, accurate gene prediction is difficult. Gene recognition in prokaryota is not straightforward, even though introns are not normally present [7]. All annotated genes are the results of running prediction programs; some of these genes encode known proteins, others do not. In some microbial genomes statistical differences between these two classes of genes suggest that many of the genes encoding hypothetical proteins are probably false. For example, the comparison of the length distribution of annotated genes with the length distribution of those matching a known protein shows that too many short genes are annotated in many genomes [8]. The annotation of protein functions itself is, by necessity, mostly based on the similarity to the characterized proteins stored in the databases available for searches. This creates a potential for the propagation of errors, especially where the annotation transfer from the preexisting database records is not critically reviewed by a scientist. Once propagated, the annotation errors tend to remain uncorrected in the archival records in EMBL/GenBank/DDBJ, due to the fact that the editorial control for a record belongs to the submitters. Even if the original data are updated and enhanced by the original sequencing group, there are not always enough resources for reannotation of the records and submission of the update.

Manually curated databases tend to represent more reliable sources of information, but the process of manual annotation is labor intensive; in many databases, there is no manual curation and the database is reliant on information provided by the submitter or transferred automatically from curated databases describing the same or similar entities. As the archival entries in EMBL/GenBank/DDBJ cannot be touched without the submitter's instructions, the solution seems to be in the creation of additional databases that take the bulk of information from the archival entries and improve the annotation by manual curation or propagating information from curated databases. The Refseq database at NCBI [9] and Genome Reviews at the European Bioinformatics Institute (EBI) are examples of such efforts, and others are available as well.

***Genome Reviews*** The goal of the Genome Reviews project is to provide an up-to-date, standardized, and comprehensively annotated view of the genomic sequence of organisms with completely deciphered genomes. The initial releases contain files for each chromosome and plasmid from all prokaryotes with completely deciphered genomes, and in later releases data will also be available for selected eukaryotic organisms. Genome Reviews files are enhanced versions of the original EMBL/GenBank/DDBJ entry in EMBL format, with additional annotation imported from other data sources (which will be described later in this chapter): the UniProt Protein Knowledgebase [10], the Proteome Analysis Database (part of the Integr8 project) [11], the Uniparc Protein Sequence Archive, the InterPro Database of protein families, domains, and functional sites [12], the Gene Ontology Annotation (GOA) database of protein annotations [13], and the HAMAP (High-Quality Automatic and Manual Annotation of Proteomes) microbial annotation project [14]. The enhancements achieved in the Genome Reviews are an increased number of cross references, from approximately 650,000 references to 9 databases in the corresponding EMBL entries to 2.5 million references to 18 databases (in release 1), and the addition of standardized gene/product names and gene/product synonyms imported from the UniProt Knowledgebase. Furthermore, systematic locus tag identifiers have been introduced for more than 99% of all genes, indicating the relative position of genes in the genome; data imported from GOA has been used to annotate coding sequences (CDS features) with functional information; and new features representing processed protein products have been added.

The Genome Reviews are presented in EMBL-like format; however, to accommodate the information from extra data sources that is normally not well compartmentalized in EMBL format, some new qualifiers were introduced in the first release of Genome Reviews (i.e. "cellular_component," "process") (see flatfile extract; Fig. 21.1).

The most obvious difference between the EMBL flatfile and the Genome Review flatfile formats is the presence of evidence tags in the qualifiers. Evidence tags are applied to feature qualifiers attached to CDS, mat_peptide, mRNA, peptide, pro_peptide,

```
FT   CDS    8238..9191
FT          /codon_start=1
FT          /gene="talB {Swiss-Prot:P30148}"
FT          /locus_tag="b0008 {Swiss-Prot:P30148}"
FT          /product="Transaldolase B {Swiss-Prot:P30148}"
FT          /EC_number="2.2.1.2 {Swiss-Prot:P30148}"
FT          /function="transaldolase activity {GO:0004801}"
FT          /function="aldolase activity {GO:0016228}"
FT          /process="pentose-phosphate shunt {GO:0006098}"
FT          /cellular_component="cytoplasm {GO:0005737}"
FT          /protein_id="AAC73119.1 {EMBL:U00096}"
FT          /db_xref="EMBL:AAB47022.1 {Swiss-Prot:P30148}"
FT          /db_xref="EMBL:AAG54308.1 {Swiss-Prot:P30148}"
FT          /db_xref="EMBL:AAN41675.1 {Swiss-Prot:P30148}"
FT          /db_xref="EMBL:AAP15554.1 {Swiss-Prot:P30148}"
FT          /db_xref="EMBL:BAA21822.1 {Swiss-Prot:P30148}"
FT          /db_xref="EMBL:BAB33431.1 {Swiss-Prot:P30148}"
FT          /db_xref="EMBL:BAB96586.1 {Swiss-Prot:P30148}"
FT          /db_xref="EcoGene:EG11556 {Swiss-Prot:P30148}"
FT          /db_xref="GO:0004801 {Swiss-Prot:P30148}"
FT          /db_xref="GO:0005737 {Swiss-Prot:P30148}"
FT          /db_xref="GO:0006098 {Swiss-Prot:P30148}"
FT          /db_xref="GO:0016228 {Swiss-Prot:P30148}"
FT          /db_xref="InterPro:IPR001585 {Swiss-Prot:P30148}"
FT          /db_xref="InterPro:IPR004730 {Swiss-Prot:P30148}"
FT          /db_xref="Swiss-Prot:P30148 {EMBL:U00096}"
FT          /db_xref="UniParc:UPI000016538F {EMBL:AAC73119}"
FT          /transl_table=11
FT          /translation="MTDKLTSLRQYTTVVADTGDIAAMKLYQPQDATTNPSLILNAAQI
FT                       PEYRKLIDDAVAWAKQQSNDRAQQIVDATDKLAVNIGLEILKLVPGRISTEVDARLSYD
FT          TEASIAKAKRLIKLYNDAGISNDRILIKLASTWQGIRAAEQLEKEGINCNLTLLFSFAQ
FT          ARACAEAGVFLISPFVGRILDWYKANTDKKEYAPAEDPGVVSVSEIYQYYKEHGYETVV
FT          MGASFRNIGEILELAGCDRLTIAPALLKELAESEGAIERKLSYTGEVKARPARITESEF
FT          LWQHNQDPMAVDKLAEGIRKFAIDQEKLEKMIGDLL"
```

**Figure 21.1** Flatfile extract of a Genome Reviews entry.

sig_peptide, transit_peptide, and tRNA features in Genome Reviews files, indicating the primary source of the information. The format for the Genome Reviews evidence tag is {database_name:evidence}, where the "database name" is the name of the database from which the information was obtained. The evidence is often the primary identifier of an entry in that source database with which the particular piece of information is associated. The initial release of Genome Reviews provides more consistent and detailed annotation than the original submissions, making comparison of data easier. Future releases will address other problems often present in the original submission, such as the incorrect gene location problem.

***Other Genome-Related Resources*** The Entrez genome database at NCBI provides nucleotide, protein, and bibliographical information for the genome scale sequence and presents both text and graphical views for a variety of completely sequenced genomes plus genetic and physical maps. The information about completed genomes is arranged on several levels: all genomes; all chromosomes for a chosen organism plus extra-chromosomal elements, if any; single chromosomes; detailed views of parts of chromosomes; and single genes. At each level there are one or more possible views of the data, summaries, links, and analyses [15]. DDBJ's Genome Information Broker (GIB) is a comparative genomics tool. GIB allows users to retrieve and display partial- and/or whole-genome sequences together with the relevant biological annotation. In the Genomic View page the genome structure map is displayed with reading frames colored by functional categories. The map is clickable to retrieve the Feature View page (a closer view). In the Feature View page the map of features in the specified region is displayed [16]. The central function of the Microbial Genome Database for Comparative Analysis (MBGD) is to create an orthologous gene classification table using precomputed all-against-all similarity relationships among genes in multiple genomes; the user interface allows the users to explore the resulting classification in detail. An automated classification algorithm allows us to create the user-specific classification table by choosing a set of organisms and parameters. This feature is useful, for example, when comparison is performed between related organisms. Users can carry out comparative analyses from various points of view, such as phylogenetic pattern analysis, gene order comparison, and detailed gene structure comparison [17].

## 21.3 PROTEOME ANALYSIS

Once the complete genome of an organism has been deciphered, new types of bioinformatic analyses become possible. In particular, it becomes possible to draw conclusions from the absence, as well as the presence, of a particular feature in a genome. Much interest in particular lies in the analysis of the complete set of proteins encoded by a particular genome. The protein sequences are stored in protein sequence databases, such as the nucleotide sequence database examples of electronic resources that can cope with the huge amount of biological sequence data, which are mostly no longer published conventionally. A comprehensive statistical analysis of all publicly available complete proteomes is available through the Integr8 resource.

### 21.3.1 Protein Sequence Databases

Protein sequence databases are sources of information on protein sequences and thus provide the underlying data for proteome compilations, sequence comparisons, and protein identification and classification. It is necessary to distinguish between universal databases, which cover proteins from all species, and specialized data collections that store information about specific families or groups of proteins or about the proteins of a specific organism. Some universal protein sequence databases are repositories that simply archive sequences, possibly redundantly, whereas others merge redundant sequences and associate additional information. For the proteomics researcher, the second group is of particular interest, because they provide the means to compare novel sequences and their features with related ones already described in the database.

The Universal Protein Resource (UniProt) [10] is a comprehensive catalog of data on protein sequence and function maintained by a collaboration of the Swiss Institute of Bioinformatics (SIB), the EBI, and the Protein Information Resource (PIR). It has three sections: (1) the UniProt Knowledgebase, which is a high-quality, comprehensive, richly annotated protein sequence knowledgebase extensively cross-referenced to other databases; (2) the UniProt sequence archive (UniParc), a comprehensive repository of all protein sequences into which new and updated sequences are loaded on a daily basis; and (3) the nonredundant sequence databases (UniRef NREFs). The UniProt Knowledgebase has two sections, UniProt/Swiss-Prot (containing manually curated entries) and UniProt/TrEMBL [computer-annotated entries derived from the translation of all coding sequences (CDS) in EMBL-Bank, except for those already included in Swiss-Prot]. Due to the unusually high quality of its annotation, parts of the UniProt Knowledgebase can also be used in new resources that are built by combining information from different databases, adding additional value by applying certain programs and/or annotation. The UniProt NREF (UniRef) databases NREF100, NREF90, and NREF50 allow more easily interpretable sequence similarity searches by providing a nonredundant sequence collection in which identical sequences and subfragments are clustered and presented as a single NREF entry (NREF100) and two collections where all records from all species with mutual sequence identity of >90% (NREF90) or >50% (NREF50) are merged into a single record.

For the model organisms, including several bacteria, there are specialized databases available, mostly gene centric but also providing information at the protein level. Examples are SubtiList [18] and NRSub [19] for *Bacillus subtilis*, TubercuList for *Mycobacterium tuberculosis*, and EchoBase [20], EcoGene [21], and EcoCyc [22] for *Escherichia coli*. A generally useful resource for all completely sequenced bacterial genomes, containing annotated protein properties, is the Comprehensive Microbial Resource (CMR) [23]. The HOBACGEN database system organizes protein sequences of different bacteria into families, allowing the selection of sets of homologous genes from bacterial species and the visualization of multiple alignments and phylogenetic trees [24].

### 21.3.2 Integr8

Integr8 is a new Web-based interface for accessing data related to complete genomes and proteomes and is a successor system to the previously available Proteome Analysis Database [11], in which currently nearly 180 organisms are represented (see Fig. 21.2). In addition to proteome analysis, Integr8 also provides access to the latest literature published about each species, general information about each species, and genomic analysis based on the Genome Reviews.

***UniProt Nonredundant Proteome Sets*** The primary data available for analysis in Integr8 comprise a number of subsets of the UniProt Knowledgebase, each representing a complete nonredundant set of proteins for each species whose sequence has been deciphered [25]. Often, when a genome is first analyzed, a set of predicted protein

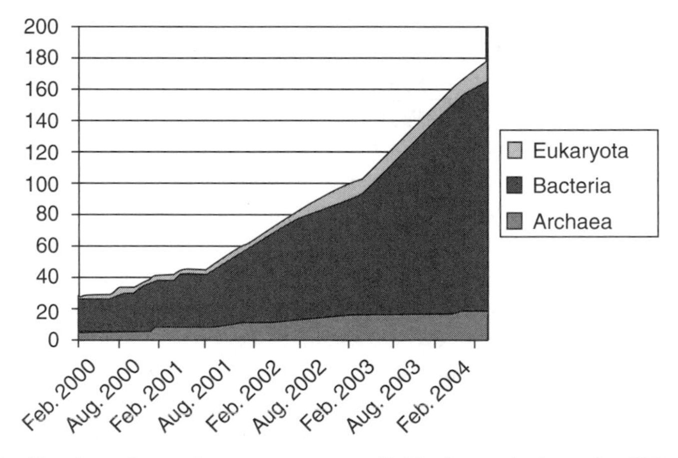


**Figure 21.2** Number of complete proteomes available for analysis at the EBI, 2000–2004.

sequences are submitted to the EMBL/GenBank/DDBJ nucleotide sequence databases complete with preliminary annotation, often derived from comparison to already-annotated sequences present in public data banks at that time. However, this annotation (and the sequences themselves) may become out of date over time. Better methods for predicting sequences or annotation may be developed, and many algorithms will anyway produce different results according to what external data are currently available to use as source information. In addition, different sequencing groups may annotate to different standards and may also represent the same information differently; in addition, integration with preexisting annotations, in cases where predicted proteins have also been experimentally characterized or predicted by others, may be poor, especially where there are differences in the reported sequences.

In the UniProt Knowledgebase, performing such integration is a high priority, and improved annotation is added to the corresponding database records through a combination of human curation and automatic annotation transfer. In the case of bacterial and hierarchical data, this occurs under the auspices of the HAMAP project, which provides automatic annotation for proteins that have no recognizable similarity to any other microbial or nonmicrobial proteins and proteins that are part of well-defined families or subfamilies. Other proteins are manually annotated. Similar processes exist to improve data derived from eukaryotic microbes. Nonredundant proteome sets can then be assembled from the data in the UniProt Knowledgebase by filtering out other predicted sequences present in the database but not yet resolved with the genomic predictions, while well-characterized proteins missed during the prediction from the genome are added to supplement the set.

***Resources for Protein Classification*** Analyses of the proteome sets are performed by classifying the individual proteins they contain in order to allow the overall composition of each proteome to be described and compared to others. The proteome analysis available in Integr8 uses a number of resources to classify proteins, principally InterPro, CluSTr, Gene Ontology (GO), and the Protein Data Bank (PDB).

InterPro [12] is a database of protein families, domains, and functional sites and the methods used to predict these de novo from primary protein sequence. InterPro is a federated database whose members use annotation present in existing UniProt entries to define characteristics of protein sequences that reliably indicate the presence of particular domains or sites within proteins or can show that a protein is a member of a particular family. They also provide algorithms for recognizing such characteristics that can then be applied to novel sequences (i.e., sequences without any preexisting annotation), allowing their classification. Different types of algorithms are provided by different members (e.g., the Pfam database [26] uses hidden Markov models, whereas the PRINTS database [27] identifies small repetitive elements that are good indicators of function in certain types of proteins). In addition, the different members have different areas of biological expertise. In InterPro, these various methods are integrated so that each entry represents a single family, domain, or functional site but may correspond to a number of different methods. Furthermore, InterPro entries are arranged in ontology such that a more general classification can be inferred from a more specific one.

CluSTr [28] is a database of protein clusters. A pairwise all-against-all protein sequence similarity comparison (using the algorithm of Smith and Waterman) has been performed on the entire contents of the UniProt Knowledgebase. Where a significant level of similarity is achieved, the result is normalized by measuring the statistical significance of the observed similarity level compared with the similarity achieved when random permutations of the same sequences are compared. Crucially, this method facilitates a determination of similarity that is wholly independent of the presence of other entries in the database (which in many approaches to normalization are used to derive a measure of significance) and allows for the incremental update of the database as new sequences become available. The matrix of protein similarities is then used to create hierarchies of clusters (from all or a chosen subset of the sequences) using the method of aggregation by single linkage. CluSTr is especially useful if used in conjunction with InterPro when it can be used to identify multidomain protein families, unclassified protein families, and singletons (entries with no close sequence homologues within the set of proteins clustered).

Gene Ontology (GO) [13] defines a hierarchical vocabulary of terms (with associated definitions) useful for the annotation of gene products (i.e., proteins and functional RNAs). In particular, it provides terms to describe the functions of a gene product, the biological processes in which it is involved, and the cellular components in which it is found. The emergence of GO as a standard used by many different groups allows the combination and comparison of annotation produced by different sources. Furthermore, the existence of (manually curated) mappings from other controlled vocabularies (that may have been used to annotate proteins) to GO allows the translation of existing annotation into the GO vocabulary. Additionally, the relevance of certain GO terms may be inferred from the fact that a protein is identified by a particular InterPro method, allowing the automatic assignment of GO terms to otherwise annotated protein sequences. As with InterPro, the hierarchical nature of GO means that, even if different proteins are annotated to different levels of specificity, a universal overview can still be provided by considering the less specific GO terms (which can be inferred from any more specific terms applied). This is especially useful when detailed manual annotation is combined with automatic predictions, which are usually less specific but which can be applied to a larger number of entries. Integrated manual and automatic annotation of UniProt is available through the GOA database [29].

***Proteome Analysis in Integr8*** InterPro provides a powerful tool for describing the physical composition of a complete proteome and for inferring the biological consequences of this. Through Integr8, it is possible to discover the most common domains, families, and functional sites present in each proteome, discover the proteins with the largest number of different InterPro classifications, and determine the overall coverage of the proteome by InterPro methods. (Unclassified proteins are usually completely uncharacterized and therefore are either novel or wrongly predicted.) The complete set of all matches between InterPro methods and sequences in each proteome set is also available for download. In addition, precomputed comparisons are available for 160 interesting combinations of species, allowing users to see the differential representation of InterPro matches between them. Additional comparisons can be customized by the user and generated interactively.

The statistical measures of protein similarity contained in the CluSTr database are used to support proteome analysis by the independent clustering of proteins in each complete proteome. The user is thus able to identify the largest and tightest clusters of proteins (grouped by similarity) in each species, singleton proteins (proteins without paralogues), and clusters with no or inconsistent InterPro annotation (which may indicate the presence of a common functional unit not yet detected by any of the InterPro methods). These results can be compared with the results of clustering proteins from all species via the

CluSTr website. The absence of proteins from a particular species in a particular multispecies cluster and the existence of loose clusters containing proteins only from a particular group of species may give clues as to the functional differentiation of species.

A GO-based analysis of each complete proteome is also available. This provides a high-level overview of the functional bias of each proteome, summarizing its composition in terms of GO. To obtain comparable data for each complete proteome, the raw data in GOA corresponding to each proteome set are mapped to a reduced set of high-level (i.e., less specific) terms, known as GO Slim. Typical GO Slim terms in each of the three GO ontologies are *transcription regulator activity* (function), *cell cycle* (process), and *external encapsulating structure* (cellular component). The overall coverage of each proteome in each of the three ontologies is provided, together with the percentage of each proteome made up by proteins to which each GO Slim term has been applied. The structure of GO allows a given lower level term to map to more than one GO Slim term (e.g., a "cellular physiological process" is both a "cellular process" and a "physiological process"); hence the cumulative proportion of proteins assigned to each high-level category may exceed 100%. In addition to this summary information, the complete set of GO terms assigned to each proteome is also available for download.

Integr8 also provides access to structural data about each proteome. The length distribution of each protein in the proteome and the cumulative amino acid usage are represented graphically. Proteins with known 3D structures (contained in the PDB [30]) from each proteome can be reviewed according to the SCOP superfamily classification [31]. Proteins with predicted 3D structures inferred from sequence similarity are also available. Coverage of the archaeal, bacterial, and eukaryotic microbial proteomes by the various classification tools is summarized in Table 21.1.

***Analysis Tools*** In addition to the precomputed analyses, a Web-based tool is also available which allows users to configure their own analyses (according to InterPro classification) of any number of selected proteomes [32]. The UniProt complete proteome sets (and corresponding data from the Genome Reviews) are also available through BioMart [33], an integrated database and query system that allows users to generate their own statistical analysis of selected proteomes using combinatorial criteria such as

**TABLE 21.1 Statistics of Data Relating to Complete Proteomes Available Through Intergr8**

| Data | Archaea | Bacteria | Eukaryotic microbes |
|---|---|---|---|
| Number of proteomes | 18 | 147 | 7 |
| Number of proteins in proteome sets | 40,118 | 437,335 | 23,491 |
| Percentage of proteins with InterPro hits | 75.9 | 78.9 | 72.3 |
| Percentage of proteins with GO "function" assignment | 48.7 | 56.2 | 49.7 |
| Percentage of proteins with GO "process" assignment | 40.6 | 49.9 | 46.7 |
| Percentage of proteins with GO "cellular component" assignment | 17.7 | 23.4 | 41.9 |
| Percentage of proteins with experimentally determined 3D structures | 0.5 | 0.7 | 1.7 |

*Note*: Data correct as of March 6, 2004. Percentages for InterPro and GO matches exclude data not yet scanned by InterPro.

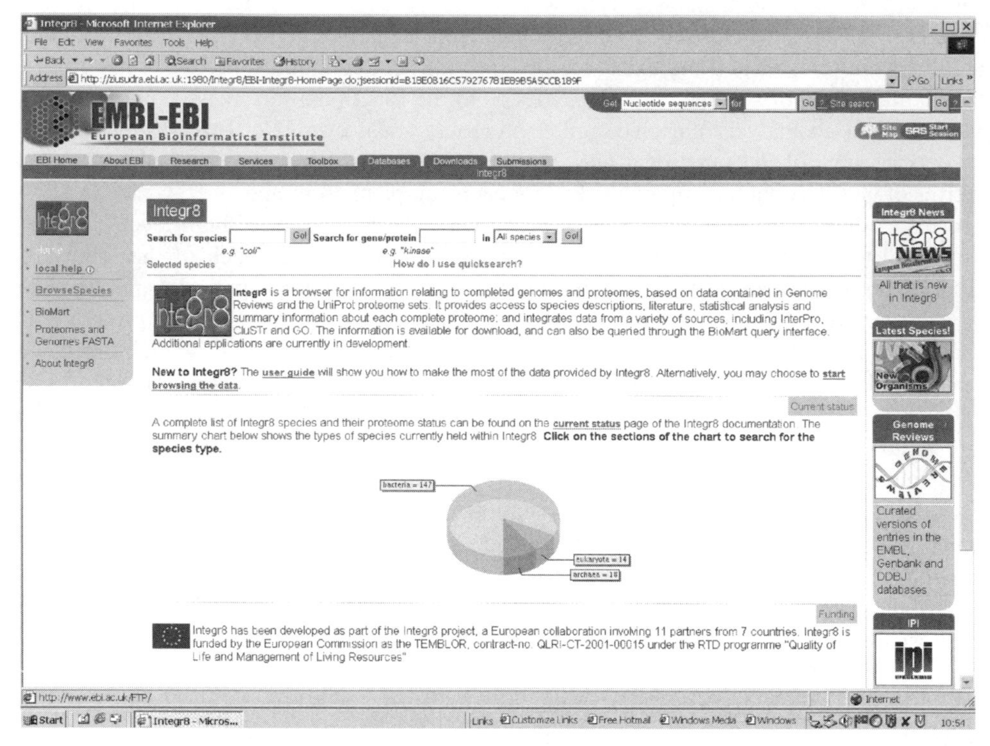


**Figure 21.3** The Integr8 web portal provides access to statistical analysis for all complete proteomes.

InterPro, CluSTr, or GO classification, UniProt annotation, or gene location. BioMart also allows users to download customized data sets based on the selected criteria. Additionally, chained queries can be specified that allow the user to link between the proteome sets and other data sets contained within BioMart (e.g., data sets from the Macromolecular Structure Database [34] and from the Ensembl eukaryotic annotation project [35] are currently available). BioMart can be queried through a Web interface or through client software installed locally. Furthermore, the entire database can be downloaded and installed locally. The Integr8 Web portal, from where all these services are available, is shown in Figure 21.3.

### 21.3.3 More Analyses of Complete Proteomes

Besides Integr8, other projects exist that deal with whole proteomes, usually with a narrower focus. The Proteome Profile Database (PPD) focuses on providing comparative analyses of protein lengths in completely sequenced prokaryotic and eukaryotic genomes [36]. Its core objective is to create protein classification tables based on the lengths of proteins by specifying a set of organisms and parameters, as protein lengths in different organisms vary considerably. The latter has been subject to research concerning functional and evolutionary aspects, showing that there are relationships between sequence length and protein conservation, expression levels, and number of functional motifs. The database of Predictions for Entire Proteomes (PEP) contains summaries of analyses of protein sequences from a range of organisms, comprising eukaryotes, prokaryotes, and archaea [37].

Proteins [derived from open reading frames (ORFs)] are identified via sequence alignments, and information about secondary structure, transmembrane helices, coiled coils, regions of low complexity, signal peptides, motifs, nuclear localization signals, and classes of cellular function is predicted on the basis of the sequence. Two more parts of the database contain information about structural domain–like fragments that have been analyzed for similar features and clusters of proteins sharing a common structural region, respectively. A database particularly useful for comparative sequence analysis, phylogeny, and molecular evolution studies is HOGENOM. Homologous proteins from fully sequenced organisms are classified into families and multiple alignments and phylogenetic trees are computed for each family.

## 21.4 ANALYZING PROTEINS: DATABASES AND TOOLS

To computationally analyze proteins in a sensible way, a whole "environment" of cooperating resources is available. These resources range from sequence databases that store protein sequences and additional information to highly specialized analysis tools. Such resources are often already integrated with computerized tools for data search and analysis. The Integr8 resource, one of the main tools for integrated proteome analysis, has already been described in detail. In the following, we will focus on some other proteomics resources, some of them sequence repositories, which are useful for the in silico discovery of protein function and protein characterization. Not all of them deal with complete proteomes, however, most focusing on single proteins and their features and functions. It can be difficult to distinguish clearly between databases and tools: In general, a database mainly is concerned with the storage of certain types of data and a tool is an instrument to analyze data and extract information from it.

### 21.4.1 "Classical" Proteomics Analyses

The classical proteomics tools deal with protein identification, translation, similarity searches, alignments, gel and data analysis, and the prediction of certain domains, posttranslational modifications, transmembrane regions, and so on. Usually, the user can enter a protein sequence, and the tool will do a comparison run against an existing data set. A comprehensive collection of such tools can be found at the Expasy server, http://ca.expasy.org/tools/. In the following, we will provide some useful examples.

***Protein Identification and Characterization*** Many different tools deal with the identification and characterization of proteins, as this is an apparent first step in dealing with an unknown amino acid sequence. Some tools also consider the isoelectric point (pI) and molecular masses for a precise characterization; others calculate masses of peptides, their posttranslational modifications, cleavage sites, and/or theoretical isotopic distribution of a peptide, protein, polynucleotide, or chemical compound.

Of special interest nowadays are software tools that enable the characterization of mass spectrometric data, as these types of data are very important in modern proteomics research. Protein-Prospector, for example, provides a collection of proteomics tools for mining sequence databases in conjunction with mass spectrometry experiments [38]. Specific programs allow a peptide mass fingerprinting that fits a user's mass spectrometry data to a protein sequence in an existing database and thus suggests the identity of the user's protein (MS-Fit), a comparison of the peptide sequence tag data contained in a user's tandem mass spectrum to a peptide sequence in an existing database (MS-Seq), and the detection of disulfide-linked peptides (MS-Bridge), among others. PROWL, developed by Rockefeller and New York Universities, is another (commercial) resource for the analysis of proteins. It consists of a number of databases and software tools which allow the analysis of mass spectrometry experiments in an interactive environment.

***Visualization and Identification of Gel Spots*** The visualization and reliable identification of gel spots are done by various techniques, including gel comparison, microsequencing, immunoblotting, amino acid composition analysis, peptide mass fingerprinting using mass spectrometry, or a combination of these techniques. To describe the identified proteins, including relevant data and their location within gels, special tools have been developed.

The SWISS-2DPAGE database is an annotated two-dimensional polyacrylamide gel electrophoresis database [39]. It contains data on proteins identified on various 2D PAGE and sodium dodecyl sulfate (SDS)–PAGE reference maps and thus enables the visualization and identification of gel spots. The current release contains 36 reference maps, mainly from eukaryotic biological samples but also from *Escherichia coli* and *Staphylococcus aeurus* (strain N315) origin. JvirGel is a collection of tools for the in silico calculation of virtual 2D protein gels. The software creates and visualizes virtual 2D protein gels based on the expected migration behavior of proteins as determined by their theoretical molecular masses in combination with their calculated pI. The utilization of all proteins of an organism of interest deduced from genes of the corresponding genome project in combination with the elimination of obvious membrane proteins permits the creation of an optimized calculated proteome map. The user can select a pI/MR range and an electrophoretic time scale in order to simulate the electrophoretic separation behavior of single proteins. The calculated pattern of protein spots is supposed to help identify unknown proteins and to localize known proteins during experimental proteomics approaches. Currently, the program provides six eukaryotic and six prokaryotic proteomes (*B. subtilis*, *E. coli*, *Helicobacter pylori*, *Listeria monocytogenes*, *Pseudomonas aeruginosa*, and *S. aureus*) [40].

### 21.4.2 Analysis of Protein–Protein Interactions

The need for, and possibility of, analyzing protein–protein interactions has arisen quite recently with the establishment of a large number of both small- and large-scale proteomics experiments, such as two-hybrid systems or tandem affinity purification. It is a major goal of proteomics to describe the protein interaction network underlying cell physiology in order to fully understand cellular functions. So once the experiments have taken place, the results have to be stored in specific databases that allow the comparison of data and further analysis. These tools do not deal with whole proteomes, but the information they are collecting is important for understanding the interactions of proteins within a proteome.

***Protein Interaction Databases*** Information about protein–protein interactions is stored in several databases, and each of these focus on certain methods to store, predict, and/or represent the data. The Biomolecular Interaction Network Database (BIND), for example, archives biomolecular interaction, molecular complex, and pathway information [41]. To capture protein function, defined at the molecular level as the set of other molecules with which a protein interacts or reacts, individual submissions, interaction data from the PDB [42], and a number of large-scale interaction and complex mapping experiments using yeast two-hybrid, mass spectrometry, genetic interactions, and phage display are stored. Molecular complexes and pathways are defined as collections of pairwise interactions between two biological objects, which can be proteins, RNAs, DNAs, molecular complexes, small molecules, photons, or genes. The Database of Interacting Proteins (DIP) also aims to integrate experimental data on protein–protein interactions and multiprotein complexes retrieved from individual research articles into a single database [43]. One focal point lies on the development of methods of quality assessment to identify the most reliable subset of interactions, as the reliability of experimental evidence can vary widely. Such a "clean" data set can then be used for evaluating the reliability of high-throughput protein–protein interaction data sets, for development of prediction methods, and in studies of the properties of protein interaction networks. The Molecular INTeraction (MINT) database stores data on functional interactions between proteins, comprising binary complexes as well as other types of functional interactions, including enzymatic modifications of one of the partners of the interaction [44]. In addition, it aims at describing the known kinetic and binding constants and the domains participating in the interactions. STRING is a database of predicted functional associations between proteins [45]. Based on the idea that protein–protein interactions are not limited to direct physical binding but that functional links between them can also often be inferred from genomic associations between the genes that encode them, STRING provides a tool to analyze these associations. Functional protein associations such as sharing a substrate in a metabolic pathway, regulating each other transcriptionally, or participating in larger multiprotein assemblies are predicted by exploiting their genomic context.

***Standards for Representation of Interaction Data*** As we have seen, many systems for the storage, representation, and analysis of protein interaction data are available. These data are of great importance for the scientific community, but unfortunately the different protein interaction data sets often show a high degree of incompatibility. Several approaches exist to help to overcome this problem by providing public repositories and data standards. The IntAct system provides a comprehensive open-source database and toolkit for the storage, presentation, and analysis of protein interactions [46]. IntAct adopted parts of the data models and some of the features of the main protein interaction databases as described above but integrated more features and controlled vocabularies. It can be locally installed and adapted to individual needs, thereby reducing development time and promoting consistency of interaction data sets through the use of uniform infrastructure and annotation system. In order to generally define community standards for data representation in proteomics and to facilitate data comparison, exchange, and verification, the Proteomics Standards Initiative (PSI) has been developed [47, 48]. This initiative is supported by the major protein interaction data providers and the Human Proteome Organization (HUPO), which does not mean that the focus is on human data alone. The standards are developed to integrate data from all species, including microorganisms. To improve the time-consuming process of reformatting and mapping of protein interaction data from different sources, the Molecular Interaction (MI) extensible markup language (XML) format has been developed as a community standard for the representation and exchange of protein interaction data.

## 21.5 CONCLUSION

The recent revolution in high-throughput techniques for molecular biology has led to the emergence of a new cooperative discipline, systems biology, whose aim is the systematic understanding of biological processes [49]. Advances in this field will enable ever more sophisticated experiments to be performed in silico, resulting in the prediction of the complete implications of perturbations to the cellular system and reducing the role of laboratory experiments to one of confirmatory checks. Microbial genomes are typically small, their proteomes simple (with less posttranscriptional modification), and their metabolisms well understood, and their single cells (in isolation from others) comprise complete living systems. It is therefore unsurprising that the greatest advances in this field are focusing on microbial organisms. This is nicely reflected in the EcoCyc database, where the entire genome of one organism is described, as well as its transcriptional regulation, transporters, and metabolic pathways. The collection of biological knowledge is accompanied by the development of mathematical tools for systematic modeling, such as E-CELL.

The knowledge necessary to assemble the whole-system picture is being provided by large-scale "-omics experiments" (genomics, transcriptomics, proteomics, metabolomics, etc.). But the interpretation of each conceptual layer of experiments is closely founded on the interpretation of the previous one. Both transcriptomic and proteomic experiments depend on the inference of an accurate set of protein predictions from the genome. The completion of metabolic modeling will in turn be dependent on the accurate description of the behavior of the proteome. A major challenge to the bioinformatics community will be to keep the annotation continuously up to date in light of our emerging knowledge of higher level function.

In this chapter we have described a variety of different databases and tools for the in silico analysis of genomes, proteins, and whole proteomes for a wide range of microbial and other organisms. Some of these tools, such as Integr8, bring together resources from many different primary sources to create an integrated picture of the composition and behavior of the genome and proteome of an organism. As the quantity of available data increases exponentially and becomes increasingly dispersed between specialized resources, the challenge of integrating these data will become increasingly important. The development of syntactic and semantic standards for data representation will be a vital component of success in this area.

## REFERENCES

1. Kaneko, T., Nakamura, Y., Sato, S., Minamisawa, K., et al., *DNA Res.* 2002, **9**, 189–197.
2. Galagan, J. E., Nusbaum, C., Roy, A., Endrizzi, M. G., et al., *Genome Res.* 2002, **12**, 532–542.
3. Koonin, E. V., and Galperin, M. Y., *Sequence – Evolution – Function Computational Approaches in Comparative Genomics*, Kluwer Academic, Norwell, MA, 2003.
4. Kulikova, T., Aldebert, P., Althorpe, N., Baker, W., et al., *Nucleic Acids Res.* 2004, **32** Database issue, D27–30.
5. Miyazaki, S., Sugawara, H., Ikeo, K., Gojobori, T., and Tateno, Y., *Nucleic Acids Res.* 2004, **32** Database issue, D31–34.
6. Benson, D. A., Karsch-Mizrachi, I., Lipman, D. J., Ostell, J., and Wheeler, D. L., *Nucleic Acids Res.* 2004, **32** Database issue, D23–26.

7. Bocs, S., Danchin, A., Medigue, C.,
8. Skovgaard, M., Jensen, L. J., Brunak, S., Ussery, D., and Krogh, A., *Trends Genet.* 2001, **17**, 425–428. *BMC Bioinformatics* 2002, 3, 5.
9. Pruitt, K. D., and Maglott, D. R., *Nucleic Acids Res.* 2001, **29**, 137–140.
10. Apweiler, R., Bairoch, A., Wu, C. H., Barker, W. C., et al., *Nucleic Acids Res.* 2004, **32** Database issue, D115–119.
11. Pruess, M., Fleischmann, W., Kanapin, A., Karavidopoulou, Y., et al., *Nucleic Acids Res.* 2003, **31**, 414–417.
12. Mulder, N. J., Apweiler, R., Attwood, T. K., Bairoch, A., et al., *Nucleic Acids Res.* 2003, **31**, 315–318.
13. Harris, M. A., Clark, J., Ireland, A., Lomax, J., et al., *Nucleic Acids Res.* 2004, **32** Database issue, D258–261.
14. Gattiker, A., Michoud, K., Rivoire, C., Auchincloss, A. H., et al., *Comput. Biol. Chem.* 2003, **27**, 49–58.
15. Tatusova, T. A., Karsch-Mizrachi, I., and Ostell, J. A., *Bioinformatics* 1999, **15**, 536–543.
16. Fumoto, M., Miyazaki, S., and Sugawara, H., *Nucleic Acids Res.* 2002, **30**, 66–68.
17. Uchiyama, I., *Nucleic Acids Res.* 2003, **31**, 58–62.
18. Moszer, I., Jones, L. M., Moreira, S., Fabry, C., and Danchin, A., *Nucleic Acids Res.* 2002, **30**, 62–65.
19. Perriere, G., Gouy, M., and Gojobori, T., *Nucleic Acids Res.* 1998, **26**, 60–62.
20. Thomas, G. H., *Bioinformatics* 1999, **15**, 860–861.
21. Rudd, K. E., *Nucleic Acids Res.* 2000, **28**, 60–64.
22. Karp, P. D., Riley, M., Saier, M., Paulsen, I. T., et al., *Nucleic Acids Res.* 2002, **30**, 56–58.
23. Peterson, J. D., Umayam, L. A., Dickinson, T., Hickey, E. K., and White, O., *Nucleic Acids Res.* 2001, **29**, 123–125.
24. Perriere, G., Duret, L., and Gouy, M., *Genome Res.* 2000, **10**, 379–385.
25. Apweiler, R., Biswas, M., Fleischmann, W., Kanapin, A., et al., *Nucleic Acids Res.* 2001, **29**, 44–48.
26. Bateman, A., Coin, L., Durbin, R., Finn, R. D., et al., *Nucleic Acids Res.* 2004, **32** Database issue, D138–141.
27. Attwood, T. K., Bradley, P., Flower, D. R., Gaulton, A., et al., *Nucleic Acids Res.* 2003, **31**, 400–402.
28. Kriventseva, E. V., Servant, F., and Apweiler, R., *Nucleic Acids Res.* 2003, **31**, 388–389.
29. Camon, E., Barrell, D., Lee, V., Dimmer, E., and Apweiler, R., *In Silico Biol.* 2004, **4**, 5–6.
30. Berman, H., Henrick, K., and Nakamura, H., *Nat. Struct. Biol.* 2003, **10**, 980.
31. Andreeva, A., Howorth, D., Brenner, S. E., Hubbard, T. J., et al., *Nucleic Acids Res.* 2004, **32** Database issue, D226–229.
32. Kanapin, A., Apweiler, R., Biswas, M., Fleischmann, W., et al., *Bioinformatics* 2002, **18**, 374–375.
33. Kasprzyk, A., Keefe, D., Smedley, D., London, D., et al., *Genome Res.* 2004, **14**, 160–169.
34. Boutselakis, H., Dimitropoulos, D., Fillon, J., Golovin, A., et al., *Nucleic Acids Res.* 2003, **31**, 458–462.
35. Birney, E., Andrews, D., Bevan, P., Caccamo, M., et al., *Nucleic Acids Res.* 2004, **32** Database issue, D468–470.
36. Sakharkar, K. R., and Chow, V. T., *In Silico Biol.* 2004, **4**, 0019.
37. Carter, P., Liu, J., and Rost, B., *Nucleic Acids Res.* 2003, **31**, 410–413.
38. Clauser, K. R., Baker, P., and Burlingame, A. L., *Anal. Chem.* 1999, **71**, 2871–2882.
39. Hoogland, C., Sanchez, J. C., Tonella, L., Binz, P. A., et al., *Nucleic Acids Res.* 2000, **28**, 286–288.

40. Hiller, K., Schobert, M., Hundertmark, C., Jahn, D., and Munch, R., *Nucleic Acids Res.* 2003, **31**, 3862–3865.
41. Bader, G. D., Betel, D., and Hogue, C. W., *Nucleic Acids Res.* 2003, **31**, 248–250.
42. Bourne, P. E., Addess, K. J., Bluhm, W. F., Chen, L., et al., *Nucleic Acids Res.* 2004, **32** Database issue, D223–225.
43. Salwinski, L., Miller, C. S., Smith, A. J., Pettit, F. K., et al., *Nucleic Acids Res.* 2004, **32** Database issue, D449–451.
44. Zanzoni, A., Montecchi-Palazzi, L., Quondam, M., Ausiello, G., et al., *FEBS Lett.* 2002, **513**, 135–140.
45. von Mering, C., Huynen, M., Jaeggi, D., Schmidt, S., et al., *Nucleic Acids Res.* 2003, **31**, 258–261.
46. Hermjakob, H., Montecchi-Palazzi, L., Lewington, C., Mudali, S., et al., *Nucleic Acids Res.* 2004, **32** Database issue, D452–455.
47. Orchard, S., Hermjakob, H., Julian, R. K., Jr., Runte, K., et al., *Proteomics* 2004, **4**, 490–491.
48. Hermjakob, H., Montecchi-Palazzi, L., Bader, G., Wojcik, J., et al., *Nat. Biotechnol.* 2004, **22**, 177–183.
49. Wolkenhauer, O., *Brief Bioinform.* 2001, **2**, 258–270.

## APPENDIX DATABASES AND TOOLS FOR IN SILICO PROTEOME ANALYSIS

| Topic | Name | Short Name | Web Address |
|---|---|---|---|
| Nucleotide sequences | DNA Data Bank of Japan | DDBJ | http://www.ddbj.nig.ac.jp/ |
| | GenBank | GenBank | http://www3.ncbi.nlm.nih.gov/Genbank/index.html |
| | EMBL Nucleotide Sequence Database | EMBL-Bank | http://www.ebi.ac.uk/embl/ |
| Genomes | Genome Reviews | Genome Reviews | http://www.ebi.ac.uk/GenomeReviews/ |
| | Entrez | Entrez | http://www.ncbi.nlm.nih.gov/Entrez/ |
| | Genome Information Broker | GIB | http://gib.genes.nig.ac.jp/ |
| | Microbial Genome Database for Comparative Analysis | MBGD | http://mbgd.genome.ad.jp/ |
| Genome annotation | Ensembl | Ensembl | http://www.ebi.ac.uk/ensembl/ |
| Protein sequences | Universal Protein Knowledgebase | UniProt | http://www.ebi.uniprot.org/index.shtml |
| Microbial sequences | SubtiList | SubtiList | http://genolist.pasteur.fr/SubtiList/ |
| | Non-Redundant *B. subtilis* Database | NRSub | http://pbil.univ-lyon1.fr/nrsub/nrsub.html |
| | TubercuList | TubercuList | http://genolist.pasteur.fr/TubercuList/ |
| | EchoBase | EchoBase | http://ecoli.bham.ac.uk/genome.html |
| | EcoGene | EcoGene | http://bmb.med.miami.edu/EcoGene/EcoWeb/ |
| | Encyclopedia of *Escherichia coli* K12 Genes and Metabolism | EcoCyc | http://ecocyc.org/ |
| | Comprehensive Microbial Resource | CMR | http://www.tigr.org/tigr-scripts/CMR2/CMRHomePage.spl |
| | Homologous Bacterial Genes Database | HOBACGEN | http://pbil.univ-lyon1.fr/databases/hobacgen.html |
| Protein classification | InterPro | InterPro | http://www.ebi.ac.uk/interpro/ |
| | CluSTr | CluSTr | http://www.ebi.ac.uk/clustr/ |
| | Gene Ontology | GO | http://www.geneontology.org/ |
| | Gene Ontology Annotation | GOA | http://www.ebi.ac.uk/GOA |

| Topic | Name | Short Name | Web Address |
|---|---|---|---|
| Proteome analysis | Proteome Analysis Database | PAD | http://www.ebi.ac.uk/proteome/index.html |
| | Integr8 | Integr8 | http://www.ebi.ac.uk/integr8/ |
| | Proteome Profile Database | PPD | http://web.bii.a-star.edu.sg/~kishore/PPD/PPD.html |
| | Predictions for Entire Proteomes | PEP | http://cubic.bioc.columbia.edu/db/PEP/ |
| | Homologous Sequences in Complete Genomes Database | HOGENOM | http://pbil.univ-lyon1.fr/databases/hogenom.html |
| Proteome annotation | High-quality automatic and Manual Annotation of Proteomes | HAMAP | http://ca.expasy.org/sprot/hamap/ |
| Protein identification | ProteinProspector | Protein Prospector | http://prospector.ucsf.edu/ |
| | PROWL | PROWL | http://prowl.rockefeller.edu/ |
| Protein visualization | SWISS-2DPAGE | SWISS-2DPAGE | http://ca.expasy.org/ch2d/ |
| | JvirGel | JvirGel | http://prodoric.tu-bs.de/proteomics.php |
| Protein structure | Protein Data Bank | PDB | http://www.rcsb.org/pdb/ |
| Protein interactions | Biomolecular Interaction Network Database | BIND | http://www.blueprint.org/bind/bind.php |
| | Database of Interacting Proteins | DIP | http://dip.doe-mbi.ucla.edu/ |
| | Molecular INTeraction Database | MINT | http://mint.bio.uniroma2.it/mint/ |
| | Search Tool for the Retrieval of Interacting Genes/Proteins | STRING | http://string.embl.de/ |

CHAPTER 22

# Interspecies and Intraspecies Comparison of Microbial Proteins: Learning about Gene Ancestry, Protein Function, and Species Life Style

BERNARD LABEDAN and OLIVIER LESPINET

Université Paris-Sud, Orsay, France

## 22.1 INTRODUCTION

With the irresistible progress of genomics, we are now approaching the 400 microbial organisms for which the genomes have been entirely sequenced, annotated, and published. Such a huge amount of data could be viewed as a gold mine, but one must be cautious that, as in real life, what is shining could just be lead. Accordingly, postgenomics approaches are emerging as a new science, achieving immense hopes but mixed with the feeling that a significant part of the data obtained by high-throughput approaches may contain fatal errors resulting in misleading inferences. Indeed, sequencing in itself appears now as a rather reliable technique. However, finding the complete set of genes present in a genome is a far more difficult task, especially for eukaryotes. Moreover, functional annotation of these putative genes is a complex and hazardous enterprise and requires caution, consistent tools, and relevant biological knowledge.

Here, we will emphasize the need for adequate concepts and powerful methodology to exploit these genomic data with pertinence. We will show how to use these concepts to study different fundamental points such as an improved understanding of protein history and how we can study genome fluidity and reconstruct minimal ancestral genomes. Furthermore, we will insist on the potential of this approach to improve functional annotation: (i) detecting (and correcting) annotated errors and (ii) assigning putative function to "conserved hypothetical proteins." Finally, we will emphasize the complexities of the relationships between sequence homology and function.

## 22.2 SOUND USE OF HOMOLOGY IS CRUCIAL IN COMPARATIVE GENOMICS

Homology is one of the most important concepts in biology, but its use—and misuse—has been dramatically accentuated since life scientists in nearly all fields entered the new era

*Microbial Proteomics: Functional Biology of Whole Organisms*, Edited by Ian Humphery-Smith and Michael Hecker. 

of genomics. Accordingly, here we will underline the importance of a few basic, essential facts about homology (with emphasis on molecules, especially proteins) when using genomic data.

### 22.2.1 Defining Homology

Although the conceptual definition of homology has been historically elaborated in a more complex way [see Fitch (2001) for discussion], we will only present the basic and widely acknowledged definition: *Two items are defined as homologues if they share a common ancestry.* Such a definition has two fundamental implications: Homology is (1) always a hypothesis and (2) all-or-none property [see Reeck et al. (1987) for discussion]. Thus, an indirect way is necessary to assess experimentally that two objects are homologues. In most cases, the level of sequence similarity is the criterion used. For example, two proteins will be labeled homologues if the number of their identical residues is higher than an imposed threshold [see, e.g., Doolittle (1981) and Altschul [1991].

### 22.2.2 Different Classes of Homology

As early as 1970 Fitch made a fundamental distinction between classes of homology. *Orthologous* genes are homologous genes which diverged by speciation. Therefore, orthologues are the pertinent objects to use to reconstruct species trees. *Paralogous* genes descend from an ancestral duplication, independently of speciation. Thus, paralogues are helpful for understanding the course of protein (gene) evolution as long as the changes to sequences over time by processes of mutation, recombination, and repair have not blurred the similarities. It has been further proposed (Solignac et al., 1995) to call *metalogues* paralogous genes which have been separated by speciation. This leads to an important point: In the case of asymmetrical loss of paralogous copies in compared species, the remaining metalogues could be erroneously interpreted as being bona fide orthologues.

### 22.2.3 Minimal Segment of Homology

In a stochastic model of protein evolution (Dayhoff et al., 1978; Schwartz and Dayhoff, 1978), the evolutionary distance separating two homologous proteins is given in PAM units (defined as the number of accepted point mutations per 100 residues separating two sequences). In two seminal papers Altschul (1991, 1993) showed, using an approach based on information theory, that 30 bits of information are necessary to distinguish an alignment from chance. He further showed that, to reach such a cutoff, the length of the segment of homology is dependent on the nature of the Dayhoff substitution matrix used. Accordingly, to be statistically significant, an alignment of sequences separated by a distance of 250 PAM units would need to have a length of at least 83 residues.

### 22.2.4 Concept of Module: Structural Segment of Homology

Riley and Labedan (1977) used the Altschul cutoffs to assess the homologous relationships when comparing *Escherichia coli* proteins. In many cases, homology was found to be limited to long structural segments with a mean size of 220 amino acids. Such segments were called modules to reflect the role they play in the mechanism of combinatorial construction of a gene from ready-made basic components. In support of this model, it is striking that, for many prokaryotic genomes the mean size of homologous proteins (~450 residues) is about twice the size of nonhomologous proteins (~250 residues), the latter being close to the module size. This would mean that many present-day proteins are the result of successive events of ancestral gene duplication and fusion of evolutionary unrelated genes which occurred at different periods and were at the origin of the present-day modules. In our eyes, the module is a new unit of evolutionary descent. Identifying a module is operationally equivalent to determining the ancestor of this gene segment.

It may be more than a coincidence that the mean size of the 220 amino acids we found for prokaryotic modules is rather close to typical fold (structural domain) size, $150 \pm 50$ determined by a completely different approach (Wheelan et al., 2000; see also Gerstein, 1998). Such a figure appears as a universal unit according to Wheelan et al. (2000), supporting our model of combinatorial construction of a protein from ready-made basic components, the modules.

Furthermore, we must emphasize that modules are conceptually different from the shorter segments of homology which have been registered as domains or motifs in various specialized databases (Bateman et al., 2002; Corpet et al., 2000; Falquet et al., 2002; Letunic et al., 2002). Domains and motifs are at a different level, being entities such as binding sites for cofactor and prosthetic groups. Domains and motifs are common features of many proteins that otherwise have no sequence similarity over a significant fraction of their total length. Although domains and motifs are important elements of the specificity of binding and the chemistry of action of a protein, as well as important features of any tertiary structure, it may be misleading to use their sequences to attempt to trace back protein history. Similar domains and motifs are found within proteins and modules that as a whole are not members of evolutionary defined families. This is the case for example of the so-called p21$^{ras}$ family of guanidine triphosphate (GTP)–binding proteins (Schulz, 1992). Besides sharing the "P-loop" functional motif, the majority of these proteins are not homologues. For example, we have shown that the adenylosuccinate synthase is clearly unrelated to the Ras protein or the elongation factor TU from *E. coli* (Bouyoub et al., 1996). It has been suggested that the structural correspondence between the adenylosuccinate synthase and the Ras protein at the level of their P loop is a consequence of convergent evolution of two distinct families of proteins which bind and hydrolyze GTP (Poland et al., 1993, 1997).

Therefore, here we emphasize how crucial is our modular approach to clearly differentiate between homology (divergent evolution) and analogy (convergent evolution)

### 22.2.5 Functional Annotation by Homology

This kind of differentiation is also very important in genomics, where homology is systematically used to annotate the coding sequences which have been identified in a newly sequenced genome, especially when no experimental evidence is available. This widely used approach is based on a two-step process: (1) detection of a homologous relationship by a pairwise sequence similarity search at the level of their primary structure and (2) inference of functional similarity from this detected homology. The public sequence databanks (GenBank/EMBL/DDBJ) are now inundated by amino acid sequences which have been annotated uniquely by this approach.

For various reasons one can however either miss or misinterpret the actual function of a protein when annotating by homology, resulting in an incorrect assumption of function by transfer. Several studies have already emphasized this point (see, e.g., Brenner, 1999;

Gerlt and Babbitt, 2000; Babbitt, 2003). We will underline later the problems inherent with this strategy which might be misleading.

After describing our approach to find all homologues when comparing exhaustively many proteomes to better understand protein history and genome evolution, we will insist below on the complexity of the relationships between sequence homology and function. Further, several cases underlining how one must be cautious when using the data obtained with the approach of exploiting homology to infer function transfer will be discussed and several controls/tools to challenge them will be proposed.

## 22.3 FINDING ALL SEGMENTS OF HOMOLOGY IN PROTEOME: COMPARATIVE GENOMICS APPROACH TO STUDY PROTEIN HISTORY

One of our main objectives is to study the evolutionary mechanisms which were used to build up the present-day proteins (and their encoding genes). The module concept, as presented in detail above, is crucial to understanding protein history by taking into account two major mechanisms occurring at the gene level: duplication and fusion.

As a matter of fact, comparison of modern-day proteins and identification of all modules will help to number the events of duplication and fusion and thus, after grouping all homologous modules in families, to trace back to the ancestral genes which were at the origin of each family.

## 22.4 SUITE OF BIOINFORMATIC TOOLS

Accordingly, in the conceptual framework based on this module approach, we are constantly updating an exhaustive comparison of whole sets of proteins (hereafter called proteomes) encoded by completely sequenced genomes of microbial species. To make this approach automatic, we have designed methodological approaches to identify all modules, to group them in families in order to number the putative ancestral genes which were at the origin of the present-day proteins, and to count the different events of gene duplication and gene fusion which helped to shape these present-day proteins. We will further focus on a few case studies illustrating how these concepts may help to illuminate various aspects of the biology of poorly known microbes.

To deal with the deluge of data released by the whole-genome sequencing programs, we have devised a suite of automatic programs which allow us to detect in a few steps the whole set of modules which constitute the proteome of any organism for which the complete DNA sequence is available (Le Bouder-Langevin et al., 2002).

Recently, this suite has been improved with new, more efficient programs corresponding to the three following main steps.

### 22.4.1 Step 1: Defining Significant Thresholds to Collect all Homologues

First, for each pair of microbial species, we compared their proteomes in order to detect homologous segments using the thresholds for both evolutionary distance (less than 250 PAM units) and alignment length (at least 80 residues), as described above.

Such an exhaustive approach allows to collect in one step all paralogous (intragenomic comparison) and all orthologous (intergenomic comparison) pairs of aligned protein sequences. To retrieve the whole set of modules, we adapted, as ready described (Riley and

Labedan, 1997; Le Bouder-Langevin et al., 2002), the Darwin *AllAll* program (Gonnet et al., 1992), which is based on a maximum-likelihood approach. We found that this *AllAll* program is very powerful in detecting all modules of interest, including distant homologues. Then, modules found in each match are classified according to their length and location inside the aligned proteins. The nomenclature used adds two suffixes to the protein name: The first one corresponds to the module structure. For example, an alignment between the first third of protein A and the last third of protein B will be interpreted as a module A_1_3 matching with a module B_3_3.

### 22.4.2 Step 2: Grouping Homologous Modules in Families to Count Their Ancestral Genes

***One Family, One Ancestral Gene*** Second, we designed a program to gather automatically into one family all modules that are related by a chain of similarities, collecting all relatives of both members of an *AllAll* pair until no further pairwise relationship is found. To meet our different experimental needs, various kinds of families were assembled, grouping either paralogues of each species or orthologues for each pair of genomes or all homologues for different groupings of species. For example, Figure 22.1

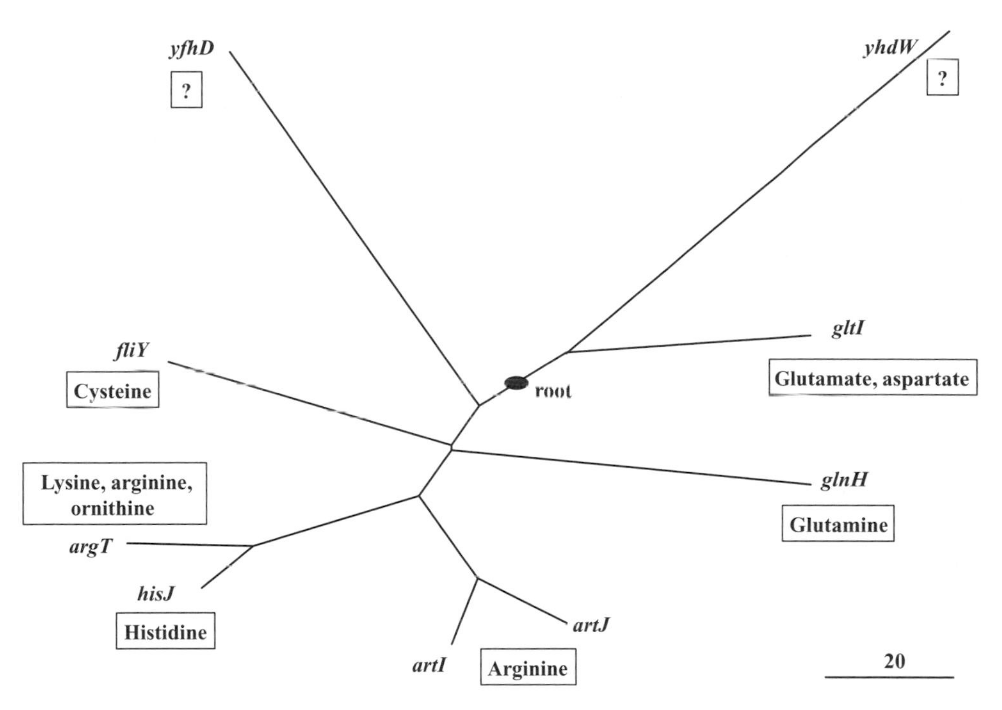


**Figure 22.1** Phylogenetic tree based on module 1_1 of nine periplasmic binding proteins of *E. coli.* The PhyloTree program of the DARWIN package (Gonnet et al., 1992) was used to make an exhaustive measure of the evolutionary distances (PAM distances) separating each sequence from all its homologues to build a multiple alignment and to derive a distance tree which is an approximation to a maximum-likelihood tree since the deduced evolutionary distances are weighted by computing the variance of the respective PAM distance when reconstructing the tree. The branch lengths (in PAM units, weighted by the respective variance) are drawn to scale. The black dot indicates the location of the putative root as it is computed by PhyloTree.

shows nine modules 1_1 (entire proteins) present in various periplasmic binding proteins of *E. coli* that are involved in transport of amino acids from the outer to the inner membrane, such as ArgT (lysine, arginine, ornithine), ArtI and ArtJ (arginine), FliY(cysteine), GlnH (glutamine), GltI (glutamate, aspartate), and HisJ (histidine). This family contains also two unknown proteins, YfhD and YhdW. Since this family looks homogeneous in terms of function, one can propose that YfhD and YhdW are putative periplasmic binding proteins for amino acids. Moreover, the more parsimonious hypothesis is that these nine modules descend from the same ancestral gene (located as the root in the tree shown in Fig. 22.1) which coded a periplasmic binding protein with a broad specificity for amino acids.

More generally, this step of collecting families allows us to number the ancestral genes which were at the origin of each present-day family.

***Distinguishing Biologically Pertinent Families in Sea of Look-Alike Modules***
It must be noted that the transitive approach we used to gather homologous modules in families is very fast and rather efficient in delivering a large array of biologically pertinent families. However, it has a major drawback. A significant *part* (which may reach 60% as soon as four or more genomes are compared) of the modules which look alike instead of forming homogeneous clusters is progressively amassed into a huge "family." This is mainly due to the presence of a limited number of modules making a simple link that bridges two unrelated clusters made of members with many links. For example, a large majority of membrane proteins are progressively and indistinctly put in the same bulky bag uniquely because, to cross the membrane, they have to harbor hydrophobic chains that look alike independently of the history and other specific features of these proteins. To prevent such a mechanical accumulation of nonhomologous modules in the same bag, we recently modified the program designed to make the families. We now require that each family member matches at least 85% of the other family members. With these new conditions, the huge artifactual family is divided into several hundreds of more pertinent small families. Moreover, a comparison between the huge transitive largest family and these more exhaustive families helps to distinguish a strong small nucleus from the weak links which are bridging them (Labedan et al., to be published).

### 22.4.3 Step 3: Intra and Intergenomic Comparisons: Four Classes of Modules

Third, for each protein with at least one homologue, we compiled all homology information in order to define its *modular structure* (identification of all events of gene duplication and gene fusion) and the *phylogenetic profiles* (listing all of the species containing at least one orthologue) for each module and for the entire protein.

Moreover, we computed to which of the four different classes previously defined (Le Bouder-Langevin et al., 2002) do each gene/protein belong: The first two categories correspond to proteins which are found in only one species (sp.) and which either have a paralogue (para-sp.) or are unique to their species (uni-sp.). The last two categories correspond to orthologous proteins which either have a paralogue (para–ortho) or are unique to their species (uni–ortho). The distribution of these different classes obtained after comparing 66 prokaryotic genomes displayed two main features as illustrated in Figure 22.2 for a set of selected organisms. (1) We observed a good correlation between the proteome size and the gene content in the two orthologous classes, para–ortho and uni–ortho. The Pearson correlation values obtained are 0.87 (associated probability

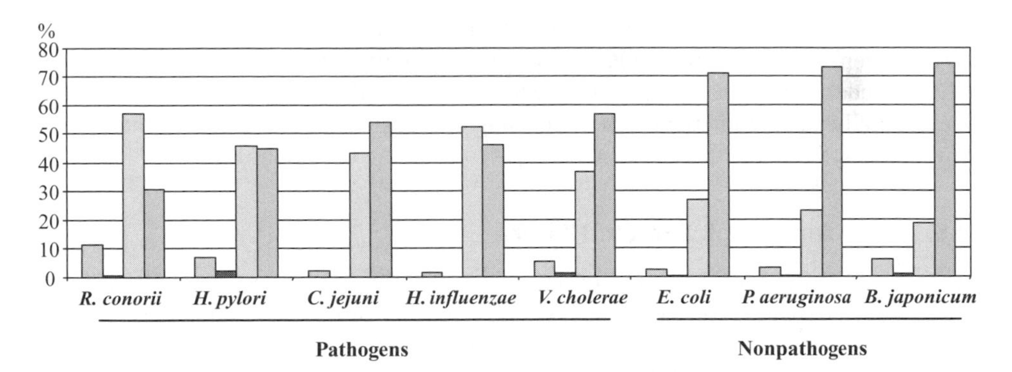


**Figure 22.2** Distribution (in percentage) of module classes of homology after inter- and intragenomic comparisons between 66 microbial genomes. Results are shown for only eight selected species (*R. conorii, H. pylori, C. jejuni, H. influenzae, V. cholera, E. coli, P. aeruginosa, B. japonicum*). From left to right, defined module classes of homology are module unique to one species, paralogous module only found in one species, paralogous modules found in several species, orthologous modules (without paralogues) found in several species.

$P = 0.005$) and $-0.92$ ($P = 0.001$), respectively. (2) The distribution of these different classes appeared to vary with the life style of each species. There is a clear difference between the pathogenic species (*Haemophilus influenzae, Vibrio cholerae, Helicobacter pylori, Campylobacter jejuni,* and *Rickettsia conorii*), which have smaller genomes, and the facultative pathogens (*E. coli, Pseudomonas aeruginosa*) and nonpathogenic species (e.g., *Bradyrhizobium japonicum*), which have larger genomes. The large majority of pathogens display a low to very low proportion of para–ortho class while the uni-sp. class is remarkably high. This is examplified in the case of *H. influenzae*, *H. pylori*, and *C. jejunii* and even more for the obligatory intracellular pathogen *R. conorii*. On the contrary, nonpathogenic bacteria display a large excess of para–ortho proteins and also a large size for the para-sp. class. These contrasted distributions confirm several of our precedent findings (De Rosa and Labedan, 1998; Le Bouder-Langevin et al., 2002; Sculo et al., 2003) which suggested that many pathogens have reduced their genome size by preferentially diminishing the size of their families of paralogous genes.

Moreover, we found that bacteria able to adapt to various environments and life conditions such as *E. coli* and *P. aeruginosa* maintain a strikingly high number (e.g., 411 in the case of *E. coli*) of small families (from two to six paralogous members) coding for putative transcriptional regulators, resistance to various substances (including antibiotics), or putative membrane proteins involved in various stages of transport of metallic ions and other rare environmental substances. These different functions may be important for survival of free-living bacteria in adverse conditions.

## 22.5 USING MODULE DATA TO RECONSTRUCT ANCESTRAL GENES AND GENOMES

After assembling all homologous modules in families (see step 2 above), we further group families which are related by a chain of neighboring unrelated homologous modules. Automatic analysis of these groups of families allows us to split them into their component parts consisting of many fused modules and/or to deduce by logic more distant modules.

All detected and inferred modules are further reassembled in refined families. These two last steps are made by SortClust, a rather complex program described at length by Le Bouder-Langevin et al. (2002). Finally, the validity of the inferred module is systematically and automatically checked using the consistency of the evolutionary tree of their deduced families.

### 22.5.1 Tracking Back Protein History

The refined families are then used to improve our study of gene and genome evolution. This is illustrated in the following analysis of the history of *E. coli* proteins (Fig. 22.3). In the first step of an exhaustive comparison, 5527 paralogous modules were detected in 2487 homologous proteins. They form 1020 families which can be separated into 307 *unique* families and 713 families which share 4586 modules. When applying SortClust to these 713 families, their modules would be reinterpreted and were then able to be assembled into only 264 refined families. Thus, the putative genome of the ancestor to *E. coli* would have been made of 1607 genes which never duplicate and 542 genes which duplicate to create the present-day 4670 paralogous modules. Moreover, 697 of the nonduplicating genes

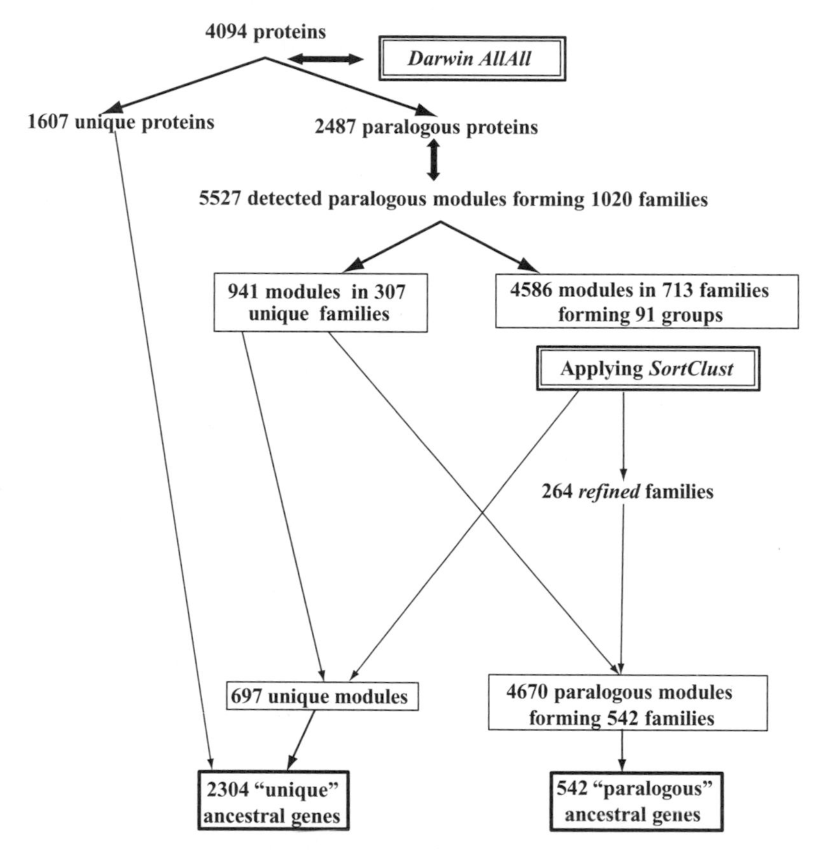


**Figure 22.3** Reconstruction of the evolutionary history of modules detected in *E. coli* proteins. The successive main steps of our automatic approach and their results are shown.

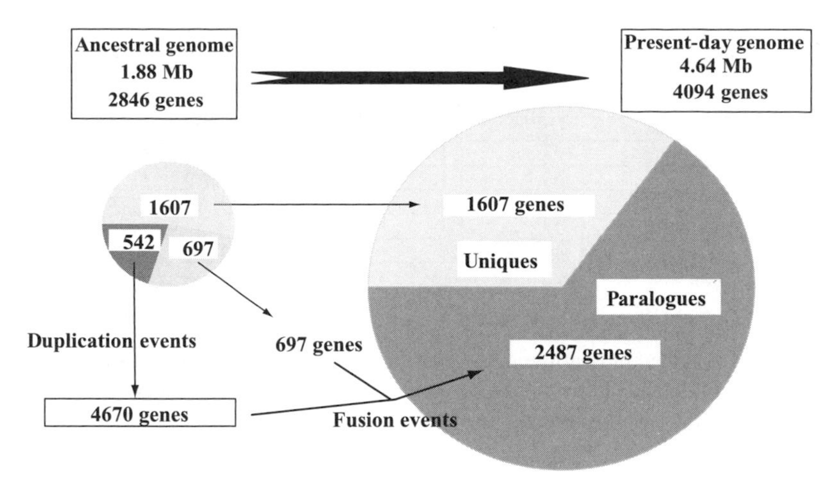


**Figure 22.4** Computing the putative ancestral *E. coli* genome based on intragenomic comparison shown in Figure 22.3.

fused in various combinations with the 4670 paralogous modules to create the 2487 paralogous present-day proteins.

In summary, the putative ancestral genome of *E. coli* (Fig. 22.4) contains a majority of genes which apparently never duplicated or, if they did, all extra copies either diverged very far or were eliminated. In contrast, only 18.6% (542) of genes duplicated often with survival of the many differentiated copies themselves duplicating again and again. Many of the products derived from the progeny of the highly duplicated minor set of ancestral genes fused either among themselves or with 697 of the unique genes in various combinations to increase the palette of functions available to the cell.

### 22.5.2 Tracing Back Genome Evolution

Although intragenomic data are helpful in disclosing many of the ancient events which created present-day proteins, they are of limited use to explore the distant history of genomes since the putative size of the ancestral genome of *E. coli* deduced by the approach summarized above appears surprisingly large (Fig. 22.4). Assuming that the mean size (220 residues) of present-day modules mirrors the size of the product encoded by the 2846 "ancestral" genes, the ancestral genome would have a size as high as 1.88 Mb. This size is clearly far larger than genomes of present-day free-living organisms.

However, intergenomic analysis helps one to go deeper into the past. Figures 22.5 and 22.6 show that the number (and type) of the compared genomes are important factors when trying to trace back protein history and genome evolution. Figure 22.5 shows the distribution of the different classes for the same set of $\gamma$-proteobacteria species (*E. coli, H. influenzae, P. aeruginosa, V. cholerae*) obtained after comparing either 66 prokaryotic genomes as already shown in Figure 22.2 or only these 4 species. It can be seen that the relative proportions of uni-sp. and para-sp. classes are decreasing and those of uni–ortho and para–ortho are increasing since the addition of more and more genomes helps confirm the existence of homologues to the genes which previously appeared as specific to their species.

From the respective distributions of the gene classes and the numbering of the families of homologous modules found after a SortClust treatment, we could tentatively

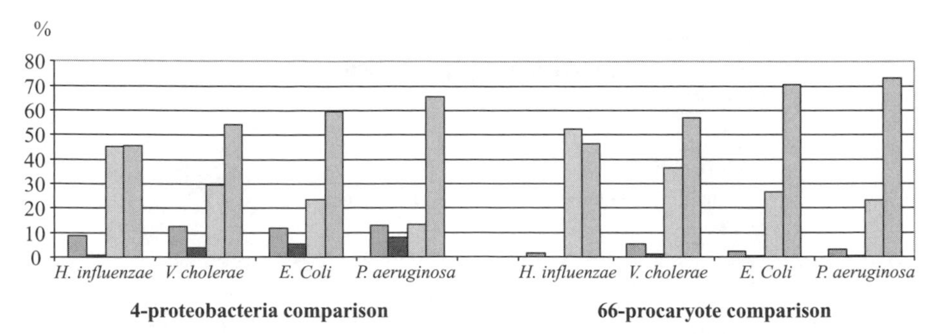


**Figure 22.5** Comparison of distribution (in percentage) of module classes of homology resulting from inter- and intragenomic comparisons of 4 proteobacteria and inter- and intragenomic comparisons of 66 microbial genomes. (See Figure 22.2 caption).

reconstitute the gene distribution in the genome of putative ancestors. Figure 22.6 shows the data obtained when comparing the four-genome set made of *E. coli, H. influenzae, P. aeruginosa*, and *V. cholerae*. The comparison with Figure 22.4 shows an increase in the relative proportions of ancestral genes which will give birth to the classes para–ortho and uni–ortho. This tendency is increased with both an increase in the number of compared genomes and the phylogenetic distance separating the compared species. In particular, there is a strong decrease in the relative proportion of ancestral genes which gave birth to uni-sp. and para-sp. when comparing a set of 66 prokaryotic species corresponding to the main branches of the bacterial and archaeal trees.

## 22.6 NUMBERING DIFFERENT EVENTS OF GENE DUPLICATION AND GENE FUSION

The comparison between a set of present-day genomes and their putative common ancestor further shed light on the way genes evolved to create contemporaneous proteins.

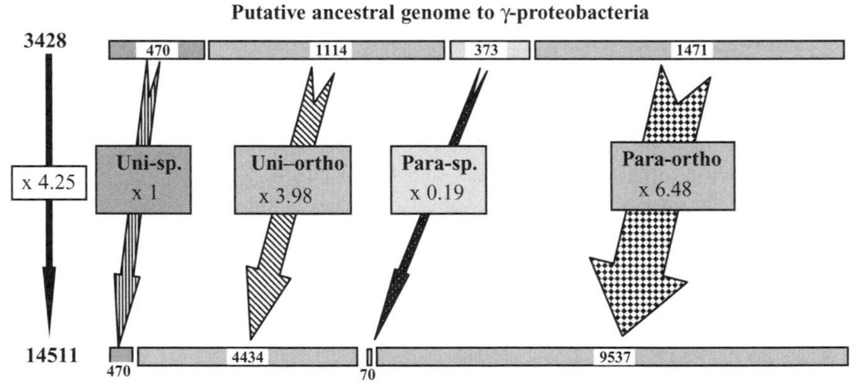


**Figure 22.6** Evolution from ancestral genes to present-day modules. The data were obtained through the intergenomic comparison of four γ-proteobacteria, *E. coli, H. influenzae, P. aeruginosa, and V. cholerae*.

Using the last common ancestor to *E. coli*, *H. influenzae*, *P. aeruginosa*, and *V. cholerae*, we can propose the following scenario for its 3428 different genes (Fig. 22.6). There are apparently two opposite behaviors. On the one hand, 45.9% of the ancestral genes never duplicated (uni-sp. and uni–ortho) or, if they did, the supplementary copy has been lost. On the other hand, 54.1% of the ancestral genes went through more or less frequent events of duplication giving birth to many paralogous orthologues. The relative proportion of paralogues specific to their species has more or less disappeared. Among the 3946 families grouping the homologous modules for these four species, we obtained only 36 para-sp. (15 for *E. coli*, 14 for *P. aeruginosa*, 8 for *V. cholerae*, and none for *H. influenzae*). The relative numbers of uni–ortho and para–ortho evolutionary units have increased by a factor of near 4 and around 6.5, respectively (Fig. 22.6). Such data are in strong support of the forecast hypothesis of Ohno (1970) that gene duplication is the main driving force to create new proteins/functions.

In this context, the thorough analysis of refined families belonging to the same group seems pertinent for elucidating how a limited number of peculiar events of gene duplication and gene fusion helped to create new proteins/functions. To illustrate this point, let us focus on the *E. coli* group 89 comprising 63 paralogous modules belonging to 33 present-day proteins (Figs 22.7*a* and 22.7*b*). This group 89 is made of 4 refined families of unequal size and contrasting groupings. The largest one, 536, is made of 1 module_1_1 (YidW) and 30 module_1_2 (Fig. 22.7*a*). The remaining modules are scattered in three different families (Fig. 22.7*b*): a two-member family of 218; a nine-member family of 510, made predominantly of aminotransferases; and a larger family, of 531, made mainly of transcriptional regulators. Thus, the most parsimonious scenario to explain the formation of these four refined families can be sketched as follows (Fig. 22.8). Four ancestral genes 536, 531, 510, and 218, are fused in various combinations to create at least three ancestral functions. The fusion 536–218 led to a protein involved in lipopolysaccharide biosynthesis. The fusion 536–510 gave a gene encoding an ancestral aminotransferase and the fusion 536–531 a gene encoding an ancestral transcriptional regulator. Then, two of these fusion products (536–510 and 536–531) duplicated once or several times to diverge in more of less large families of more specialized biochemical functions. Moreover, several of these bimodular proteins may be bifunctional, as has been demonstrated for the recent crystallized MalY protein (Clausen et al., 2000). The presence of MalY_1_2 in family 536 and of MalY_2_2 in family 510 appears consistent with the demonstration by Clausen et al. (2000) that MalY is capable of a direct protein–protein interaction with MalT, the central transcriptional activator of the maltose system composed of a pyridoxal 5′-phoshate-binding domain and a structural domain similar to aminotransferases.

## 22.7 USING MODULE DATA TO DETERMINE CORE OF ESSENTIAL GENES IN EACH GENOME ANALYZED

### 22.7.1 Comparing close and distant species

In a next step, we focused our comparison on the complete genomes of 26 proteobacteria with a special focus on the 2 ε because they look far from the α, β, and γ phyla available at that time (Sculo, unpublished data). We have observed two opposite trends with the steady increase of genomes to be compared.

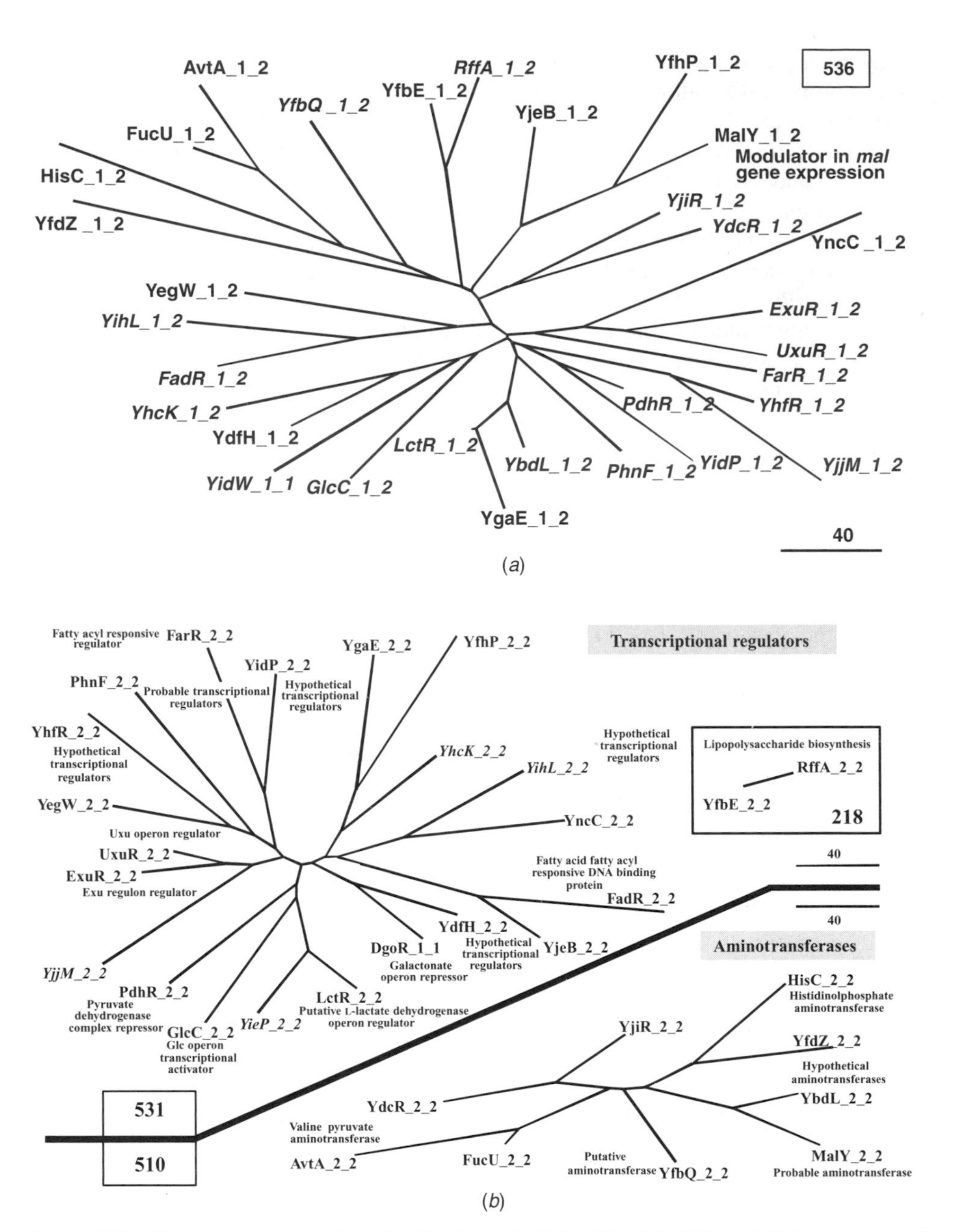


**Figure 22.7** Phylogenetic trees of (*a*) family 536 and (*b*) families 218, 510, and 531 reconstructed as described in legend to Figure 22.1.

1. Exhaustive comparison of the 89,555 proteobacterial proteins gave 303,529 modules present in 78,659 homologous proteins (87.8% of the total proteins). The probability of finding a homologue for a gene increased with the number of genomes, that is, with the total number of genes. It seems that we are approaching a sort of plateau due to the presence in each genome of a significant proportion of orphans (genes we called uni-sp. in our nomenclature).

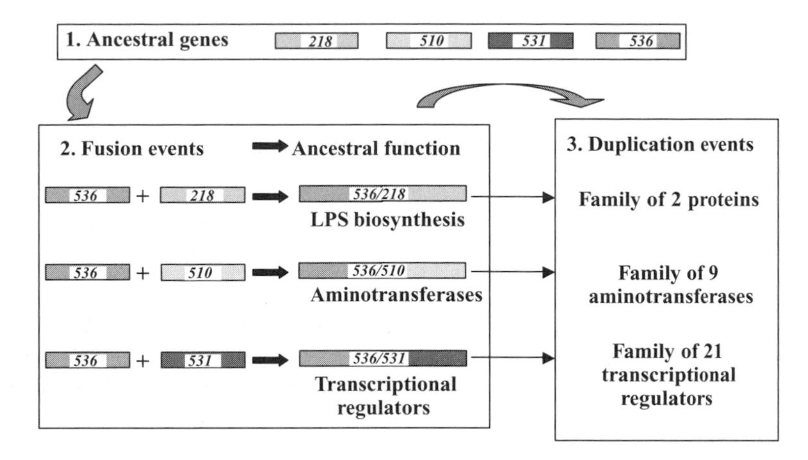


**Figure 22.8** Proposed scenario for evolutionary history of genes belonging to families 218, 510, 531, and 536 (see Figs. 22.7*a*, *b*).

2. Among the homologous modules, only a limited set present in a genome has been found to have orthologues in all 25 other proteobacteria. The number of these omnipresent genes is decreasing very rapidly with the increase in the number and nature of the compared genomes, as shown on Table 22.1.

Therefore, such universal genes would define a core of essential genes and would be representative of the basic set of genes present in the common ancestor to all proteobacteria. Moreover, these figures are species dependent and fall in two main categories. As many as 233 modules in *C. jejuni* and 201 in *H. pylori* correspond to a unique segment of homology, the huge majority of them being entire proteins. The minor category corresponds to more complex cases where a protein appears to be universal because, besides having homologues in a large majority of the other proteobacteria, only one of its modules has orthologues in the missing species. An example of this minor category is the *dnaA* gene present in *C. jejuni*: It aligns along its full length with the other *dnaA* genes present in 24 proteobacterial species except in the case of *Wigglesworthia brevipalpis*, where the homology is limited to the module 1_2. Another, more complex case is *rplC* (encoding a protein of the large ribosomal subunit), with a full-length alignment limited to 11 species and a homology limited to the module 2_2 in the 14 remaining species (*Agrobacterium tumefaciens, Buchnera, Caulobacter crescentus, E. coli, H. influenzae, Mesorhizobium loti, Neisseria. meningitidis,*

**TABLE 22.1 Effect of Increasing Number of Genomes**[a]

| | Four γ-Proteobacteria | Twenty-Six Proteobacteria | Sixty-Six Prokaryotes |
|---|---|---|---|
| *E. coli* | 1834 | 306 | 141 |
| *H. influenzae* | 1133 | 241 | 107 |
| *P. aeruginosa* | 2147 | 367 | 163 |
| *V. cholerae* | 1532 | 333 | 161 |

[a]Defining the core of ubiquitous genes is improved when increasing the number of compared genomes.

*Pasteurella mulcida, P. aeruginosa, Rickettsia prowazekii, Salmonella enterica, W. brevipalpis, Xylella fastidiosa, Yersinia. pestis*).

### 22.7.2 Determining Minimal Genome at Recent or More Distant Ancestors

We further showed that more than 90% of these universal genes play essential roles in basic processes such as DNA replication, transcription, and translation; cell wall biosynthesis; and cell division, metabolism, or active transport. This is strikingly reminiscent of the known universal genes which have been described by different approaches as being the core of bacterial life (see, e.g., Kobayashi et al., 2003). Thus, our approach may help to define the whole set of genes encoding essential functions, including those which deserve to be further ascertained by experimental studies.

Indeed, we have found in the case of ε-proteobacteria that a significant proportion of these omnipresent genes have no known function and have been annotated as putative or hypothetical in *C. jejuni* and *H. pylori*. Therefore, we can infer that these universal genes for which we do not have any information probably encode important functions. We will explain below how we have proposed putative functions for the reannotation of these universal genes.

## 22.8 USING MODULE FAMILIES DATA TO UNDERSTAND RELATIONSHIPS BETWEEN SEQUENCE AND FUNCTION

The relationships between primary sequence, tertiary structure, and function of a protein appear to be very complex when considering the large set of families we have progressively assembled with our approach of exhaustive comparison. We have already described an interesting case above where a group of homologous bimodular proteins present in *E. coli* display three different unrelated functions at the level of their second module: lipopolysaccharide biosynthesis, aminotransferase, and transcriptional regulator. To illustrate this complexity, we now present three example case studies. Moreover, we underline how these relationships are helpful in reannotating newly sequenced genomes.

### 22.8.1 Case 1: Homology of Sequence = Homology of Function

Figure 22.9 shows the genealogical tree of a family of 16 kinases of various sugars present in the set of the four γ-proteobacteria and two unknown open reading frames present in *E. coli* (YdeV and YgcE). As can be seen, this family looks rather homogeneous. The recent events of duplication are discernible, and they always occurred before the speciation events which separated the different orthologues descending from the ancestral paralogues (e.g., *glpK*). The topology of this tree is rather typical of many midsize families, with a center having both a bad resolution and distances probably underestimated due to saturation phenomena. However, this tree strongly suggests that the ancestral module at the origin of this family and which extends along the entire length of these 18 present-day proteins was already a sugar kinase with broad specificity. Moreover, it must be underlined that other paralogous kinases of sugar having different sizes are

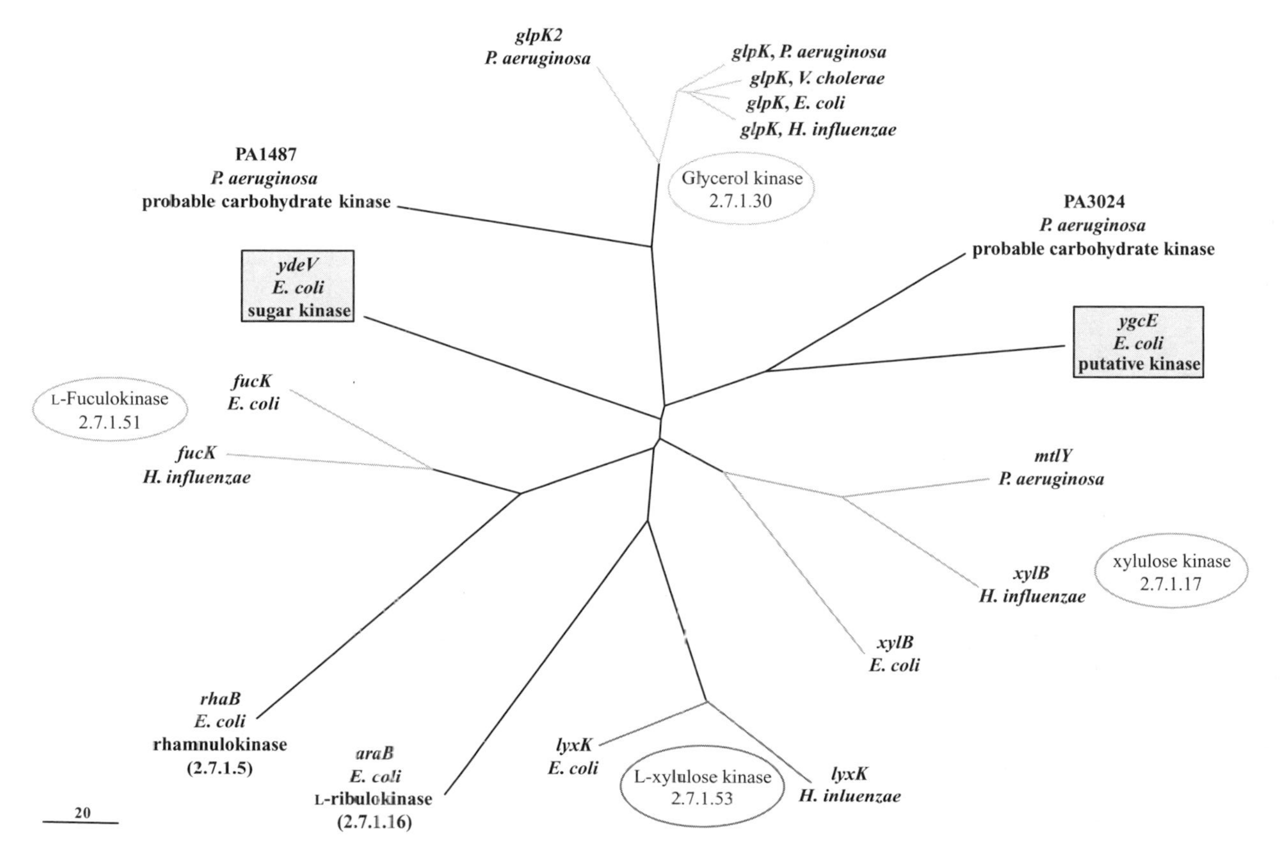


**Figure 22.9** Phylogenetic tree of homogeneous family of kinases reconstructed as described in legend to Figure 22.1.

present in other families showing different evolutionary history to obtain this kind of functional protein.

### 22.8.2 Case 2: Homology of Sequence But Different Functions

Figure 22.10 shows an example of the genealogical tree of a nonhomogeneous family. We have evolutionary relationships between repressors of type LacI which group on the left of the tree and several periplasmic binding proteins (PBPs) of transport systems which group on the right of the tree (red branches). The PBPs which have kept significant sequence similarities to one or several repressors are underlined. This is the case, for example, of the pair made of RbsB which participates in the transport of ribose and RbsR, the repressor of the ribose operon. An explanation, already proposed by Mauzy and Hermodson (1992), might be that differentiation of an ancestral protein specifically recognizing ribose gave rise to a ribose-specific repressor on the one hand and a ribose-specific transport protein on the other hand. More generally, the ancestor which is probably located at the connection point of the two subtrees could have been a protein able to bind to a small molecule and which after gene duplication and divergence of the copies progressively gained substrate specificity. This tree topology further suggests that one ancestral copy modified its cellular location first, before gaining substrate specificity. The addition of a signal sequence would have been a crucial step for such a transfer from cytoplasm to periplasm.

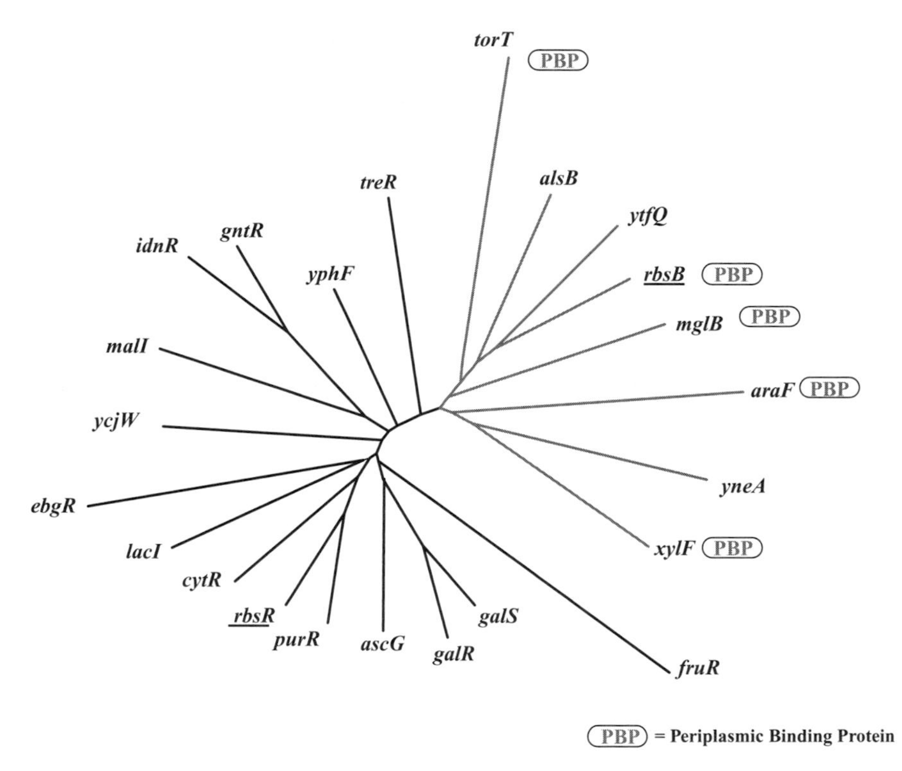


**Figure 22.10** Phylogenetic tree of heterogeneous family made of paralogous transcriptional regulators and periplasmic binding proteins present in *E. coli*. The tree was reconstructed as described in legend to Figure 22.1.

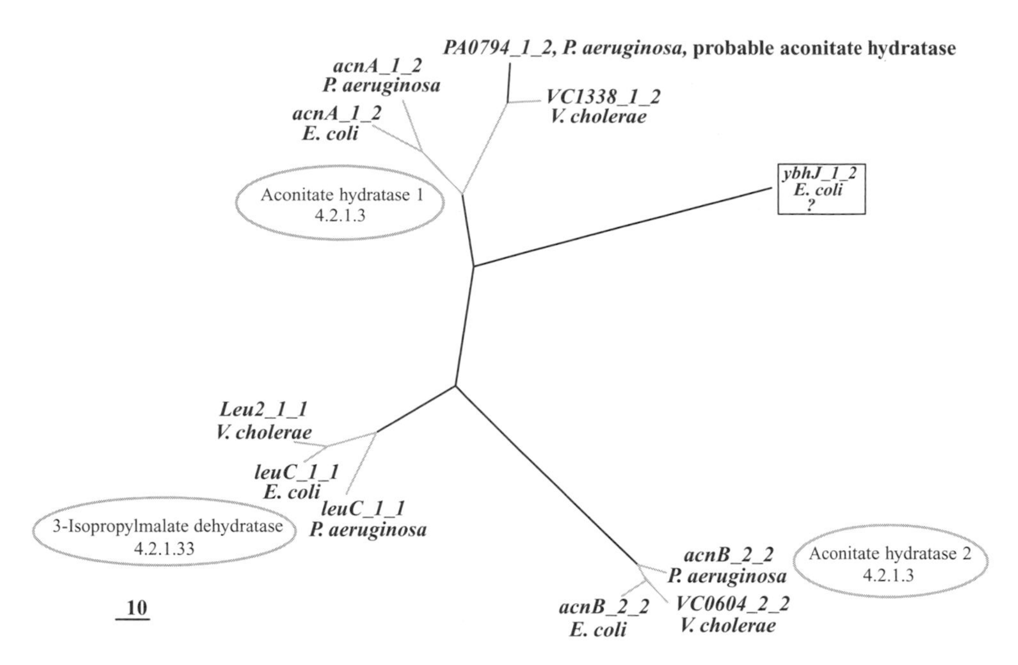


**Figure 22.11** Grouping modules which are not homologues but display identical function. The tree was reconstructed as described in legend to Figure 22.1.

### 22.8.3 Case 3: Nonhomology of Sequence But Identity of Function

Figure 22.11 shows the genealogical tree of an 11-member family obtained after comparing the set of the four $\gamma$-proteobacteria. These modules correspond to two isoenzymes (aconitate hydratases 1 and 2, EC 4.2.1.3), the large subunit (LeuC) of a functionally similar enzyme, the 3-isopropylmalate dehydratase (EC 4.2.1.33), and one unknown open reading frame (YbhJ), respectively.

Several interesting features specific to this family must be underlined. (i) Surprisingly, the two isoenzymes are very distant and their similarities of sequence are too low for our criteria of homology. They are found belonging to the same group only because both aconitate hydratases display significant sequence similarities to LeuC and YbhJ. (ii) The module of this family is about the size of the entire LeuC protein. The corresponding segment of homology is located on the N-terminal side in the aconitate hydratase 1 and the same thing is true for YbhJ, but it is located on the C-terminal side in the aconitate hydratase 2. This striking difference in location suggests, at least in this case, that module location has no effect on protein activity. It also confirms that the evolutionary histories of both aconitate hydratases are significantly different. (iii) The long segment present in the N-terminal side of aconitate hydratase 2 has no homologue known in another protein beside the orthologues present in species closely related. On the other hand, we have detected on the C-terminal side of aconitate hydratase 1 a short module homologue to the corresponding segment present in the second subunit (LeuD) of the 3-isopropylmalate dehydratase. Thus, it seems that during their evolutionary history the two similar enzymes aconitate hydratase 1 and 3-isopropylmalate dehydratase have followed two different paths: In one case two independent modules remain as separate subunits (encoded by two adjacent genes inside the *leu* operon) which form a multimeric active complex. In the other case a similar function is made by a monomer which corresponds to the fusion of the two ancestral modules. Note also that YbhJ is apparently too short to contain this entire 64-amino-acid module, and this may predict absence of functionality.

## 22.9 POWERFUL TOOL TO ANNOTATE UNKNOWN GENES

The phylogenetic approach described above may help to transfer a function from known proteins to unknown homologues. Unknown open reading frames which are highly similar to several proteins coding similar function may be assigned this function with good probability. This could be made, for example, in the case of the kinases of sugar (YgcE and YdeV in Fig. 22.9). This correlation also strongly suggests that the ancestral sequence of such a family which appears homogeneous for function was already coding for a function similar to that of its present-day descendants, although its specificity was probably broader.

## 22.10 ANOTHER IMPORTANT POINT ABOUT RELATIONSHIPS BETWEEN STRUCTURE AND FUNCTION: REANNOTATING CONSERVED HYPOTHETICAL PROTEINS

To go a step further and illustrate how helpful might be our approach in improving the annotation of such essential proteins, we focused on genes which are universal to the 26 proteobacteria and which remain annotated as hypothetical (no putative function suggested) proteins in *C. jejuni*.

Figure 22.12 shows that the unknown protein Cj0495 affords complex homology relationships with various partners. We detected significant matches only between its

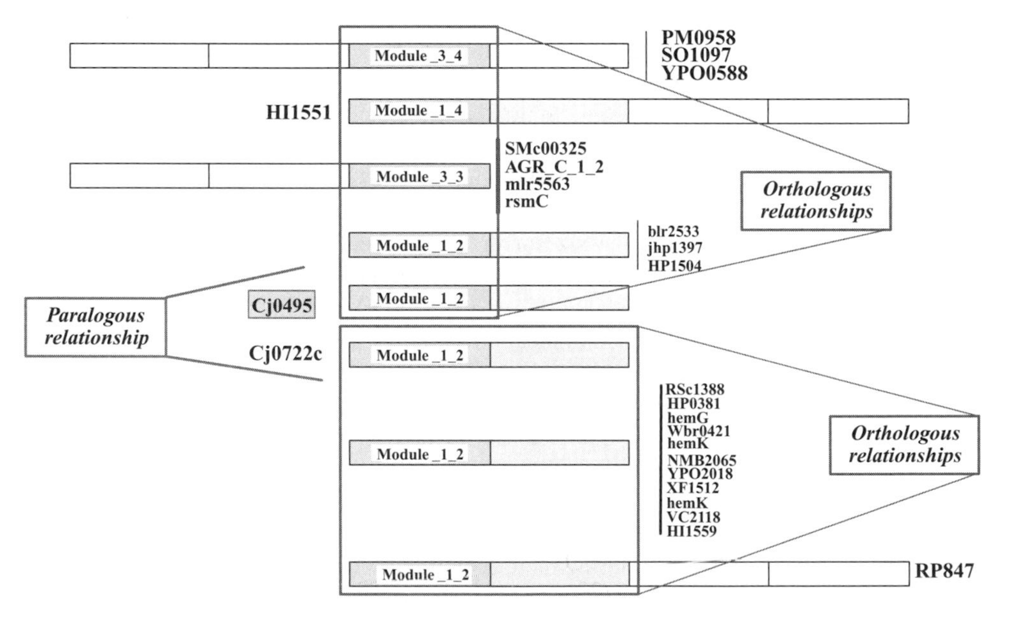


**Figure 22.12** Differentiating homologous modules in a variety of multimodular proteins. The complexity of the paralogous and orthologous relationships is underlined.

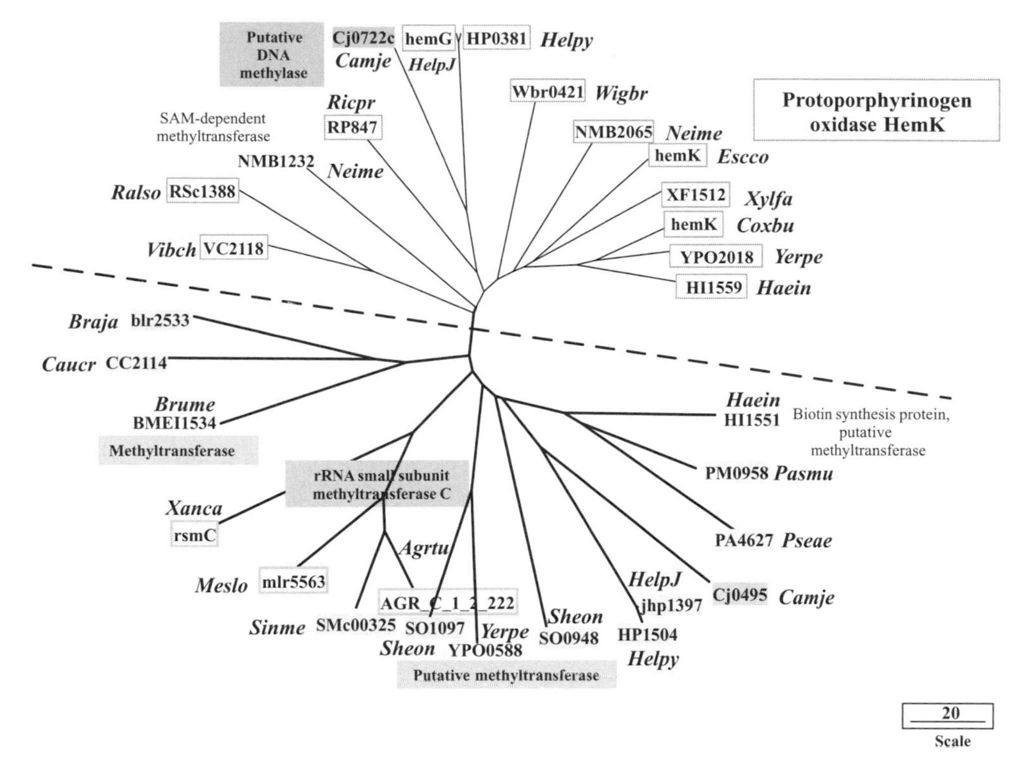


**Figure 22.13** Phylogenetic tree of the homologous modules identified in Figure 22.12. The tree was reconstructed as described in legend to Figure 22.1.

module 1_2 and either the module 1_2 of its paralogue Cj0722c or different modules of various proteobacterial proteins. Moreover, the Cj0722c protein, annotated as a putative DNA methylase, is aligning along its full length with the putative protoporphyrinogen oxidase HemK of many proteobacteria. To better understand the complex relationships occurring between Cj0495 and its different orthologous and paralogous relatives, we aligned its module 1_2 with the homologous modules (all the yellow modules in Fig. 22.12) in order to reconstruct an evolutionary tree. Figure 22.13 shows that this tree is made of two well-separated subtrees. In the subtree located in the upper part and made essentially of the HemK family and containing also the SAM-dependent methyltransferase NMB1232 of *N. meningitidis*, we found the putative DNA methyltransferase Cj0722c branching on a node common with the *H. pylori* HemK. In the subtree located in the lower part, we find Cj0495 branching on a node common with its two unknown *H. pylori* orthologues. This lower subtree contains a set of proteins which are annotated as either hypothetical or predicted methyltransferases. Note that the three ribosomal RNA small-subunit methyltransferases C (rsmC) present in Xanca, Meslo, and Agrtu form a clade, strongly suggesting that the *Sinorhizobium meliloti* SMc00325 is also an rRNA small-subunit methyltransferase C. Thus, Fig. 22.13 shows a clear evolutionary separation between the two paralogues Cj0722c and Cj0495 and their respective partners, which seems to concur with some functional separation between the different kinds of methyltransferases. This suggests that Cj0495 would be a methyltransferase but is probably not a HemK protein since it is found to be closer to the two *H. pylori* HP1504 and jhp1397 proteins than the two HemK-like HP0381 and hemG.

## 22.11 CONCLUSIONS

We have described how our conceptual and methodogical approach is powerful in finding many homology relationships between proteins of any organism. From these data, we have deduced fundamental conclusions on the way genes (proteins) are evolving by duplication and/or fusion to diversify function already existing or create de novo new function. Data obtained on gene evolution are also crucial to understand the way genomes are evolving and new species are created.

Essential (universal) genes which are supposed to be present in the last universal common ancestor to all extant life are nearly always unimodular. This seems to imply that the conservation of both copies issued from a gene duplication event has over time been retained only for genes which were necessary to conquer new environments and to adapt to new life conditions. Thus, it appears that ancestral organisms accepted to expand their genome beyond maintaining basic life conditions. Accordingly, they appear to have progressively transformed their superfluous genes to those most necessary. For example, photosynthesis is not crucial to many forms of life but, as soon as it was acquired, it was a source of tremendous potentialities for the future of life on Earth.

## REFERENCES


Altschul, S. F. (1991). Amino acid substitution matrices from an information theoretic perspective. *J. Mol. Biol.* **219**:555–565.

Altschul, S. F. (1993). A protein alignment scoring system sensitive at all evolutionary distances. *J. Mol. Evol.* **36**:290–300.

Babbit, P. C. (2003). Definitions of enzyme function for the structural genomics era. *Curr. Opin. Chem. Biol.* **2**:230–237.

Bateman, A., Birney, E., Cerruti, L., Durbin, R., Etwiller, L., Eddy, S. R., Griffiths-Jones, S., Howe, K. L., Marshall, M., and Sonnhammer, E. L. (2002). The Pfam protein families database. *Nucleic Acids Res.* **30**:276–280.

Bouyoub, A., Barbier, G., Forterre, P., and Labedan, B. (1996). The adenylosuccinate synthetase from the hyperthermophilic archaeon *Pyrococcus* species display unusual structural features. *J. Mol. Biol.* **261**:144–154.

Brenner, S. E. (1999). Errors in genome annotation. *Trends Genet.* **15**:132–133.

Clausen, T., Schlegel, A., Peist, R., Schneider, E., Steegborn, C., Chang, Y.-S., Haase, A., Bourenkov, G. P., Bartunik, H. D., and Boos, W. (2000). X-ray structure of MalY from *Escherichia coli*: A pyridoxal 5′-phosphase-dependent enzyme acting as a modulator in *mal* gene expression. *EMBO J.* **19**:831–842.

Corpet, F., Servant, F., Gouzy, J., and Kahn, D. (2000). ProDom and ProDom-CG: Tools for protein domain analysis and whole genome comparisons. *Nucleic Acids Res.* **28**:267–269.

Dayhoff, M. O., Schwartz, R. M., and, Orcutt, B. C. (1978). A model for evolutionary change. In M. O. Dayhoff (Ed.), *Atlas of Protein Sequence and Structure*, Vol. 5, Suppl. 3. Washington, DC: National Biomedical Research Foundation, pp. 345–352.

De Rosa, R., and Labedan, B. (1998). The evolutionary relationships between the two bacteria *Escherichia coli* and *Haemophilus influenzae* and their putative last common ancestor. *Mol. Biol. Evol.* **15**:17–27.

Doolittle, R. F. (1981). Similar amino acid sequences: Chance or common ancestry? *Science* **214**:149–159.

Falquet, L., Pagni, M., Bucher, P., Hulo, N., Sigrist, C. J., Hofmann, K., and Bairoch, A. (2002). The PROSITE database, its status in 2002. *Nucleic Acids Res.* **30**:235–238.

Fitch, W. M. (1970). Distinguishing homologous from analogous proteins. *Systematic Zool.* **19**:99–113.

Fitch, W. M. (2001). Homology: A personal view on some of the problems. *Trends Genet.* **16**:227–231.

Gerlt, J. A., and Babbitt, P. C. (2000). Can sequence determine function? *Genome Biol.* **1**: Reviews 0005.10

Gerstein, M. (1998). How representative are the known structures of the proteins in a complete genome? A comprehensive structural census. *Folds Des* **3**:497–512.

Gonnet, G. H., Cohen, M. A., and Benner, S. A. (1992). Exhaustive matching of the entire protein sequence database. *Science* **256**:1443–1445.

Kobayashi, K., Ehrlich, S. D., Albertini, A., Amati, G., Andersen, K. K., Arnaud, M., Asai, K., Ashikaga, S., Aymerich, S., Bessieres, P., Boland, F., Brignell, S. C., Bron, S., Bunai, K., Chapuis, J., Christiansen, L. C., Danchin, A., Debarbouille, M., Dervyn, E., Deuerling, E., Devine, K., Devine, S. K., Dreesen, O., Errington, J., Fillinger, S., Foster, S. J., Fujita, Y., Galizzi, A., Gardan, R., Eschevins, C., Fukushima, T., Haga, K., Harwood, C. R., Hecker, M., Hosoya, D., Hullo, M. F., Kakeshita, H., Karamata, D., Kasahara, Y., Kawamura, F., Koga, K., Koski, P., Kuwana, R., Imamura, D., Ishimaru, M., Ishikawa, S., Ishio, I., Le Coq, D., Masson, A., Mauel, C., Meima, R., Mellado, R. P., Moir, A., Moriya, S., Nagakawa, E., Nanamiya, H., Nakai, S., Nygaard, P., Ogura, M., Ohanan, T., O'Reilly, M., O'Rourke, M., Pragai, Z., Pooley, H. M., Rapoport, G., Rawlins, J. P., Rivas, L. A., Rivolta, C., Sadaie, A., Sadaie, Y., Sarvas, M., Sato, T., Saxild, H. H., Scanlan, E., Schumann, W., Seegers, J. F., Sekiguchi, J., Sekowska, A., Seror, S. J., Simon, M., Stragier, P., Studer, R., Takamatsu, H., Tanaka, T., Takeuchi, M., Thomaides, H. B., Vagner, V., van Dijl, J. M., Watabe, K., Wipat, A., Yamamoto, H., Yamamoto, M., Yamamoto, Y., Yamane, K., Yata, K., Yoshida, K., Yoshikawa, H., Zuber, U., and Ogasawara, N. (2003). Essential *Bacillus subtilis* genes. *Proc. Natl. Acad. Sci. USA* **100**:4678–4683.

Le Bouder-Langevin, S., Capron-Montaland, I., De Rosa, R., and Labedan, B. (2002). A strategy to retrieve the whole set of protein modules in microbial proteomes. *Genome Res.* **12**:1961–1973.

Letunic, I., Goodstadt, L., Dickens, N. J., Doerks, T., Schultz, J., Mott, R., Ciccarelli, F., Copley, R. R., Ponting, C. P., and Bork, P. (2002). Recent improvements to the SMART domain-based sequence annotation resource. *Nucleic Acids Res.* **30**:242–244.

Mauzy, C. A., and Hermodson, M. A. (1992). Structural homology between *rbs* repressor and ribose binding protein implies functional similarity. *Protein Sci.* **1**:843–849.

Ohno, S. (1970). *Evolution by Gene Duplication*. New York: Springer-Verlag.

Poland, B. W., Bruns, C., Fromm, H. J., and Honzatko, R. B. (1997). Entrapment of 6-thiophosphoryl-IMP in the active site of crystalline adenylosuccinate sythetase from *Escherichia coli*. *J Biol Chem.* **272**:15200–15205.

Poland, B. W., Silva, M. M., Serra, M. A., Cho, Y., Kim, K. H., Harris, E. M., and Honzatko, R. B. (1993). Crystal structure of adenylosuccinate synthetase from *Escherichia coli*. Evidence for convergent evolution of GTP-binding domains. *J. Biol. Chem.* **268**:25334–25342.

Reeck, G. R., de Haen, C., Teller, D. C., Doolittle, R. F., Fitch, W. M., Dickerson, R. E., Chambon, P., McLachlan, A. D., Margoliash, E., Jukes, T. H., and Zuckerkandl, E. (1987). "Homology" in proteins and nucleic acids: A terminology muddle and a way out of it. *Cell* **50**:667.

Riley, M., and Labedan, B. (1997). Protein sequences: Introducing the notion of a structural segment of homology, the module. *J. Mol. Biol.* **268**:857–868.

Schulz, G. E. (1992). Binding of nucleotides by proteins. *Curr. Opin. Struct. Biol.* **2**:61–67.

Schwartz, R., and Dayhoff, M. O. (1978). Matrices for detecting distant relationships. In M. O. Dayhoff (Ed.), *Atlas of Protein Sequence and Structure*, Vol. 5, Suppl. 3. Washington, DC: National Biomedical Research Foundation, pp. 353–358.

Sculo, Q., Lespinet, O., and Labedan, B. (2003). Retreiving the whole set of protein modules of *Campylobacter jejuni* and *Helicobacter pylori. Genome Lett.* **2**:2–9.

Solignac, M., Periquet, C., Anzolabehere, D., and Petit, C. (1995). *Genetique et Evolution*. Paris:Hermann.

Wheelan, S. J., Marchler-Bauer, A., and Bryant, S. H. (2000). Domain size distributions can predict domain boundaries. *Bioinformatics* **16**:613–618.

CHAPTER 23

# Cellular Kinetic Modeling of the Microbial Metabolism

[1]IGOR I. GORYANIN, [2]GALINA V. LEBEDEVA, [2]EKATERINA A. MOGILEVSKAYA, [2]EUGENIY A. METELKIN, and [2]OLEG V. DEMIN

[1] The University of Edinburgh, Edinburgh, United Kingdom
[2]A.N. Belozersky Institute of Physico-Chemical Biology, Moscow State University, Moscow, Russia

## 23.1 INTRODUCTION

The last several years have seen substantial progress in molecular biology and genetic research (Perna et al., 2001). Sequence information on genomes of more than 100 different organisms has stimulated the emergence of functional genomics, a discipline that sets out to understand the meaning of sequenced data using high-throughput gene and protein expression data. Life scientists have transformed old style protein chemistry into proteomics and traditional biochemistry into metabolomics. These new fields provide essential clues to the underlying metabolic, signaling, and gene regulation networks that operate in different organisms under different conditions.

Cellular metabolism, the integrated interconversion of thousands of metabolic substrates through enzyme-catalysied biochemical reactions, is the most investigated complex of intracellular molecular interactions. When one has knowledge of most, or all, of the major biological entities and stoicheometry of their interactions, an illusion can appear that this voluminous knowledge will enable us to predict the whole cell behaviors for the purposes of mechanistic understanding and bioengineering control. Indeed, in some specific cases, it is possible to make plausible predictions based on "static" information without relying on kinetic data (Edwards et al., 2001). Unfortunately, this is not the general case, overall cellular behavior is determined not only by what biological entities are available but also mainly by their dynamic interactions and their individual properties. Activities of most, if not all, of the enzymes involved in cellular metabolism are regulated by end products and intermidiates of corresponding pathways. This complex system of positive and negative feedbacks as well as genetic regulation of expression level of many proteins provide flexible adaptation of the metabolic network to a changing external environment, and it is the overall dynamic nature of the cell that determines not only its present properties, but its future ones as well. That is a regulatory system that is

*Microbial Proteomics: Functional Biology of Whole Organisms*, Edited by Ian Humphery-Smith and Michael Hecker. 

responsible for maintenance of cellular homeostasis as well as for transitions between different available steady states. Therefore when we try to describe cellular metabolism with a model, it seems to be extremely important to include into consideration main regulatory properties (effects) known for corresponding metabolic pathways. From our point of view, these requirements could be fullfilled in the framework of kinetic modeling.

The main purpose of this contribution is to describe a new approach for constructing large-scale kinetic network models. In the framework of this approach, we suggest a novel way to collect and mine large-scale experimental data and use them to build and verify kinetic models. This approach is illustrated by the example of histidine pathway model development. Then, we apply this approach to develop and analyze the kinetic model of the phosphotransferase glycolysis system in *Escherichia coli* and study the regulatory properties of the cell. From these predictions from the kinetic models, we generate new functional biological knowledge about the phenotypes of the cell based on the expression levels of the corresponding enzymes.

## 23.2 BASIC PRINCIPLES OF KINETIC MODEL CONSTRUCTION

The term "kinetic model" is used in two senses, one biological and one mathematical. In the biological sense, it is used to indicate that the network interactions between the different biological entities are always changing and the network always includes them, even if the fluxes may become zero. It is important to consider two entities as connected, if, for any time, there is ever a concentration flux connecting them. Thus, there is only static network with temporally changing concentrations as the cell undergoes genotypic and phenotypic changes. In the mathematical sense, it refers to a system of mechanistic ordinary differential equations that determine the temporal state of the corresponding system of biochemical reactions. In these equations, there is mass conservation between the production and the consumption of each species:

$$dX/dt = V_{\text{production}} - V_{\text{consumption}} \tag{23.1}$$

where $V_{\text{production}}$ and $V_{\text{consumption}}$ are the respective rates of production and consumption of species $X$.

The development of kinetic models for metabolic systems, gene regulation, and signaling networks is accomplished in several steps. We will describe each stage of the process in details and illustrate it by several examples.

### 23.2.1 Development of System of Ordinary Differential Equations Describing Dynamics of Selected Biochemical System

The first step is to elucidate a static model of the system, i.e., to find out all cellular players, intermediates, enzymes, small molecules, cofactors, and all non-enzymatic processes in the cellular network. The result is a network (i.e., a directed bond graph) of all interactions connecting all species. For the network to be proper, each species must exist in at least one reaction or behave as a cofactor. Disconnected fragments, resulting from incomplete knowledge, can, optionally, be considered as part of the network, although for all practical purposes they will be treated as separate networks. We will illustrate all details of static model development by the example of the pathway of histidine biosynthesis in *E. coli* (Umbarger,

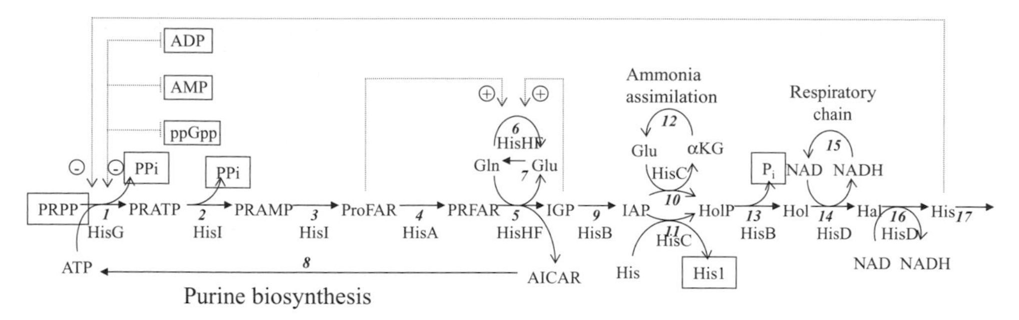


**Figure 23.1** Scheme of the histidine biosynthesis pathway in *Escherichia coli*.

1996). This pathway is depicted in Figure 23.1. List of intermediates of the pathway is shown in Table 23.1. Chemical equations of reactions of the pathway are summarized in Table 23.2. Taking into account this information, we create a stoichiometric matrix:

$$
\begin{array}{c}
\begin{array}{ccccccccccccccccc}
v_1 & v_2 & v_3 & v_4 & v_5 & v_6 & v_7 & v_8 & v_9 & v_{10} & v_{11} & v_{12} & v_{13} & v_{14} & v_{15} & v_{16} & v_{17}
\end{array} \\
\left(\begin{array}{ccccccccccccccccc}
-1 & 0 & 0 & 0 & 0 & 0 & 0 & 1 & 0 & 0 & 0 & 0 & 0 & 0 & 0 & 0 & 0 \\
1 & -1 & 0 & 0 & 0 & 0 & 0 & 0 & 0 & 0 & 0 & 0 & 0 & 0 & 0 & 0 & 0 \\
0 & 1 & -1 & 0 & 0 & 0 & 0 & 0 & 0 & 0 & 0 & 0 & 0 & 0 & 0 & 0 & 0 \\
0 & 0 & 1 & -1 & 0 & 0 & 0 & 0 & 0 & 0 & 0 & 0 & 0 & 0 & 0 & 0 & 0 \\
0 & 0 & 0 & 1 & -1 & 0 & 0 & 0 & 0 & 0 & 0 & 0 & 0 & 0 & 0 & 0 & 0 \\
0 & 0 & 0 & 0 & -1 & -1 & 1 & 0 & 0 & 0 & 0 & 0 & 0 & 0 & 0 & 0 & 0 \\
0 & 0 & 0 & 0 & 1 & 1 & -1 & 0 & 0 & -1 & 0 & 1 & 0 & 0 & 0 & 0 & 0 \\
0 & 0 & 0 & 0 & 1 & 0 & 0 & 0 & -1 & 0 & 0 & 0 & 0 & 0 & 0 & 0 & 0 \\
0 & 0 & 0 & 0 & 1 & 0 & 0 & -1 & 0 & 0 & 0 & 0 & 0 & 0 & 0 & 0 & 0 \\
0 & 0 & 0 & 0 & 0 & 0 & 0 & 0 & 1 & -1 & -1 & 0 & 0 & 0 & 0 & 0 & 0 \\
0 & 0 & 0 & 0 & 0 & 0 & 0 & 0 & 0 & 0 & -1 & 0 & 0 & 0 & 0 & 1 & -1 \\
0 & 0 & 0 & 0 & 0 & 0 & 0 & 0 & 0 & 1 & 1 & 0 & -1 & 0 & 0 & 0 & 0 \\
0 & 0 & 0 & 0 & 0 & 0 & 0 & 0 & 0 & 1 & 0 & -1 & 0 & 0 & 0 & 0 & 0 \\
0 & 0 & 0 & 0 & 0 & 0 & 0 & 0 & 0 & 0 & 0 & 0 & 1 & -1 & 0 & 0 & 0 \\
0 & 0 & 0 & 0 & 0 & 0 & 0 & 0 & 0 & 0 & 0 & 0 & 0 & -1 & 1 & -1 & 0 \\
0 & 0 & 0 & 0 & 0 & 0 & 0 & 0 & 0 & 0 & 0 & 0 & 0 & 1 & -1 & 1 & 0 \\
0 & 0 & 0 & 0 & 0 & 0 & 0 & 0 & 0 & 0 & 0 & 0 & 0 & 1 & 0 & -1 & 0
\end{array}\right)
\begin{array}{l}
ATP \\ PRATP \\ PRAMP \\ \mathrm{Pr}oFAR \\ PRFAR \\ G\ln \\ Glu \\ IGP \\ AICAR \\ IAP \\ His \\ HolP \\ \alpha KG \\ Hol \\ NAD \\ NADH \\ Hal
\end{array}
\end{array}
$$

The columns of the matrix correspond to reactions of the pathway, and the rows correspond to metabolites. The construction of a stoichiometric matrix of a selected biochemical system corresponds to development of its "static" model. The next stage on the way of development of a kinetic model of the system consists in that using a static model (or, in other words, stoichiometric matrix), we can write out a system of differential equations describing the dynamics of the selected pathway:

$$\frac{d\mathbf{x}}{dt} = \mathbf{N} \cdot \mathbf{v}, \qquad \mathbf{x(0)} = \mathbf{x_0} \tag{23.2}$$

Here, $\mathbf{x} = [\mathrm{x}_1, \ldots, \mathrm{x}_\mathrm{m}]^\mathrm{T}$ is a vector of intermediate concentrations, $\mathbf{x_0} = [\mathrm{x}_{10}, \ldots, \mathrm{x}_{\mathrm{m}0}]^\mathrm{T}$ is a vector of initial concentrations of intermediates, $\mathbf{v} = [\mathrm{v}_1, \ldots, \mathrm{v}_\mathrm{n}]^\mathrm{T}$ is a vector of reaction rates, and $\mathbf{N}$ is a stoichiometric matrix that has $n$ columns and $m$ rows. In the case

**TABLE 23.1 Intermediates of the Histidine Biosynthesis Pathway in *Escherichia coli***

| Intermediate Designation | Chemical Name of Intermediate |
|---|---|
| PRATP | N1-(5′-phosphoribosyl)-ATP |
| PRAMP | N1-(5′-phosphoribosyl)-AMP |
| ProFAR | Pro-phosphoribosyl-formimino-5-aminoimidazole-4-carboxamide ribonucleotide |
| PRFAR | Phosphoribosyl-formimino-5-aminoimidazole-4-carboxamide ribonucleotide |
| IGP | Imidazoleglycerol phosphate |
| AICAR | 5-aminoimidazole-4-carboxamide ribonucleotide |
| IAP | Imidazoleacetol phosphate |
| HolP | L-Histidinol phosphate |
| Hol | L-Histidinol |
| Hal | L-Histidinal |
| His | L-Histidine |
| ATP | Adenosine triphosphate |
| Gln | Glutamine |
| αKg | αketoglutarate |
| Glu | Glutamate |
| NAD | Nicotineamide adenine dinucleotide phosphate oxidized |
| NADH | Nicotineamide-adenine-dinucleotide phosphate reduced |

**TABLE 23.2 Reactions of the Histidine Biosynthesis Pathway in *Escherichia coli***

| Reaction Number | Chemical Equation | Enzyme | Gene |
|---|---|---|---|
| 1 | PRPP + ATP = PRATP + PPi | ATP-phosphoribosyltransferase | HisG |
| 2 | PRATP = PPi + PRAMP | Phosphoribosyl-ATP-pyrophosphohydrolase: Phosphoribosyl-AMP cyclohydrolase | *HisI* |
| 3 | PRAMP = ProFAR | Phosphoribosyl-ATP-pyrophosphohydrolase: Phosphoribosyl-AMP cyclohydrolase | *HisI* |
| 4 | ProFAR = PRFAR | Phosphoribosyl-formimino-5-amino-1-phosphoribosyl-4-imidazole-carboxamide isomerase | *HisA* |
| 5 | PRFAR + Gln = Glu + IGP + AICAR | IGP synthase | *HisHF* |
| 6 | Gln = Glu | IGP synthase | *HisHF* |
| 7 | Glu = Gln | Glutamine-synthatase | |
| 8 | AICAR = ATP | Purine biosynthesis | |
| 9 | IGP = IAP | IGP dehydratase | *HisB* |
| 10 | IAP + Glu = HolP + αKG | Histidinol phosphate aminotransferase | *HisC* |
| 11 | IAP + His = HolP + His1 | Histidinol phosphate aminotransferase | *HisC* |
| 12 | αKG = Glu | Ammonia assimilation | |
| 13 | HolP = Hol + Pi | Histidinol phosphatase | *HisB* |
| 14 | Hol + NAD = Hal + NADH | Histidinoldehydrogenase | *HisD* |
| 15 | NADH = NAD | Respiratory chain | |
| 16 | Hal + NAD = His +NADH | Histidinaldehydrogenase | *HisD* |
| 17 | His → | Histidine consumption | |

of a pathway of histidine biosynthesis, both $m$ and $n$ are equal to 17 and vectors of intermediate concentrations and initial conditions are as follows:

$$\mathbf{x} = [\mathrm{ATP}, \mathrm{PRATP}, \mathrm{PRAMP}, \mathrm{ProFAR}, \mathrm{PRFAR}, \mathrm{Gln}, \mathrm{Glu}, \mathrm{IGP}, \mathrm{AICAR}, \mathrm{IAP}, \mathrm{His}, \mathrm{HolP}, \alpha\mathrm{KG}, \mathrm{Hol}, \mathrm{NAD}, \mathrm{NADH}, \mathrm{Hal}]^{\mathrm{T}}$$

$$\mathbf{x_0} = [\mathrm{ATP}_0, \mathrm{PRATP}_0, \mathrm{PRAMP}_0, \mathrm{ProFAR}_0, \mathrm{PRFAR}_0, \mathrm{Gln}_0, \mathrm{Glu}_0, \mathrm{IGP}_0, \mathrm{AICAR}_0, \mathrm{IAP}_0, \mathrm{His}_0, \mathrm{HolP}_0, \alpha\mathrm{KG}_0, \mathrm{Hol}_0, \mathrm{NAD}_0, \mathrm{NADH}_0, \mathrm{Hal}_0]^{\mathrm{T}}$$

Before going into methods, peculiarities, and details of rate equation derivation, we will focus on what the static model enables us to understand about the system of interest. It turns out that the stoichiometric matrix allows us, first, to derive relationships between steady-state fluxes and, second, to find out the number of conservation laws and to write out their expressions. Indeed, solving the system of linear algebraic equations

$$\mathbf{N} \cdot \mathbf{v} = \mathbf{0} \tag{23.3}$$

we find that any steady-state reaction rate (steady state flux), $\mathrm{v_i}$, $\mathrm{i} = 1, \ldots n$, can be expressed as a linear combination of $s$ independent steady-state rates. Number $s$, equal to a dimension of kernel of matrix $\mathbf{N}$, and coefficients of relationships expressing any steady-state rate in terms of $s$ selected independent rates are fully determined by a stoichiometric matrix. As an example, we consider relationships between the steady-state rates of the histidine biosynthesis pathway (see Fig. 23.1):

$$\begin{aligned}
&\mathrm{v}_{16} = \mathrm{v}_{14} = \mathrm{v}_{13} = \mathrm{v}_9 = \mathrm{v}_8 = \mathrm{v}_5 = \mathrm{v}_4 = \mathrm{v}_3 = \mathrm{v}_2 = \mathrm{v}_1, \\
&\mathrm{v}_7 = \mathrm{v}_1 + \mathrm{v}_6, \\
&\mathrm{v}_{10} = \mathrm{v}_{17} = \mathrm{v}_{12}, \\
&\mathrm{v}_{11} = \mathrm{v}_1 - \mathrm{v}_{12}, \\
&\mathrm{v}_{15} = 2 \cdot \mathrm{v}_1
\end{aligned}$$

From these relationships, it follows that any steady-state rate can be expressed in terms of the three independent rates $\mathrm{v}_1$, $\mathrm{v}_6$, and $\mathrm{v}_{12}$, i.e., $s$ is equal to 3. Conservation laws are the first, linear integrals of the system of differential equations [Eq. (23.2)] describing kinetics of the selected biochemical system. As a simplest example of the conservation law valid for the pathway of histidine biosynthesis, we can consider the following algebraic expression:

$$\mathrm{NAD} + \mathrm{NADH} = \mathrm{const}_1 \tag{23.4}$$

This relationship results from summing up and integration of differential equations describing how concentrations NAD and NADH change with time. The meaning of Eq. (23.3) consists in that the sum of concentrations of NAD and NADH does not change with time. It is easy to show that the number of conservation laws of kinetic model describing biochemical system consisting of $m$ intermediates connected with $n$ reactions is given by following formula:

$$(\text{number of conservation laws}) = m - n + s \tag{23.5}$$

In the case of the histidine biosynthesis pathway, both *m* and *n* are equal to 17 and *s* is equal to 3. Consequently, in accordance with Eq. (23.5), we find that the number of conservation laws of the kinetic model of the histidine biosynthesis pathway shown in Figure 23.1 is equal to 3. Relationship (23.4) is one of the three conservation laws. Two others are given by the following expressions:

$$\mathrm{PRATP} + \mathrm{PRAMP} + \mathrm{ProFAR} + \mathrm{PRFAR} + \mathrm{AICAR} + \mathrm{ATP} = \mathrm{const}_2$$
$$\mathrm{Glu} + \mathrm{Gln} + \alpha\mathrm{KG} = \mathrm{const}_3 \tag{23.6}$$

As Eqs. (23.4) and (23.6) are true for any moment of time, including time equal to zero, then values of parameters $\mathrm{const}_i$, $i = 1, 2, 3$, are completely determined by initial conditions:

$$\mathrm{const}_1 = \mathrm{NAD}_0 + \mathrm{NADH}_0$$
$$\mathrm{const}_2 = \mathrm{PRATP}_0 + \mathrm{PRAMP}_0 + \mathrm{ProFAR}_0 + \mathrm{PRFAR}_0 + \mathrm{AICAR}_0 + \mathrm{ATP}_0$$
$$\mathrm{const}_3 = \mathrm{Glu}_0 + \mathrm{Gln}_0 + \alpha\mathrm{KG}_0$$

### 23.2.2 Basic Principles of Kinetic Description of Enzymatic Reactions Using *in vitro* Experimental Data

Once an appropriate static network has been chosen, the second stage is to generate rate equations to describe the dependence of each reaction rate against concentrations of intermediates involved in the selected pathway. To make the models scalable and comparable with different kinds of experimental data, we have developed both detailed and reduced descriptions for every biochemical process in the model. The detailed reaction description includes the exact molecular mechanism of the biomolecular reaction (i.e., enzyme catalytic cycle) and takes into account all possible states of the protein, including possible non-active states (i.e., phosphorylated) or dead-end inhibitor complexes. Usually, the detailed description comprises a set of differential algebraic equations from the ordinary differential flux equations and nonlinear algebraic equations (if steady-state or conservation constraint assumptions are made) simultaneously.

The *reduced* description represents the reaction rates as an explicit analytic function of the substrates and products. We identify from the literature or hypothesize the catalytic cycle based on three-dimensional structures and other relative biological information for each active protein involved in the model (i.e., enzyme with catalytic function). To derive the corresponding rate equations from the catalytic cycle, we have used quasi-steady-state and rapid equilibrium approaches. The catalytic cycle of each enzyme is described by nonlinear differential equations. Initially, concentrations of substrates, products, and effectors (inhibitors and activators) are assumed to be buffered, i.e., do not change with time.

The quasi-steady state of the system is calculated as a function of substrates, products, inhibitors, activators, total protein concentrations, and all kinetic constants of the processes. The rate law for every process is derived as a flux from the catalytic cycle for this quasi-steady state. Finally, the rate law depends on temporal changes of the total concentration of the protein, concentrations of the effectors (activators, inhibitors, agonists, and antagonists), substrates, products, and the values of the kinetic parameters ($K_m$, $K_i$, $K_d$, and elementary rate constants). Although derived from a quasi-steady-state approach, these rate laws will be used in simulations based on the full differential equations with any such simplifications made.

In the framework of approach suggested in this article, the level of detailed elaboration of the catalytic cycle of selected enzyme and subsequent derivation of rate equation are fully determined by available experimental data on structural and functional organization of the enzyme. Indeed, if catalytic cycle of the enzyme is established and proved experimentally, then we use it to derive the rate equation. If the mechanism underlying the enzyme operation is unknown, we suggest a "minimal" catalytic cycle, which

(1) Satisfies all structural and stoichiometric data available from literature
(2) Allows us to derive the rate equation describing available kinetic experimental data
(3) Is the simplest catalytic cycle of all possible ones satisfying the first two clauses

Another important feature of development of the enzyme kinetic description based on *in vitro* data consists in that kinetic experimental data published in different literature sources are measured under different conditions (pH, temperature). This means that we should construct such catalytic cycle and derive such rate equation that

(4) To satisfy available experimental data describing dependence of enzyme operation on pH, temperature and other experimental conditions
(5) Mechanism describing dependence of reaction rate on pH and temperature should be taken into account in the catalytic cycle of the enzyme in the simplest of all possible ways

Parameter estimation is the third stage of model development. To estimate the kinetic parameter values, we use the following sources:

1) Literature data on values of $K_m$, $K_i$, $K_d$, rate constants, pH optimum, etc.
2) Electronic databases; only a few databases with specific kinetic content are available at the moment, in particular EMP (Selkov et al., 1996) and BRENDA (Schomburg et al., 2002)
3) Experimentally measured dependencies of the initial reaction rates on concentrations of substrates, products, inhibitors, and activators
4) Time series data from enzyme kinetics

However, many processes (i.e., enzyme reactions) have not been studied kinetically. Many kinetic parameters cannot be estimated from the literature or databases due to a lack of available experimental data. One remedy is to express these unknown or "free" parameters in terms of other measured kinetic parameters. The result is the establishment of functional relationships between "free" parameters and measured kinetic parameters. Each parameter value, of course, is constrained by physicochemical properties and any other information available from other organisms or related processes. The more constraints available, the more dimensionality reduction can occur.

To illustrate the basic principles of construction of the catalytic cycle and derivation of the rate equation described above, we consider two enzymes of the *E. coli* histidine biosynthesis pathway (see Fig. 23.1): imidazole glycerol phosphate synthase and histidinol dehydrogenase. Development of the kinetic model of imidazole glycerol phosphate synthase exemplifies how available data on enzyme effectors can be taken into account to construct the catalytic cycle, derive the rate equation, and estimate the values of kinetic parameters. On the other hand, development of the kinetic model of histidinol dehydrogenase illustrates another basic principle of kinetic description of enzymatic reactions sited above, namely, how kinetic data measured at different conditions (pH, temperature, and others) can be taken all together into account to construct a quantitative description predicting the kinetic behavior of the enzyme at any conditions.

To develop and analyze kinetic models, we use the software package DBSolve (Goryanin et al., 1999).

### 23.2.3 Development of the Kinetic Model of Imidazole Glycerol Phosphate Synthase from *E. coli*

In this section, we develop a kinetic model of imidazole glycerol phosphate synthase and illustrate how available experimental data on structure, kinetics, and effectors of the enzyme can be taken into account to construct the catalytic cycle, derive the rate equation, and estimate the values of kinetic parameters.

Imidazologlycerol-phosphate synthetase (IGPS) belongs to the group of aminotransferases that catalyze the transfer of the amido group of glutamine onto various acceptor molecules—intermediates in biosynthesis of purine and pyrimidine nucleotides and amino acids (Zalkin, 1993). IGPS from *E. coli* is a key enzyme in the biosynthesis of histidine and structurally is a heterodimer encoded by *hisH* and *hisF* (Zalkin and Smith, 1998). Similar to other glutamine aminotransferases, imidazologlycerol-phosphate synthetase catalyzes two reactions (Zalkin, 1993, Zalkin and Smith, 1998):

$$\mathrm{PRFAR} + \mathrm{Gln} = \mathrm{AICAR} + \mathrm{IGP} + \mathrm{Glu} \tag{23.7}$$

$$\mathrm{Gln} = \mathrm{Glu} + \mathrm{NH_3} \tag{23.8}$$

The essence of the first reaction is transfer of the amido group of glutamine conjugated with decomposition of 5′-phosphoribulosylformimino-5-aminoimidazolo-4-carboxamideribonucleotide (PRFAR) into 5-aminoimidazolo-4-carboxamido-1-β-D-ribofuranosyl 5′-monophosphate (AICAR) and imidazologlycerol phosphate (IGP) and synthesis of the imidazole ring. This reaction is catalyzed due to the synthetase activity of IGPS. In the second reaction, glutamine (Gln) is decomposed to glutamate (Glu). The rate of this reaction is governed by the glutaminase activity of IGPS.

In this section, we develop a kinetic model of the catalytic cycle of imidazologlycerol-phosphate synthetase, derive the rate equations for glutaminase and synthetase activities of the enzyme, and evaluate the unknown parameters characterizing the kinetic properties of this enzyme. Based on the rate equations and parameter values obtained, we studied how concentrations of the substrates and effectors determine the contributions of glutaminase and synthetase activities into the enzyme functioning.

*Available experimental data.* In this study, we use the following available data on structural and functional properties of IGPS:

1 The enzyme has two catalytic sites: one for binding to glutamine and another to PRFAR (Zalkin H. 1993; Zalkin and Smith, 1998).

2 Binding to the substrates and dissociation of the products occur accidentally (Zalkin, 1993; Zalkin and Smith, 1998).
3 The amido group cleaved from glutamine is transferred on the catalytic site binding PRFAR without emerging into solution, that is, an intramolecular transfer of the amido group occurs (Zalkin, 1993; Zalkin and Smith, 1998).
4 The glutaminase activity of imidazologlycerol-phosphate synthetase manifests itself essentially only with excess glutamine and depletion of PRFAR, the second substrate (Klem and Davisson, 1993; Chittur et al., 2001).
5 Whereas IGP, one of the products of the synthetase reaction catalyzed by imidazologlycerol-phosphate synthetase, catalyzes the glutaminase reaction, another product, AICAR, does not affect the latter (Klem and Davisson, 1993; Chittur et al., 2001).
6 On the pathway of biosynthesis of histidine, the reaction catalyzed by IGPS is preceded by another reaction catalyzed by ProFAR isomerase, its substrate, 5′-phosphoribosylformimino-5-aminoimidazolo-4-carboxamidoribonucleotide (ProFAR), being an activator of the glutaminase activity of IGPS (Klem and Davisson, 1993; Chittur et al., 2001).

Some kinetic properties of the enzyme functioning as synthetase, i.e., catalyzing only reaction (1) were investigated earlier (Klem and Davisson, 1993). The Michaelis constants for Gln and PRFAR:

$$K^{S}_{m,Gln} = 240\,\mu M, \quad K^{S}_{m,PRFAR} = 1.5\,\mu M \tag{23.9}$$

and the maximal number of cycles of the enzyme:

$$k^{S}_{cat} = 8.5\,s^{-1} \tag{23.10}$$

were determined at pH 8 and 25°C. In the same work, Klem and Davisson evaluated the Michaelis constant for Gln and the maximal number of cycles of the enzyme functioning as glutaminase, i.e., catalyzing only reaction (23.8):

$$K^{g}_{m,Gln} = 4800\,\mu M, \quad k^{g}_{cat} = 0.07\,s^{-1} \tag{23.11}$$

To characterize quantitatively the effect of ProFAR and IGP on the glutaminase activity of IGPS, Klem and Davisson evaluated the apparent Michaelis constant for Gln and apparent maximal number of cycles of the enzyme functioning as glutaminase in the presence of the activators ProFAR and IGP:

$$[\mathrm{ProFAR}]_0 = 2000\,\mu M, \quad K^{g,ProFAR}_{m,app,Gln} = 2800\,\mu M, \quad k^{g,ProFAR}_{cat,app} = 2.6\,s^{-1} \tag{23.12}$$

$$[\mathrm{IGP}]_0 = 9000\,\mu M, \quad K^{g,IGP}_{m,app,Gln} = 1900\,\mu M, \quad k^{g,IGP}_{cat,app} = 2.7\,s^{-1} \tag{23.13}$$

Kinetic characteristics of the enzyme functioning, that is, changes in concentrations of the substrates and products in the presence of 9-nM IGPS with time, were also monitored in Klem and Davisson (1993).

*Construction of catalytic cycle.* The experimental data mentioned in items (1)–(6) of the previous section were accounted for on development of a kinetic model of the catalytic

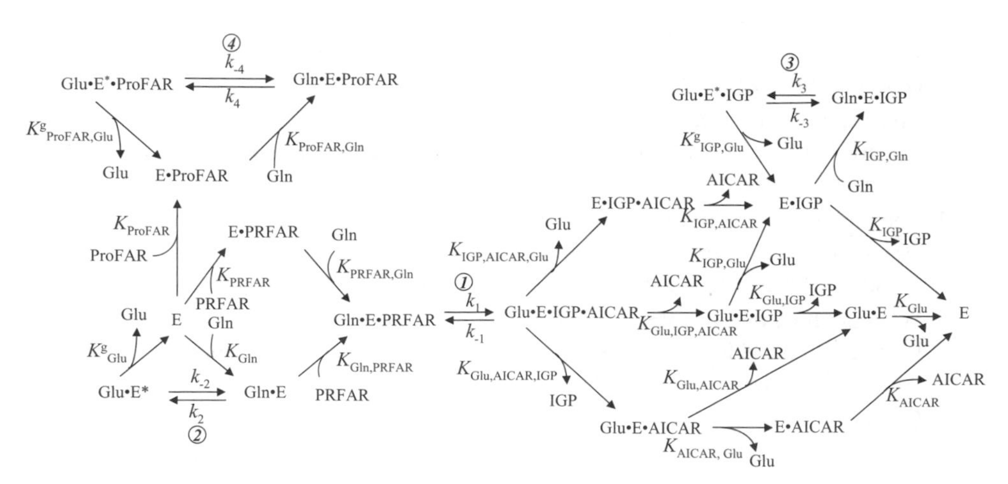


**Figure 23.2** The catalytic cycle of imidazologlycerol-phosphate synthetase. The dissociation constants of the substrate–product complexes and the rate constants are presented near the corresponding stages of the catalytic cycle.

cycle of imidazologlycerol-phosphate synthetase presented in Figure 23.2. We suggested that the enzyme functions via the Random Bi Ter mechanism according to the Cleland classification (Cleland, 1963); this agrees with items (1) and (2) of the previous section. Transfer of the triple enzyme–substrate complex Gln·E·PRFAR into the enzyme–product complex Glu·E·IGP·AICAR corresponds with an intermolecular transfer of the amido group mentioned in item (3) of the previous section and subsequent transformation of PRFAR into AICAR and IGP. As a result of the accidental order of the substrate binding, in the absence of PRFAR, a complex of the enzyme with glutamine Gln·E is formed and glutaminase reaction (23.2) catalyzed by free enzyme proceeds ($E \rightarrow Gln\cdot E \rightarrow Glu\cdot E^* \rightarrow E$); this corresponds with item (4) of the previous section. Activation of the glutaminase reaction in the presence of ProFAR and IGP described in items (5) and (6) of the previous section is accounted for due to inclusion of the additional reaction cycles ($E \rightarrow E\cdot ProFAR \rightarrow Gln\cdot E\cdot ProFAR \rightarrow Glu\cdot E^*\cdot ProFAR \rightarrow E\cdot ProFAR$) and($E\cdot IGP \rightarrow Gln\cdot IGP \rightarrow Glu\cdot E^*\cdot IGP \rightarrow E\cdot IGP$) into the standard Random Bi Ter mechanism. In this work, we did not study how the rate of glutaminase reaction depends on concentration of one of its products ($NH_3$); that is why $NH_3$ is not presented directly in Figure 23.2.

*Derivation of rate equations.* According to the scheme of the catalytic cycle of IGPS presented in Figure 23.1, the rate equations for synthetase and glutaminase reactions can be written as follows:

$$v_s = k_1[\mathrm{Gln\cdot E\cdot PRFAR}] - k_{-1}[\mathrm{Glu\cdot E\cdot IGP\cdot AICAR}] \tag{23.14}$$

$$\begin{aligned} v_g = {} & (k_2[\mathrm{Gln\cdot E}] - k_{-2}[\mathrm{Glu\cdot E^*\cdot IGP}]) + (k_3[\mathrm{Gln\cdot E\cdot IGP}] - k_{-3}[\mathrm{Glu\cdot E^*\cdot IGP}]) \\ & + (k_4[\mathrm{Gln\cdot E\cdot ProFAR}] - k_{-4}[\mathrm{Glu\cdot E^*\cdot ProFAR}]), \end{aligned} \tag{23.15}$$

Deriving the equations that describe the dependence of the rates of synthetase and glutaminase reactions on concentrations of the substrates, products, and effectors, we suggested that the rates of all reactions of the substrate binding and dissociation of the products are significantly higher than the rates of catalytic reactions designated as 1, 2, 3, and 4 in

Figure 23.2. This suggestion allowed us to obtain analytical expressions for concentrations of the enzyme states being the terms of Eqs. (23.14) and (23.15) (Gln·E·PRFAR, Glu·E·IGP·AICAR, Gln·E, Glu·E$^*$, Gln·E·IGP, Glu·E$^*$·IGP, Gln·E·ProFAR) and, Glu·E$^*$·ProFAR) and, thus, derive dependence of the rates of synthetase and glutaminase reactions on concentrations of the substrates, products, and effectors (see Appendix 1):

$$v_s = \frac{[\text{HisHF}]}{\Delta}\left(k_1 \frac{[\text{Gln}]}{K_{\text{PRFAR,Gln}}}\frac{[\text{PRFAR}]}{K_{\text{PRFAR}}} - k_{-1}\frac{[\text{AICAR}]}{K_{\text{Glu,IGP,AICAR}}}\frac{[\text{IGP}]}{K_{\text{Glu,IGP}}}\frac{[\text{Glu}]}{K_{\text{Glu}}}\right)$$

$$v_g = \frac{[\text{HisHF}]}{\Delta}\left(\frac{k_2}{K_{\text{Gln}}} + \frac{k_3}{K_{\text{IGP,Gln}}}\frac{[\text{IGP}]}{K_{\text{IGP}}} + \frac{k_4}{K_{\text{ProFAR,Gln}}}\frac{[\text{ProFAR}]}{K_{\text{ProFAR}}}\right)\left([\text{Gln}] - \frac{\text{k}_{-2}}{\text{k}_2}\frac{\text{K}_{\text{Gln}}}{\text{K}^{\text{g}}_{\text{Glu}}}[\text{Glu}]\right)$$

$$\begin{aligned}\Delta = 1 &+ \frac{[\text{PRFAR}]}{K_{\text{PRFAR}}} + \frac{[\text{Gln}]}{K_{\text{PRFAR,Gln}}}\frac{[\text{PRFAR}]}{K_{\text{PRFAR}}} + \frac{[\text{Gln}]}{K_{\text{Gln}}} + \frac{[\text{Glu}]}{K^{g}_{\text{Glu}}} + \frac{[\text{ProFAR}]}{K_{\text{ProFAR}}} \\ &\times\left(1 + \frac{[\text{Glu}]}{K^{g}_{\text{ProFAR,Glu}}} + \frac{[\text{Gln}]}{K_{\text{ProFAR,Gln}}}\right) + \frac{[\text{IGP}]}{K_{\text{IGP}}}\left(1 + \frac{[\text{Glu}]}{K^{g}_{\text{IGP,Glu}}} + \frac{[\text{Gln}]}{K_{\text{IGP,Gln}}}\right) + \frac{[\text{AICAR}]}{K_{\text{AICAR}}} \\ &+ \frac{[\text{Glu}]}{K_{\text{AICAR,Glu}}}\frac{[\text{AICAR}]}{K_{\text{AICAR}}} + \frac{[\text{Glu}]}{K_{\text{Glu}}} + \frac{[\text{IGP}]}{K_{\text{Glu,IGP}}}\frac{[\text{Glu}]}{K_{\text{Glu}}} + \frac{[\text{AICAR}]}{K_{\text{IGP,AICAR}}}\frac{[\text{IGP}]}{K_{\text{IGP}}} \\ &+ \frac{[\text{AICAR}]}{K_{\text{Glu,IGP,AICAR}}}\frac{[\text{IGP}]}{K_{\text{Glu,IGP}}}\frac{[\text{Glu}]}{K_{\text{Glu}}}\end{aligned} \qquad (23.16)$$

where [HisHF] is the total concentration of imidazologlycerol-phosphate synthetase; $K_{\text{A}}$ and $K^{\text{g}}{}_{\text{A}}$ are the dissociation constants of the substrate (or product) A from free enzyme; $K_{\text{BA}}$ and $K^{\text{g}}{}_{\text{BA}}$ are the dissociation constants of the substrate (or product) A from the enzyme complex with the substrate (or product) B; $K_{\text{CBA}}$ are the dissociation constants of the substrate (or product) A from the enzyme complexes with the substrates (or products) B and C; and $k_{\text{i}}$, $k_{-\text{i}}$ (i = 1, 2, 3, and 4) are the rate constants of catalytic stages of the enzyme cycle. The dissociation and rate constants are presented in Figure 23.1 near the corresponding reactions.

In Eqs. (23.16), there are 22 parameters—the rate and dissociation constants characterizing the kinetic properties of certain stages of the catalytic cycle presented in Figure 23.2. However, in the enzymatic kinetics, the rate equations are usually written using parameters, which characterize kinetic properties of the enzyme as a whole. The Michaelis constants, the maximal number of the enzyme cycles, and the equilibrium constant are used as such kinetic parameters (Cornish-Bouden, 2001). To describe kinetic properties of imidazologlycerol-phosphate synthetase, we used 17 kinetic parameters of this kind. The synthetase reaction was characterized by the maximal number of cycles of the enzyme catalyzing this reaction ($k^{\text{s}}_{\text{cat}}$), the equilibrium constant of synthetase reaction ($K^{\text{s}}_{\text{eq}}$), and the Michaelis constants of the substrates and products of this reaction ($K^{\text{s}}_{\text{m,PRFAR}}$, $K^{\text{s}}_{\text{m,Gln}}$, $K^{\text{s}}_{\text{m,Glu}}$, $K^{\text{s}}_{\text{m,IGP}}$, and $K^{\text{s}}_{\text{m,AICAR}}$). The glutaminase reaction in the absence of activators was characterized by the maximal number of cycles of the enzyme catalyzing this reaction ($K^{\text{g}}_{\text{cat}}$), the equilibrium constant of glutaminase reaction ($K^{\text{g}}_{\text{eq}}$), and the Michaelis constants of the substrates and products of this reaction ($K^{\text{g}}{}_{\text{m,Gln}}$ and $K^{\text{g}}_{\text{m,Glu}}$). To account for the effect of activators on the glutaminase reaction, we entered the following parameters: the maximal number of the enzyme cycles at an infinitely high concentration of IGP ($k^{\text{g,IGP}}_{\text{cat}}$) or ProFAR ($k^{\text{g,ProFAR}}_{\text{cat}}$) and the Michaelis constants of the substrates and products of the glutaminase reaction at an infinitely high concentration of IGP ($K^{\text{g,IGP}}_{\text{m,Gln}}$, $K^{\text{g,IGP}}_{\text{m,Glu}}$) or ProFAR

$(K_{\mathrm{m,Gln}}^{\mathrm{g,IGP}}, K_{\mathrm{m,Glu}}^{\mathrm{g,IGP}})$. We found functional interrelations between thus defined kinetic parameters of the imidazologlycerol-phosphate synthetase and parameters characterizing kinetic properties of certain stages of its catalytic cycle. Using this interrelation, we expressed the parameters of the catalytic cycle via kinetic parameters (see Appendix 2) and wrote the rate equations (23.16) using traditional kinetic parameters:

$$v_s = \frac{[\mathrm{HisHF}]}{\Delta} \frac{k_{\mathrm{cat}}^{\mathrm{s}}}{K_{\mathrm{m,PRFAR}}^{\mathrm{s}} K_{\mathrm{m,Gln}}^{\mathrm{g}}} ([\mathrm{PRFAR}][\mathrm{Gln}] - [\mathrm{AICAR}][\mathrm{IGP}][\mathrm{Glu}]/K_{\mathrm{eq}}^{\mathrm{s}})$$

$$v_g = \frac{[\mathrm{HisHF}}{\Delta} \left( \frac{k_{\mathrm{cat}}^{\mathrm{g}}}{K_{\mathrm{m,Gln}}^{\mathrm{g}}} + \frac{k_{\mathrm{cat}}^{\mathrm{g,IGP}}}{K_{\mathrm{m,Gln}}^{\mathrm{g,IGP}}} \frac{[\mathrm{IGP}]}{K_{\mathrm{IGP}}} + \frac{k_{\mathrm{cat}}^{\mathrm{g,ProFAR}}}{K_{\mathrm{m,Gln}}^{\mathrm{g,ProFAR}}} \frac{[\mathrm{ProFAR}]}{K_{\mathrm{ProFAR}}} \right) ([\mathrm{Gln}] - [\mathrm{Glu}]/K_{\mathrm{eq}}^{\mathrm{g}})$$

$$\begin{aligned}
\Delta = {} & 1 + \frac{[\mathrm{Gln}]}{K_{\mathrm{m,Gln}}^{\mathrm{g}}} + \frac{[\mathrm{PRFAR}]}{K_{\mathrm{m,PRFAR}}^{\mathrm{s}}} \left( \frac{K_{\mathrm{m,Gln}}^{\mathrm{s}}}{K_{\mathrm{m,Gln}}^{\mathrm{g}}} + \frac{[\mathrm{Gln}]}{K_{\mathrm{m,Gln}}^{\mathrm{g}}} \right) + \frac{[\mathrm{ProFAR}]}{K_{\mathrm{ProFAR}}} \left( 1 + \frac{[\mathrm{Glu}]}{K_{\mathrm{m,Glu}}^{\mathrm{g,ProFAR}}} + \frac{[\mathrm{Gln}]}{K_{\mathrm{m,Gln}}^{\mathrm{g,ProFAR}}} \right) \\
& + \frac{[\mathrm{IGP}]}{K_{\mathrm{IGP}}} \left( 1 + \frac{[\mathrm{Glu}]}{K_{\mathrm{m,Glu}}^{\mathrm{g,IGP}}} + \frac{[\mathrm{Gln}]}{K_{\mathrm{m,Gln}}^{\mathrm{g,IGP}}} \right) + \frac{[\mathrm{Glu}]}{K_{\mathrm{m,Glu}}^{\mathrm{g}}} + \frac{[\mathrm{AICAR}]}{K_{\mathrm{AICAR}}} + \frac{[\mathrm{Glu}]}{K_{\mathrm{Glu}}} \frac{[\mathrm{IGP}]}{K_{\mathrm{Glu,IGP}}} + \left( 1 + \frac{K_{\mathrm{IGP}} K_{\mathrm{m,Glu}}^{\mathrm{g,IGP}}}{K_{\mathrm{Glu,IGP}} K_{\mathrm{Glu}}} \right) \\
& \times \left( \frac{[\mathrm{AICAR}]}{K_{\mathrm{m,AICAR}}^{\mathrm{s}}} \frac{[\mathrm{Glu}]}{K_{\mathrm{m,Glu}}^{\mathrm{g,IGP}}} \frac{K_{\mathrm{m,IGP}}^{\mathrm{s}}}{K_{\mathrm{IGP}}} + \frac{[\mathrm{AICAR}]}{K_{\mathrm{m,AICAR}}^{\mathrm{s}}} \frac{[\mathrm{IGP}]}{K_{\mathrm{IGP}}} \frac{K_{\mathrm{m,Glu}}^{\mathrm{s}}}{K_{\mathrm{m,Glu}}^{\mathrm{g,IGP}}} + \frac{[\mathrm{AICAR}]}{K_{\mathrm{m,AICAR}}^{\mathrm{s}}} \frac{[\mathrm{IGP}]}{K_{\mathrm{IGP}}} \frac{[\mathrm{Glu}]}{K_{\mathrm{m,Glu}}^{\mathrm{g,IGP}}} \right)
\end{aligned} \tag{23.17}$$

In Eqs. (23.17), there are 22 parameters: 17 are kinetic parameters, and the other 5 are the dissociation constants of the products and effectors ($K_{\mathrm{AICAR}}$, $K_{\mathrm{Glu}}$, $K_{\mathrm{IGP}}$, $K_{\mathrm{Glu,IGP}}$, and $K_{\mathrm{ProFAR}}$); as shown in Appendix 2, the dissociation constant of glutamate from the free enzyme must be higher than the Michaelis constant for glutamate in the glutaminase reaction:

$$K_{\mathrm{m,Glu}}^{\mathrm{g}} < K_{\mathrm{Glu}} \tag{23.18}$$

*Estimation of kinetic parameters of the rate equations using in vitro experimental data.* To find the values of parameters, which appear in Eqs. (23.17), we used experimental data from [3, 4]. As mentioned in subsection "*Available experimental data*," these authors evaluated Michaelis constants for Gln and PRFAR (23.9) and the maximal number of cycles (23.10) of the enzyme catalyzing only the synthetase reaction (23.7). The Michaelis constant for Gln and the maximal number of cycles (23.11) of the enzyme functioning as glutaminase, that is, catalyzing only reaction (23.8) were also evaluated in [3, 4]. In addition, Klem and Davisson (1993) [3] found the values of apparent Michaelis constants for glutamine and apparent maximal number of cycles of the enzyme catalyzing only the glutaminase reaction in the presence of 2-mM ProFAR (23.12) or 9-mM IGP (23.13). We evaluated five kinetic parameters appearing in Eqs. (23.17) using Eqs. (23.9)–(23.11). The values of the apparent constants (23.12)–(23.13) were used to obtain four relationships among the remaining 17 parameters; this allowed reducing the number of unknown parameters to 13. These four relationships are algebraic expressions for the apparent constants, whose values are given in (23.12)–(23.13) via the parameters of Eqs. (23.17).

To derive expressions for the apparent Michaelis constant for glutamine and the apparent maximal number of cycles of the enzyme catalyzing only the glutaminase reaction in the presence of 2-mM ProFAR, we rewrote the rate equation of the glutaminase reaction accounting for conditions of the experiment, which gave evaluation of these apparent constants. In fact, in this experiment, we measured the initial rate of imidazologlycerol-phosphate synthetase depending on the glutamine concentration at various concentrations of the substrates, products, and effectors: [PRFAR] = [Glu] = [IGP] = [AICAR] = 0, and [ProFAR] = $[\mathrm{ProFAR}]_o$ = 2 mM. Substitution of these concentration values into the rate equation for glutaminase reaction (23.17) gives

$$v_g = \frac{[\mathrm{HisHF}]\left(\dfrac{k_{cat}^{g}}{K_{m,Gln}^{g}} + \dfrac{k_{cat}^{g,ProFAR}}{K_{m,Gln}^{g,ProFAR}}\dfrac{[\mathrm{ProFAR}]_0}{K_{ProFAR}}\right)[\mathrm{Gln}]}{1 + \dfrac{[\mathrm{Gln}]}{K_{m,Gln}^{g}} + \dfrac{[\mathrm{ProFAR}]_0}{K_{ProFAR}}\left(1 + \dfrac{[\mathrm{Gln}]}{K_{m,Gln}^{g,ProFAR}}\right)} \tag{23.19}$$

From Eq. (23.19), we obtained the following expressions for the apparent Michaelis constant for glutamine and apparent maximal number of cycles of the enzyme:

$$K_{m,app,Gln}^{g,ProFAR} = \frac{1 + \dfrac{[\mathrm{ProFAR}]_0}{K_{ProFAR}}}{\dfrac{1}{K_{m,Gln}^{g}} + \dfrac{1}{K_{m,Gln}^{g,ProFAR}}\dfrac{[\mathrm{ProFAR}]_0}{K_{ProFAR}}}$$

$$k_{cat,app}^{ProFAR} = \frac{\dfrac{k_{cat}^{g}}{K_{m,Gln}^{g}} + \dfrac{k_{cat}^{g,ProFAR}}{K_{m,Gln}^{g,ProFAR}}\dfrac{[\mathrm{ProFAR}]_0}{K_{ProFAR}}}{\dfrac{1}{K_{m,Gln}^{g}} + \dfrac{1}{K_{m,Gln}^{g,ProFAR}}\dfrac{[\mathrm{ProFAR}]_0}{K_{ProFAR}}} \tag{23.20}$$

These two relationships allowed expressing $K_{ProFAR}$ and $K_{cat}^{g,ProFAR}$ as functions of the other kinetic parameters of Eq. (23.17), the apparent Michaelis constant, the apparent catalytic constant, and $[\mathrm{ProFAR}]_o$ value at which these constants were measured:

$$K_{ProFAR} = [\mathrm{ProFAR}]_0 \frac{K_{m,Gln}^{g}\left(K_{m,app,Gln}^{g,ProFAR} - K_{m,Gln}^{g,ProFAR}\right)}{K_{m,Gln}^{g,ProFAR}\left(K_{m,Gln}^{g} - K_{m,app,Gln}^{g,ProFAR}\right)} \tag{23.21}$$

$$k_{cat}^{g,ProFAR} = k_{cat,app}^{g,ProFAR} + \left(k_{cat,app}^{g,ProFAR} - k_{cat}^{g}\right)\frac{K_{m,app,Gln}^{g,ProFAR} - K_{m,Gln}^{g,ProFAR}}{K_{m,Gln}^{g} - K_{m,app,Gln}^{g,ProFAR}} \tag{23.22}$$

As $K_{ProFAR}$ and $k_{cat}^{g,ProFAR}$ are to be always positive, the values of parameters of the right parts of Eqs. (23.21) and (23.22) should satisfy the following inequalities:

$$K_{m,Gln}^{g} > K_{m,app,Gln}^{g,ProFAR} > K_{m,Gln}^{g,ProFAR}, \qquad k_{cat,app}^{g,ProFAR} > k_{cat}^{g} \tag{23.23}$$

Using the data on the apparent kinetic parameters of the glutaminase reaction activated by IGP (23.13) and repeating all reasons and manipulations presented above, we obtained relationships that allowed us expressing $K_{IGP}$ and $k_{cat}^{g,IGP}$ as functions of other kinetic parameters of Eqs. (23.17), the apparent Michaelis constant, the apparent catalytic constant, and $[\mathrm{IGP}]_o$ value at which these constants were measured:

$$K_{\mathrm{IGP}} = [\mathrm{IGP}]_0 \frac{K^{g}_{\mathrm{m,Gln}}\left(K^{g,\mathrm{IGP}}_{\mathrm{m,app,Gln}} - K^{g,\mathrm{IGP}}_{\mathrm{m,Gln}}\right)}{K^{g,\mathrm{IGP}}_{\mathrm{m,Gln}}\left(K^{g}_{\mathrm{m,Gln}} - K^{g,\mathrm{IGP}}_{\mathrm{m,app,Gln}}\right)}$$

$$k^{g,\mathrm{IGP}}_{\mathrm{cat}} = k^{g,\mathrm{IGP}}_{\mathrm{cat,app}} + \left(k^{g,\mathrm{IGP}}_{\mathrm{cat,app}} - k^{g}_{\mathrm{cat}}\right)\frac{K^{g,\mathrm{IGP}}_{\mathrm{m,app,Gln}} - K^{g,\mathrm{IGP}}_{\mathrm{m,Gln}}}{K^{g}_{\mathrm{m,Gln}} - K^{g,\mathrm{IGP}}_{\mathrm{m,app,Gln}}} \tag{23.24}$$

$$K^{g}_{\mathrm{m,Gln}} > K^{g,\mathrm{IGP}}_{\mathrm{m,app,Gln}} > K^{g,\mathrm{IGP}}_{\mathrm{m,Gln}}, \qquad k^{g,\mathrm{IGP}}_{\mathrm{cat,app}} > k^{g}_{\mathrm{cat}} \tag{23.25}$$

Thus, accounting for the values of kinetic parameters (23.9)–(23.11) and using Eqs. (23.21)–(23.23), we reduced the number of unknown parameters involved in the rate equation for IGPS from 22 to 13, and these 13 parameters should satisfy inequalities (23.18), (23.23), and (23.25).

To obtain the values of the remaining 13 parameters, we used kinetic data for imidazologlycerol-phosphate synthetase obtained at pH 8.0 and 25°C (Klem and Davisson, 1993). Experimental conditions were as follows: to 90-μM PRFAR and 5-mM glutamine imidazologlycerol-phosphate synthetase were added to the final concentration of 9-nM, and changes in PRFAR, AICAR, and Glu concentrations with time were monitored. Kinetic model describing this experiment can be depicted as the kinetic scheme presented in Figure 23.3; this model is given by a system of differential and algebraic equations:

$$\begin{aligned} d[\mathrm{PRFAR}]/dt &= -v_s \\ d[\mathrm{Glu}]/dt &= v_s + v_g \\ [\mathrm{Gln}] + [\mathrm{Glu}] &= 500\ \mu\mathrm{M} \\ [\mathrm{PRFAR}] + [\mathrm{AICAR}] &= 90\ \mu\mathrm{M} \\ [\mathrm{PRFAR}] + [\mathrm{IGP}] &= 90\ \mu\mathrm{M} \end{aligned} \tag{23.26}$$

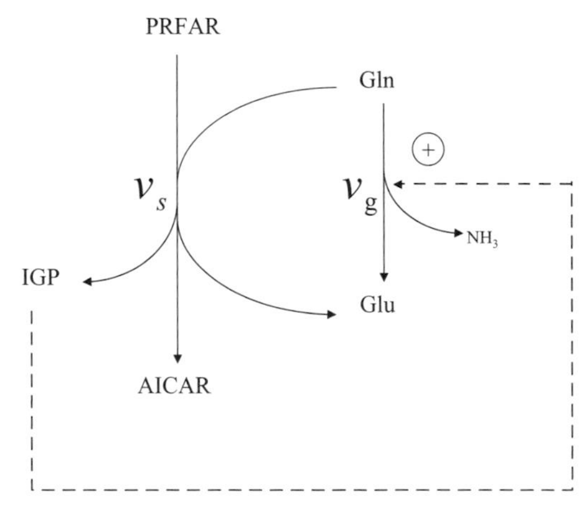


**Figure 23.3** Kinetic scheme of the model corresponding with the experimental data on the kinetics of imidazologlycerol-phosphate synthetase [3]. Solid arrows specify synthetase ($v_s$) and glutaminase ($v_g$) activities of the enzyme, and broken arrow shows that IGP activates the glutaminase reaction.

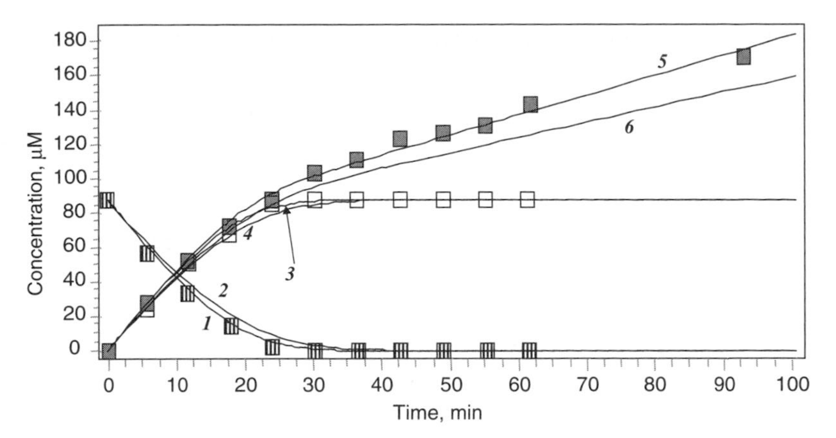


**Figure 23.4** Time dependencies of PRFAR (curves *1* and *2* and hatched squares), AICAR (curves *3* and *4* and open squares), and Glu (curves *5* and *6* and gray squares) concentrations, experimentally measured in [3] (hatched, open, and gray squares) and obtained from the kinetic model based on this experiment (curves *1–6*).

The initial values of variables in this model coincide with the initial experimental concentrations: [PRFAR] = 90 μM, [Gln] = 5000 μM, and [Glu] = [IGP] = [AICAR] = 0. Eleven of 13 kinetic parameters of the rate equations of the synthetase and glutaminase reactions were fitted so that the time dependencies of PRFAR, AICAR, and Glu concentrations obtained as numerical solution of system (23.26) best coincided with corresponding experimental dependencies (Klem and Davisson, 1993). The values of the two parameters $K_{m,Gln}^{g,ProFAR}$ and $K_{m,Glu}^{g,ProFAR}$ characterizing the kinetic properties of the enzyme in the presence of ProFAR (activator of the glutaminase reaction) could not be evaluated from the experimental data (Klem and Davisson, 1993) because of zero concentration of this activator in the considered experiment. Figure 23.4 presents experimental data (Klem and Davisson, 1993) (squares) and model results (solid lines). Approximation of the experimental data (Fig. 23.4, curves *2*, *4*, and *6*) seemed to be impossible with a sufficient range of accuracy by changing values of 11 parameters, which remained nonevaluated (that is, free) after accounting for the experimental data [Klem and Davisson, (1993), Chittur et al., 2001] by the values of true and apparent catalytic and Michaelis constants in our model. However, if we suppose that the values of these true and apparent catalytic and Michaelis constants from expressions (23.9)–(23.13) are measured with 30% accuracy and that is why we can vary these values within this 30% interval together with 11 nonevaluated parameters, the results of modeling will exactly coincide with experimental data (Fig. 23.4, curves *1*, *3*, and 5). The thus fitted parameter values are given in the Table 23.3.

The remaining two parameters $K_{m,Gln}^{g,ProFAR}$ and $K_{m,Glu}^{g,ProFAR}$ characterizing kinetic properties of imidazologlycerol-phosphate synthetase in the presence of ProFAR, an activator of the glutaminase reaction, were evaluated by the data of Jurgens et al., (2000); in this work, Jurgens and coauthors studied the properties of ProFAR isomerase, a precursor of IGPS in the pathway of biosynthesis of histidine and a catalyst of the isomerization reaction of ProFAR into PRFAR. Experimental conditions were as follows: pH 7.5 and 25°C. To 0.25-μM ProFAR isomerase, 1-μM imidazologlycerol-phosphate synthetase, and 5-mM glutamine ProFAR was added to the final concentration 20 μM, and the time dependence of its concentration was monitored. The kinetic model describing this experiment can be depicted as the kinetic scheme presented in Figure 23.5; this model is given by

**TABLE 23.3 Kinetic Parameters of Imidazole Glycerol-Phosphate Synthase and ProFAR Isomerase**

| Parameter | Experimental Data [ref] | Values Obtained as a Result of Identification of the Model with Experimental Data (Klem and Davisson, 1993) and (Jurgens et al., 2000) |
|---|---|---|
| $k^{s}_{cat}$ | 510 $min^{-1}$(Klem and Davisson, 1993) | 585 $min^{-1}$ |
| $K^{s}_{eq}$ | | 4,85 mM |
| $K^{s}_{m,PRFAR}$ | 1,5 μM (Klem and Davisson, 1993) | 1,86 μM |
| $K^{s}_{m,Gln}$ | 240 μM (Klem and Davisson, 1993) | 180 μM |
| $K^{s}_{m,Glu}$ | | 0,01 μM |
| $K^{s}_{m,IGP}$ | | 0,14 μM |
| $K^{s}_{m,AICAR}$ | | 20,6 mM |
| $k^{g}_{cat}$ | 4.2 $min^{-1}$ (Klem and Davisson, 1993) | 5,28 $min^{-1}$ |
| $K^{g}_{eq}$ | | 1,03 mM |
| $K^{g}_{m,Gln}$ | 4,8 mM (Klem and Davisson, 1993) | 5,96 mM |
| $K^{g}_{m,Glu}$ | | 51,4 mM |
| $k^{g,IGP}_{cat}$ | | 210 $min^{-1}$ |
| $k^{g,ProFAR}_{cat}$ | | 156 $min^{-1}$ |
| $K^{g,IGP}_{m,Gln}$ | | 1,32 mM |
| $K^{g,IGP}_{m,Glu}$ | | 10,4 mM |
| $K^{g,ProFAR}_{m,Gln}$ | | 946 μM |
| $K^{g,ProFAR}_{m,Glu}$ | | 46,2 mM |
| $K_{AICAR}$ | | 31,7 mM |
| $K_{Glu}$ | | 51,4 mM |
| $K_{IGP}$ | | 61,3 μM |
| $K_{Glu,\ IGP}$ | | 30,9 mM |
| $K_{ProFAR}$ | | 4,7 μM |
| $k^{HisA}_{cat}$ | 42 $min^{-1}$ (Jurgens et al., 2000) | 42 $min^{-1}$ |
| $K^{HisA}_{m,ProFAR}$ | 0,6 μM (Jurgens et al., 2000) | 0,6 μM |
| $K^{HisA}_{m,PRFAR}$ | | 0,03 μM |
| $k^{g,IGP}_{cat,app}$ | 162 $min^{-1}$ (Klem and Davisson, 1993) | 210 $min^{-1}$ |
| $K^{g,IGP}_{m,app,Gln}$ | 1,9 mM (Klem and Davisson, 1993) | 1,33 mM |
| $k^{g,ProFAR}_{cat,app}$ | 156 $min^{-1}$ (Klem and Davisson, 1993) | 156 $min^{-1}$ |
| $K^{g,ProFAR}_{m,app,Gln}$ | 2,8 mM (Klem and Davisson, 1993) | 2,8 mM |
| $K^{1}_{H}$ | | 0,01 μM |
| $K^{2}_{H}$ | | 0,93 μM |

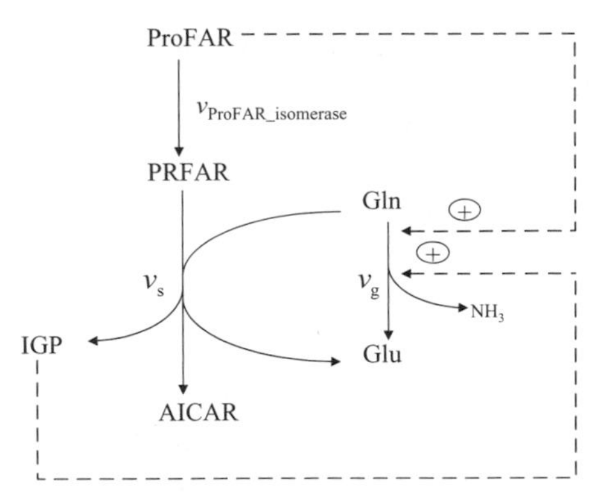


**Figure 23.5** Kinetic scheme of a model corresponding to the kinetic study of imidazologlycerol-phosphate synthetase–ProFAR isomerase system [8]. Solid arrows designate reactions catalyzed by ProFAR isomerase ($v_{\text{ProFAR\_isomerase}}$) and the synthetase ($v_s$) and glutaminase ($v_g$) activities of IGPS. Broken arrows show that IGP and ProFAR stimulate the glutaminase activity of IGPS.

a system of differential and algebraic equations:

$$\begin{aligned} d[\text{ProFAR}]/dt &= -v_{\text{ProFAR-isomerase}} \\ d[\text{PRFAR}]/dt &= v_{\text{ProFAR-isomerase}} - v_s^{\text{pH}} \\ d[\text{Glu}]/dt &= ***v_s^{\text{pH}} + ***v_g^{\text{pH}} \\ [\text{Gln}] + [\text{Glu}] &= 5000\,\mu\text{M}, \\ [\text{ProFAR}] + [\text{PRFAR}] + [\text{AICAR}] &= 20\,\mu\text{M} \\ [\text{ProFAR}] + [\text{PRFAR}] + [\text{IGP}] &= 20\,\mu\text{M} \end{aligned} \tag{23.27}$$

The initial values of variables in the model coincide with the initial experimental concentrations: [ProFAR] = 20 μM, [Gln] = 5000 μM, and [Glu] = [PRFAR] = [IGP] = [AICAR] = 0. As most imidazologlycerol-phosphate synthetase parameters were evaluated from experimental data obtained at pH 8.0 [(Klem and Davisson, (1993)], for evaluation of the remaining parameters from experimental data obtained at pH 7.5 (Jurgens et al., 2000), it is necessary to account for the pH dependence of the activity of this enzyme. Following the method for describing pH dependence of the rate of enzymatic reaction (Cornish-Bouden, 2001), we suggested that a catalytic site (one or several amino acid residues directly participating in catalysis) can be deprotonated and protonated singly or doubly, the singly protonated form of the enzyme being catalytically active. We also suggested that the values of the proton dissociation constants do not depend on the enzyme state. These suggestions do not contradict experimental data on kinetics of this enzyme; as a result, the pH dependence of the rate of IGPS functioning is given as follows:

$$\begin{aligned} v_s^{\text{pH}} &= v_s \frac{1 + \dfrac{K_H^1}{10^{6-\text{pH}_1}} + \dfrac{10^{6-\text{pH}_1}}{K_H^2}}{1 + \dfrac{K_H^1}{10^{6-\text{pH}}} + \dfrac{10^{6-\text{pH}}}{K_H^2}} \\ v_g^{\text{pH}} &= v_g \frac{1 + \dfrac{K_H^1}{10^{6-\text{pH}_1}} + \dfrac{10^{6-\text{pH}_1}}{K_H^2}}{1 + \dfrac{K_H^1}{10^{6-\text{pH}}} + \dfrac{10^{6-\text{pH}}}{K_H^2}} \end{aligned} \tag{23.28}$$

where $v_s$ and $v_g$ are given by Eqs. (23.17) describing the synthetase and glutaminase activities of imidazologlycerol-phosphate synthetase; $K_H^1$, $K_H^2$ are the proton dissociation constants from non-protonated and singly protonated forms of the enzyme, respectively; and $pH_1 = 8.0$ coincides with pH in the experiment [Klem and Davisson, (1993)] used for evaluation of most of the enzyme constants. As the data (Jurgens et al., 2000) were obtained at pH 7.5, in the rate equations for imidazologlycerol-phosphate synthetase involved in kinetic model (27), which describes these data, the pH should be also 7.5.

The rate equation for ProFAR isomerase can be written as follows:

$$v_{\text{ProFAR\_isomerase}} = \frac{k_{\text{cat}}^{\text{HisA}}[\text{HisA}][\text{ProFAR}]}{K_{\text{m,ProFAR}}^{\text{HisA}} + [\text{ProFAR}] + \frac{K_{\text{m,ProFAR}}^{\text{HisA}}}{K_{\text{m,PRFAR}}^{\text{HisA}}} \cdot [\text{PRFAR}]} \tag{23.29}$$

The catalytic and Michaelis constants for ProFAR evaluated from the data obtained at pH 7.5 and 25°C were taken from Jurgens et al. (2000): $k_{\text{cat}}^{\text{HisA}} = 0,7\,\text{c}^{-1}$, $K_{\text{m,ProFAR}}^{\text{HisA}} = 0,6\,\text{мкM}$. The remaining unknown parameters of model (23.21) ($K_{\text{m,Gln}}^{\text{g,ProFAR}}$, $K_{\text{m,Glu}}^{\text{g,ProFAR}}$ $u$ $K_{\text{m,PRFAR}}^{\text{HisA}}$, $K_H^1$, $K_H^2$) were fitted so that the time dependence of ProFAR concentration obtained as a result of a numerical solution of system (23.27) best coincided with the corresponding experimentally obtained dependence (Jurgens et al., 2000). As shown in Figure 23.6, the thus found parameters provide a precise coincidence of the results of modeling (solid line) with experimental data (Jurgens et al., 2000) (squares). The values of fitted parameters are given in the Table 23.3.

*How the synthetase and glutaminase activities of imidazologlycerol-phosphate synthetase depend on concentrations of the substrates and effectors.* As shown in Zalkin; (1993) and Zalkin and Smith (1998), imidazologlycerol-phosphate synthetase can catalyze two different processes: synthesis of imidazologlycerol phosphate (23.7) and decomposition of glutamine (23.8). To study what activity dominates and what is an accompanying one and how it depends on the concentrations of the substrates and effectors, we plotted (Figure 23.7) the rates of the glutaminase and synthetase reactions versus the substrate (PRFAR) concentration at various concentrations of effectors (IGP and ProFAR ), zero concentrations of the products, pH 8.0, and a fixed concentration of glutamine ([Gln] = 1 mM),

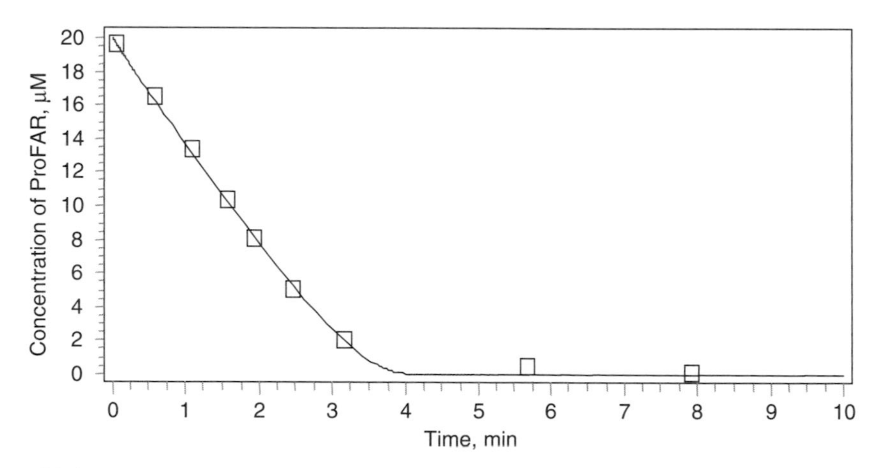


**Figure 23.6** Time dependence of ProFAR concentrations: experimentally measured in [8] (open squares) and obtained from the kinetic model based on this experiment (solid line).

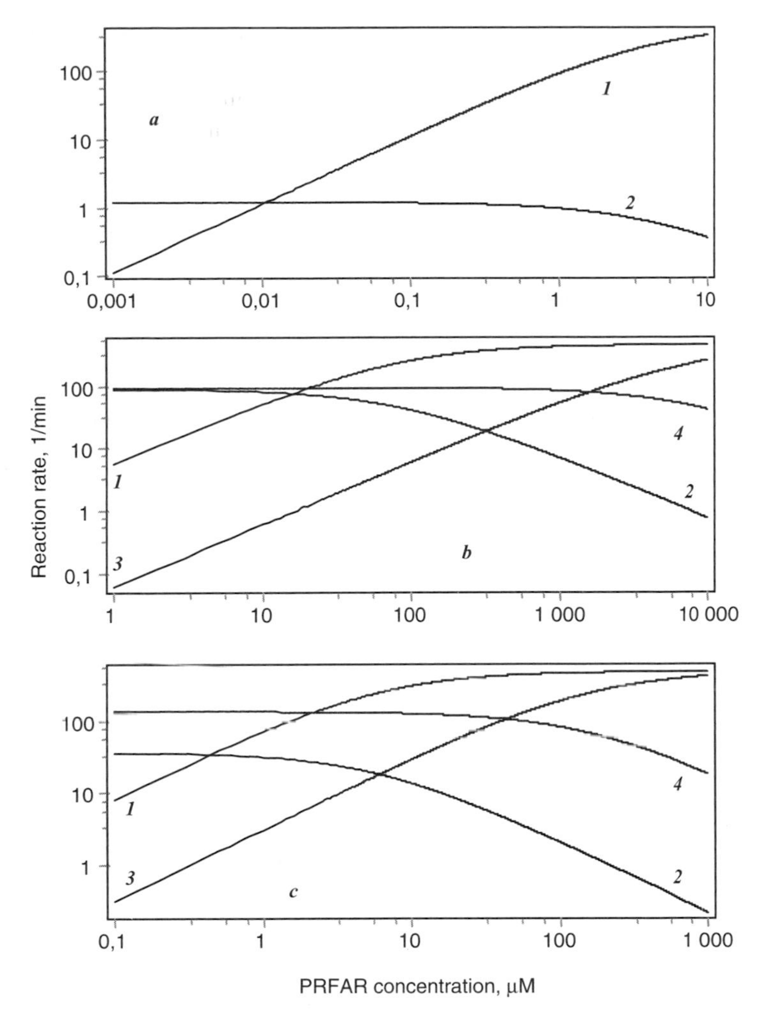


**Figure 23.7** Rates of the synthetase (curves *1* and *3*) and glutaminase (curves *2* and *4*) reactions catalyzed by imidazologlycerol-phosphate synthetase versus PRFAR concentration: (a) in the absence of activators of the glutaminase reaction; (b) at various ProFAR concentrations (10-μM ProFAR, curves *1* and *2*; 1-mM ProFAR, curves *3* and *4*); and (c) at various IGP concentrations (10-μM IGP, curves *1* and *2*; 1-mM IGP, curves *3* and *4*). All curves were obtained at zero concentrations of the products and 1-mM Gln.

using the results of the previous section (the rate equations (23.17) and the values of parameters from the table). It seemed that in the absence of effectors the synthetase activity can be considered as dominating. In fact, as shown in Figure 23.7a, at PRFAR concentrations exceeding 10 nM, the rate of the synthetase reaction is higher than the rate of the glutaminase reaction. Moreover, beginning from ~100-nM PRFAR, the rate of the synthetase reaction is more than ten times higher than that of the glutaminase reaction, and this difference grows with an increase in PRFAR concentration, achieving 1000 times at [PRFAR] = 10 μM. In the presence of effector (ProFAR), the situation is not so definite. As shown in Figure 23.7b, at 10-μM ProFAR, the rate of the synthetase reaction (curve *1*) becomes equal to the rate of the glutaminase reaction (curve *2*) at a PRFAR concentration

~13 μM. At lower PRFAR concentrations, the rate of the glutaminase reaction is higher than the rate of the synthetase reaction, whereas at higher PRFAR concentrations, the synthetase activity dominates. At higher ProFAR concentrations, the interval of PRFAR concentrations corresponding to domination of the glutaminase reaction expands. In fact, at 1-mM ProFAR, the rate of the synthetase reaction (curve *3*) becomes equal to the rate of the glutaminase reaction (curve *4*) at a PRFAR concentration ~1300 μM. Moreover, at PRFAR concentrations lower than 130 μM, the rate of the glutaminase reaction is more than ten times higher than the rate of the synthetase reaction; that is, the glutaminase activity becomes dominating. Analogous conclusions are true also for another effector—IGP (Fig. 23.7c). At 1-mM IGP, the rate of the synthetase reaction (curve *3*) becomes equal to the rate of the glutaminase reaction (curve *4*) at a PRFAR concentration ~23 μM. As shown in Figure 23.7c, an increase in IGP concentration also results in expansion of the interval of PRFAR concentrations at which the glutaminase activity dominates.

As follows from Figure 23.6, independent of the presence of the effectors—ProFAR and IGP—an increase in the substrate (PRFAR) concentration results in a decrease in glutaminase and an increase in synthetase activities of imidazologlycerol-phosphate synthetase.

### 23.2.4 Development of Kinetic Model of Histidinol Dehydrogenase from *Escherichia coli*

In this section we develop a kinetic model of histidinol dehydrogenase and illustrate how kinetic data measured at different conditions (pH, temperature, and others) can be taken all together into account to construct the catalytic cycle of the enzyme, derive the rate equation, and estimate its kinetic parameters. The method developed in the section allows us to predict kinetic behavior of the enzyme at any conditions.

Histidinol dehydrogenase (EC 1.1.1.23), encoded by gene *hisD*, can catalyze two separate reactions (Umbarger, 1996): oxidation of histidinol to histidinal (histidinoldehydrogenase activity)

$$\mathrm{Hol} + \mathrm{NAD} = \mathrm{Hal} + \mathrm{NADH} \tag{23.30}$$

and oxidation of histidinal to histidine (histidinaldehydrogenase activity)

$$\mathrm{Hal} + \mathrm{NAD} = \mathrm{His} + \mathrm{NADH} \tag{23.31}$$

*Available experimental data.* To construct the catalytic cycle of histidinol dehydrogenase, we take into account the following data on structural and functional features of the enzyme:

(i) When histidinol is used as a substrate of the enzyme, one histidine molecule is formed and two NAD molecules are reduced per one histidinol molecule consumed, but histidinal accumulation is not experimentally detected (Loper et al., 1965; Adams, 1955).

(ii) When histidinal is used as a substrate of the enzyme, one histidine molecule is formed and one NAD molecule is reduced per one histidinal molecule consumed (Loper et al., 1965).

(iii) The enzyme has one catalytic site (Loper et al., 1965; Adams, 1955); in other words, different substrates of histidinoldehydrogenase and histidinaldehydrogenase reactions compete with each other.

(iv) Binding of substrates as well as dissociation of products proceeds in random order (Umbarger, 1996; Loper et al., 1965; Adams, 1955).

(v) The pH optimum of histidinoldehydrogenase reaction differs from that of histidinaldehydrogenase reaction by two units of pH (Loper et al., 1965).

Kinetic properties of the enzyme operating as histidinoldehydrogenase, i.e., catalyzing reaction (30) only, have been partly studied (Loper et al., 1965), where the turnover number and Michaelis constants for Hol and NAD have estimated at different pH values:

$$\begin{aligned} K_{m,Hol}^{pH=7.7} &= 0.029\ mM, \quad pH = 7.7 \\ K_{m,Hol}^{pH=9.3} &= 0.012\ mM, \quad pH = 9.3 \\ K_{m,NAD}^{Hol} &= 1.53\ mM, \quad pH = 7.5 \\ K_{m,NAD}^{Hol} &= 1.26\ mM, \quad pH = 9.3 \\ k_{cat,Hol} &= 1073\ \min^{-1}, \quad pH = 9.4 \end{aligned} \tag{23.32}$$

From Eqs. (23.32), it follows that an increase in pH from 7.7 to 9.3 decreases the Michaelis constant for histidinol by almost three times but changes the Michaelis constant for NAD ($K_{m,NAD}^{Hol}$) by less than 20 percentages, i.e., in the range of experimental error. Thus, we can conclude that the Michaelis constant for NAD does not depend on pH.

Kinetic properties of the enzyme operating as histidinaldehydrogenase, i.e., catalyzing reaction (23.31) only, has been partly studied (Loper et al., 1965), where the turnover number and Michaelis constants for Hal and NAD have estimated at different pH values:

$$\begin{aligned} K_{m,Hal} &= 0.0078\ mM, \quad pH = 7.7 \\ K_{m,NAD}^{Hal} &= 0.21\ mM, \quad pH = 7.5 \\ k_{cat,Hal} &= 1073\ \min^{-1}, \quad pH = 7.6 \end{aligned} \tag{23.33}$$

In addition, dependencies of maximal activity of histidinoldehydrogenase and histidinaldehydrogenase on pH have been measured in Loper et al. (1965) and the time dependence of NADH accumulation at a presence of 2.7 mM of histidinol dehydrogenase and pH equal to 8.9 has been followed in Adams (1955).

*Construction of catalytic cycle.* To construct the catalytic cycle of histidinol dehydrogenase (see Fig. 23.2), we used experimental data described in clauses (1)–(4) of the previous section. Indeed, we assumed that the enzyme operated in accordance with a Random Bi Bi mechanism of the Cleland classification (Cleland, 1963) that conformed to clause (4) of the previous section. The suggested mechanism has four ternary complexes: Hol°E°NAD, Hal°E°NADH, Hal°E°NAD, and His°E°NADH. The transition of the ternary enzyme-substrate complex Hol°E°NAD to a complex of enzyme with products Hal°E° NADH corresponds to histidinoldehydrogenase activity of the enzyme but irreversible transition (Loper et al., 1965) of the ternary complex Hal°E°NAD to complex His°E° NADH corresponds to histidinaldehydrogenase activity of the enzyme.

In accordance with clause (3), histidinol should compete with histidinal for a catalytic site of the enzyme. To take this fact into account, we introduce into the catalytic cycle, an additional reaction of binding of histidinal to the enzyme state with NAD bound in the catalytic site (E°NAD $\rightarrow$ Hal°E°NAD).

If histidinol and NAD are used as initial substrates, then in accordance with the suggested catalytic cycle, one molecule of histidine is formed and two molecules of NAD are reduced per one histidinol molecule consumed. This stoichiometry corresponds to successive transitions via the following stages of the catalytic cycle: E → E°NAD → Hol° E°NAD → Hal°E°NADH → Hal°E → Hal°E°NAD → His°E°NADH → His°E → E. Note that the histidinal molecule being intermediate on the way of histidine formation from histidinol is not released during cycling through these stages of catalytic cycle but remains in the catalytic site of the enzyme and is oxidized to histidine. Therefore, it is not possible to detect free histidinal experimentally if the enzyme mainly operates via a selected cycle of stages of the catalytic cycle. That corresponds to clause (1) of the previous section. If histidinal and NAD are used as initial substrates, then one molecule of histidine is formed and one molecule of NAD is reduced per one histidinal molecule consumed. This stoichiometry corresponds to successive transitions via the following stages of the catalytic cycle: E → E°NAD → Hal°E°NAD → His°E°NADH → His°E → E and is in agreement with clause (2) of the previous section.

The next stage of catalytic cycle construction consists in taking into account the possibility of proton binding to different states of enzyme, which means that we should modify the catalytic cycle depicted in Figure 23.8a in such a way that it enables us to derive the rate equation, which pH dependence is in line with following the experimental observations:

1. pH optimums of histidinoldehydrogenase activity differs from that of the histidinaldehydrogenase one (that is in agreement with clause (5) of the previous section).
2. The Michaelis constant with respect to histidinol depends on pH.

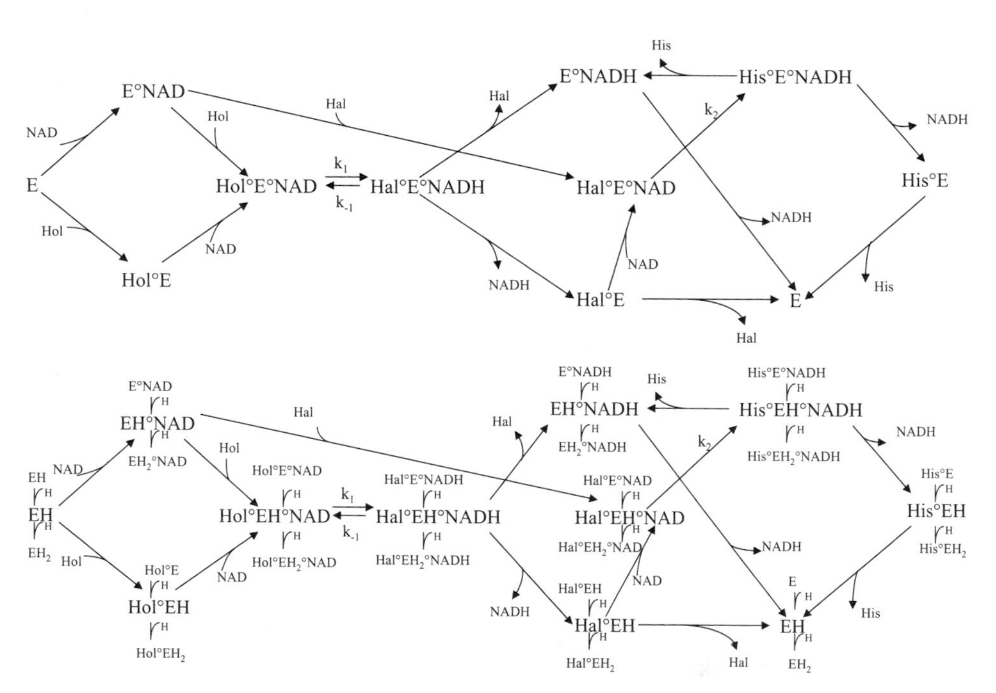


**Figure 23.8** Catalytic cycle of histidinol dehydrogenase without (a) and with (b) mechanism assigning pH dependence.

3. The Michaelis constant with respect to NAD of the histidinoldehydrogenase reaction does not depend on pH.

To introduce pH dependence of enzyme operation and to satisfy the requirements formulated above, we, following the approach described in (Cornish-Bouden, 2001), suggested that enzyme (or, in other words, one or several aminoacid residues of catalytic cycle directly participating in catalysis) can be indeprotonated, once protonated, and twice protonated states. Assuming that the protonated enzyme is catalytically active only once, we modified the catalytic cycle in the manner presented in Figure 23.8b.

*Derivation of rate equations.* Using the scheme of the catalytic cycle depicted in Figure 23.8b, the rates of the histidinoldehydrogenase and histidinaldehydrogenase reactions can be written out in following manner:

$$\mathrm{V_{Hol}} = \mathrm{k_1 \cdot Hol^\circ EH^\circ NAD - k_{-1} \cdot Hal^\circ EH^\circ NADH} \tag{23.34}$$

$$\mathrm{V_{Hal}} = \mathrm{k_2 \cdot Hal^\circ EH^\circ NAD} \tag{23.35}$$

To derive rate equations describing the dependence of rates of histidinoldehydrogenase and histidinaldehydrogenase reactions on concentrations of substrates, products, and effectors, we assume that reactions of substrate binding, product dissociation, and enzyme protonation are much more faster than catalytic reactions designated by numbers 1 and 2 in Figure 23.8b. Consequently, we can consider all fast reactions as quasi-equilibrium and obtain analytical expressions of concentrations of enzyme states (Hol°EH°NAD, Hal°EH° NADH, Hal°EH°NAD) included in the right-hand sides of Eqs. (23.34) and (23.35). Now we can derive dependencies of rates of histidinoldehydrogenase and histidinaldehydrogenase reactions on substrates, products, and proton concentrations:

$$v_{Hol} = \frac{HisD}{\Delta} \cdot \left( k_1 \cdot \frac{Hol}{K_{NAD,Hol}} \cdot \frac{NAD}{K_{NAD}} - k_{-1} \cdot \frac{NADH}{K_{Hal,NADH}} \cdot \frac{Hal}{K_{Hal}} \right)$$

$$v_{Hal} = \frac{HisD \cdot k_2 \cdot NAD \cdot Hal / K_{Hal,NAD} / K_{Hal}}{\Delta}$$

$$\begin{aligned}\Delta = {} & h_E + h_{Hol \circ E} \cdot \frac{Hol}{K_{Hol}} + h_{E \circ NAD} \cdot \frac{NAD}{K_{NAD}} + h_{Hol \circ E \circ NAD} \cdot \frac{Hol}{K_{NAD,Hol}} \cdot \frac{NAD}{K_{NAD}} \\ & + h_{Hal \circ E} \cdot \frac{Hal}{K_{Hal}} + h_{Hal \circ E \circ NAD} \cdot \frac{NAD}{K_{Hal,NAD}} \cdot \frac{Hal}{K_{Hal}} + h_{E \circ NADH} \cdot \frac{NADH}{K_{NADH}} \\ & + h_{Hal \circ E \circ NADH} \cdot \frac{NADH}{K_{Hak,NADH}} \cdot \frac{Hal}{K_{Hal}} + h_{His \circ E} \cdot \frac{His}{K_{His}} + h_{His \circ E \circ NADH} \cdot \frac{His}{K_{NADH,His}} \cdot \frac{NADH}{K_{NADH}}\end{aligned} \tag{23.36}$$

where functions $h_X$ define the level of once protonated enzyme state $X$, $X \in \{\mathrm{E, E^\circ NAD, Hol^\circ E^\circ NAD, etc.}\}$:

$$h_X = h_X(pH) = 1 + \frac{K_X^1}{H} + \frac{H}{K_X^2} \tag{23.37}$$

Here, $H$ is proton concentration; $K_X^1$ and $K_X^2$ are dissociation constants describing proton dissociation from once and twice protonated enzyme state $X$, correspondently. HisD is total enzyme concentration: $K_A$ is dissociation constant of substrate (or product) $A$ from free enzyme; $K_{BA}$ is dissociation constant of substrate (or product) $A$ from a complex of enzyme with substrate (or product) $B$; and $k_1, k_{-1}, k_2$ are rate constants of catalytic steps of the catalytic cycle. Dissociation constants of substrates/products and rate constants are shown in Figure 23.8b near the corresponding reactions.

Equations (23.36) involves 32 parameters: 12 parameters are rate and dissociation constants describing kinetic properties of individual steps of the catalytic cycle, depicted in Figure 23.8b and 20 parameters characterize proton binding to different enzyme states. In accordance with the approach suggested in this chapter, we will introduce several assumptions that do not conflict with the experimental data cited above on the one hand and simplify the mechanism of pH dependence of enzyme operation on the other hand. This method enables us to reduce the number of unknown parameters. Indeed, let us assume that

1) Ten enzyme states of catalytic cycle, depicted in Figure 23.8a can be subdivided into three groups:
   a. Group 1: E, E°NAD, E°NADH
   b. Group 2: Hol°E, Hol°E°NAD
   c. Group 3: Hal°E°NADH, Hal°E°NAD, His°E°NADH, Hal°E, His°E
2) Protonation of amino acid residues of catalytic site of enzyme states included in one group is described by identical dissociation constants:

$$h_E = h_{E\circ NAD} = h_{E\circ NADH} = h_0 = 1 + \frac{K_0^1}{H} + \frac{H}{K_0^2} \tag{23.38}$$

$$h_{Hol\circ E} = h_{Hol\circ E\circ NAD} = h_{Hol} = 1 + \frac{K_{Hol}^1}{H} + \frac{H}{K_{Hol}^2} \tag{23.39}$$

$$\begin{aligned} h_{Hal\circ E} &= h_{Hal\circ E\circ NAD} = h_{Hal\circ E\circ NADH} = h_{His\circ E} = h_{His\circ E\circ NADH} \\ &= h_{Hal,His} = 1 + \frac{K_{Hal,His}^1}{H} + \frac{H}{K_{Hal,His}^2} \end{aligned} \tag{23.40}$$

These assumptions allows us to reduce the number of unknown parameters describing pH dependence of enzyme operation from 20 to 6 and to rewrite equations (23.36) in the following manner:

$$\begin{aligned} v_{Hol} &= \frac{HisD}{\Delta} \cdot \left( k_1 \cdot \frac{Hol}{K_{NAD,Hol}} \cdot \frac{NAD}{K_{NAD}} - k_{-1} \cdot \frac{NADH}{K_{Hal,NADH}} \cdot \frac{Hal}{K_{Hal}} \right) \\ v_{Hal} &= \frac{HisD \cdot k_2 \cdot NAD \cdot Hal / K_{Hal,NAD} / K_{Hal}}{\Delta} \\ \Delta &= h_0 \cdot \left( 1 + \frac{NAD}{K_{NAD}} + \frac{NADH}{K_{NADH}} \right) + h_{Hol} \left( \frac{Hol}{K_{Hol}} + \frac{Hol}{K_{NAD,Hol}} \cdot \frac{NAD}{K_{NAD}} \right) \\ &+ h_{Hal,His} \left( \frac{Hal}{K_{Hal}} + \frac{NAD}{K_{Hal,NAD}} \cdot \frac{Hal}{K_{Hal}} + \frac{NADH}{K_{Hal,NADH}} \cdot \frac{Hal}{K_{Hal}} + \frac{His}{K_{His}} + \frac{His}{K_{NADH,His}} \cdot \frac{NADH}{K_{NADH}} \right) \end{aligned} \tag{23.41}$$

It is easy to see that this simplified mechanism of pH dependence completely satisfies experimentally established facts (1), (2) and (3) cited in the previous section. Moreover, it is not difficult to prove that further simplification of mechanism of pH dependence (such as subdivision of all enzyme states to two groups instead of three) does not allow us to derive the rate equation describing these experimental facts correctly. It means that the suggested mechanism of pH dependence of enzyme operation is a minimal one of all possible mechanisms able to describe experimental facts (1), (2), and (3).

*Estimation of kinetic parameters of the rate equations using in vitro experimental data.* To reduce the number of unknown parameters of Eqs. (23.41), we used experimental data measured in Loper et al., (1965) and described above in Eqs. (23.32) and (23.33). As was mentioned, Loper et al. (1965) estimated turnover numbers and Michaelis constants with respect to substrates of histidinoldehydrogenase and histidinaldehydrogenase reactions at different pH values. We obtained analytical expressions for these kinetic parameters in terms of parameters of the catalytic cycle, i.e., dissociation and rate constants. Then, using these relationships between kinetic parameters and parameters of the catalytic cycle, we expressed seven parameters of Eqs. (23.41) in terms of experimentally measured kinetic parameters (23.32) and (23.33) and 11 remaining parameters of the catalytic cycle:

$$k_1 = k_{cat,Hol} \cdot h_{Hol}(pH = 9.4), \quad k_2 = k_{cat,Hal} \cdot h_{Hal,His}(pH = 7.6), K_{Hal,NAD} = K_{m,NAD}^{Hal}$$

$$K_{NAD} = \frac{K_{Hol} \cdot K_{m,NAD}^{Hol}}{K_{m,Hol}^{pH=9.3}} \cdot \frac{h_0(pH = 9.3)}{h_{Hol}(pH = 9.3)}, \quad K_{NAD,Hol} = K_{m,Hol}^{pH=9.3} \cdot \frac{h_{Hol}(pH = 9.3)}{h_0(pH = 9.3)}$$

$$K_0^1 = \frac{\dfrac{K_{m,Hol}^{pH=9.3}}{K_{m,Hol}^{pH=7.7}} \cdot \dfrac{h_{Hol}(pH = 9.3)}{h_{Hol}(pH = 7.7)} \cdot \left(1 + \dfrac{10^{-4.7}}{K_0^2}\right) - \left(1 + \dfrac{10^{-6.3}}{K_0^2}\right)}{\dfrac{1}{10^{-6.3}} - \dfrac{1}{10^{-4.7}} \cdot \dfrac{K_{m,Hol}^{pH=9.3}}{K_{m,Hol}^{pH=7.7}} \cdot \dfrac{h_{Hol}(pH = 9.3)}{h_{Hol}(pH = 7.7)}},$$

$$K_{Hal} = \frac{K_{Hol} \cdot K_{m,NAD}^{Hol} \cdot K_{m,Hal}}{K_{m,NAD}^{Hal} \cdot K_{m,Hol}^{pH=9.3}} \cdot \frac{h_0(pH = 9.3)}{h_{Hol}(pH = 9.3)} \cdot \frac{h_{Hal,His}(pH = 7.7)}{h_0(pH = 7.7)} \tag{23.42}$$

where

$$h_0(pH) = 1 + \frac{K_0^1}{10^{-3+pH}} + \frac{10^{-3+pH}}{K_0^2}, \quad h_{Hol}(pH) = 1 + \frac{K_{Hol}^1}{10^{-3+pH}} + \frac{10^{-3+pH}}{K_{Hol}^2}$$

$$h_{Hal,His}(pH) = 1 + \frac{K_{Hal,His}^1}{10^{-3+pH}} + \frac{10^{-3+pH}}{K_{Hal,His}^2} \tag{23.43}$$

To estimate the values of the remaining 11 parameters, we used experimental data (Loper et al., 1965) on pH dependence of maximal velocity of histidinoldehydrogenase and histidinaldehydrogenase activity of the enzyme as well as experimental data (Adams, 1955) on the time dependence of NADH accumulation at a presence of 2.7 mM of histidinol dehydrogenase and pH equal to 8.9.

It is easy to find that pH dependence of maximal velocity of histidinoldehydrogenase activity of the enzyme normalized to its value at optimal pH is determined by following expression:

$$V_{\max,norm}^{Hol}(pH) = \frac{1 + 2 \cdot \sqrt{\dfrac{K_{Hol}^{1}}{K_{Hol}^{2}}}}{h_{Hol}(pH)} \tag{23.44}$$

In a similar way, it obtains that pH dependence of maximal velocity of histidinaldehydrogenase activity of the enzyme normalized to its value at optimal pH is determined by the following expression:

$$V_{\max,norm}^{Hal}(pH) = \frac{1 + 2 \cdot \sqrt{\dfrac{K_{Hal,His}^{1}}{K_{Hal,His}^{2}}}}{h_{Hal,His}(pH)} \tag{23.45}$$

The values of the four parameters of Eqs. (23.44) and (23.45) have been chosen in such a way that pH dependencies of maximal velocities of histidinol dehydrogenase given by these two equations have coincided with corresponding pH dependencies experimentally measured in Loper et al. (1965). Figure 23.9 shows pH dependencies of maximal velocities calculated from Eqs. (23.44) and (23.45) (solid lines) and experimentally measured ones (symbols). The values of estimated parameters are listed in Table 23.4.

The remaining seven parameters of Eq. (23.41) have been estimated on the basis of experimental data (Adams, 1955) obtained in the kinetic experiment that has been conducted in the following manner: The reaction has started by adding 2.7 μM of histidinol to a cuvette containing 500 μM of NAD and 2.7 μM of histidinol dehydrogenase at pH 8.9, and accumulation of NADH has been followed in time. The kinetic model corresponding

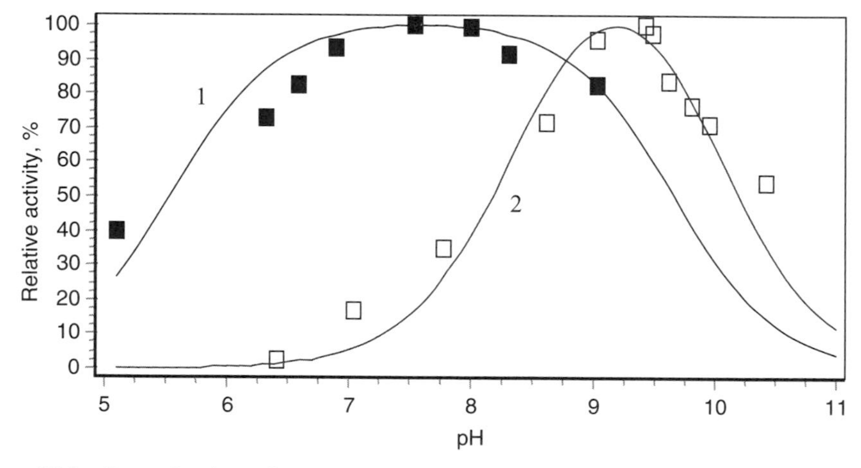


**Figure 23.9** Dependencies of experimentally measured (symbols; Loper et al., 1965) and calculated from model (solid lines) histidinoldehydrogenase (empty squares and line 2) and histidinaldehydrogenase (filled squares and line 1) activities of histidinol dehydrogenase on pH.

**TABLE 23.4 Values of Kinetic Parameters of Histidinol Dehydrogenase**

| Parameter | Values Resulted from Model Identification Against Experimental Data Published in Loper et al. (1965) and Adams (1955) | Parameter | Values Resulted from Model Identification Against Experimental Data Published in Loper et al. (1965) and Adams (1955) |
|---|---|---|---|
| $k_1$ | $1073\ \text{min}^{-1}$ | $K_{NAD}$ | 47.2 mM |
| $k_{-1}$ | $645\ \text{min}^{-1}$ | $K_{Hol}$ | 0.208 mM |
| $k_2$ | $1073\ \text{min}^{-1}$ | $K_{NAD,Hol}$ | $6.7 \cdot 10^{-3}$ mM |
| $K_0^1$ | $3.8 \cdot 10^{-7}$ mM | $K_{Hal}$ | 1.8 mM |
| $K_0^2$ | $8.5 \cdot 10^{-7}$ mM | $K_{Hal,NAD}$ | 0.21 mM |
| $K_{Hol}^1$ | $10^{-7}$ mM | $K_{NADH}$ | 0.63 mM |
| $K_{Hol}^2$ | $4.2 \cdot 10^{-6}$ mM | $K_{His}$ | 0.2 mM |
| $K_{Hal,His}^1$ | $2.2 \cdot 10^{-7}$ mM | $K_{Hal,NADH}$ | $2.8 \cdot 10^{-4}$ mM |
| $K_{Hal,His}^2$ | $2.8 \cdot 10^{-3}$ mM | $K_{NADH,His}$ | 2.1 mM |

to this experiment can be presented as a kinetic scheme depicted in Figure 23.10 and consists of the following set of differential and algebraic equations:

$$\begin{aligned} dNADH/dt &= v_{Hol} + v_{Hal} \\ NAD + NADH &= 500\,\mu M \\ Hol + Hal + His &= 20\,\mu M \\ NADH + 2 \cdot Hol + Hal &= 40\,\mu M \end{aligned} \qquad (23.46)$$

Initial values of model variables correspond to initial concentrations of substrates in experiment:

$$\begin{aligned} Hol &= 20\,\mu M, \\ NAD &= 500\,\mu M \\ NADH &= Hal = His = 0 \end{aligned} \qquad (23.47)$$

Seven unknown parameters of rate equations of histidinodehydrogenase and histidinadehydrogenase reactions have been chosen in such a way that the time dependence of NADH accumulation resulted from a numerical solution of system (23.46), (23.47) (solid line in Fig. 23.11) has coincided with a corresponding time dependence measured experimentally (symbols in Fig. 23.11) in Adams (1955). The values of the estimated parameters are listed in Table 23.4.

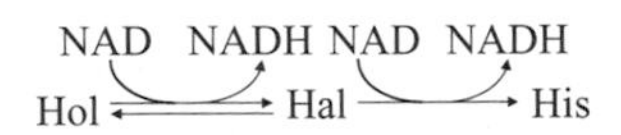


**Figure 23.10** Kinetic scheme of experiment described in Adams (1955).

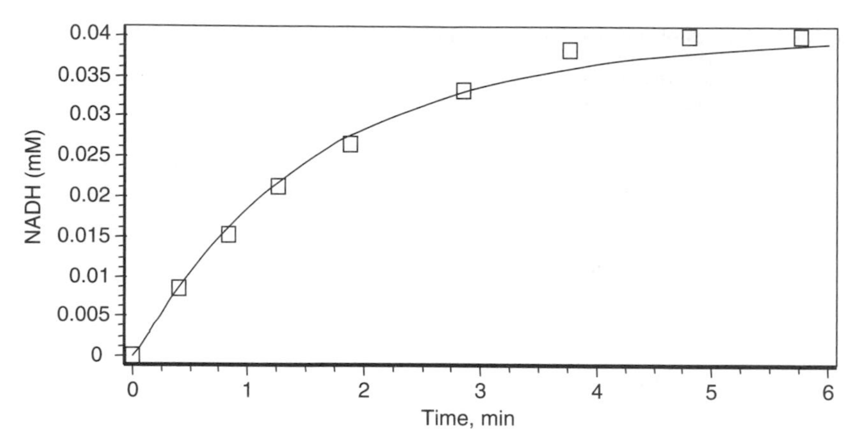


**Figure 23.11** Esperimentally measured (empty squares; Adams, 1955) and calculated from model (solid line) time dependence of NADH accumulation.

## 23.3 KINETIC MODEL OF PTS AND GLYCOLYSIS IN *E. COLI*. EXPRESSION LEVEL OF GLUCOKINASE CONTROLS THE *E. COLI* METABOLISM

The phosphoenolpyruvate-dependent phosphotransferase system (PTS) is one of the main mechanisms responsible for substrate transport in *E. coli* (Postma et al., 1996). Glucose PTS consists of four proteins localized in both the cytoplasm and the inner bacterial membrane (Postma et al., 1996). PTS is responsible for the transport and phosphorylation of glucose. Glucose PTS is tightly coupled with glycolysis, and therefore, the PTS glycolysis in *E. coli* is a cyclic metabolic pathway. Using PEP to phosphorylate glucose, PTS couples upstream glycolytic stages (activation of glucose via its phosphorylation) with downstream reactions of substrate phosphorylation. Indeed, both glucokinase and PTS can catalyze phosphorylation of glucose to form G6P. Similarly, both pyruvate kinase and PTS can dephosphorylate PEP. One can question what are the implications of this redundancy and how the cell can regulate its metabolism under different external conditions by using this particular pathway organization.

### 23.3.1 The Kinetic Model

As a preliminary step to study the regulatory properties of carbohydrate catabolic mechanisms and to understand possible control schemes, we have developed kinetic models of the following metabolic pathways:

1. Glucose PTS.
2. Glucokinase.
3. Pyruvate kinase.
4. Reaction of the production of two molecules of PEP and one molecule of ATP from one molecule of glucose-6-phosphate (G6P) and one molecule of ADP. This reaction mimics the part of glycolysis from G6P isomerization up to PEP formation.
5. ATP consumption process.

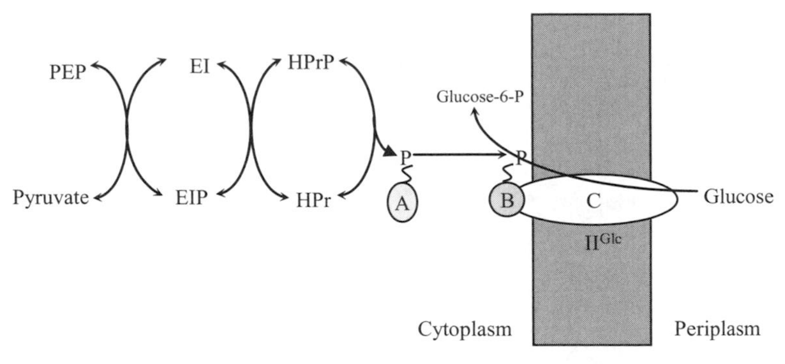


**Figure 23.12** Structural and functional organization of glucose PTS of *Escherichia coli*.

Glucose PTS consists of the following proteins:

Cytoplasmic protein Enzyme I (EI)
Cytoplasmic protein HPr
Cytoplasmic protein Enzyme $IIA^{glc}$ ($EIIA^{glc}$)
Protein of inner bacterial membrane Enzyme $IICB^{glc}$ ($EIICB^{glc}$)

The transfer (see Fig. 23.12) of the phosphate group from PEP to glucose is coupled with the transport of glucose molecules inside the cell. First, EI is auto-phosphorylated by PEP as a phosphate donor; then phosphorylated Enzyme I (EIP) donates a phosphate group to HPr and forms HPrP, which in turn phosphorylates $EIIA^{glc}$. Phosphorylated Enzyme $IIA^{glc}$, $EIIA^{glc}P$, associates with the membrane protein $EIICB^{glc}$ and phosphorylates it. Finally, complex $EIIA^{glc}$ and $EIICB^{glc}P$ catalyze the transfer and phosphorylation of glucose [9].

*Enzyme I.* EI catalyzes the transfer of a phosphate group from PEP to HPr. The catalytic cycle, depicted in Fig. 23.2 reflects the peculiarities of the way Enzyme I functions. Monomers of Enzyme I can associate to form a dimer, $EI_2$ (reaction 1), which is the catalytically active form. The level of dimerization strongly depends on $Mg^{2+}$ concentrations and, as was shown in (Postma et al., 1996), approaches 90% at physiological concentrations of $Mg^{2+}$. Therefore, we assume that all Enzyme I proteins are present in the dimeric form. Upon dimer formation, each monomer is phosphorylated on histidine 189 (reactions 2–7). Only doubly phosphorylated dimer is able to transfer one phosphate group to protein HPr [9]. Mechanism of HPr phosphorylation (see Fig. 23.13) is determined by reactions 8–10. Dynamics of the Enzyme I catalytic cycle is described by the following system of ordinary differential equations:

$$\begin{aligned}
\mathrm{d}\,EI_2/\mathrm{dt} &= -\mathrm{v}_2 \\
\mathrm{d}\,EI_2 \circ PEP/\mathrm{dt} &= \mathrm{v}_2 - \mathrm{v}_3 \\
\mathrm{d}\,EI_2P \circ Pyr/\mathrm{dt} &= \mathrm{v}_3 - \mathrm{v}_4 \\
\mathrm{d}\,EI_2P/\mathrm{dt} &= \mathrm{v}_4 + \mathrm{v}_{10} - \mathrm{v}_5 \\
\mathrm{d}\,EI_2P \circ PEP/\mathrm{dt} &= \mathrm{v}_5 - \mathrm{v}_6 \\
\mathrm{d}\,EI_2P_2 \circ Pyr/\mathrm{dt} &= \mathrm{v}_6 - \mathrm{v}_7 \\
\mathrm{d}\,EI_2P_2/\mathrm{dt} &= \mathrm{v}_7 - \mathrm{v}_8 \\
\mathrm{d}\,EI_2P_2 \circ H\,\mathrm{Pr}/\mathrm{dt} &= \mathrm{v}_8 - \mathrm{v}_9 \\
\mathrm{d}\,EI_2P \circ H\,\mathrm{Pr}\,P/\mathrm{dt} &= \mathrm{v}_9 - \mathrm{v}_{10}
\end{aligned} \tag{23.48}$$

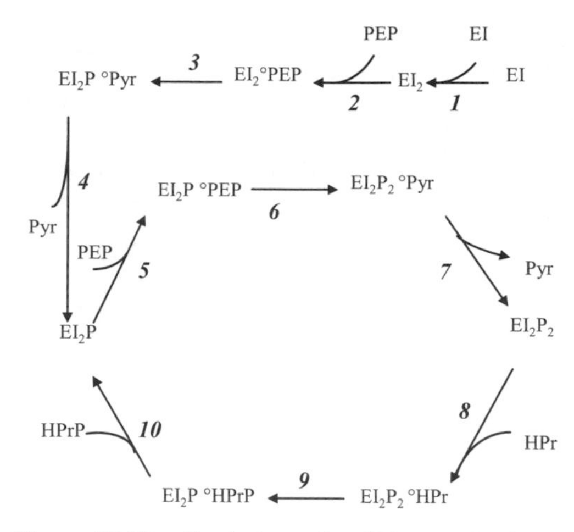


**Figure 23.13** Catalytic cycle of Enzyme I of PTS.

where

$$
\begin{aligned}
v_2 &= a_2 \cdot (EI_2 \cdot \mathrm{PEP}/K_{d,0}^{PEP} - EI_2 \circ PEP) \\
v_3 &= a_3 \cdot (EI_2 \circ PEP - EI_2P \circ Pyr/K_{t,0}) \\
v_4 &= a_4 \cdot (EI_2P \circ Pyr - EI_2P \cdot Pyr/K_{d,1}^{Pyr}) \\
v_5 &= a_5 \cdot (EI_2P \cdot PEP/K_{d,1}^{PEP} - EI_2P \circ PEP) \\
v_6 &= k_1 \cdot EI_2P \circ PEP - k_{-1} \cdot EI_2P_2 \circ Pyr \\
v_7 &= a_7 \cdot (EI_2P_2 \circ Pyr - EI_2P_2 \cdot \mathrm{Pyr}/K_{d,2}^{Pyr}) \\
v_8 &= a_8 \cdot (EI_2P_2 \cdot \mathrm{HPr}/K_d^{H\,\mathrm{Pr}} - EI_2P_2 \circ H\ \mathrm{Pr} \\
v_9 &= k_2 \cdot EI_2P_2 \circ H\,\mathrm{Pr} - k_{-2} \cdot EI_2P \circ H\ \mathrm{Pr}\ P \\
v_{10} &= a_{10} \cdot (EI_2P \circ H\ \mathrm{Pr}\ P - EI_2P \cdot \mathrm{HPrP}/K_d^{H\,\mathrm{Pr}\,P})
\end{aligned}
\tag{23.49}
$$

Here, $EI_2P \circ Pyr$, $EI_2 \circ PEP$, $EI_2$, $EI_2P \circ PEP$, $EI_2P_2 \circ Pyr$, $EI_2P$, $EI_2P_2$, $EI_2P_2 \circ H$ Pr, and $EI_2P \circ H$ Pr $P$ are the concentrations of Enzyme I's different states. $K_{d,0}^{PEP}$, $K_{d,1}^{Pyr}$, $K_{t,0}$, $K_{d,1}^{PEP}$, $K_d^{H\,\mathrm{Pr}\,P}$, $K_{d,2}^{Pyr}$, and $K_d^{H\,\mathrm{Pr}}$ are the corresponding dissociation constants; $k_1$, $k_{-1}$, $k_2$, $k_{-2}$, and $a_i$, i = 2,3,4,5,7,8,10 are rate constants.

We assume that the stages of the catalytic cycle corresponding to the association of substrates (PEP, HPr; reactions 2, 5, and 8) and dissociation of products (Pyr, HPrP; reactions 4, 7, and 10) are much faster than the processes of the intra-molecular transfer of phosphate (reactions 6 and 9). This means that any of $a_i$, i = 2,3,4,5,7,8,10, is much larger than any of $k_1$, $k_{-1}$, $k_2$ $k_{-2}$. From this assumption, it follows that stages 2, 3, 4, 5, 7, 8, and 10 are at quasi-equilibrium; i.e., the following relationships hold:

$$
\begin{aligned}
K_{d,0}^{PEP} &= \frac{PEP \cdot EI_2}{EI_2 \circ PEP} \\
K_{t,0} &= \frac{EI_2P \circ Pyr}{EI_2 \circ PEP}
\end{aligned}
$$

$$
\begin{aligned}
K_{d,1}^{Pyr} &= \frac{Pyr \cdot EI_2P}{EI_2P \circ Pyr} \\
K_{d,1}^{PEP} &= \frac{PEP \cdot EI_2P}{EI_2P \circ PEP} \\
K_d^{H\,\mathrm{Pr}\,P} &= \frac{H\ \mathrm{Pr}\ P \cdot EI_2P}{EI_2P \circ H\ \mathrm{Pr}\ P} \\
K_{d,2}^{Pyr} &= \frac{Pyr \cdot EI_2P_2}{EI_2P_2 \circ Pyr} \\
K_d^{H\,\mathrm{Pr}} &= \frac{H\ \mathrm{Pr} \cdot EI_2P_2}{EI_2P_2 \circ H\ \mathrm{Pr}}
\end{aligned}
\tag{23.50}
$$

and the system of differential equations (23.48) is reduced to the following two-component system:

$$
\begin{aligned}
\mathrm{dX/dt} &= \mathrm{v}_9 - \mathrm{v}_6 \\
\mathrm{dY/dt} &= \mathrm{v}_6 - \mathrm{v}_9
\end{aligned}
\tag{23.51}
$$

where X and Y stands for the sums of Enzyme I states:

$$
\begin{aligned}
X &= EI_2 + EI_2 \circ PEP + EI_2 \circ Pyr + EI_2P + EI_2P \circ PEP + EI_2P \circ H\ \mathrm{Pr}\ P \\
Y &= EI_2P_2 \circ Pyr + EI_2P_2 + EI_2P_2 \circ H\ \mathrm{Pr}
\end{aligned}
\tag{23.52}
$$

Using Eqs. (50) and (52), rate equations $\mathrm{v}_6$ and $\mathrm{v}_9$ can be rewritten as follows:

$$
\begin{aligned}
\mathrm{v}_6 &= \rho_1 \cdot \mathrm{PEP} - \rho_{-1} \cdot \mathrm{Pyr} \\
\mathrm{v}_9 &= \rho_2 \cdot \mathrm{HPr} - \rho_{-2} \cdot \mathrm{HPrP}
\end{aligned}
$$

where

$$
\rho_1 = \frac{K_{t,0} \cdot \dfrac{PEP}{K_{d,1}^{PEP}} \cdot \dfrac{PEP}{K_{d,0}^{PEP}}}{\dfrac{Pyr}{K_{d,1}^{Pyr}} + \dfrac{PEP}{K_{d,0}^{PEP}} \cdot \dfrac{Pyr}{K_{d,1}^{Pyr}} + K_{t,0} \cdot \dfrac{PEP}{K_{d,0}^{PEP}} \cdot \dfrac{Pyr}{K_{d,1}^{Pyr}} + K_{t,0} \dfrac{PEP}{K_{d,0}^{PEP}} + K_{t,0} \dfrac{PEP}{K_{d,1}^{PEP}} \cdot \dfrac{PEP}{K_{d,0}^{PEP}} + K_{t,0} \dfrac{H\,\mathrm{Pr}\,P}{K_d^{H\,\mathrm{Pr}\,P}} \cdot \dfrac{PEP}{K_{d,0}^{PEP}}}
$$

$$
\rho_{-1} = \frac{\dfrac{Pyr}{K_{d,2}^{Pyr}}}{1 + \dfrac{Pyr}{K_{d,2}^{Pyr}} + \dfrac{H\,\mathrm{Pr}}{K_d^{H\,\mathrm{Pr}}}}, \quad \rho_2 = \frac{\dfrac{H\,\mathrm{Pr}}{K_d^{H\,\mathrm{Pr}}}}{1 + \dfrac{Pyr}{K_{d,2}^{Pyr}} + \dfrac{H\,\mathrm{Pr}}{K_d^{H\,\mathrm{Pr}}}}
$$

$$
\rho_{-2} = \frac{K_{t,0} \cdot \dfrac{H\,\mathrm{Pr}\,P}{K_d^{H\,\mathrm{Pr}\,P}} \cdot \dfrac{PEP}{K_{d,0}^{PEP}}}{\dfrac{Pyr}{K_{d,1}^{Pyr}} + \dfrac{PEP}{K_{d,0}^{PEP}} \cdot \dfrac{Pyr}{K_{d,1}^{Pyr}} + K_{t,0} \cdot \dfrac{PEP}{K_{d,0}^{PEP}} \cdot \dfrac{Pyr}{K_{d,1}^{Pyr}} + K_{t,0} \dfrac{PEP}{K_{d,0}^{PEP}} + K_{t,0} \dfrac{PEP}{K_{d,1}^{PEP}} \cdot \dfrac{PEP}{K_{d,0}^{PEP}} + K_{t,0} \dfrac{H\,\mathrm{Pr}\,P}{K_d^{H\,\mathrm{Pr}\,P}} \cdot \dfrac{PEP}{K_{d,0}^{PEP}}}
$$

Equations (23.51) describe the dynamics of the "reduced" two-component catalytic cycle of Enzyme I, which is depicted in Figure 23.14.

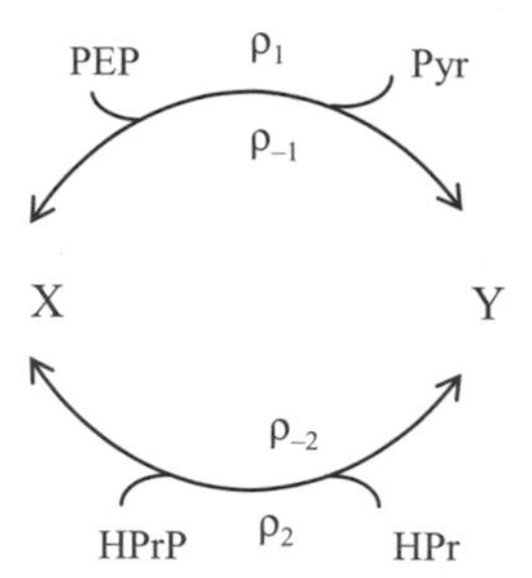


**Figure 23.14** "Reduced" catalytic cycle of Enzyme I of PTS. $\rho_1$, $\rho_{-1}$, $\rho_2$, and $\rho_{-2}$ designate apparent rate constants.

Applying the "reduced" catalytic cycle quasi-steady-state approach, we obtain the following equation:

$$v = \frac{EI_{tot} \cdot PEP \cdot \dfrac{K_{t,0} \cdot k_1 \cdot k_2}{K_{d,0}^{PEP} \cdot K_{d,1}^{PEP} \cdot K_d^{H\,\mathrm{Pr}}} \cdot \left[ PEP \cdot H\,\mathrm{Pr} - \dfrac{Pyr \cdot H\,\mathrm{Pr}\,P \cdot K_{d,1}^{PEP} \cdot K_d^{H\,\mathrm{Pr}} \cdot k_{-1} \cdot k_{-2}}{K_{d,2}^{Pyr} \cdot K_d^{H\,\mathrm{Pr}\,P} k_1 \cdot k_2} \right]}{\left\{ \begin{array}{l} K_{t,0} \cdot \dfrac{PEP}{K_{d,0}^{PEP}} \cdot \left( k_1 \cdot \dfrac{PEP}{K_{d,1}^{PEP}} + k_{-2} \cdot \dfrac{H\,\mathrm{Pr}\,P}{K_d^{H\,\mathrm{Pr}\,P}} \right) \cdot \left( 1 + \dfrac{Pyr}{K_{d,2}^{Pyr}} + \dfrac{H\,\mathrm{Pr}}{K_d^{H\,\mathrm{Pr}}} \right) + \\ \left( k_{-1} \dfrac{Pyr}{K_{d,2}^{Pyr}} + \dfrac{H\,\mathrm{Pr}}{K_d^{H\,\mathrm{Pr}}} \cdot k_2 \right) \cdot \left( \begin{array}{l} \dfrac{Pyr}{K_{d,1}^{Pyr}} + \dfrac{PEP}{K_{d,0}^{PEP}} \cdot \dfrac{Pyr}{K_{d,1}^{Pyr}} + K_{t,0} \cdot \dfrac{PEP}{K_{d,0}^{PEP}} \cdot \dfrac{Pyr}{K_{d,1}^{Pyr}} + \\ K_{t,0} \cdot \dfrac{PEP}{K_{d,0}^{PEP}} + K_{t,0} \cdot \dfrac{PEP}{K_{d,1}^{PEP}} \cdot \dfrac{PEP}{K_{d,0}^{PEP}} + K_{t,0} \cdot \dfrac{H\,\mathrm{Pr}\,P}{K_d^{H\,\mathrm{Pr}\,P}} \cdot \dfrac{PEP}{K_{d,0}^{PEP}} \end{array} \right) \end{array} \right\}} \tag{23.53}$$

Equation (23.8) contains 12 parameter. To estimate these parameters, we have used experimentally measured values of the total enzyme concentration (Rohwer et al., 2000), $EI_{tot}$, the equilibrium constant of Enzyme I (Postma et al., 1996), $K_{eq}^{EI}$, the rate constants of the overall process (Hoving et al., 1981), $k_b$ and $k_f$, $K_m$'s for PEP and HPr (Waygood et al., 1980), as listed in Table 23.5.

**TABLE 23.5 Kinetic Parameter Values of the Kinetic Model of Glycolysis**

| Reaction Number | Parameter | Experimentally Measured Value (refs) (mM/min) | Values Estimated from Fitting (refs) (mM/min) | Values of "Free" Parameters (mM/min) |
|---|---|---|---|---|
| 1 | $EI_{tot}$ | 0.03 (Postma et. al., 1993) | | |
| | $k_b$ | 800 (Hoving et. al., 1981) | | |
| | $k_f$ | 30000 (Hoving et. al., 1981) | | |
| | $K_m^{PEP}$ | 0.18 (Waygood and Steeves, 1980) | | |
| | $K_m^{HPr}$ | 0.009 (Waygood and Steeves, 1980) | | |
| | $K_{eq}^{EI}$ | 11 (Postma et al., 1996) | | |
| | $\alpha$ | | | 1.1 |
| | $\beta$ | | | 1 |
| | $\gamma$ | | | 2 |
| | $\varepsilon$ | | | 0.001 |
| | $K_{d,1}^{Pyr}$ | | | 5 |
| | $K_d^{HPrP}$ | | | 0.1 |

**TABLE 23.5** (*Continued*)

| Reaction Number | Parameter | Experimentally Measured Value (refs) (mM/min) | Values Estimated from Fitting (refs) (mM/min) | Values of "Free" Parameters (mm/min) |
|---|---|---|---|---|
| 2 | $k_{HPrP\_A}$ | | 128700 (Meadow and Roseman, 1996) | |
| | $K_d^{HPrP\_A}$ | | 0.000156 (Meadow and Roseman, 1996) | |
| 3 | $k_{HPrP\_transfer}$ | | 75600 (Meadow and Roseman, 1996) | |
| | $K_{eq}^{HPrP\_transfer}$ | | 0.84 (Meadow and Roseman, 1996) | |
| 4 | $k_{HPr\_AP}$ | | 2040 (Meadow and Roseman, 1996) | |
| | $K_d^{HPr\_AP}$ | | 0.000289 (Meadow and Roseman, 1996) | |
| 5–7 | $EIICB^{glc}{}_{tot}$ | | | 1 |
| | $K_d^A$ | 0.02 (Postma et al., 1996) | | |
| | $K_d^{AP}$ | 0.002 (Postma et al., 1996) | | |
| | $K_d^{G6P}$ | 0.001 (Postma et al., 1996) | | |
| | $K_d^{out}$ | 0.002 (Postma et al., 1996) | | |
| | $K_t$ | 1 (Postma et al., 1996) | | |
| | $k_{Glc}$ | | 60000 (Lolkema et al., 1993) | |
| | $k_{-Glc}$ | | 6 (Lolkema et al., 1993) | |
| | $k_{10}$ | | 17000 (Lolkema et al., 1993) | |
| | $k_{-10}$ | | 560 (Lolkema et al., 1993) | |
| | $k_8$ | | 52000 (Lolkema et al., 1993) | |
| | $k_{-8}$ | | 0.1 (Lolkema et al., 1993) | |
| | $k_{14}$ | | 1000000 (Lolkema et al., 1993) | |
| | $k_{11}$ | | 470 (Lolkema et al., 1993) | |
| | $k_{19}$ | | 80 (Lolkema et al., 1993) | |
| | $k_{-19}$ | | 3 (Lolkema et al., 1993) | |
| 8 | $V_{max}^{glk}$ | | | Varied |
| | $K_{eq}^{glk}$ | | | 10 |
| | $K_{m,Glc}^{glk}$ | 0.15 (Meyer et al., 1997) | | |
| 9 | $k_9$ | | | 50 |
| | $K_9^{glycolysis}$ | | | 10 |
| 10 | $k_{10}$ | | | 50 |
| | $K_{10}^{glycolysis}$ | | | 10 |
| 11 | $k_{11}$ | | | 50 |
| Conservation laws | Total $EIIA^{glc}$ | | | 1 |
| | Total HPr | 0.3 (Postma et al., 1996) | | |
| | Total ADP + ATP | 4 (Postma et al., 1996) | | |
| Buffered concentrations | $Glc_{out}$ | | | Varied |
| | Pyr | 0.4 (Postma et al., 1996) | | |

The kinetic parameters of Eq. (23.8) can be expressed in terms of these measured parameters in the following manner:

$$
\begin{aligned}
K_{d,1}^{PEP} &= \alpha \cdot K_{d,0}^{PEP} \\
K_{d,2}^{Pyr} &= \beta \cdot K_{d,1}^{Pyr} \\
k_1 &= \gamma \cdot k_{-1} \cdot K_{t,0}, K_{d,0}^{PEP} = \frac{K_m^{PEP} + \varepsilon}{\alpha} \\
k_{-1} &= k_b \cdot K_{d,1}^{Pyr} \\
k_{-2} &= \frac{\beta \cdot \gamma^2 \cdot k_f^2 \cdot K_d^{H\,\mathrm{Pr}\,P} \cdot K_m^{PEP}}{\alpha^2 \cdot k_b \cdot K_{eq}^{EI} \cdot K_m^{H\,\mathrm{Pr}}} \\
k_2 &= \frac{\gamma \cdot k_f \cdot K_m^{PEP} \cdot K_{d,1}^{Pyr}}{\varepsilon} \\
k_{t,0} &= \frac{k_f \cdot (K_m^{PEP} + \varepsilon)}{\alpha \cdot k_b \cdot K_{d,1}^{Pyr}} \\
K_d^{H\,\mathrm{Pr}} &= \frac{\alpha \cdot K_m^{H\,\mathrm{Pr}} \cdot K_{d,1}^{Pyr}}{\varepsilon}, \varepsilon > 0
\end{aligned} \tag{23.54}
$$

Equation (23.54) allows us to reduce the number of "free" parameters of Eq. (23.53) from 12 to just 6: $\alpha$, $\beta$, $\varepsilon$, $\gamma$, $K_{d,1}^{Pyr}$, $K_d^{H\,PrP}$. The values of these "free" parameters are listed in Table 23.5.

*Enzyme IIA$^{glc}$ and protein HPr.* The doubly phosphorylated dimer of Enzyme I phosphorylates the protein HPr on histidine 15 (Postma et al., 1993). Then, the phosphorylated HPr reacts with EIIA$^{\mathrm{glc}}$ resulting in the transfer of a phosphate group from HPrP to histidine 90 of EIIA$^{\mathrm{glc}}$ (Postma et al., 1993). The phosphorylated Enzyme IIA$^{\mathrm{glc}}$, EIIA$^{\mathrm{glc}}$P, reacts with Enzyme IICB$^{\mathrm{glc}}$. The reaction rates of binding HPrP with EIIA$^{\mathrm{glc}}$ ($\mathrm{v_{HPrP\_A}}$), the phosphate transfer ($\mathrm{v_{HPrP\_transfer}}$), and the dissociation of HPr from EIIA$^{\mathrm{glc}}$P ($\mathrm{v_{HPr_AP}}$) are described by mass action:

$$
\begin{aligned}
\mathrm{v_{HPrP\_A}} &= \mathrm{k_{HPrP\_A}} \cdot (\mathrm{HPrP} \cdot \mathrm{EIIA^{glc}}/K_d^{H\,\mathrm{Pr}\,P\_A} - \mathrm{HPrP\_A}) \\
\mathrm{v_{HPrP\_transfer}} &= \mathrm{k_{HPrP\_transfer}} \cdot (\mathrm{HPrP\_A} - \mathrm{HPr\_AP}/K_{eq}^{H\,\mathrm{Pr}\,P\_transfer}) \\
\mathrm{v_{HPr\_AP}} &= \mathrm{k_{HPr\_AP}} \cdot (\mathrm{HPr\_AP} - \mathrm{HPr} \cdot \mathrm{EIIA^{glc}P}/K_d^{H\,\mathrm{Pr}\,\_AP})
\end{aligned} \tag{23.55}
$$

The values of the constants, $\mathrm{k_{HPrP\_A}}$, $\mathrm{k_{HPrP\_transfer}}$, $\mathrm{k_{HPr\_AP}}$, $K_{eq}^{H\,\mathrm{Pr}\,P\_transfer}$, $K_d^{H\,\mathrm{Pr}\,\_AP}$, $K_d^{H\,\mathrm{Pr}\,P\_A}$, have been fitted (see Table 23.5) to describe the experimental data published in (Meadow et al., 1996). In this experiment, EIIA$^{\mathrm{glc}}$ phosphorylation has been measured upon addition of HPrP (Fig. 23.15, open cycles). To describe the experiment and to estimate the kinetic parameters of Eq. 23.6, we have developed a kinetic model for transfer of phosphate group from HPrP to EIIA$^{\mathrm{glc}}$:

$$
\begin{aligned}
\mathrm{dHPrP/dt} &= -\mathrm{v_{HPrP\_A}} \\
\mathrm{dEIIA^{glc}/dt} &= -\mathrm{v_{HPrP\_A}} \\
\mathrm{dHPrP\_A/dt} &= \mathrm{v_{HPrP\_A}} - \mathrm{v_{HPrP\_transfer}} \\
\mathrm{dHPr\_AP/dt} &= \mathrm{v_{HPrP\_transfer}} - \mathrm{v_{HPr\_AP}}
\end{aligned}
$$

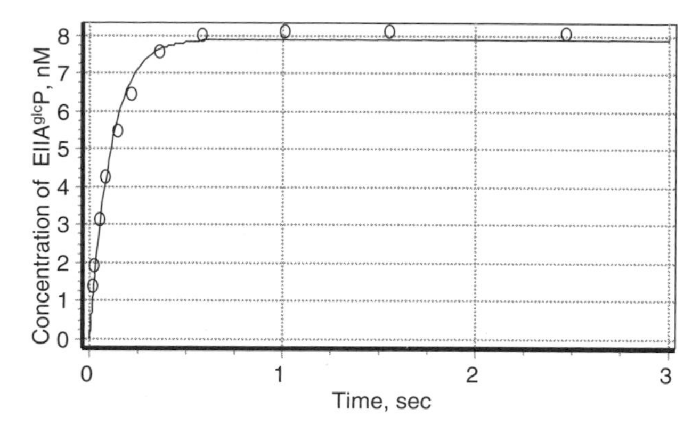


**Figure 23.15** Time dependence of $EIIA^{glc}$ phosphorylation level measured experimentally in [13] (open circles) and resulted from kinetic model (solid line).

$$dHPr/dt = v_{HPr\_AP}$$
$$dEIIA^{glc}P/dt = v_{HPr\_AP}$$
$$HPrP = 9.4\,nM$$
$$EIIA^{glc} = 86\,nM$$
$$HPrP\_A = HPr\_AP = HPr = EIIA^{glc}P = 0$$

The kinetic model corresponds to the kinetic scheme depicted in Figure 23.16. The initial concentrations are taken from Meadow et al. (1996). Figure 23.15 shows the good quality of the correlation between the resulting simulation results (solid line) with the experimental data (open cycles) Meadow et al. (1996).

*Enzyme IICB$^{glc}$*. Two domains, B and C, of the enzyme $IICB^{glc}$ are incorporated into the inner membrane of *E. coli*. The transmembrane domain C has a glucose binding site that can be exposed to either the periplasm or the cytoplasm. Domain B is exposed to the cytoplasm and carries cystein 421 that can be phosphorylated with $EIIA^{glc}P$. Enzyme $IICB^{glc}$ catalyzes the transport and phosphorylation of glucose from the periplasm to the cytoplasm. The $EIICB^{glc}$ catalytic cycle depicted in Figure 23.17 consists of 13 states designated as $x_i$, $i = 1,2,3$; $y_j$, $z_j$, $j = 1, \ldots, 5$. Each state is determined by the substates of the glucose binding site belonging to domain C, substates of cysteine 421 situated on domain B, as well as substates of histidine 90 of $EIIA^{glc}$, which can form a catalytically active complex with $EIICB^{glc}$. The glucose binding site can be exposed either to the periplasm (states $y_5$ and $z_5$) or to the cytoplasm (states $x_i$, $i = 1,2,3$; $y_j$, $z_j$, $j = 1, \ldots, 4$). Cystein 421 can be either phosphorylated (states $y_j$, $z_j$, $j = 4,5$) or non-phosphorylated (states $x_i$, $i = 1,2,3$; $y_j$, $z_j$, $j = 1,2,3$). $EIICB^{glc}$ can be free (states $x_2$, $y_2$, $z_2$) or in the form of complex with $EIIA^{glc}$ (states $x_i$, $i = 1,3$; $y_j$, $z_j$, $j = 1,3,\ldots 5$), which histidine 90, in its turn, can be either phosphorylated (states $x_3$, $y_3$, $z_3$) or non-phosphorylated (states $x_1$, $y_j$, $z_j$, $j = 1,4,5$).

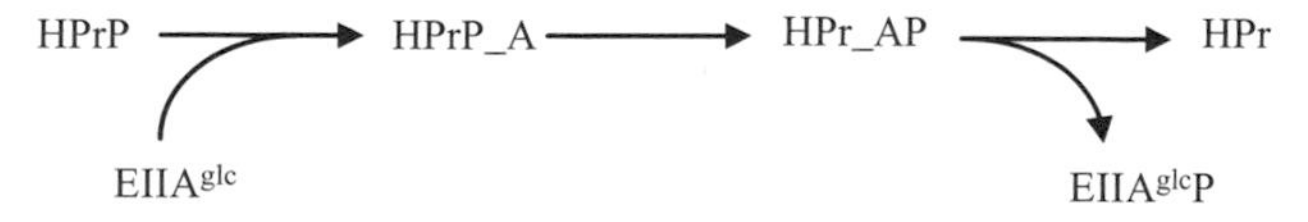


**Figure 23.16** Kinetic scheme of processes of $EIIA^{glc}$ phosphorylation.

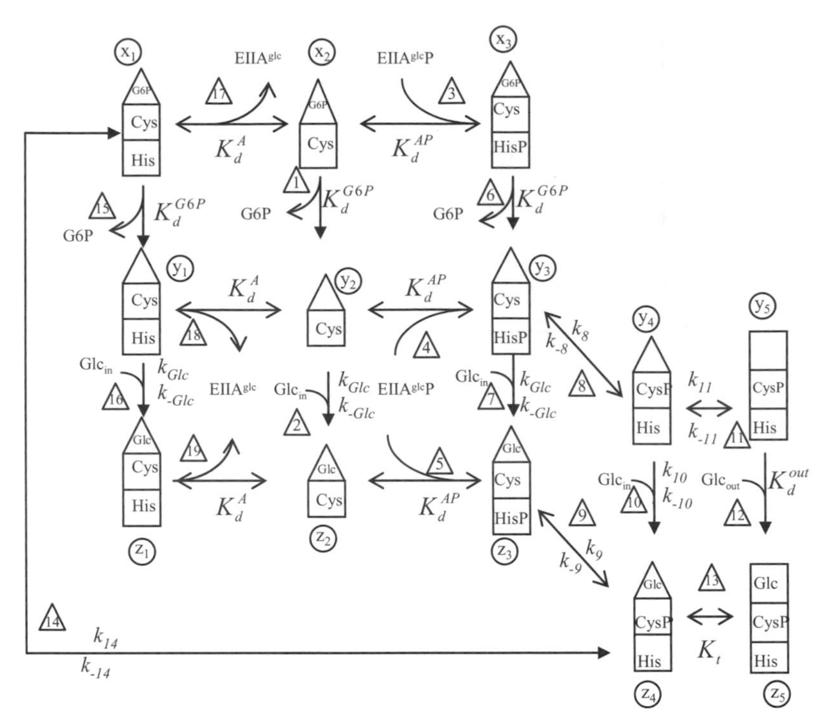


**Figure 23.17** Catalytic cycle of Enzyme IICB$^{glc}$ of PTS. Each state of EIICB$^{glc}$ is presented by two designations: structural and operational. Structural designation shows how the state of EIICB$^{glc}$ is determined by states of Cys 421, glucose, and EIIA binding sites and His 90 of EIIA. Operational designations of EIICB$^{glc}$ state are used in the main text. Each operational designation x[i], i = 1,2,3, y[j], z[j], $j = 1, \ldots, 5$, is shown in empty cycles situated near a corresponding structural designation of a corresponding EIICB$^{glc}$ state. Each reaction of the catalytic cycle is numbered by numbers in empty triangles. Dissociation or rate constants is situated near each corresponding reaction.

It was found (Lolkema et al., 1993) that free Enzyme IICB$^{glc}$ is always in the state where glucose binding site is exposed to cytoplasm (states $x_2$, $y_2$, $z_2$). When this occurs, EIICB$^{glc}$ can bind cytoplasmic glucose, Glc$_{in}$ (reaction 2) or G6P (reaction 1) using the glucose binding site or EIICB$^{glc}$ can form a complex with EIIA$^{glc}$P (reactions 3–5). The complex of EIICB$^{glc}$ with EIIA$^{glc}$P can also bind Glc$_{in}$ (reaction 7 in Fig. 23.17) or G6P (reaction 6) using the glucose binding site. After formation of the complex EIICB$^{glc}$ with EIIA$^{glc}$P, a phosphate group is transferred from histidine 90 of EIIA$^{glc}$ to cysteine 421 of EIICB$^{glc}$ (reactions 8 and 9). As a result of this intramolecular phosphate transfer, the complex becomes catalytically active—it can transport and phosphorylate glucose. If the glucose binding site is empty, then the phosphorylation of cysteine 421 leads to conformational changes in domain C resulting in the glucose binding site being exposed to the periplasm (reaction 11) where periplasmic glucose, Glc$_{out}$, binds (reaction 12). The EIICB$^{glc}$ conformation changes how the glucose binding site (with glucose) is exposed to the cytoplasm (reaction 13). The EIICB$^{glc}$P and EIIA$^{glc}$ complex can be also attained due to binding of the cytoplasmic glucose (reaction 10). Furthermore, a phosphate is transferred from cysteine 421 to the glucose (reaction 14) to form G6P and inactivate the EIICB$^{glc}$. Then the dissociation of G6P and the subsequent binding of Glc$_{in}$ (reactions 15 and 16) or the dissociation of $EIIA^{glc}$ from $EIICB^{glc}$ (reactions 17–19) proceeds in random order resulting in the regeneration of free $EIICB^{glc}$ and the completion of the catalytic cycle.

Enzyme $IICB^{glc}$ can function in three modes:

1. Transport of periplasmic glucose to the cytoplasm with its concomitant phosphorylation to form G6P (transition via y2 → y3 → y4 → y5 → z5 → z4 → x1 → y1 → y2)
2. Phosphorylation of cytoplasmic glucose to form G6P (transition via y2 → y3 → y4 → z4 → x1 → y1 → y2)
3. Transport of periplasmic glucose to cytoplasm without its phosphorylation (transition via y4 → y5 → z5 → z4 → y4)

The contribution of each mode to the overall Enzyme $IICB^{glc}$ reaction rate is determined by the concentrations of G6P, $Glc_{in}$, $Glc_{out}$, PEP, and Pyr.

We assume that the following processes of the catalytic cycle are at quasi-equilibrium:

- $Glc_{out}$ association (reaction 12)
- Dissociation of G6P (reactions 1, 6, and 15),
- Association/dissociation of $EIICB^{glc}$ (reactions 3–5 and 17–19)
- Transition of the glucose binding site from the periplasm to the cytoplasm (reaction 13)

Applying the strategy described for derivation of Enzyme I rate equation, the following relationships have been produced:

$$
\begin{aligned}
K_d^A &= \frac{EIIA^{glc} \cdot x_2}{x_1} \\
K_d^A &= \frac{EIIA^{glc} \cdot y_2}{y_1} \\
K_d^A &= \frac{EIIA^{glc} \cdot z_2}{z_1} \\
K_d^{AP} &= \frac{EIIA^{glc}P \cdot x_2}{x_3} \\
K_d^{AP} &= \frac{EIIA^{glc}P \cdot y_2}{y_3} \\
K_d^{AP} &= \frac{EIIA^{glc}P \cdot z_2}{z_3} \\
K_d^{G6P} &= \frac{G6P \cdot y_1}{x_1} \\
K_d^{out} &= \frac{Glc_{out} \cdot y_5}{z_5} \\
K_t &= \frac{z_4}{z_5}
\end{aligned}
\qquad (23.56)
$$

In Scheme depicted in Figure 23.6, $K_d^A$, $K_d^A$, $K_d^A$, $K_d^{AP}$ $K_d^{AP}$, $K_d^{AP}$, $K_d^{G6P}$, $K_d^{out}$, and $K_t$ stand for corresponding dissociation constants; $k_{Glc}$, $k_{-Glc}$, $k_i$, $k_{-i}$, $i = 8, \ldots 11, 14$ are rate constants. Using Eqs. 23.56, the number of states of the catalytic cycle of Enzyme $IICB^{glc}$ can be reduced to four ("reduced" catalytic cycle depicted in Figure 23.18). Each state of the

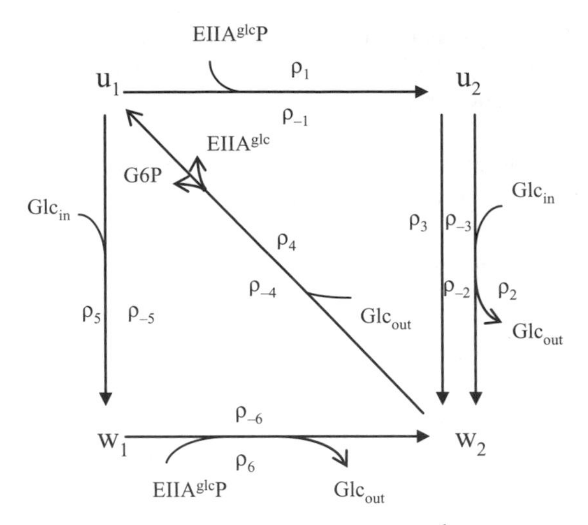


**Figure 23.18** "Reduced" catalytic cycle of Enzyme IICB$^{glc}$ of PTS. $\rho_1$, $\rho_{-1}$, $\rho_2$, and $\rho_{-2}$ designate apparent rate constants.

"reduced" catalytic cycle corresponds to the sum of states of the unreduced catalytic cycle:

$$
\begin{aligned}
u_1 &= x_1 + x_2 + x_3 + y_1 + y_2 + y_3 \\
u_2 &= y_4 \\
w_1 &= z_1 + z_2 + z_3 \\
w_2 &= y_5 + z_4 + z_5
\end{aligned}
$$

As shown in Figure 23.18, four states of the "reduced" catalytic cycle are connected with six reactions with rate constants given by following expressions:

$$\rho_1 = \frac{k_8 \cdot K_d^{G6P} \cdot K_d^A}{\left(K_d^{G6P} + G6P\right) \cdot \left\{K_d^A \cdot \left(EAP + K_d^{AP}\right) + EA \cdot K_d^{AP}\right\}}$$

$$\rho_{-1} = k_{-8} \qquad \rho_2 = k_{10}$$

$$\rho_{-2} = \frac{k_{-10}}{(1 + K_t) \cdot Glc_{out} + K_t \cdot K_d^{out}}$$

$$\rho_3 = k_{11}$$

$$\rho_{-3} = \frac{k_{-11} \cdot K_t \cdot K_d^{out}}{K_t \cdot K_d^{out} + (1 + K_t) \cdot Glc_{out}}$$

$$\rho_4 = \frac{k_{14}}{Glc_{out}(1 + K_t) + K_t \cdot K_d^{out}}, \qquad \rho_{-4} = \frac{k_{-14} \cdot K_d^{AP}}{(K_d^{G6P} + G6P) \cdot \{K_d^A \cdot (EAP + K_d^{AP}) + EA \cdot K_d^{AP}\}}$$

$$\rho_5 = \frac{k_{Glc} \cdot K_d^{G6P}}{K_d^{G6P} + G6P}$$

$$\rho_{-5} = k_{-Glc}$$

$$\rho_6 = \frac{k_9 \cdot K_d^A}{K_d^A \cdot (K_d^{AP} + EAP) + EA \cdot K_d^{AP}}$$

$$\rho_{-6} = \frac{k_{-9}}{K_t \cdot K_d^{out} + (1 + K_t) \cdot Glc_{out}}$$

As mentioned, $EIICB^{glc}$ can function in three different modes. In terms of states of the "reduced" catalytic cycle (see Fig. 23.18), these three modes can correspond to the following transitions through states $u_1$, $u_2$, $w_1$, $w_2$:

1. $u_1 \rightarrow u_2 \rightarrow w_2 \rightarrow u_1$: transport of periplasmic glucose to cytoplasm with its concomitant phosphorylation to form G6P ($\nu_{G6P}^{out}$)
2. $u_1 \rightarrow w_1 \rightarrow w_2 \rightarrow u_1$: phosphorylation of cytoplasmic glucose to form G6P ($\nu_{G6P}^{in}$)
3. $u_2 \rightarrow w_2 \rightarrow u_2$: transport of periplasmic glucose to cytoplasm without its phosphorylation ($\nu_{Glc}^{influx}$)

The rate equations derived from the reduced catalytic cycle are as follows:

$$v_{Glc_{out}} = \frac{EIICB_{tot}\left\{\begin{array}{l}(\rho_{-5}+\rho_6\cdot EIIA^{glc}P)\cdot\rho_1\cdot\rho_4\left[\begin{array}{r}\rho_3\cdot\left(EIIA^{glc}P\cdot Glc_{out}-\dfrac{EA\cdot G6P}{K_{eq}^{EIICB}}\right)\\ +\rho_2\cdot Glc_{out}\cdot\left(EIIA^{glc}P\cdot Glc_{in}-\dfrac{EIIA^{glc}\cdot G6P}{K_{eq}^{EIICB}}\right)\end{array}\right]\\ +\rho_1\cdot\rho_3\cdot\rho_{-5}\cdot\rho_{-6}\cdot EIIA^{glc}P\cdot(Glc_{out}-Glc_{in})\end{array}\right\}}{(G6P)^2\cdot\left\{\begin{array}{l}K_d^A\cdot\left[EIIA^{glc}P\cdot(K_d^{G6P}+G6P)+K_d^{AP}\cdot(K_d^{G6P}+G6P)\right]\\ +EIIA^{glc}\cdot K_d^{AP}(K_d^{G6P}+G6P)\end{array}\right\}}$$

$$v_{Glc_{in}} = \frac{EIICB_{tot}\left\{\begin{array}{l}(\rho_{-1}+Glc_{in}\cdot\rho_2+\rho_3)\cdot\rho_4\cdot Glc_{out}\cdot\left(Glc_{in}\cdot EIIA^{glc}P-\frac{EIIA^{glc}\cdot G6P}{K_{eq}^{EIICB}}\right)\\ +\rho_{-1}\cdot\rho_{-3}\cdot EIIA^{glc}P\cdot(Glc_{in}-Glc_{out})\end{array}\right\}}{(G6P)^2\cdot\left\{\begin{array}{l}K_d^A\cdot\left[EIIA^{glc}P\cdot(K_d^{G6P}+G6P)+K_d^{AP}\cdot(K_d^{G6P}+G6P)\right]\\ +EIIA^{glc}\cdot K_d^{AP}(K_d^{G6P}+G6P)\end{array}\right\}}$$

$$v_{influx} = \frac{EIICB_{tot}\left\{\begin{array}{l}(\rho_{-5}+\rho_6\cdot EIIA^{glc}P)\cdot\rho_1\cdot\rho_2\cdot\rho_4\cdot Glc_{out}\cdot\left(\dfrac{EIIA^{glc}\cdot G6P}{K_{eq}^{EIICB}}-EIIA^{glc}P\cdot Glc_{in}\right)\\ +[(\rho_{-5}+\rho_6\cdot EIIA^{glc}P)\cdot(\rho_1\cdot EIIA^{glc}P+\rho_{-4}\cdot EIIA^{glc}\cdot G6P)+\rho_5\cdot\rho_6\cdot Glc_{in}\cdot EIIA^{glc}P]\\ \cdot\rho_2\cdot\rho_{-3}\cdot(Glc_{out}-Glc_{in})\end{array}\right\}}{(G6P)^2\cdot\left\{\begin{array}{l}K_d^A\cdot[EIIA^{glc}P\cdot(K_d^{G6P}+G6P)+K_d^{AP}\cdot(K_d^{G6P}+G6P)]\\ +EIIA^{glc}\cdot K_d^{AP}(K_d^{G6P}+G6P)\end{array}\right\}} \tag{23.57}$$

These equations describe the dependencies of reaction rates of Enzyme $IICB^{glc}$ functioning in different modes on substrates and product concentrations. Eqs. 23.57 contain 17 parameters. Some of them have been estimated in Lolkema et al. (1993). Others (listed in Table 23.5) were chosen to provide a good fit of simulation results (solid line, Fig. 23.19) to the experimental data (Lolkema et al., 1993) (open cycles, Fig. 23.19).

*Glycolysis*. All processes for model construction are shown in Figure 23.20. For the purposes of illustration, we have included only three processes to this model of glycolysis. The first process is phosphorylation of $Glc_{in}$ catalyzed by glucokinase (reaction 8 in Fig. 23.20). As $K_m$ for ATP is five-fold greater than $K_m$ for $Glc_{in}$ (Meyer et al., 1997), the rate equation can be written as follows:

$$v_8 = V_{max}^{glk}\cdot(Glc_{out}\cdot ATP - G6P\cdot ADP/K_{eq}^{glk})/(K_{m,Glc}^{glk}+GlC_{out}) \tag{23.58}$$

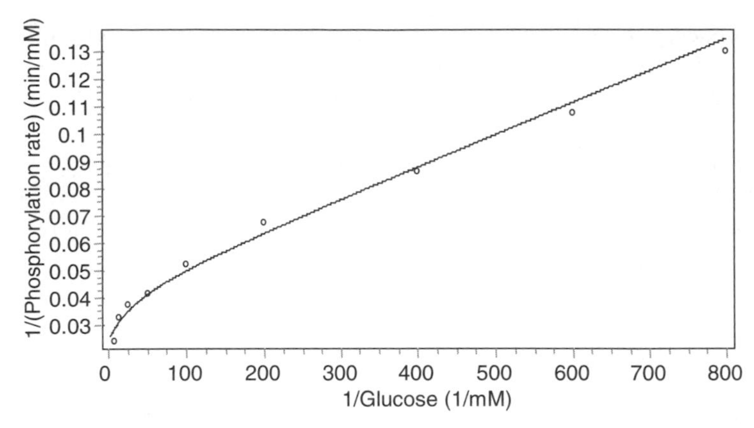


**Figure 23.19** Experimentally measured (open circles, [14]) and generated with kinetic model (solid line) dependence of rate of reaction catalyzed by Enzyme IICB$^{glc}$ on glucose concentration.

where $V_{\max}^{glk}$, $K_{eq}^{glk}$, and $K_{m,Glc}^{glk}$ are the kinetic constants of glucokinase (values presented in Table 23.5).

The second process (reaction 9 in Fig. 23.20) mimics the glycolysis reactions from G6P isomerization up to PEP formation. This reaction describes the production of two molecules of PEP and one ATP from one molecule of glucose-6-phosphate (G6P) and one molecule of ADP. A reaction rate can be written from the corresponding mass action law:

$$v_9 = k_9 \cdot (G6P \cdot ADP - ATP \cdot PEP^2 / K_9^{glycolysis}) \tag{23.59}$$

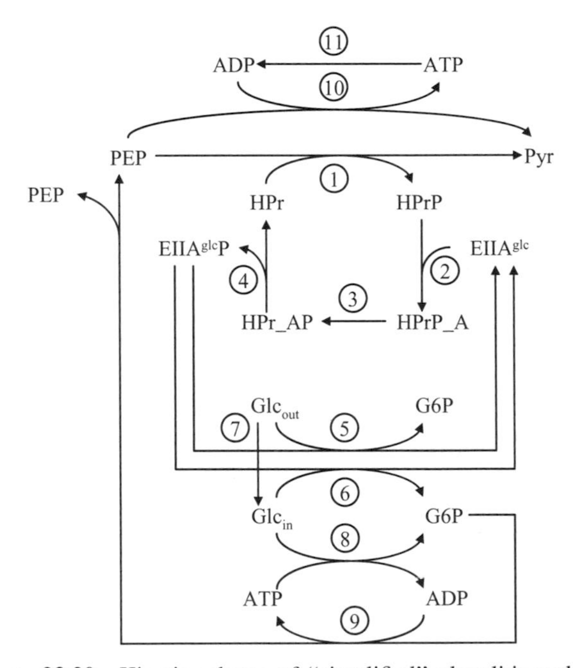


**Figure 23.20** Kinetic scheme of "simplified" glycolitic pathway.

The third process is the reaction of PEP dephosphorylation catalyzed by pyruvate kinase (reaction 10). The reaction rate is derived from mass action:

$$\mathrm{v}_{10} = k_{10} \cdot (\mathrm{PEP} \cdot \mathrm{ADP} - \mathrm{ATP} \cdot \mathrm{Pyr}/K_{10}^{glycolysis}) \tag{23.60}$$

We also consider the process of ATP consumption (reaction 11), which is described according to the mass action law:

$$\mathrm{v}_{11} = k_{11} \cdot \mathrm{ATP} \tag{23.61}$$

The values of the kinetic parameters in Eqs. (23.59)–(23.61) are listed in Table 23.5. The kinetic model of PTS and glycolysis shown in Figure 23.20 is described by:

$$\begin{aligned}
\mathrm{dHPrP/dt} &= \mathrm{v}_1 - \mathrm{v}_2 \\
\mathrm{dEIIA^{glc}/dt} &= \mathrm{v}_5 + \mathrm{v}_6 - \mathrm{v}_2 \\
\mathrm{dHPrP_A/dt} &= \mathrm{v}_2 - \mathrm{v}_3 \\
\mathrm{dHPr_AP/dt} &= \mathrm{v}_3 - \mathrm{v}_4 \\
\mathrm{dHPr/dt} &= \mathrm{v}_4 - \mathrm{v}_1 \\
\mathrm{dEIIA^{glc}P/dt} &= \mathrm{v}_4 - \mathrm{v}_5 - \mathrm{v}_6 \\
\mathrm{dGlc_{in}/dt} &= \mathrm{v}_7 - \mathrm{v}_6 - \mathrm{v}_8 \\
\mathrm{dG6P/dt} &= \mathrm{v}_6 + \mathrm{v}_8 - \mathrm{v}_9 \\
\mathrm{dPEP/dt} &= 2 \cdot \mathrm{v}_9 - \mathrm{v}_1 - \mathrm{v}_{10} \\
\mathrm{dADP/dt} &= v_8 + v_{11} - v_9 - v_{10} \\
\mathrm{dATP/dt} &= v_9 + v_{10} - v_8 - v_{11}
\end{aligned} \tag{23.62}$$

Here, $\mathrm{Glc_{out}}$ and Pyr are buffered i.e., their concentrations do not change with time and $\mathrm{v_i}$, $\mathrm{i} = 1, \ldots 7$, are as follows:

$$\begin{aligned}
\mathrm{v}_1 &= \mathrm{v_{EI}} \\
\mathrm{v}_2 &= \mathrm{v_{HPrP_A}} \\
\mathrm{v}_3 &= \mathrm{v_{HPrP_transfer}} \\
\mathrm{v}_4 &= \mathrm{v_{HPr_AP}} \\
\mathrm{v}_5 &= \mathrm{v_{Glcout}} \\
\mathrm{v}_6 &= \mathrm{v_{Glcin}} \\
\mathrm{v}_7 &= \mathrm{v_{influx}}
\end{aligned}$$

### 23.3.2 Determining Quasi-Steady States in Reduced Kinetic Models

We use the following method to determine the number of quasi-steady states in the kinetic model. Consider a system of differential equations corresponding to a kinetic model and assume that one of the intermediates $X_i$ is constant. The assumption eliminates the specific differential equation for $X_i$. Yet, it does not change the functional dependencies of the other rate equations on $X_i$. There are still reactions producing and consuming $X_i$, because $X_i$ is a concentration of an intermediate. Because $X_i$ is an intermediate and assumed to be in steady state (hence, the $\mathrm{dxi/dt} = 0$ constancy assumption), we do not expect the dynamics to change much.

From Eq. (23.1), we designate the production and consumption fluxes for $X_i$ as $v_{X_i}^{prod}$ and $v_{X_i}^{cons}$. Consider a *reduced* system where $X_i$ is a constant so that [from Eq. (23.1)], we obtain two functional relationships based on $X_i$:

$$v_{X_i}^{prod}(X_i) = v_{X_i}^{cons}(X_i)$$

This effective quasi-steady state can only depend on that set of the $X_i$ values that the roots of this nonlinear algebraic equation. Clearly there can be one root, many roots, or no bounded positive solutions. Where possible we strive for analytic solutions, but if that is not feasible, a numerical root finder can always be used. The resulting quasi-steady states can be further characterized by their stability characteristics as being either stable or unstable. Such characterization is important to understand, which are biologically meaningful and are artifacts of the model. Usually, we are doing stability analysis numerically calculating eigenvalues and other relative parameters.

### 23.3.3 The Number of Quasi-Steady States Depends on the Expression Level of Glucokinase

We applied the previously described method for finding quasi-steady states to the kinetic model of PTS and glycolysis in *E. coli* (Fig. 23.20) with PEP as the intermediate held constant in the *reduced* system. We consider quasi-steady-state dependencies of the rate of PEP consumption and production as a function of PEP.

$$v_{PEP}^{prod}(PEP) = 2v_9(PEP)$$
$$v_{PEP}^{cons}(PEP) = v_1(PEP) + v_{10}(PEP)$$

Varying PEP over a reasonable range from 0 to 0.2 mM we have found the quasi-steady states at different levels of glucokinase and periplasmic glucose, $Glc_{out}$. Figure 23.21a shows the dependencies of $v_{PEP}^{prod}$ and ${v_{cons}}^{PEP}$ on PEP and the three quasi-steady states that result for the original system (Fig. 23.20).

We found that two of these quasi-steady states are stable (indicated as "stable 0" and "stable 1" in Fig. 23.21) and one is unstable. Quasi-steady-state "stable 1" is a physiological quasi-steady state; i.e., the values of fluxes and concentrations at this quasi-steady state are comparable with those measured experimentally. The "stable 0" state is not physiologically meaningful because it is just the trivial null point of all zero fluxes and represents a "dead" cell. The unstable quasi-steady state cannot correspond to any meaningful biological state because any fluctuation, however small, will result in a transition to one of the other stable quasi-steady states. The expression level of glucokinase given by $V_{max}^{glk}$ value determines the number of quasi-steady states and their stability with a low glucose concentration ($Glc_{out} = 50\,\mu M$). When glucokinase expression is low ($V_{max}^{glk}$ is equal to 0.1 mM/min), then our kinetic model has three quasi-steady states (see Fig. 23.21a), only one of which is a nontrivial and stable. Increasing the glucokinase expression level with low glucose results in a decrease of the distance between the physiological and unstable steady states and, finally, to a transition to the quasi-steady state corresponding to cell death. Indeed, when the glucokinase expression reaches a critical level of $V_{max}^{glk}$ equal to 120 mM/min (this value depends on all other parameters of the model), the physiological quasi-steady-state "stable 1" and unstable states merge and the resulting quasi-steady state will be unstable (Fig. 23.21b). Further increases in the glucokinase expression level to $V_{max}^{glk}$ equal to 500 mM/min lead to the disappearance of this unstable quasi-steady state entirely (Fig. 23.21c). Figure 23.22 shows how the level of glucokinase expression determines the number of quasi-steady states and their stability when glucose concentration is not less than 1 mM. Indeed, when the glucokinase

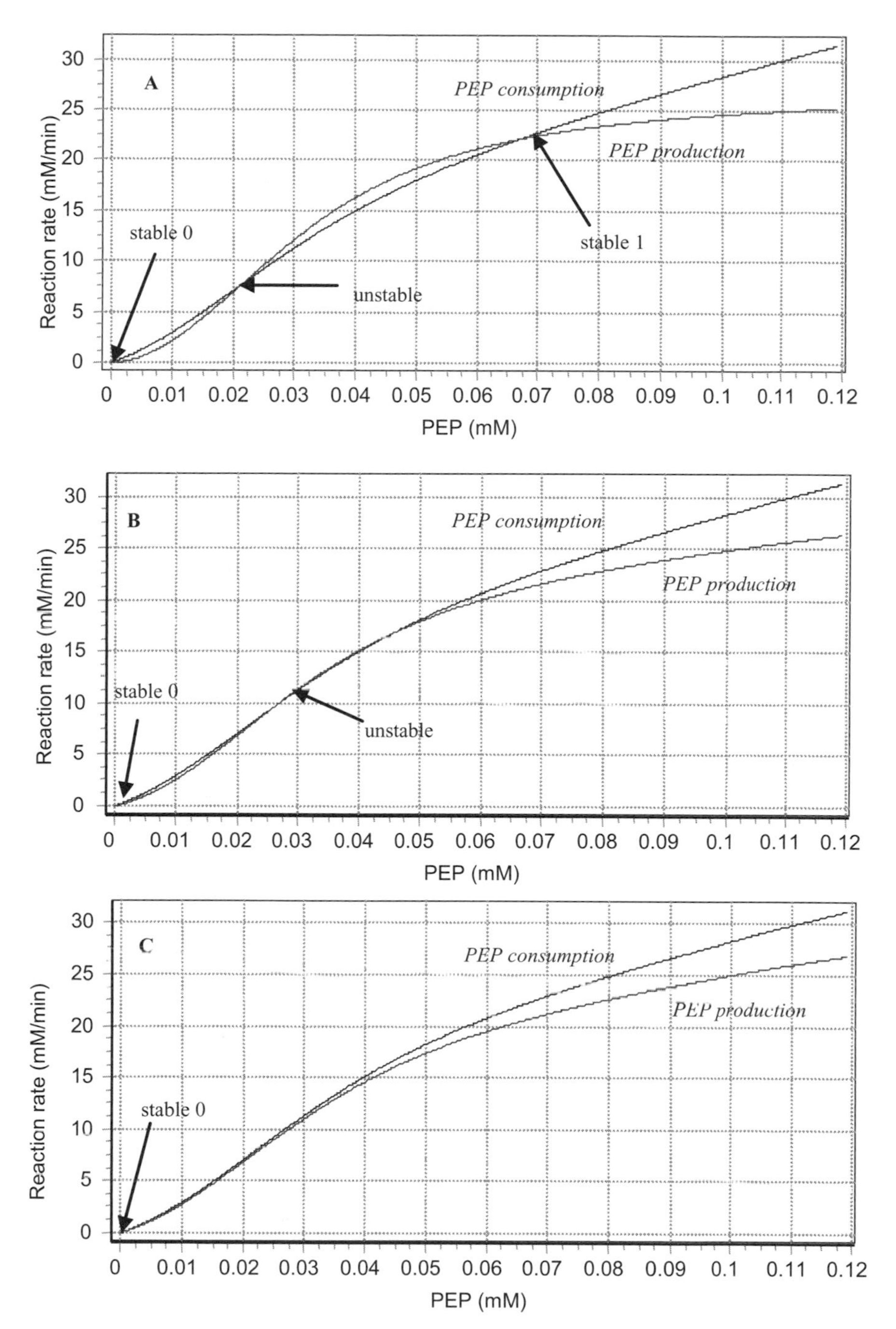


**Figure 23.21** Number of steady states of kinetic model of glycolysis at $Glc_{out} = 0.05$ mM and different activities of glucokinase: (a) correponds to $V_{max} = 0.1$ mM/min, (b) corresponds to $V_{max} = 120$ mM/min, and (c) corresponds to $V_{max} = 500$ mM/min.

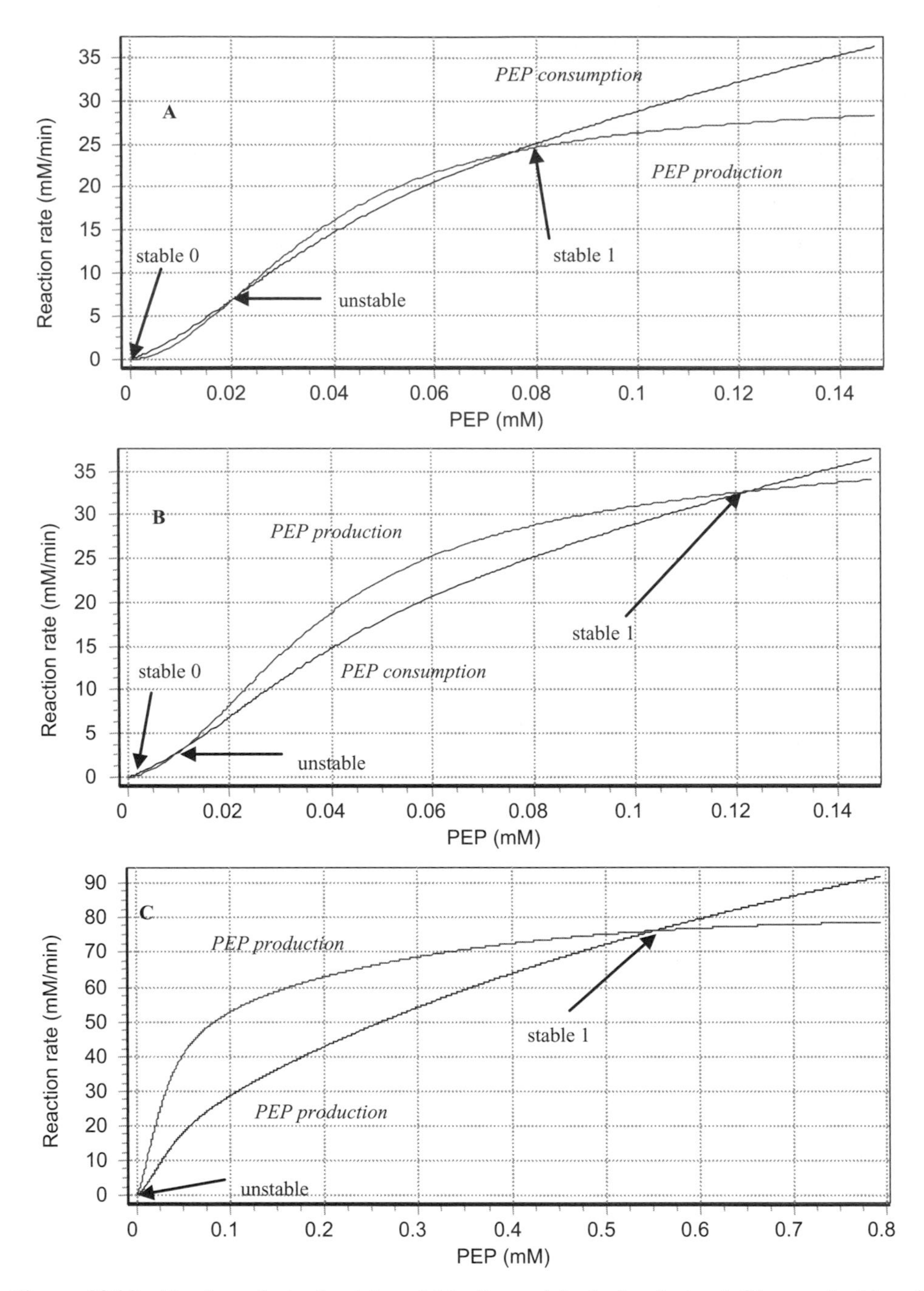


**Figure 23.22** Number of steady states of kinetic model of glycolysis at $Glc_{out} = 1$ mM and different activities of glucokinase: (a) correponds to $V_{max} = 1$ mM/min, (b) corresponds to $V_{max} = 8$ mM/min, and (c) corresponds to $V_{max} = 100$ mM/min.

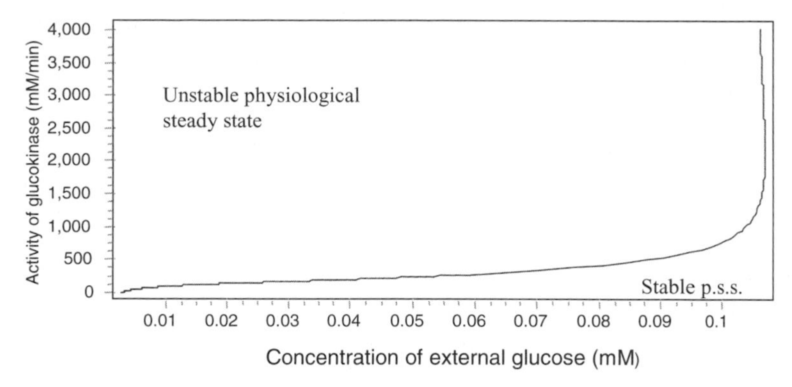


**Figure 23.23** Bifurcational diagram in two-dimensional parametric space, reflecting regions of stability of physiological steady-state "stable 1."

expression reaches a critical level of $V_{max}^{glk}$ equal to 8 mM/min (this value depends on the values of all other parameters of the model), the non-physiological quasi-steady state and unstable quasi-steady state merge. The resulting quasi-steady state will be unstable (Fig. 23.22b). This analysis of the stability properties of the quasi-steady states as a function of glucokinase expression leads to a very simple biological interpretation: The glucokinase expression level can lead to either cell death or keep the cell alive depending on substrate availability. By using a bifurcation diagram, such as shown in Figure 23.23, an estimate can be made of the glucose concentrations and glucokinase expression levels that correspond to physiological quasi-steady states.

## 23.4 CONCLUSION REMARKS

In this chapter, we have tried to show that mathematical modeling in biology in general, and metabolic kinetic models in particular, have been moving from pure academic exercises to a powerful predictive technology. We are only at the beginning of a new era, where such tools can be applied to a variety of challenging practical problems in medical and pharmaceutical research. As systems biology and other technologies deliver more and more data, and more is known about biological pathways, such techniques will be critical. The complexity of biological systems, even with our current understanding, is no longer intuitive and only mathematical approaches will allow the fine dissection of behavior. Indeed, it is both the hypothesis testing and the hypothesis generation capabilities of mathematical modeling that gives it such wide practability. In this article, we have tried to provide some examples of both.

Acknowledgments: This work was financially supported by GlaxoSmithKline Company.

## REFERENCES

1. Aulabaugh, A., and Schloss, J.V. (1990). Oxalyl hydroxamates as reaction-intermediate analogs for ketol-acid reductoisomerase. *Biochemistry* **29**:2824–2830.
2. Barak, Z., Chipman, D.M., and Gollop, N. (1987). Physiological implications of the specificity of acetohydroxy acid synthase isozymes of enteric bacteria. *J. Bacteriol.* **169**:3750–3756.

3. Bar-Ilan, A., Balan, V., Tittmann, K., Golbik, R., Vyazmensky, M., Hubner, G., Barak, Z., Chipman, D.M. (2001). Binding and activation of thiamin diphosphate in acetohydroxyacid synthase. *Biochemistry* **40**:11946–11954.
4. Chassagnole, C., Noisommit-Rizzi, N., Schmid, J.W., Klaus Mauch, K., and Reuss, M. (2002). Dynamic modeling of the central carbon metabolism of Escherichia coli. *Biotechnol. Bioeng.* **79**:53–73.
5. Chunduru, S.K., Mrachko, G.T., and Calvo, K.C. (1998). Mechanism of ketol acid reductoisomerase. Steady-state analysis and metal ion requirement. *Biochemistry* **28**:486–493.
6. Cleland, W.W. (1963). The kinetics of enzyme-catalysed reactions with two or more substrates or products. *Biochim. Biophys. Acta.* **67**:104–137.
7. Cornish-Bouden, A. (2001). Fundamentals of Enzyme Kinetic. Cambridge: Portland Press.
8. Chittur, S.V., Klem, T.J., Shafer, C.M., and Davisson, V.J. (2001). *Biochemistry* **40**:876–887.
9. Edwards, J.S., Ibarra, R.U., and Palsson, B.O. (2001). *In silico* predicrions of *Escherichia coli* metabolic capabilities are consistent with experimental data. *Nat. Biotechnol.* **19**:125–130.
10. Adams, A. (1955). I-histidinal, a biosynthetic precursor of histidine. *J. Biol. Chem.* **217**:325–344.
11. Engel, S., Vyazmensky, M., Barak, Z., Chipman, D.M., and Merchuk, J.C. (2000). Determination of the dissociation constant of valine from acetohydroxy acid synthase by equilibrium partition in an aqueous two-phase system. *J. Chromatogr. B* **743**:225–229.
12. Eoyang, L., and Silverman, P.M. (1984). Purification and subunit composition of acetohydroxyacid synthase I from Escherichia coli K-12. *J. Bacteriol.* **157**:184–189.
13. Goryanin, I., Hodgman, T.C., and Selkov, E. (1999). Mathematical simulation and analysis of cellular metabolism and regulation. *Bioinformatics* **15**:749–758.
14. Hall, T.R., Wallin, R., Reinhart, G.D., and Hutson, S.M. (1993). Branched chain aminotransferase isoenzymes. Purification and characterization of the rat brain isoenzyme. *J. Biol. Chem.* **268**:3092–3098.
15. Hill, C.M., and Duggleby, R.G. (1998). Escherichia coli acetohydroxyacid synthase II mutants. *Biochem. J.* **335**:653–661.
16. Hill, C.M., Pang, S.S., and Duggleby, R.G. (1997). Escherichia coli acetohydroxyacid synthase II. *Biochem. J.* **327**:891–898.
17. Holms, W.H. (1986). The central metabolic pathways of Escherichia coli: relationship between flux and control at a branch point, efficiency of conversion to biomass, and excretion of acetate. *Curr. Top. Cell Regul.* **28**:69–104.
18. Hoving, H., Lolkema, J.S., and Robillard, G.T. (1981). Escherichia coli phosphoenolpyruvate-dependent phosphotransferase system: equilibrium kinetics and mechanism of enzyme i phosphorylation. *Biochemistry* **20**(1):87–93.
19. Inoue, K., Kuramitsu, S., Aki, K., Watanabe, Y., Takagi, T., Nishigai, M., Ikai, A., and Kagamiyama, H. (1988). Branched-chain amino acid aminotransferase of Escherichia coli: overproduction and properties. *J. Biochem.* **104**:777–784.
20. Ivanitzky, G.R., Krinsky, V.I., and Selkov, E.E. (1978). Mathematical Biophysics of the Cell. Nauka: Moscow.
21. Jurgens, C., Strom, A., Wegener, D., Hettwer, S., Wilmanns, M., and Sterner, R. (2000). *Proc. Natl. Acad. Sci. USA.*, **97**:9925–9930.
22. Klem, T.J., and Davisson, V.J. (1993). *Biochemistry* **32**: 5177–5186.
23. Lee-Peng, F.C., Hermodson, M.A., and Kohlhaw, G.B. (1979). Transaminase B from *Escherichia coli*: quaternary structure, amino-terminal sequence, substrate specificity, and absence of a separate valine-α-ketoglutarate activity. *J. Bacteriol.* **139**(2):339–345.
24. Limberg, G., Klaffkc, W., and Thiem, J. (1995). Conversion of aldonic acids to their corresponding 2-keto-3-deoxy-analogs by the non-carbohydrate enzyme dihydroxy acid dehydratase (DHAD). *Bioorg. Med. Chem.* **3**:487–494.

25. Loper, J., and Adams, E. (1965). Purification and properties of histidinol dehydrogenase from *Salmonella typhimurium*. *J. Biol. Chem.* **240**:788–795.

26. Lolkema, J.S., ten Hoeve-Duurkens, R.H., and Robillard, G.T. (1993). Quasi-steady state kinetics of mannitol phosphorylation catalyzed by enzyme IImtl of the Escherichia coli phosphoenolpyruvate-dependent phosphotransferase system. *J. Biol. Chem.* **268**(24): 17844–17849.

27. Myers, J.W. (1961). Dihydroxy acid dehydrase: an enzyme involved in the biosynthesis of isoleucine and valine. *J. Biol. Chem.* **236**:1414–1418.

28. Meadow, N.D., and Roseman, S. (1996). Rate and equilibrium constants for phosphoryltransfer between active site histidines of Escherichia coli HPr and the signal transducing protein IIIGlc. *J. Biol. Chem.* **271**(52):33440–33445.

29. Meyer, D., Schneider-Fresenius, C., Horlacher, R., Peist, R., and Boos, W. (1997). Molecular characterization of glucokinase from Escherichia coli K-12. *J. Bacteriol.* **179**(4): 1298–1306.

30. Perna, N.T., Plunkett, G. 3rd, Burland, V., Mau, B., Glasner, J.D., Rose, D.J., Mayhew, G.F., Evans, P.S., et al. (2001). Genome sequence of enterohaemorrhagic Escherichia coli O157:H7. *Nature* **409**:529–533.

31. Postma, P.W., Lengeler, J.W., and Jacobson, J.R. (1996). *Escherichia coli and Salmonella typhimurium: Cellular and Molecular Biology.* Vol. 1, Washington, D.C.: American Society for Microbilogy, pp. 1149–1174.

32. Postma, P.W., Lengeler, J.W., and Jacobson, G.R. (1993). Phosphoenolpyruvate: Carbohydrate phosphotransferase systems of bacteria. *Microbiol. Rev.* **57**(3):543–594.

33. Rane, M.J., and Calvo, K.C. (1997). Reversal of the nucleotide specificity of ketol acid reductoisomerase by site-directed mutagenesis identifies the NADPH Binding Site1. *Arch. Biochem. Biophys.* **338**:83–89.

34. Rohwer, J.M., Meadow, N.D., Roseman, S., Westerhoff, H.V., and Postma, P.W. (2000). Understanding glucose transport by the bacterial phosphoenolpyruvate: Glycose phosphotransferase system on the basis of kinetic measurements in vitro. *J. Biol. Chem.* **275**(45):34909–34921.

35. Selkov, E., Basmanova, S., Gaasterland, T., Goryanin, I., Gretchkin, Y., Maltsev, N., Nenashev, V., Overbeek, R., Panyushkina, E., Pronevitch, L., and Yunis, I. (1996). The metabolic pathway collection from EMP: The enzymes and metabolic pathways database. *Nucleic Acids Res.* **24**:26–28.

36. Shomburg, I., Chang, A., and Shomburg, D. (2002). BRENDA, enzyme data and metabolic information. *Nucleic Acids Res.* **30**:47–49.

37. Umbarger, H.E. (1996). *Escherichia coli and Salmonella: Cellular and molecular biology.* Washington, D.C. ASM Press, pp. 442–458.

38. Vyazmensky, M., Sella, C., Barak, Z., and Chipman, D.M. (1996). Isolation and characterization of subunits of acetohydroxy acid synthase isozyme III and reconstitution of the holoenzyme. *Biochemistry* **35**:10339–10346.

39. Wessel, P.M., Graciet, E., Douce, R., and Dumas, R. (2000). Evidence for two distinct effector-binding sites in threonine deaminase by site-directed mutagenesis, kinetic, and binding experiments. *Biochemistry* **39**:15136–151143.

40. Waygood, E.B., and Steeves, T. (1980). Enzyme I of the phosphoenolpyruvate: Sugar phosphotransferase system of Escherichia coli. Purification to homogeneity and some properties. *Can. J. Biochem.* **58**(1):40–48.

41. Zalkin, H. (1993). *Adv. Enzymol. Relat. Areas Mol. Biol.*, **66**:203–309.

42. Zalkin, H., and Smith, J. L. (1998). *Adv. Enzymol. Relat. Areas Mol. Biol.*, **72**:87–144.

## APPENDIX 1

To deduce expressions defining the rates of the synthetase and glutaminase reactions, let us suppose that all reactions of the substrate addition and dissociation of the products are much faster than the catalytic reactions designated by numbers 1, 2, 3, and 4 in Figure 23.1. Thus each of these "fast" reactions can be considered as a quasi-equilibrium one (the dissociation constants are given near the corresponding reactions in Fig. 23.1); thus, we can write the following relationships:

$$K_{\mathrm{PRFAR}} = \frac{[\mathrm{PRFAR}][\mathrm{E}]}{[\mathrm{E} \bullet \mathrm{PRFAR}]}, \quad K_{\mathrm{PRFAR,Gln}} = \frac{[\mathrm{Gln}][\mathrm{E} \bullet \mathrm{PRFAR}]}{[\mathrm{Gln} \bullet \mathrm{E} \bullet \mathrm{PRFAR}]}, \quad K_{\mathrm{Gln}} = \frac{[\mathrm{Gln}][\mathrm{E}]}{[\mathrm{Gln} \bullet \mathrm{E}]},$$

$$K^{g}_{\mathrm{Glu}} = \frac{[\mathrm{Glu}][\mathrm{E}]}{[\mathrm{Glu} \bullet \mathrm{E}^{*}]}, \quad K_{\mathrm{ProFAR}} = \frac{[\mathrm{ProFAR}][\mathrm{E}]}{[\mathrm{E} \bullet \mathrm{ProFAR}]}, \quad K_{\mathrm{ProFAR,Gln}} = \frac{[\mathrm{Gln}][\mathrm{E} \bullet \mathrm{ProFAR}]}{[\mathrm{Gln} \bullet \mathrm{E} \bullet \mathrm{ProFAR}]},$$

$$K^{g}_{\mathrm{ProFAR,Gln}} = \frac{[\mathrm{Glu}][\mathrm{E} \bullet \mathrm{ProFAR}]}{[\mathrm{Glu} \bullet \mathrm{E}^{*} \bullet \mathrm{ProFAR}]}, \quad K_{\mathrm{AICAR}} = \frac{[\mathrm{AICAR}][\mathrm{E}]}{[\mathrm{E} \bullet \mathrm{AICAR}]},$$

$$K_{\mathrm{AICAR,Glu}} = \frac{[\mathrm{Glu}][\mathrm{E} \bullet \mathrm{AICAR}]}{[\mathrm{Glu} \bullet \mathrm{E} \bullet \mathrm{AICAR}]}, \quad K_{\mathrm{Glu}} = \frac{[\mathrm{Glu}][\mathrm{E}]}{[\mathrm{Glu} \bullet \mathrm{E}]}, \quad K_{\mathrm{Glu,IGP}} = \frac{[\mathrm{IGP}][\mathrm{Glu} \bullet \mathrm{E}]}{[\mathrm{Glu} \bullet \mathrm{E} \bullet \mathrm{IGP}]},$$

$$K_{\mathrm{Glu,IGP,AICAR}} = \frac{[\mathrm{AICAR}][\mathrm{Glu} \bullet \mathrm{E} \bullet \mathrm{IGP}]}{[\mathrm{Glu} \bullet \mathrm{E} \bullet \mathrm{IGP} \bullet \mathrm{AICAR}]}, \quad K_{\mathrm{IGP}} = \frac{[\mathrm{IGP}][\mathrm{E}]}{[\mathrm{E} \bullet \mathrm{IGP}]},$$

$$K_{\mathrm{IGP,AICAR}} = \frac{[\mathrm{AICAR}][\mathrm{E} \bullet \mathrm{IGP}]}{[\mathrm{E} \bullet \mathrm{IGP} \bullet \mathrm{AICAR}]}, \quad K_{\mathrm{IGP,Gln}} = \frac{[\mathrm{Gln}][\mathrm{E} \bullet \mathrm{IGP}]}{[\mathrm{Gln} \bullet \mathrm{E} \bullet \mathrm{IGP}]},$$

$$K^{g}_{\mathrm{IGP,Glu}} = \frac{[\mathrm{Glu}][\mathrm{E} \bullet \mathrm{IGP}]}{[\mathrm{Glu} \bullet \mathrm{E}^{*} \bullet \mathrm{IGP}]} \tag{23.A1}$$

For concentrations of the enzyme states, the following conservation law is also fulfilled:

$$\begin{aligned}
&[\mathrm{E}] + [\mathrm{E} \bullet \mathrm{PRFAR}] + [\mathrm{Gln} \bullet \mathrm{E} \bullet \mathrm{PRFAR}] + [\mathrm{Gln} \bullet \mathrm{E}] + [\mathrm{Glu} \bullet \mathrm{E}^{*}] + [\mathrm{E} \bullet \mathrm{ProFAR}] \\
&\quad + [\mathrm{Glu} \bullet \mathrm{E}^{*} \bullet \mathrm{ProFAR}] + [\mathrm{Gln} \bullet \mathrm{E} \bullet \mathrm{ProFAR}] + [\mathrm{E} \bullet \mathrm{AICAR}] + [\mathrm{Glu} \bullet \mathrm{E} \bullet \mathrm{AICAR}] \\
&\quad + [\mathrm{Glu} \bullet \mathrm{E}] + [\mathrm{Glu} \bullet \mathrm{E} \bullet \mathrm{IGP}] + [\mathrm{E} \bullet \mathrm{IGP}] + [\mathrm{E} \bullet \mathrm{IGP} \bullet \mathrm{AICAR}] + [\mathrm{Glu} \bullet \mathrm{E}^{*} \bullet \mathrm{IGP}] \\
&\quad + [\mathrm{Gln} \bullet \mathrm{E} \bullet \mathrm{IGP}] + [\mathrm{Glu} \bullet \mathrm{E} \bullet \mathrm{IGP} \bullet \mathrm{AICAR}] = [\mathrm{HisHF}]
\end{aligned} \tag{23.A2}$$

where [HisHF] is the total concentration of imidazologlycerol-phosphate synthetase. Solving the system of linear (relative to concentrations of the enzyme states) algebraic equations (23.A1) and (23.A2), we obtain the following expressions for the enzyme forms:

$$[\mathrm{Gln} \bullet \mathrm{E} \bullet \mathrm{PRFAR}] = \frac{[\mathrm{Gln}]}{K_{\mathrm{PRFAR,Gln}}} \frac{[\mathrm{PRFAR}]}{K_{\mathrm{PRFAR}}} \frac{[\mathrm{HisHF}]}{\Delta}, \; [\mathrm{Gln} \bullet \mathrm{E}] = \frac{[\mathrm{Gln}]}{K_{\mathrm{Gln}}} \frac{[\mathrm{HisHF}]}{\Delta},$$

$$[\mathrm{Glu} \bullet \mathrm{E}^{*}] = \frac{[\mathrm{Glu}]}{K^{g}_{\mathrm{Glu}}} \frac{[\mathrm{HisHF}]}{\Delta}, \; [\mathrm{Gln} \bullet \mathrm{E} \bullet \mathrm{IGP}] = \frac{[\mathrm{IGP}]}{K_{\mathrm{IGP}}} \frac{[\mathrm{Gln}]}{K_{\mathrm{IGP,Gln}}} \frac{[\mathrm{HisHF}]}{\Delta},$$

$$[\text{Glu} \bullet \text{E}^* \bullet \text{IGP}] = \frac{[\text{IGP}]}{K_{\text{IGP}}} \frac{[\text{Glu}]}{K^{\text{g}}_{\text{IGP,Glu}}} \frac{[\text{HisHF}]}{\Delta}, [\text{Gln} \bullet \text{E} \bullet \text{ProFAR}] = \frac{[\text{ProFAR}]}{K_{\text{ProFAR}}} \frac{[\text{Gln}]}{K_{\text{ProFAR,Gln}}} \frac{[\text{HisHF}]}{\Delta},$$

$$[\text{Glu} \bullet \text{E}^* \bullet \text{IGP}] = \frac{[\text{ProFAR}]}{K_{\text{ProFAR}}} \frac{[\text{Glu}]}{K^{\text{g}}_{\text{ProFAR,Glu}}} \frac{[\text{HisHF}]}{\Delta},$$

$$[\text{Gln} \bullet \text{E} \bullet \text{PRFAR}] = \frac{[\text{AICAR}]}{K_{\text{Glu,IGP,AICAR}}} \frac{[\text{IGP}]}{K_{\text{Glu,IGP}}} \frac{[\text{Glu}]}{K_{\text{Glu}}} \frac{[\text{HisHF}]}{\Delta},$$

$$\begin{aligned}\Delta = 1 &+ \frac{[\text{PRFAR}]}{K_{\text{PRFAR}}} + \frac{[\text{Gln}]}{K_{\text{PRFAR,Gln}}} \frac{[\text{PRFAR}]}{K_{\text{PRFAR}}} + \frac{[\text{Gln}]}{K_{\text{Gln}}} + \frac{[\text{Glu}]}{K^{\text{g}}_{\text{Glu}}} + \frac{[\text{ProFAR}]}{K_{\text{ProFAR}}} \\ &\times \left(1 + \frac{[\text{Glu}]}{K^{\text{g}}_{\text{ProFAR,Glu}}} + \frac{[\text{Gln}]}{K_{\text{ProFAR,Gln}}}\right) + \frac{[\text{IGP}]}{K_{\text{IGP}}} \left(1 + \frac{[\text{Glu}]}{K^{\text{g}}_{\text{IGP,Glu}}} + \frac{[\text{Gln}]}{K_{\text{IGP,Gln}}}\right) \\ &+ \frac{[\text{AICAR}]}{K_{\text{AICAR}}} + \frac{[\text{Glu}]}{K_{\text{AICAR,Glu}}} \frac{[\text{AICAR}]}{K_{\text{AICAR}}} + \frac{[\text{Glu}]}{K_{\text{Glu}}} + \frac{[\text{IGP}]}{K_{\text{Glu,IGP}}} \frac{[\text{Glu}]}{K_{\text{Glu}}} \\ &+ \frac{[\text{AICAR}]}{K_{\text{IGP,AICAR}}} \frac{[\text{IGP}]}{K_{\text{IGP}}} + \frac{[\text{AICAR}]}{K_{\text{Glu,IGP,AICAR}}} \frac{[\text{IGP}]}{K_{\text{Glu,IGP}}} \frac{[\text{Glu}]}{K_{\text{Glu}}}\end{aligned} \tag{23.A3}$$

It should be noted that the equilibrium constant of the glutaminase reaction does not depend on what of the three forms of the enzyme (free enzyme, enzyme bound to ProFAR or to IGP) catalyzes the reaction; consequently, the following relationships are correct:

$$K^{\text{g}}_{\text{eq}} = \frac{K^{\text{g}}_{\text{Glu}}}{K_{\text{Gln}}} \frac{k_2}{k_{-2}} = \frac{K^{\text{g}}_{\text{ProFAR,Glu}}}{K_{\text{ProFAR,Gln}}} \frac{k_4}{k_{-4}} = \frac{K^{\text{g}}_{\text{IGP,Glu}}}{K_{\text{IGP,Gln}}} \frac{k_3}{k_{-3}} \tag{23.A4}$$

which make it possible to express $k_{-3}$ and $k_{-4}$ via $k_2$ and $k_{-2}$:

$$k_{-4} = k_4 \frac{K_{\text{ProFAR,Gln}}}{K^{\text{g}}_{\text{ProFAR,Glu}}} \frac{K^{\text{g}}_{\text{Glu}}}{K_{\text{Gln}}} \frac{k_2}{k_{-2}}, \quad k_{-3} = k_3 \frac{K_{\text{IGP,Gln}}}{K^{\text{g}}_{\text{IGP,Glu}}} \frac{K^{\text{g}}_{\text{Glu}}}{K_{\text{Gln}}} \frac{k_2}{k_{-2}} \tag{23.A5}$$

Substitution of (23.A3) and (23.A5) into Eqs. (23.8) and (23.9) of the main text gives the following equations for the rates of the synthetase and glutaminase reactions:

$$v_s = \frac{[\text{HisHF}]}{\Delta} \left( k_1 \frac{[\text{Gln}]}{K_{\text{PRFAR,Gln}}} \frac{[\text{PRFAR}]}{K_{\text{PRFAR}}} - k_{-1} \frac{[\text{AICAR}]}{K_{\text{Glu,IGP,AICAR}}} \frac{[\text{IGP}]}{K_{\text{Glu,IGP}}} \frac{[\text{Glu}]}{K_{\text{Glu}}} \right)$$

$$v_g = \frac{[\text{HisHF}]}{\Delta} \left( \frac{k_2}{K_{\text{Gln}}} + \frac{k_3}{K_{\text{IGP,Gln}}} \frac{[\text{IGP}]}{K_{\text{IGP}}} + \frac{k_4}{K_{\text{ProFAR,Gln}}} \frac{[\text{ProFAR}]}{K_{\text{ProFAR}}} \right) \left( [\text{Gln}] - \frac{k_{-2}}{k_2} \frac{K_{\text{Gln}}}{K^{\text{g}}_{\text{Glu}}} [\text{Glu}] \right)$$

## APPENDIX 2

Let (Eq. 23.B1) be the rate equation of the enzyme functioning in which parameters of the catalytic cycle (the rate and dissociation constants of certain stages) are the terms:

$$v = [\text{E}]_{\text{tot}} \cdot f(\text{S}_1, \ldots, \text{S}_i, \ldots, \text{S}_n, \text{P}_1, \ldots, \text{P}_j, \ldots, \text{P}_m, \text{M}_1, \ldots, \text{M}_h, \ldots, \text{M}_q) \tag{23.B1}$$

where $[E]_{tot}$ is the total enzyme concentration and $S_i$ (i = 1, ..., n), $P_j$ (j = 1, ..., m), and $M_h$ (h = 1, ..., q) are concentrations of the substrates, products, and modifiers (inhibitors and activators), respectively. Using biochemical definitions of conventional parameters of enzymatic kinetics (the Michaelis constants of the substrates and products, the equilibrium constants, the maximal number of enzyme cycles in forward reaction in the presence and in the absence of activators and inhibitors), let us find how to express the kinetic parameters via parameters of the catalytic cycle. By definition, the maximal number of the enzyme cycles in the forward reaction is the ratio of the maximal rate of enzyme functioning to the total enzyme concentration at saturating concentrations of all substrates and zero concentrations of all products and modifiers. Thus, for calculation of the maximal number of enzyme cycles in the forward reaction, the following expression should be used:

$$k_{cat}^{f} = \lim_{\substack{S_i \to \infty, i=1,\ldots,n \\ P_j = 0, j=1,\ldots,m \\ M_h = 0, h=1,\ldots,q}} f(S_1, \ldots, S_n, P_1, \ldots, P_m, M_1, \ldots, M_q) \tag{23.B2}$$

Analogously the maximal number of enzyme cycles at the saturating concentration of modifier (inhibitor or activator) $M_r$ can be calculated as follows:

$$k_{cat}^{f,M_r} = \lim_{\substack{S_i \to \infty, i=1,\ldots,n \\ M_r \to \infty \\ P_j = 0, h=1,\ldots,m \\ M_h = 0, h=1,\ldots r+1,\ldots,q}} f(S_1, \ldots, S_n, P_1, \ldots, P_m, M_1, \ldots, M_q) \tag{23.B3}$$

The equilibrium constant can be found from the following expression:

$$K_{eq} = \prod_{j=1}^{m} P_j^{eq} \Big/ \prod_{i=1}^{n} S_i^{eq} \tag{23.B4}$$

where the equilibrium concentrations of the substrates $S_i^{eq}$ ($i$ = 1, ..., n) and products $P_j^{eq}$ ($j$ = 1, ..., m) are solutions of the following equation:

$$f(S_1^{eq}, \ldots, S_i^{eq}, \ldots, S_n^{eq}, P_1^{eq}, \ldots, P_j^{eq}, \ldots, P_m^{eq}, M_1, \ldots, M_h, \ldots, M_q) = 0 \tag{23.B5}$$

By definition, the Michaelis constant of the enzyme with some substrate is the concentration of the considered substrate at which the rate of the enzyme functioning is a half of its maximal rate under the conditions when the products and modifiers (inhibitors and activators) are absent and the concentrations of all other substrates are saturating. In accord with this definition, $K_{m,St}$, the Michaelis constant of substrate $S_t$, is a solution of the following equation:

$$\frac{k_{cat}^{f}}{2} = F_{S_t}(K_{m,S_t}), \text{где} \qquad F_{S_t}(S_t) = \lim_{\substack{S_i \to \infty, i=1,\ldots,t-1,t+1,\ldots,n \\ P_j = 0, j=1,\ldots,m \\ M_h = 0, h=1,\ldots,q}} f(S_1, \ldots, S_n, P_1, \ldots, P_m, M_1, \ldots, M_q) \tag{23.B6}$$

Analogously, the Michaelis constant of substrate $S_t$ at the saturating concentration of modifier (inhibitor or activator) $M_r$:

$$\frac{k_{cat}^{f,M_r}}{2} = F_{S_t}^{M_r}\left(K_{m,S_t}^{M_r}\right)$$

where

$$F_{S_t}^{M_r}(S_t) = \lim_{\substack{S_i \to \infty, i=1,\ldots,t-1,t+1,\ldots,n \\ M_r \to \infty \\ P_j=0, j=1,\ldots,m \\ M_h=0, h=1,\ldots,r-1,r+1,\ldots,q}} f(S_1, \ldots, S_n, P_1, \ldots, P_m, M_1, \ldots, M_q) \qquad (23.B7)$$

and the Michaelis constant of product $P_t$ is calculated as follows:

$$\frac{k_{cat}^{b}}{2} = F_{P_t}(K_{m,P_t})$$

where

$$F_{P_t}(P_t) = \lim_{\substack{P_j \to \infty, j=1,\ldots,t-1,t+1,\ldots,m \\ S_i=0, i=1,\ldots,n \\ M_h=0, h=1,\ldots,q}} (-f(S_1, \ldots, S_n, P_1, \ldots, P_m, M_1, \ldots, M_q)) \qquad (23.B8)$$

In this expression, the maximal number of enzyme cycles in the reverse reaction, $k_{cat}^{b}$, is calculated as follows:

$$k_{cat}^{b} = \lim_{\substack{P_j \to \infty, j=1,\ldots,m \\ S_i=0, i=1,\ldots,n \\ M_h=0, h=1,\ldots,q}} (-f(S_1, \ldots, S_n, P_1, \ldots .P_m, M_1, \ldots, M_q)) \qquad (23.B9)$$

Applying formulas (23.B1)–(23.B9) to the rate equations of the synthetase and glutaminase reactions [Eqs. (23.10) of the main text], we deduced the following relationships between the parameters of the catalytic cycle and the conventional parameters of enzymatic kinetics:

$$k_{cat}^{s} = k_1; K_{m,Gln}^{s} = K_{PRFAR,Gln}; K_{m,PRFAR}^{s} = K_{PRFAR}\frac{K_{PRFAR,Gln}}{K_{Gln}}$$

$$K_{eq}^{s} = \frac{k_1}{k_{-1}}\frac{K_{Glu,IGP,AICAR}K_{Glu,IGP}K_{Glu}}{K_{PRFAR,Gln}K_{PRFAR}}; \quad K_{m,AICAR}^{s} = K_{Glu,IGP,AICAR}\left(1 + \frac{K_{Glu,IGP}K_{Glu}}{K_{IGP,Glu}^{g}K_{IGP}}\right);$$

$$K_{m,Glu}^{s} = K_{Glu,IGP,AICAR}\frac{K_{Glu,IGP}K_{Glu}}{K_{IGP,AICAR}K_{IGP}}; \quad K_{m,IGP}^{s} = K_{Glu,IGP,AICAR}\frac{K_{Glu,IGP}K_{Glu}}{K_{AICAR,Glu}K_{AICAR}};$$

$$k_{cat}^{g} = k_2; \quad K_{m,Gln}^{g} = K_{Gln}; \quad K_{eq}^{g} = \frac{k_2}{k_{-2}}\frac{K_{Glu}^{g}}{K_{Gln}}; \quad K_{m,Glu}^{g} = K_{Glu}\frac{K_{Glu}^{g}}{K_{Glu} + K_{Glu}^{g}};$$

$$k_{cat}^{g,IGP} = k_3; \quad K_{m,Gln}^{g,IGP} = K_{IGP,Gln}; \quad K_{m,Glu}^{g,IGP} = K_{IGP,Glu}^{g}; \quad k_{cat}^{g,ProFAR} = k_4;$$

$$K_{m,Gln}^{g,ProFAR} = K_{ProFAR,Gln}; \quad K_{m,Glu}^{g,ProFAR} = K_{ProFAR,Glu}^{g}$$

These relationships allow expressing parameters of the catalytic cycle via kinetic parameters:

$$k_1 = k^{s}_{cat}; \quad K_{PRFAR,Gln} = K^{s}_{m,Gln}; \quad k_2 = k^{g}_{cat}; K_{Gln} = K^{g}_{m,Gln}; \quad k_3 = k^{g,IGP}_{cat};$$

$$K_{IGP,Gln} = K^{g,IGP}_{m,Gln}; \quad K^{g}_{IGP,Glu} = K^{g,IGP}_{m,Glu}; \quad k_4 = k^{g,ProFAR}_{cat}; \quad K_{ProFAR,Gln} = K^{g,ProFAR}_{m,Gln};$$

$$K^{g}_{ProFAR,Glu} = K^{g,ProFAR}_{m,Glu}; \quad K_{PRFAR} = K^{s}_{m,PRFAR}\frac{K^{g}_{m,Gln}}{K^{s}_{m,Gln}};$$

$$K_{Glu,IGP,AICAR} = \frac{K^{s}_{m,AICAR}K^{g,IGP}_{m,Glu}K_{IGP}}{K_{Glu}K_{Glu,IGP} + K_{IGP}K^{g,IGP}_{m,Glu}}; \quad K^{g}_{Glu} = K^{g}_{m,Glu}\frac{K_{Glu}}{K_{Glu} - K^{g}_{m,Glu}}, K_{Glu} > K^{g}_{m,Glu};$$

$$k_{-2} = \frac{k^{g}_{cat}}{K^{g}_{eq}K^{g}_{m,Gln}}\frac{K^{g}_{m,Glu}K_{Glu}}{K_{Glu} - K^{g}_{m,Glu}}, \quad K_{Glu} > K^{g}_{m,Glu};$$

$$K_{IGP,AICAR} = \frac{K^{s}_{m,AICAR}K^{g,IGP}_{m,Glu}K_{Glu}K_{Glu,IGP}}{K^{s}_{m,Glu}\left(K_{Glu}K_{Glu,IGP} + K_{IGP}K^{g,IGP}_{m,Glu}\right)};$$

$$K_{AICAR,Glu} = \frac{K^{s}_{m,AICAR}K^{g,IGP}_{m,Glu}K_{Glu}K_{Glu,IGP}K_{IGP}}{K^{s}_{m,IGP}K_{AICAR}\left(K_{Glu}K_{Glu,IGP} + K_{IGP}K^{g,IGP}_{m,Glu}\right)};$$

$$k_{-1} = \frac{k^{s}_{cat}}{K^{s}_{eq}K^{s}_{m,PRFAR}K^{g}_{m,Gln}}\frac{K^{s}_{m,AICAR}K^{g,IGP}_{m,Glu}K_{Glu}K_{Glu,IGP}K_{IGP}}{K_{Glu}K_{Glu,IGP} + K_{IGP}K^{g,IGP}_{m,Glu}}$$

Substitution of these relationships into Eqs. (23.10) of the main text gives Eqs. (23.11).

# INDEX



*Microbial Proteomics: Functional Biology of Whole Organisms*, Edited by Ian Humphery-Smith and Michael Hecker.